STATISTICS THROUGH APPLICATIONS

STATISTICS THROUGH APPLICATIONS

SECOND EDITION

DAREN S. STARNES
THE LAWRENCEVILLE SCHOOL

DANIEL S. YATES
STATISTICS CONSULTANT

DAVID S. MOORE
PURDUE UNIVERSITY

W. H. FREEMAN AND COMPANY / NEW YORK

Executive Publisher: Craig Bleyer
Publisher: Ruth Baruth
Executive Editor: Ann Heath
Senior Developmental Editor: Shona Burke
Executive Marketing Manager: Cindi Weiss
Marketing Manager: Nicole Sheppard Klophaus
Media Editor: Laura Capuano
Associate Editor: Katrina Wilhelm
Editorial Assistant: Catriona Kaplan
Project Editor: Jane O'Neill
Text and Cover Designer: Blake Logan
Photo Editor: Cecilia Varas
Photo Researcher: Elyse Rieder
Senior Illustration Coordinator: Bill Page
Illustrations: Network Graphics
Production Coordinator: Paul Rohloff
Composition: Aptara
Printing and Binding: RR Donnelley

Library of Congress Control Number: 2009936387
ISBN-13: 9-781-4292-1974-7
ISBN-10: 1-4292-1974-2

Printed in the United States of America

First Printing

W. H. Freeman and Company
41 Madison Avenue
New York, NY 10010
Houndmills, Basingstoke RG21 6XS, England
www.whfreeman.com

CONTENTS

ABOUT THE AUTHORS

DAREN S. STARNES holds the endowed Master Teacher chair in Mathematics at The Lawrenceville School near Princeton, New Jersey. He earned his MA in Mathematics from the University of Michigan. In 1997, he received a GTE GIFT Grant to integrate AP Statistics and AP Environmental Science. He was named a Tandy Technology Scholar in 1999. Daren has led numerous one-day and weeklong AP Statistics institutes for new and experienced AP teachers, and he has been a reader, table leader, and question leader for the AP Statistics exam. In 2001–2002, he served as coeditor of the Technology Tips column in the NCTM journal *The Mathematics Teacher*. Since 2004, Daren has served on the ASA/NCTM Joint Committee on the Curriculum in Statistics and Probability (which he chaired in 2009). While on the committee, he edited the *Guidelines for Assessment and Instruction in Statistics Education (GAISE) Pre-K–12 Report* and coauthored (with Roxy Peck) *Making Sense of Statistical Studies,* a capstone module in statistical thinking for high school students. Daren is also coauthor of the popular text *The Practice of Statistics,* second and third editions.

DANIEL S. YATES has taught AP Statistics in the Electronic Classroom (a distance learning facility affiliated with Henrico County Public Schools in Richmond, Virginia). Before he was a high school teacher, Dan was on the mathematics faculty at Virginia Tech and Randolph-Macon College. He has a PhD in Mathematics Education from Florida State University. He has served as President of the Greater Richmond Council of Teachers of Mathematics and the Virginia Council of Teachers of Mathematics. Named a Tandy Technology Scholar in 1997, Dan is a 2000 recipient of the College Board/Siemens Foundation Advanced Placement Teaching Award. Although recently retired from classroom teaching, he stays in step with trends in teaching by frequently conducting College Board workshops for new and experienced AP Statistics teachers and by monitoring and participating in the AP Statistics electronic discussion group. Dan is coauthor of *The Practice of Statistics,* first, second, and third editions.

DAVID S. MOORE is Shanti S. Gupta Distinguished Professor of Statistics at Purdue University and 1998 President of the American Statistical Association. David is an elected fellow of the American Statistical Association and of the Institute of Mathematical Statistics and an elected member of the International Statistical Institute. He has served as program director for statistics and probability at the National Science Foundation. David has devoted his attention to the teaching of statistics. He was the content developer for the Annenberg/Corporation for Public Broadcasting college-level telecourse *Against All Odds: Inside Statistics* and for the series of video modules *Statistics: Decisions Through Data,* intended to aid the teaching of statistics in schools. He is the author of influential articles on statistics education and of several leading textbooks, including *Introduction to the Practice of Statistics* (written with George P. McCabe and Bruce Craig), *The Basic Practice of Statistics, Statistics: Concepts and Controversies, The Practice of Business Statistics, The Practice of Statistics in the Life Sciences* (written with Brigitte Baldi), and most recently *Essential Statistics*.

PREFACE

Statistics Through Applications, second edition (*STA* 2e), is designed to support a first course in statistics that emphasizes statistical thinking. The focus of this text is on statistical ideas and reasoning and on their relevance to such fields as medicine, education, environmental science, business, psychology, sports, politics, and entertainment.

In its publication *Principles and Standards for School Mathematics,* the National Council of Teachers of Mathematics (NCTM) identifies "Data Analysis and Probability" as one of five content standards. Here is the precise statement of this NCTM standard:

> Instructional programs from prekindergarten through grade 12 should enable all students to—
>
> - formulate questions that can be addressed with data and collect, organize, and display relevant data to answer them;
> - select and use appropriate statistical methods to analyze data;
> - develop and evaluate inferences and predictions that are based on data;
> - understand and apply basic concepts of probability.[1]

More recently, the American Statistical Association (ASA) published its *Guidelines for Assessment and Instruction in Statistics Education (GAISE) Pre-K–12 Report* to elaborate on the NCTM Data Analysis and Probability standard. The joint curriculum committee of the ASA and the Mathematical Association of America (MAA) recommends that any first course in statistics "emphasize the elements of statistical thinking" and feature "more data and concepts, fewer recipes and derivations."[2] *STA* 2e incorporates the recommendations made by the NCTM, ASA, and MAA by taking a conceptual and verbal approach rather than a methods-oriented one.

Statistics Through Applications is ideally suited for a non-AP-level introduction to statistics for high school students. It may be used effectively in either a one- or a two-semester course.

THE NATURE OF *STA* 2e

Statistics Through Applications is written to be read by students. It is somewhat informal, with thought-provoking stories ("nuggets") in the margins and cartoons and photographs interspersed throughout. Activities, Applications, and Data Explorations give students an opportunity to investigate, discuss, and make use of statistical ideas and methods. Examples and exercises have been carefully selected to pique students' interest and curiosity. Although the text has a fairly relaxed style and demands only an algebra background, its emphasis on ideas and reasoning asks more of the student than many texts that emphasize procedures over concepts.

Students learn to think about data by working with data. Consequently, we have included many elementary graphical and numerical techniques in *STA* 2e. We have not, however, allowed techniques to dominate concepts. Our intention is to invite discussion and even argument about statistical ideas rather than to focus exclusively on computation (though some computation remains essential). The coverage in *STA* 2e is considerably broader than that of some traditional statistics texts, as the table of contents reveals.

THE STRUCTURE OF *STA* 2e

STA 2e contains 10 chapters organized in four parts: analyzing data, producing data, chance, and inference. Each chapter consists of two or three sections, each devoted to a single coherent set of ideas. Sections are further divided into small subsections of material that can be fully addressed in a standard 45- or 50-minute high school period. Exercises appear at the end of each subsection, section, and chapter. Summaries at the end of each section and chapter help students distill what they have learned.

THE SECOND EDITION

In revising *Statistics Through Applications,* we have attempted to build on the elements that made the first edition successful:

- **Clear explanations** of statistical ideas and terminology are presented.
- Hands-on **Activities** that allow students to explore statistical concepts begin each section.
- Abundant **examples** from a variety of fields illustrate important statistical ideas.
- Varied **exercises** (with descriptive titles) reflect the usefulness of statistics in many different subject areas.
- Numerous **cartoons** enliven the pages.
- More than 30 short **marginal nuggets** offer interesting and often amusing anecdotes about statistics and its impact.

- **Applications** at the end of each section require students to put what they have learned to use in a real-world setting.
- A short **Summary** reviews the big ideas at the end of each section. Each chapter concludes with a **Chapter Review** that lists the skills students should have acquired by that point and **Chapter Review Exercises** to check understanding.

WHAT'S NEW IN THE SECOND EDITION?

Combining our classroom experience, input from users of the first edition, and a careful review of NCTM, ASA, and MAA recommendations, we have made some important changes in content and features. Here is a brief summary.

Content Changes

In response to users' overwhelming preference to get students involved in doing data analysis more quickly, we have reorganized the first six chapters. In addition, we have expanded the coverage of probability concepts and techniques in Chapters 7 and 8.

- The revised **Chapter 1** gives students an overview of the statistical problem-solving process—from defining a research question to producing, analyzing, and drawing conclusions from data.
- **Chapters 2, 3, and 4** provide a foundation in exploratory data analysis—graphical and numerical methods for describing one- and two-variable data.
- The details of sampling, surveys, and experiments can now be found in **Chapters 5 and 6.**
- We have significantly expanded the coverage of probability in **Chapters 7 and 8** to include conditional probability, basic counting techniques (permutations and combinations), and binomial distributions.
- In **Chapter 10,** we added the chi-square test for goodness of fit and inference for means using the *t* distributions.
- Former Section 1.2, "Measuring," and Section 5.2, "The Consumer Price Index and Government Statistics," have been removed from the textbook. They are still available on a CD and at the *STA* 2e Web site, **www.whfreeman.com/sta2e.**

Features

- **Four-color design** and **photographs** add interest value and visual appeal.
- Each chapter begins with a brief **opening story** that gives an overview of issues that will be addressed in the pages ahead.
- Many of the **Activities** that open each section have been revised to engage students in "doing statistics," and a number of the **Applications** have been updated to ask students to critique and interpret results from published statistical studies.
- New **Data Exploration** boxes put students in the role of data detectives as they use graphical and numerical tools to investigate variation in data.
- The four-step **statistical problem-solving process,** introduced on page 19 and first used in Example 1.10, provides a framework for helping students to think through the process of asking and answering good statistical questions. This four-step process is used as a learning tool to aid in interpreting and analyzing data in selected examples and exercises.
- **Calculator Corners** provide step-by-step instructions with screen shots for performing important statistical functions on the TI-84 and the TI-Nspire.
- **Solutions** to odd-numbered exercises in the back of the book now include figures that are part of the answers.
- Expanded use of **technology**—Java applets, computer software, and Web links—helps enrich student learning. Look for the new Applet icon and Web icon.
- **Output from Fathom, Minitab, and Excel** is integrated in examples and exercises throughout the textbook.

SUPPLEMENTS

A full range of supplements is available to help teachers and students use *STA* 2e.

Teacher's Resource Binder (TRB), 1-4292-5111-5

This indispensable resource compiled by Daren Starnes, Dan Yates, and Michael Legacy (The Greenhill School) gives teachers a multitude of useful tools for planning and managing their courses, assessing student progress, and enhancing day-to-day instruction. The TRB offers the following:

- Two sample quizzes for every section of every chapter
- Two sample chapter tests for every chapter
- Extensive teaching suggestions, including additional examples for in-class use, for each chapter
- Additional Activities and Applications
- New worksheets that focus on acquisition of vocabulary and basic skills, many written by Sally Miller of T. C. Williams High School

- Suggestions for short-term and long-term projects, with sample grading rubrics
- Expanded technology tips for using the TI-84, TI-Nspire, Fathom software, and Java applets
- An extensive list of references to books, journal articles, video series, data sources, interesting statistics-related Web sites, and more

Instructor's Solutions Manual (ISM), 1-4292-5110-7

Written by Sarah Streett and Duane Hinders, the ISM has been thoroughly revised and updated to reflect the content changes made to the second edition. It contains fully worked out solutions to every problem in the book. Where appropriate, solutions now follow the four-step statistical problem-solving process outlined in the text.

Instructor's Resource CD, 1-4292-5107-7

The CD contains the following:

- Microsoft Word files of the contents of the Teacher's Resource Binder
- The Teacher's Solutions Manual in Adobe .pdf format
- All text figures in an exportable presentation format, JPEG for Windows users and PICT for Macintosh users
- PowerPoint Slides that may be used directly or customized to fit your needs
- Applets that allow students to manipulate data and see the corresponding results graphically
- Data sets from the text in a variety of formats
- Sections 1.2 and 5.2 from the first edition in .pdf format
- Multiple-choice questions from the Test Bank as Microsoft Word files to allow teachers to add, edit, and resequence test items

Printed Test Bank, 1-4292-5108-5

Tim Brown (The Lawrenceville School) has updated the printed Multiple-Choice Test Bank to reflect the changes to the second edition and has added 150 new problems, bringing the total number to 1000.

Student Companion Web Site (www.whfreeman.com/sta2e)

The *STA* 2/e Student Companion Web Site offers a variety of resources to help students succeed in their high school statistics course. This site is open to all students; no registration is required.

Resources include the following:

- Online quizzes with instant feedback to help students master material and prepare for testing
- Interactive statistical applets that allow students to manipulate data and see the corresponding results graphically
- Calculator program files and data sets
- Key tables and formulas summary sheet
- All tables from the text in .pdf format for quick, easy reference
- Optional content (S.1, "Measuring," and S.2, "The Consumer Price Index and Government Statistics")
- Statistics on the Web: links to useful sites and resources, such as the Census Bureau, Gallup Poll, Nielsen, etc.

Interactive eBook (Access code or online purchase required)

This online resource integrates a complete version of the text with its media resources for ease of use. Students may quickly search the text and personalize the eBook with notes, highlights, and bookmarks just as they would with the print version.

- Allows students and teachers to bookmark pages, take notes, and highlight content
- Provides answers to the odd-numbered problems with a click of the "Short Answer" button, allowing students to check their answers immediately without having to flip to the back of the book
- Applets are linked directly to the supporting activity via hot-linked icons that encourage students to explore and experience hands-on simulations
- Data sets and calculator programs are accessed easily when they are needed
- Definitions to key terms are easily found through the online glossary
- EESEE Case Studies developed by The Ohio State University Statistics Department provide additional support to teach students to apply their statistical skills by exploring actual case studies, using real data, and answering questions about the study

ACKNOWLEDGMENTS

To Craig Bleyer and Ruth Baruth, I offer sincere gratitude for their unwavering support of *STA* as it continues to evolve. My heartfelt thanks go to Shona Burke for shepherding the manuscript through development, and especially for her timely responses to my occasionally frantic emails and phone calls. Jennifer Battista did yeoman's work coordinating the external reviews of the manuscript and consolidating disparate opinions into coherent recommendations. Thanks to Ann Heath, who came in late in the game as our "closer" and pitched a shutout with her careful attention to the design of the new edition and the supplements and saw us through production. Pamela Bruton did her usual superb copyediting, leaving no stone unturned and no grammar rule violated. Jane O'Neill capably fulfilled her role as project editor, and Blake Logan worked tirelessly to help us develop and refine the design and cover.

All of us involved in the project appreciate the thoughtful comments and suggestions made by our colleagues who reviewed draft chapters of the *STA* 2e manuscript:

Matthew J. Adam, McHenry High School—West Campus, IL
Peter R. Apple, Badger High School, WI
Barbara Aylworth, Alexander High School, GA
Virginia Blake, Lincoln-Sudbury Regional High School, MA
Denise Bogues, Hartland High School, MI
Kevin Brenner, Cloquet Senior High School, MN
Timothy Brown, Lawrenceville School, NJ
Katie Bruck, Secondary Technical Education Program, MN
Robert Bunio, Anoka High School, MN
Julie A. Burke, North Springs Charter School, GA
Joani Byrum-Wright, Louisville Male High School, KY
Lyn Davies, Denver School of the Arts, CO
Diane DeSantis, Algonquin Regional High School, MA
Allan DeToma, Pinewood Preparatory School, SC
Melissa E. DosSantos, Bay Shore High School, NY
Erika Draiss, Celina High School, OH
Karen Eggebrecht, Charlotte Catholic High School, NC
Vickie Fosdick, David W. Butler High School, NC
Terry C. French, Lake Braddock Secondary School, VA
Eva Galdo, Robinson Secondary School, VA
Adam Goudge, Wheat Ridge High School, CO
Steve Haldeman, Hempfield High School, PA
Katherine Harris, Patrick Henry High School, VA
Karen Huff, Spencer–Van Etten High School, NY
Cheryl Hughes, Providence School, FL
Jean-Marie Johnson, Shrewsbury High School, MA
Melissa Johnson, Robinson Secondary School, VA
John Thomas Matsik, West Mecklenburg High School, NC
Sally C. Miller, T. C. Williams High School, VA
Jim Mydloski, Monroe High School, MI
Marcia Prill, Madison High School, NJ
Sue Ravalese, Concord-Carlisle Regional High School, MA
Gerry Reynolds, Wayne Valley High School, NJ
Linda Ruppenthal, Hancock High School, MD
Colleen A. Russell, Warwick Valley High School, NY
Dianne Saugier, Granada High School, CA
Tom Shafer, Leslie High School, MI
Holly K. Shrader, Edison High School, OH
Allison Stanley, McHenry High School—East Campus, IL
Matt Townsley, Solon High School, IA
Natalie Tucker, Concord High School, NC
Doug Tyson, Central York High School, PA
Melissa Warfield, Boulder High School, CO
Lisa Weiss, Lincoln-Sudbury Regional High School, MA
David White, Ponderosa High School, CA
Robert Wickwire, Simi Valley High School, CA
Jodi N. Wilson, West Carrollton High School, OH
Jeannette Wilt, Brandywine High School, DE
Christine Wittenberg, Warwick Valley High School, NY

Special thanks go to Tim Brown (The Lawrenceville School) and Michael Legacy (The Greenhill School) for their detailed reviews of the chapters that involved major content changes in the new edition and for their efforts in revising the *Test Bank* and *Teacher's Resource Binder,* respectively. I also owe a debt of gratitude to Eric Zagorski from The Lawrenceville School, who proved his mettle as one of my AP Statistics students and as a student reviewer for *STA* 2e. Eric offered some remarkably insightful feedback at several critical stages in the writing process. Finally, I would like to thank Sarah Streett for her thorough and careful work in preparing the *Instructor's Solutions Manual* and Ann Cannon (Cornell College) for her fastidiousness in checking the book and solutions for accuracy.

As the saying goes, behind every good man, there is a great woman. For those of you who have had the pleasure of meeting my wife and soul mate, Judy, you know that the adage is true. I cannot possibly thank her enough for the sacrifices she has made to ensure that *STA* 2e got done well and on time. She inspires me every day.

Statistics Through Applications was Dan Yates's brainchild. He articulated a clear vision for *STA* as a text that would show students the applicability of statistics in their daily lives. Following his successful launch of *The Practice of Statistics* as the first book written specifically for high school AP Statistics, Dan persuaded the good folks at W. H. Freeman and Company that a textbook for non-AP high school statistics students was an absolute necessity. With the publication of *STA* in 2005, Dan's vision became reality. His pioneering efforts have touched the lives of thousands of students and teachers.

As I started work on the second edition, Dan provided invaluable advice on several potential changes. He read every manuscript chapter, offering meaningful suggestions that enhanced the quality of the finished product. Dan's influence can be seen throughout the pages of *STA* 2e, from his conviction that students learn statistics best by doing statistics (hence all the Activities, Applications, and Data Explorations) to the many cartoons pulled from his extensive collection.

To my mentor, colleague, and friend, Dan Yates, I dedicate *STA* 2e as a tribute to his innumerable contributions to statistics education.

Daren S. Starnes

CHAPTER 1

Statistics: The Art and Science of Data

1.1 Where Do Data Come From?
1.2 Drawing Conclusions from Data

Alamy

Improving health with the Wii?

In the past, video game players spent hours sitting idly in front of a television or computer screen. Critics argued that excessive video game play was a major contributor to the increase in obesity. With the growing popularity of the Nintendo Wii, however, the days of passive video gaming may be over. More importantly, the Wii may actually lead to better health for some individuals.

Some doctors have begun using the Wii to help stroke victims regain movement in their arms and legs. Patients who find traditional exercise too boring or too difficult are motivated to play the motion-intensive Wii games. "Wii therapy" has been used successfully with recovering military patients, too. The Wii has also caught on among senior citizens, who enjoy the social and competitive nature of games like bowling, tennis, and golf.

Research has even been conducted on whether playing Wii games regularly can help prevent childhood obesity. A study at Liverpool's John Moores University compared teenagers' energy use when playing with a traditional video game console and when using the Wii console. The 12 participants in the study increased their energy expenditure by an average of 60% over resting values when playing the traditional video game. When playing with the Wii, the teens increased their energy expenditure by an average of 156% over resting values. However, as Professor Tim Cable, the lead researcher for this study points out, "Parents should encourage other physical activities and outdoor pursuits in order for their children to lead well-balanced lives."[1]

1.1 Where Do Data Come From?

ACTIVITY 1.1

The "real facts"

Materials: Computer with Internet access

Do you drink Snapple? If so, then you have probably noticed a "Real Fact" printed on each bottle cap. Here is one we found: *Licking a stamp burns 10 calories.* How did they determine this? Does this "fact" seem reasonable? In this Activity, you will start to think carefully about how the data were produced.

1. Consider the following "Real Facts." For each one, (a) discuss whether the result makes sense, and (b) suggest how you might find data to support or oppose the "fact."
   - The average American walks 18,000 steps a day.
   - August has the highest percent of births.
   - The average person spends about 2 years on the phone in a lifetime.
   - Termites eat through wood 2 times faster when listening to rock music.
   - 27% of Americans have spent at least one night in jail. (This one is actually an incorrect choice in a previous Real Facts game at the Snapple Web site. Do you think the correct percent is higher or lower? Why?)

2. On each Snapple bottle cap, you are encouraged to "Get all the Real Facts at snapple.com." Go to the Snapple Web site (**www.snapple.com**) and locate the Real Facts page.

3. Find two "Real Facts" that are different from the ones above and that involve a numerical result. Record each fact. Then tell how you think the data were produced and whether the result makes sense.

You can hardly go a day without meeting data and statistical studies. Have you ever wondered how the numbers were produced? Consider these examples.

- In the year 2007, 12.5% of drivers involved in fatal crashes were 15 to 20 years old, according to the National Highway Traffic Safety Administration.[2]
- According to researchers in the Netherlands, people who are 30 or more pounds overweight could lose up to 7 years from their life expectancy.[3]
- The Gallup Poll reports (November 2007) that 51% of U.S. adults engage in "vigorous exercise" at least once a week.
- A major medical study concluded that taking aspirin regularly reduces the risk of a heart attack.

Where do these data come from? Why can we trust them? Or maybe we can't trust them. You can observe a lot just by watching. But you can't discover just by watching that 51% of adults exercise vigorously or that aspirin reduces heart attack risk. Good data are the product of intelligent human effort. Bad data result from laziness or lack of understanding, or even the desire to mislead others. "Where do the data come from?" is the first question you should ask when someone throws a number at you.

Data beat personal experiences

Suppose you want data on a question that interests you. Do cell phones cause cancer? What percent of people recycle? Does listening to music while studying improve student learning? Why did pollsters incorrectly predict the results of a recent election? Can a new drug help people quit smoking? How will you get relevant data to help answer your question?

It is tempting to base conclusions on your own experiences. You probably know many people who use cell phones. Do they have cancer? And polls aren't always right, are they? If one of your friends swears that listening to music helps him concentrate better, you might be persuaded to try it. But our own experiences may not be typical. In fact, the incidents that stick in our memory are often unusual. We are much safer relying on carefully produced data.

Example 1.1 Do cell phones cause brain cancer?

In August 2000, Dr. Chris Newman appeared as a guest on CNN's *Larry King Live*. Dr. Newman had developed brain cancer. He was also a frequent cell phone user. Dr. Newman's physician suggested that the brain tumor may have been caused by cell phone use. So Dr. Newman decided to sue the cell phone maker, Motorola, and the phone company that provided his service, Verizon. As people heard Dr. Newman's sad story, they began to worry about whether their own cell phone use might lead to cancer.

Since 2000, several statistical studies have investigated the link between cell phone use and brain cancer. One of the largest was conducted by the Danish Cancer Society. Over 400,000 Denmark residents who regularly used cell phones were included in the study. Researchers compared the brain cancer rate for the cell phone users with the rate in the general population. The result: no difference.[4] In fact, most studies have produced similar conclusions. In spite of the evidence, many people are still convinced that Dr. Newman's experience is typical.

Talking about data: individuals and variables

Statistics is the art and science of dealing with data. Good judgment, good math, and even good taste make good statistics. A big part of good judgment is deciding what to measure in order to produce data that help answer your questions. Measurements are made on **individuals** and organized in **variables**.

Individuals and variables

Individuals are the objects described by a set of data. Individuals may be people, animals, or things.

A **variable** is any characteristic of an individual. A variable can take different values for different individuals.

For example, here is a small part of a data set from the Cyber Stat Corporation:

	A	B	C	D	E	F
1	Name	Job Type	Age	Gender	Race	Salary
2	Cedillo, Jose	Technical	27	Male	White	52,300
3	Chambers, Tonia	Management	42	Female	Black	112,800
4	Childers, Amanda	Clerical	39	Female	White	27,500
5	Chen, Huabang	Technical	51	Male	Asian	83,600
6						
7						
8						

Ready NUM

The individuals are company employees. In addition to each employee's name, there are five variables. The first says what type of job the employee holds. The second variable gives the age, the third records gender, and the fourth reports race. The fifth variable records employee salaries.

Statistics deals with numbers, but not all variables are numerical. Of the five variables in the company's data set, only age and salary have numbers as values. It makes sense to talk about the average age or the average salary of the company's employees. However, it wouldn't make sense to ask about the average job type, average race, or average gender. For nonnumerical variables like these, we can use *counts or percents* to summarize the data. We might give the percent of employees who work in management, for example, or the number who are Asian.

When we examine data, it is helpful to distinguish these two types of variables.

Categorical and quantitative variables

A **categorical variable** places an individual into one of several groups or categories.

A **quantitative variable** takes numerical values for which arithmetic operations such as adding and averaging make sense.

For the data from Cyber Stat Corporation, age and salary are quantitative variables. Job type, gender, and race are categorical variables.

Deciding what variables to measure and how to measure them isn't easy. Bad judgment in choosing variables can result in data that cost lots of time and money to produce but that don't tell us much. Here is an example that shows some challenges in deciding what and how to measure.

Example 1.2 Who recycles?

Do wealthier people recycle more than poorer people? Researchers spent lots of time and money weighing the stuff put out for recycling in two neighborhoods in a California city, call them Upper Crust and Lower Mid. The *individuals* here are households,

Jupiterimages

because trash and recycling pickup are done for residences. The *variable* measured was the weight in pounds of the curbside recycling basket each week.

The Upper Crust households contributed more pounds per week on average than those in Lower Mid. Can we say that the rich are more serious about recycling? No. Someone noticed that Upper Crust recycling baskets contained lots of heavy glass wine bottles. In Lower Mid, they put out lots of light plastic soda bottles and light metal beer and soda cans. Weight tells us little about commitment to recycling.[5]

Exercises

1.1 Making the grade Here are a few lines from a statistics teacher's grade book:

Name	S	Homeroom	Gr	Calculator #	Test 1
Hsu, Danny	M	Blair	Sr.	B319	81
Iris, Francine	F	Kingsley	Sr.	B298	92
Ruiz, Ricardo	M	Alfonso	Jr.	B304	87

(a) What individuals does this data set describe?

(b) For each individual, what variables are given? Identify each variable as categorical or quantitative.

1.2 Hotel choices A high school's debate team is planning to go to Salt Lake City for a three-day competition. Team members will be responsible for sharing the cost of staying in a hotel for two nights. The competition's sponsor provides a list of available hotels, along with some information about each hotel. Here is a chart that summarizes the hotel options.

Hotel	Pool	Exercise room?	Internet ($/day)	Restaurants	Distance to competition	Room service?	Room rate ($/day)
Comfort Inn	Out	Y	0	1	8.2 mi	Y	149
Fairfield Inn & Suites	In	Y	0	1	8.3 mi	N	119
Baymont Inn & Suites	Out	Y	0	1	3.7 mi	Y	60
Chase Suite Hotel	Out	N	15	0	1.5 mi	N	139
Courtyard	In	Y	0	1	0.2 mi	Dinner	114
Hilton	In	Y	10	2	0.1 mi	Y	156
Marriott	In	Y	9.95	2	0.0 mi	Y	145

(a) What individuals does this data set describe?

(b) Clearly identify each of the variables. Which are quantitative?

(c) If you were a debate team member, which hotel would you recommend? Why?

1.3 Who recycles more? In Example 1.2, weight is not a good measure of which of the two neighborhoods recycles more. What variables would you measure instead?

1.4 Choosing a college Popular magazines rank colleges and universities on their "academic quality." Describe four variables that you would like to see measured for each college if you were choosing where to apply. Identify each as categorical or quantitative. Give reasons for your choices.

WWW **1.5 More "Real Facts"** According to the Real Facts page at **www.snapple.com**, Americans on average eat 18 acres of pizza every day.

(a) Tell how you think the data were produced.

(b) Discuss whether this result makes sense.

WWW **1.6 Chart toppers** Visit the Billboard Music Web site at **www.billboard.com.**

(a) What are the top five songs on the Billboard Hot 100 chart?

(b) How do you think they determined these rankings?

Observational studies

Sometimes all you can do is watch. To learn how chimpanzees in the wild behave, watch. To study how a teacher and young children interact in a schoolroom, watch. It helps if the watcher knows what to look for. The chimpanzee expert may be interested in how males and females interact, in whether some chimps in the troop are dominant, in whether the chimps hunt and eat meat. In fact, chimps were thought to be vegetarians until Jane Goodall watched them carefully in Gombe National Park, Tanzania. Now it is clear that meat is a natural part of the chimpanzee diet.

At first, the observer may not know what to record. Eventually patterns seem to emerge. Then we can decide what variables we want to measure. How often do chimpanzees hunt? Alone or in groups? How large are hunting groups? Males alone, or both males and females? How much of the diet is meat? Observation that is organized and measures clearly defined variables is more convincing than just watching. A carefully designed **observational study** is far better than educated guesswork.

Observational study

An **observational study** observes individuals and measures variables of interest but does not attempt to influence the responses. The purpose of an observational study is to describe some group or situation.

Here is an example of highly organized observation.

Example 1.3 Pain and suffering

An Associated Press news article begins, "people hurt in traffic accidents actually recover more quickly when they cannot collect money for their pain and suffering, researchers say in a new study."[6] The Canadian province of Saskatchewan changed its insurance laws. The old system allowed lawsuits for "pain and suffering." The new no-fault system paid for medical costs and lost work but not for subjective suffering. The study looked at insurance claims filed between July 1 of the year before the change and December 31 of the year after the change. Under the new system there were not just fewer claims of whiplash neck injuries but faster recovery with less pain for the people who filed claims. This is a **comparative observational study**.

It is possible that the effect of the change in the insurance system is mixed up with some other difference between the two time periods. The study gives reasonably convincing evidence that people report less pain when they can't collect money for it. But the study's design prevents us from concluding that the change in insurance law *causes* a decrease in pain.

Sampling

You don't have to drink the whole pitcher of lemonade to know that it's too sweet. A few sips from a well-mixed pitcher will tell you. That's the idea of sampling: to gain information about the whole by examining only a part. Observational studies that use sampling are called **sample surveys**. They attempt to measure characteristics of some group of individuals (the **population**) by studying only some of its members (the **sample**). The individuals in the sample are selected not because they are of special interest but because they represent the larger group. Here's the vocabulary we use to discuss sampling.

Populations and samples

The **population** in a statistical study is the entire group of individuals about which we want information.

A **sample** is a part of the population from which we actually collect information, which is then used to draw conclusions about the whole.

Notice that the *population* is the group we want to study. If we want information about all U.S. high school students, that is our population even if students at only one high school are available for sampling. To make sense of any sample result, you must know what population the sample represents. Did a preelection poll, for example, ask the opinions of all adults? Citizens only? Registered voters only? Democrats only? The *sample* consists of the people we actually have information about. If the poll can't contact some of the people it selected, those people aren't in the sample.

The distinction between population and sample is basic to statistics. The following examples illustrate this distinction and also introduce some major uses of sampling. These brief descriptions also indicate the variables measured for each individual in the sample.

Example 1.4 Public opinion polls

Polls such as those conducted by Gallup and many news organizations ask people's opinions on a variety of issues. The variables measured are responses to questions about public issues. Though most noticed at election time, these polls are conducted on a regular basis throughout the year. For a typical opinion poll:

Population: U.S. residents 18 years of age and over. Noncitizens and even illegal immigrants are included.

Sample: Between 1000 and 1500 people interviewed by telephone.

www For more information about Gallup polls, visit **www.gallup.com.**

Example 1.5 The Current Population Survey

Government economic and social data come from large sample surveys of a nation's individuals, households, or businesses. The monthly Current Population Survey (CPS) is the most important government sample survey in the United States. Many of the variables recorded by the CPS concern the employment or unemployment of everyone over 16 years old in a household. The government's monthly unemployment rate comes from the CPS. The CPS also records many other economic and social variables. For the CPS:

Population: The more than 110 million U.S. households. Notice that the individuals are households rather than people or families. A household consists of all people who share the same living quarters, regardless of how they are related to each other.

Sample: About 50,000 households interviewed each month.

www For more information about the Current Population Survey, visit **www.bls.gov/cps/.**

Example 1.6 TV ratings

Market research is designed to discover what consumers want and what products they use. One example of market research is the television-rating service of Nielsen Media Research. The Nielsen ratings influence how much advertisers will pay to sponsor a program and whether or not the program stays on the air. For the Nielsen national TV ratings:

Population: The over 110 million U.S. households that have a television set.

Sample: About 25,000 households that agree to use a "people meter" to record the TV viewing of all people in the household.

The variables recorded include the number of people in the household and their ages and gender, whether the TV set is in use at each time period, and, if so, what program is being watched and who is watching it.

www For more information on Nielsen ratings, visit **www.nielsenmedia.com** and search for "television rankings."

Example 1.7 The General Social Survey

Social science research makes heavy use of sampling. The General Social Survey (GSS), carried out every second year by the National Opinion Research Center at the University of Chicago, is the most important social science sample survey. The variables cover the participant's personal and family background, experiences and habits, and attitudes and opinions on subjects from abortion to war.

Population: Adults (aged 18 and over) living in households in the United States. The population does not include adults in institutions such as prisons and college dormitories. It also does not include persons who cannot be interviewed in English.

Sample: About 3000 adults interviewed in person in their homes.

For more information on the General Social Survey, visit **www.norc.org.**

We reserve the term "sample survey" for studies that use an organized plan to choose a sample that represents some specific population, like those in Examples 1.4 to 1.7. By our definition, the population in a sample survey can consist of people, animals, or things. Some people use the terms "survey" or "sample survey" to refer only to studies in which people are asked one or more questions, like the opinion poll of Example 1.4. We'll avoid this more restrictive terminology.

Most statistical studies use samples in the broad sense. Consider the people who had traffic accidents in the year following the change in insurance laws from Example 1.3 (page 6). They are supposed to represent all accident victims (not just those from that particular year) following the change in insurance laws. Expert judgment says they are probably typical, but we can't be sure. A sample survey doesn't rely on judgment: it starts with an entire population and *chooses* a sample to represent it. Chapter 5 discusses the art and science of sample surveys.

Exercises

1.7 Populations and samples, I For each of the following sampling situations, identify the population and the sample as exactly as possible.

(a) A furniture maker buys hardwood in large batches. The supplier is supposed to dry the wood before shipping (wood that is not dry won't hold its size and shape). The furniture maker chooses five pieces of wood from each batch and tests their moisture content. If any piece exceeds 12% moisture content, the entire batch is sent back.

(b) An insurance company wants to monitor the quality of its procedures for handling loss claims from its auto insurance policyholders. Each month the company selects a sample from all auto insurance claims filed that month to examine the accuracy and promptness with which they were handled.

1.8 Populations and samples, II For each of the following sampling situations, identify the population and the sample as exactly as possible.

(a) A business school researcher wants to know what factors affect the survival and success of small businesses. She selects a sample of 150 restaurants from those listed in the phone book for a large city.

(b) Your local television station wonders if its viewers would rather watch a local college basketball team play or an NBA game scheduled at the same time. It announces that it will show the NBA game and receives 89 calls asking that it show the local game instead.

1.9 Bad apples A truckload of apples arrives at an apple juice production plant. The plant's quality control team selects three large buckets of apples from various locations within the truck. These apples are inspected carefully. Based on inspection results, the entire truckload is either accepted or rejected by the plant. Identify the population, sample, individuals, and variable(s) in this setting.

AP Photo/Bizuayehu Tesfaye

1.10 J. K. Rowling's words Different types of writing can sometimes be distinguished by the lengths of the words used. A student interested in this fact wants to study the lengths of words used by J. K. Rowling in her Harry Potter books. She opens a Harry Potter book at random and records the lengths of the first 50 words on the page.

(a) What is the population in this study? What is the sample?

(b) What variable does the student measure?

1.11 Nielsen ratings Find a current copy of the Nielsen television ratings. (Many local newspapers report these ratings weekly.)

(a) What individuals are being measured?

(b) What variables are recorded? For each quantitative variable, identify the units of measurement.

(c) How were the data produced?

1.12 Gallup Poll Go to the Gallup Poll Web site, **www.gallup.com.** Locate the results of a sample survey on a topic that interests you.

(a) What questions did the survey include?

(b) Who were the individuals in the sample?

(c) Summarize one or two of the important results from the survey. Be sure to cite source information for your article, including the title, date, and Web URL.

Census

A sample survey looks at only a part of the population. Why not look at the entire population? A **census** tries to do this.

> **Census**
>
> A **census** is a sample survey that attempts to include the entire population in the sample.

The U.S. Constitution requires a census of the American population every 10 years. A census of so large a population is expensive and takes a long time.

Is a census old-fashioned?

The United States has taken a census every 10 years since 1790. Technology marches on, however, and replacements for a national census look promising. Denmark has no census, and France plans to eliminate its census. Denmark has a national register of all its residents, who carry identification cards and change their register entry whenever they move. France will replace its census by a large sample survey that rotates among the nation's regions. The U.S. Census Bureau has a similar idea: the American Community Survey has already started and will replace the census "long form."

Even the federal government, which can afford a census, uses samples such as the Current Population Survey to produce timely data on employment and many other variables. If the government asked every adult in the country about his or her employment, this month's unemployment rate wouldn't be available until next year!

So time and money favor samples over a census. Samples can have other advantages as well. If you are testing fireworks or fuses, the individuals in the sample are destroyed. Moreover, a sample can produce more accurate data than a census. A careful sample of an inventory of spare parts will almost certainly give more accurate results than asking the employees to count all 500,000 parts in the warehouse. Bored people don't count accurately.

Until 2000, even the every-10-years census included a sample survey. A census "long form" that asked many more questions than the basic census form was sent to a sample of one-sixth of all households. Starting with the 2010 census, the long form will be replaced by the American Community Survey (ACS).

Example 1.8 Census undercounts and overcounts

A census can only attempt to sample the entire population. The Census Bureau estimates that the 1990 census missed 1.6% of the American population. These missing persons included an estimated 4.4% of the black population, largely in inner cities.[7] In 2000, the census overcounted the U.S. population by about 1.3 million. Millions of people who live in two places—like college students—were counted twice. About 1.8% of blacks and 0.7% of Hispanics were not counted at all. A census is not foolproof, even with the resources of the government behind it.

Why take a census at all? The government needs block-by-block population figures to create election districts with equal populations. The main function of the U.S. census is to provide this local information.

What kind of information can we get from the U.S. census? You'll find out in the following Data Exploration.

DATA EXPLORATION

Sampling from the Census

We used Fathom software to take a *random sample* of 100 individuals from the 2000 U.S. census. For each individual, the values of several variables were recorded. The table on the following page displays data for the first 10 people selected in the sample. (The full data set is available on the book's Web site, **www.whfreeman.com/sta2e.**) www

DATA EXPLORATION *(continued)*

DATA EXPLORATION *(continued)*

Sex	Age	Race	Siblings	Marital status	Education
Male	52	White	0	Married_ spouse present	1 to 3 years of college
Male	50	White	0	Married_ spouse present	1 to 3 years of college
Female	8	White	3	Never married/single (N/A)	Grade 1_ 2_ 3_ or 4
Female	18	Black/Negro	2	Never married/single (N/A)	Grade 11
Female	7	White	1	Never married/single (N/A)	None or preschool
Female	60	White	0	Married_ spouse present	Grade 12
Female	11	White	1	Never married/single (N/A)	Grade 5_ 6_ 7_ or 8
Male	33	White	0	Never married/single (N/A)	1 to 3 years of college
Female	45	White	0	Never married/single (N/A)	1 to 3 years of college
Male	79	White	0	Married_ spouse present	Grade 12

1. Describe the population and the sample as exactly as possible.

2. What variables were recorded? Identify each as categorical or quantitative.

In Chapter 2, you will learn about graphical and numerical tools for analyzing data. Here is a brief preview of what you'll see. (You may find that you have seen, or even used, some of these methods in previous math courses.)

Let's begin with a categorical variable, like sex. We have constructed a **bar graph** (often called a bar chart) and a summary table that display information about this variable for our sample.

3. What *percent* of the individuals in the sample are male?

4. What percent of the *population* is male? Justify your answer.

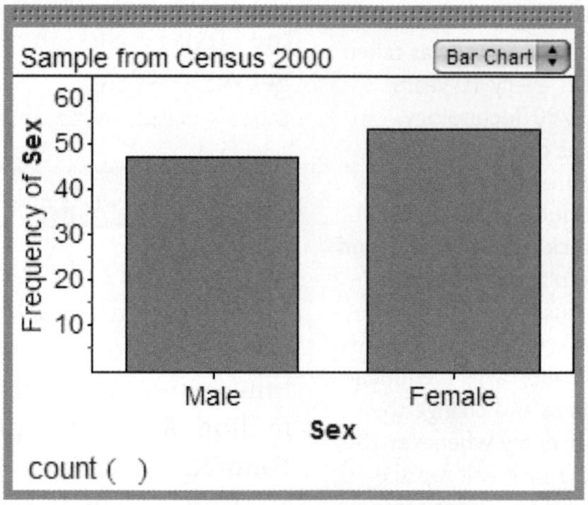

Sex	Count
Male	47
Female	53
Total	100

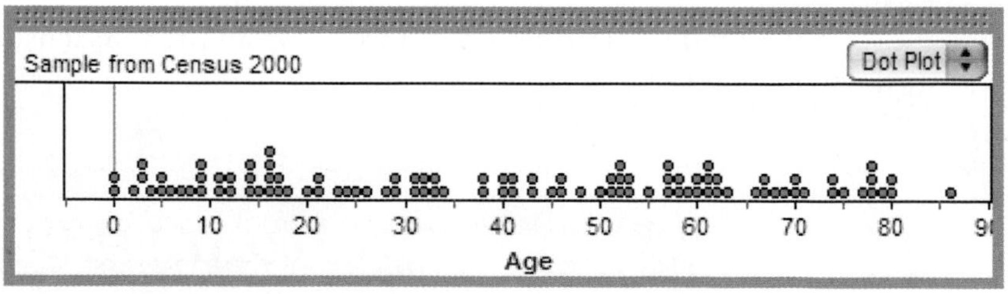

Now let's look at the quantitative variable age. We have constructed a *dotplot* to display the ages of the individuals in our sample.

5. How old is the youngest person in the sample? The oldest person?

6. About what percent of the sample is 65 or older?

7. About what percent of the population is 65 or older? Justify your answer.

"Now eat that banana. The nice statistician is watching us"

Experiments

Our goal in choosing a sample is a picture of the population, disturbed as little as possible by the act of gathering information. All observational studies share the principle "observe but don't disturb." When Jane Goodall first began observing chimpanzees in Tanzania, she set up a feeding station where the chimps could eat bananas. She later said that was a mistake, because it might have changed the apes' behavior.

In **experiments,** on the other hand, we want to change behavior. In doing an experiment, we don't just observe individuals or ask them questions. We actively impose some *treatment* in order to observe the response. Experiments can answer questions such as "Does aspirin reduce the chance of a heart attack?" and "Do a majority of high school students prefer Pepsi to Coke when they taste both without knowing which they are drinking?"

Experiments

An **experiment** deliberately imposes some treatment on individuals in order to observe their responses. The purpose of an experiment is to study whether the treatment causes a change in the response.

Example 1.9 Curing a cold

It has been claimed that large doses of vitamin C will prevent colds. An experiment to test this claim was performed in Toronto during the winter months. About 500 volunteer *subjects* were assigned "at random" to each of two groups. Group 1 received 1 gram per day of vitamin C and 4 grams per day at the first sign of a cold. (This is a large amount of vitamin C; the recommended daily allowance of this vitamin for adults is only 60 milligrams, or 60/1000 of a gram.) Group 2 served as a *control group* and received a *placebo* pill—identical in appearance to the vitamin C capsules but with no active ingredient. Both groups were regularly checked for illness during the winter.

Some of the subjects dropped out of the experiment for various reasons, but 818 completed at least two months. Groups 1 and 2 were very similar in age, occupation, smoking habits, and other background variables. At the end of the winter, 26% of the subjects in Group 1 had not had a cold, compared with 18% in Group 2. Thus, vitamin C did appear to prevent colds better than the placebo, but not much better.[8]

The vitamin C example illustrates the big advantage of experiments over observational studies: *in principle, experiments can give good evidence for cause and effect.* If we design the experiment properly, we start with two very similar groups of subjects. The *individual* subjects differ from each other in age, occupation, smoking habits, and other respects. But the two *groups* resemble each other when we look at those variables for all subjects in each group. During the experiment, the subjects' lives differ, but there is only one *systematic* difference between the two groups: whether they take vitamin C or a placebo. So we should be able to say whether taking vitamin C reduces the likelihood of getting a cold.

The fact that experiments can give good evidence that a treatment causes a response is one of the big ideas of statistics. A big idea needs a big caution: statistical conclusions hold "on the average" for groups of individuals. They don't tell us much about one individual. *On the average,* the subjects taking vitamin C had fewer colds than those who were taking a placebo. That says vitamin C was somewhat effective. It doesn't say everyone who takes vitamin C will be healthy. And a big idea may also raise big questions: if we think vitamin C will prevent colds, is it ethical to offer it to some and not to others? Chapter 6 explains how to design good experiments and looks at ethical issues.

Reprinted with special permission of King Features Syndicate

Exercises

1.13 Tasty muffins Before a new variety of frozen muffin is put on the market, it is subjected to extensive taste testing. People are asked to taste the new muffin and a competing brand and to say which they prefer. Is this an observational study or an experiment? Explain your answer.

1.14 Cell phones and cancer One study of cell phones and the risk of brain cancer looked at a group of 469 people who have brain cancer. The investigators matched each cancer patient with a person of the same sex, age, and race who did not have brain cancer, then asked about use of cell phones.[9] Result: "Our data suggest that use of hand-held cellular telephones is not associated with risk of brain cancer."

(a) Is this an observational study or an experiment? Why?

(b) What individuals are measured, and what variables are recorded?

1.15 Choose your study purpose Give an example of a question about high school students, their behavior, or their opinions that would best be answered by

(a) an observational study.

(b) an experiment.

1.16 Child care In 2001, researchers announced that "children who spend most of their time in child care are three times as likely to exhibit behavioral problems in kindergarten as those who are cared for primarily by their mothers."[10]

(a) Was this likely an observational study or an experiment? Why?

(b) Can we conclude from this study that child care causes behavior problems? Why or why not?

1.17 Teenage drivers Go to the National Highway Traffic Safety Administration Web site: **www.nhtsa.dot.gov.** Locate a traffic safety fact about young drivers that includes a supporting chart or graph. If possible, print the image.

(a) Summarize your safety fact with a few well-written sentences in your own words. Be sure to describe what the chart or graph tells you. Cite the title, date, and Web URL of the article from which this information comes.

(b) How did the NHTSA obtain these data? Justify your answer.

1.18 Whiter teeth In an effort to make their teeth whiter, some people use whitening strips or bleaching gels. Are new "light-activated bleaching techniques" any more effective? Researchers carried out a study to help answer this question. They gathered a small group of dental patients who were willing to participate. On one side of patients' mouths, the researchers used bleaching gel alone. On the other side, they used light plus bleaching gel. The results? The patients' teeth got whiter on both sides of their mouths; there was no real difference in whiteness between the two sides.

(a) Was this study an experiment or an observational study? Explain.

(b) Why did the researchers apply both possible treatments to each patient, rather than using bleaching gel on only half of the patients and light plus bleaching gel on the other half?

APPLICATION 1.1

Making sense of the census

The U.S. Census Bureau Web site contains a vast amount of data about the U.S. population. In this Application, you will take a closer look at where these data come from.

1. Go to the Census Bureau's site at **www.census.gov.**

2. Use the "Population Clocks" section of the site to answer these questions:

- What are the current U.S. and world populations, according to the Population Clocks?
- How is the U.S. population figure obtained?
- When did the world population reach 6 billion?

APPLICATION 1.1 *(continued)*

APPLICATION 1.1 *(continued)*

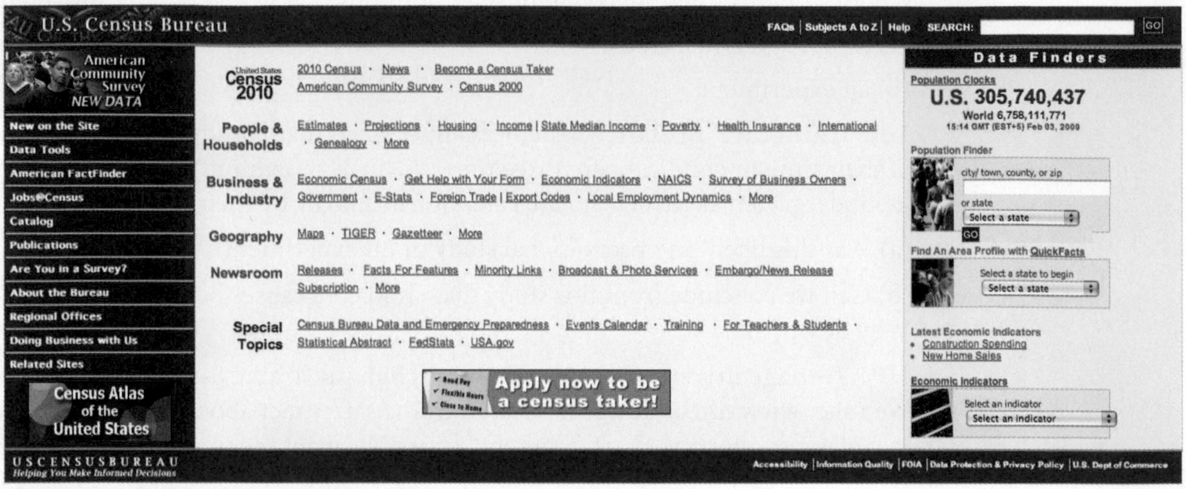

U.S. Census Bureau/www.census.gov

3. Click on the link for the most recent census to answer these questions:
- How large was the U.S. population?
- What percent of the U.S. population was male? How many people were aged 15 to 19?
- What percent of the U.S. population was Hispanic? How does this compare with the Hispanic population in the previous census?
- How did the Census Bureau obtain their data?

4. Go to the "QuickFacts" section of the site. For the most recent census:
- What was your state's population?
- For your state, what was the average travel time to work? How do you think this value was obtained?

5. Click on the American Community Survey (ACS) link.
- Why does the Census Bureau conduct surveys, such as the ACS?
- How many people does the ACS survey, and how are they selected?

6. Click on "Subjects A to Z." For the most recent census:
- Computer use and ownership: What percent of homes had a computer? What percent had Internet access? Where did these data come from?
- Educational attainment: How do males and females compare in terms of bachelor's, master's, and higher degrees earned? Where do these data come from?

Section 1.1 Summary

Any statistical study records data about some **individuals** (people, animals, or things) by giving the value of one or more **variables** for each individual. **Quantitative variables,** such as age and income, take numerical values. **Categorical variables,** such as occupation and gender, do not. Be sure the variables in a study really do tell you what you want to know.

The most important fact about any statistical study is how the data were produced. **Observational studies** try to gather information without disturbing the scene they are observing. **Sample surveys** are an important kind of observational study. A sample survey chooses a **sample** from a specific **population** and uses the sample to get information about the entire population. A **census** attempts to measure every individual in a population. **Experiments** actually do something to individuals in order to see how they respond. The goal of an experiment is usually to learn whether some treatment actually causes a certain response.

Section 1.1 Exercises

1.19 Public housing To study the effect of living in public housing on family stability in low-income households, researchers obtain a list of all applicants for public housing in Chicago last year. Some applicants were accepted, while others were turned down by the housing authority. The researchers interview both groups and compare them. Is this an experiment or an observational study? Explain your answer.

1.20 Baking bread A flour company wants to know what fraction of Minneapolis households bake some or all of their own bread. A sample of 500 residential addresses is taken, and interviewers are sent to those addresses. All of the interviews take place during regular working hours on weekdays.

(a) What population is the flour company interested in? What is the sample?

(b) Do you think this sample will provide accurate information to the flour company? Why or why not?

1.21 Fit to lead? Is there a relationship between physical fitness and leadership ability? To answer this question, researchers recruit 100 students who are willing to take part in an exercise program. The volunteers are divided into a low-fitness group and a high-fitness group on the basis of a physical examination. All students then take a test designed to measure leadership, and the results for the two groups are compared.

(a) Is this an experiment or an observational study? Explain your answer.

(b) What population does it appear that the investigators were interested in? What variables did they measure?

(c) Explain why we might prefer a sample survey to using students who volunteer for a fitness program.

1.22 Cool cars You and your friends want to find out which student at your school has the "coolest" car.

(a) What variables would you record? Which are quantitative?

(b) What are the individuals in your data set?

(c) What method would you use to produce your data? Justify your choice.

1.23 Angry bees My grandmother once told me that the color red makes bees angry. Here's a method I've designed to test her claim. I'll select half of my students (by drawing names from a hat) to wear red clothes and the other half to wear white clothes. Then I'll turn a bunch of bees loose in our classroom and record how many times each student is stung.

(a) Is this an observational study or an experiment? Why?

(b) What variables are recorded?

(c) If students wearing red clothes are stung much more often than students wearing white, can we conclude that the color red *causes* bees to sting more? Why or why not?

(d) Comment on any flaws you see with my methods.

1.24 Fast food Many people eat fast food as a regular part of their diet. Is fast food unhealthy? The best answer may be "It depends on what you eat." Here are some nutritional data about four popular fast-food burgers:

Burger	Restaurant	Calories	Fat	Cholesterol	Sodium
Quarter Pounder with cheese	McDonald's	430	30	95	1310
Classic Single with everything	Wendy's	410	19	70	890
Whopper with cheese	Burger King	706	43	113	1164
Cheeseburger	In-N-Out	480	27	60	1000

(a) What individuals and variables are recorded in this data set?
(b) In what units is each quantitative variable measured?
(c) Are fast-food chicken sandwiches healthier than burgers? Collect some data for yourself. (Most fast-food restaurants will provide nutritional information on request. You can also search the Web.)

1.2 Drawing Conclusions from Data

ACTIVITY 1.2

See no evil, hear no evil?

Materials: Index cards (two per student, prepared by your teacher); clock, watch, or stopwatch to measure 30 seconds; coin (one per pair of students)

Confucius said, "I hear and I forget. I see and I remember. I do and I understand." Do people really remember what they see better than what they hear?[11] In this Activity, you will perform an experiment to try to find out.

1. Your teacher will give each student two index cards with 10 pairs of numbers on them. Do not look at the numbers until it is time for you to do the experiment.

2. With your partner, decide who will perform the experiment first and who will be the timer.

The experiment consists of two tasks; let's call them Task A and Task B.

3. Flip a coin to determine which task you will perform first: heads = Task A first; tails = Task B first.

4. Flip the coin again to decide which of the two index cards will be used for the first task.
Task A: Study the pairs of numbers on one of your index cards for 30 seconds. Then turn the card over. Recite the alphabet aloud (A, B, C, etc.). Then tell your partner what you think the numbers on the card are.
Task B: Your partner will read the pairs of numbers on one of your index cards aloud three times slowly. Next, you will recite the alphabet aloud (A, B, C, etc.). Then you will tell your partner what you think the numbers on the card are.

5. Perform both tasks in the order specified by the coin toss. Your partner will record how many pairs of numbers you recalled correctly for each task.

6. Switch roles, and repeat Steps 3 to 5.

7. Your teacher will make a chart on the board like the one shown below. Record the data for each partner on a separate row of the chart.

Student number	First task?	"See" correct	"Hear" correct	Difference
1	A	5	3	2
2	B	4	5	−1

8. As a class, make a dotplot (see page 12) of the difference values on the board.

9. Did students in your class remember numbers better when they saw them or when they heard them? Give appropriate evidence to support your answer.

10. Based on the results of this experiment, can we conclude that people in general remember better when they see than when they hear? Why or why not?

The statistical problem-solving process

Activity 1.2 outlines the steps in the **statistical problem-solving process.**[12] We begin with a question of interest. Then, we produce data using an observational study or experiment. Next, we analyze the data using graphs and numerical summaries. Finally, we interpret the results of the analysis and draw conclusions about our original question.

Statistical problem-solving process

I. Ask a question of interest A statistics question involves some characteristic that varies from individual to individual.

II. Produce data The methods of choice are observational studies and experiments.

III. Analyze data Graphs and numerical summaries are the tools for describing patterns in the data, as well as any deviations from those patterns.

IV. Interpret results The results of the data analysis should help answer the question of interest.

You have already seen some examples of the kinds of questions that statistics can help answer. Here are a few more. How common is texting while driving? Can yoga help dogs live longer? How well do SAT Writing scores predict the GPAs of college freshmen? If a casino's roulette wheel comes up red five times in a row, is the casino cheating? Will a low-fat diet or a low-carbohydrate diet reduce weight more effectively in the long run? Do angry people have more heart disease? How much time per day does the average teen spend on social-networking sites? Do employers discriminate against job candidates based on their names?

What do all of these questions have in common? Variation. Individuals vary. Chance outcomes—like spins of a roulette wheel or tosses of a coin—vary. Data vary. Some would say that statistics is really the study of variation. The following example illustrates how the statistical problem-solving process can be used to make sense of variation.

Example 1.10 Who has tattoos?

I. Ask a question of interest What percent of U.S. adults have one or more tattoos?
II. Produce data The Harris Poll conducted an online survey of 2302 adults during January 2008. According to the published report, "Respondents for this survey were selected from among those who have agreed to participate in Harris Interactive surveys."[13]
III. Analyze data The *pie chart* below summarizes the responses from those who were surveyed: 14% said they had one or more tattoos. Did responses differ by gender? The accompanying *bar graph* shows that males and females responded similarly (13% of females said Yes, compared with 15% of males).

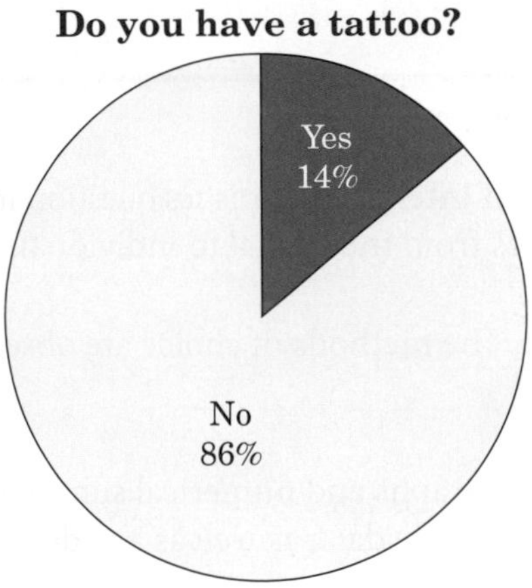

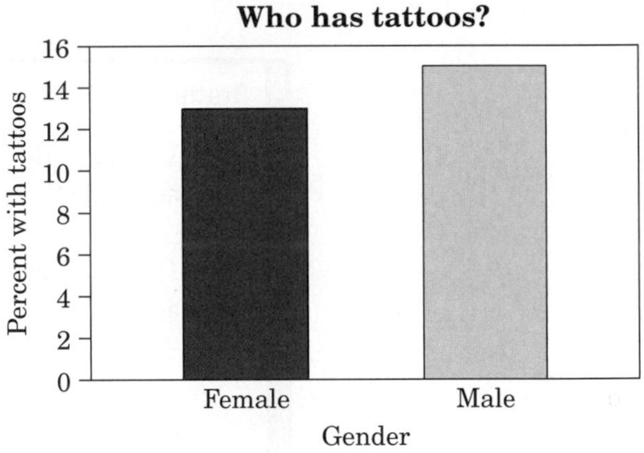

IV. Interpret results In this sample of adults, 14% said they had one or more tattoos. Of course, a different sample of 2302 adults might yield different results. That's the idea of **sampling variability.** If this sample is truly representative of the adult population, then we would expect about 14% of the population to say that they had at least one tattoo if asked.

www You can find reports about various sample surveys—like this one on tattoos—at the Harris Poll's Web site: **www.harrisinteractive.com/harris_poll/.**

Interesting questions can come from almost anywhere—something you read in the newspaper, see while surfing the Internet, or hear on television or the radio. As one of the authors learned, interesting questions can even emerge during a family lunch. In this case, Mr. Starnes's mother mentioned that several people she knew had begun filling their car tires with nitrogen instead of compressed air. Her friends were convinced that nitrogen-filled tires would lead to a smoother ride, steadier air pressure, and better gas mileage. After lunch, Mr. Starnes took to the Internet to research this issue. To learn what he discovered, keep reading.

Example 1.11 Nitrogen in tires—a lot of hot air?

Courtesy Judy Starnes

Most automobile tires are inflated with compressed air, which consists of about 78% nitrogen. Aircraft tires are filled with nitrogen, which is safer than air in case of fire.

I. Ask a question of interest Does filling automobile tires with nitrogen improve safety, performance, or both?

II. Produce data Consumers Union designed a study to test whether nitrogen-filled tires would maintain pressure better than air-filled tires. They obtained pairs of tires from each of several brands and then filled one tire in each pair with air and one with nitrogen. All tires were inflated to a pressure of 30 pounds per square inch and then placed outside for a year. At the end of the year, Consumers Union measured the pressure in each tire.[14]

III. Analyze data The amount of pressure lost (in pounds per square inch) during the year for the air-filled and nitrogen-filled tires of each brand is shown in the following table.

Brand	Air	Nitrogen	Brand	Air	Nitrogen
BF Goodrich Traction T/A HR	7.6	7.2	Pirelli P6 Four Seasons	4.4	4.2
Bridgestone HP50 (Sears)	3.8	2.5	Sumitomo HTR H4	1.4	2.1
Bridgestone Potenza EL400	2.1	1.0	Yokohama Avid H4S	4.3	3.0
Bridgestone Potenza G009	3.7	1.6	BF Goodrich Traction T/A V	5.5	3.4
Bridgestone Potenza RE950	4.7	1.5	Bridgestone Potenza RE950	4.1	2.8
Continental Premier Contact H	4.9	3.1	Continental ContiExtreme Contact	5.0	3.4
Cooper Lifeliner Touring SLE	5.2	3.5	Continental ContiProContact	4.8	3.3
Dayton Daytona HR	3.4	3.2	Cooper Lifeliner Touring SLE	3.2	2.5
Falken Ziex ZE-512	4.1	3.3	General Exclaim UHP	6.8	2.7
Fuzion Hrl	2.7	2.2	Hankook Ventus V4 H105	3.1	1.4
General Exclaim	3.1	3.4	Michelin Energy MXV4 Plus	2.5	1.5
Goodyear Assurance Tripletred	3.8	3.2	Michelin Pilot Exalto A/S	6.6	2.2
Hankook Optimo H418	3.0	0.9	Michelin Pilot HX MXM4	2.2	2.0
Kumho Solus KH16	6.2	3.4	Pirelli P6 Four Seasons	2.5	2.7
Michelin Energy MXV4 Plus	2.0	1.8	Sumitomo HTR+	4.4	3.7
Michelin Pilot XGT H4	1.1	0.7			

A brief inspection of the data on pressure loss reveals a lot of variation. What are some possible sources of this variation? Some tires were filled with compressed air, while others were filled with nitrogen. So part of the variation in pressure loss is due to what the tires were filled with. Some brands of tires might maintain pressure better than others. Thus, some of the variation we see is due to differences among the brands of tires that were used in this study. Even two tires of the same brand are not completely identical. As a result, some of the variation in the data may be due to differences in individual tires of the same brand. How can we separate the effects of these sources of variation?

We'll continue the data analysis process shortly.

Exercises

1.25 A misleading graph When we entered the results of the tattoo survey of Example 1.10 (page 20) into Excel, the program produced the default bar graph that you see below. What's wrong with this picture?

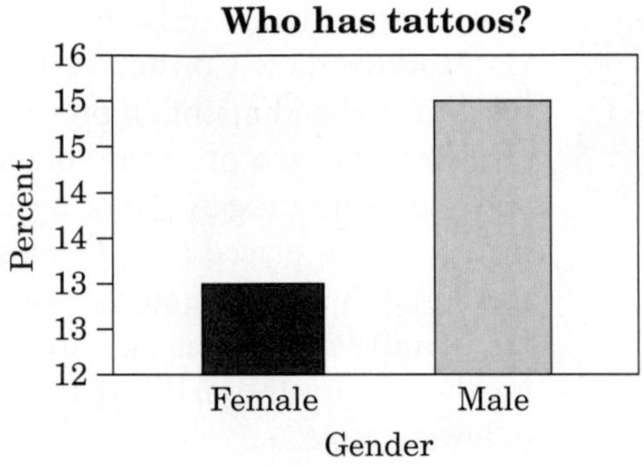

1.26 Tattoos across the United States An Excel bar graph displaying more results from the tattoo survey of Example 1.10 (page 20) is shown below. Write a couple of sentences describing what the graph tells you.

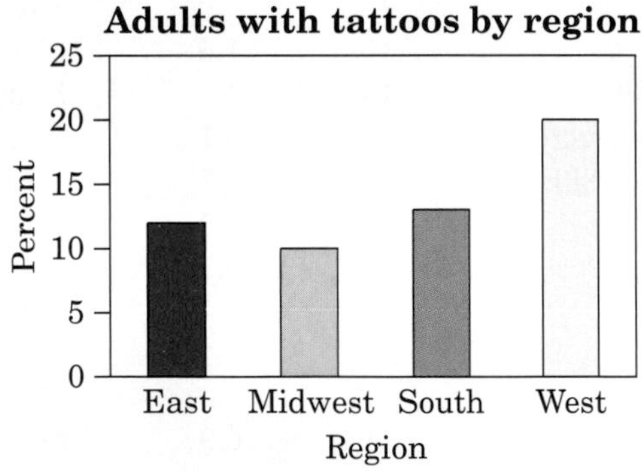

1.27 Variation in tire pressure In Example 1.11, we discussed three possible sources of variation in tire pressure loss. Give specific evidence to show that each of these three types of variation is present in the experiment.

1.28 Nitrogen and gas mileage Can filling car tires with nitrogen instead of compressed air increase gas mileage? Describe how you would design a study to try to answer this question. Be as specific as you can about the details of your study.

1.29 Teens and tattoos What percent of teens have tattoos? Describe how you would design a study to try to answer this question. Be as specific as you can about the details of your study.

1.30 Who talks more—women or men? According to Louann Brizendine, author of *The Female Brain,* women say nearly three times as many words per day as men. Skeptical researchers devised a study to test this claim. They used electronic devices to record the talking patterns of 396 university students from Texas, Arizona, and Mexico. The device was programmed to record 30 seconds of sound every 12.5 minutes without the carrier's knowledge. What were the results?

According to a published report of the study in *Scientific American*, "Men showed a slightly wider variability in words uttered. . . . But in the end, the sexes came out just about even in the daily averages: women at 16,215 words and men at 15,669."[15]

(a) Did researchers conduct an observational study or an experiment? Explain.

(b) Why was it important for the recording device to operate without the knowledge of the carrier?

(c) To what population can the results of this study be generalized? Justify your answer.

Statistics in action: from tattoos to tires

The air versus nitrogen study of Example 1.11 is an experiment, because researchers deliberately imposed *treatments*—filling with air or filling with nitrogen—on the individuals (tires) in the study. A common goal of experiments is to draw conclusions about cause and effect. Consumers Union hoped that their experiment would provide convincing evidence about whether filling tires with nitrogen rather than air causes a reduction in pressure loss. Did they achieve their goal? Let's examine the data more closely before we try to answer this question.

DATA EXPLORATION

Nitrogen in Tires

www The nitrogen in tires data set from Example 1.11 is available at the book's Web site, **www.whfreeman.com/sta2e.** If you have a TI-Nspire graphing calculator, you can download the file *tires.tns* from your teacher or the book's Web site.

1. The accompanying TI-Nspire screen shot shows a comparative dotplot of the pressure loss data. What does this graph tell you about pressure loss for air-filled versus nitrogen-filled tires?

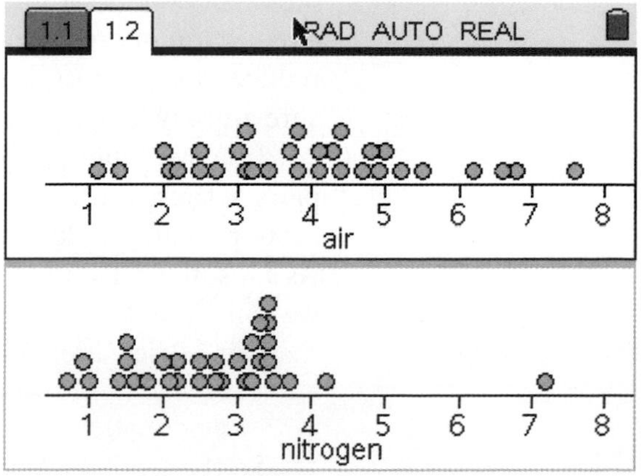

DATA EXPLORATION *(continued)*

DATA EXPLORATION *(continued)*

2. Quick calculations reveal that the average pressure loss for the air-filled tires was 3.9 pounds per square inch (psi), compared with 2.7 psi for the nitrogen-filled tires. That seems like a pretty large difference. Can we conclude that this difference was due to what the tires were filled with? Or is there another possible explanation?

3. Did you realize that our examination of the tire pressure data in Questions 1 and 2 ignored the fact that the data values are *paired* by brand of tire? Since the data are paired, it makes sense to examine the *differences* in pressure loss for the air-filled and nitrogen-filled tires in each pair. For example, for the BF Goodrich Traction T/A HR tires, the air-filled tire lost 7.6 psi in a year. The nitrogen-filled tire lost 7.2 psi over that same period. The difference in pressure loss (air − nitrogen) for this brand is 7.6 − 7.2 = 0.4 psi.

A dotplot of the differences in pressure loss is shown in the TI-Nspire screen shot below.

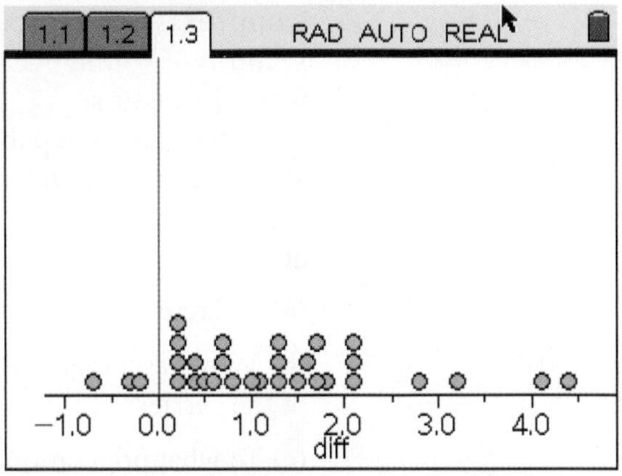

Are most of the difference values positive or negative? The average difference in pressure loss is 1.2 psi. What do these results suggest?

What have we learned so far from the tire experiment? On average, a nitrogen-filled tire of a given brand lost 1.2 psi less pressure than an air-filled tire of the same brand over a one-year period. For 28 of the 31 pairs of tires tested, the nitrogen-filled tire lost less pressure than the air-filled tire. Are we safe concluding that these observed differences in tire pressure loss are due to the different effects of filling with air and filling with nitrogen? Or could such results have happened "just by chance"?

Example 1.12 Nitrogen in tires—a lot of hot air? *(continued)*

IV. Interpret results In the tire pressure experiment, the air-filled tire lost more pressure than the nitrogen-filled tire for 28 of the 31 tire pairs. There are two possible explanations for this result: (1) filling tires with nitrogen rather than air caused a reduction in pressure loss, or (2) filling tires with nitrogen rather than air did not reduce pressure loss, but just by chance, the tire that was going to lose more pressure anyway happened to get filled with air in 28 of the 31 pairs.

Imagine flipping a coin 31 times to simulate assigning air and nitrogen to all 31 pairs of tires. There is virtually no chance of selecting the tire that will end up losing more pressure in 28 of those pairs to fill with air just by the luck of the coin toss. As a result, explanation (2) is not plausible. Based on the results of this experiment, we conclude that there is convincing evidence that filling tires with nitrogen rather than air caused a reduction in pressure loss.

So there you have it: a brief introduction to how statistics can be used to draw conclusions from data. Whether it's an observational study about who has tattoos or an experiment about filling tires with nitrogen instead of air, the statistical problem-solving process can help us make sense of the world around us. In the chapters that follow, you'll learn more about the last three steps in this process—producing data, analyzing data, and interpreting results.

Statistics Through Applications presents the important ideas of statistics in four parts:

Part A: Analyzing Data (Chapters 2 to 4) concerns strategies for exploring, organizing, and describing data using graphs and numerical summaries. You can learn to look at data intelligently even with very simple tools, as you have seen in this chapter.

Part B: Producing Data (Chapters 5 and 6) describes methods for data production that can give clear answers to puzzling questions. As you learned in this chapter, where the data come from is really important. Basic concepts about how to select samples and design experiments are the most fundamental principles in statistics.

Part C: Chance (Chapters 7 and 8) gives us the language we need to describe variation, risk, and probability. Because variation is present in any statistical problem, we need a deeper understanding of the role that chance plays in contributing to that variation.

Part D: Inference (Chapters 9 and 10) moves beyond the data in hand to draw conclusions about some wider universe. Done properly, inference will allow us to estimate some characteristic of a population based on results from a sample. Other inference methods help us answer the question "Could this result have happened just by chance?"

APPLICATION 1.2

What is "normal" body temperature?

If a thermometer under your tongue reads higher than 98.6°F, do you have a fever? Maybe not. People vary in their "normal" temperature. Your own temperature also varies—it is lower around 6 A.M. and higher around 6 P.M.

Researchers designed an observational study to investigate the long-accepted value of 98.6°F, which was established in 1878 by German doctor Carl Wunderlich. In the study, 148 healthy men and women aged 18 to 40 had their temperatures taken orally several times over a three-day period. A total of 700 temperature readings were produced.[16]

When researchers published their results, they did not provide the original data. This is unfortunately a common practice. By working backward from graphs and numerical information given in the article, however, Allen Shoemaker from Calvin College produced a data set with many of the same characteristics as the original temperature readings.[17]

QUESTIONS

1. Explain why this is an observational study and not an experiment.

2. To what population does the conclusion about normal body temperature apply? Justify your answer.

3. A dotplot of Shoemaker's temperature data is shown below. We have added a vertical line at 98.6°F for reference. What does this graph tell you about "normal" body temperature?

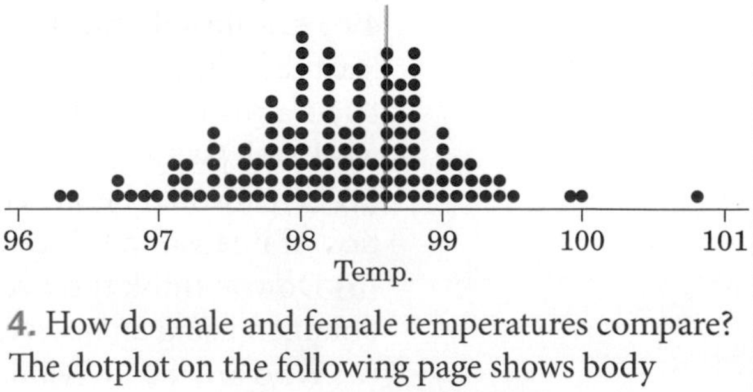

4. How do male and female temperatures compare? The dotplot on the following page shows body

APPLICATION 1.2 *(continued)*

APPLICATION 1.2 *(continued)*

temperature readings by gender. Write a few sentences comparing normal body temperatures for males and females in the study.

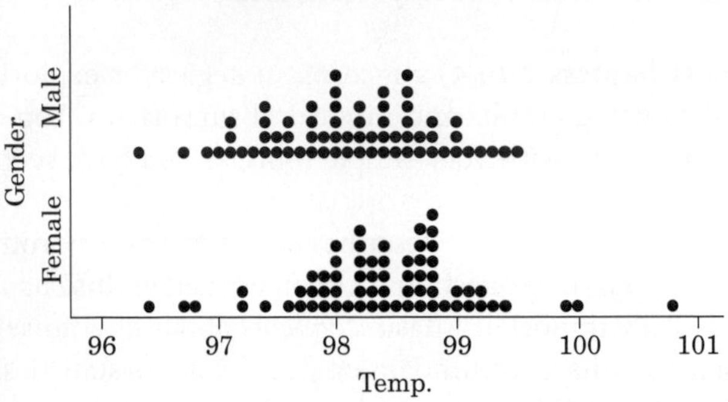

5. List and describe as many possible sources of variation in this study as you can.

The average of the 700 temperature readings was 98.2°F. For the men in the study, the average temperature reading was 98.1°F. For the women in the study, the average temperature reading was 98.4°F. Only 8% of the temperatures recorded were equal to 98.6 degrees.

Researchers used a statistical test to determine that an average temperature of 98.2°F in the sample of 700 measurements was extremely unlikely if the true average temperature in the population was 98.6°F. They concluded that Carl Wunderlich's value of 98.6°F is not a reasonable value for "normal" body temperature.

Section 1.2 Summary

Statistics is the study of variation—among individuals, groups, and measurements. The **statistical problem-solving process** can help us make sense of the variation we see. The process begins with some question of interest about the world around us—a question that involves variation. To answer the question, we need some relevant data. The most common ways of obtaining data are through an **observational study** or an **experiment.** With data in hand, we proceed to analysis. Graphs and numerical summaries help us make sense of observed variation in the data. Finally, we interpret the results of our analysis and answer the original question of interest. You'll learn more about each of these aspects of the statistical problem-solving process in later chapters.

Section 1.2 Exercises

1.31 Meat eaters According to a *USA Today* Snapshot, about 3% of Americans never eat meat. A pie chart presented in the Snapshot shows that 49% eat meat daily, 34% eat meat at least three times a week, 8% eat meat twice a week, and 6% eat meat once per week or less.

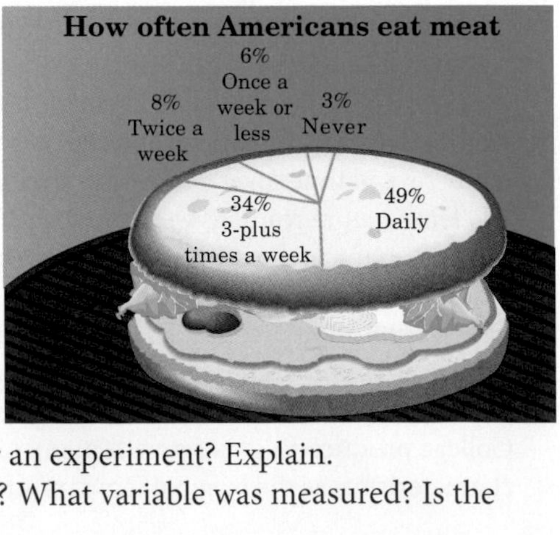

(a) What would be a reasonable question of interest in this case?

(b) Do you think these results were obtained using an observational study or an experiment? Explain.

(c) Who are the individuals in this study? What variable was measured? Is the variable categorical or quantitative?

1.32 TV and obesity According to the National Institute on Media and the Family, a preschoolers' risk of obesity jumps 6% for every hour of television watched per day. The risk increases by 31% if the TV is in their bedroom.[18]
(a) What would be a reasonable question of interest in this case?
(b) Do you think these results were obtained using an observational study or an experiment? Why?
(c) The actual study that produced these results involved 2761 low-income adults in New York with children aged 1 to 4 years. Who are the individuals in this study? What variable(s) were measured?

1.33 Throw in the towel! Which paper towel brand absorbs best? Samantha, a 12-year-old, carried out a science project to compare the absorbency of four brands of paper towels—Bounty, Scott, Sparkle, and Western Family. She started with four individual sheets of each brand. One at a time, Samantha placed a sheet of paper towel in a container with 200 milliliters of water. After 15 seconds, she removed the paper towel from the water and squeezed out as much water as she could into a graduated cylinder. Samantha then determined the amount of water that had been absorbed.[19] Here are the data she recorded:

Bounty	33	31	49	32
Scott	42	34	60	36
Sparkle	53	36	47	33
Western Family	47	26	41	31

(a) Identify two sources of variation in this study. Give specific evidence to show that each of these types of variation is present.
(b) Write a brief report describing Samantha's study. Follow the four-step statistical problem-solving process.

1.34 Healthy cereal? Researchers collected data on 77 brands of cereal at a local supermarket.[20] For each brand, the values of several variables were recorded, including sugar (grams per serving), calories per serving, and the shelf in the store on which the cereal was located (1 = bottom, 2 = middle, 3 = top). Here is a dotplot of the data on sugar content and shelf:

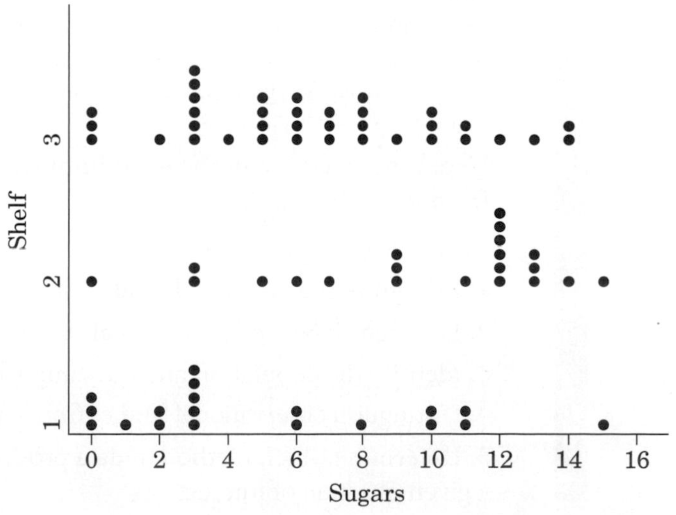

Critics claim that supermarkets tend to put sugary kids' cereal on lower shelves, where the kids can see it. Do the data from this study support this claim? Follow the four-step statistical problem-solving process.

1.35 Creative artwork Does competing for a prize improve children's artistic creativity? A psychologist carried out a study with about 100 third-graders that was designed to help answer this question. In the study, the children were divided into two groups and instructed to make a "silly" collage using materials that were provided. Before they started, the children in one group were told that their collages would be judged by experts and that the winners would receive prizes. The children in the other group were told that they would share their collages in an art party. In fact, expert judges rated the creativity of all the collages.[21]

(a) Was this an observational study or an experiment? Justify your answer.

(b) The judges were not aware of which students' collages came from which group. Why is this important?

(c) Describe two possible sources of variation in this study.

(d) Children in the group who believed they were competing for prizes received much lower average creativity ratings than children in the art party group. Can we conclude that the difference in creativity scores was caused by the difference in what the two groups of children were told? Why or why not?

1.36 Find your own study Locate a published report about an experiment or an observational study.

(a) Print the report if possible. Be sure to record all relevant source information.

(b) Describe carefully how the study reflects the four-step statistical problem-solving process.

CHAPTER 1 REVIEW

The first and most important question to ask about any statistical study is "Where did the data come from?" The distinction between observational and experimental data is a key part of the answer. Good statistics starts with good designs for producing data. Then, we measure the characteristics of interest to obtain numbers we can work with. With data in hand, we turn to analysis. Graphs and numerical summaries help us explore variation in the data. Finally, we interpret the results of our data analysis in light of the original question of interest. Ask a question, produce data, analyze data, interpret results—these are the four steps in the statistical problem-solving process.

Here is a review list of the most important skills you should have developed from your study of this chapter.

A. DATA PRODUCTION

1. Recognize the individuals and variables in a statistical study.
2. Distinguish between categorical and quantitative variables.
3. Identify the population and the sample in a statistical study.
4. Distinguish observational studies from experiments.
5. Determine which method of data production is most appropriate for answering a given question of interest.

B. DRAWING CONCLUSIONS FROM DATA

1. Identify the four steps in the statistical problem-solving process.
2. Describe possible sources of variation in a statistical study.
3. Interpret the results of a statistical study from basic graphical displays.
4. Recognize limitations in the kinds of conclusions that can be drawn from a statistical study based on how the data were produced.

CHAPTER 1 REVIEW EXERCISES

1.37 Multi-airline flights A passenger who takes a trip involving two or more airlines pays the first carrier, which then owes the other carrier a portion of the ticket cost. It is too expensive for the airlines to calculate exactly how much they owe each other. Instead, a sample of about 12% of tickets sold is examined and accounts are settled on that basis.

(a) What is the population that the airlines want information about? What is the sample?
(b) Will the sample results definitely, probably, probably not, or definitely not yield the correct amount of money for each airline? Explain.

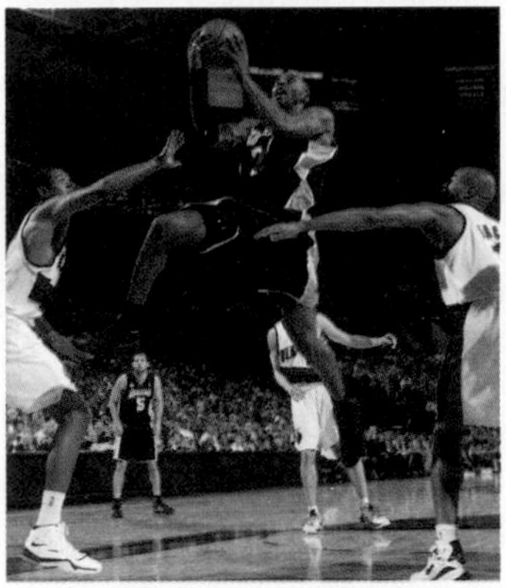

Sam Forencich/NBAE via Getty Images

1.38 NBA All-Stars In the National Basketball Association (NBA), fans vote to decide which players get to play in the NBA All-Star game. Kevin Garnett of the Boston Celtics led all players, with 2,399,148 votes for the 2008 All-Star game. The table below provides data about the top five vote-getters that year:[22]

Player	Team	Votes	Points per game	Height	Position
Kevin Garnett	BOS	2,399,148	18.8	83	F
LeBron James	CLE	2,108,831	30.0	80	F
Dwight Howard	ORL	2,066,991	20.7	83	C
Kobe Bryant	LAL	2,004,940	28.3	78	G
Carmelo Anthony	DEN	1,723,701	25.7	80	F

(a) What individuals are measured?
(b) What variables are recorded? Which are quantitative? In what *units* is each quantitative variable recorded?
(c) Which variable or variables do you think most influence the number of votes received by an individual player?

1.39 Cow-a-bunga! According to scientists, a typical cow burps about every 40 seconds. When cows burp, they release methane gas that is built up while digesting their food. Some estimates suggest that livestock produce over 25% of methane emissions each year. Researchers in Germany have developed a pill that is designed to reduce the amount of methane that cows produce.

(a) What would be a reasonable question of interest in this case?
(b) Describe how you would design a study to answer the question of interest from (a). Be as specific as you can about the details of your design.

1.40 Noisy snowmobiles In spite of a study that predicted harmful environmental effects, the government lifted a ban on snowmobiles in two national parks in November 2002. The study found that air pollutants would be far higher with snowmobiles than with the snowcoaches that were currently being used. Noise from snowmobiles would be audible over half the time in the parks.
(a) Was this an observational study or an experiment? Why?
(b) What individuals were measured, and what variables were recorded?
(c) Why do you think the government decided to allow snowmobiles, in spite of the study results?

1.41 How much allowance? According to a *USA Today* Snapshot, about 24% of children aged 5 to 12 receive more than $20 a month in allowance. A pie chart presented in the Snapshot shows that 13% receive $5 or less, 23% get $6 to $10, and 39% receive $11 to $20.

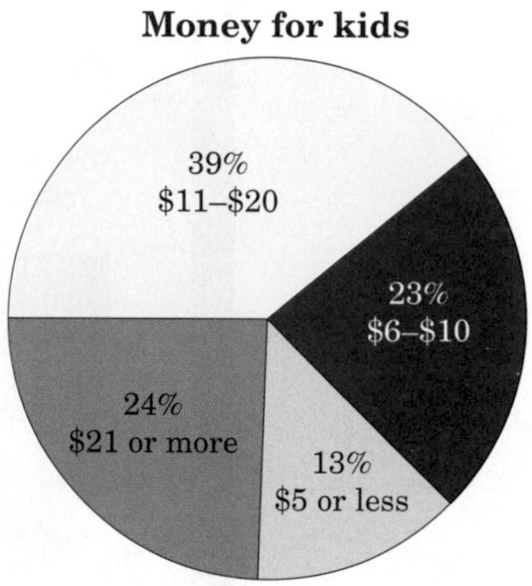

(a) What would be a reasonable question of interest in this case?
(b) Do you think these results were obtained using an observational study or an experiment? Why?
(c) Who are the individuals in this study? What variable was measured?
(d) Think: why don't the four percents given add up to 100%?

1.42 Creative writing Do external rewards—things like money, praise, fame, and grades—promote creativity? Not according to the study in Exercise 1.35 (page 28). Children who competed for prizes generally produced less creative collages than children who simply shared their artwork. Will similar results hold for college students? The psychologist of Exercise 1.35 designed another study—this time involving 47 experienced creative writers who were college students. Students were divided into two groups using a chance process (like drawing names from a hat). The students in one group were given a list of statements about external reasons for writing, like public recognition, making money, or pleasing their parents. Students in the other group were given a list of statements about internal reasons for writing, such as expressing yourself and enjoying working with words. Both groups were then instructed to write a poem about laughter. Each student's poem was rated separately by 12 different poets using a creativity scale.[23]
(a) Was this an observational study or an experiment? Justify your answer.
(b) Why were students divided into the two groups using a chance process?
(c) None of the poets was aware of which students were in which group. Why is this important?

1.43 More creative writing Refer to Exercise 1.42. The 12 poets' ratings of each student's poem were averaged to obtain an overall creativity score. Parallel dotplots of the two groups' creativity scores are shown on the facing page.

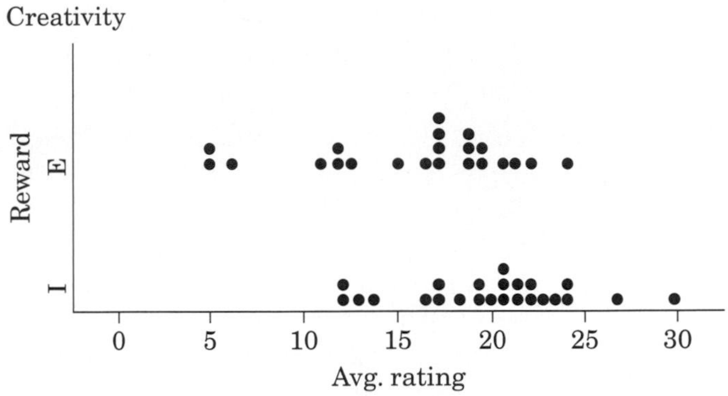

What do you conclude? Follow the four-step problem-solving process.

1.44 Find another study Locate a published report about a sample survey.
(a) Print the report if possible. Be sure to record all relevant source information.
(b) Describe carefully how the study reflects the four-step statistical problem-solving process.

PART A

ANALYZING DATA

"Tonight we're going to let the statistics speak for themselves."

Words alone don't tell a story. A writer organizes words into sentences and organizes the sentences into a story. If the words are badly organized, the story isn't clear. Data also need organizing if they are to tell a clear story. Too many words can make a subject confusing. Vast amounts of data are even harder to digest—we often need a brief summary to highlight essential facts. How to organize, summarize, and present data are our topics in this part of the book.

As the old saying goes, a picture is worth a thousand words. So we start by looking at graphical methods for presenting data. Then we add appropriate numerical summaries—like counts or percents for categorical variables, and the mean and standard deviation for quantitative variables. The final step is to interpret what the graphs and numbers tell us about the data.

Organizing and summarizing a large body of facts opens the door to distortions, both unintentional and deliberate. This is also the case when the facts take the form of numbers rather than words. We'll point out some common pitfalls to avoid when presenting data graphically and numerically. Mostly, though, we'll show you how to present data clearly and accurately.

CHAPTER 2

Describing Distributions of Data

Does education pay?

People with more education earn more, on average, than people with less education. How much more? A simple way to answer this question is to compare *median* incomes, the amount such that half of a group of people earn more and half earn less. Here are the median incomes of adults aged 25 and older with four different levels of education:

High school graduate only	Some college, no degree	Bachelor's degree	Advanced degree
$24,800	$30,000	$45,000	$62,000

These numbers come from 113,320 people interviewed by the Current Population Survey (CPS) in March 2008.[1] Every March, the CPS asks detailed questions about income. Boiling down many incomes into four numbers helps us get a quick picture of the relationship between education and income. The median college graduate, for example, earns about twice as much as the median high school graduate.

We know that incomes vary a lot. In fact, the highest income reported by the 33,943 people in the CPS sample with only a high school diploma was $692,270. The medians compare the centers of the four income distributions. Can we also describe the spreads with just a few numbers? The largest and smallest incomes in a group of people don't tell us much. Instead, let's give the range covered by the middle half of the incomes in each group. Here they are:

High school graduate only	Some college, no degree	Bachelor's degree	Advanced degree
$11,000 to $40,000	$14,645 to $48,000	$23,545 to $72,050	$37,550 to $100,000

The message of the 113,320 incomes is now clear: going to college but not getting a degree doesn't raise income much. People with bachelor's degrees, however, earn quite a bit more, and an advanced degree gives income another healthy boost. But there's a lot of variation among individuals—some very rich people never went to college. Finally, remember that this observational study says nothing about cause and effect. People who get advanced education are often smart, ambitious, and well-off to start with, so they might earn more even without their degrees.

2.1 Displaying Distributions with Graphs

ACTIVITY 2.1A

Did you know?

How much do you know about your classmates? Without looking, what is the most common eye color in your class? Without asking, how many siblings does a "typical" classmate have? In this Activity, you'll begin by collecting some data to help you answer these questions. Then, you'll work with a few teammates to analyze the data.

1. Have each member of the class record his or her eye color and number of siblings on the board or overhead projector.

2. Work with your teammates to produce a graph that displays the eye color data effectively. Add any numerical information that you think helps summarize the data.

3. Share your team's analysis with the rest of the class. What is the most common eye color?

4. Now work with your teammates to produce a graph that displays the siblings data effectively. Add any numerical information that you think helps summarize the data.

5. Share your team's analysis with the rest of the class. How many siblings does a "typical" student have?

Bar graphs and pie charts

How well educated are young adults in the United States? Table 2.1 presents some data for people aged 25 to 34 years. This table illustrates some good practices for data tables. It is clearly *labeled* so that we can see the subject of the data at once. The table title gives the general subject of the data and the date because these data will change over time. Labels within the table identify the variables and state the *units* in which they are measured. Notice, for example, that the counts are in thousands. The *source* of the data appears at the bottom of the table. Note that these data come from our old friend, the Current Population Survey.

Table 2.1 starts with the **counts** of young adults with each level of education. **Rates** (percents or proportions) are often clearer than counts—it is more helpful

Table 2.1 Level of Education of People 25 to 34 Years Old in the United States, 2007

Level of education	Number of persons (thousands)	Percent
Less than high school	5,126	12.9
High school graduate	11,408	28.6
Some college	10,961	27.5
Bachelor's degree	9,076	22.8
Advanced degree	3,299	8.3
Total	39,868	100

Source: U.S. Census Bureau, Current Population Survey, 2007.

to hear that 12.9% of young adults did not finish high school than to hear that there are 5,126,000 such people. The percents also appear in Table 2.1. The last two columns of the table present the **distribution** of the variable "level of education" in two forms. Each column gives information about what values the variable takes and how often it takes each value.

Distribution of a variable

The **distribution** of a variable tells us what values the variable takes and how often it takes these values.

Did you check Table 2.1 for consistency? The total number of people should be

$$5{,}126 + 11{,}408 + 10{,}961 + 9{,}076 + 3{,}299 = 39{,}870 \text{ (thousands)}$$

The table gives the total as 39,868. What happened? Each entry is rounded to the nearest thousand. The rounded entries don't quite add to the total, which is rounded separately. Such **roundoff errors** will be with us from now on as we do more arithmetic.

The distribution in Table 2.1 is quite simple because the categorical variable "level of education" has only five possible values. To picture this distribution, we might use a **bar graph.**

Example 2.1 Education level in the United States: a bar graph

Figure 2.1 (on the next page) shows a bar graph of the data. The height of each bar shows the percent of young adults who have the level of education that is marked at the bar's base. The bar graph makes it clear that there are more people with only a high school diploma than there are people who also have some college—the "HS grad." bar is taller.

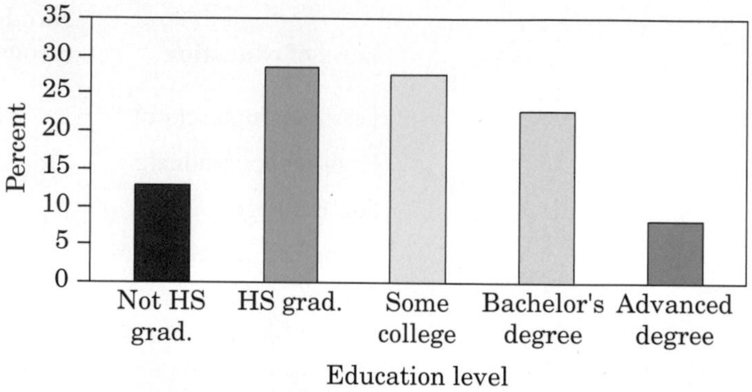

Figure 2.1 Bar graph of the distribution of level of education for persons aged 25 to 34 in the United States in 2007.

We can also display the distribution of education level with a **pie chart.**

Example 2.2 Education level in the United States: a pie chart

Figure 2.2 is a pie chart of the level of education of young adults. Pie charts show how a whole is divided into parts. Because it is hard to see this from the wedges in the pie chart, we added labels with the actual percents.

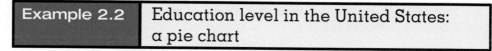

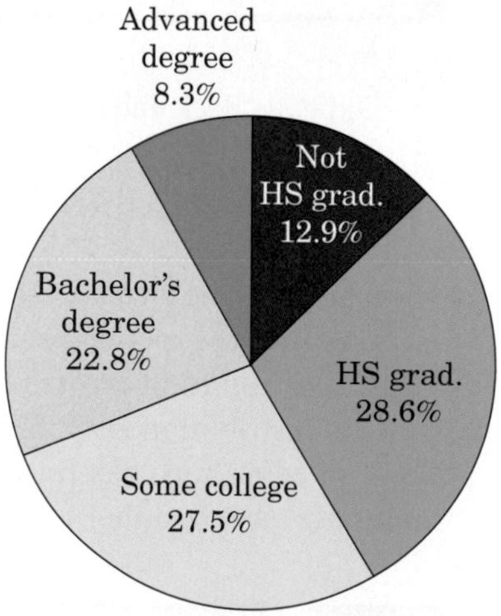

Figure 2.2 Pie chart of the distribution of education level among persons aged 25 to 34 in the United States in 2007.

To make a pie chart, first draw a circle. The circle represents the whole—in this case, all people in the United States aged 25 to 34 years. Wedges within the circle represent the parts. The angle spanned by each wedge is proportional to the size of that part. For example, 22.8% of young adults have a bachelor's degree but not an advanced degree. Because there are 360 degrees in a circle, the "bachelor's degree" wedge spans an angle of $0.228 \times 360 = 82$ degrees.

You can calculate the angles for the other wedges in the same way and then use a protractor to help you draw an accurate pie chart. Our advice is to let technology draw the graph for you.

Pie charts force us to see that the parts do make a whole. But because angles are harder to compare than lengths, a pie chart is not the best way to compare the sizes of the various parts of the whole. A bar graph is easier to draw than a pie chart unless a computer is doing the drawing for you.

When we think about graphs, it is helpful to distinguish between *quantitative variables,* like height in centimeters and SAT scores, and *categorical variables,* such as sex, eye color, and level of education. Pie charts and bar graphs are useful for displaying distributions of categorical variables. Although both pie charts and bar graphs can show the distribution (either counts or percents) of a categorical variable such as level of education, bar graphs have other uses, too.

Example 2.3 Hang up and drive!

The Harris Poll conducted an online survey of 2085 U.S. adults about cell phone use while driving.[2] Over half (56%) of those who responded said that it is dangerous or very dangerous for someone to use a cell phone while driving. One survey question asked, "How often do you talk on a cell phone while you are driving?" Respondents could choose from the options "Always," "Sometimes," or "Never." Figure 2.3 shows the percent who chose "Always" or "Sometimes" by region of the United States in which they live.

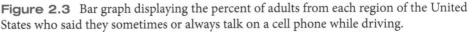

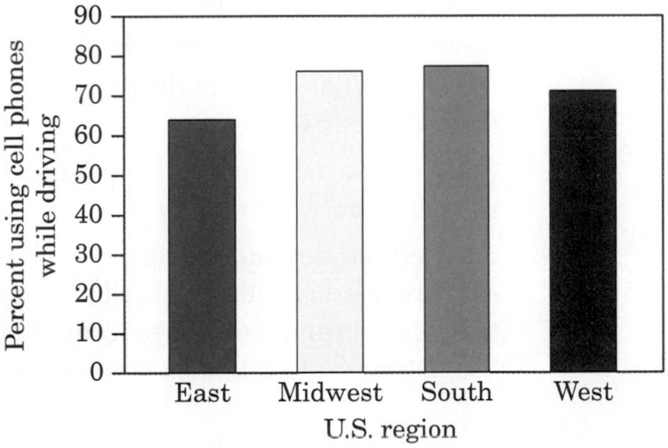

Figure 2.3 Bar graph displaying the percent of adults from each region of the United States who said they sometimes or always talk on a cell phone while driving.

We cannot replace Figure 2.3 by a pie chart, because it compares four separate quantities, not the parts of some whole. A pie chart can compare only parts of a whole. Bar graphs can compare quantities that are not parts of a whole.

Exercises

2.1 Marital status In the *Statistical Abstract of the United States* we find these data on the marital status of adult American women as of 2007:

Marital status	Count (thousands)
Never married	25,262
Married	65,128
Widowed	11,208
Divorced	13,210
Total	**114,807**

(a) How many women were not married in 2007?

(b) Make a bar graph to show the distribution of marital status.

(c) Would it also be correct to use a pie chart? If so, make a pie chart for these data.

2.2 Consistency? Refer to the previous exercise. What is the sum of the counts for the four marital status categories? Why is this sum not equal to the total given in the table?

2.3 College freshmen A survey of college freshmen asked what field they planned to study. The results: 25.2% arts and humanities, 19.3% business, 7.1% education, 16.6% engineering and science, 7.8% professional, and 15.3% social science.[3]

(a) What percent plan to study fields other than those listed?

(b) Make a graph that compares the percents of college freshmen planning to study various fields.

2.4 Girls excel Is it true that girls perform better than boys in the study of languages and so-called soft sciences? Here are several Advanced Placement subjects and the percent of examinations taken by female candidates in 2007: English Language/Composition, 63%; French Language, 70%; Spanish Language, 64%; and Psychology, 65%.[4]

(a) Explain clearly why we cannot use a pie chart to display these data, even if we know the percent of exams taken by girls for every subject.

(b) Make a bar graph of the data. Order the bars from tallest to shortest; this will make comparisons easier.

(c) Do these data answer the question about whether girls perform better in these subject areas? Why or why not?

2.5 Cell phones and driving The Harris Poll survey of Example 2.3 (page 39) also provided results about cell phone use and driving by age group. Here is a table that summarizes the percent of each age group who said they "Always" or "Sometimes" use cell phones while driving

Age group			
Gen. Y	**Gen. X**	**Baby boomers**	**Matures**
86%	79%	76%	48%

(a) Would it be appropriate to use a pie chart to display these data? If so, do it. If not, explain why not.

(b) Make a bar graph of the data. Describe what you see.

2.6 Lottery sales States sell lots of lottery tickets. Table 2.2 shows where the money comes from. Make a bar graph that shows the distribution of lottery sales by type of game. Is it also proper to make a pie chart of these data?

Table 2.2 State Lottery Sales by Type of Game, 2007

Game	Sales (millions of dollars)
Instant games	29,736
Three-digit games	5,586
Four-digit games	3,499
Lotto	10,014
Other games	3,579
Total	52,414

Source: 2009 *Statistical Abstract of the United States.*

Displaying quantitative variables: dotplots

ACTIVITY 2.1B

How good is your balance?

Materials: Stopwatch, plus open space with a desk, table, or wall nearby for each pair of students.

Can you stand on one foot without hopping or touching anything? Try it. (You might need to practice!) In this Activity, you and a partner will take turns trying to balance on one foot—with your eyes closed!

1. With your partner, decide who will perform the activity first (the "doer") and who will be the timer. Your teacher may give you some practice time before data collection begins.
2. The doer should get in position—standing on one foot with eyes closed—and then say, "Go." The timer starts the stopwatch on "Go" and stops it when the doer hops, opens his or her eyes, or touches anything. Record the doer's time to the nearest second.
3. Switch roles and repeat Step 2.
4. While you are collecting data, your teacher will put a blank data table on the board or overhead projector. When both team members have finished, record your individual times in the data table.
5. Once all teams have finished, make a dotplot of the class data. Then describe what you see.

In Chapter 1, we used a **dotplot** to display:

- ages of a random sample of 100 people from the U.S. census
- pressure loss values in an experiment involving 31 brands of tires
- temperature readings of 148 healthy adults in an observational study about normal body temperature
- sugar content of 77 brands of cereal

What do all of these variables—age, pressure loss, temperature, and sugar content—have in common? They're all quantitative variables. A dotplot is the simplest graph for displaying the distribution of a quantitative variable.

Example 2.4 Are you driving a gas guzzler?

The Environmental Protection Agency (EPA) is in charge of how fuel economy ratings for cars are determined and reported (think of those large window stickers on a new car). For years, consumers complained that their actual gas mileages were noticeably lower than the values reported by the EPA. It seems that the EPA's tests—all of which are done on computerized devices to ensure consistency—did not consider things like outdoor temperature, use of the air conditioner, or realistic acceleration and braking by drivers. In 2008, the EPA changed the method for measuring a vehicle's fuel economy to try to give more accurate estimates.

Table 2.3 displays the EPA estimates of highway gas mileage for 24 model year 2009 midsize cars. Let's construct a dotplot that shows the distribution of EPA highway mileage ratings.

Table 2.3 Highway Gas Mileages for Model Year 2009 Midsize Cars

Model	mpg	Model	mpg
Acura RL	22	Lexus GS 350	26
Audi A6 Quattro	23	Lincoln MKZ	28
Bentley Arnage	14	Mazda 6	29
BMW 528I	28	Mercedes-Benz E350	24
Buick Lacrosse	28	Mercury Milan	29
Cadillac CTS	25	Mitsubishi Galant	27
Chevrolet Malibu	33	Nissan Maxima	26
Chrysler Sebring	30	Rolls-Royce Phantom	18
Dodge Avenger	30	Saturn Aura	33
Hyundai Elantra	33	Toyota Camry	31
Jaguar XF	25	Volkswagen Passat	29
Kia Optima	32	Volvo S80	25

Source: Fuel Economy Guide 2009, from the U.S. Environmental Protection Agency's Web site at **www.fueleconomy.gov**.

Step 1: *Label your axis and title your graph.* Draw a horizontal line and label it with the variable, highway gas mileage (mpg). Title your graph.

Step 2: *Scale and number the axis based on the values of the variable.* For these data, the smallest value is 14 and the largest is 33.

Step 3: *Mark a dot above the number on the horizontal axis corresponding to each data value.*

Figure 2.4 displays the dotplot.

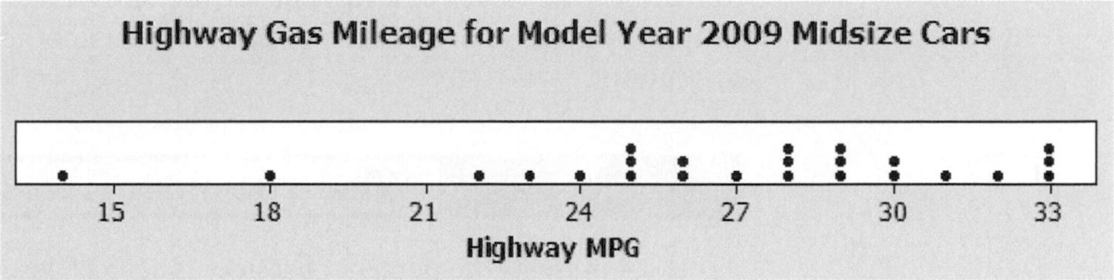

Figure 2.4 Minitab dotplot of EPA highway gas mileage ratings for 24 model year 2009 midsize cars.

Current information on vehicle fuel economy ratings is available at **www.epa.gov/fueleconomy**.

Making a statistical graph is not an end in itself. The purpose of the graph is to help us understand the data. After you (or your computer) make a graph, always ask, "What do I see?" Here is a general strategy for looking at graphs.

Pattern and deviations

In any graph of data, look for an **overall pattern** and also for striking **deviations** from that pattern.

In the dotplot of Figure 2.4, we can see three distinct *clusters* of values: cars that get around 25 mpg, cars that get about 28–30 mpg, and cars that get around 33 mpg. We also see two cars with unusually low gas mileage ratings—the Bentley Arnage and the Rolls-Royce Phantom. These cars are potential **outliers.**

Outliers

An **outlier** in any graph of data is an individual observation that falls outside the overall pattern of the graph.

Whether an observation is an outlier is to some extent a matter of judgment. Both the Arnage and the Phantom are clearly separated from the main body of observations, which seems to put them "outside the overall pattern." In Section 2.2, we'll establish a numerical procedure for identifying outliers. For now, once you have spotted potential outliers, look for an explanation. Many outliers are due to mistakes, such as typing 4.0 as 40. Other outliers point to the special nature of some observations. Explaining outliers usually requires some background information. Does it surprise you that the Bentley and Rolls-Royce midsize cars have much lower gas mileages than the other models?

Displaying quantitative variables: stemplots

Sometimes the values of a variable are too spread out for us to make a reasonable dotplot by hand. In these cases, we can consider another simple graphical display—a **stemplot.**

Example 2.5 Where do older folks live?

Table 2.4 presents the percent of residents aged 65 years and over in each of the 50 states. Let's make a stemplot to display this distribution.

Table 2.4 Percent of Residents Aged 65 and Over in the States, 2007

State	Percent	State	Percent
Alabama	13.5	Montana	14.0
Alaska	7.0	Nebraska	13.3
Arizona	12.9	Nevada	11.1
Arkansas	14.0	New Hampshire	12.6
California	11.0	New Jersey	13.1
Colorado	10.1	New Mexico	12.7
Connecticut	13.5	New York	13.2
Delaware	13.6	North Carolina	12.2
Florida	17.0	North Dakota	14.6
Georgia	9.9	Ohio	13.5
Hawaii	14.3	Oklahoma	13.3
Idaho	11.7	Oregon	13.1
Illinois	12.1	Pennsylvania	15.2
Indiana	12.5	Rhode Island	13.9
Iowa	14.7	South Carolina	13.0
Kansas	13.0	South Dakota	14.3
Kentucky	13.0	Tennessee	12.9
Louisiana	12.2	Texas	10.0
Maine	14.8	Utah	8.9
Maryland	11.8	Vermont	13.6
Massachusetts	13.3	Virginia	11.8
Michigan	12.7	Washington	11.7
Minnesota	12.2	West Virginia	15.5
Mississippi	12.5	Wisconsin	13.1
Missouri	13.4	Wyoming	12.2

Source: 2008 *Statistical Abstract of the United States.*

Step 1(a): *Separate each observation into a* ***stem,*** *consisting of all but the final (rightmost) digit, and a* ***leaf,*** *the final digit. Stems may have as many digits as needed, but each leaf contains only a single digit.* For the "65 and over" percents in Table 2.4, the whole-number part of the observation is the stem and the final digit (tenths) is the leaf. The Alabama entry, 13.5, has stem 13 and leaf 5.

Step 1(b): *Write the stems in a vertical column with the smallest at the top, and draw a vertical line at the right of this column.* For these data, the stems run from 7 to 17.

Step 2: *Write each leaf in the row to the right of its stem.* For Alabama, we put the leaf 5 on the 13 stem. For Alaska, we add a 0 leaf to the 7 stem. We continue in this way for the other 48 states in alphabetical order.

Step 3: *Sort the leaves in increasing order as they move out from the stem.*

Figure 2.5 shows the steps in making a stemplot for the data in Table 2.4.

Step 1: Stems	Step 2: Add leaves (stem)	Step 2: Add leaves (leaves)	Step 3: Order leaves (stem)	Step 3: Order leaves (leaves)
7	7	0	7	0
8	8	9	8	9
9	9	9	9	9
10	10	10	10	01
11	11	078187	11	017788
12	12	915272567292	12	122225567799
13	13	55600343125319061	13	00011123334555669
14	14	0378063	14	0033678
15	15	25	15	25
16	16		16	
17	17	0	17	0

Step 1: Stems **Step 2: Add leaves** **Step 3: Order leaves**

Figure 2.5 Making a stemplot of the data in Table 2.4. Whole percents form the stems, and tenths of a percent form the leaves.

To describe the distribution in Figure 2.5, it is easiest to begin with deviations from the overall pattern. Which states stand out as unusual? Florida has 17% of its residents over age 65, which is 1.5% higher than for any other state. There's a gap in the stemplot that calls our attention to Florida. On the other end of the stemplot, only 7% of Alaska's residents are 65 or older. That's 1.9% lower than for any other state. So Alaska and Florida are clear outliers. Is Utah, with 8.9% of its population over 65, an outlier? It's not as obvious from the stemplot that Utah is an outlier, but it is. As we'll see in Section 2.2, so is Georgia at 9.9%. It is not surprising that Florida, with its many retired people, has many residents over 65 and that Alaska, the northern frontier, has few. How would you explain Utah and Georgia?

To see the **overall pattern** of a distribution, ignore any outliers. Here is a simple way to organize your thinking.

Overall pattern of a distribution

To describe the overall pattern of a distribution:

- Give the **center** and the **spread.**
- See if the distribution has a simple **shape** that you can describe in a few words.

We will learn how to describe center and spread numerically in Section 2.2. For now, we can describe the center of a distribution by its *midpoint,* the value with half the observations taking smaller values and half taking larger values. We can describe the spread of a distribution by giving the *smallest and largest values,* ignoring any outliers.

Example 2.6 Where do older folks live? *(continued)*

Look again at the stemplot in Figure 2.5. **Shape:** The distribution has a *single peak.* It is roughly *symmetric*—that is, the pattern is similar on both sides of the peak. **Center:** The midpoint of the distribution is close to the single peak, at about 13%. **Spread:** The spread is about 10% to 15.5% if we ignore the outliers.

Exercises

2.7 Gooooaaal! The number of goals scored by each team in the first round of the California high school soccer playoffs is shown below.

5	0	1	0	7	2	1	0	4	0	3	0	2	0
3	1	5	0	3	0	1	0	1	0	2	0	3	1

(a) Make a dotplot of the data.

(b) Describe the distribution of the variable "number of goals scored." Be sure to discuss the overall pattern (shape, center, and spread) and any deviations from that pattern.

2.8 Poverty in the states, I Table 2.5 shows the percent of people living below the poverty line in the 26 states east of the Mississippi.

(a) Make a stemplot of these data.

(b) Describe the distribution of the variable "percent living in poverty." Be sure to discuss the overall pattern (shape, center, and spread) and any deviations from that pattern.

Table 2.5 Percent of State Residents Living in Poverty, 2007

State	Percent	State	Percent	State	Percent
Alabama	16.9	Maryland	8.3	Pennsylvania	11.6
Connecticut	7.9	Massachusetts	9.9	Rhode Island	12.0
Delaware	10.5	Michigan	14.0	South Carolina	15.0
Florida	12.1	Mississippi	20.6	Tennessee	15.9
Georgia	14.3	New Hampshire	7.1	Vermont	10.1
Illinois	11.9	New Jersey	8.6	Virginia	9.9
Indiana	12.3	New York	13.7	West Virginia	16.9
Kentucky	17.3	North Carolina	14.3	Wisconsin	10.8
Maine	12.0	Ohio	13.1		

Source: U.S. Census Bureau, Current Population Survey, Annual Social and Economic Supplements.

2.9 Driving in town In Example 2.4, we examined data on highway gas mileages of model year 2009 midsize cars. The Minitab dotplot below shows the EPA estimates of city gas mileages for these same 24 car models. Describe the overall pattern of the distribution and any deviations from that pattern.

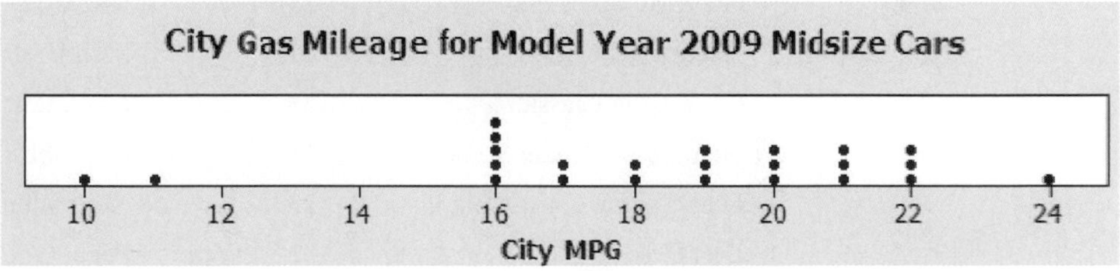

2.10 Fuel efficiency Refer to the previous exercise. The Minitab dotplot below shows the difference (Highway—City) in EPA mileage ratings for each of the 24 car models from Example 2.4. What does the graph tell us about fuel economy in the city versus on the highway for these car models? Be specific.

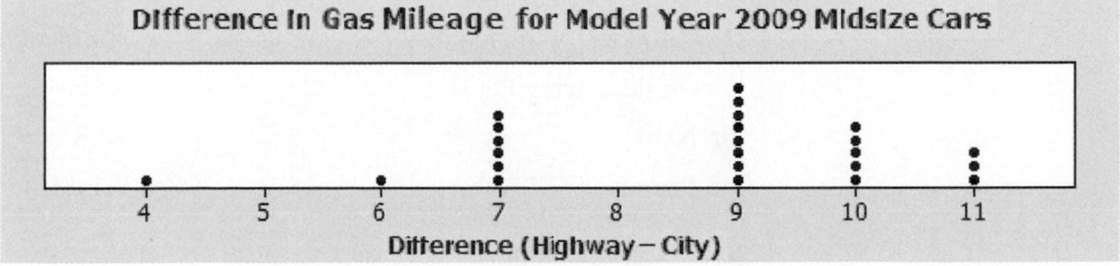

11	44
11	66778
12	0134
12	666778888
13	0000001111444
13	7788999
14	0044
14	567
15	11
15	
16	0

Figure 2.6 Stemplot of the percent of each state's residents who are 25 to 34 years old.

2.11 Where do the young live? Figure 2.6 is a stemplot of the percent of residents aged 25 to 34 in each of the 50 states. As in Figure 2.5 (page 45) for older residents, the stems are whole percents and the leaves are tenths of a percent. This time, each stem has been split in two, with values having leaves 0 through 4 placed on one stem and values ending in 5 through 9 placed on another stem.

(a) Utah has the highest percent of residents aged 25 to 34. What is that percent? Why does Utah have an unusually high percent of residents in this age group?

(b) Describe the shape, center, and spread of the distribution, ignoring Utah.

(c) Is the distribution for young adults more or less spread out than the distribution in Figure 2.5 for older adults? Justify your answer.

2.12 Watch that caffeine! The U.S. Food and Drug Administration limits the amount of caffeine in a 12-ounce can of carbonated beverage to 72 milligrams. That translates to a maximum of 48 milligrams of caffeine per 8-ounce serving. Data on the caffeine content of popular soft drinks is provided in Table 2.6. How does the caffeine content of these drinks compare to the USFDA's limit?

Table 2.6 Caffeine Content (in Milligrams) for an 8-Ounce Serving of Popular Soft Drinks

Brand	Caffeine (mg per 8-oz serving)	Brand	Caffeine (mg per 8-oz serving)
A&W Cream Soda	20	IBC Cherry Cola	16
Barq's Root Beer	15	Kick	38
Cherry Coca-Cola	23	KMX	36
Cherry RC Cola	29	Mello Yello	35
Coca-Cola Classic	23	Mountain Dew	37
Diet A&W Cream Soda	15	Mr. Pibb	27
Diet Cherry Coca-Cola	23	Nehi Wild Red Soda	33
Diet Coke	31	Pepsi One	37
Diet Dr. Pepper	28	Pepsi-Cola	25
Diet Mello Yello	35	RC Edge	47
Diet Mountain Dew	37	Red Flash	27
Diet Mr. Pibb	27	Royal Crown Cola	29
Diet Pepsi-Cola	24	Ruby Red Squirt	26
Diet Ruby Red Squirt	26	Sun Drop Cherry	43
Diet Sun Drop	47	Sun Drop Regular	43
Diet Sunkist Orange Soda	28	Sunkist Orange Soda	28
Diet Wild Cherry Pepsi	24	Surge	35
Dr. Nehi	28	TAB	31
Dr. Pepper	28	Wild Cherry Pepsi	25

Source: National Soft Drink Association.

(a) You could construct a dotplot of the caffeine content data, but a stemplot might be preferable. Explain why.

(b) Construct a stemplot of the data using the first digit as the stem and the second digit as the leaf. What problem do you see with this display?

(c) Figure 2.7 shows a stemplot with "split stems" for the caffeine content data. This time, values having leaves 0 through 4 are placed on one stem, while values ending in 5 through 9 are placed on another stem. Describe the shape, center, and spread of the distribution.

1	5 5 6
2	0 3 3 3 4 4
2	5 5 6 6 7 7 7 8 8 8 8 8 9 9
3	1 1 3
3	5 5 5 6 7 7 7 8
4	3 3
4	7 7

Key: 2|8 means the soft drink contains 28 mg of caffeine per 8-ounce serving.

Figure 2.7 Stemplot showing the caffeine content (in milligrams per 8-ounce serving) of various soft drinks.

Displaying quantitative variables: histograms

ACTIVITY 2.1C

The nutritional value of cereals

APPLET

What makes a "healthy" breakfast cereal? One without much sugar? A cereal with lots of vitamins? You probably didn't think about how much salt the cereal contains. In this Activity, you will examine data on the sodium content (in milligrams) for 77 brands of breakfast cereal. In the process, you will meet another kind of graph for displaying the distribution of a quantitative variable—a *histogram*.

Your teacher will give you a data set called CEREALS that provides nutritional information for the 77 brands of breakfast cereal.[5] For each brand of cereal, we show the number of calories per serving, milligrams of sodium, grams of sugar, and grams of fat.

1. Launch the *One-Variable Statistical Calculator* applet at the book's Web site, **www.whfreeman.com/sta2e.** In the Data Sets tab, choose Cereals. Click on the Data tab to look at the values of sodium content.

2. Click on the Stemplot tab to view a stemplot of the sodium data. Try the "Split stems" option. Which graph displays the data more effectively?

3. Now click on the Histogram tab. Let's explore this new type of graph. The variable of interest, sodium content, is on the horizontal axis. The applet divides the values on the horizontal axis into "classes" of equal width. The vertical axis uses a frequency scale that indicates how many values fall in each class interval.

4. You can adjust the width of the classes by clicking and dragging with your mouse. Try this for yourself. As you do this, do the features of the histogram change your impression of what the data say? Explain.

5. Once you have found an interval width that shows the distribution well, describe the shape, center, and spread of the distribution. Are there any outliers?

6. Write a few sentences describing what you have learned about the sodium content of these breakfast cereals.

We can use a dotplot or a stemplot to display the distribution of quantitative variables that have relatively few values. What about variables such as the SAT scores of students admitted to a college or the income of families? These variables, like the sodium content of cereals in Activity 2.1C, take so many values that a graph of the distribution is clearer if nearby values are grouped together. The most common graph of the distribution of a quantitative variable is a **histogram.**

Example 2.7 College costs

Tonya is a student at Grand Valley State University in her home state of Michigan. Grand Valley State charges in-state students \$8196 in tuition and fees. Tonya wonders how Grand Valley State's tuition compares with the amount charged by other Michigan colleges and universities.

There are 90 colleges and universities in Michigan. Their tuition and fees for the 2008–2009 school year run from \$1873 at Oakland Community College to \$30,823 at Kalamazoo College.[6] It isn't easy to compare Grand Valley State with a list of 89 other colleges. So let's make a graph. The *histogram* in Figure 2.8 shows that many colleges charge less than \$3000, but the distribution also extends out to the right. At the upper extreme, 8 colleges charge more than \$24,000. Grand Valley State is one of 19 colleges in the \$6000 to \$9000 group.

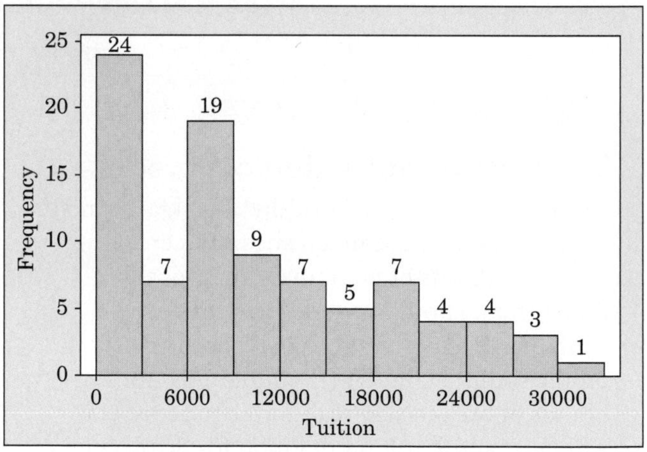

Figure 2.8 Minitab histogram of the tuition and fees charged by 90 Michigan colleges and universities in the 2008–2009 academic year.

The distribution of tuition and fees at Michigan colleges has a distinct **shape.** There is a strong *peak* at the lowest cost class. And even though most colleges charge less than \$9000, there is a long right "tail" extending up to \$33,000. We call a distribution with a long tail on one side *skewed.* The **center** is roughly \$8500 (half the colleges charge less than this). The **spread** is large, from less than \$3000 to more than \$30,000. There are no outliers—the colleges with the highest tuition just continue the long right tail that is part of the overall pattern.

□

When you describe a distribution, concentrate on the main features. Look for major peaks, not for minor ups and downs in the bars of the histogram like those in Figure 2.8. Look for clear outliers, not just for the smallest and largest observations. Look for rough **symmetry** or clear **skewness.**

Symmetric and skewed distributions

A distribution is **symmetric** if the right and left sides of the graph are approximately mirror images of each other.

A distribution is **skewed to the right** if the right side of the graph (containing the half of the observations with larger values) extends much farther out than the left side. It is **skewed to the left** if the left side of the graph extends much farther out than the right side.

In mathematics, symmetry means that the two sides of a figure like a histogram are exact mirror images of each other. Data are almost never exactly symmetric, so we were willing to call the stemplot in Figure 2.5 (page 45) roughly symmetric as an overall description. The tuition distribution in Figure 2.8, on the other hand, is clearly skewed to the right.

Example 2.8 How to make a histogram

Table 2.4 (page 44) presents the percent of residents aged 65 years and over in each of the 50 states. We made a stemplot of the data in Example 2.5. To make a histogram of this distribution, proceed as follows.

Step 1: *Divide the range of the data into classes of equal width.* The data in Table 2.4 range from 7.0 to 17.0, so we choose as our classes

$$7.0 \le \text{percent over 65} < 8.0$$
$$8.0 \le \text{percent over 65} < 9.0$$
$$\vdots$$
$$17.0 \le \text{percent over 65} < 18.0$$

Be sure to specify the classes precisely so that each individual falls into exactly one class. A state with 7.9% of its residents aged 65 or older would fall into the first class, but 8.0% falls into the second.

Step 2: *Count the number of individuals in each class.* Here are the counts:

Class	Count	Class	Count	Class	Count
7.0 to 7.9	1	11.0 to 11.9	6	15.0 to 15.9	2
8.0 to 8.9	1	12.0 to 12.9	12	16.0 to 16.9	0
9.0 to 9.9	1	13.0 to 13.9	17	17.0 to 17.9	1
10.0 to 10.9	2	14.0 to 14.9	7		

Step 3: *Draw the histogram.* Put the variable whose distribution you are displaying on the horizontal axis. That's "percent of residents aged 65 and over" in this example. The scale runs from 6 to 19 because that range spans the classes we chose. The vertical axis contains the scale of counts. Each bar represents a class. The base of the bar covers the class, and the bar height is the class count. *There is no horizontal space between the bars unless a class is empty,* so that its bar has height zero. Figure 2.9 (on the next page) is our histogram.

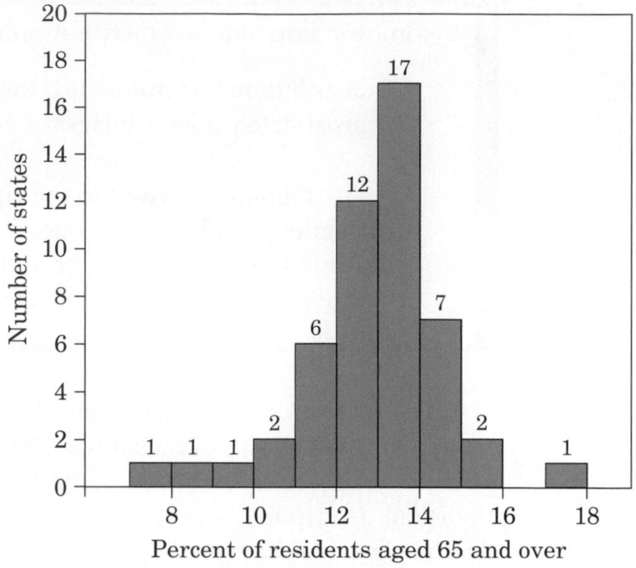

Figure 2.9 Histogram of the percent of residents aged 65 and older in the 50 states.

Just as with bar graphs, our eyes respond to the area of the bars in a histogram. Be sure that the classes for a histogram have equal widths. There is no one right choice of the classes. Too few classes will give a "skyscraper" histogram, with all values in a few classes with tall bars. Too many classes will produce a "pancake" graph, with most classes having one or no observations. Neither choice will give a good picture of the shape of the distribution. You must use your judgment in choosing classes to display the shape. Technology will usually choose classes for you. The computer's or calculator's choice is usually a good starting point, but you can change the settings if you want, as you'll see in the following Calculator Corner.

CALCULATOR CORNER Entering data and making histograms

The TI-84 and TI-Nspire will make histograms for you. Here's how to do it.

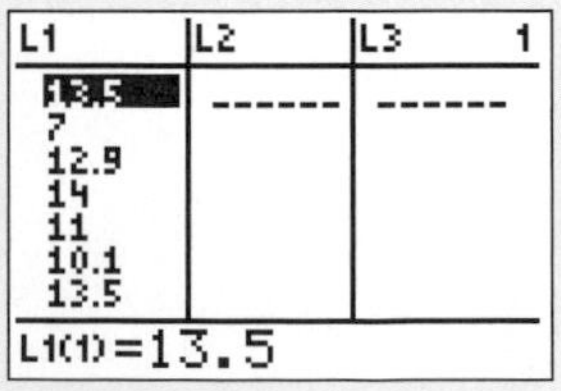

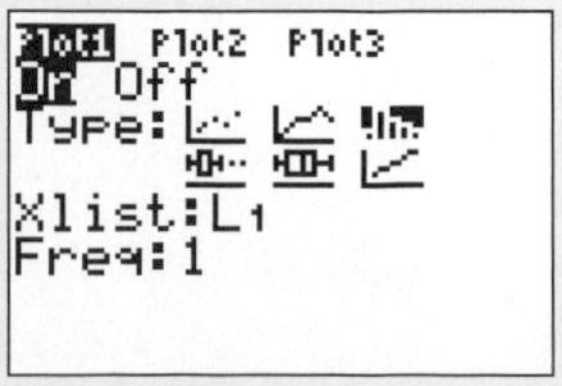

TI-84

1. Enter the data from Table 2.4 (page 44).
- Press [STAT] and choose `1:Edit...`
- Type the values into list L1.

2. Set up a histogram in the statistics plots menu.
- Press [2nd][Y=] (STAT PLOT)
- Press [ENTER] or [1] to go into Plot 1.
- Adjust your settings as shown.

3. Get a "quick graph" using ZoomStat.
- Press [ZOOM] and choose `9:ZoomStat`.
- Press [TRACE] to look at the class intervals chosen by the calculator.

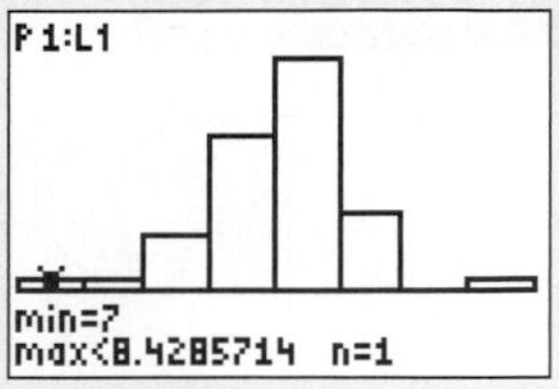

4. Set the window to match the class intervals chosen in Example 2.8.
- Press [WINDOW].
- Enter the values shown.

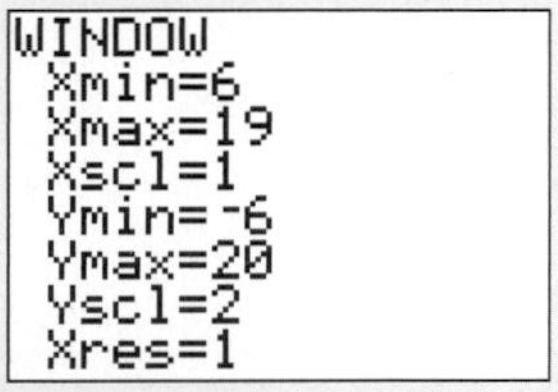

5. Graph the histogram.
- Press [GRAPH]. Then press [TRACE].
- Compare with Figure 2.9.

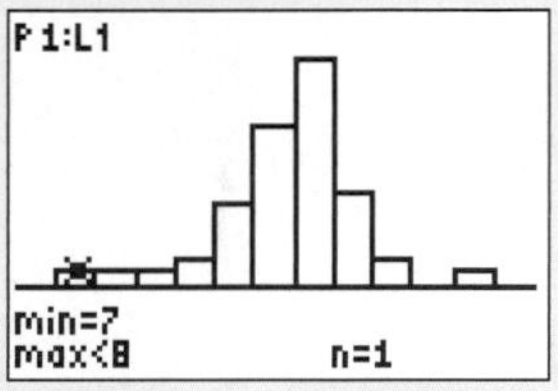

TI-Nspire

1. Enter the data from Table 2.4 (Page 44). First, open a new document with a Lists & Spreadsheets page.
- Press (home) and choose `6:New Document`.
- Press (menu); choose `3:Add Lists & Spreadsheet`.
- In the name box at the top of column A, type "pctolder."

Second, type the values into list pctolder.

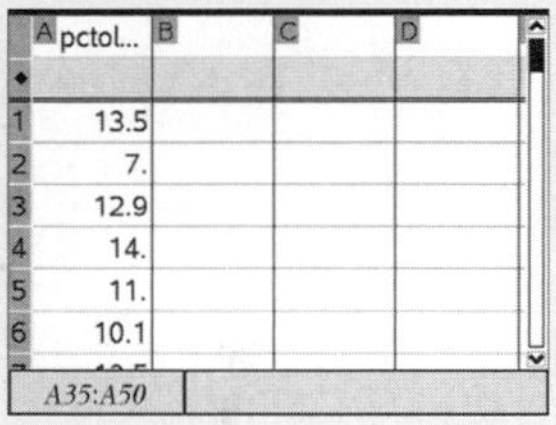

2. Make a "quick graph" of the data.
- Press (menu); choose `3:Data` and `5:Quick Graph`. You should see a dotplot of the data.
- Press (menu); choose `1:Plot Type` and `3:Histogram` to change the graph to a histogram.

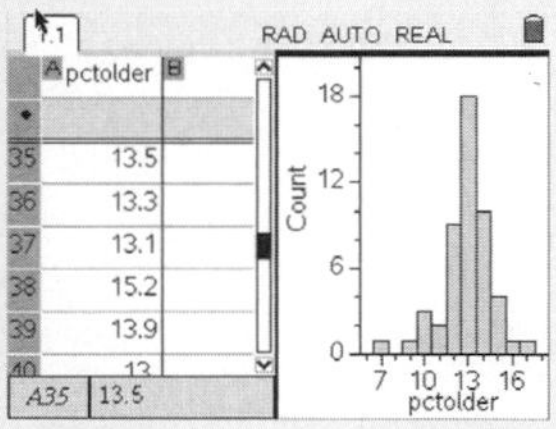

3. Change the horizontal scale to match the class intervals chosen in Example 2.8.
- Press (menu); choose `5:Window/Zoom` and `1:Window Settings`.
- Change Xmin to 6, Xmax to 19, and Ymax to 20, and click OK.
- Press (menu); choose `2:Plot Properties` and `2:Bin Settings`.
- Change Width to 1, Alignment to 6, and click OK.

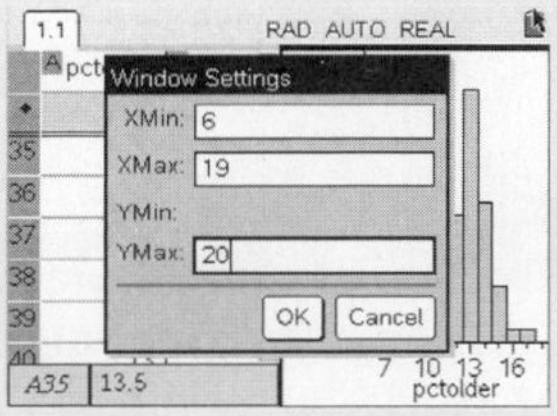

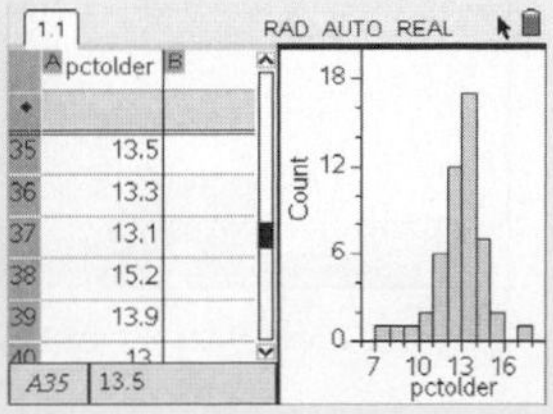

Exercises

2.13 Minority students in engineering Figure 2.10 is a histogram of the number of minority students (black, Hispanic, Native American) who earned doctorate degrees in engineering from each of 115 universities over a 5-year period.[7] Briefly describe the shape, center, and spread of this distribution.

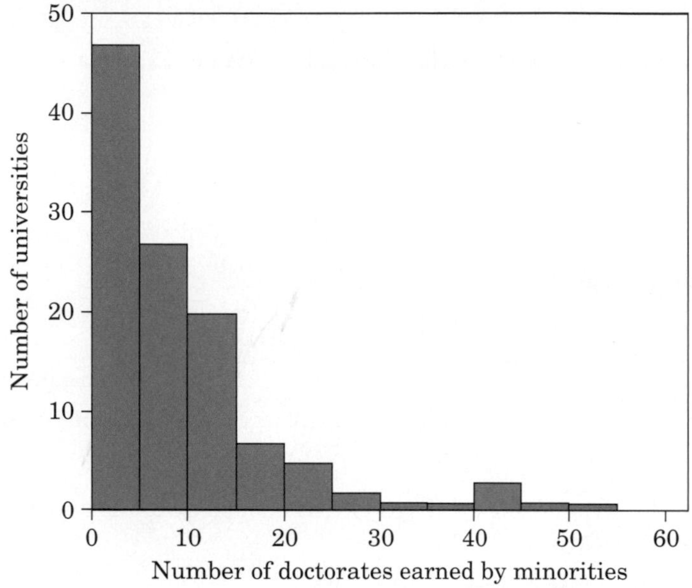

Figure 2.10 The distribution of number of engineering doctorates earned by minority students at 115 universities over a 5-year period.

2.14 Lightning flashes Figure 2.11 comes from a study of lightning storms in Colorado.[8] It shows the distribution of the hour of the day during which the first lightning flash for that day occurred. Describe the shape, center, and spread of this distribution. Are there any outliers?

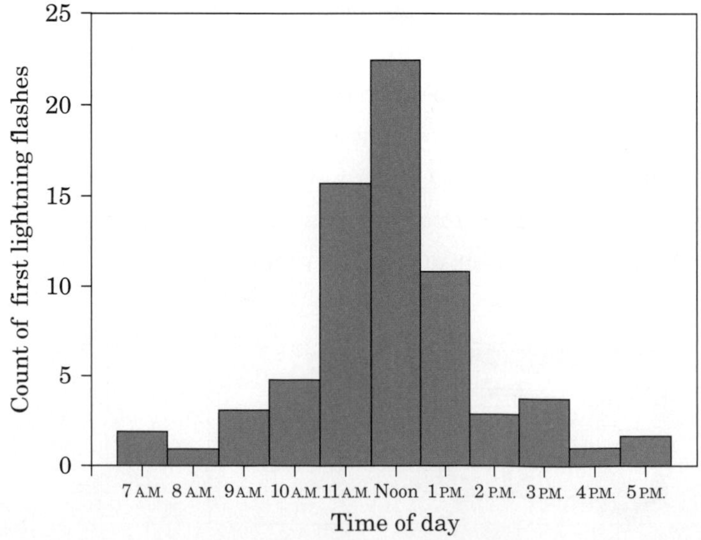

Figure 2.11 Histogram of time of day at which the day's first lightning flash occurred.

2.15 Poverty in the states, II Refer to Exercise 2.8 (page 46). Joe made the histogram shown below from the poverty data in Table 2.5. Write a sentence or two clearly explaining to Joe why his histogram is not legitimate.

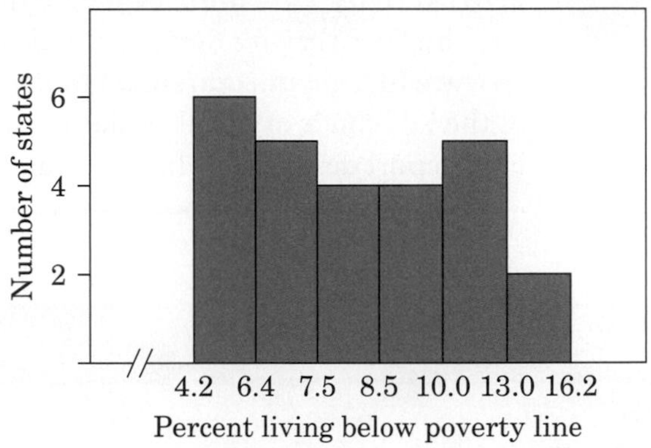

2.16 Poverty in the states, III Refer to Exercise 2.8 (page 46). Make a histogram for the data in Table 2.5. Use seven classes of width 2, beginning with 7–9. If you made a stemplot of these data in Exercise 2.8, how does your histogram compare?

www

2.17 Yankee money Table 2.7 gives the salaries of the players on the New York Yankees baseball team for the 2008 season.

(a) Make a histogram of these data.

Table 2.7 Salaries of the New York Yankees, 2008

Alex Rodriguez	$28,000,000	Morgan Ensberg	$1,750,000
Jason Giambi	$23,428,571	Andrew Brackman	$1,184,788
Derek Jeter	$21,600,000	Wilson Betemit	$1,165,000
Bobby Abreu	$16,000,000	Brian Bruney	$725,000
Andy Petite	$16,000,000	Billy Traber	$500,000
Mariano Rivera	$15,000,000	Melky Cabrera	$461,200
Jorge Posada	$13,100,000	Phil Hughes	$406,350
Johnny Damon	$13,000,000	Shelley Duncan	$398,300
Hideki Matsui	$13,000,000	Sean Henn	$397,448
Mike Mussina	$11,071,029	Ian Kennedy	$394,275
Carl Pavano	$11,000,000	Jeffrey Karstens	$393,300
Kyle Farnsworth	$5,916,666	John Albaladejo	$393,225
Chien-Ming Wang	$4,000,000	Ross Ohlendorf	$391,425
LaTroy Hawkins	$3,750,000	Joba Chamberlain	$390,000
Robinson Cano	$3,000,000	Humberto Sanchez	$390,000
Jose Molina	$1,875,000		

Source: USA Today Salary Database at **http://content.usatoday.com/sports/baseball/salaries/default.aspx**.

(b) Is the distribution of Yankees' salaries roughly symmetrical, skewed to the left, or skewed to the right? Explain briefly.

(c) What is the spread of the distribution?

APPLET

2.18 Activity 2.1C follow-up Use the CEREALS data from Activity 2.1C and the *One Variable Statistical Calculator* applet on the book's Web site (**www.whfreeman.com/sta2e**) to investigate the calories, sugar, or fat content of the 77 brands of cereal. Make a histogram of the variable you chose. Write a brief report describing what you learned.

APPLICATION 2.1

Mercury in tuna

Do you like to eat tuna? Many people do. Unfortunately, some of the tuna that people eat may contain high levels of mercury. (Yes, mercury, like the stuff found in glass thermometers—yuck!) Exposure to mercury can be especially hazardous for pregnant women and small children. How much mercury is safe to consume? That depends on your body weight. According to the Environmental Protection Agency (EPA), a 180-pound man would be safe eating one 6-ounce can of tuna per week if the mercury content was less than 0.31 parts per million (ppm). For a 45-pound child to be safe, the same can of tuna would have to contain less than 0.08 ppm of mercury. The Food and Drug Administration will take action (such as removing the product from store shelves) if the mercury concentration in a 6-ounce can of tuna is 1.00 ppm or higher.

What is the typical mercury concentration in cans of tuna sold in stores? A study conducted by Defenders of Wildlife set out to answer this question. Defenders collected a sample of 164 cans of tuna from stores across the United States. They sent the selected cans to a laboratory that is often used by the EPA for mercury testing. The table below shows data on the brand, country of origin, type of tuna (light or albacore), and mercury content (ppm) for 8 of the 164 cans.[9] You can get the entire set of data from your teacher or from the book's Web site, **www.whfreeman.com/sta2e.**

Ron Sumners/FeaturePics

Can	Brand	Country	Type	Mercury (ppm)
11	Dave's Ahi Tuna	United States	Light	0.790
12	Van Triunfo	Ecuador	Light	0.760
13	Whole Foods	Thailand	Albacore	0.730
14	Tuna Real	Ecuador	Light	0.720
15	Starkist	United States	Albacore	0.710
16	Tuna Real	Ecuador	Light	0.700
17	Tuna Real	Ecuador	Light	0.660
18	Tuny	Mexico	Light	0.640

QUESTIONS

1. What is the population of interest? What is the sample?

2. List the variables that were measured. Identify each as categorical or quantitative.

3. The Fathom bar graph below displays information about the country of origin for the whole sample. About what percent of the cans in the sample originated in the United States? Is this value a good estimate for the percent of all cans of tuna that originate in the United States? Why or why not?

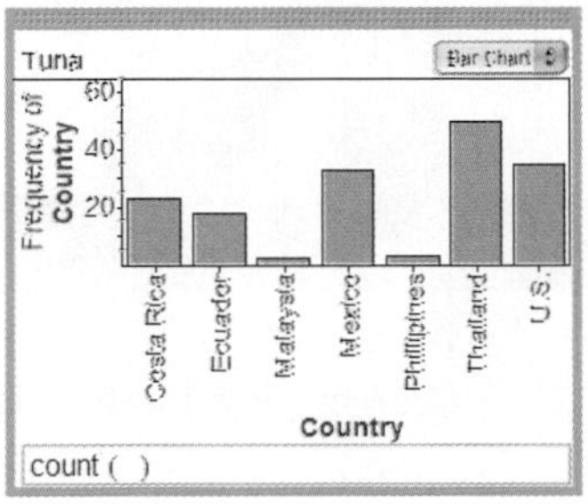

4. The Fathom histogram below displays the mercury concentration in the sampled cans. Write a few sentences describing this distribution.

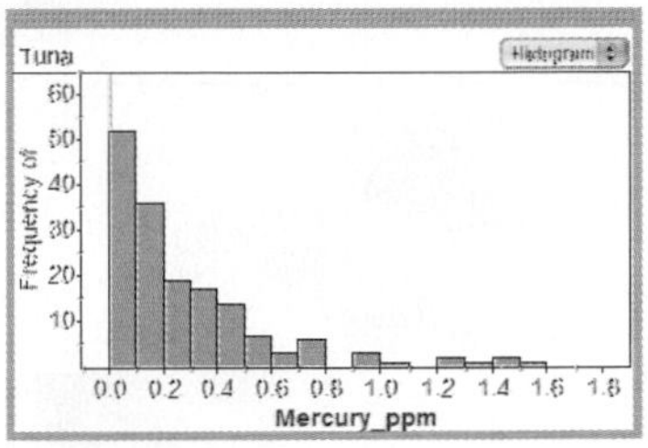

We'll return to the tuna data a little later, so keep your work handy.

Section 2.1 Summary

To see what data say, start with graphs. The choice of graph depends on the type of data. The **distribution** of a variable tells us what values the variable takes and how often it takes each value. To display the distribution of a categorical variable, use a **pie chart** or a **bar graph.** Pie charts always show the parts of some whole, but bar graphs can compare any set of numbers measured in the same units. To display the distribution of a quantitative variable, use a **dotplot, stemplot,** or **histogram.** We usually favor dotplots or stemplots when we have a small number of observations. For large amounts of data, histograms are usually better.

When you look at a graph, look for an **overall pattern** and for **deviations** from that pattern, such as **outliers.** We can describe the overall pattern of a dotplot, stemplot, or histogram by giving its **shape, center,** and **spread.** Some distributions have simple shapes such as **symmetric** or **skewed,** but others are too irregular to describe by a simple shape.

Section 2.1 Exercises

2.19 Feeling sleepy? Students in a college statistics class responded to a survey designed by their teacher. One of the survey questions was "How much sleep did you get last night?" Here are the data (in hours):

9	6	8	6	8	8	6	6.5	6	7	9	4	3	4
5	6	11	6	3	6	6	10	7	8	4.5	9	7	7

(a) Make a dotplot to display the data.
(b) Describe the overall pattern of the distribution and any deviations from that pattern.

2.20 Bad habits According to the National Household Survey on Drug Abuse, 31.8% of adolescents aged 12 to 17 years used alcohol in 2007, 12.5% used marijuana, 1.5% used cocaine, and 15.7% used cigarettes. Explain why it is *not* correct to display these data in a pie chart.

2.21 Low birth weights Figure 2.12 (on the next page) shows the distribution of percent of low birth weights for the 26 states east of the Mississippi.[10] Notice that the vertical scale in Figure 2.12(a) is the *number* of states in each class. In Figure 2.12(b), we have calculated and displayed the *percent* of states in each low-birth-weight class.

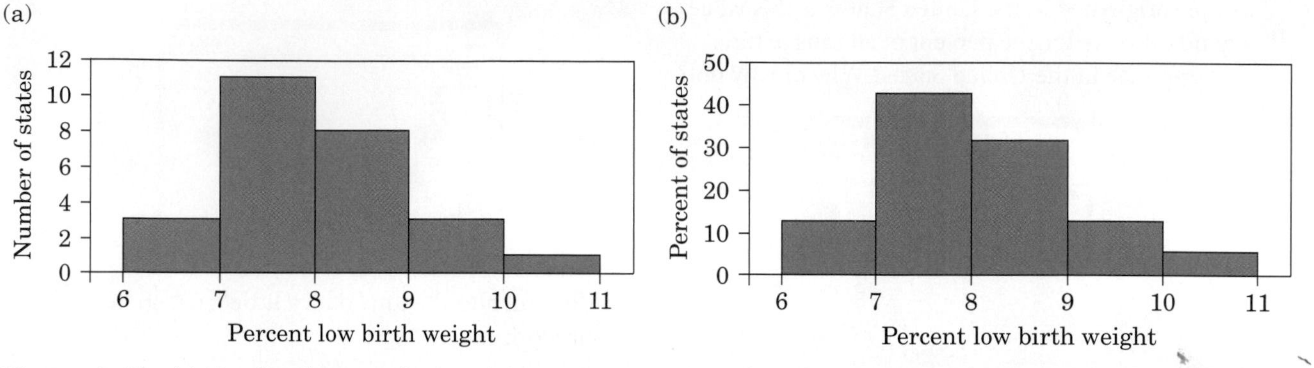

Figure 2.12 (a) The distribution of infants born with low birth weight in the 26 states east of the Mississippi. (b) The heights of the bars are the same, but the vertical scale has been changed to show the percent of states in each class.

(a) Describe the shape, center, and spread of the distribution.
(b) When might it be preferable to use percents rather than counts on the vertical axis of a histogram?
(c) Mississippi had the highest percent of low birth weight, 10.1. Why do you think this state has such a high percent of low-birth-weight infants?

2.22 Skewed left
(a) Sketch a histogram for a distribution that is skewed to the left.
(b) Suppose that you and your friends emptied your pockets of coins and recorded the year marked on each coin. The distribution of dates would be skewed to the left. Explain why.

Exercises 2.23 to 2.26 refer to the following setting. CensusAtSchool is an international project that collects data about primary and secondary school students using surveys. As of early 2009, students in Australia, Canada, New Zealand, South Africa, and the United Kingdom have participated in the project. Data from CensusAtSchool surveys are available from the project's Web site, **www.censusatschool.com.** We used the "Random Data Selector" to choose a simple random sample of 50 Canadian students who completed the survey in 2007–2008. The data set CASCAN0708 can be found on the book's Web site, **www.whfreeman.com/sta2e.**

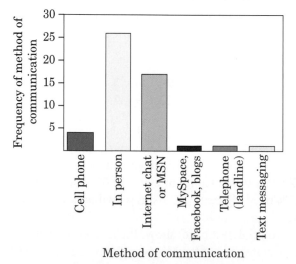

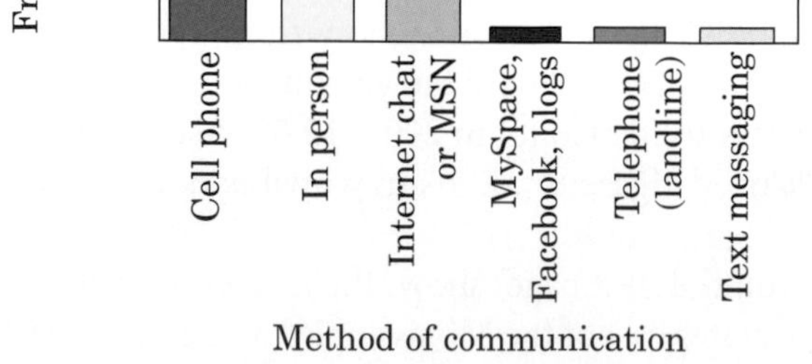

www **2.23 Let's chat** The bar graph on the left displays data on students' responses to the question "Which of these methods do you most often use to communicate with your friends?"
(a) Would it be appropriate to make a pie chart for these data? If so, do it. If not, explain why not.
(b) Summarize what the bar graph tells you about students' communication preferences.

www **2.24 Travel time** The dotplot on the facing page displays data on students' responses to the question "How long does it usually take you to travel to school?"

(a) Make a well-labeled histogram of the data.
(b) Describe the shape, center, and spread of the distribution. Are there any outliers?

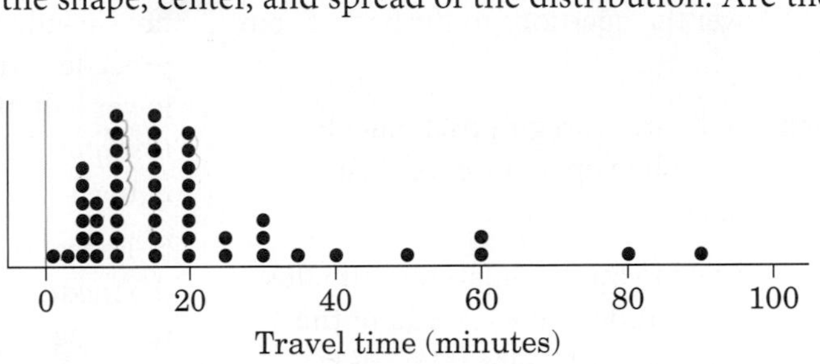

2.25 Who's left-handed? Students were asked, "Are you right-handed, left-handed or ambidextrous?" The responses of the 50 randomly selected Canadian students are shown below (R = right-handed; L = left-handed; A = ambidextrous).

R	R	R	R	R	R	R	R	R	R	R	L	R	R
R	R	R	R	R	R	R	R	R	R	R	R	R	A
R	R	R	R	A	R	R	L	R	R	R	R	L	A
R	R	R	R	R	R	R	R						

(a) Make an appropriate graph to display these data.
(b) Over 10,000 Canadian high school students took the CensusAtSchool survey in 2007–2008. What percent of this population would you estimate is left-handed? Justify your answer.

2.26 How tall are you? Here are the heights (in centimeters) of the 50 randomly selected Canadian students who participated in CensusAtSchool in 2007–2008.

166.5	170	178	163	150.5	169	173	169	171	166	190	183	178	161
171	170	191	168.5	178.5	173	175	160.5	166	164	163	174	160	174
182	167	166	170	170	181	171.5	160	178	157	165	187	168	157.5
145.5	156	182	168.5	177	162.5	160.5	185.5						

Make a stemplot of these data. Describe the shape, center, and spread of the distribution. Are there any outliers?

2.2 Describing Distributions with Numbers

ACTIVITY 2.2A

Do you know your geography?

How well do you know the U.S. state capitals? The locations of major rivers and mountain chains? The populations of states and cities? In this Activity, you'll have a chance to demonstrate your geographic knowledge by estimating the population of a capital city.[11]

1. Your teacher will give each student a slip of paper with two questions about the population of a U.S.

ACTIVITY 2.2A *(continued)*

ACTIVITY 2.2A *(continued)*

capital city. Answer the questions to the best of your ability.

2. Your teacher will form two groups. Students in each group should line up in order of their estimates, from smallest to largest.

3. Find the *median* of your population estimates as follows. Have the student at each end of the line begin counting off with "One." The next student in line at each end should call out "Two." Continue until you locate the middle person in line (if the last person is the only one calling out a number) or the middle two people in line. In the latter case, average the estimates of the two middle students.

4. The median from Step 3 divides the line into two halves. (Note that the median itself is not included in either of these halves.) Use the procedure from Step 3 to find the median of the lower half of the estimates. This value is called the *first quartile.* (Why do you think we give it that name?)

5. Repeat the process from Step 3 again to find the median of the upper half of the estimates. The resulting value is called the *third quartile.*

6. Record the following five values for the estimates in your line: the minimum, first quartile, median, third quartile, and maximum. This is known as the *five-number summary.*

7. Your teacher will provide the capital city's actual population. How close was your estimate?

8. Which group did better at estimating the city's population? How can you tell?

Measuring center: the median

A natural way to describe the center of a distribution is to use the "middle value" in a histogram or stemplot. That is, find the number such that half the observations are smaller and the other half are larger. This is the **median** of the distribution. We will call the median M for short. Although the idea of the median as the midpoint of a distribution is simple, we need a precise rule for putting the idea into practice.

Example 2.9 Finding the median

To find the median of the numbers

8 4 9 1 3

arrange them in increasing order:

1 3 **4** 8 9

The boldface 4 is the center observation, because there are 2 observations to its left and 2 to its right. When the number of observations n is odd, there is always one observation in the center of the ordered list. This is the median, $M = 4$.

If n is even, there is no one middle observation. But there is a middle pair, and we take the median to be the mean of this middle pair, the point halfway between them. So the median of

8 4 1 9 1 5

is found by arranging these numbers in increasing order,

1 1 **4** **5** 8 9

and averaging the middle pair,

$$M = \frac{4 + 5}{2} = 4.5$$

Here is our rule for finding medians.

The median M

The **median** M is the midpoint of a distribution, the number such that half the observations are smaller and the other half are larger. To find the median of a distribution:

1. Arrange all the observations in order of size, from smallest to largest.
2. If the number of observations n is odd, the median M is the center observation in the ordered list.
3. If the number of observations n is even, the median M is the average of the two center observations in the ordered list.

Example 2.10 How many text messages?

In a September 28, 2008, article titled "Letting Our Fingers Do the Talking," the *New York Times* reported that Americans now send more text messages than they make phone calls. According to a study by Nielsen Mobile, "Teenagers ages 13 to 17 are by far the most prolific texters, sending or receiving 1,742 messages a month." Mr. Brown, a high school statistics teacher, was skeptical of the texting result stated in the article. So he collected data from his first-period statistics class on the number of text messages they had sent and received in the past 24 hours. Here are the data:

0 7 1 29 25 8 5 1 25 98 9 0 26 8 118 72 0 92 52 14 3 3 44 5 42

Let's find the median number of text messages sent and received in the past 24 hours. We begin by sorting the values from lowest to highest:

0 0 0 1 1 3 3 5 5 7 8 8 **9** 14 25 25 26 29 42 44 52 72 92 98 118

Since there are 25 values, the median is the bold 9 in the center of the ordered list.

Measuring spread with quartiles

In Example 2.10, the median number of text messages in the past 24 hours for Mr. Brown's students was 9. That information is helpful but incomplete. Did most students have close to 9 text messages in the previous 24 hours, or were the data values very spread out?

The simplest useful description of a distribution consists of both a measure of *center* and a measure of *spread*. If we choose the median (the midpoint) to describe center, the **quartiles** give us a natural way to measure spread. Again, the idea is clear: find the points one-quarter and three-quarters up the ordered list of observations. Here's an example.

Example 2.11 How many text messages? *(continued)*

The first half of the ordered list of number of text messages sent and received in the past 24 hours is

0 0 0 1 1 **3** **3** 5 5 7 8 8

There are 12 data values *less than* the median, and we have put the two middle numbers in boldface type. The first quartile, Q_1, is the average of the middle two numbers, which is 3. The 12 data values *greater than* the median are

14 25 25 26 29 **42** **44** 52 72 92 98 118

The third quartile, Q_3, is the average of the two middle numbers, also in bold.

$$Q_3 = \frac{42 + 44}{2} = 43$$

We can measure the spread of the values around the median by calculating the distance between Q_1 and Q_3:

$$Q_3 - Q_1 = 43 - 3 = 40$$

This is known as the *interquartile range* (*IQR* for short).

Note: If there is an odd number of observations less than or greater than the median, then the middle number of that half of the data is Q_1 or Q_3. Remember that in finding the quartiles, you never include the median, whether it is a data value or not.

The quartiles get their name because, with the median, they divide the observations into quarters—one-quarter of the observations lie below the first quartile, half the observations lie below the median, and three-quarters lie below the third quartile. The **interquartile range** (*IQR*) measures the range of the middle 50% of the data. That's the idea. Again, we need a rule to make the idea precise. The rule for calculating the quartiles and the *IQR* uses the rule for the median.

The quartiles Q_1 and Q_3; the interquartile range (*IQR*)

To calculate the **quartiles:**

1. Arrange the observations in increasing order and locate the median M in the ordered list of observations.
2. The **first quartile Q_1** is the median of the observations whose position in the ordered list is to the left of the location of the overall median.
3. The **third quartile Q_3** is the median of the observations whose position in the ordered list is to the right of the location of the overall median.

The **interquartile range** (*IQR*) is defined as

$$IQR = Q_3 - Q_1$$

The five-number summary and boxplots

The smallest and largest observations tell us little about the distribution as a whole, but they give information about the tails of the distribution that is missing if we know only the median and the quartiles. To get a quick summary of both center and spread, combine all five numbers.

The five-number summary

The **five-number summary** of a distribution consists of the smallest observation, the first quartile, the median, the third quartile, and the largest observation, written in order from smallest to largest. In symbols, the five-number summary is

Minimum $\quad Q_1 \quad M \quad Q_3 \quad$ Maximum

These five numbers offer a reasonably complete description of center and spread. We can use them to make a new type of graphical display, a **boxplot.**

Example 2.12 How many text messages? A boxplot

The five-number summary for the text message data of Examples 2.10 and 2.11 is

$$0 \quad 3 \quad 9 \quad 43 \quad 118$$

Figure 2.13 (on the next page) shows a *boxplot* of the distribution.

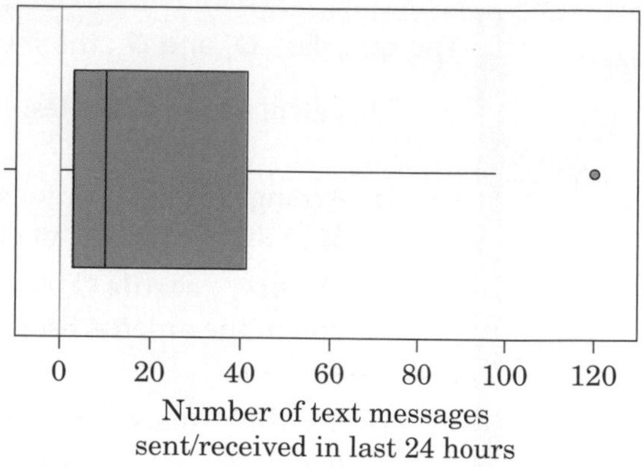

Figure 2.13 A boxplot of the text message data for Mr. Brown's students from Examples 2.10 and 2.11.

Here are the main properties of boxplots.

Boxplot

A boxplot is a graph of the five-number summary:

- A central box is drawn from the first quartile (Q_1) to the third quartile (Q_3).
- A line in the box marks the median.
- Lines extend from the box out to the smallest and largest observations that are not outliers.

We will adopt the convention of identifying outliers as isolated points in the boxplot. In Example 2.12, the student who sent and received 118 text messages in the past 24 hours seems like a clear outlier. What about the student with 98 text messages? This observation is not considered an outlier, because it isn't "far enough" from the other values in the data set. How far is "far enough?" Here's the rule that's most commonly used.

Identifying outliers

Any observation that is more than 1.5 times the interquartile range (IQR) above Q_3 or below Q_1 is considered an outlier.

For the text message data of Example 2.12, $IQR = 43 - 3 = 40$ and $1.5(IQR) = 1.5(40) = 60$. Any observation that is *above*

$$Q_3 + 1.5(IQR) = 43 + 60 = 103$$

or *below*

$$Q_1 - 1.5(IQR) = 3 - 60 = -57$$

would be identified as an outlier. So the student with 98 text messages is not an outlier, but the one with 118 text messages is.

Exercises

2.27 Median income You read that the median income of U.S. households in 2007 was $50,233. Explain in plain words what "the median income" is.

2.28 Acing the first test Here are the scores of Mrs. Liao's students on their first statistics test:

93	93	87.5	91	94.5	72	96	95	93.5	93.5	73	82	45	88	80
86	85.5	87.5	81	78	86	89	92	91	98	85	82.5	88	94.5	43

(a) Find the median. Explain what this number means in this setting.

(b) Find the quartiles and the *IQR*. Interpret the meaning of the *IQR* in this setting.

(c) Are there any outliers? Justify your answer with appropriate calculations.

(d) Make a boxplot of the test score data.

2.29 Phone calls, I After hearing about the text message study of Example 2.10, Mrs. Krebs asked the students in her statistics class how many phone calls they had made or received in the past 24 hours. Here are their responses:

3 0 12 10 8 4 2 7 10 11 6 7 0 35 12 3 10 2 7 9 15 5 14 15 6

(a) Find the median. Explain what this number means in this setting.

(b) Find the quartiles and the *IQR*. Interpret the meaning of the *IQR* in this setting.

(c) Determine whether there are any outliers. Show your work.

2.30 Phone calls, II Refer to the previous exercise. Make a boxplot of the phone call data from Mrs. Krebs's class. How would you describe the distribution of the variable "number of phone calls made or received in the past 24 hours"?

2.31 Texting or calling, I After hearing about the text message study of Example 2.10, Mr. Williams collects data from each student in his math class on the number of cell phone texts and calls sent or received in the past 24 hours. Figure 2.14 is a boxplot of the difference (texts − calls) in the number of text messages and calls for each student.

(a) Estimate the five-number summary for these data from the boxplot.

(b) Interpret the value of Q_1, M, and Q_3 in this setting.

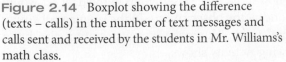

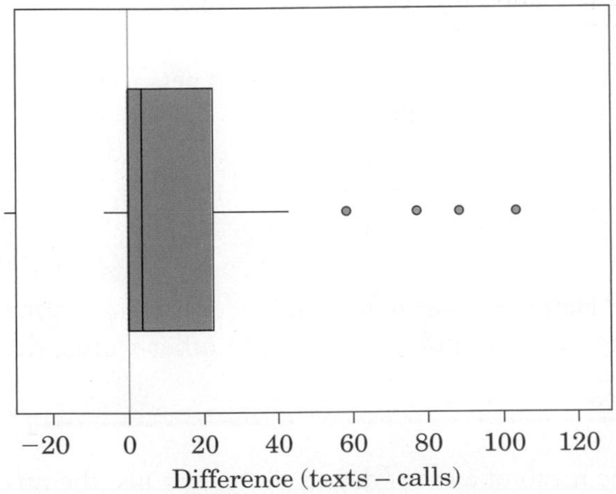

Figure 2.14 Boxplot showing the difference (texts – calls) in the number of text messages and calls sent and received by the students in Mr. Williams's math class.

2.32 Texting or calling, II Refer to the previous exercise.

(a) Do Mr. Williams's students seem to prefer texting or talking on the phone? Give appropriate evidence to support your answer.

(b) Can we draw any conclusion about the preferences of all students in the school based on the data from Mr. Williams's math class? Why or why not?

Measuring center: the mean

The five-number summary is not the most common numerical description of a distribution. That distinction belongs to the combination of the **mean** to measure center and the **standard deviation** to measure spread. The mean is familiar—it is the ordinary average of the observations. We'll meet the standard deviation shortly.

The mean $\bar{x}$

The **mean,** denoted by $\bar{x}$ (pronounced "x-bar"), of a set of observations is their average. To find the mean of n observations, add the values and divide by n:

$$\bar{x} = \frac{\text{sum of the observations}}{n}$$

Let's consider a simple example involving the mean.

Example 2.13 Do you have pets?

Jonathan Brizendine/iStockphoto

Nine elementary school children were asked how many pets they had. Here are their responses, arranged from lowest to highest:[12]

1 3 4 4 4 5 7 8 9

Since there are 9 observations, the median is the middle number: 4 pets. What's the mean number of pets for this group of children? It's

$$\bar{x} = \frac{\text{sum of observations}}{n} = \frac{1+3+4+4+4+5+7+8+9}{9} = 5$$

But what does that number tell us? Here's one way to look at it: if every child in the group had the same number of pets, each would have 5 pets. In other words, the mean is the "fair share" value.

As we saw in Example 2.13, the mean of a set of data is more than just the numerical average. The mean tells us how large each observation in the data set would be if the total were split equally among all the observations. In the language of young children, the mean is the "fair share" value. The mean of a distribution also has a physical interpretation, as the following Activity shows.

ACTIVITY 2.2B

What does the mean mean?

Materials: One pad of Post-it Notes

In this Activity, you'll use the pet data from Example 2.13 to explore an interesting property of the mean.

1. Draw a number line on the board. Label the horizontal axis from 0 to 9, with the tick marks spaced out by at least the width of a Post-it Note.

2. Make a sticky-note dotplot with all nine values equal to 5. Mark the location of the mean.

3. Change the value of one observation from 5 to 9 by moving a sticky note. Move one other sticky note so that the mean of the distribution becomes 5 again.

4. Return the two sticky notes to their original location at 5. Now change the value of one observation from 5 to 3. Move two other sticky notes to the right so that the mean becomes 5 again.

5. Start with all nine sticky notes above 5 on the dotplot. Change the value of one observation to 9, the maximum number of pets among the children. Now move another sticky note to keep the mean at 5. Change the value of one of the 5s to a 3. Move another sticky note from the stack above 5 to make the mean 5 again. Finally, change three of the 5s to 4s. Move one of the remaining sticky notes from above 5 to return the mean of the distribution to 5.

6. Do you see why the mean is sometimes called the "balance point" of a distribution?

The mean and median give two different ways to describe the center of a distribution. The median is right in the middle of the ordered data. But the median ignores the size of the values at either end of the distribution. As a result, we say that the median is **resistant.** The mean, on the other hand, incorporates every value in the data set. As a result, extreme values such as outliers can have a large effect on the mean.

Example 2.14 The median is resistant; the mean is not

In Example 2.10 (page 61), we examined data on the number of text messages sent and received in one day by Mr. Brown's statistics class. The median number of texts was 9. What about the mean? If we add the number of text messages and divide by 25, we get

$$\bar{x} = \frac{0 + 7 + 1 + 29 + \ldots + 42}{25} = \frac{687}{25} = 27.48$$

That's a lot more than 9. Notice that the 12 values below the median are much less spread out than the 12 values above the median. Since the mean is the balance point of the distribution, it is pulled far to the right by the several large values above the median.

In Example 2.14, the mean is pulled far above the median by extreme skewness and an outlier. Because of this, the mean is the preferred measure of center only when a distribution is roughly symmetric with no outliers. When a distribution is skewed or has outliers, the median does a much better job of summarizing its center.

You can compare the behavior of the mean and median by using the *Mean and Median* applet at the book's Web site, **www.whfreeman.com/sta2e.**

Measuring spread: the standard deviation

When we use the median to measure the center of a distribution, the interquartile range (*IQR*) is our corresponding measure of spread. If we summarize the center of a distribution with the mean, then the standard deviation is used to describe the spread of observations around the mean. The idea of the standard deviation is to give the average distance of observations from the mean.

Example 2.15 Do you have pets? *(continued)*

In Example 2.13, we found the mean number of pets for the group of children to be $\bar{x} = 5$. Let's look at where the observations in the data set are relative to the mean. The table below shows the deviation (value − mean) of each value in the data set from the mean. Add up the deviations from the mean. Did you get 0? You should, because the mean is the balance point of the distribution. Since the sum of the deviations from the mean will be 0 for any set of data, we need another way to calculate spread around the mean.

Value:	1	3	4	4	4	5	7	8	9
Value − mean:	1 − 5	3 − 5	4 − 5	4 − 5	4 − 5	5 − 5	7 − 5	8 − 5	9 − 5
Deviation:	−4	−2	−1	−1	−1	0	2	3	4

How can we fix the problem of the positive and negative deviations canceling each other out? We could take the absolute value of each deviation. Or we could square the deviations. For mathematical reasons beyond the scope of this book, statisticians choose to square rather than to use absolute values.

The table below shows the pet data and the deviations from the mean once again. We have added a row at the bottom that shows the square of each deviation. Add up the squared deviations. Did you get 52?

Value:	1	3	4	4	4	5	7	8	9
Deviation:	−4	−2	−1	−1	−1	0	2	3	4
Squared deviation:	16	4	1	1	1	0	4	9	16

Now we compute the average squared deviation—sort of. Instead of dividing by the number of observations n, we divide by $n - 1$:

$$\text{“average” squared deviation} = \frac{16 + 4 + 1 + 1 + 1 + 0 + 4 + 9 + 16}{9 - 1} = \frac{52}{8} = 6.5$$

Because we squared all the deviations, our units are in "squared pets." That's no good. We'll take the square root to get back to the correct units—pets. The resulting value is the *standard deviation*:

$$\text{standard deviation} = \sqrt{\text{"average" squared deviation}} = \sqrt{6.5} = 2.55$$

This 2.55 is "sort of like" the average distance of the values in the data set from the mean.

As you saw in Example 2.15, the "average distance" in the standard deviation is found in a rather unexpected way. You may want to just think of the standard deviation as "average distance from the mean" and leave the details to your calculator.

The standard deviation s

The **standard deviation** s measures the average distance of the observations from their mean. It is calculated by finding an average of the squared distances and then taking the square root. To find the standard deviation of n observations:

1. Find the distance of each observation from the mean and square each of these distances.
2. Average the distances by dividing their sum by $n - 1$. This average squared distance is called the **variance.**
3. The standard deviation s is the square root of this average squared distance.

In practice, you can key the data into your calculator and get the mean and standard deviation by asking for one-variable statistics. (See the Calculator Corner on page 72.) Or you can enter the data into a spreadsheet or other software to find $\bar{x}$ and s. It is usual, for good but somewhat technical reasons, to average the squared distances by dividing their total by $n - 1$ rather than by n. Many calculators report two standard deviations, giving you a choice between dividing by n and dividing by $n - 1$. The former is usually labeled σ, the symbol for the standard deviation of a population. If your data set consists of the entire population, then it's appropriate to use σ. More often, the data we're examining come from a sample. In that case, we should use s and divide by $n - 1$.

More important than the details of the calculation are the properties that show how the standard deviation measures spread.

Properties of the standard deviation s

- s measures spread about the mean $\bar{x}$. Use s to describe the spread of a distribution only when you use $\bar{x}$ to describe the center.
- $s = 0$ only when there is *no variability*. This happens only when all observations have the same value. So standard deviation zero means no spread at all. Otherwise, $s > 0$. As the observations become more spread out about their mean, s gets larger.

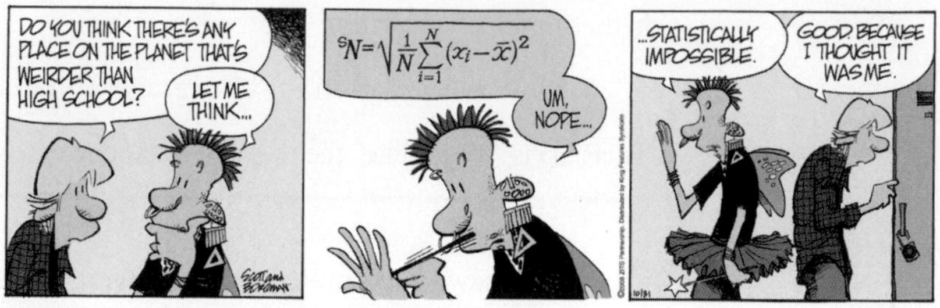

Exercises

2.33 Metabolism, I A person's metabolic rate is the rate at which the body consumes energy. Metabolic rate is important in studies of weight gain, dieting, and exercise. Here are the metabolic rates of seven men who took part in a study of dieting. (The units are calories per 24 hours. These are the same calories used to describe the energy content of foods.)

1792	1666	1362	1614	1460	1867	1439

Use the formula to calculate the mean. Interpret this value.

2.34 Metabolism, II Refer to the previous exercise.

(a) Use the method in Example 2.15 to calculate the deviation of each observation from the mean. Show that the sum of the deviations is 0.

(b) Calculate the standard deviation. Show your work.

2.35 Mean income You read that the mean income of U.S. households in 2007 was $67,609. Recall from Exercise 2.27 (page 65) that the median U.S. household income in 2007 was $50,233. Explain why the mean household income is so much higher than the median household income.

2.36 Feeling sleepy? The first four students to arrive for a first-period statistics class were asked how much sleep (to the nearest hour) they got last night. Their responses were 7, 7, 9, and 9.

(a) Use the formula to calculate the mean. Interpret this value.

(b) Use the method in Example 2.15 to calculate the deviation of each observation from the mean. Then compute the standard deviation. Explain what this value means.

2.37 A matchup Match the summary statistics with the histograms. Explain how you made your decision.

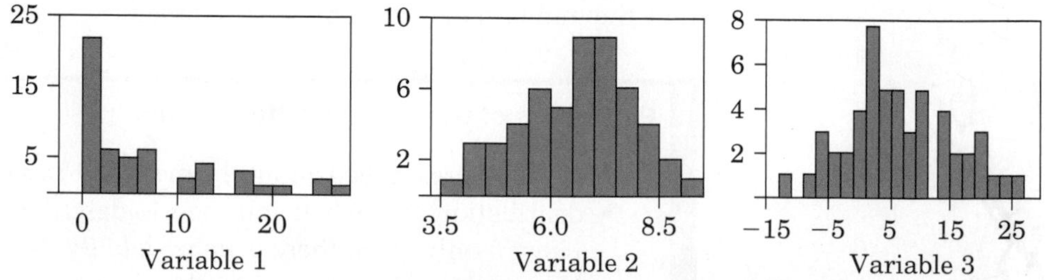

(a) mean = 6.6, median = 6.8, standard deviation = 1.3, variable = ______.

(b) mean = 6.6, median = 6.0, standard deviation = 8.65, variable = ______.

(c) mean = 6.6, median = 3.75, standard deviation = 7.4, variable = ______.

2.38 Properties of the standard deviation

(a) Juan says that, if the standard deviation of a list is zero, then all the numbers on the list are the same. Is Juan correct? Explain your answer.

(b) Letishia alleges that, if the means and standard deviations of two different lists of numbers are the same, then all of the numbers in the two lists are the same. Is Letishia correct? Explain your answer.

CALCULATOR CORNER Boxplots and summary statistics

You can use the TI-84 or TI-Nspire to make boxplots and calculate numerical summaries. We'll show you how using the text message data from Example 2.10 (page 61).

I. MAKING A BOXPLOT

TI-84

1. Enter the texting data: Press STAT and choose 1:`Edit`... Then type the values into list L1.

2. Define Plot 1 as a boxplot of the data in L1.

- Press 2nd Y= (STAT PLOT).
- Press ENTER or 1 to go into Plot 1.
- Adjust your settings as shown.

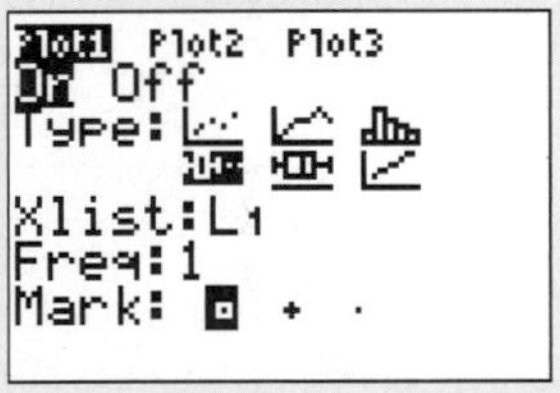

3. Use the calculator's zoom feature to display the boxplot.

- Press ZOOM and choose `9:ZoomStat`.
- To see the five-number summary for the texting data, press the TRACE key.

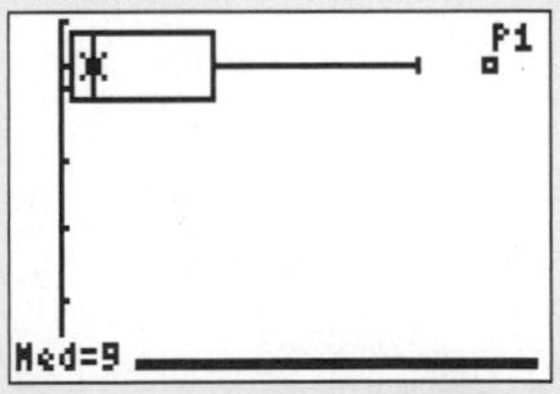

TI-Nspire

1. Enter the texting data in a Lists & Spreadsheet page.

- Press (home) and choose `6:New Document.`
- Press (menu); choose `3:Add Lists & Spreadsheet.`
- In the name box at the top of column A, type "texts."
- Type the values into list "texts."

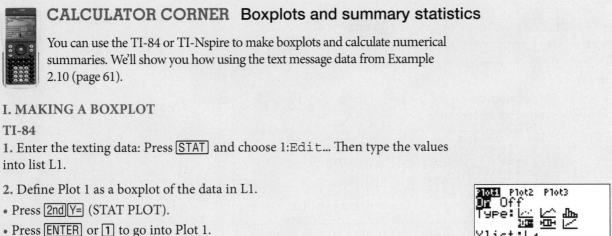

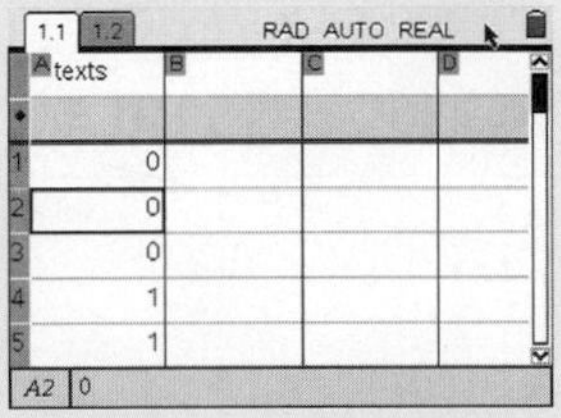

2. Insert a new Data & Statistics page. Press (ctrl)(I) and choose `5:Add Data & Statistics.`

CALCULATOR CORNER *(continued)*

CALCULATOR CORNER *(continued)*

3. Now make your boxplot.

- Use the "NavPad" to move your cursor over the "Click to add variable" box at the bottom of the screen.
- Click the mouse [click]. You should see the variable "texts" highlighted. Press [enter] or click [click] to make a dotplot.
- Press [menu]; choose `1:Plot Type` and `2:Box Plot` to change the graph to a boxplot.
- Move your cursor over the boxplot to see the five-number summary.

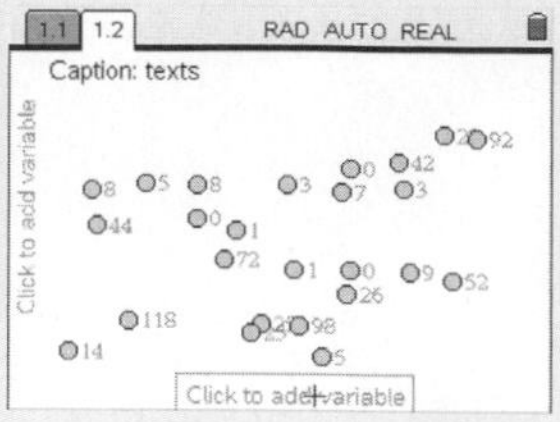

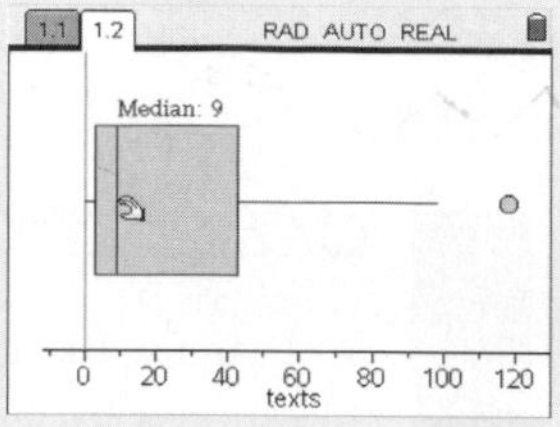

II. CALCULATING SUMMARY STATISTICS

TI-84

- Press [2nd][MODE] `(QUIT)` to go to the home screen.
- Press [STAT], arrow-right [▶] to `CALC`, and choose `1:1-Var Stats`.
- Complete the command `1-Var Stats L1` and press [ENTER]. (Press [2nd][1] to get `L1`.)
- Notice the down-arrow on the left side of the display. Press the down-arrow key [▼] several times to see more.

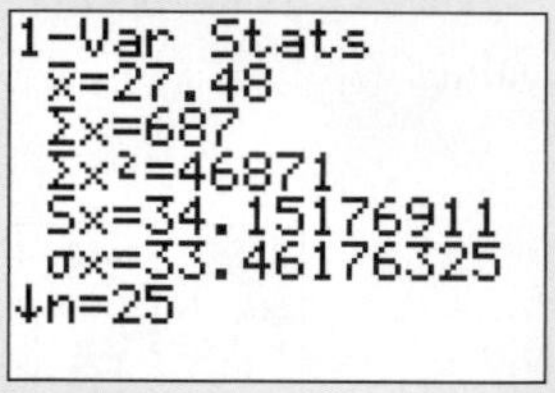

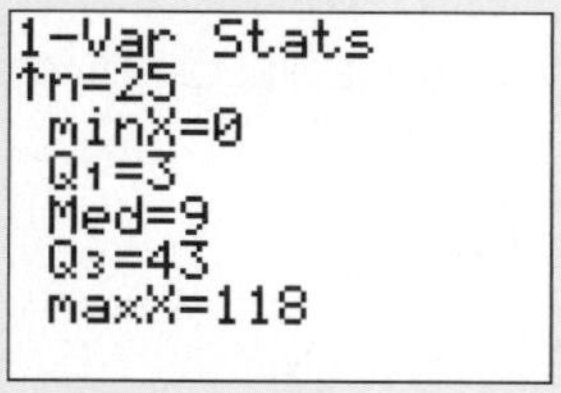

TI-NSPIRE

- Press [ctrl]◀ to return to the Lists & Spreadsheet page.
- Press [menu]; choose `4:Statistics`, `1:Stat Calculations...`, and `1:One-Variable Statistics`.
- When the dialog box appears, press [enter]. The One-Variable Statistics dialog box will appear.
- Use the drop-down menu for `X1 List:` to choose "texts." Then press [enter].
- Use the NavPad to scroll down the values in column C.

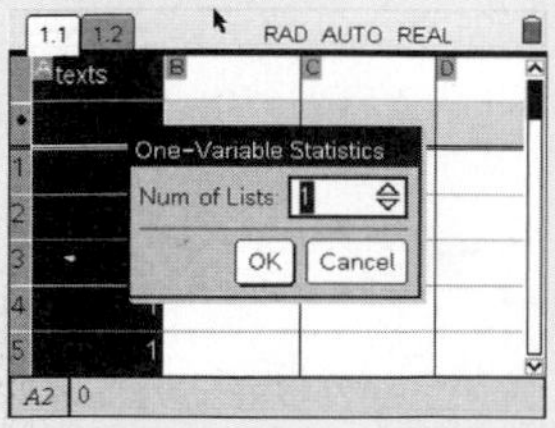

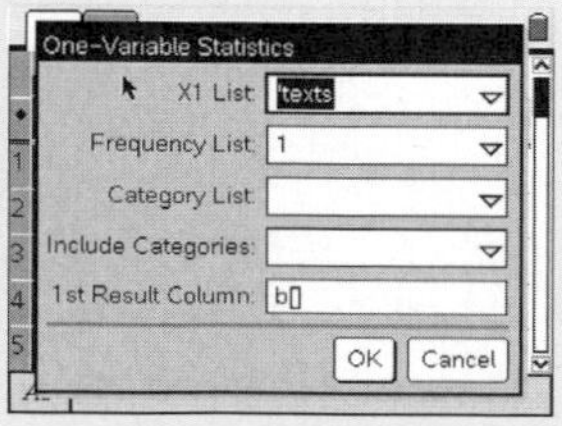

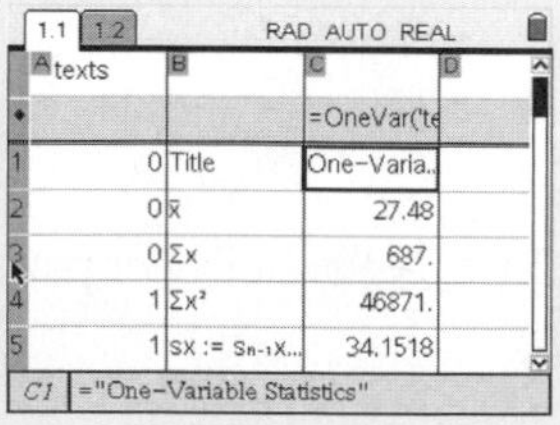

Choosing numerical descriptions

The five-number summary is easy to understand and is the best short description for most distributions. The mean and standard deviation are harder to understand but are more common. How can we decide which of these two descriptions of center and spread we should use?

Example 2.16 First-test scores

Figure 2.15 shows a dotplot of students' scores on Mrs. Liao's first test (see Exercise 2.28, page 65). The median is $M = 87.75$; the mean is $\bar{x} = 84.8$. Which of these values more accurately summarizes how a "typical" student did on this test? The mean is clearly affected by the two extremely low scores. Only 9 of the 30 students in the class earned a score below 84.8, so the median does a better job of describing a typical student's performance.

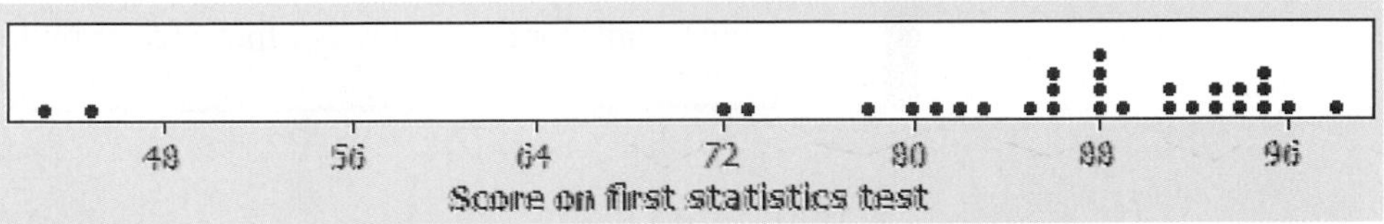

Figure 2.15 Minitab dotplot of students' scores on their first statistics test.

To investigate the effect of the two low outliers on the summary statistics, we calculated the mean, standard deviation, and five-number summary with and without these two scores. The table below summarizes the results. When the outliers are removed, the mean increases to be almost equal to the median. Also notice the large effect that the two extreme scores have on the standard deviation. What about the *IQR*? With all 30 observations, $IQR = 93 - 82 = 11$. With the two outliers removed, $IQR = 93.25 - 83.75 = 9.5$. The *IQR* is fairly *resistant* to the extreme scores.

	n	$\bar{x}$	s	Min.	Q_1	M	Q_3	Max.
All data	30	84.8	12.857	43	82	87.75	93	98
No outliers	28	87.7	6.735	72	83.75	88	93.25	98

The mean and median of a symmetric distribution are close to each other. In fact, $\bar{x}$ and M are exactly equal if the distribution is exactly symmetric. In skewed distributions, however, the mean runs away from the median toward the long tail. Many distributions of monetary values—incomes, house prices, wealth—are strongly skewed to the right. The mean may be much larger than the median. For example, we saw at the beginning of the chapter that the median income of people with advanced degrees in the Current Population Survey sample is $62,000. The distribution is skewed to the right, and the long right tail pulls the mean up to $82,729. Because monetary data often have a few extremely high observations, descriptions of these distributions usually employ the median.

The standard deviation is pulled up by outliers or the long tail of a skewed distribution even more strongly than the mean. The standard deviation of students' test scores in Mrs. Liao's class is $s = 12.857$ for all 30 students and only $s = 6.735$ when the two outliers are removed. The quartiles are much less sensitive to a few extreme observations.

There is another reason to avoid the standard deviation in describing skewed distributions. Because the two sides of a strongly skewed distribution have different spreads, no single number such as s describes the spread well. The five-number summary, with its two quartiles and two extremes, usually does a better job. In most situations, it is wise to use $\bar{x}$ and s only for distributions that are roughly symmetric.

Poor New York?

Is New York a rich state? New York's mean income per person ranks fourth among the states, right up there with its rich neighbors Connecticut and New Jersey, which rank first and second. But while Connecticut and New Jersey rank seventh and second in median household income, New York stands at twenty-ninth, well below the national average. What's going on? Just another example of mean versus median. New York has many very highly paid people who pull up its mean income per person. But it also has a higher proportion of poor households than do New Jersey and Connecticut, and this brings the median down. New York is not a rich state—it's a state with extremes of wealth and poverty.

Choosing a summary

The mean and standard deviation are strongly affected by outliers and by the long tail of a skewed distribution. The median and quartiles are less affected.

The median and *IQR* are usually better than the mean and standard deviation for describing a skewed distribution or a distribution with outliers. Use $\bar{x}$ and s only for reasonably symmetric distributions that are free of outliers.

Why do we bother with the mean and standard deviation at all? One answer appears in the next chapter: the mean and standard deviation are the natural measures of center and spread for an important kind of symmetric distribution, called a *normal distribution.*

Transforming data

After looking at the distribution of test scores in Example 2.16, Mrs. Liao is considering "adjusting" the scores. She has two ideas in mind:

1. Add 5 points to each student's test score.
2. Double each student's score and grade the test out of 200 points.

Before she implements either idea, she wants to know how each possible adjustment would affect the shape, center, and spread of the test score distribution.

ACTIVITY 2.2C

First-test scores: making an adjustment

In this Activity, you will investigate the effect of Mrs. Liao's two ideas for adjusting her class's test scores on the mean, standard deviation, and five-number summary. Here are the data again:

93	93	87.5	91	94.5	72	96	95	93.5	93.5	73	82	45	88	80
86	85.5	87.5	81	78	86	89	92	91	98	85	82.5	88	94.5	43

1. On your paper, make a chart like the one shown below to record your results.

	n	$\bar{x}$	s	Min.	Q_1	M	Q_3	Max.
Original scores	30	84.8	12.857	43	82	87.75	93	98
Our guess: add 5	30							
Our guess: multiply by 2	30							
Actual result: add 5	30							
Actual result: multiply by 2	30							

2. Discuss with a partner: what do you think will happen to each of the numerical summary values in the table for Mrs. Liao's two ideas? Record your educated guesses in the appropriate rows.

3. Start with the original data and add 5 to each student's test score. Use `One-Variable Statistics` on your calculator or the *One Variable Statistical Calculator* applet at the book's Web site (**www.whfreeman.com/sta2e**) to fill in the values in the fourth row of the table. How do the results compare with your educated guesses?

4. Start over with the original data. This time, multiply each student's score by 2. Proceed as in Step 3 to complete the final row of the table. How do the results compare with your educated guesses?

5. Summarize what you have learned about the effects of (i) adding a constant and (ii) multiplying by a constant on our popular measures of center and spread.

6. *Investigate:* What happens to the shape of the distribution with each of these two transformations?

Activity 2.2C shows what happens to the shape, center, and spread of a distribution if the original values are adjusted in a particular way. Adding a constant shifts the center of the distribution by that amount but doesn't affect the shape or spread. Multiplying all of the original values by a constant multiplies the mean, standard deviation, and every value in the five-number summary by that constant. Again, the shape of the distribution stays the same.

Comparing distributions

Some of the most interesting statistics questions involve comparing two or more groups. Which of two popular diets leads to greater long-term weight loss? How does income relate to educational level? Who texts more—males or females? Does seat belt use differ across regions of the United States? To make such comparisons, start with a graph. For categorical variables, use a bar graph or a pie chart. For quantitative variables, make a dotplot, stemplot, or histogram. Then add numerical summaries and a boxplot.

Example 2.17 Habits of mind for college freshmen

The 2007 CIRP Freshman Survey asked first-year college students about their "habits of mind"—specific behaviors that college faculty have identified as being important for student success. One question asked students, "How often in the past year did you revise your papers to improve your writing?" Another asked, "How often in the past year did you seek feedback on your academic work?" Figure 2.16 (on the next page) is a bar graph comparing male and female responses to these two questions.[13]

What does the bar graph tell us about the habits of mind of male and female college freshmen? Over half of females surveyed said that they had frequently revised their papers to improve their writing as high school seniors. Only 37% of the males who were surveyed answered "frequently." Nearly half of females said that they had frequently sought feedback on their work in the past year, compared to about 38% of males who were surveyed. It appears that females engaged

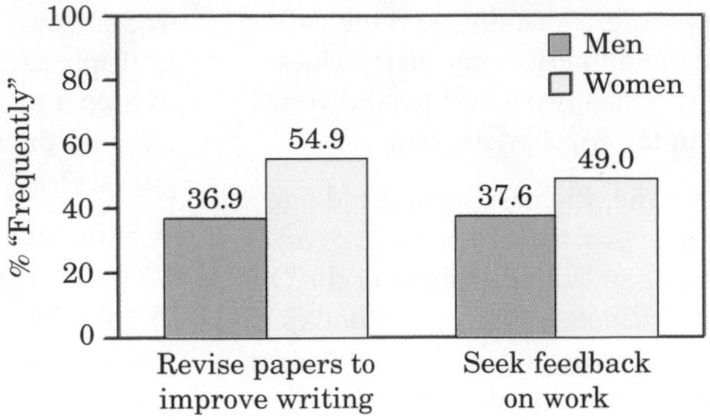

Figure 2.16 Comparative bar graph of male and female students' responses to the 2007 CIRP Freshman Survey about the frequency of specific learning behaviors in the past year.

in these two specific learning behaviors more often than males did as high school seniors.

Example 2.17 shows you how to compare distributions of a categorical variable. The following Data Exploration involves a quantitative variable.

DATA EXPLORATION

Mercury in tuna (continued)

In Application 2.1 (page 56), you examined data from a study of mercury concentration in canned tuna. You should reread the details of that study before you answer the questions that follow.

Is there a difference in the mercury concentration (parts per million) in light tuna and albacore tuna? The accompanying parallel boxplots and summary chart provide information to help you answer this question.

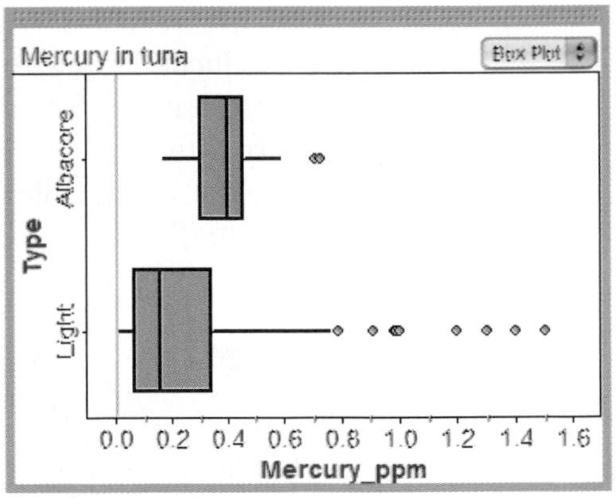

Type	Count	$\bar{x}$	s	Min.	Q_1	M	Q_3	Max.
Albacore	20	0.401	0.152	0.17	0.295	0.4	0.46	0.73
Light	144	0.269	0.312	0.012	0.0595	0.16	0.345	1.5

1. Explain why the mean mercury content is much higher than the median for the light tuna but not for the albacore tuna.

2. Confirm that the outliers shown in the boxplots are identified by the 1.5(IQR) rule.

3. Write a few sentences comparing the distributions of mercury concentration for these two types of canned tuna.

As the Data Exploration suggests, you should always discuss shape, center, spread and any unusual values whenever you compare distributions of a quantitative variable.

Exercises

2.39 Music CDs How many music CDs do students own? The 24 members of a college statistics class provided data in response to this question.[14] Figure 2.17 is a dotplot of the data. Which would better summarize the center and spread of the distribution: (i) the mean and standard deviation or (ii) the median and *IQR*? Justify your answer.

Figure 2.17 Minitab dotplot showing the number of music CDs owned by students in a college statistics class.

www

2.40 Domain names When it comes to Internet domain names, is shorter better? According to one ranking of Web sites in 2008, the top 8 sites (by number of "hits") were **yahoo.com**, **google.com**, **youtube.com**, **live.com**, **msn.com**, **myspace.com**, **wikipedia.org**, and **facebook.com**. These familiar sites certainly have short domain names. Figure 2.18 is a histogram of the domain name lengths for the 500 most popular Web sites.

(a) Estimate the mean and median of the distribution. Explain your method clearly.

(b) If you wanted to argue that shorter domain names were more popular, which measure of center would you choose—the mean or the median? Justify your answer.

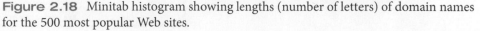

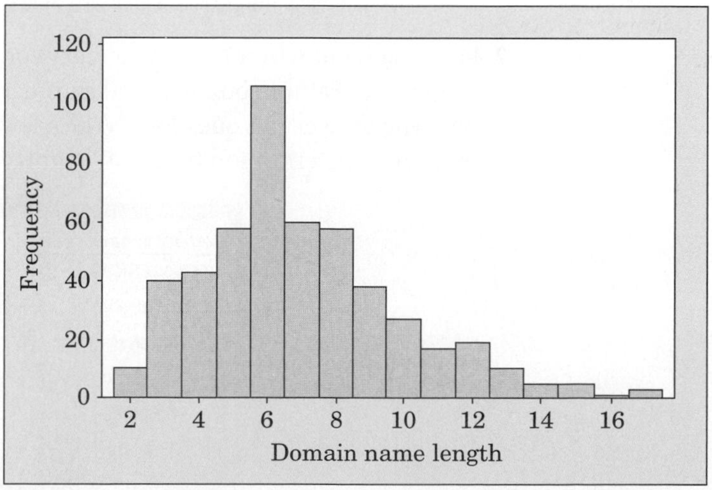

Figure 2.18 Minitab histogram showing lengths (number of letters) of domain names for the 500 most popular Web sites.

2.41 Tall or short, I Mr. Walker measures the heights (in inches) of the students in one of his classes. He uses a computer to calculate the following numerical summaries:

Mean	Std. dev.	Min.	Q_1	Med.	Q_3	Max.
69.188	3.20	61.5	67.75	69.5	71	74.5

Next, Mr. Walker has his entire class stand on their chairs, which are 18 inches off the ground. Then he measures the distance from the top of each student's head to the floor.

(a) Find the mean and median of these measurements. Show your work.

(b) Find the standard deviation and *IQR* of these measurements. Show your work.

2.42 Tall or short, II Refer to the previous exercise. Mr. Walker converts his students' original heights from inches to feet.

(a) Find the mean and median of the students' heights in feet. Show your work.

(b) Find the standard deviation and *IQR* of the students' heights in feet. Show your work.

2.43 Cool car colors Table 2.8 gives information about the most popular colors of vehicles purchased in 2007. Make a bar graph that compares the color distributions for full size/intermediate cars and SUVs/trucks. Describe any similarities and differences between the two distributions.

Table 2.8 Most Popular Vehicle Colors in 2007

Color	Full size/intermediate cars	SUVs/trucks
Silver	21%	14%
Blue	14%	11%
White	14%	26%
Black	13%	16%
Red	13%	13%
Gray	12%	12%
Beige/brown	7%	4%

Source: 2009 *World Almanac and Book of Facts.*

2.44 Tuna from where? Does mercury content in canned tuna differ by country of origin? The Fathom boxplots and summary chart that follow provide information to help you answer this question. Write a few sentences comparing the distributions of mercury concentration based on country of origin.

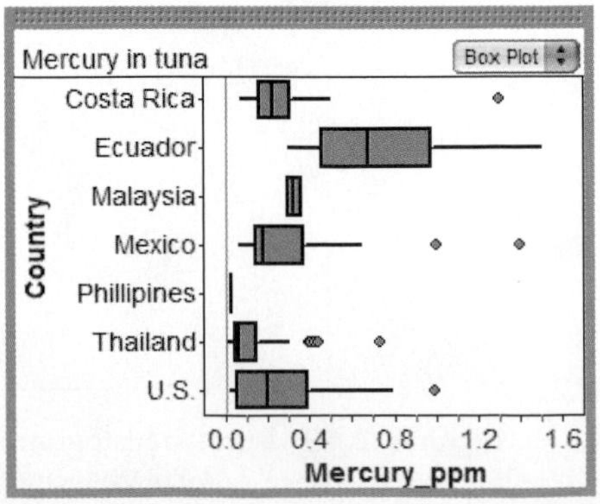

Country	Count	$\bar{x}$	s	Min.	Q_1	M	Q_3	Max.
Costa Rica	23	0.281	0.243	0.079	0.16	0.23	0.32	1.3
Ecuador	18	0.754	0.367	0.3	0.45	0.68	0.98	1.5
Malaysia	2	0.33	0.057	0.29	0.29	0.33	0.37	0.37
Mexico	33	0.31	0.285	0.064	0.14	0.18	0.38	1.4
Phillipines	3	0.031	0.006	0.025	0.025	0.034	0.035	0.035
Thailand	50	0.126	0.148	0.012	0.043	0.065	0.16	0.73
United States	35	0.268	0.242	0.023	0.052	0.2	0.4	0.99

APPLICATION 2.2

Did Mr. Starnes stack his class?

Mr. Starnes teaches AP Statistics, but he also does the class scheduling for the high school. There are two AP Statistics classes—one taught by Mr. Starnes and one taught by Ms. McGrail. The two teachers give the same first test to their classes and grade the test together. Mr. Starnes's students earned an average score that was 8 points higher than the average for Ms. McGrail's class. Ms. McGrail wonders whether Mr. Starnes might have "adjusted" the class rosters from the computer scheduling program. In other words, she thinks he might have "stacked" his class with better students. He denies this, of course.

To help resolve the dispute, the teachers collected data on the cumulative grade point averages and SAT Math scores of their students. Mr. Starnes provided the GPA data from his computer. The students reported their SAT Math scores. The table below shows the data for each student in the two classes. Note that the two data values in each row come from a single student.

Did Mr. Starnes stack his class? Give appropriate graphical and numerical evidence to support your conclusion.

Starnes GPA	Starnes SAT-M	McGrail GPA	McGrail SAT-M
2.9	670	2.9	620
2.86	520	3.3	590
2.6	570	3.98	650
3.6	710	2.9	600
3.2	600	3.2	620
2.7	590	3.5	680
3.1	640	2.8	500
3.085	570	2.9	502.5
3.75	710	3.95	640
3.4	630	3.1	630
3.338	630	2.85	580
3.56	670	2.9	590
3.8	650	3.245	600
3.2	660	3.0	600
3.1	510	3.0	620
		2.8	580
		2.9	600
		3.2	600

Section 2.2 Summary

To describe a set of data, always start with graphs. Then add well-chosen numbers that summarize specific aspects of the data. If we have data on a single quantitative variable, we start with a dotplot, stemplot, or histogram to display the distribution. Then we add numbers to describe the **center** and **spread** of the distribution.

There are two common descriptions of center and spread: **the five-number summary** and the **mean and standard deviation.** The five-number summary consists of the **median** to measure center and the **interquartile range** (*IQR*) to describe spread, along with the minimum and maximum. We can use a **boxplot** to display the five-number summary in graphical form. The median is the midpoint of the observations. The **mean** is the average of the observations. The **standard deviation** measures spread as a kind of average distance from the mean, so use it only with the mean.

The mean and standard deviation are not **resistant**—they can be changed a lot by a few outliers. The mean and median agree for symmetric distributions, but the mean moves farther toward the long tail of a skewed distribution. In general, use the five-number summary to describe most distributions and the mean and standard deviation only for roughly symmetric distributions.

Section 2.2 Exercises

2.45 Electoral votes, I To become president of the United States, a candidate does not have to receive a majority of the popular vote. The candidate does, however, have to win a majority of the 538 electoral votes that are cast in the Electoral College. Figure 2.19 is a stemplot of the number of electoral votes for each of the 50 states and the District of Columbia.

(a) Find the five-number summary for this distribution.

(b) Are there any outliers? Justify your answer with a calculation.

(c) Make a boxplot for these data.

```
0 | 3333333344444
0 | 55555666777788999
1 | 0000111123
1 | 5557
2 | 011
2 | 7
3 | 14
3 |
4 |
4 |
5 |
5 | 5
```

Figure 2.19 Stemplot of the number of electoral votes for the 50 states and the District of Columbia.

2.46 Electoral votes, II Refer to the previous exercise. Which measure of center and spread would you use to summarize the distribution in Figure 2.19: (i) the mean and standard deviation or (ii) the median and *IQR*? Justify your answer.

2.47 Which is easier—AP Calculus AB or AP Statistics? The table below gives the distribution of grades earned by students taking the AP Calculus AB and AP Statistics exams in 2008.[15]

		Grade				
	No. of exams	**5**	**4**	**3**	**2**	**1**
AP Calculus AB	222,835	22.1%	21.2%	17.9%	15.2%	23.7%
AP Statistics	108,284	12.9%	22.7%	23.7%	18.8%	21.8%

(a) Make an appropriate graphical display to compare the grade distributions for AP Calculus AB and AP Statistics.

(b) Write a few sentences comparing the two distributions of exam grades. Can we tell which exam is easier? Explain why or why not.

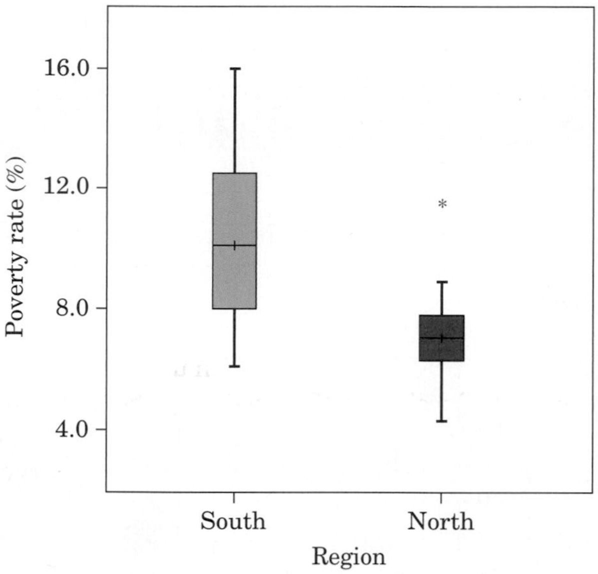

Figure 2.20 Boxplots comparing the poverty rates of southern and northern states east of the Mississippi in the same year.

"Yup, Old Bob drowned due to being Ignorant of statistics. He thought it was enough to know the average depth of the river."

2.48 Mean or median? You are planning a party and want to know how many cans of soda to buy. A genie offers to tell you either the mean number of cans guests will drink or the median number of cans. Which measure of center should you ask for? Why? To make your answer concrete, suppose that there will be 30 guests and the genie will tell you either $\bar{x} = 5$ cans or $M = 3$ cans. How many cans should you have on hand?

2.49 Poverty in the eastern states The poverty rates for states east of the Mississippi have been divided into northern and southern states, according to the geographic divisions used by the Census Bureau. Figure 2.20 shows boxplots of the poverty rates for the states in each region. Write a few sentences comparing the distributions.

2.50 Phosphate levels The level of various substances in the blood influences our health. Here are measurements of the level of phosphate in the blood of a patient, in milligrams of phosphate per deciliter of blood, made on six consecutive visits to a clinic:

5.6	5.2	4.6	4.9	5.7	6.4

(a) Find the mean from its definition. Show your work.
(b) Find the standard deviation from its definition. Show your work.
(c) Now enter the data into your calculator and use it to obtain $\bar{x}$ and s. Do the results agree with your hand calculations?

2.51 Teacher raises, I A school system employs teachers at salaries between \$18,000 and \$40,000. The teachers' union and the school board are negotiating the form of next year's increase in the salary scale.
(a) If every teacher is given a \$1000 raise, what will this do to the mean salary? To the median salary? Explain your answers.
(b) What would an across-the-board \$1000 raise do to the extremes and quartiles of the salary distribution? To the standard deviation of teachers' salaries? Explain your answers.

2.52 Teacher raises, II Refer to the previous exercise. If each teacher receives a 5% raise instead of a \$1000 raise, the amount of the raise will vary from \$900 to \$2000, depending on the present salary.
(a) What will this do to the mean salary? To the median salary? Explain your answers.
(b) Will a 5% raise increase the *IQR*? Will it increase the standard deviation? Explain your answers.

2.3 Use and Misuse of Statistics

ACTIVITY 2.3

Who buys iMacs?

After Apple Computer introduced the iMac, the company wanted to know whether this new computer was expanding Apple's market share. Was the iMac mainly being bought by previous Macintosh owners, or was it being purchased by first-time computer buyers and by previous PC users who were switching over? To find out, Apple hired a firm to conduct a survey of 500 iMac customers. Each customer was categorized as a new computer purchaser, a previous PC owner, or a previous Macintosh owner. The table below summarizes the survey results.[16]

Previous ownership	Count	Percent
None	85	17.0
PC	60	12.0
Macintosh	355	71.0
Total	500	100.0

Discuss each of the following questions with two or three classmates. Be prepared to present your answers to the class.

1. Here's a clever graph of the data that uses pictures instead of the more traditional bars. How is this graph misleading?

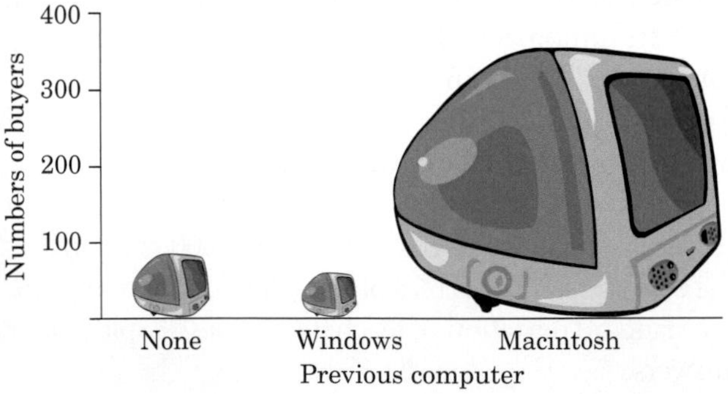

2. Two possible bar graphs of the data are shown below. Which one could be considered deceptive? Why?

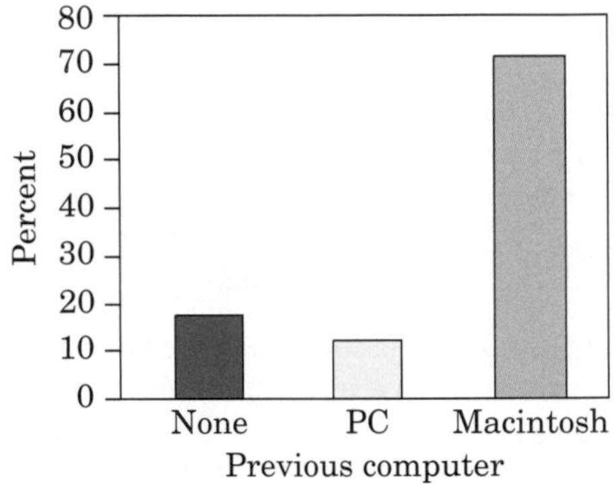

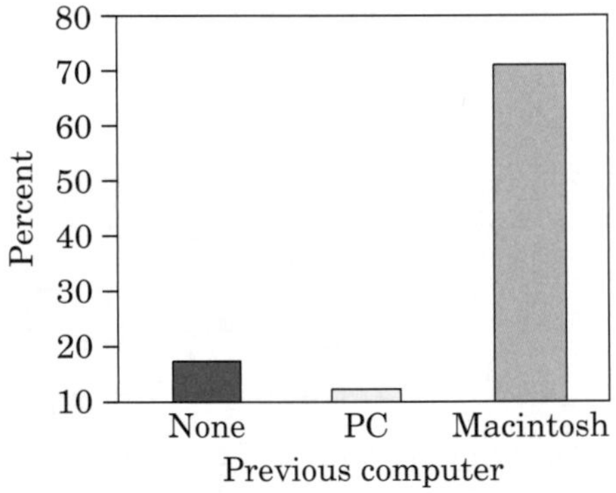

3. A clever marketing consultant wants to advertise the results of the survey as: "Over 40% of new iMac owners are former PC users." The consultant got this figure as follows: $60/145 = 0.414$, or 41.4%. Explain how this value distorts the truth.

In his *Chapters from My Autobiography,* Mark Twain wrote: "There are three kinds of lies: lies, damned lies, and statistics." Several popular books have been written that build on this theme. There's Darryl Huff's classic *How to Lie with Statistics* and Joel Best's *Damned Lies and Statistics.* Sometimes the "word on the street" about statistics focuses mainly on how people can make data say anything they want. Activity 2.3 gives you a few illustrations of this kind of deceptive statistics.

In this section, we'll introduce you to some of the most common ways that graphs and numbers can be used to deceive people. We'll also show you how to use graphs and numbers to display data honestly and effectively.

Change over time: line graphs

Many quantitative variables are measured at intervals over time. We might, for example, measure the height of a growing child or the price of a stock at the end of each month. In these examples, our main interest is change over time. To display change over time, make a **line graph.**

Line graph

A **line graph** of a variable plots each observation against the time at which it was measured. Always put time on the horizontal scale of your plot and the variable you are measuring on the vertical scale. Connect the data points by lines to display the change over time.

Example 2.18 The price of gasoline

How has the price of gasoline at the pump changed over time? Figure 2.21 is a line graph of the average price of regular unleaded gasoline each month from January 2000 to January 2009. For reference, the point for January 2000 is at 128.9 on the vertical scale.[17]

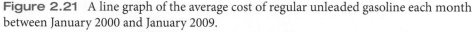

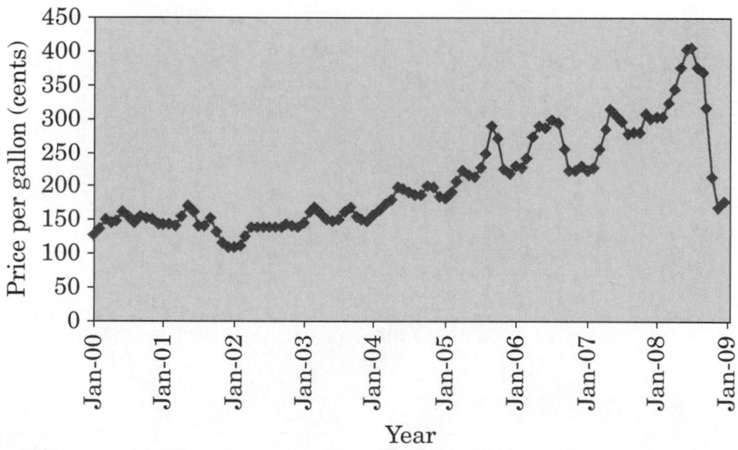

Figure 2.21 A line graph of the average cost of regular unleaded gasoline each month between January 2000 and January 2009.

It would be difficult to see patterns in a long table of monthly prices. Figure 2.21 makes the patterns clearer. What should we look for?

First, look for an overall pattern. For example, a **trend** is a long-term upward or downward movement over time. There was an overall upward trend in gasoline prices over this time period. This upward trend began as OPEC countries drove the price up by reducing their oil production. A major strike in Venezuela, the U.S. war with Iraq, and unrest in Nigeria are just a few of the events that have contributed to the upward trend.

Next, look for striking deviations from the overall pattern. The 2001 price drop was due to weakening global demand. In fall 2008, a serious financial crisis in the United States led to decreased demand and lower prices.

Change over time often has a regular pattern of **seasonal variation** that repeats each year. Gasoline prices are usually highest during the summer driving season and lowest in the winter, when there is less demand for gasoline. You can see this up-in-summer then down-in-the-fall pattern for many years in Figure 2.21.

Watch those scales!

Because graphs speak so strongly, they can mislead the unwary. The careful reader of a line graph looks closely at the scales marked off on the axes.

Example 2.19 Living together

The number of unmarried couples living together has increased in recent years, to the point that some people say that cohabitation is delaying or even replacing marriage. Figure 2.22 presents two line graphs of the number of unmarried-couple households in the United States. The data once again come from the Current Population Survey. The graph on the left suggests a steady but moderate increase. The right-hand graph says that cohabitation is rapidly increasing.

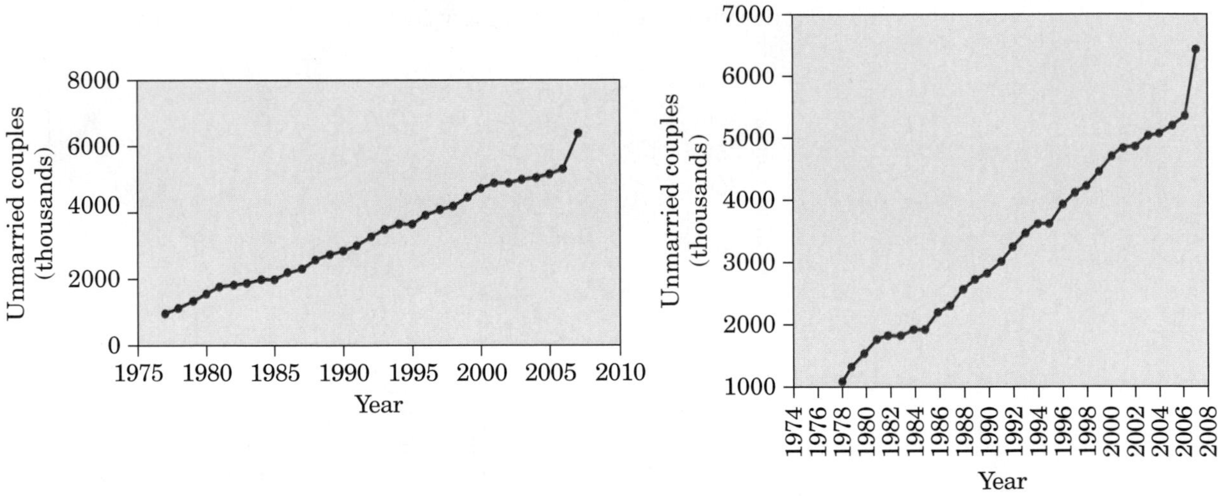

Figure 2.22 The effect of changing the scales in a line graph. Both graphs plot the same data, but the right-hand graph makes the increase appear much more rapid.

The secret is in the scales. You can transform the left-hand graph into the right-hand graph by stretching the vertical scale, squeezing the horizontal scale, and cutting off the vertical scale just above and below the values to be plotted. Now you know how to either exaggerate or play down a trend in a line graph.

Which graph is correct? Both are accurate graphs of the data, but both have scales chosen to create a specific effect. Because there is no one "right" scale for a line graph, correct graphs can give different impressions by their choices of scale. Watch those scales!

Did you notice the sudden increase in the number of unmarried couples living together from 2006 to 2007? In previous years, the Current Population Survey only counted couples that included the head of household. Starting in 2007, the CPS changed its survey question to "Do you have a boyfriend/girlfriend or partner in the household?" They found about 1.1 million more unmarried couples with this direct question.

Beware the pictogram

Bar graphs compare several quantities by comparing the heights of bars that represent the quantities. Our eyes, however, react to the *area* of the bars as well as to their height. When all bars have the same width, the area (width × height) varies in proportion to the height and our eyes receive the right impression. When you draw a bar graph, make the bars equally wide. Artistically speaking, bar graphs are a bit dull. It is tempting to replace the bars with pictures for greater eye appeal.

Example 2.20 Favorite pets: a misleading graph

All 900 students in an elementary school responded to the following survey question: "What is your favorite animal to have as a pet?" Most students said "dog," "cat," or "horse," but a few named another animal, like a guinea pig or a rabbit. Rather than use a bar graph to display the data, one of the students suggested using the **pictogram** shown in Figure 2.23. Do you see how this graph is misleading?

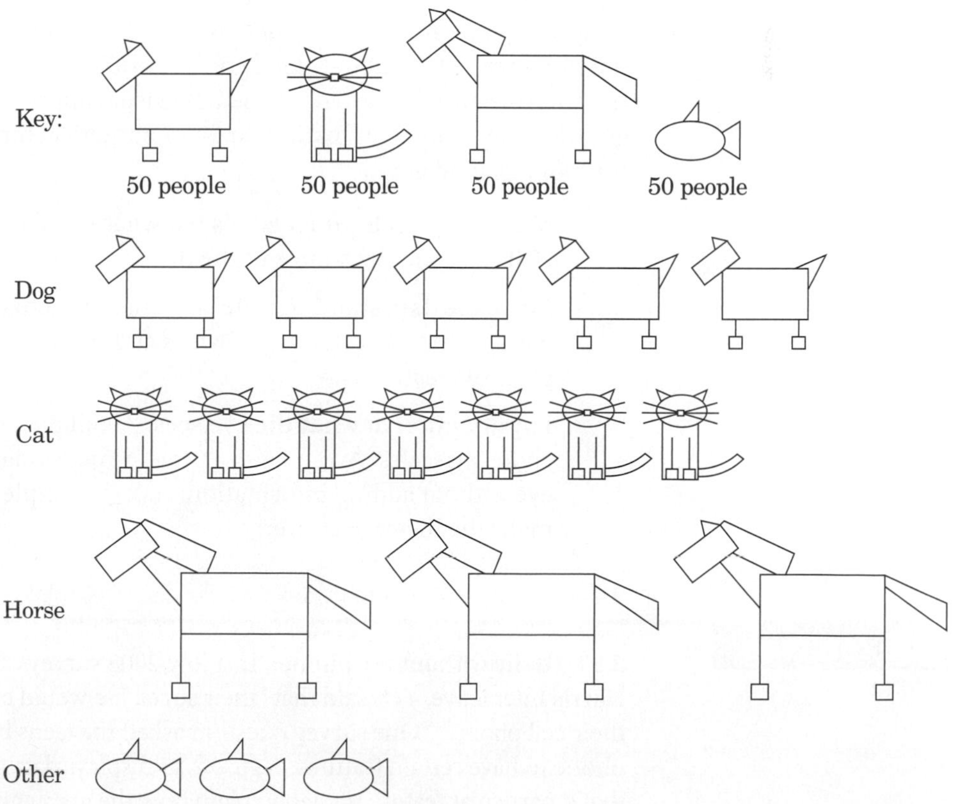

Figure 2.23 Pictogram showing the favorite pets of students in an elementary school.

The pictogram makes it look like horses are the most popular pets. In fact, only 150 students chose "horse" as their favorite pet. But the pictures of the horses in Figure 2.23 are so large that even three of them extend past the images for all the other pet categories. To make a pictogram that isn't deceptive, we need to use pictures of the animals that are all the same size. Alternatively, we could just make a bar graph, like the one shown in Figure 2.24. Now we see that cats were actually the most popular pet for students at this elementary school.

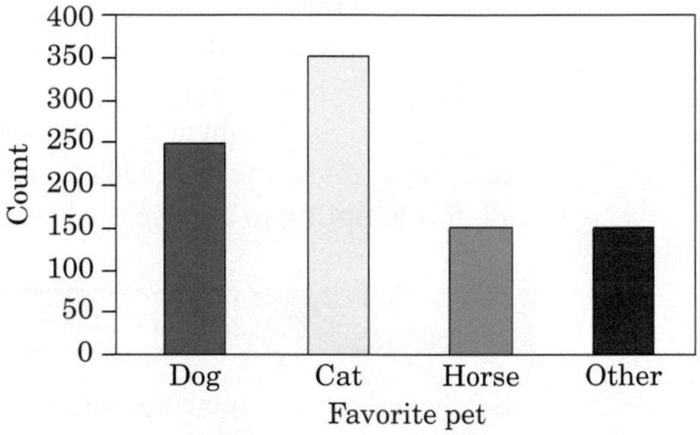

Figure 2.24 Bar graph that displays the data in Figure 2.23 more accurately.

Making good graphs

Graphs are the most effective way to communicate using data. A good graph frequently reveals facts about the data that would be difficult or impossible to detect from a table. In addition, the immediate visual impression of a graph is much stronger than the impression made by data in numerical form. Here are some principles for making good graphs:

- Make sure **labels and legends** tell what variables are plotted, what their units are, and the source of the data.
- **Make the data stand out.** Be sure that the actual data, not labels, grids, or background art, catch the viewer's attention. You are drawing a graph, not a piece of creative art.
- **Pay attention to what the eye sees.** Avoid pictograms and be careful choosing scales. Avoid fancy "three-dimensional" effects that confuse the eye without adding information. Ask if a simple change in a graph would make the message clearer.

Exercises

2.53 I'd die without my phone! In a July 2008 survey of over 2000 U.S. teenagers by Harris Interactive, 47% said that "their social life would end or be worsened without their cell phone."[18] One survey question asked the teens how important it is for their phone to have certain features. Figure 2.25 displays data on the percent who indicated that a particular feature is vital. Explain how the graph gives a misleading impression.

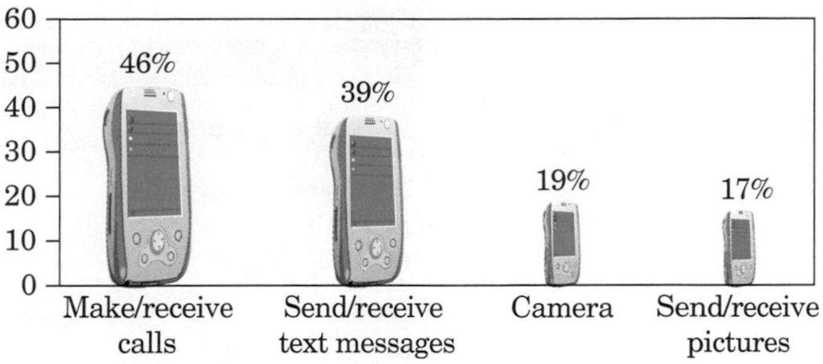

Figure 2.25 A graph showing the percent of teens who said that a particular feature is essential to have on their cell phones in a July 2008 survey.

2.54 The cost of fresh oranges Figure 2.26 is a line graph of the average cost of fresh oranges each month from January 2000 to December 2008. These data, from the Bureau of Labor Statistics monthly survey of retail prices, are "index numbers" rather than prices in dollars and cents. That is, they give each month's price as a percent of the price in a base period (in this case, the years 1982 to 1984). So 250 means "250% of the base price."

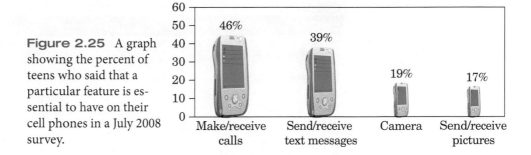

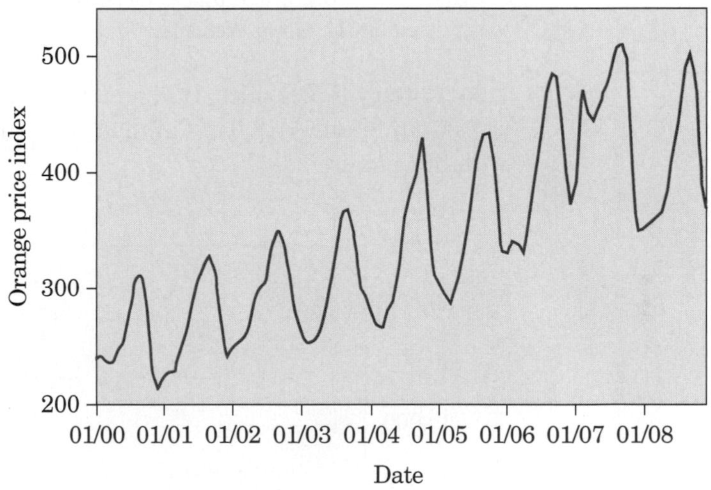

Figure 2.26 The price (index numbers; base period 1982 to 1984) of fresh oranges, January 2000 to December 2008.

(a) The graph shows strong seasonal variation. How is this visible in the graph? Why would you expect the price of fresh oranges to show seasonal variation?

(b) What is the overall trend in orange prices during this period, after we take account of the seasonal variation?

2.55 Support the court? In 2005, CNN reported the results of a survey about a Florida court's decision to remove the feeding tube from coma patient Terry Schiavo, effectively ending her life. Figure 2.27 (on the next page) shows a bar graph of the data that CNN initially posted on its Web site.

(a) What visual impression does the graph give about support for the court's decision?

(b) Make your own graph that displays the data in a less misleading way.

(*Note:* When notified about the misleading nature of their graph, CNN posted a corrected version.)

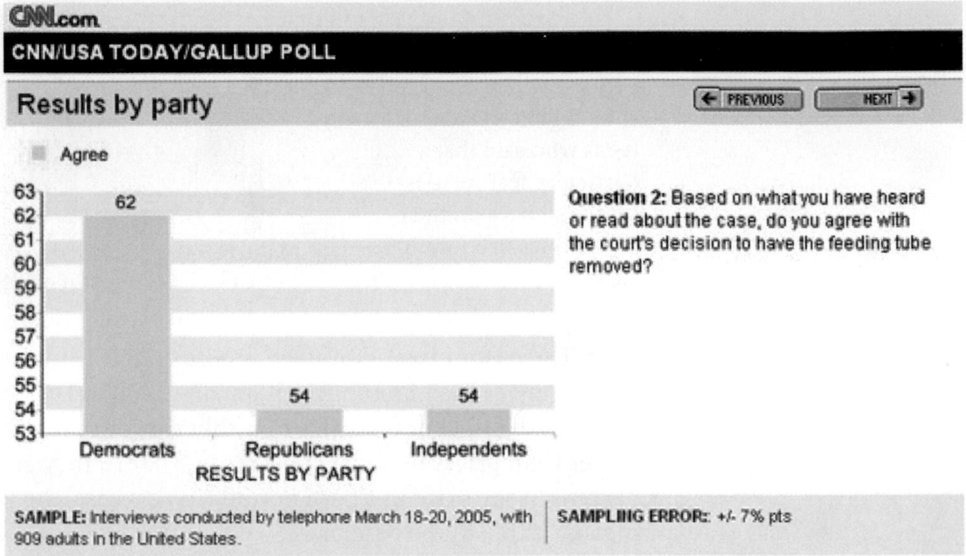

Figure 2.27 Bar graph that displays the results of a survey about Terry Schiavo and was posted on the CNN Web site.

2.56 Lottery ticket sales, I Figure 2.28 is a line graph of the total lottery ticket sales (in millions of dollars) in the United States from 1980 to 2007. What's wrong with this picture?

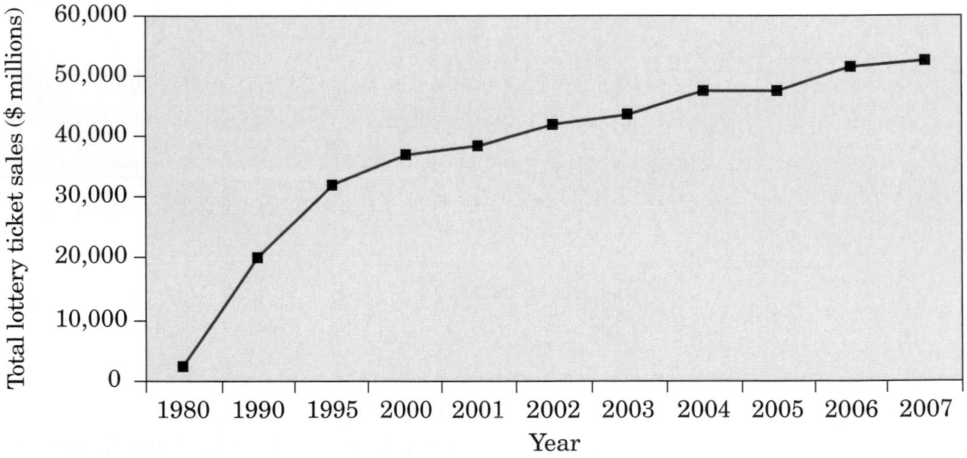

Figure 2.28 Line graph of lottery ticket sales in the United States, in millions of dollars, from 1980 to 2007.

2.57 Lottery ticket sales, II Refer to the previous exercise. Draw a new line graph of the data displayed in Figure 2.28 that isn't deceptive. Describe what you see.

2.58 Getting to school Students in a high school statistics class were given data about the primary method of transportation to school for a group of 30 students. They produced the pictograph shown on the facing page.

(a) How is this graph misleading?

(b) Make a graph of the data that is not misleading.

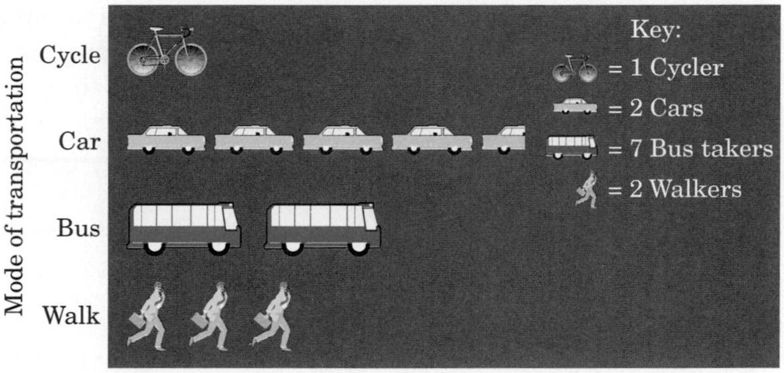

Do the numbers make sense?

According to the July 21 to 28, 2008, edition of the *Nation,* "Roughly one-third of all eligible Americans, 64 million people, are not registered to vote. This percentage is even higher for African-Americans (30 percent) and Hispanics (40 percent)." Did you notice the error in that quote? If not, read it again now.

Some people lack the skills needed to use numbers carefully. Others use data to argue a cause and care more for the cause than for the accuracy of the data. Some people feed us fake data. We know that we should always ask:

- How were the data produced?
- What exactly was measured?

To help develop number sense, we will look at how bad data, or good data used wrongly, can trick the unwary.

What didn't they tell us? The most common way to mislead with data is to cite correct numbers that don't tell the whole story. The numbers are not made up, so the fact that the information is a bit incomplete may be an innocent oversight. Or maybe not.

Example 2.21 Snow! Snow! Snow!

Crested Butte attracts skiers by advertising that it has the highest average snowfall of any ski town in Colorado. That's true. But skiers want snow on the ski slopes, not in the town—and many other Colorado resorts get more snow on the slopes.[19]

Example 2.22 We attract really good students

Colleges know that many prospective students look at popular guidebooks to decide where to apply for admission. The guidebooks print information supplied by the colleges themselves. Surely no college would simply lie about, say, the average SAT score of its entering students. But they do want their scores to look good. How about leaving out the scores of international and remedial students? Northeastern University did this, making the average SAT score of its freshman class 50 points higher than if all students were included.

"Sure your patients have 50% fewer cavities. That's because they have 50% fewer teeth!"

If the school admits economically disadvantaged students under a special program sponsored by the state, surely no one will complain if the school leaves their SAT scores out of the average. New York University did this.[20]

The point of these examples is that numbers have a context. If you don't know the context, the lonely, isolated, naked number doesn't tell you much.

Are the numbers plausible? You can often detect suspicious numbers simply because they don't seem plausible. Sometimes you can check an implausible number against data in reliable sources like the *Statistical Abstract of the United States*. Sometimes, as the next example illustrates, you can do a calculation to show that a number can't be right.

Example 2.23 The abundant melon field

The very respectable journal *Science,* in an article on insects that attack plants, mentioned a California field that produces 750,000 melons per acre. A reader responded, "I learned as a farm boy that an acre covers 43,560 square feet, so this remarkable field produces about 17 melons per square foot. If these are cantaloupes, with each fruit covering about 1 square foot, I guess they must grow in a stack 17 deep." Here is the calculation the reader did:

$$\text{melons per square foot} = \frac{\text{melons per acre}}{\text{square feet per acre}} = \frac{750{,}000}{43{,}560} = 17.2$$

The editor, a bit embarrassed, replied that the correct figure was about 11,000 melons per acre.[21]

Oh, those percents! Rates and percents seem to cause particular trouble, as the following example illustrates.

Example 2.24 Something doesn't add up!

The *Canberra Times* reported, "Of those aged more than 60 living alone, 34% are women and only 15% are men." That's 49% of those living alone. What about the other 51%?

Calculating the percent increase or decrease in some quantity seems particularly prone to mistakes. The percent change in a quantity is found by

$$\text{percent change} = \frac{\text{amount of change}}{\text{starting value}} \times 100$$

Example 2.25 Stocks go up, stocks go down

In 2008, the NASDAQ composite index of stock prices dropped from 2653.91 to 1577.03. What percent decrease was this?

$$\begin{aligned}\text{percent change} &= \frac{\text{amount of change}}{\text{starting value}} \times 100\\ &= \frac{1577.03 - 2653.91}{2653.91} \times 100\\ &= \frac{-1076.88}{2653.91} \times 100 = -0.406 \times 100 = -40.6\%\end{aligned}$$

That's a big drop! Remember to always use the *starting* value, not the smaller value, in the denominator of your fraction.

A quantity can increase by any amount—a 100% increase just means it has doubled. But nothing can go down more than 100%—it has then lost 100% of its value, and 100% is all there is.

APPLICATION 2.3

You can't fool me!

In this section, you have seen examples of proper and improper use of statistics. This Application gives you a chance to show what you have learned.

QUESTIONS

1. The two graphs that follow display information about the percent of a Web site's traffic that comes from each of three search engines: Google, Yahoo, and MSN.[22] Explain how each graph is misleading. Make a graph of these data that isn't misleading.

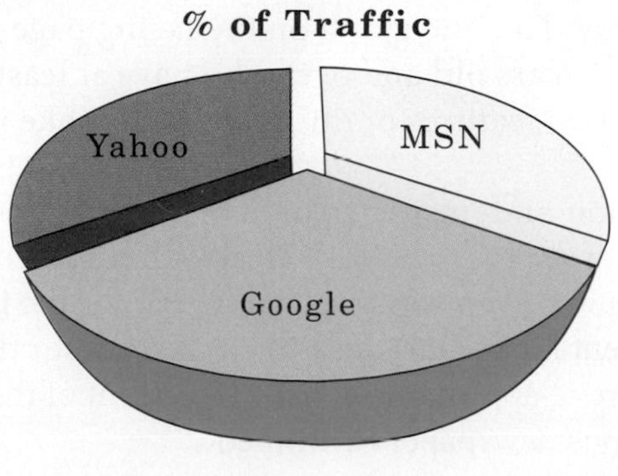

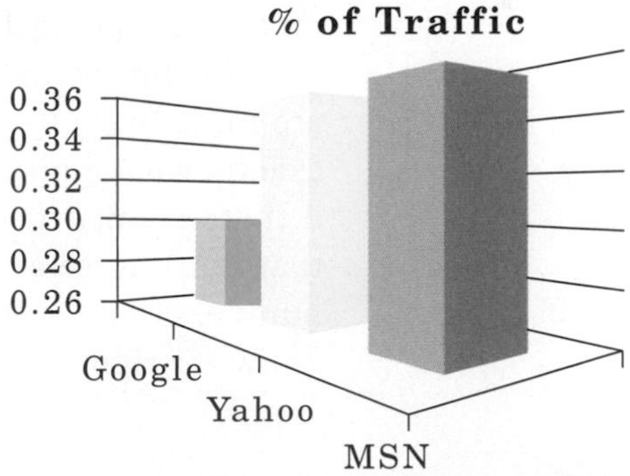

2. Continental Airlines once advertised that it had "decreased lost baggage by 100% in the past six months." Do you believe this claim?

3. Here's a quotation from a book review in a scientific journal:

> *. . . a set of 20 studies with 57 percent reporting significant results, of which 42 percent agree on one conclusion while the remaining 15 percent favor another conclusion, often the opposite one.*[23]

APPLICATION 2.3 *(continued)*

APPLICATION 2.3 *(continued)*

Do the numbers given in this quotation make sense? Explain.

4. The pointy-haired boss in the *Dilbert* cartoon once noted that 40% of all sick days taken by the staff are Fridays and Mondays. Is this evidence that the staff are trying to take long weekends?

5. An article in a midwestern newspaper about flight delays at major airports said:

> *According to a Gannett News Service study of U.S. airlines' performance during the past five months, Chicago's O'Hare Field scheduled 114,370 flights. Nearly 10 percent, 1,136, were canceled.*[24]

Check the newspaper's arithmetic. What percent of scheduled flights from O'Hare were actually canceled?

6. A report on the problem of vacation cruise ships polluting the sea by dumping garbage overboard said:

> *On a seven-day cruise, a medium-size ship (about 1,000 passengers) might accumulate 222,000 coffee cups, 72,000 soda cans, 40,000 beer cans and bottles, and 11,000 wine bottles.*[25]

Are these numbers plausible? Do some arithmetic to back up your conclusion.

Section 2.3 Summary

To show how a quantitative variable changes over time, use a **line graph** that plots values of the variable (vertical scale) against time (horizontal scale). Look for an upward or downward **trend** and for any **seasonal variation.** Graphs can mislead the eye. Look at the scales of a line graph to see if they have been stretched or squeezed to create a particular impression. Avoid **pictograms,** which replace the bars of a bar graph with pictures whose height and width both change. Avoid clutter that makes the data hard to see.

The aim of statistics is to provide insight by means of numbers. Numbers are most likely to yield their insights to those who examine them closely. Ask exactly what a number measures. Look for the context of the numbers and ask if there is important **missing information.** Look for **inconsistencies,** numbers that don't agree as they should, and check for **incorrect arithmetic.** Compare numbers that are **implausible**—surprisingly large or small—with numbers you know are right.

Section 2.3 Exercises

2.59 The rise in college education The line graph on the facing page shows the rise in the percent of women 25 years old and over who have at least a bachelor's degree. Identify at least two features of this graph that make it confusing or misleading.

2.60 Greeks on campus The question-and-answer column of a college campus newspaper was asked what percent of the campus was "Greek" (that is, members of fraternities or sororities). The answer given was that "the figures for the fall semester are approximately 13 percent for the girls and 15–18 percent for the guys, which produces a 'Greek' figure of approximately 28–31 percent of the undergraduates."[26] Discuss the campus newspaper's arithmetic.

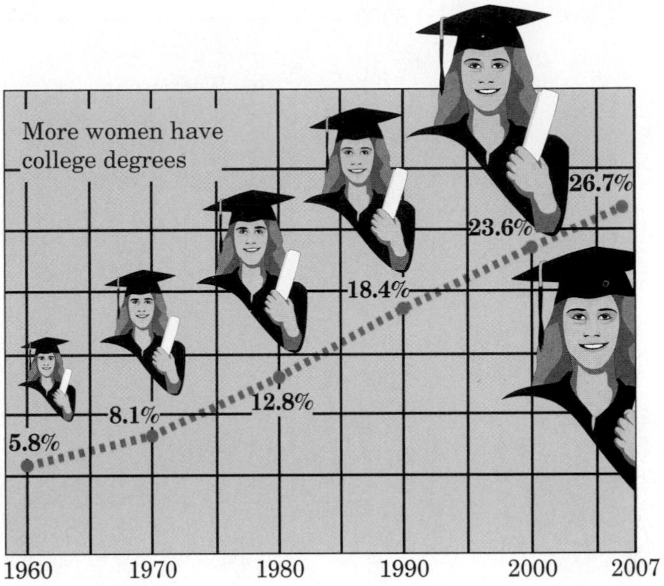

2.61 Oatmeal and cholesterol Does eating Quaker Oatmeal reduce cholesterol? An advertisement included the following graph as evidence that the answer is "Yes."

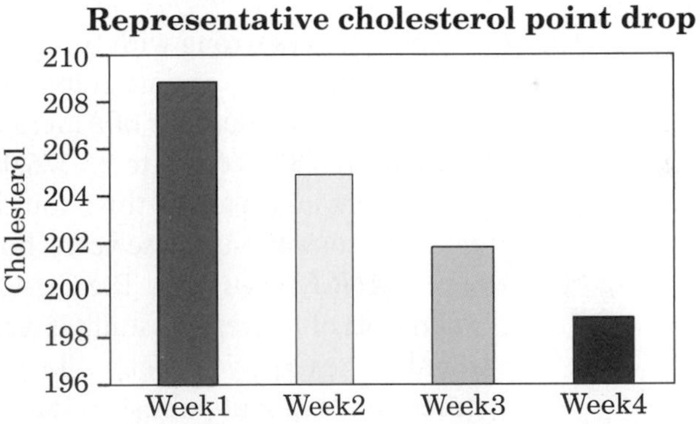

(a) How is this graph misleading?
(b) Make a new graph that isn't misleading. What do you conclude about the effect of eating Quaker Oats on cholesterol reduction?

2.62 Deer in the suburbs Westchester County is a suburban area covering 438 square miles immediately north of New York City. A garden magazine claimed that the county is home to 800,000 deer.[27] Do a calculation that shows this claim to be implausible.

2.63 We can read, but can we count? The Census Bureau once gave a simple test of literacy in English to a random sample of 3400 people. The *New York Times* printed some of the questions under the headline "113% of Adults in U.S. Failed This Test."[28] Why is the percent in the headline clearly wrong?

2.64 Online card games Figure 2.29 (on the next page) shows the number of people playing card games at the Yahoo Web site on a Sunday and on a Wednesday in the same week.

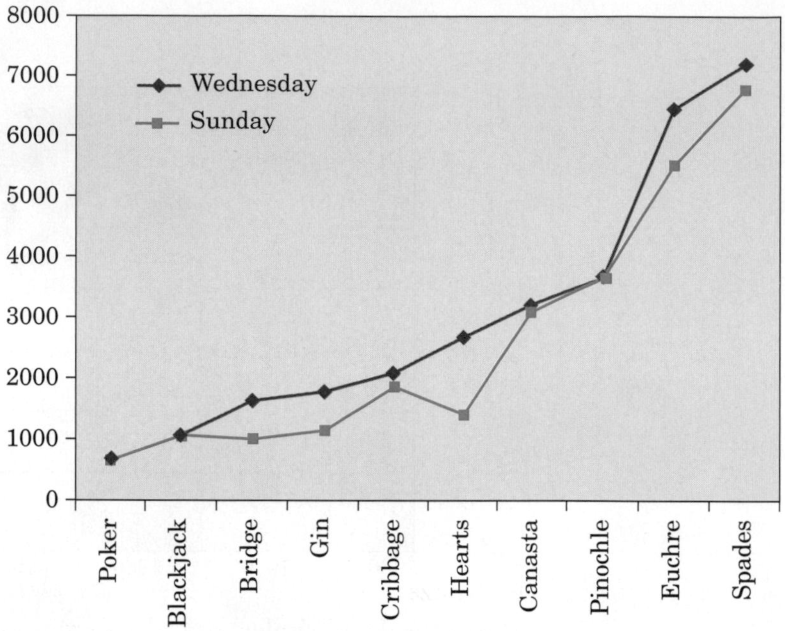

Figure 2.29 A graph showing the number of people playing various card games at the Yahoo Web site on Sunday and Wednesday of the same week.

(a) Explain what's wrong with this graph.
(b) Make a more appropriate graph to display these data.

2.65 Poverty The number of Americans living below the official poverty line increased from 32,476,000 to 37,276,000 in the 10 years between 1998 and 2007. What percent increase was this? You should not conclude from this that poverty grew more common in these years, however. Why not?

www **2.66 *CHANCE News* wiki** The CHANCE Web site at Dartmouth College contains lots of interesting stuff (at least if you are interested in statistics). In particular, the *CHANCE News* wiki at **http://chance.dartmouth.edu/chancewiki** features articles about published statistical results in the media, including deceptive statistics. Go to the site and find an example of one of the following: leaving out essential information, implausible numbers, or faulty arithmetic. Try looking in the "Forsooth" column of a recent issue.

CHAPTER 2 REVIEW

Data analysis is the art of describing data using graphs and numerical summaries. The purpose of data analysis is to help us see and understand the most important features of a set of data. Section 2.1 focused on graphical displays: pie charts, bar graphs, dotplots, stemplots, and histograms. Section 2.2 showed how to describe shape, center, and spread for distributions of quantitative variables. Section 2.3 examined proper and improper uses of graphs (line graphs, pictograms) and numerical summaries. Figure 2.30 organizes the big ideas. We plot our data, and then we describe their center and spread using either the mean and standard deviation or the median and *IQR*.

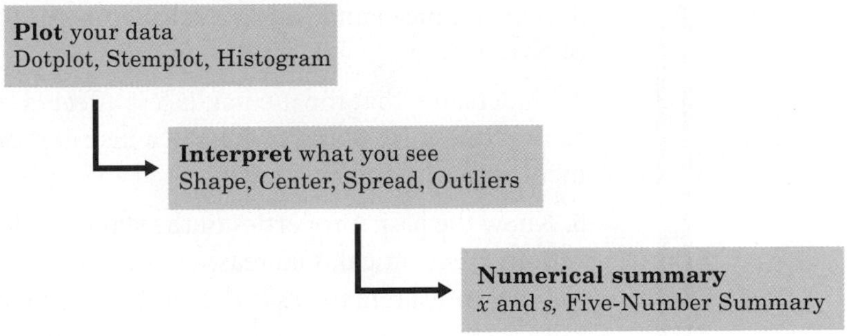

Figure 2.30 To describe the distribution of a quantitative variable, plot, interpret, and summarize.

Here is a review list of the most important skills you should have developed from your study of this chapter.

A. DISPLAYING DISTRIBUTIONS

1. Recognize when a pie chart can and cannot be used.
2. Make a bar graph of the distribution of a categorical variable or, in general, to compare related quantities.
3. Interpret pie charts and bar graphs.
4. Make a line graph of a quantitative variable over time. Recognize patterns such as trends and seasonal variation in line graphs.
5. Be aware of graphical abuses, especially in line graphs and pictograms.
6. Make a dotplot or stemplot of the distribution of a small set of observations.
7. Make a histogram of the distribution of a quantitative variable.

B. DESCRIBING DISTRIBUTIONS (QUANTITATIVE VARIABLE)

1. Look for the overall pattern of a dotplot, stemplot, or histogram and for major deviations from the pattern.
2. Assess from a dotplot, stemplot, or histogram whether the shape of a distribution is roughly symmetric, distinctly skewed, or neither. Assess whether the distribution has one or more major peaks.
3. Describe the overall pattern by giving numerical measures of center and spread in addition to a verbal description of shape.
4. Decide which measures of center and spread are more appropriate: the mean and standard deviation (especially for symmetric distributions) or the five-number summary (especially for skewed distributions).
5. Identify outliers and give plausible explanations for them.
6. Compare distributions of categorical or quantitative variables.

C. NUMERICAL SUMMARIES OF DISTRIBUTIONS

1. Find the median M and the quartiles Q_1 and Q_3 for a set of observations.
2. Give the five-number summary and draw a boxplot; use boxplots to compare distributions.

3. Find the mean and (using a calculator) the standard deviation *s* for a small set of observations.

4. Understand that the median is less affected by extreme observations than the mean. Recognize that skewness in a distribution moves the mean away from the median toward the long tail.

5. Know the basic properties of the standard deviation: $s = 0$ only when all observations are identical; *s* increases as the spread increases; *s* has the same units as the original measurements; *s* is pulled strongly up by outliers or skewness.

CHAPTER 2 REVIEW EXERCISES

2.67 Longevity of presidents Table 2.9 shows the ages at death of U.S. presidents.

Table 2.9 Age at Death of U.S. Presidents

Washington	67	Fillmore	74	T. Roosevelt	60	J. Adams	90
Pierce	64	Taft	72	Jefferson	83	Buchanan	77
Wilson	67	Madison	85	Lincoln	56	Harding	57
Monroe	73	A. Johnson	66	Coolidge	60	J. Q. Adams	80
Grant	63	Hoover	90	Jackson	78	Hayes	70
F. Roosevelt	63	Van Buren	79	Garfield	49	Truman	88
Harrison	68	Arthur	56	Eisenhower	78	Tyler	71
Cleveland	71	Kennedy	46	Polk	53	Harrison	67
L. Johnson	64	Taylor	65	McKinley	58	Nixon	81
Ford	93	Reagan	93				

(a) Make a stemplot of these data. Did you decide to split the stems?
(b) Now make a histogram. Describe the shape, center, and spread of the distribution. Are there any outliers?
(c) Which plot is better at displaying this distribution: a stemplot or a histogram? Why?

2.68 Be creative! Make up a list of numbers of which only 10% are above the average (that is, above the mean). What percent of the numbers in your list fall above the median?

2.69 Smoking-related deaths Below is a table from *Smoking and Health Now,* a report of the British Royal College of Physicians. It shows the number and percent of deaths among men aged 35 and over from the chief diseases related to smoking. One of the entries in the table is incorrect and an erratum slip was inserted to correct it. Which entry is wrong, and what is the correct value?

	Lung cancer	Chronic bronchitis	Coronary heart disease	All causes
Number	26,973	24,976	85,892	312,537
Percent	8.6%	8.0%	2.75%	100%

2.70 Who sells cars? Figure 2.31 is a pie chart of the percent of passenger car sales in a given year by various manufacturers.
(a) The artist has tried to make the graph more interesting by using the wheel of a car for the "pie." Is the graph still a correct display of the data? Explain your answer.
(b) Make a bar graph of the data. What advantage does your new graph have over the pie chart in Figure 2.31?

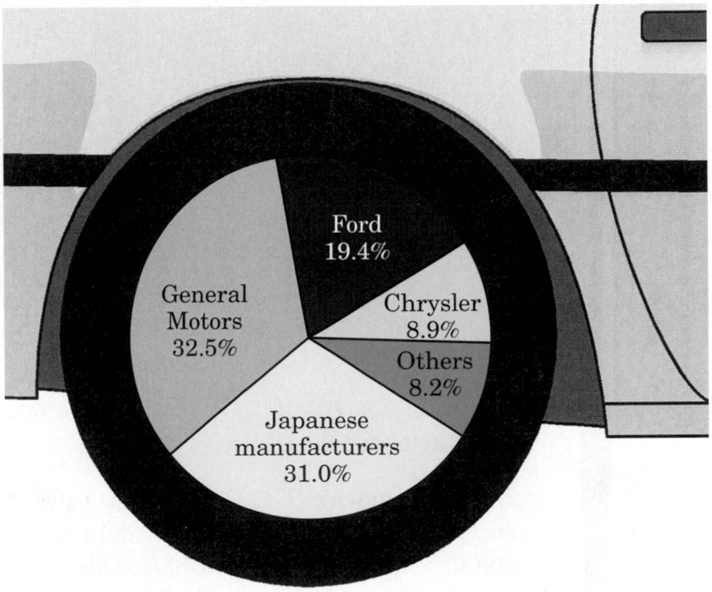

Figure 2.31 Passenger car sales by several manufacturers in the same year.

2.71 We pay high interest Figure 2.32 shows a graph taken from an advertisement for an investment that promises to pay a higher interest rate than bank accounts and other competing investments. Is this graph a correct comparison of the four interest rates? Explain your answer.

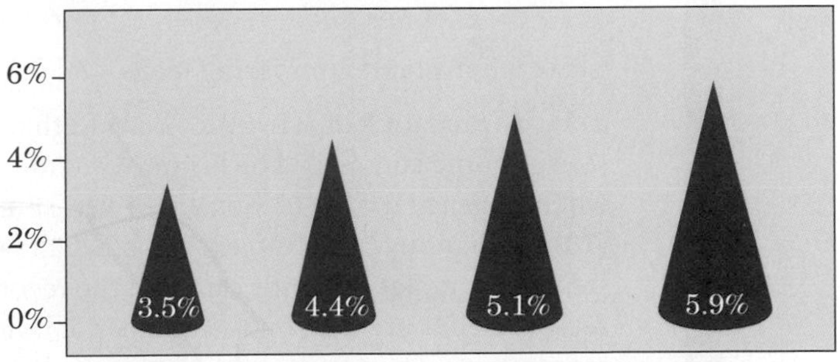

Figure 2.32 Pictogram comparing interest rates.

2.72 State SAT scores Figure 2.33 (on the next page) is a histogram of the average scores on the mathematics part of the SAT exam for students in the 50 states and the District of Columbia.[29] The distinctive overall shape of this distribution implies that a single measure of center such as the mean or the median is of little value in describing the distribution. Explain why.

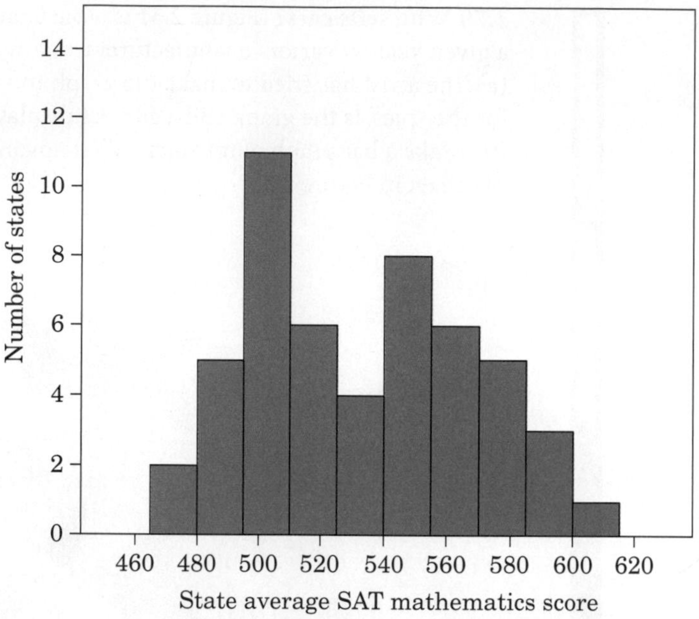

Figure 2.33 Histogram of the average scores on the SAT Math exam for students in the 50 states and the District of Columbia.

2.73 Are you wired? In late 2006, the Pew Internet and American Life Project conducted a telephone survey of 935 adults and their children aged 12 to 17. One question they asked was "Do you, personally, happen to have . . . a desktop computer? A cell phone? An iPod or other MP3 player? A laptop computer? A PDA like a Palm Pilot or Blackberry?" The table below summarizes the responses.

	Desktop	Cell phone	iPod/MP3	Laptop	PDA
Parent and teen both own one	64%	60%	22%	18%	1%
Parent owns one, teen doesn't	19%	29%	7%	19%	12%
Teen owns one, parent doesn't	8%	3%	29%	7%	7%
Neither parent nor teen owns one	9%	8%	42%	56%	80%

Make a bar graph comparing teens' and parents' ownership of these devices.

2.74 Home run king In 1927, Babe Ruth broke Major League Baseball's single-season home run record by hitting 60 home runs. The Babe's record stood until 1961, when Roger Maris hit 61 homers in a season. Almost 40 years later, Mark McGwire (70) and Sammy Sosa (68) excited baseball fans by smashing Maris's record in the same season. Barry Bonds captured the record in 2001 by hitting 73 home runs in a season. Below are data on the number of home runs hit by Bonds and McGwire each season during the prime of their careers. Who is the better home run hitter? Make comparative boxplots and provide numerical evidence to support your answer.

Bonds

16 25 24 19 33 25 34 46 37 33 42 40 37 34 49 73

McGwire

49 32 33 39 22 42 39 52 58 70

2.75 Getting more sleep An experiment was carried out with 10 patients to investigate the effectiveness of a drug that was designed to increase sleep time. The data below show the number of additional hours of sleep gained by each subject after taking the drug.[30] (A negative value indicates that the subject got less sleep after taking the drug.) Do these data provide sufficient evidence to conclude that the drug was effective? Follow the four-step problem-solving process from Chapter 1 in answering this question.

1.9 0.8 1.1 0.1 −0.1 4.4 5.5 1.6 4.6 3.4

2.76 Sleep: hours or minutes? Refer to the previous exercise. Suppose the sleep increase data are converted from hours to minutes. How will this affect the mean, median, standard deviation, and *IQR*? Explain.

CHAPTER 3

Modeling Distributions of Data

AP Photo/Joseph Kaczmarek

Do you sudoku?

The sudoku craze has officially swept the globe. Here's what Will Shortz, crossword puzzle editor for the *New York Times,* said about sudoku in August 2005:

> *As humans we seem to have an innate desire to fill up empty spaces. This might explain part of the appeal of sudoku, the new international craze, with its empty squares to be filled with digits. Since April, when sudoku was introduced to the United States in the* New York Post, *more than half the leading American newspapers have begun printing one or more sudoku a day. No puzzle has had such a fast introduction in newspapers since the crossword craze of 1924–25.*[1]

Since then, millions of people have made sudoku part of their daily routines.

www One of the authors played an online game of sudoku at **www.websudoku.com.** The graph below provides information about how well he did. (His time is marked with an arrow.)

In this chapter, you'll learn more about how to describe the location of an individual observation—like the author's sudoku time—within a distribution.

3.1 Measuring Location in a Distribution

ACTIVITY 3.1

Where do I stand?

Materials: Masking tape to mark a number line scale; rope (if desired) to make human boxplot

In this Activity, you and your classmates will explore ways to describe where you stand (literally!) within a distribution.

1. Your teacher will mark out a number line on the floor with a scale running from about 58 to 78 inches.

2. Make a human dotplot. Each member of the class should stand above the appropriate location along the number line scale based on height (to the nearest inch).

3. Count the number of people in the class that have heights less than or equal to your height. Remember this value—you'll use it shortly.

4. Your teacher will make a copy of the dotplot on the board for your reference.

5. What percent of the students in the class have heights equal to or less than yours? This is your *percentile* in the distribution of heights.

6. Work with a partner to calculate the mean and standard deviation of the class's height distribution from the dotplot.

7. Where does your height fall relative to the mean: above or below? How far above or below the mean is it? How many standard deviations above or below the mean is it? This last number is the *z*-score corresponding to your height.

www Want to know more about where you stand—in terms of height, weight, or even body mass index? Go to the National Center for Health Statistics Web site, **www.cdc.gov/growthcharts/**, and click on the "Clinical Growth Charts" link.

Mrs. Navard's statistics class has just completed the first three steps of Activity 3.1. Figure 3.1 shows a dotplot of the class's height distribution, along with summary statistics from computer output. Lynette, who is a student in the class, is 65 inches tall. Is she tall or short relative to her classmates?

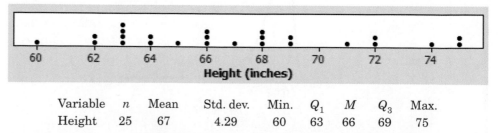

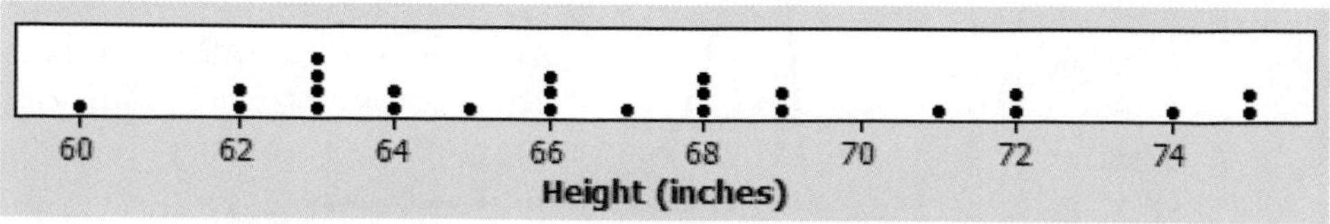

Variable	n	Mean	Std. dev.	Min.	Q_1	M	Q_3	Max.
Height	25	67	4.29	60	63	66	69	75

Figure 3.1 Dotplot and summary statistics for the heights of Mrs. Navard's statistics students.

Measuring location: percentiles

One way to describe Lynette's location within the distribution of heights is to tell what percent of students in the class are her height or shorter. That is, we can calculate Lynette's **percentile.**

Percentile

The ***p*th percentile** of a distribution is the value with *p* percent of the observations less than or equal to it.

Using the dotplot in Figure 3.1, we see that Lynette's 65-inch height is tenth from the bottom. Since 10 of the 25 observations (40%) are at or below her height, Lynette is at the 40th percentile in the class's height distribution.

Example 3.1 Where do they stand: percentiles?

For Brett, who is 74 inches tall, only two students in the class are taller than he is. Brett's percentile is computed as follows: $23/25 = 0.92$, or 92%. So Brett's height puts him at the 92nd percentile in the class's height distribution.

The two students whose heights are 75 inches are by definition at the 100th percentile, since 25 out of 25 students are that height or less. **Note that some people define the *p*th percentile of a distribution as the value with *p* percent of observations *below* it.** Using this alternative definition of percentile, it is never possible for an individual to fall at the 100th percentile. That is why you never see a standardized test score reported above the 99th percentile.

Consider the three students who are 66 inches tall. Thirteen of the 25 students in the class are this height or shorter, which, by our definition of percentile, places these three students at the 52nd percentile. If we used the alternative definition of percentile, these three students would fall at the 40th percentile (10 of the 25 heights are *below* 66 inches). But we can see from the computer output in Figure 3.1 that 66 inches is the median height for the class. So we'd prefer to think of these heights as being at the 50th percentile. Calculating percentiles is not an exact science!

Measuring location: *z*-scores

Where does Lynette's height of 65 inches fall relative to the mean of this distribution? Since the mean height of the class is 67 inches, we can see that Lynette's height is "below average." But how much below average is it?

We can describe Lynette's location within the class's height distribution by telling how many standard deviations above or below the mean her height is. Since the mean is 67 inches and the standard deviation is a little over 4 inches, Lynette's height is about one-half standard deviation below the mean. Converting observations like this from original values to standard deviation units is known as **standardizing.** To standardize a value, subtract the mean of the distribution and then divide by the standard deviation.

Standardized values and *z*-scores

If x is an observation from a distribution that has known mean and standard deviation, the **standardized value** of x is

$$z = \frac{x - \text{mean}}{\text{standard deviation}}$$

A standardized value is often called a ***z*-score.**

A z-score tells us how many standard deviations from the mean the original observation falls, and in what direction. Observations larger than the mean have positive z-scores; observations smaller than the mean have negative z-scores.

Example 3.2 Where do they stand: z-scores?

Lynette's height is 65 inches. Her *standardized* height is

$$z = \frac{x - \text{mean}}{\text{standard deviation}} = \frac{65 - 67}{4.29} = -0.47$$

In other words, her height is about one-half of a standard deviation below the mean height of the class.

Brett, who is 74 inches tall, has a corresponding z-score of

$$z = \frac{x - \text{mean}}{\text{standard deviation}} = \frac{74 - 67}{4.29} = 1.63$$

His height is 1.63 standard deviations above the mean height of the class.

We can also use z-scores to compare the locations of individuals in different distributions, as the following example illustrates.

Example 3.3 SAT versus ACT

Sofia scores 660 on the SAT Math test. The distribution of SAT scores in the population is roughly symmetric and single-peaked with mean 500 and standard deviation 100. Jim takes the ACT Math test and scores 26. ACT scores follow a symmetric, single-peaked distribution with mean 18 and standard deviation 6. Assuming that both tests measure the same kind of ability, who did better?

Sofia's standardized score on the SAT Math test is

$$z = \frac{x - \text{mean}}{\text{standard deviation}} = \frac{660 - 500}{100} = 1.60$$

So her score is 1.6 standard deviations above the mean. Jim's 26 on the ACT Math test converts to a z-score of

$$z = \frac{x - \text{mean}}{\text{standard deviation}} = \frac{26 - 18}{6} = 1.33$$

That is, Jim's score is 1.33 standard deviations above the mean. Sofia performed slightly better on the SAT than Jim did on the ACT.

We often standardize observations from symmetric distributions to express them on a common scale. We might, for example, compare the heights of two children of different ages by calculating their z-scores. The standardized scores tell us where each child stands (pun intended!) in the distribution for his or her age group.

Exercises

3.1 Interpreting percentiles

(a) Mrs. Munson is concerned about how her daughter's height and weight compare with those of other girls her age. She uses an online calculator to determine that her daughter is at the 87th percentile for weight and the 67th percentile for height. Explain to Mrs. Munson what this means.

(b) According to the *Los Angeles Times,* speed limits on California highways are set at the 85th percentile of vehicle speeds on those stretches of road. Explain to someone who knows little statistics what that means.

3.2 Comparing performance: percentiles Peter is a star runner on the track team, and Molly is one of the best sprinters on the swim team. Both athletes qualify for the league championship meet based on their performance during the regular season.

(a) In the track playoffs, Peter records a time that would fall at the 80th percentile of all his race times that season. But his performance places him at the 50th percentile in the league championship meet. Explain how this is possible.

(b) Molly swims a bit slowly for her in the league swim meet, recording a time that would fall at the 50th percentile of all her meet times that season. But her performance places Molly at the 80th percentile in this event at the league meet. Explain how this could happen.

3.3 Measuring bone density Individuals with low bone density have a high risk of broken bones (fractures). Physicians who are concerned about low bone density (osteoporosis) in patients can refer them for specialized testing. Currently, the most common method for testing bone density is dual-energy X-ray absorptiometry (DEXA). A patient who undergoes a DEXA test usually gets bone density results in grams per square centimeter (g/cm^2) and also in standardized units.

Francine, who is 25 years old, has her bone density measured using DEXA. Her results indicate a bone density in the hip of 948 g/cm^2, which converts to a standardized score of $z = -1.45$. In the reference population of 25-year-old women like Francine,[2] the mean bone density in the hip is 956 g/cm^2.

(a) Francine has not taken a statistics class in a few years. Explain to her in simple language what the standardized score tells her about her bone density.

(b) Use the information provided to calculate the standard deviation of bone density in the reference population.

3.4 Comparing bone density Refer to the previous exercise. One of Francine's friends, Louise, has the bone density in her hip measured using DEXA. Louise is 35 years old. Her bone density is also reported as 948 g/cm^2, but her standardized score is $z = 0.50$. The mean bone density in the hip for the reference population of 35-year-old women is 944 g/cm^2.

(a) Whose bones are healthier—Francine's or Louise's? Justify your answer.

(b) Calculate the standard deviation of bone density in Louise's reference population. How does this compare with your answer to Exercise 3.3(b)? Does this make sense to you?

Exercises 3.5 and 3.6 refer to information found in Table 3.1 and Figure 3.2 about the salaries of the 2008 World Champion Philadelphia Phillies baseball team.

Table 3.1 shows the salaries for each member of the Phillies baseball team on the opening day of the 2008 season. Figure 3.2 gives a dotplot and summary statistics for the salary data.

Table 3.1 Opening-Day Salaries for the Philadelphia Phillies, 2008

Player	Salary ($)	Player	Salary ($)	Player	Salary ($)
Burrell, Pat	14,250,000	Romero, J. C.	3,250,000	Kendrick, Kyle	445,000
Howard, Ryan	10,000,000	Feliz, Pedro	3,000,000	Dobbs, Greg	440,000
Myers, Brett	8,583,333	Helms, Wes	2,400,000	Ruiz, Carlos	425,000
Rollins, Jimmy	8,000,000	Werth, Jayson	1,700,000	Condrey, Clayton	420,000
Eaton, Adam	7,958,333	Madson, Ryan	1,400,000	Coste, Chris	415,000
Utley, Chase	7,785,714	Durbin, Chad	900,000	Rosario, Francisco	395,000
Lidge, Brad	6,350,000	Taguchi, So	900,000	Zagurski, Mike	392,500
Moyer, Jamie	6,000,000	Bruntlett, Eric	600,000	Lahey, Tim	390,000
Gordon, Tom	5,500,000	Hamels, Cole	500,000	Mathieson, Scott	390,000
Jenkins, Geoff	5,000,000	Victorino, Shane	480,000		

Source: USA Today online salary data base, **http://content.usatoday.com/sports/baseball/salaries/default.aspx.**

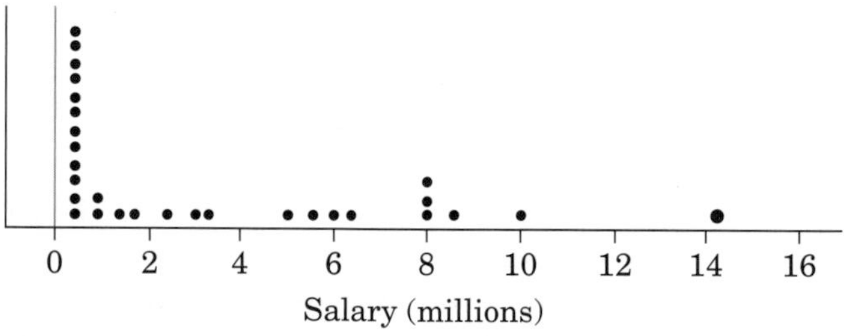

Variable	n	Mean	Std. dev.	Min.	Q_1	M	Q_3	Max.
Salary	29	3388617	3767484	390000	440000	1400000	6000000	14250000

Figure 3.2 Dotplot of salaries for the 2008 Philadelphia Phillies baseball team, along with summary statistics for these data.

Elsa/Getty Images

3.5 Baseball salaries, I Brad Lidge played a crucial role as the Phillies' "closer"; that is, he pitched the end of many games throughout the season.

(a) Find the percentile corresponding to Lidge's salary. Explain what this value means.

(b) Find the z-score corresponding to Lidge's salary. Explain what this value means.

3.6 Baseball salaries, II Did Ryan Madson have a high salary or a low salary compared with the rest of the team? Justify your answer using Madson's percentile and z-score.

Density curves

We now have a kit of graphical and numerical tools for describing distributions. What's more, we have a clear strategy for exploring data on a single quantitative variable:

1. Always plot your data: make a graph, usually a histogram or a stemplot.
2. Look for the overall pattern (shape, center, spread) and for striking deviations such as outliers.
3. Choose either the five-number summary or the mean and standard deviation to briefly describe center and spread in numbers.

Here is one more step to add to this strategy:

4. Sometimes the overall pattern of a large number of observations is so regular that we can describe it by a smooth curve.

Example 3.4 Histograms and smooth curves

Figure 3.3 is a histogram of the scores of all 947 seventh-grade students in Gary, Indiana, on the vocabulary part of the Iowa Test of Basic Skills.[3] Scores on this national test have a very regular distribution.

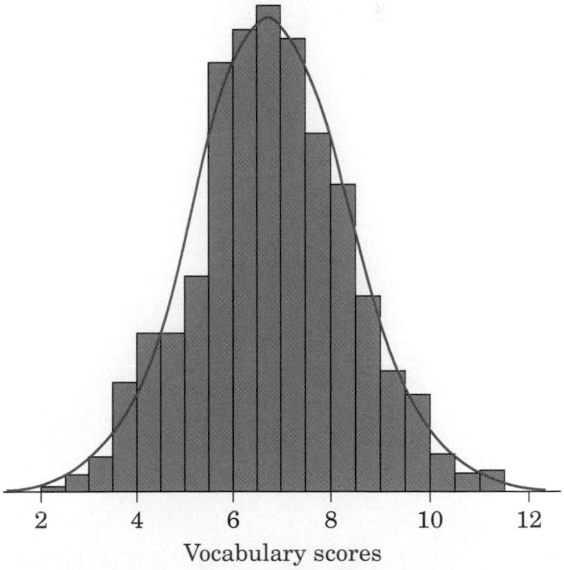

Figure 3.3 Histogram of the vocabulary scores of all seventh-grade students in Gary, Indiana.

The histogram is symmetric, and both tails fall off smoothly from a single center peak. There are no large gaps or obvious outliers. The smooth curve drawn through the tops of the histogram bars in Figure 3.3 is a good description of the overall pattern of the data. It is called a *density curve*.

Figure 3.3 shows a curve used in place of a histogram to picture the overall shape of a distribution of data. You can think of drawing a curve through the tops of the bars in a histogram and smoothing out the irregular ups and downs of the bars. There is one important distinction between histograms and these curves. Most histograms show the *counts* of observations in each class by the heights of their bars and therefore by the areas of the bars. We set up curves to show the *proportion* of observations in any region by areas under the curve. To do that, we choose the scale so that the total area under the curve is exactly 1. We then have a **density curve.**

Example 3.5 Seventh-grade vocabulary scores and density curves

Our eyes respond to the areas of the bars in a histogram. The bar areas represent proportions of the observations. Figure 3.4(a) is a copy of Figure 3.3 with the left-most bars shaded dark blue.

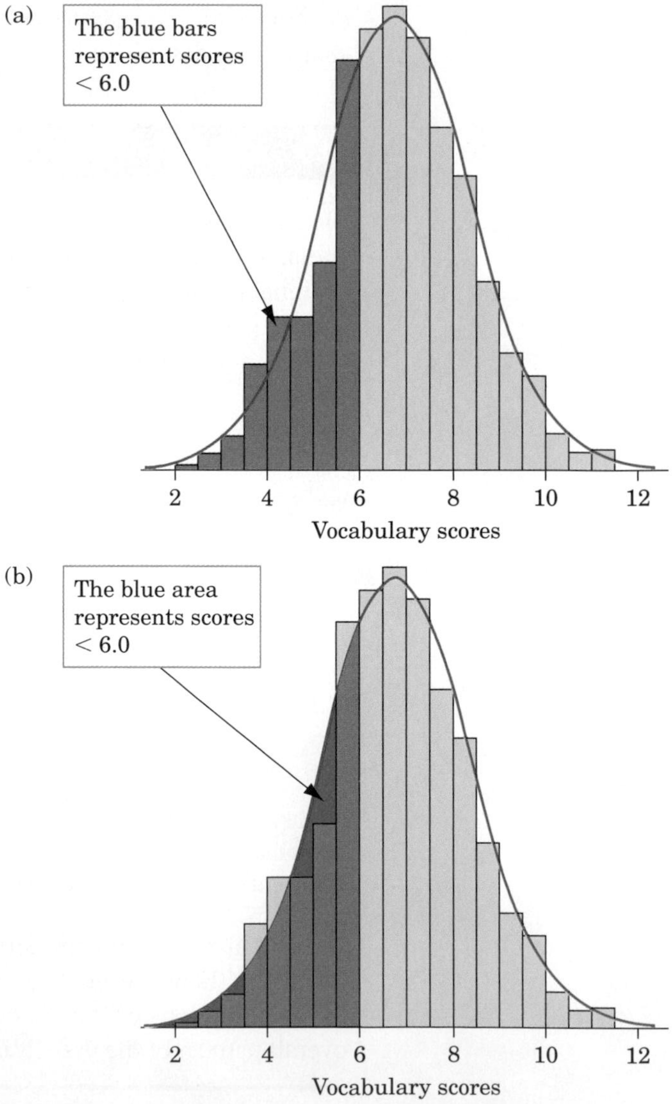

Figure 3.4 (a) The proportion of scores less than 6.0 from the histogram is 0.303. (b) The proportion of scores less than 6.0 from the density curve is 0.293.

The area of the dark blue bars in Figure 3.4(a) represents the students with vocabulary scores lower than 6.0. There are 287 such students, who make up the proportion $287/947 = 0.303$ of all Gary seventh-graders. In other words, a score of 6.0 corresponds to about the 30th percentile.

Now concentrate on the curve drawn through the bars. In Figure 3.4(b), the area under the curve to the left of 6.0 is shaded. Adjust the scale of the graph so that the total area under the curve is exactly 1. This area represents the proportion 1, that is, all the observations. Areas under the curve then represent proportions of the observations. The curve is now a *density curve*. The dark blue area under the density curve in Figure 3.4(b) represents the proportion of students with scores lower than 6.0. This area is 0.293, only 0.010 away from the histogram result. So our estimate based on the density curve is that a score of 6.0 falls at about the 29th percentile. You can see that areas under the density curve are quite good approximations of areas given by the histogram.

The area under the density curve in Figure 3.4(b) is not exactly equal to the true proportion, because the curve is an idealized picture of the distribution. For example, the curve is exactly symmetric but the actual data are only approximately symmetric. Because density curves are smoothed-out, idealized pictures of the overall shapes of distributions, they are most useful for describing large numbers of observations.

The density curve in Figures 3.3 and 3.4 is a *Normal curve*. The Normal curve has a distinctive, symmetric, single-peaked bell shape. The Normal curve is easy to work with and does not require clever software. We will see in Section 3.2 that Normal curves have special properties that help us use them and think about them. Only some kinds of data fit Normal curves, however, so keep the technology handy for those that don't.

The center and spread of a density curve

Density curves help us better understand our measures of center and spread. The median and quartiles are easy. Areas under a density curve represent proportions of the total number of observations. The median is the point with half the observations on either side. So *the median of a density curve is the equal-areas point,* the point with half the area under the curve to its left and the remaining half of the area to its right. The quartiles divide the area under the curve into quarters. One-fourth of the area under the curve is to the left of the first quartile, and three-fourths of the area is to the left of the third quartile. You can roughly locate the median and quartiles of any density curve by eye by dividing the area under the curve into four equal parts.

Because density curves are idealized patterns, a symmetric density curve is exactly symmetric. The median of a symmetric density curve is therefore at its center. Figure 3.5(a) (on the next page) shows the median of a symmetric curve. Even on a skewed curve like that in Figure 3.5(b), we can roughly locate the equal-areas point by eye.

What about the mean? The mean of a set of observations is their arithmetic average. If we think of the observations as weights stacked on a seesaw, the mean is

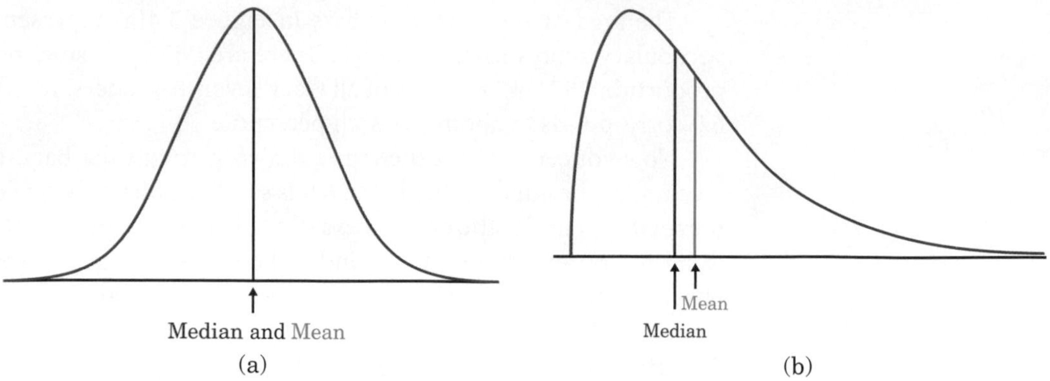

Figure 3.5 The median and mean for two density curves: a symmetric Normal curve and a curve that is skewed to the right.

the point at which the seesaw would balance. This fact is also true of density curves. *The mean is the point at which the curve would balance if made of solid material.* Figure 3.6 illustrates this fact about the mean. A symmetric curve balances at its center because the two sides are identical. The mean and median of a symmetric density curve are equal, as in Figure 3.5(a). We know that the mean of a skewed distribution is pulled toward the long tail. Figure 3.5(b) shows how the mean of a skewed density curve is pulled toward the long tail more than is the median.

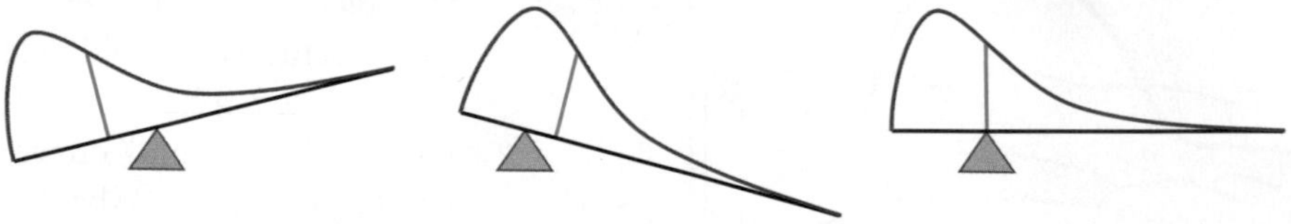

Figure 3.6 The mean of a density curve is the point at which it would balance.

Median and mean of a density curve

The **median** of a density curve is the equal-areas point, the point that divides the area under the curve in half.

The **mean** of a density curve is the balance point, the point at which the curve would balance if made of solid material.

The median and mean are the same for a symmetric density curve. They both lie at the center of the curve. The mean of a skewed curve is pulled away from the median in the direction of the long tail.

Exercises

3.7 A uniform distribution Figure 3.7 (on the facing page) shows the density curve for a **uniform distribution.** This curve has height 1 over the interval from 0 to 1 and is zero outside that range.

Use this density curve to answer the following questions.

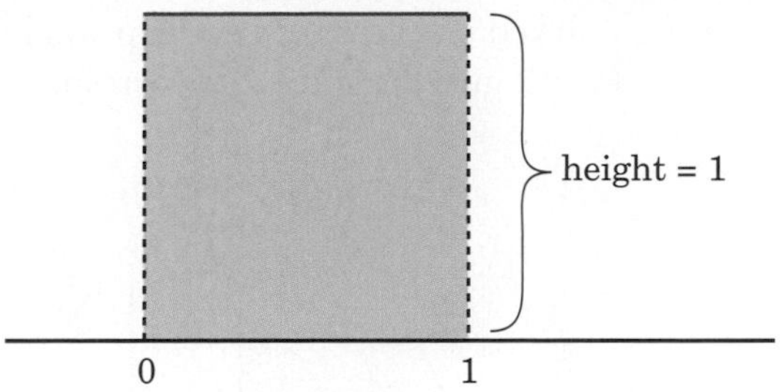

Figure 3.7 The density curve of a uniform distribution.

(a) Why is the height of the curve equal to 1?

(b) Recall that the mean of a density curve is the "balance point." What is the value of the mean of this uniform distribution?

(c) What is the median? What property of this density curve tells you that the mean and median are related?

(d) What percent of the observations lie between 0 and 0.4? Explain.

3.8 Another uniform distribution Refer to the previous exercise. Now let's consider a uniform distribution over the interval from 0 to 2.

(a) Sketch a graph of the density curve. What is the height of the curve? Why?

(b) What percent of the observations lie between 1 and 1.4? Explain.

(c) Find the median and the quartiles for this distribution. Show your work.

3.9 From histogram to density curve, I Copy the distribution in Figure 3.8 onto your paper. Then sketch a smooth curve that describes the distribution well. Mark your best guess for the mean and median of the distribution.

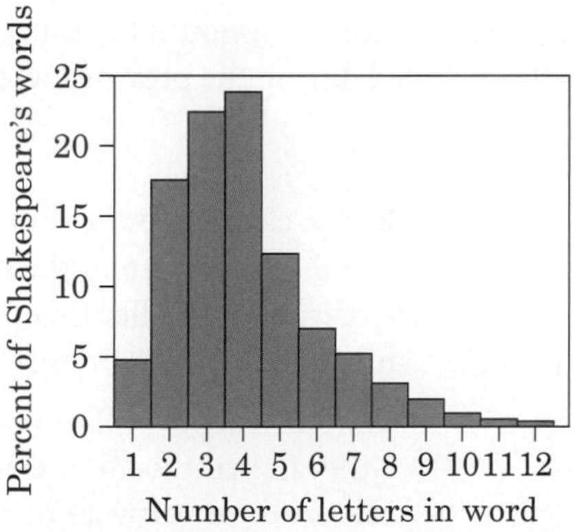

Figure 3.8 A histogram of the distribution of lengths of words used in Shakespeare's plays.

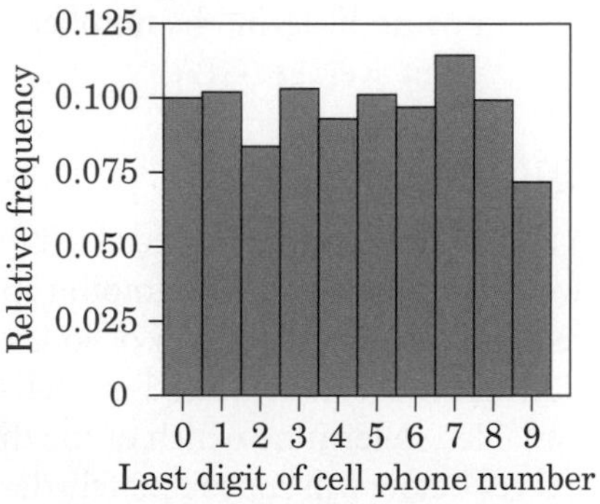

Figure 3.9 A histogram of the distribution of last digits in students' cell phone numbers.

3.10 From histogram to density curve, II Copy the distribution in Figure 3.9 onto your paper. Then sketch a smooth curve that describes the distribution well. Mark your best guess for the mean and median of the distribution.

3.11 Mean and median Figure 3.10 (on the next page) shows two density curves. Briefly describe the overall shape of each distribution. Two points are marked on each curve to help you locate the mean and the median. For each curve, give the letter that corresponds to (i) the median and (ii) the mean.

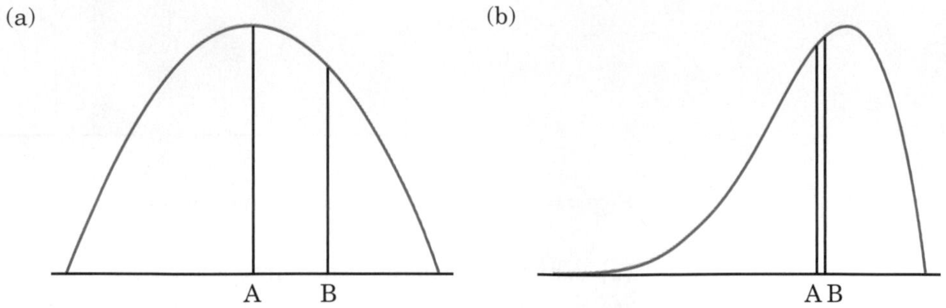

Figure 3.10 Two density curves of different shapes.

3.12 Percentiles and density curves Joey told his dad that he scored at the 98th percentile on a national standardized test. The scores on this test are approximately Normally distributed. Sketch a Normal density curve and show the approximate location of Joey's score in the distribution.

APPLICATION 3.1

A student survey

Recently, students in an introductory statistics class at a large state university responded to a survey. The first question instructed students to roll a six-sided die and record the value on the up-face. Other questions on the survey asked students to report their gender, birth month, the amount they spent on textbooks at the beginning of the term, the number of sodas they consumed during the previous week, and their high school grade point average (GPA).

QUESTIONS

1. Three of the variables recorded in the survey were outcome of die roll, amount spent on textbooks ($), and number of sodas consumed during the previous week. For each of the three variables, determine which of the three density curves below will approximately describe the distribution. Explain your choices.

2. For each of your choices in Question 1, decide whether the mean of the distribution is probably larger than, smaller than, or about the same as the median. Justify your answers.

3. Your teacher will provide you with the data, either in print form or electronically. Use your calculator or software to draw separate histograms for the three variables in Question 1 from the

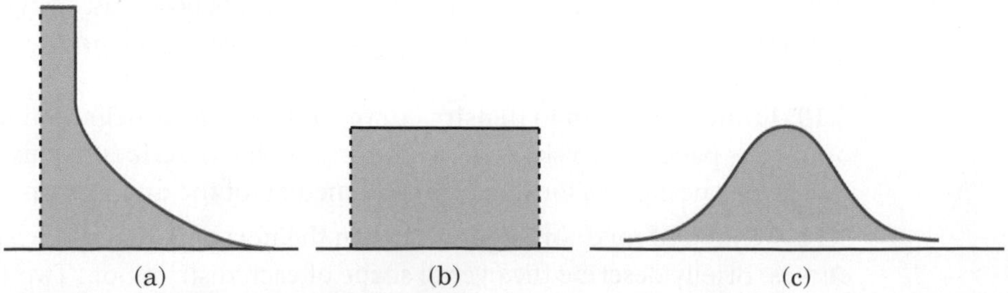

actual data. How do the histograms compare with what you expected?

4. Look at the three histograms for die roll, textbook cost, and soda consumption. Guess the value of the median, the mean, and the standard deviation for each distribution. Record your values in a table like the one below.

Variable	My guess at the median	My guess at the mean	My guess at the standard deviation
Die roll			
Textbooks			
Sodas			

5. Use your calculator or software to compute the actual values of the median, mean, and standard deviation for the three distributions. Record your results in a table like the one that follows.

Variable	Actual median	Actual mean	Actual standard deviation
Die roll			
Textbooks			
Sodas			

6. Compare the actual values with your guesses. Which variable's distribution was most difficult to judge? Explain.

7. One student reported spending $300 on books and consuming 12 sodas in the previous week.
(a) Find the z-score for each of these values.
(b) Find the percentile corresponding to each of these values.
(c) In which distribution is this student's value more surprising? Justify your answer.

Section 3.1 Summary

Two ways of describing an individual's location within a distribution are **percentiles** and ***z*-scores.** An observation's percentile is the percent of the distribution that is at or below the value of that observation. To **standardize** any observation x, subtract the mean of the distribution and then divide by the standard deviation. The resulting z-score

$$z = \frac{x - \text{mean}}{\text{standard deviation}}$$

says how many standard deviations x lies from the distribution mean. We can also use z-scores and percentiles to compare the relative standing of individuals in different distributions.

Stemplots, histograms, and boxplots all describe the distributions of quantitative variables. **Density curves** are another kind of graph that serves the same purpose. A density curve is a curve with area exactly 1 underneath it whose shape describes the overall pattern of a distribution. An area under the curve gives the proportion of the observations that fall in an interval of values. You can roughly locate the median (equal-areas point) and the mean (balance point) by eye on a density curve.

The mean and median are equal for symmetric density curves. The mean of a skewed curve is located farther toward the long tail than is the median.

Section 3.1 Exercises

3.13 Finding means and medians Figure 3.11 (on the next page) displays three density curves, each with three points indicated. At which of these points on each curve do the mean and the median fall?

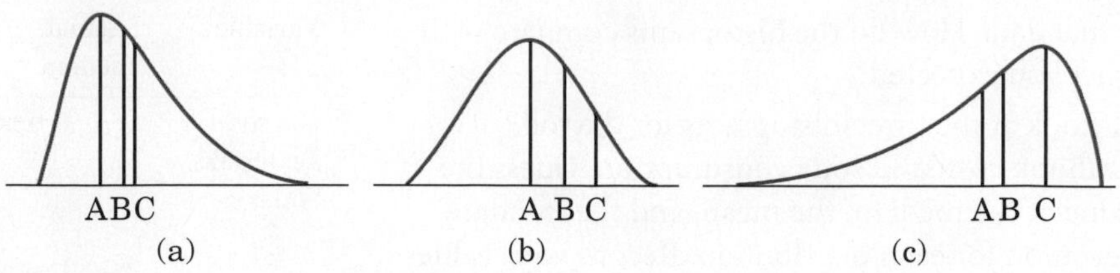

Figure 3.11 Three density curves: can you locate the mean and median?

3.14 Cholesterol: good or bad? Martin came home very excited after a visit to his doctor. He announced proudly to his wife, "My doctor says my cholesterol level is at the 90th percentile among men like me. That means I'm better off than about 90% of similar men." How should his wife, who is a statistician, respond to Martin's statement?

3.15 Unemployment in the states, I Each month the Bureau of Labor Statistics announces the unemployment rate for the previous month. Unemployment rates are economically important and politically sensitive. Unemployment may vary greatly among the states because types of work are unevenly distributed across the country. Table 3.2 presents the unemployment rates for each of the 50 states in September 2008.

Table 3.2 Unemployment Rates by State, September 2008

State	Percent	State	Percent	State	Percent
Alabama	5.3	Louisiana	5.2	Ohio	7.2
Alaska	6.8	Maine	5.6	Oklahoma	3.8
Arizona	5.9	Maryland	4.6	Oregon	6.4
Arkansas	4.9	Massachusetts	5.2	Pennsylvania	5.7
California	7.7	Michigan	8.7	Rhode Island	8.8
Colorado	5.2	Minnesota	5.9	South Carolina	7.3
Connecticut	6.1	Mississippi	7.8	South Dakota	3.2
Delaware	4.8	Missouri	6.4	Tennessee	7.2
Florida	6.6	Montana	4.6	Texas	5.1
Georgia	6.5	Nebraska	3.5	Utah	3.5
Hawaii	4.5	Nevada	7.3	Vermont	5.2
Idaho	5.0	New Hampshire	4.1	Virginia	4.3
Illinois	6.9	New Jersey	5.8	Washington	5.8
Indiana	6.2	New Mexico	4.0	West Virginia	4.5
Iowa	4.2	New York	5.8	Wisconsin	5.0
Kansas	4.8	North Carolina	7.0	Wyoming	3.3
Kentucky	7.1	North Dakota	3.6		

Source: Bureau of Labor Statistics Web site, **www.bls.gov.**

(a) Make a histogram of these data. Be sure to label and scale your axes.
(b) Calculate numerical summaries for this data set. Describe the shape, center, and spread of the distribution of unemployment rates.

(c) Determine the percentile for Illinois. Explain in simple terms what this says about the unemployment rate in Illinois relative to the other states.
(d) Which state is at the 40th percentile? Calculate the z-score for this state.

3.16 Unemployment in the states, II Refer to the previous exercise. The December 2000 unemployment rates for the 50 states had a symmetric, single-peaked distribution with a mean of 3.47% and a standard deviation of about 1%. The unemployment rate for Illinois that month was 4.5%. There were 42 states with lower unemployment rates than Illinois.
(a) Write a sentence comparing the actual rates of unemployment in Illinois in December 2000 and September 2008.
(b) Compare the percentiles for the Illinois unemployment rate in these same two months in a sentence or two.
(c) Compare the z-scores for the Illinois unemployment rate in these same two months in a sentence or two.

Exercises 3.17 to 3.20 refer to the following information. According to the National Center for Health Statistics, the distribution of heights for 15-year-old males is modeled well by a Normal density curve. For this distribution, a z-score of 0 corresponds to a height of 170 centimeters (cm), and a z-score of 1 corresponds to a height of 177.5 cm. The distribution of heights for 16-year-old females is also modeled well by a Normal density curve. For this distribution, a z-score of 0 corresponds to a height of 162.5 cm, and a z-score of 1 corresponds to a height of 169 cm.

Stone/Getty Images

3.17 Male heights Consider the height distribution for 15-year-old males.
(a) Find its mean and standard deviation. Show your method clearly.
(b) What height would correspond to a z-score of 2.5? Show your work.

3.18 Female heights Consider the height distribution for 16-year-old females.
(a) Find its mean and standard deviation. Show your method clearly.
(b) What height would correspond to a z-score of -1.5? Show your work.

3.19 Is Paul tall? Paul is 15 years old and 175 cm tall.
(a) Find the z-score corresponding to Paul's height. Explain what this value means.
(b) Paul's height puts him at the 75th percentile among 15-year-old males. Explain what this means to someone who knows no statistics.

3.20 Is Miranda taller? Miranda is 16 years old and 170 cm tall.
(a) Find the z-score corresponding to Miranda's height. Explain what this value means.
(b) Miranda's height puts her at the 88th percentile among 16-year-old females. Explain what this means to someone who knows no statistics.
(c) Refer to Exercise 3.19. Who is taller relative to their peers—Miranda or Paul? Justify your answer.

3.2 Normal Distributions

ACTIVITY 3.2A

Tennis, anyone?

Materials: Tennis ball and ruler with centimeter scale for each pair of students

What's the diameter of a tennis ball? In this Activity, you and your classmates will make some measurements to try to answer this question. Your teacher will provide you and a partner with a tennis ball and a ruler.

1. Each of you should measure the ball's diameter to the nearest millimeter. Do not tell your partner what value you got until both of you have measured.

2. Compare results with your partner. If your measurements are very different, discuss possible reasons for this. Then make two new measurements of the tennis ball's diameter. Continue until your two measurements are similar (but not necessarily the same). Record these values.

3. Your teacher will draw a number line on the board scaled in millimeters. Put your final two values on the class dotplot.

4. Once the dotplot is complete, discuss the shape, center, and spread of the distribution of measurements with your classmates. Are there any outliers?

5. As a class, identify as many possible sources of variation in this Activity as you can.

6. If you had to report a single value for the diameter of a tennis ball based on your class's results, what value would you choose? Explain.

Normal distributions as density curves

One particularly important class of density curves has already appeared in Figures 3.3 (page 107) and 3.5(a) (page 110). These density curves are symmetric, single-peaked, and bell-shaped. They are called **Normal curves,** and they describe **Normal distributions.** Normal distributions play a large role in statistics, but they are rather special and not at all "normal" in the sense of being average or natural. We capitalize "Normal" to remind you that these curves are special.

Why are the Normal distributions important in statistics? Here are three reasons. First, Normal distributions are good descriptions for some distributions of *real data.* Distributions that are often close to Normal include

- scores on tests taken by many people (such as SAT exams and IQ tests)
- repeated careful measurements of the same quantity (like the diameter of a tennis ball)
- characteristics of biological populations (such as yields of corn and lengths of animal pregnancies)

From Gaussian to Normal

Normal curves were first applied to data by the great mathematician Carl Friedrich Gauss (1777–1855), who used them to describe the small errors made by astronomers and surveyors in repeated careful measurements of the same quantity. You will sometimes see Normal distributions labeled "Gaussian" in honor of Gauss. For much of the nineteenth century Normal curves were called "error curves" because they were first used to describe the distribution of measurement errors. As it became clear that the distributions of some biological and psychological variables were at least roughly Normal, the "error curve" terminology was dropped. The curves were first called "normal" by Francis Galton in 1889. Galton, a cousin of Charles Darwin, pioneered the statistical study of inheritance.

Second, Normal distributions are good approximations to the results of many kinds of *chance outcomes*, such as tossing a coin many times. Third, and most important, we will see that many *statistical inference* procedures are based on Normal distributions.

Figure 3.12 presents two more Normal density curves. Normal curves are symmetric, single-peaked, and bell-shaped. Their tails fall off quickly, so that we do not expect outliers. Because Normal distributions are symmetric, the mean and median lie together at the peak in the center of the curve.

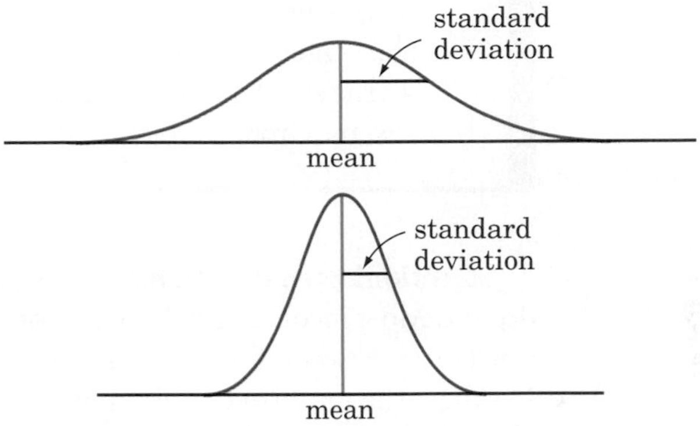

Figure 3.12 Two Normal curves. The standard deviation fixes the spread of a Normal curve.

Normal curves also have the special property that we can locate the standard deviation of the distribution by eye on the curve. This isn't true for most other density curves. Here's how to do it. Imagine that you are skiing down a mountain that has the shape of a Normal curve. At first, you descend at an ever-steeper angle as you go out from the peak:

Fortunately, before you find yourself going straight down, the slope begins to grow flatter rather than steeper as you go out and down.

The points at which this change of curvature takes place are located one standard deviation on either side of the mean. The standard deviations are marked on the two curves in Figure 3.12. You can feel the change as you run a pencil along a Normal curve, and so find the standard deviation.

Normal curves have the special property that giving the mean and the standard deviation completely specifies the curve. The mean fixes the center of the curve, and the standard deviation determines its shape. Changing the mean of a Normal distribution does not change its shape, only its location on the axis. Changing the standard deviation does change the shape of a Normal curve, as Figure 3.12 illustrates. The distribution with the smaller standard deviation is less spread out and more sharply peaked. Here is a summary of basic facts about Normal curves.

Normal density curves

The **Normal curves** are symmetric, bell-shaped curves that have these properties:

- A specific Normal curve is completely described by giving its mean and its standard deviation.
- The mean determines the center of the distribution. It is located at the center of symmetry of the curve.
- The standard deviation determines the shape of the curve. It is the distance from the mean to the change-of-curvature points on either side.

Caution! Even though many sets of data follow a Normal distribution, many do not. Most income distributions, for example, are skewed to the right and so are not Normal. Some distributions are symmetric but not Normal or even close to Normal. The uniform distribution of Exercise 3.7 (page 110) is one such example. Non-Normal data, like nonnormal people, not only are common but also are sometimes more interesting than their Normal counterparts.

The 68–95–99.7 rule

ACTIVITY 3.2B

The *Normal Curve* applet

APPLET ***Directions:*** Go to the texbook's Web site, **www.whfreeman.com/sta2e,** and select the *Normal Curve* applet. Drag the green flags to have the applet compute the part of the Normal curve that is in the bright yellow shaded region you create. If you drag one flag past the other, the applet will show the area under the curve between the two flags. When the "2-Tail" box is checked, the applet calculates symmetric areas around the mean.

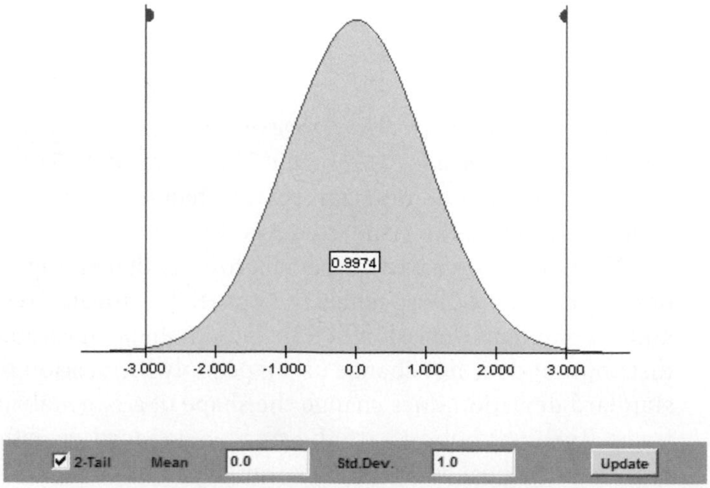

Think:

Answer the following without using the applet.

1. If you were to put one flag at the extreme left of the curve and the second flag exactly in the middle, what would be the proportion reported by the applet?

2. If you were to place the two flags exactly one standard deviation on either side of the mean, what would the applet say is the area between them?

3. As you can see from the figure, about 99.7% of the area under the Normal density curve lies within three standard deviations of the mean. Does this mean that about 99.7%/2 = 49.85% will lie within one and a half standard deviations?

Verify:

4. Use the applet to test your answers to the three questions above. Did the proportion of shaded area under the curve behave as you thought it would in each case?

APPLET TIP: The direction in which the green flags point is the side of the flag poles that is shaded.

Experiment:

5. What percent of the area under the Normal curve lies within two standard deviations of the mean?

6. Change the mean to 100 and the standard deviation to 15. Then click "Update." What percent of the area under this Normal density curve lies within one, two, and three standard deviations of the mean?

7. Change the mean to 500 and the standard deviation to 100. Answer the question in Step 6.

Summarize:

8. Complete the following sentence: "For any Normal density curve, the area under the curve within one, two, and three standard deviations of the mean is about ___%, ___%, and ___%."

There are many Normal curves, each described by its mean and standard deviation. All Normal curves share many properties. In particular, the standard deviation is the natural unit of measurement for Normal distributions. This fact is reflected in the following rule.

The 68–95–99.7 rule

In any Normal distribution, approximately

- **68%** of the observations fall within one standard deviation of the mean
- **95%** of the observations fall within two standard deviations of the mean
- **99.7%** of the observations fall within three standard deviations of the mean

Figure 3.13 (on the next page) illustrates the 68–95–99.7 rule. (Some people refer to this result as the "empirical rule.") By remembering these three numbers, you can think about Normal distributions without constantly making detailed calculations. Remember also, though, that no set of data is exactly described by a Normal curve. The 68–95–99.7 rule will be only approximately true for SAT scores or the lengths of crickets, for example.

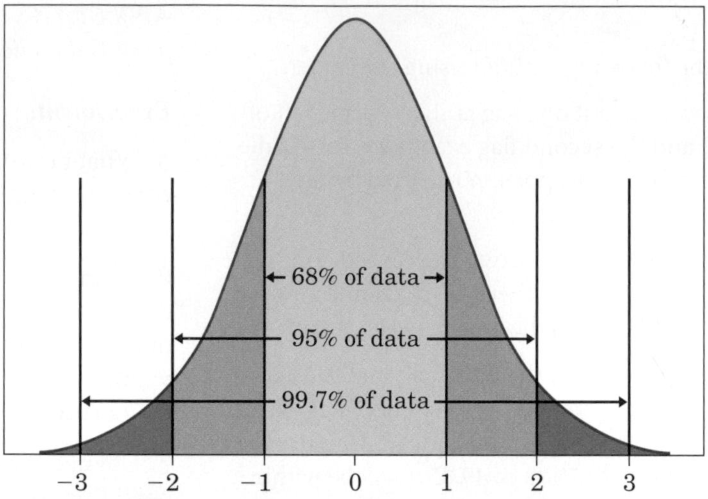

Figure 3.13 The 68–95–99.7 rule for Normal distributions.

Here's an example that shows how we can use the 68–95–99.7 rule to find a percentile.

Example 3.6 Performing well on the SAT

iStockphoto

Jennie scored 600 on the Critical Reading section of the SAT Reasoning test. How good a score is this? That depends on where a score of 600 lies in the distribution of all scores. The SAT exams are scaled so that scores on the Critical Reading section should roughly follow the Normal distribution with mean 500 and standard deviation 100.

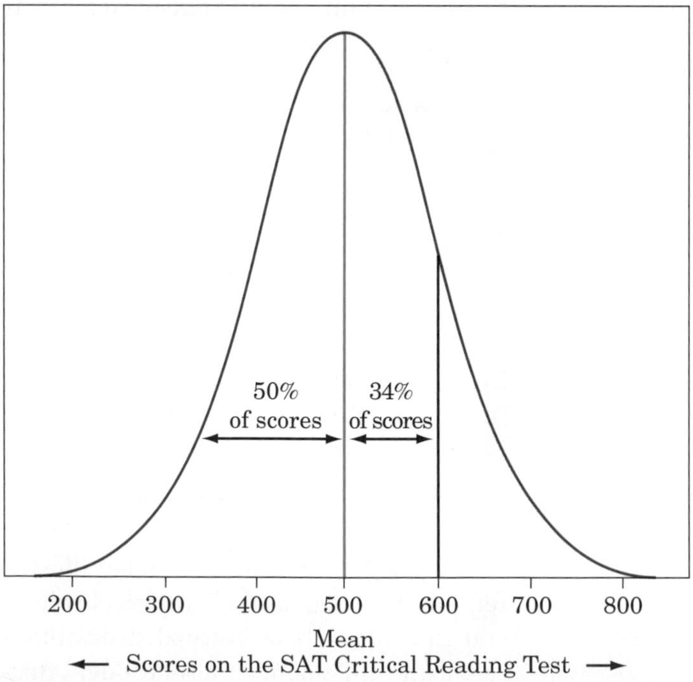

Figure 3.14 The 68–95–99.7 rule shows that 84% of any Normal distribution lies to the left of the point one standard deviation above the mean. Here, this fact is applied to SAT scores.

Jennie's 600 is one standard deviation above the mean. The 68–95–99.7 rule now tells us just where she stands (see Figure 3.14 on the facing page). Half of all scores are below 500, and another 34% are between 500 and 600. So Jennie did better than 84% of the students who took the SAT. Her score report not only will say that she scored 600 but will add that this is at the "84th percentile." As you learned in Section 3.1, that's statistics speak for "You did better than 84% of those who took the test."

Because the standard deviation is the natural unit of measurement for Normal distributions, we restated Jennie's score of 600 as "one standard deviation above the mean." As you learned in Section 3.1, observations expressed in standard deviations above or below the mean of a distribution are called z-scores or standardized scores. To obtain the standardized score for a particular observation, we compute

$$z = \frac{x - \text{mean}}{\text{standard deviation}}$$

The usual notation for the **mean of a density curve** is μ (the Greek letter "mu"). We write the **standard deviation of a density curve** as σ (the Greek letter "sigma"). These are the same symbols we use to represent the mean and standard deviation of a population distribution. This makes sense because density curves are quite often used to describe population distributions.

With our new symbols, the standardized score formula becomes

$$z = \frac{x - \mu}{\sigma}$$

For Jennie's SAT Critical Reading score, $x = 600$, $\mu = 500$, and $\sigma = 100$. Her z-score is

$$z = \frac{600 - 500}{100} = 1$$

Exercises

```
 8 | 6 9
 9 | 0 1 3 3
 9 | 6 7 7 8
10 | 0 0 2 2 3 3 3 3 4 4
10 | 5 5 5 6 6 6 7 7 7 7 8 9
11 | 0 0 0 0 1 1 1 1 2 2 2 2 3 3 3 4 4 4 4
11 | 5 5 6 8 8 9 9 9
12 | 0 0 3 3 4 4
12 | 6 7 7 8 8 8
13 | 0 2
13 | 6
```

Figure 3.15 Stemplot of the IQ scores of 74 seventh-grade students, for Exercises 3.21 to 3.23.

Use the distribution in Figure 3.15 and the 68–95–99.7 rule for Exercises 3.21 to 3.23. Figure 3.15 is a stemplot of the IQ test scores of 74 seventh-grade students.[4] This distribution is very close to Normal with mean 111 and standard deviation 11. It includes all the seventh-graders in a rural Midwest school. Take the Normal distribution with mean 111 and standard deviation 11 as a description of the IQ test scores of all rural Midwest seventh-grade students.

3.21 IQ test scores, I Between what values do the IQ scores of the middle 95% of *all* rural Midwest seventh-graders lie? Explain.

3.22 IQ test scores, II What percent of IQ scores for *all* rural Midwest seventh-graders are greater than 100? Explain. How does this compare with the percent in our sample?

3.23 IQ test scores, III What percent of all students have IQ scores of 144 or higher? Explain. None of the 74 students in our sample school had scores this high. Are you surprised at this? Why?

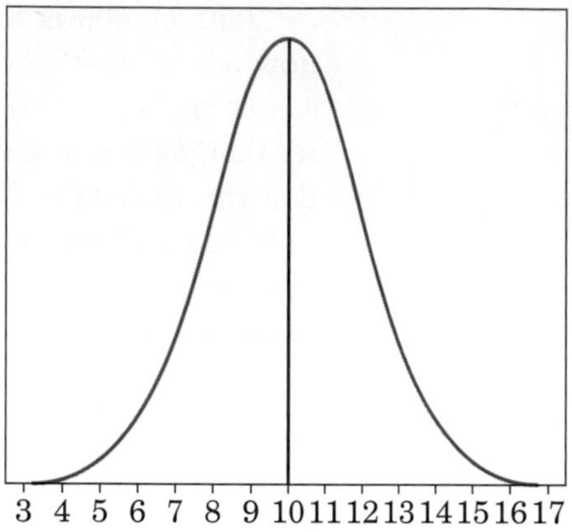

Figure 3.16 What are the mean and standard deviation of this Normal density curve?

3.24 A Normal curve Estimate the mean and standard deviation of the Normal curve in Figure 3.16.

3.25 Horse pregnancies Bigger animals tend to carry their young longer before birth. The length of horse pregnancies from conception to birth varies according to a roughly Normal distribution with mean 336 days and standard deviation 3 days. Use the 68–95–99.7 rule to answer the following questions.

(a) Almost all (99.7%) horse pregnancies fall in what range of lengths?

(b) What percent of horse pregnancies are longer than 339 days? Show your work.

3.26 Eggs A truck is loaded with cartons of eggs that weigh an average of 2 pounds each with a standard deviation of 0.1 pound. A histogram of these weights looks very much like a Normal distribution. Use the 68–95–99.7 rule to answer the following questions.

(a) What percent of the cartons weigh less than 2.1 pounds? Show your work.

(b) What percent of the cartons weigh less than 1.8 pounds? Show your work.

(c) What percent of the cartons weigh more than 1.9 pounds? Show your work.

DATA EXPLORATION

Presidents' Ages at Inauguration

How can we decide whether a given data set follows a Normal distribution? We can start by using graphical and numerical methods from Chapter 2 to analyze the data in Table 3.3.

Table 3.3 Ages of Presidents at Inauguration

President	Age	President	Age	President	Age
Washington	57	J. Q. Adams	57	Polk	49
J. Adams	61	Jackson	61	Taylor	64
Jefferson	57	Van Buren	54	Fillmore	50
Madison	57	W. Harrison	68	Pierce	48
Monroe	58	Tyler	51	Buchanan	65

President	Age	President	Age	President	Age
Lincoln	52	T. Roosevelt	42	L. Johnson	55
A. Johnson	56	Taft	51	Nixon	56
Grant	46	Wilson	56	Ford	61
Hayes	54	Harding	55	Carter	52
Garfield	49	Coolidge	51	Reagan	69
Arthur	51	Hoover	54	G. H. Bush	64
Cleveland	47	F. Roosevelt	51	Clinton	46
B. Harrison	55	Truman	60	G. W. Bush	54
Cleveland	55	Eisenhower	61	Obama	47
McKinley	54	Kennedy	43		

1. Plot a histogram of the data. Your teacher will decide whether you may use technology.

2. Describe the distribution. Include the shape, center, spread, and any deviations from the pattern.

3. Which units of center and spread would be more appropriate: mean and standard deviation or median and interquartile range? Explain your choice.

4. There are 44 data points. What percent of them are within one standard deviation of the mean? Use the One-Variable Statistics command on your calculator or the *One-Variable Statistical Calculator* applet to compute the mean and standard deviation. Count the number of observations that fall within one standard deviation of the mean and convert to a percent.

5. Use the method in Step 4 to find the percent of presidents' ages at inauguration within two standard deviations of the mean and the percent within three standard deviations of the mean.

Commentary: Although the distribution of presidents' ages at inauguration does not follow a Normal distribution, we can say that it is *approximately Normal.* Why? First, the histogram you created in Step 1 is single-peaked, roughly symmetric, and looks bell-shaped. Second, the percents of observations within one, two, and three standard deviations of the mean fall fairly close to those specified by the 68–95–99.7 rule for Normal distributions. When both of these conditions are met, we can describe the distribution as approximately Normal.

The standard Normal distribution

As the 68–95–99.7 rule suggests, all Normal distributions are the same if we measure in units of size σ about the mean μ as center. Changing to these units requires us to standardize:

$$z = \frac{x - \mu}{\sigma}$$

If the variable we standardize has a Normal distribution, then so does the new variable z. This new distribution is called the **standard Normal distribution.**

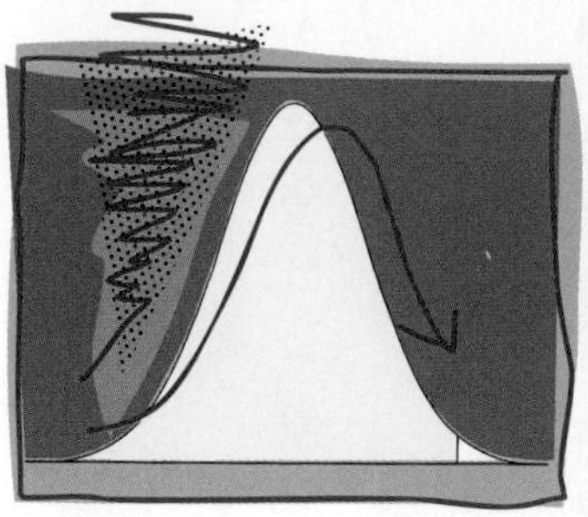

The bell curve?

Does the distribution of human intelligence follow the "bell curve" of a Normal distribution? Scores on IQ tests do roughly follow a Normal distribution. That is because a test score is calculated from a person's answers in a way that is designed to produce a Normal distribution. To conclude that intelligence follows a bell curve, we must agree that the test scores directly measure intelligence. Many psychologists don't think there is one human characteristic called "intelligence" that can be measured by a single test score.

Standard Normal distribution

The **standard Normal distribution** is the Normal distribution with mean 0 and standard deviation 1 (Figure 3.17).

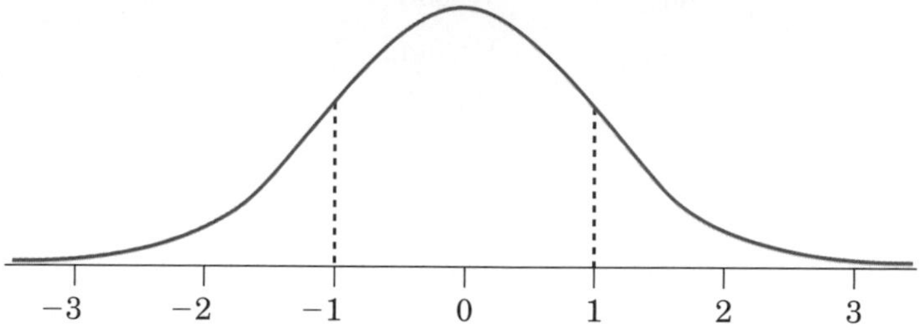

Figure 3.17 Standard Normal distribution.

If a variable x has any Normal distribution with mean μ and standard deviation σ, then the standardized variable

$$z = \frac{x - \mu}{\sigma}$$

has the standard Normal distribution.

An area under a density curve is a proportion of the observations in a distribution. Any question about what proportion of observations lie in some range of values can be answered by finding an area under the curve. In a standard Normal distribution, the 68–95–99.7 rule tells us that about 68% of the observations fall between and $z = -1$ and $z = 1$ (that is, within one standard deviation of the mean). What if we want to find the percent of observations that fall between $z = -1.25$ and $z = 1.25$? The 68–95–99.7 rule can't help us.

Because all Normal distributions are the same when we standardize, we can find areas under any Normal curve from a single table, a table that gives areas under the curve for the standard Normal distribution. Table A, the **standard Normal table,** gives areas under the standard Normal curve. You can find Table A at the back of the book.

The standard Normal table

Table A is a table of areas under the standard Normal curve. The table entry for each value z is the area under the curve to the left of z.

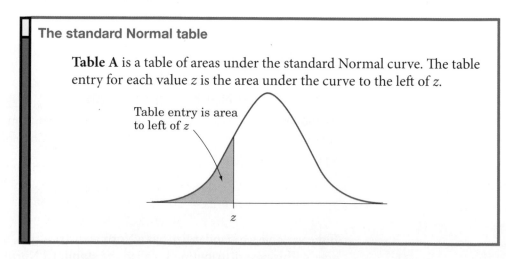

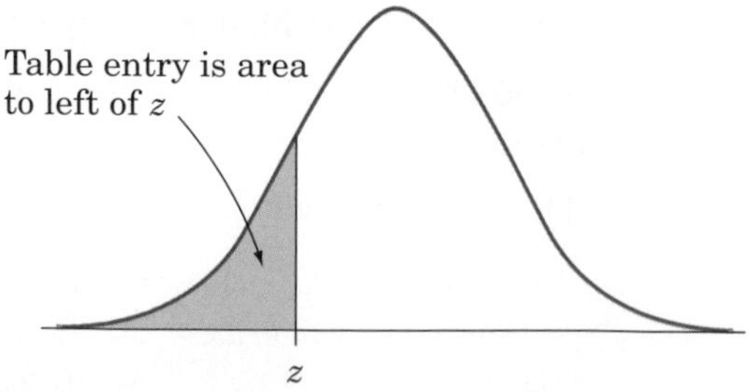

The next few examples show you how to use Table A.

Example 3.7 Using the standard Normal table, I

Problem: Find the proportions of observations from the standard Normal distribution that are (a) less than -1.25 and (b) greater than 0.81.

Solution:
(a) To find the area to the left of -1.25, locate -1.2 in the left-hand column of Table A, then locate the remaining digit 5 as .05 in the top row. The entry opposite -1.2 and under .05 is .1056. This is the area we seek. Here is a reproduction of the relevant portion of Table A, which appears on the inside back cover of this book:

z	**.00**	**.01**	**.02**	**.03**	**.04**	**.05**
-1.3	.0968	.0951	.0934	.0918	.0901	.0885
-1.2	.1151	.1131	.1112	.1093	.1075	.1056
-1.1	.1357	.1335	.1314	.1292	.1271	.1251

Figure 3.18 illustrates the relationship between the value $z = -1.25$ and the area 0.1056.

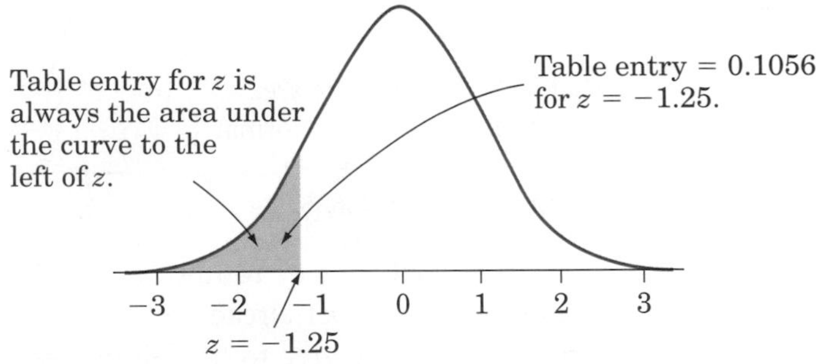
Table entry for z is always the area under the curve to the left of z.
Table entry = 0.1056 for $z = -1.25$.
−3
−2
−1
0
1
2
3
$z = -1.25$

Figure 3.18 The area under the standard Normal curve to the left of the point $z = -1.25$ is 0.1056.

z	.00	**.01**	**.02**
0.7	.7580	.7611	.7642
0.8	.7881	.7910	.7939
0.9	.8159	.8186	.8212

(b) To find the area to the right of $z = 0.81$, locate 0.8 in the left-hand column of Table A, then locate the remaining digit 1 as .01 in the top row. The entry opposite 0.8 and under .01 is .7910. This is the area *to the left* of $z = 0.81$. To find the area *to the right* of $z = 0.81$, we use the fact that the total area under the standard Normal density curve is 1. So the desired proportion is $1 - 0.7910 = 0.2090$. The relevant portion of Table A is reproduced in the margin.

Figure 3.19 (on the next page) illustrates the relationship between the value $z = 0.81$ and the area 0.2090.

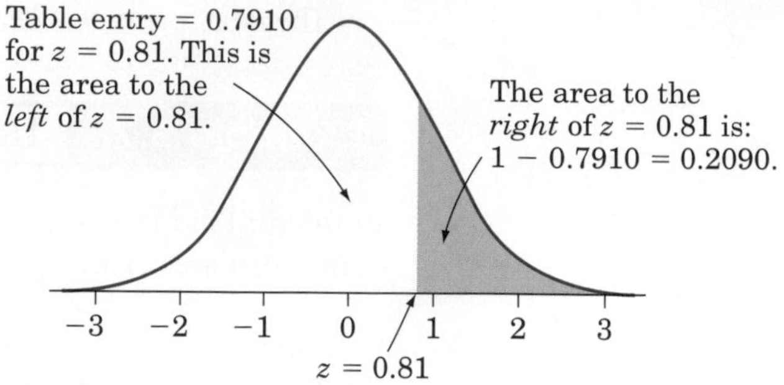

Figure 3.19 The area under the standard Normal curve to the right of the point $z = 0.81$ is 0.2090.

Example 3.8 Using the standard Normal table, II

Problem: Find the proportion of observations from the standard Normal distribution that are between −1.25 and 0.81.

Solution: In Example 3.7, we found the area to the left of $z = -1.25$ and the area to the right of $z = 0.81$ under the standard Normal curve. Now we want to find the area *between* these two z-scores. Figure 3.20 shows one way to do this. From Example 3.7, the area to the left of $z = -1.25$ is 0.1056 and the area to the right of $z = 0.81$ is 0.2090. So the area *between* these two z-scores is

$$1 - (0.1056 + 0.2090) = 1 - 0.3146 = 0.6854$$

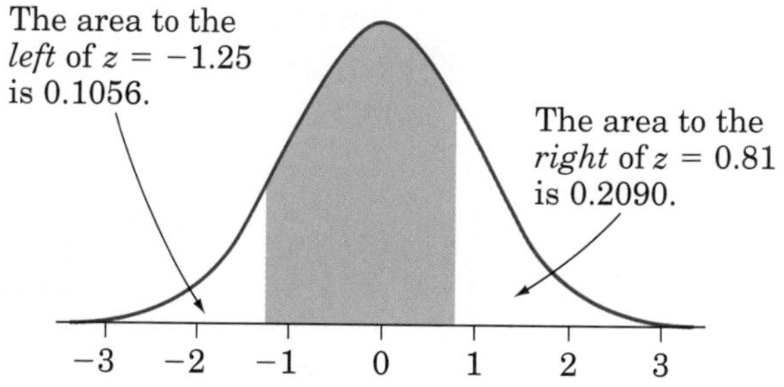

Figure 3.20 The area under the standard Normal curve between $z = -1.25$ and $z = 0.81$ is 0.6854.

Here's another way to find the shaded area in Figure 3.20. This is especially helpful if you haven't already found the areas in the two tails of the distribution. From Table A, the area to the left of $z = 0.81$ is 0.7910 and the area to the left of $z = -1.25$ is 0.1056. So the area under the standard Normal curve between these two z-scores is $0.7910 - 0.1056 = 0.6854$. Figure 3.21 shows why this approach works.

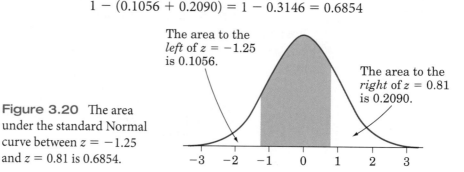

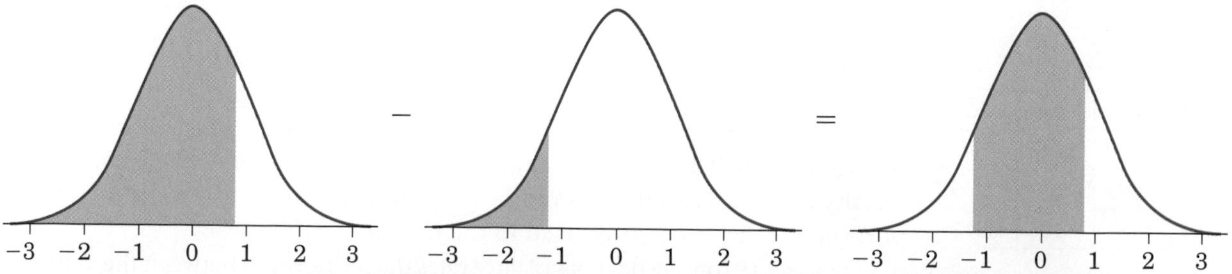

Figure 3.21 Another way to find the area between $z = -1.25$ and $z = 0.81$.

Caution! A common student mistake is to look up a z-value in Table A and report the entry corresponding to that z-value, regardless of whether the problem asks for the area to the left or to the right of that z-value. Always sketch the standard Normal curve, mark the z-value, and shade the area of interest. And before you finish, make sure your answer is reasonable in the context of the problem.

In Examples 3.7 and 3.8, we used Table A to find areas under the standard Normal curve from z-scores. What if we want to find the z-score that corresponds to a particular area?

Example 3.9 Finding z-scores from percentiles

Problem: Find the z-score that corresponds to the 90th percentile of the standard Normal curve.

Solution: We're looking for the z-score that has 90 percent of the area to its left, as shown in Figure 3.22. Since Table A gives areas to the left of a specified z-score, all we need to do is find the value closest to 0.90 in the middle of the table. From the reproduced portion of Table A, you can see that the desired z-score is $z = 1.28$. That is, the area to the left of $z = 1.28$ is as close to 0.90 as the table gets.

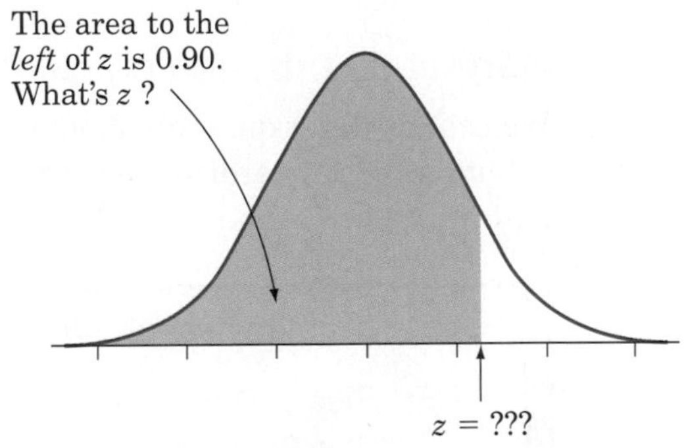

Figure 3.22 The z-score with area 0.90 to its left under the standard Normal curve.

z	.07	.08	.09
1.1	.8790	.8810	.8830
1.2	.8980	.8997	.9015
1.3	.9147	.9162	.9177

Exercises

For Exercises 3.27 to 3.30, use Table A to find the proportion of observations from the standard Normal distribution that satisfies each of the following statements. In each case, sketch a standard Normal curve and shade the area under the curve that is the answer to the question. Use the *Normal Curve* applet to check your answers.

3.27 Table A practice, I

(a) z is less than -0.37 **(b)** z is greater than -0.37

(c) z is less than 2.15 **(d)** z is greater than 2.15

3.28 Table A practice, II

(a) z is less than -1.58 **(b)** z is greater than 1.58

(c) z is greater than -0.46 **(d)** z is less than 0.93

3.29 Table A practice, III

(a) z is between -1.33 and 1.65 **(b)** z is between 0.50 and 1.79

3.30 Table A practice, IV

(a) z is between -2.05 and 0.78 **(b)** z is between -1.11 and -0.32

APPLET

For Exercises 3.31 and 3.32, use Table A to find the value z from the standard Normal distribution that satisfies each of the following conditions. (Use the value of z from Table A that comes closest to satisfying the condition.) In each case, sketch a standard Normal curve with your value of z marked on the axis. Use the *Normal Curve* applet to check your answers.

3.31 Working backward, I

(a) The 20th percentile of the standard Normal distribution.

(b) 45% of all observations are greater than z.

3.32 Working backward, II

(a) The 63rd percentile of the standard Normal distribution.

(b) 75% of all observations are greater than z.

Normal distribution calculations

We can answer a question about proportions of observations in *any* Normal distribution by standardizing and then using the standard Normal table. Here is an outline of the method for finding the proportion of the distribution in any region.

> **Solving problems involving Normal distributions**
>
> **Step 1:** *State the problem* in terms of the observed variable x.
>
> **Step 2:** *Standardize and draw a picture.* Standardize x to restate the problem in terms of a standard Normal variable z. Draw a picture to show the area of interest under the standard Normal curve.
>
> **Step 3:** *Use the table.* Find the required area under the standard Normal curve using Table A and the fact that the total area under the curve is 1.
>
> **Step 4:** *Conclusion.* Write your conclusion in the context of the problem.

Here's an example of the method at work.

David Cannon/Getty Images

Example 3.10 Tiger on the range

On the driving range, Tiger Woods practices his swing with a particular club by hitting many, many balls. When Tiger hits his driver, the distance the ball travels follows a Normal distribution with mean 304 yards and standard deviation 8 yards. What percent of Tiger's drives travel at least 290 yards?

Step 1: *State the problem.* Let $x =$ the distance that Tiger's ball travels. The variable x has a Normal distribution with $\mu = 304$ and $\sigma = 8$. We want the proportion of

Tiger's drives with $x \geq 290$. Figure 3.23(a) is a picture of the distribution with the area of interest shaded.

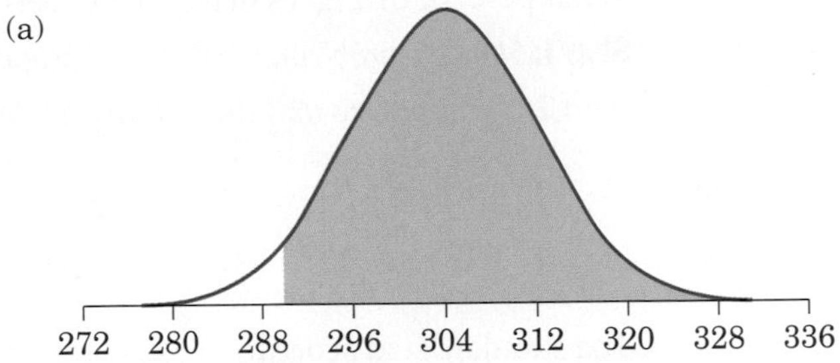

Figure 3.23(a) Distance traveled by Tiger Woods's drives on the range.

Step 2: *Standardize and draw a picture.* For $x = 290$, we have

$$z = \frac{x - \mu}{\sigma} = \frac{290 - 304}{8} = -1.75$$

So $x \geq 290$ corresponds to $z \geq -1.75$ under the standard Normal curve. Figure 3.23(b) shows the standard Normal distribution with the area of interest shaded.

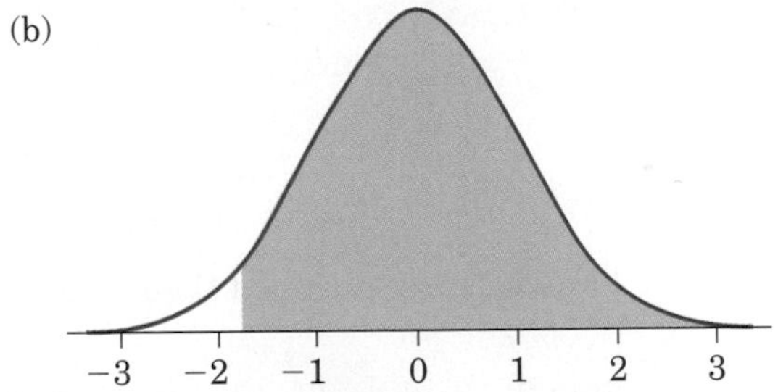

Figure 3.23(b) Area under the standard Normal curve corresponding to drives of length 290 yards or more in Figure 3.23(a).

Step 3: *Use the table.* From Table A, we see that the proportion of observations less than -1.75 is 0.0401. About 4% of Tiger's drives travel less than 290 yards. The area to the right of -1.75 is therefore $1 - 0.0401 = 0.9599$. This is about 0.96, or 96%.

Step 4: *Conclusion.* About 96% of Tiger Woods's drives on the range travel at least 290 yards.

In a Normal distribution, the proportion of observations with $x \geq 290$ is the same as the proportion with $x > 290$. There is no *area* under the curve exactly above the point 290 on the horizontal axis, so the areas under the curve with $x \geq 290$ and $x > 290$ are the same. This isn't true of the actual data. Tiger may hit a drive exactly 290 yards. The Normal distribution is just an easy-to-use approximation, not a description of every detail in the actual data.

The key to doing a Normal calculation is to sketch the area you want, then match that area with the area that the table gives you. Here is another example.

Example 3.11 Tiger on the range *(continued)*

What percent of Tiger's drives travel between 305 and 325 yards?

Step 1: *State the problem.* We want the proportion of Tiger's drives with $305 \leq x \leq 325$.

Step 2: *Standardize and draw a picture.* When $x = 305$,

$$z = \frac{305 - 304}{8} = 0.13$$

The standardized score for a 325-yard drive ($x = 325$) is

$$z = \frac{325 - 304}{8} = 2.63$$

Figure 3.24 shows the standard Normal curve with the area of interest shaded.

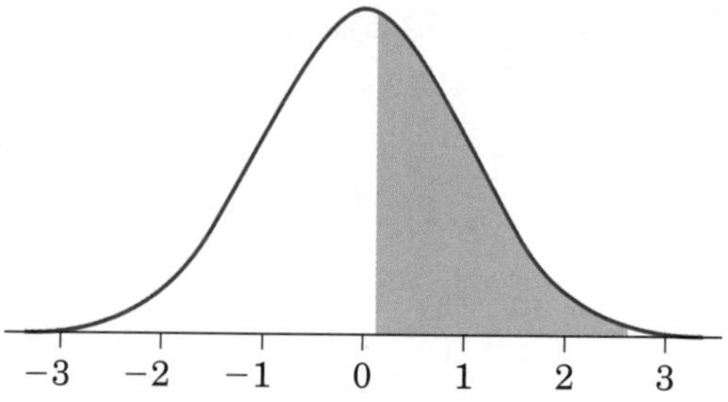

Figure 3.24 Standard Normal curve showing the area between $z = 0.13$ and $z = 2.63$.

Step 3: *Use the table.* The area between $z = 0.13$ and $z = 2.63$ is the area to the left of 2.63 minus the area to the left of 0.13. Look at Figure 3.21 to see why this makes sense. From Table A,

$$\begin{aligned}\text{area between 0.13 and 2.63} &= \text{area to the left of 2.63} - \text{area to the left of 0.13}\\ &= 0.9957 - 0.5517 = 0.4440\end{aligned}$$

Step 4: *Conclusion.* About 44% of Tiger's drives travel between 305 and 325 yards.

What if we meet a z that falls outside the range covered by Table A? For example, the area to the left of $z = -4$ does not appear in the table. But since -4 is less than -3.4, this area is smaller than the entry for $z = -3.40$, which is 0.0003. There is very little area under the standard Normal curve outside the range covered by Table A. You can take this area to be approximately zero with little loss of accuracy.

Finding a value, given a proportion

Examples 3.10 and 3.11 illustrate the use of Table A to find what proportion of the observations satisfies some condition, such as "Tiger's drive travels between

305 and 325 yards." Sometimes, we may want to find the observed value that corresponds to a given percentile. To do this, use Table A backward. Find the given proportion in the body of the table, read the corresponding z from the left column and top row, then "unstandardize" to get the observed value. Here is an example.

Example 3.12 Tiger on the range: using Table A in reverse

What distance would a ball have to travel to be at the 80th percentile of Tiger's drive lengths?

Step 1: *State the problem and draw a picture.* We want to find the drive distance x with area 0.8 to its left under the Normal curve with mean $\mu = 304$ and standard deviation $\sigma = 8$. Figure 3.25 poses the question in graphical form.

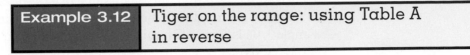

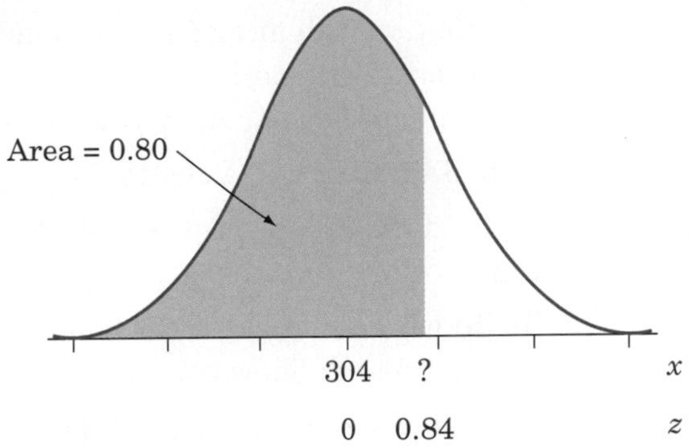

Figure 3.25 Locating the point on a Normal curve with area 0.80 to its left.

Step 2: *Use the table.* Look in the body of Table A for the entry closest to 0.8. It is 0.7995. This is the entry corresponding to $z = 0.84$. So $z = 0.84$ is the standardized value with area 0.8 to its left.

Step 3: "*Unstandardize*" to transform the solution from a z-score to a value of x. We know that the standardized value of the unknown x is $z = 0.84$. So x satisfies the equation

$$\frac{x - 304}{8} = 0.84$$

Solving this equation for x gives

$$x = 304 + (0.84)(8) = 310.7$$

This equation should make sense: it finds the x that lies 0.84 standard deviations above the mean on this particular Normal curve. That is the "unstandardized" meaning of $z = 0.84$.

Step 4: *Conclusion.* A ball would have to travel 310.7 yards to be at the 80th percentile of Tiger's drive length distribution.

Exercises

For Exercises 3.33 to 3.38, follow the four-step method of Examples 3.10 to 3.12 to answer the questions.

3.33 Tiger prowls After hitting plenty of balls on the practice range, Tiger Woods heads out to the first tee to begin a golf tournament. A large creek crosses the fairway 317 yards from the tee. Assume that the distance traveled by Tiger's ball follows a Normal distribution with $\mu = 304$ yards and $\sigma = 8$ yards, as in Example 3.10. Is Tiger safe hitting his driver? Show your work.

3.34 High IQ scores Scores on the Wechsler Adult Intelligence Scale for 20- to 34-year-olds are approximately Normally distributed with mean 110 and standard deviation 25. How high must a person score to be in the top 25% of all scores? Show your work.

3.35 SAT scores, I The average performance of females on the SAT, especially the Math section, is lower than that of males. The reasons for this gender gap are controversial. In 2007, female scores on the SAT Math test followed a Normal distribution with mean 500 and standard deviation 111. Male scores had a mean of 533 and a standard deviation of 118. What percent of females scored higher than the male mean? Show your work.

3.36 SAT scores, II Refer to the previous exercise.

(a) Find the 85th percentile of the SAT Math score distribution for males. Show your work.

(b) To what percentile in the female score distribution does your answer to (a) correspond? Show your work.

3.37 Potatoes Bags of potatoes in a shipment averaged 10 pounds with a standard deviation of 0.5 pounds. A histogram of these weights followed a Normal curve quite closely.

(a) What percent of the bags weighed less than 10.25 pounds? Show your work.

(b) What percent weighed between 9.5 and 10.25 pounds? Show your work.

3.38 Brush your teeth The amount of time Ricardo spends brushing his teeth follows a Normal distribution with unknown mean μ and standard deviation $\sigma = 20$ seconds. Ricardo spends less than 60 seconds brushing his teeth about 40% of the time. Use this information to determine the mean of this distribution. Show your method clearly.

CALCULATOR CORNER Normal curve areas

The TI-84 and TI-Nspire can be used to find the area under a Normal curve above any interval on the horizontal axis. We'll show you how to use the `normalcdf` (`normCdf` on the TI-Nspire) command to find the percent of Tiger Woods's drives that travel at least 290 yards. Recall from Example 3.10 (page 128) that $\mu = 304$ yards and $\sigma = 8$ yards.

We want to find the area to the right of 290 under a Normal curve with mean 304 and standard deviation 8. The screen shot shows the desired area. Notice that we chose an "upper bound" of 400 yards on Tiger's drive lengths, which is more than 10 standard deviations above the mean. To find this area, we'll execute the calculator command `normalcdf(290,400,304,8)`, where 290 is the "lower bound," 400 is the "upper bound," 304 is the mean, and 8 is the standard deviation.

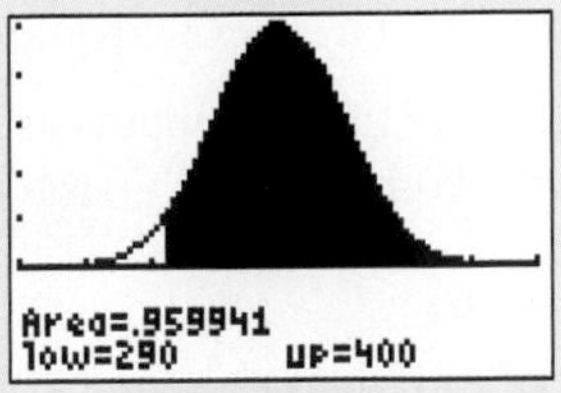

TI-84

- Find the command `normalcdf` by pressing [2nd][VARS] `[DISTR]` and choosing `2:normalcdf(`.
- Complete the command `normalcdf(290,400,304,8)` and press [ENTER]. The result should be 0.9599, after rounding to four decimal places.

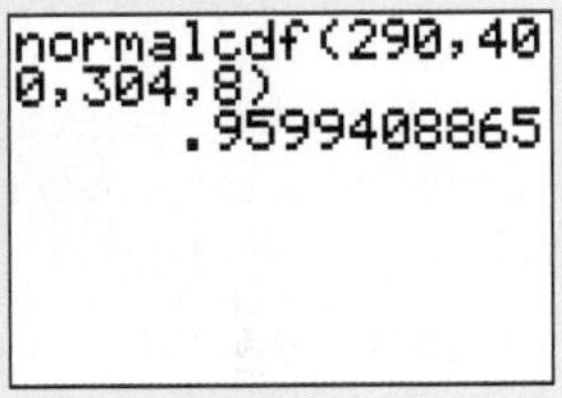

TI-Nspire

- Open a Calculator page. Go into the catalog (press [catalog]) and then press (N). Use the NavPad to arrow down to `normCdf` and click the mouse ([click]) or press [enter] to select the command. A dialog box should appear.
- Enter these values in the dialog box: Lower Bound, 290; Upper Bound, 400; μ, 304; σ, 8. (Use the (tab) key to move between the fields in the dialog box.) Then press [enter].
- Compare your results with the screen shot shown.

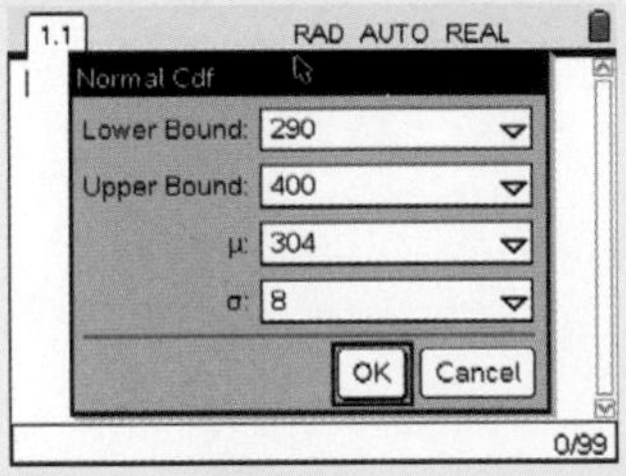

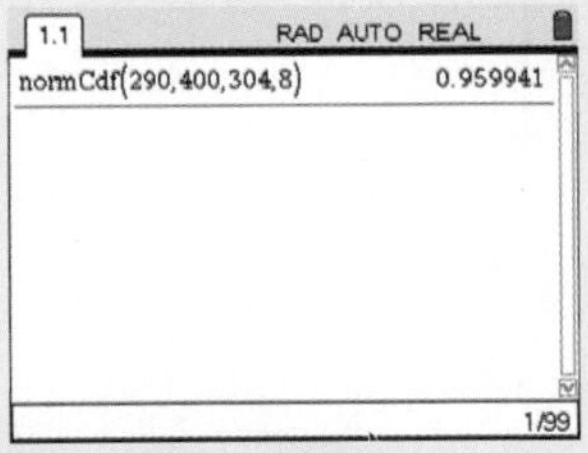

These results agree with what we got by using Table A earlier. The first two numbers that you entered (290 and 400) specify the interval corresponding to the shaded area under the curve. The last two numbers (304 and 8) identify the particular Normal curve by specifying the mean and standard deviation.

Now see if you can calculate the area in Example 3.11 (page 130).

CALCULATOR CORNER "Backward" Normal calculations

You can use the TI-84's or the TI-Nspire's `invNorm` command to find the value with a specified area to its left under a Normal curve. To determine the 80th percentile of Tiger's drive distances from Example 3.12 (page 131), we need to execute the command `invNorm(0.8,304,8)`.

CALCULATOR CORNER *(continued)*

CALCULATOR CORNER *(continued)*

Notice that the inputs are the area to the *left* of our desired value (which is 0.8 for the 80th percentile), the mean (304), and the standard deviation (8).

TI-84

- Find the command `invNorm` by pressing [2nd][VARS] `[DISTR]` and choosing `3:invNorm(`.
- Complete the command `invNorm(0.8,304,8)` and press [ENTER]. The result should be about 310.73.

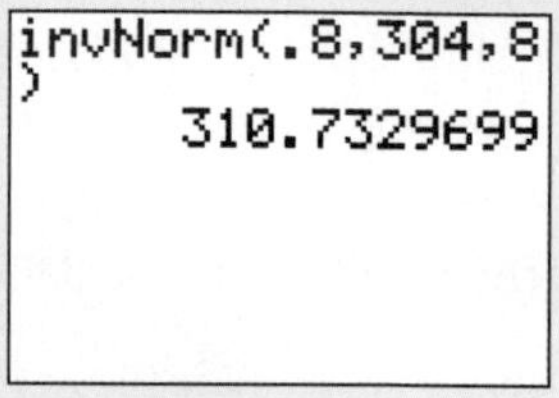

TI-Nspire

- In a Calculator page, go into the catalog (press [catalog]) and then press [I]. Use the NavPad to arrow down to `invNorm` and click the mouse ([click]) or press [enter] to select the command. A dialog box should appear.
- Enter these values in the dialog box: Area, 0.8; μ, 304; σ, 8. (Use the [tab] key to move between the fields in the dialog box.) Then press [enter].
- Compare your results with the screen shots shown.

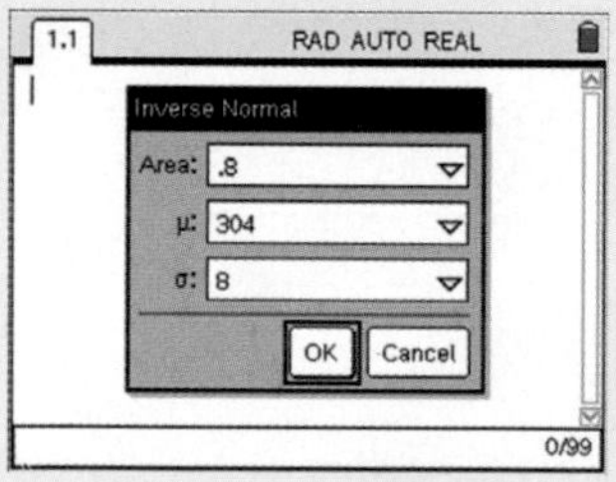

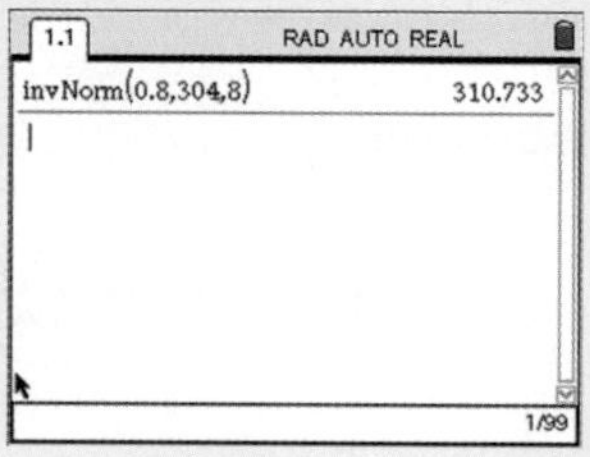

APPLICATION 3.2

The vending machine problem

Have you ever purchased a hot drink from a vending machine? The intended sequence of events runs something like this. You insert your money into the machine and select your preferred beverage. A cup falls out of the machine, landing upright. Liquid pours out until the cup is nearly full. You reach in, grab the piping hot cup, and drink happily.

Sometimes, things go wrong. The machine might take your money and not give you anything in return. Or the cup might fall over. More frequently, everything goes smoothly until the liquid begins to flow. It might stop flowing when the cup is only half full. Or the liquid might keep coming until your cup overflows. Neither of these results leaves you satisfied.

The vending machine company wants to keep customers happy. So they have decided to hire you as a statistical consultant. They provide you with the following summary of important facts about the vending machine:

- Cups will hold 8 ounces.
- The amount of liquid dispensed varies according to a Normal distribution centered at the mean μ that is set in the machine.
- $\sigma = 0.2$ ounces

QUESTIONS

1. Suppose the company sets the machine's mean at $\mu = 7.5$ ounces. What percent of cups would overflow? Show your work.

2. If a cup contains too much liquid, a customer may get burned from a spill. This could result in an expensive lawsuit for the company. Explain to the company president why setting the machine's mean at 7.0 ounces might be smarter than setting it at 7.5 ounces.

3. Customers may be irritated if they get a cup with too little liquid from the machine. At what value should the machine's mean be set to ensure that 90% of cups contain at least 7 ounces?

4. Given the issues raised in the previous two questions, what mean setting for the machine would you recommend? Write a brief report to the vending machine company president that explains your answer.

Section 3.2 Summary

Normal curves are a special kind of density curve that describe the overall pattern of some sets of data. Normal curves are symmetric and bell-shaped. A specific Normal curve is completely described by its mean and standard deviation. You can locate the mean (center point) and the standard deviation (distance from the mean to the change-of-curvature points) on a Normal curve. All **Normal distributions** obey the **68–95–99.7 rule,** which describes what percent of observations lie within one, two, and three standard deviations of the mean.

All Normal distributions are the same when measurements are standardized. If x follows a Normal distribution with mean μ and standard deviation σ, we can standardize using

$$z = \frac{x - \mu}{\sigma}$$

The variable z has the **standard Normal distribution** with mean 0 and standard deviation 1.

Table A gives percentiles for the standard Normal curve. By standardizing, we can use Table A to determine the percentile for a given score or the score corresponding to a given percentile in any Normal distribution. You can use the *Normal Curve* applet or your calculator to perform Normal calculations quickly.

Section 3.2 Exercises

3.39 Heights of young women The distribution of the heights of young women aged 18 to 24 is approximately Normal with mean 65 inches and standard deviation 2.5 inches. Sketch a picture of a Normal curve and then use the 68–95–99.7 rule to show what the rule states about these women's heights.

3.40 Normal curve properties, I Figure 3.26 is a Normal density curve. Estimate the mean and the standard deviation of this distribution.

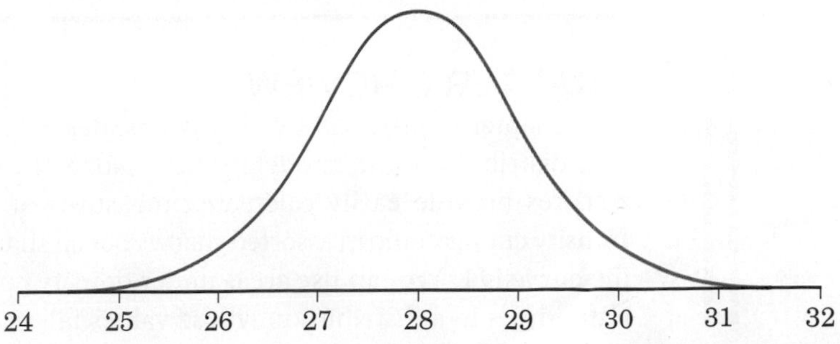

Figure 3.26 A Normal density curve.

3.41 Normal curve properties, II Explain why the point that is one standard deviation below the mean in a Normal distribution is always the 16th percentile. Explain why the point that is two standard deviations above the mean is the 97.5th percentile.

3.42 More Table A practice, I Use Table A to find the proportion of observations from a standard Normal distribution that satisfies each of the following statements. In each case, sketch a standard Normal curve and shade the area under the curve that is the answer to the question. Use an applet or your calculator to check your answers.

(a) $z > -1.81$ (b) $z < 2.29$

(c) $-1.81 \leq z \leq -0.47$ (d) $-1.02 \leq z \leq 0.65$

3.43 More Table A practice, II Use Table A to find the value z from a standard Normal distribution that satisfies each of the following conditions. In each case, sketch a standard Normal curve with your value of z marked on the axis. Use an applet or your calculator to check your answers.

(a) The point z with 32% of the observations falling to its left.

(b) The point z with 40% of the observations falling to its right.

(c) The 10th percentile of the standard Normal distribution.

(d) The 83rd percentile of the standard Normal distribution.

3.44 NCAA rules for athletes The National Collegiate Athletic Association (NCAA) requires Division I athletes to score at least 820 on the combined Mathematics and Critical Reading parts of the SAT exam in order to compete in their first college year. (Higher scores are required for students with poor high school grades.) In 2007, the combined scores of the millions of students taking the SATs were approximately Normal with mean 1017 and standard deviation 211. What percent of all students had scores less than 820? Show your work.

3.45 Are we getting smarter? When the Stanford-Binet "IQ test" came into use in 1932, it was adjusted so that scores for each age group of children followed roughly the Normal distribution with mean 100 and standard deviation 15. The test is readjusted from time to time to keep the mean at 100. If present-day American children took the 1932 Stanford-Binet test, their mean score would be about 120. The reasons for the increase in IQ over time are not known but probably include better childhood nutrition and more experience in taking tests.[5]

(a) IQ scores above 130 are often called "very superior." What percent of children had very superior scores in 1932?

(b) If present-day children took the 1932 test, what percent would have very superior scores? (Assume that the standard deviation 15 does not change.)

3.46 Get smart Refer to the previous exercise. What IQ score was at the 98th percentile in 1932? Show your work.

CHAPTER 3 REVIEW

This chapter focused on two big issues: describing an observation's location within a distribution and modeling with Normal distributions. Both percentiles and z-scores provide easily calculated measures of relative standing for individuals. Density curves come in assorted shapes, but all share the property that the area beneath the curve is 1. We can use areas under density curves to estimate the proportion of individuals in a distribution whose values fall in a specified range.

Sometimes the overall pattern of a large number of observations is so regular that we can describe it by a smooth curve, called a Normal curve. Normal curves are handy models because they have properties that make them easy to use and think about. When data are approximately Normally distributed, the mean and standard deviation are natural measures of center and spread. In such cases, we can use the standard Normal curve and Table A to calculate areas.

Here is a review list of the most important skills you should have developed from your study of this chapter.

A. MEASURING LOCATION IN A DISTRIBUTION

1. Use percentiles to locate individual values within a distribution of data.

2. Find the standardized value (z-score) of an observation. Interpret z-scores in context.

B. DENSITY CURVES

1. Know that areas under a density curve represent proportions of all observations and that the total area under a density curve is 1.

2. Approximately locate the median (equal-areas point) and the mean (balance point) on a density curve.

C. NORMAL DISTRIBUTIONS

1. Recognize the shape of Normal curves and estimate by eye both the mean and the standard deviation from such a curve.

2. Use the 68–95–99.7 rule and symmetry to state what percent of the observations from a Normal distribution fall between two points when both points lie at the mean or one, two, or three standard deviations on either side of the mean.

3. Use Table A to find the percentile of a value from any Normal distribution and the value that corresponds to a given percentile.

CHAPTER 3 REVIEW EXERCISES

3.47 Textbook costs Students taking a college introductory statistics class reported spending an average of \$305 on textbooks that quarter with a standard deviation of \$90. Here's a rough sketch of a Normal density curve that fitted the histogram well:

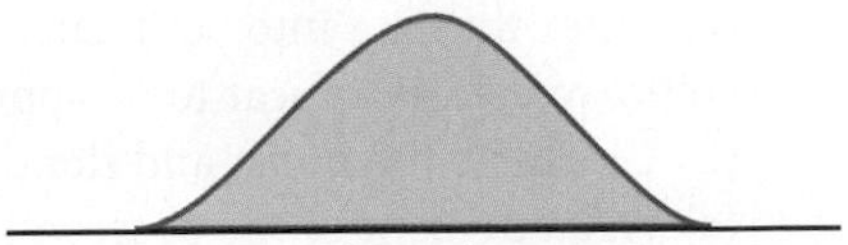

(a) Approximately what percent of the students spent between \$215 and \$395 on textbooks that quarter? Explain how you got this answer.
(b) One student spent \$287 on textbooks. What was her standard score? What percent of the students spent less than she did on textbooks that quarter?

3.48 Standardized test scores as percentiles Joey received a report that he scored in the 97th percentile on a national standardized math test but in the 72nd percentile on the reading portion of the test.

(a) Explain to Joey's grandmother, who knows little statistics, what these numbers mean.
(b) Can we determine Joey's *z*-scores for his math and reading performance? Why or why not?

3.49 Finding areas Use Table A to find the proportion of observations from a standard Normal distribution that falls in each of the following regions. In each case, sketch a standard Normal curve and shade the area representing the region. Use an applet or your calculator to check your answers.
(a) $z \geq -2.25$ **(b)** $-2.25 < z < 1.77$
(c) $z < 2.89$ **(d)** $1.44 < z < 2.89$

3.50 Finding *z*-scores Use Table A to find the value *z* from a standard Normal distribution that satisfies each of the following conditions. In each case, sketch a standard Normal curve with your value of *z* marked on the axis. Use an applet or your calculator to check your answers.
(a) The point *z* with 70% of the observations falling below it.
(b) The point *z* such that 90% of all observations are greater than *z*.
(c) The 46th percentile of the standard Normal distribution.

3.51 Low-birth-weight babies Researchers in Norway analyzed data on the birth weights of 400,000 newborns over a 6-year period. The distribution of birth weights is approximately Normal with a mean of 3668 grams and a standard deviation of 511 grams.[6] Babies that weigh less than 2500 grams at birth are classified as "low birth weight."
(a) What percent of babies will be identified as low birth weight? Show your work.
(b) Find the quartiles of the birth weight distribution. Show your work.

3.52 Bacteria in milk A study of bacterial contamination in milk counted the number of coliform bacteria per milliliter in 100 specimens of milk purchased in East Coast grocery stores. The U.S. Public Health Service recommends no more than 10 coliform bacteria per milliliter. Here are the data:

5	8	6	7	8	3	2	4	7	8	6	4	4	8	8	8	6	10	6	5
6	6	6	6	4	3	7	7	5	7	4	5	6	7	4	4	4	3	5	7
7	5	8	3	9	7	3	4	6	6	8	7	4	8	5	7	9	4	4	7
8	8	7	5	4	10	7	6	6	7	8	6	6	6	0	4	5	10	4	5
7	9	8	9	5	6	3	6	3	7	1	6	9	6	8	5	2	8	5	3

(a) Enter the data into your calculator and plot a histogram. Does the distribution of coliform counts appear to be approximately Normal?
(b) Calculate the mean and standard deviation for the distribution.
(c) What percent of the observations fall within one, two, and three standard deviations of the mean?

3.53 Density curves Figure 3.27 (on the facing page) shows density curves of two different shapes. Briefly describe the overall shape of each distribution. Two or three points are marked on each curve. The mean and the median are among these points. For each curve, which point is the median and which is the mean?

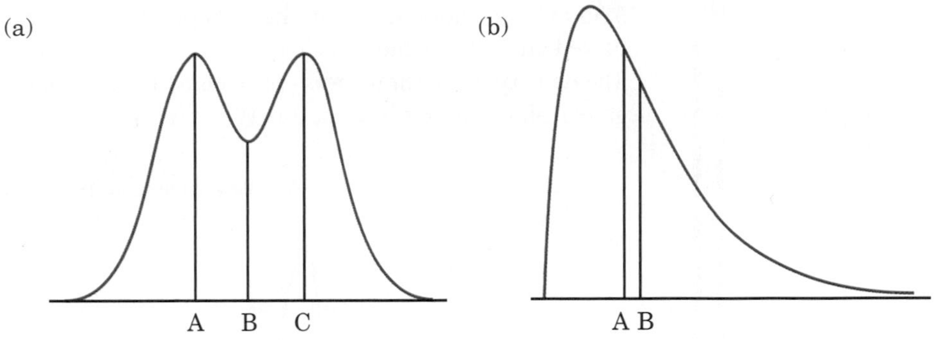

Figure 3.27 Two density curves of different shapes.

3.54 Helmet sizes The army reports that the distribution of head circumference among soldiers is approximately Normal with mean 22.8 inches and standard deviation 1.1 inches. Helmets are mass-produced for all except the smallest 5% and the largest 5% of head sizes. Soldiers in the smallest or largest 5% get custom-made helmets. What head sizes get custom-made helmets? Show your work.

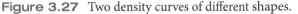

3.55 The stock market The annual rate of return on stock indexes (which combine many individual stocks) is very roughly Normal. Since 1945, the Standard & Poor's 500 index has had a mean yearly return of 12%, with a standard deviation of 16.5%. Take this Normal distribution to be the distribution of yearly returns over a long period.

(a) In what range do the middle 95% of all yearly returns lie? Explain.

(b) The market is down for the year if the return on the index is less than zero. In what proportion of years is the market down? Show your work.

(c) In what proportion of years does the index gain between 15% and 25%? Show your work.

3.56 Do you sudoku? In the chapter opening story (page 101), one of the authors played an online game of sudoku. At the end of his game, this graph was displayed. The density curve shown was constructed from a histogram of times from 4,000,000 games played in one week at this Web site.

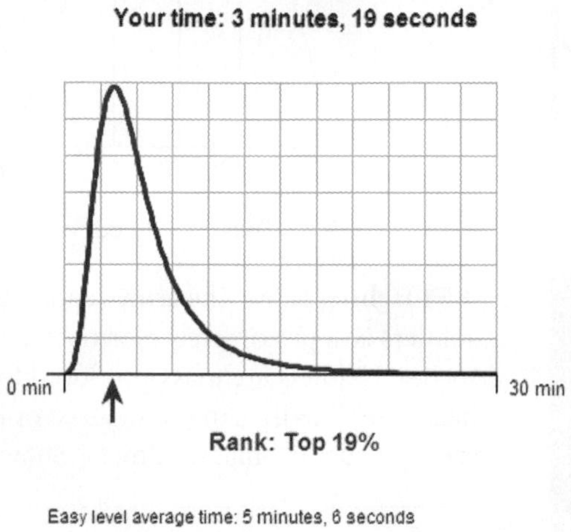

(a) How would you describe the shape of the density curve? Explain why this shape makes sense in this setting.
(b) Use what you have learned in this chapter to describe the author's performance in a few sentences.

CHAPTER 4

Describing Relationships

4.1 Scatterplots and Correlation
4.2 Regression and Prediction

How faithful Is Old Faithful?

A geyser is a natural phenomenon: a type of hot spring that shoots hot water and steam into the air. One of the best-known geysers is Old Faithful, located in Yellowstone National Park in Wyoming. Accurate predictions of eruption times allow the National Park Service to inform visitors of the approximate time of the next eruption. Visitors can then adjust their schedules accordingly.[1] How can park geologists make predictions with any certainty? To see how, let's look at some data.

Figure 4.1 (on the following page) shows the distribution of intervals between eruptions of Old Faithful in July 1995. The shortest interval was 47 minutes and the longest was 113 minutes. That's quite a range! The distribution has an unusual shape: it has two clear peaks—one at about 60 minutes and the other at about 90 minutes. Let's put this in context. On many occasions, the time between eruptions of Old Faithful is about an hour. Even more often, the interval between eruptions is around 90 minutes. Can park geologists make predictions based on this double-peaked distribution?

If geologists predict a 60-minute gap between eruptions, but the actual interval is 90 minutes, some visitors will become impatient. If they predict a 90-minute interval, but the gap is closer to 60 minutes, some visitors may go elsewhere in the park and return after the eruption has occurred. What should the geologists do?

Later in the chapter, we'll answer this question. For now, keep this in mind: to understand one variable, you often have to look at how it is related to other variables.

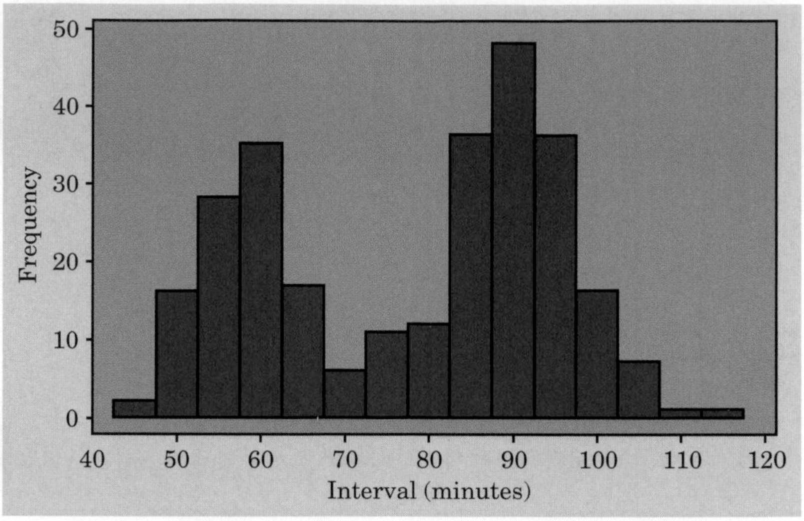

Figure 4.1 Histogram of the interval (in minutes) between eruptions of the Old Faithful geyser in July 1995.

4.1 Scatterplots and Correlation

ACTIVITY 4.1

Tedious tasks take time

Materials: Graphing calculator, graph paper, and stopwatch for each pair of students

Many jobs require employees to perform repetitive tasks. Fast-food restaurants rely on their employees to fill customers' orders quickly and accurately. Bookkeepers and accountants enter pages of numerical data into Excel spreadsheets. Elementary school students practice their multiplication facts. Cashiers at retail stores determine the total price of goods that are purchased. Quality control inspectors examine a sample of the items produced. In each of these situations, it is important to know how long the task will take. This Activity will give you a chance to see how long a tedious task takes you.

1. **TI-84:** Clear your lists. Go into your Statistics/List Editor (press STAT, then choose `1:Edit...`). Clear list L1 by highlighting L1 with your cursor, then pressing CLEAR ENTER. Clear lists L2 through L6 in the same way.

TI-Nspire: Open a new Lists & Spreadsheet page (press ⌂ and choose `3:Lists & Spreadsheet`).

2. Your first task is to enter the digits 0, 1, 2, . . . , 9 in list L1/column A. Your partner will time how long it takes you to complete the task. When your partner says "Go!" enter the digits one at a time into the list. As soon as you have finished, say "Done!" so your partner will stop timing. If you entered all of the numbers correctly, record your time to the nearest second. If you made a mistake, repeat the task.

3. Your second task is to enter the numbers 10, 11, 12, . . . , 19 in list L2/column B. As in the previous task, begin when your partner says "Go!" and say "Done!" when you are finished. If you entered all the numbers correctly, record your time to the nearest second. Otherwise, repeat the task.

4. You can probably guess what the third, fourth, fifth, and sixth tasks are now.

Task 3: Enter the numbers 100, 101, 102, . . . , 109 in list L3/column C.
Task 4: Enter the numbers 1000, 1001, 1002, . . . , 1009 in list L4/column D.
Task 5: Enter the numbers 10000, 10001, 10002, . . . , 10009 in list L5/column E.
Task 6: Enter the numbers 100000, 100001, 100002, . . . , 100009 in list L6/column F.

For each task, start typing when your partner says "Go!" and be sure to say "Done!" when you finish. Once you enter all the numbers correctly for each task, record your time to the nearest second.

5. Switch roles with your partner and repeat Steps 2 through 4.

6. Summarize the data for your six tasks in a table like this one:

Task no.	Total no. of digits typed	Time to nearest second
1	10	
2	20	
. . .		

7. Make a *scatterplot* of your data as follows:

- On graph paper, draw x and y axes. Label the x axis "Total digits typed" and the y axis "Time." Scale the horizontal axis by adding tick marks at 0, 10, 20, . . . , 70. Scale the vertical axis in half-second increments beginning with 0.
- Plot each of your six data points on the graph.

8. Describe what the graph tells you about the relationship between the two variables.

9. Compare graphs with your partner. What similarities and differences do you see?

A medical study finds that short women are more likely to have heart attacks than women of average height, while tall women have the fewest heart attacks. An insurance group reports that heavier cars have fewer deaths per 10,000 vehicles registered than lighter cars do. These and many other statistical studies look at the relationship between two variables. To understand such a relationship, we must often examine other variables as well. To conclude that shorter women have higher risk of heart attacks, for example, the researchers had to eliminate the effect of other variables such as weight and exercise habits.

Our topic in this chapter is relationships between variables. One of our main themes is that the relationship between two variables can be strongly influenced by other variables that are lurking in the background. Most statistical studies examine data on more than one variable. Fortunately, statistical analysis of several-variable data builds on the tools we used to examine individual variables. The principles that guide our work also remain the same:

- First plot the data, then add numerical summaries.
- Look for overall patterns and deviations from those patterns.
- When the overall pattern is quite regular, there is sometimes a way to describe it very briefly.

Scatterplots

The most common way to display the relationship between two quantitative variables is a **scatterplot.**

Scatterplot

A **scatterplot** shows the relationship between two quantitative variables measured on the same individuals. The values of one variable appear on the horizontal axis, and the values of the other variable appear on the vertical axis. Each individual in the data appears as a point in the plot fixed by the values of both variables for that individual.

Here's an example that shows how a scatterplot can illuminate the relationship between two variables.

Example 4.1 Waiting on Old Faithful

In July 1995, the Starnes family visited Yellowstone in hopes of seeing Old Faithful erupt. They had only about four hours available in the park before they had to drive on to their hotel for the night. When they pulled into the parking lot at Old Faithful, a large crowd of people was headed back to their cars from the geyser. Old Faithful had just finished erupting.

How long would it be until the next eruption? From the histogram in the opening story, Figure 4.1 (page 142), we see that the Starnes family might have to wait anywhere from about 50 to about 100 minutes. Can we narrow it down a bit? Figure 4.2 helps us do that.

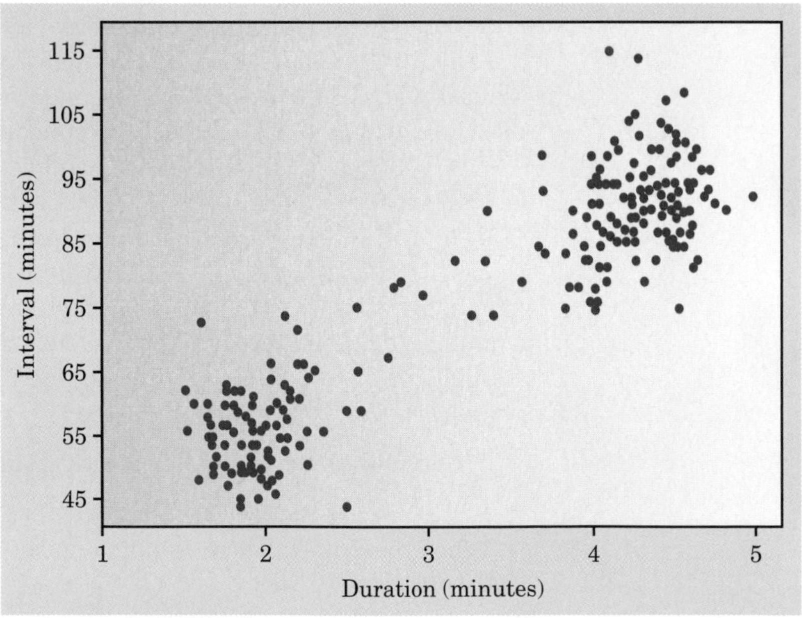

Figure 4.2 Scatterplot of the interval between eruptions of Old Faithful against the duration of the previous eruption.

Figure 4.2 is a scatterplot that plots the interval between successive eruptions of Old Faithful on the vertical axis against the duration of the previous eruption on the horizontal axis. The plot clearly shows two groups of eruptions. In one group, the duration of the eruptions is about 2 minutes. In the other group of eruptions, most of them last at least 4 minutes. The plot as a whole shows that the interval between eruptions increases as the duration of the previous eruption increases.

Since a long eruption depletes more water from a geyser's reservoir than a short eruption, we would expect that it would take more time to refill the reservoir after a longer eruption. Consequently, we would expect a longer time interval until the next eruption. Thus, the physical properties of the geyser suggest a relationship between the duration of an eruption and the time until the next eruption. Such a relationship may be used to predict geyser eruption times.

The scatterplot in Figure 4.2 shows how the interval between eruptions of Old Faithful is related to the duration of the previous eruption. We think that "duration" will help explain "interval." That is, "duration" is the **explanatory variable** and "interval" is the **response variable.** We want to see how the interval changes when duration changes, so we put duration (the explanatory variable) on the horizontal axis. We can then see that as the duration of the previous eruption increases, the interval between eruptions increases. Each point on the plot represents one eruption. For one eruption that lasted 1.6 minutes, there was an interval of 74 minutes until the next eruption. Find 1.6 on the x (horizontal) axis to locate this eruption.

Explanatory and response variables

A **response variable** is a variable that measures an outcome or result of a study.

An **explanatory variable** is a variable that we think explains or causes changes in the response variable.

Always plot the explanatory variable, if there is one, on the horizontal axis (the x axis) of a scatterplot. As a reminder, we usually call the explanatory variable x and the response variable y. If there is no explanatory-response distinction, either variable can go on the horizontal axis.

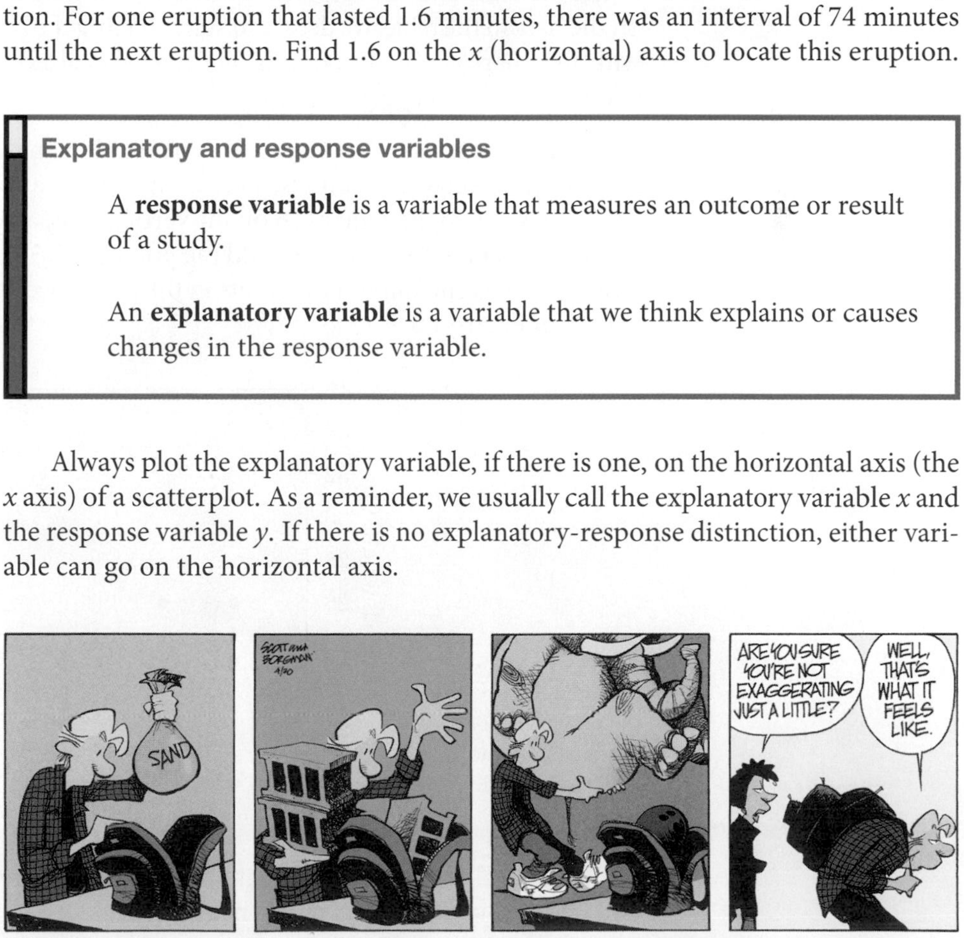

Example 4.2 Heavy backpacks

Ninth-grade boys at the Webb School go on a backpacking trip each fall. Students are divided into hiking groups of size 8 by selecting names from a hat. Before leaving, students and their backpacks are weighed. Here are data from one hiking group in a recent year:

Body weight (lb):	120	187	109	103	131	165	158	116
Backpack weight (lb):	26	30	26	24	29	35	31	28

Let's make a scatterplot to investigate the relationship between body weight and pack weight. To construct a scatterplot:

Step 1: *Identify explanatory and response variables.* The amount of weight a student can comfortably carry depends on his body weight. Therefore, we'll use body weight as the explanatory variable and backpack weight as the response variable.

Step 2: *Label and scale axes.* We put the explanatory variable, body weight, on the horizontal axis and the response variable, pack weight, on the vertical axis. Based on the data in the table, we decide to start labeling the horizontal axis at 100 pounds, with tick marks every 10 pounds. For the vertical axis, we start at 20 pounds and add tick marks every 2 pounds.

Step 3: *Plot individual data values.* The first student in the group weighs 120 pounds and his pack weighs 26 pounds. We plot this point directly above 120 on the horizontal axis and to the right of 26 on the vertical axis, as shown in Figure 4.3. For the second student in the group, we add the point (187, 30) to the graph. By adding the points for the remaining six students in the group, we get the completed scatterplot shown in Figure 4.3.

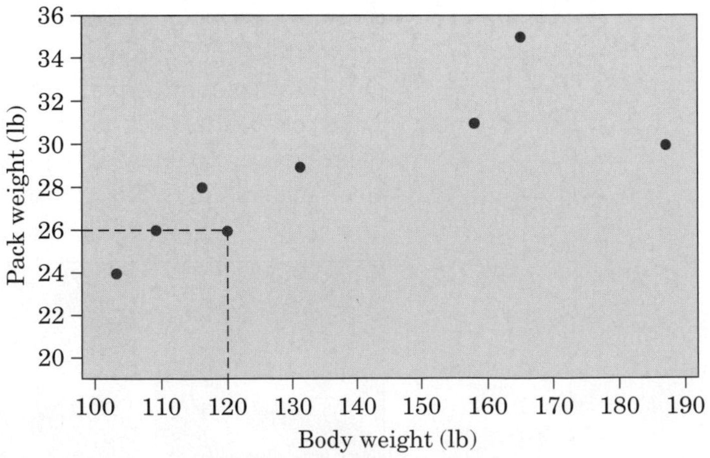

Figure 4.3 Completed scatterplot of body weight and backpack weight for the 8 students.

In general, it appears as though lighter students are carrying lighter backpacks and heavier students are carrying heavier packs.

Exercises

4.1 Explanatory or response, I In each of the following situations, is it more reasonable to simply explore the relationship between the two variables or to view one of the variables as an explanatory variable and the other as a response variable? In the latter case, which is the explanatory variable and which is the response variable?

(a) The amount of time spent studying for a statistics exam and the grade on the exam

(b) The weight in kilograms and height in centimeters of a person

(c) Inches of rain in the growing season and the yield of corn in bushels per acre

(d) A student's scores on the SAT Math test and the SAT Critical Reading test

4.2 Explanatory or response, II In each of the following situations, is it more reasonable to simply explore the relationship between the two variables or to view one of the variables as an explanatory variable and the other as a response variable? In the latter case, which is the explanatory variable and which is the response variable?

(a) A family's income and the years of education their eldest child completes

(b) Price of a house and square footage of the house

(c) The arm span and height of a person

(d) Amount of snow in the Colorado mountains and the volume of water in area rivers

4.3 IQ and GPA Figure 4.4 is a scatterplot of school grade point average versus IQ score for all 78 seventh-grade students in a rural midwestern school.

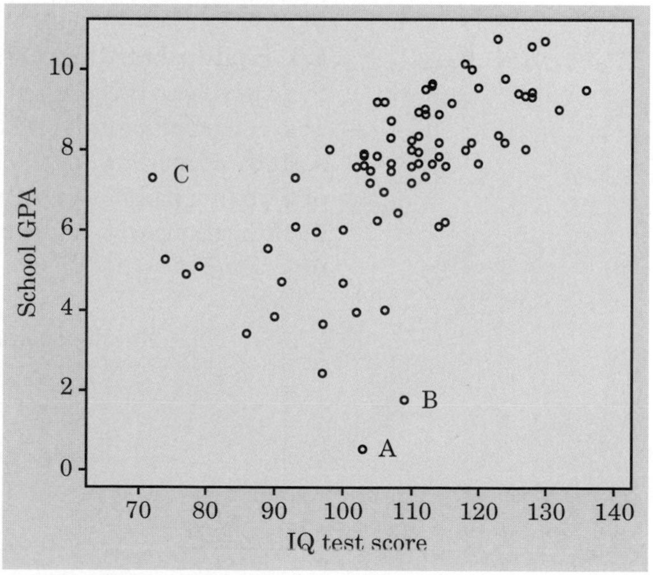

Figure 4.4 School grade point averages and IQ test scores for 78 seventh-grade students.

(a) Describe the overall pattern of the relationship in words. (Points A, B, and C might be called *outliers.*)

(b) About what are the IQ and GPA for Student A?

(c) For each point A, B, and C, say how it is unusual (for example, "low GPA but a moderately high IQ score").

4.4 SAT scores by state More than 1.5 million students took the SAT during the 2007–2008 school year. For the entire group of test takers, the mean score on the Math test was 515. If we look at the data on a state-by-state basis, we see quite a different picture. Figure 4.5 is a scatterplot of the percent of high school graduates who took the SAT and the mean Math test score in each state.

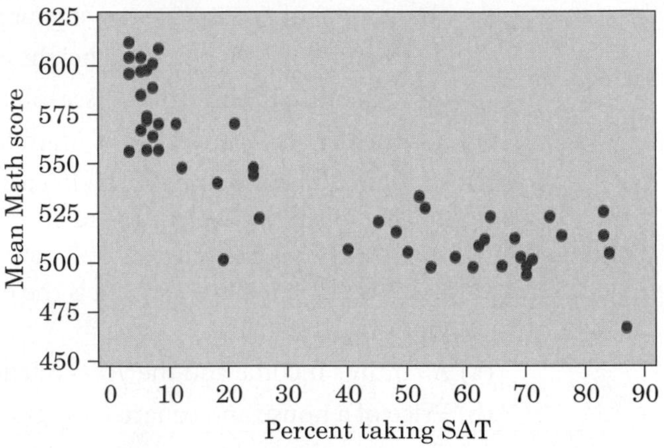

Figure 4.5 Scatterplot showing the mean SAT Math score and the percent of high school graduates who took the SAT in each state during the 2007–2008 school year.

(a) Describe the overall relationship between the two variables. Why are there two distinct clusters of points?

(b) Two states stand out in Figure 4.5: West Virginia at (19, 501) and Maine at (87, 466). In what way is each of these states "unusual"?

4.5 Healthy breeding Often the percent of an animal species in the wild that survive to breed again is lower following a successful breeding season. This is part of nature's self-regulation, tending to keep population size stable. A study of merlins (small falcons) in northern Sweden observed the number of breeding pairs in an isolated area and the percent of males (banded for identification) who returned the next breeding season. Here are the data for nine years:[2]

Breeding pairs	Percent of males returning
28	82
29	83, 70, 61
30	69
32	58
33	43
38	50, 47

(a) Why is the response variable the *percent* of males that return rather than the *number* of males that return?

(b) Make a well-labeled scatterplot.

(c) Describe the pattern you see. Do the data support the theory that a smaller percent of birds survive following a successful breeding season?

4.6 Fast cars Interested in a sporty car? Worried that it might use too much gas? The Environmental Protection Agency lists most such vehicles in its "minicompact" or "two-seater" categories. Table 4.1 gives city and highway gas mileages (in miles per gallon) for all model year 2009 cars in these two groups.

Table 4.1 Gas Mileages (mpg) for Model Year 2009 Cars

Mini/subcompact cars			Two-seater cars		
Model	**City**	**Highway**	**Model**	**City**	**Highway**
Audi TT coupe	23	31	Aston Martin DBS Coupe	17	24
BMW 328CI Convertible	18	27	Aston Martin V8 Vantage	13	19
BMW 335CI Convertible	17	26	Audi TT Roadster	22	30
BMW M3 Convertible	14	20	Cadillac XLR	14	23
Jaguar XK Convertible	16	25	Chevrolet Corvette	15	25
Jaguar XKR Convertible	15	23	Dodge Viper	13	22
Mercedes-Benz CLK350	17	25	Ferrari GTB Fiorano	11	15
Mercedes-Benz CLK550	15	22	Ferrari F430	11	16
Mitsubishi Eclipse Spyder	19	26	Honda S2000	18	25
Porsche 911 Carrera	18	26	Lamborghini Gallardo Coupe	14	20
Porsche 911 Turbo	15	23			
			Mercedes-Benz SL500	13	21
			Mercedes-Benz SL600	11	18
			Mercedes-Benz SLK320	19	26
			Nissan 350Z Roadster	17	23
			Pontiac Solstice	19	27
			Porsche 911 GT2	16	23
			Saturn Sky	19	27
			Smart fortwo convertible	33	41
			Spyker C8	13	18

Source: Environmental Protection Agency, *2008 Fuel Economy Guide*, **www.fueleconomy.gov.**

(a) Make a scatterplot that shows the relationship between city and highway mileage for minicompact cars using city mileage as the explanatory variable. Be sure to label your axes.

(b) On the same graph, make a scatterplot that shows the relationship between city and highway mileage for two-seater cars. Use a different color or plotting symbol.

(c) Describe what you see in the scatterplot. Is the form of the relationship similar for the two types of car? What is the most important difference between the two types?

Interpreting scatterplots

To interpret a scatterplot, apply the usual strategies of data analysis.

> **Examining a scatterplot**
>
> As in any graph of data, look for the **overall pattern** and for striking **deviations** from that pattern.
>
> You can describe the overall pattern of a scatterplot by the **direction, form,** and **strength** of the relationship.
>
> An important kind of deviation is an **outlier,** an individual value that falls outside the overall pattern of the relationship.

Example 4.3 Health and wealth

www

The Gapminder Web site, **www.gapminder.org**, provides loads of data on the health and well-being of the world's inhabitants. Figure 4.6 is a scatterplot of data from Gapminder.[3] The individuals are all the world's nations for which data are available. The explanatory variable is a measure of how rich a country is: the income per person, also known as the gross domestic product (GDP) per person. GDP is the total value of the goods and services produced in a country, converted into dollars. The response variable is life expectancy at birth.

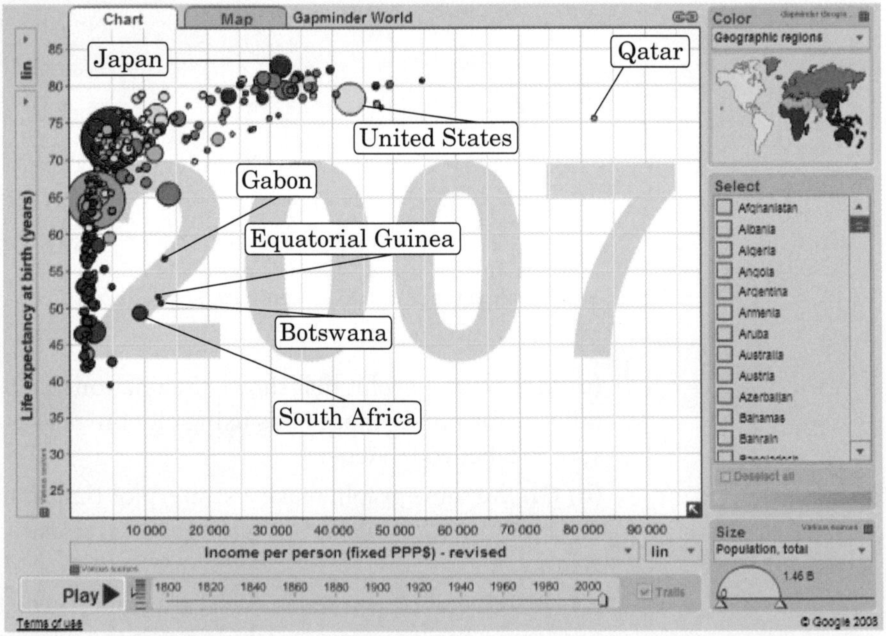

Figure 4.6 Scatterplot of the life expectancy of people in many nations against each nation's gross domestic product per person.

We expect people in richer countries to live longer. The overall pattern of the scatterplot does show this, but the relationship has an interesting shape. Life expectancy rises very quickly as GDP increases and then levels off. People in very rich countries such as the United States live no longer than people in poorer but not extremely poor nations. In some less wealthy countries, such as Japan, people live longer than in the United States.

Four African nations are outliers. Their life expectancies are similar to those of their neighbors but their GDP is higher. Gabon and Equatorial Guinea produce oil, and South Africa and Botswana produce diamonds. It may be that income from mineral exports goes mainly to a few people and so pulls up GDP per person without much effect on either the income or the life expectancy of ordinary citizens. That is, GDP per person is a mean, and we know that mean income can be much higher than median income.

Both Figures 4.2 (page 144) and 4.6 have a clear **direction:** the time between eruptions goes up as the length of the previous eruption increases, and life expectancy generally goes up as GDP increases. We say that Figures 4.2 and 4.6 show a **positive association** between the variables.

Positive association, negative association

Two variables are **positively associated** when above-average values of one tend to accompany above-average values of the other and below-average values also tend to occur together. The scatterplot slopes upward as we move from left to right.

Two variables are **negatively associated** when above-average values of one tend to accompany below-average values of the other, and vice versa. The scatterplot slopes downward from left to right.

Each of our scatterplots has a distinctive **form.** Figure 4.2 shows two *clusters* of points, and Figure 4.6 shows a *curved relationship.* The **strength** of a relationship in a scatterplot is determined by how closely the points follow a clear form. The relationships in Figures 4.2 and 4.6 are not strong. Eruptions of similar lengths show quite a bit of scatter in times until the next eruption, and nations with similar GDPs can have quite different life expectancies. Here is an example of two variables with a strong negative relationship and a simple form.

Example 4.4 How much natural gas does a household use?

Joan is concerned about the amount of energy she uses to heat her home in the Midwest. She keeps a record of the natural gas she consumes each month over one year's heating season. Because the months are not equally long, she divides each month's consumption by the number of days in the month to get the average number of cubic feet of gas used per day. Demand for heating is strongly influenced by the outside temperature. From local weather records, Joan obtains the average temperature for each month, in degrees Fahrenheit. Here are Joan's data:

Month:	Oct.	Nov.	Dec.	Jan.	Feb.	Mar.	Apr.	May
Temperature, x:	49.4	38.2	27.2	28.6	29.5	46.4	49.7	57.1
Gas consumed, y:	520	610	870	850	880	490	450	250

A scatterplot of these data appears in Figure 4.7. Temperature is the explanatory variable x (plotted on the horizontal axis) because outside temperature helps explain gas consumption. The scatterplot shows *a strong negative straight-line association.* The straight-line form is important because it is common and simple. The association is strong because the points lie close to a line. It is negative because, as temperature increases, gas consumption goes down because less gas is used for heating.

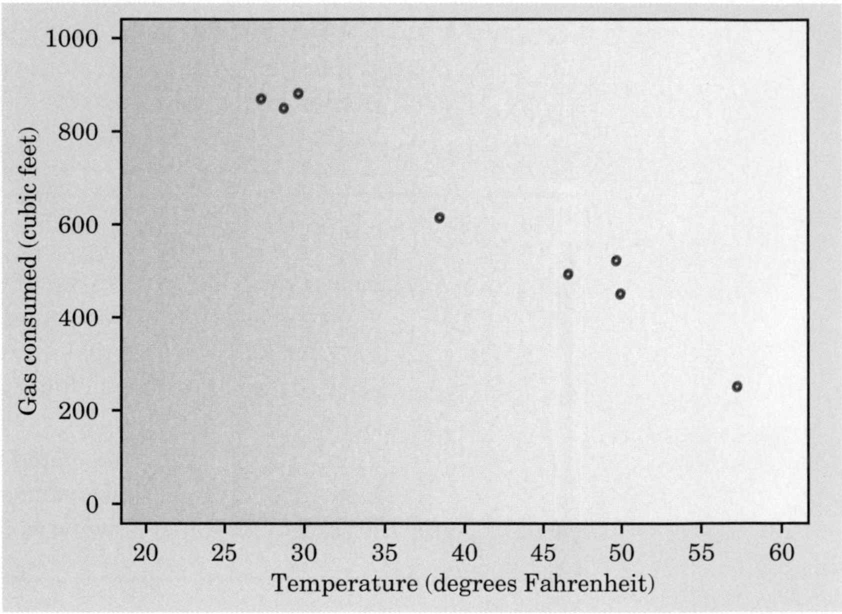

Figure 4.7 Home consumption of natural gas versus outdoor temperature.

Making scatterplots with technology is much easier than constructing them by hand.

CALCULATOR CORNER How to make a scatterplot

In this Calculator Corner, you will make a scatterplot of the natural-gas consumption data from Example 4.4.

1. Enter the data in two lists.

TI-84

- Go into your Statistics/List Editor (press STAT, then choose `1:Edit...`). Clear lists L1 and L2 as you did in Activity 4.1.
- Enter the temperature data in list L1 and the corresponding gas consumption values in list L2.

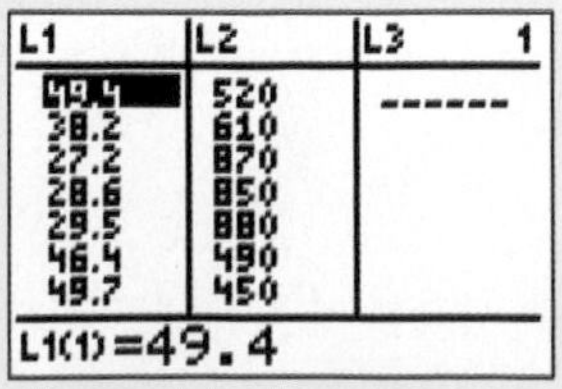

TI-Nspire

• Open a new Lists & Spreadsheet page (Press ⌂ and choose `3:Lists & Spreadsheet`).

• Name column A "Temp" and column B "Gas." Enter the temperature data in column A and the corresponding gas consumption values in column B.

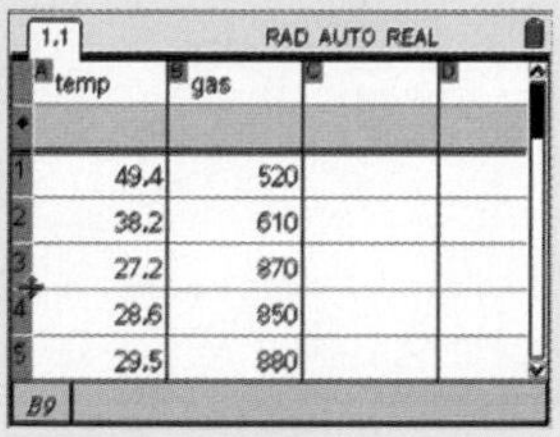

2. Make a scatterplot.

TI-84

• Go to the statistics plots menu (press [2nd][Y=]) and select `1:Plot1`. Adjust your settings as shown below.

• Make sure that Plot2 and Plot3 are Off and that all functions are deactivated in the Y = menu. Then use `ZoomStat` (press [ZOOM] and choose `9:ZoomStat`) to produce the graph.

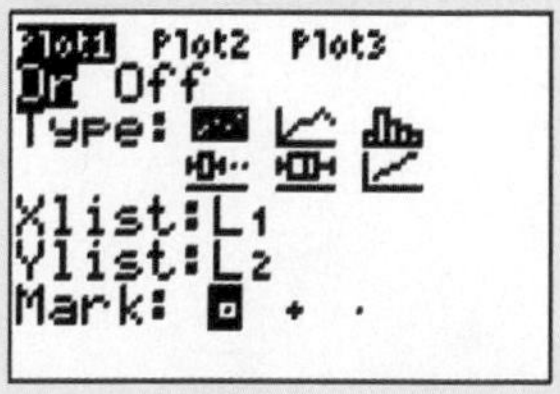

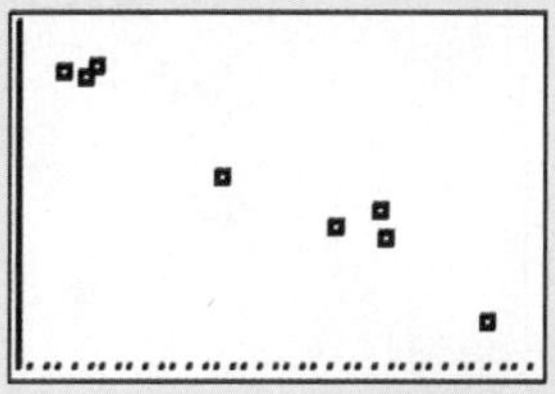

TI-Nspire

• Insert a Data & Statistics page (press ctrl I and choose `5:Add Data & Statistics`).

• Use the NavPad to move your cursor to "Click to add variable" at the bottom of the screen, and press the "mouse" button (☝). From the drop-down menu, choose Temp as the explanatory variable. A dotplot should appear. To choose Gas as the response variable, move your cursor to the left-hand side of the screen until you see "Click to add variable," press ☝, and choose Gas from the drop-down menu.

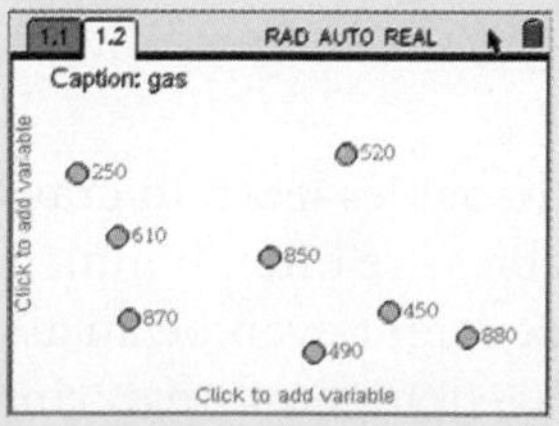

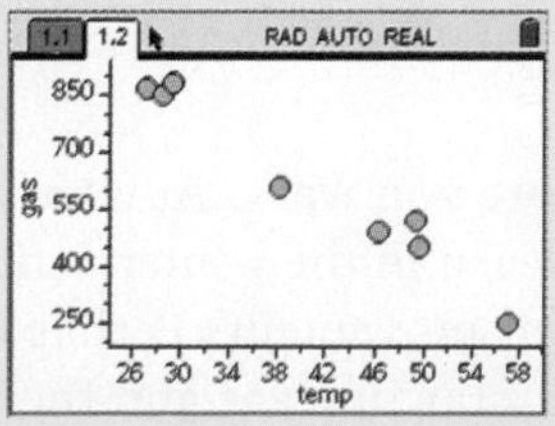

3. Adjust window settings as desired.

CALCULATOR CORNER *(continued)*

CALCULATOR CORNER *(continued)*

TI-84

- Press [WINDOW] and modify the settings to values that seem reasonable. Press [GRAPH] to obtain the updated scatterplot.

TI-Nspire

- Press (menu), choose `5:Window/Zoom, 1:Window Settings`, enter desired values, and choose OK. The graph should update automatically.

Exercises

4.7 Rich states, poor states One measure of a state's prosperity is the median income of its households. Another measure is the mean personal income per person in the state. Figure 4.8 is a scatterplot of these two variables, both measured in dollars.

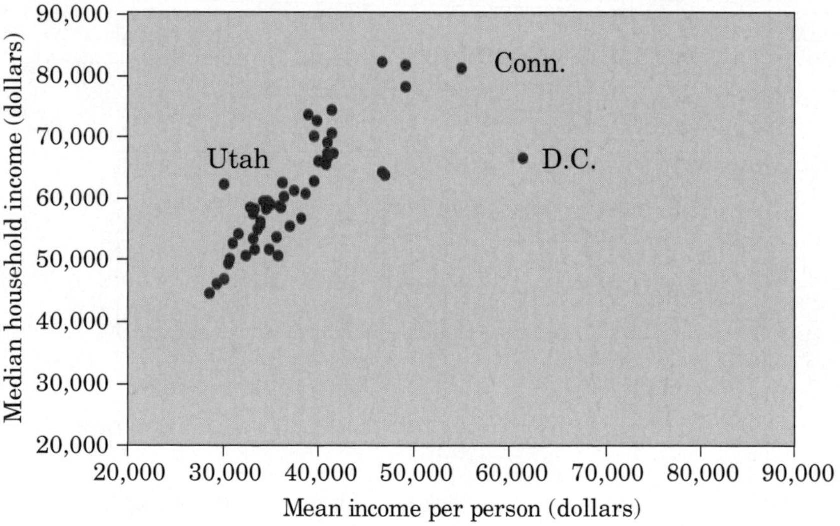

Figure 4.8 Median household income and mean income per person in the states.

(a) Explain why you expect a positive association between these variables. Also, explain why you expect household income to be generally higher than income per person.

(b) Describe the overall pattern of the plot, ignoring the outliers.

(c) We have labeled some interesting states on the graph. Describe what's unusual about each one in terms of mean income per person and median household income.

4.8 Crawl before you walk At what age do babies learn to crawl? Does it take longer to learn in the winter, when babies are often bundled in clothes that restrict their movement? Perhaps there might even be an association between babies' crawling age and the average temperature during the month they first try to crawl (around six months after birth). Data were collected from parents who brought their babies into the University of Denver Infant Study Center to participate in one of a number of experiments. Parents reported the birth month and the age at which their child was first able to creep or crawl a distance of four feet within one minute. Information was obtained on 414 infants, 208 boys and 206 girls. Crawling age is given in weeks, and the average temperature (in °F) is for the month that is six months after the birth month.[4]

Birth month	Average crawling age (wk)	Average temp. (°F) 6 mo after birth
January	29.84	66
February	30.52	73
March	29.70	72
April	31.84	63
May	28.58	52
June	31.44	39
July	33.64	33
August	32.82	30
September	33.83	33
October	33.35	37
November	33.38	48
December	32.32	57

(a) Make a scatterplot that displays the relationship between average crawling age and average temperature. Be sure to think about which is the explanatory variable.

(b) Describe the direction, form, and strength of the relationship. What do you conclude about when babies learn to crawl?

4.9 Interested in stocks? When interest rates are high, investors may avoid stocks because they can get high returns with less risk. Figure 4.9 plots the annual returns on U.S. common stocks over a 50-year period against the returns on Treasury bills for the same years.[5] (The interest paid by Treasury bills is a measure of how high interest rates were that year.)

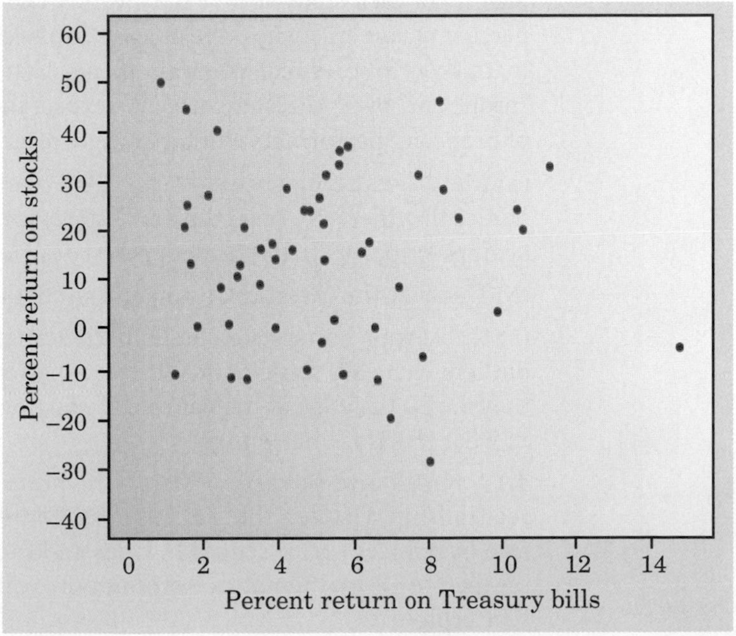

Figure 4.9 Percent returns on Treasury bills and common stocks over a 50-year period.

(a) Approximately what were the highest and lowest annual percent returns on stocks during this period? What were the highest and lowest returns for Treasury bills?

(b) Describe the pattern you see. Are high interest rates bad for stocks? Is the relationship between interest rates and stock returns strong or weak?

4.10 Teen drivers Thirty students in a high school statistics class were asked to report the age (in years) and odometer reading of their primary vehicle. Here are their data:[6]

Age	Mileage	Age	Mileage	Age	Mileage
3	40,300	6	85,000	8	110,000
2	11,912	4	20,000	4	30,323
3	30,000	3	30,000	7	100,000
4	40,000	2	17,000	10	98,000
8	98,000	3	25,000	2	12,000
3	48,000	7	150,000	5	53,000
8	120,000	2	10,000	7	40,000
11	185,000	10	110,000	4	76,000
4	40,000	5	103,000	2	3,000
1	1,050	5	66,610	7	75,000

(a) Enter the data into your calculator lists. Refer to the Calculator Corner (page 152).

(b) Make a scatterplot of the data. Adjust the viewing window, then transfer your graph to paper.

(c) Describe what the scatterplot tells you about the direction, form, and strength of the relationship. Are there any outliers?

4.11 NAEP for fourth-graders The National Assessment of Educational Progress (NAEP) assesses what students know in several subject areas based on large representative samples. Table 4.2 (on the facing page) reports some findings of the NAEP year 2007 Mathematics Assessment for fourth-graders in the United States. For each state we give the mean NAEP math score (out of 500) and also the percent of students who were at least "proficient" in the sense of being able to use math skills to solve real-world problems. Nationally, about 25% of students are "proficient" by NAEP standards. We expect that average performance and percent of proficient performers will be strongly related.

(a) Make a scatterplot, using mean NAEP scores as the explanatory variable. Notice that there are several pairs of states with identical values. Use a different symbol for points that represent two states.

(b) Describe the direction, form, and strength of the relationship.

(c) Circle your home state's point in the scatterplot. Although there are no clear outliers, there are some points that you may consider interesting, perhaps because they are on the edge of the pattern. Choose one such point: which state is this, and in what way is it interesting?

4.12 NAEP and poverty Refer to the previous exercise. What is the relationship between the percent of students in a state who were eligible for free or reduced-price school lunches and the percent who scored proficient on the NAEP mathematics test? Support your answer with a well-labeled scatterplot.

Table 4.2 State Performance on the 2007 NAEP Mathematics Assessment of Fourth-Graders

State	Mean NAEP score	Percent proficient	Percent free/ reduced lunch	State	Mean NAEP score	Percent proficient	Percent free/ reduced lunch
AL	229	23	55	MT	244	39	38
AK	237	32	44	NE	238	33	39
AZ	232	27	52	NV	232	27	45
AR	238	32	57	NH	249	44	19
CA	230	25	53	NJ	249	42	29
CO	240	35	40	NM	228	22	67
CT	243	37	31	NY	243	37	49
DE	242	36	39	NC	242	35	48
FL	242	34	48	ND	245	41	32
GA	235	28	52	OH	245	39	37
HI	234	29	42	OK	237	30	55
ID	241	35	44	OR	236	31	44
IL	237	31	44	PA	244	40	35
IN	245	40	41	RI	236	31	40
IA	243	38	34	SC	237	31	53
KS	248	42	41	SD	241	37	36
KY	235	27	53	TN	233	26	49
LA	230	22	70	TX	242	35	55
ME	242	36	36	UT	239	35	37
MD	240	32	34	VT	246	42	31
MA	252	47	27	VA	244	35	30
MI	238	32	38	WA	243	37	39
MN	247	41	30	WV	236	30	50
MS	228	20	69	WI	244	40	34
MO	239	33	42	WY	244	40	36

Source: National Center for Education Statistics Web site, **nces.ed.gov/nationsreportcard/**.

Correlation

A scatterplot displays the direction, form, and strength of the relationship between two variables. Straight-line relations are particularly important because a straight line is a simple pattern that is quite common. A straight-line relation is strong if the points lie close to a straight line, and it is weak if they are widely scattered about a line. Our eyes are not good judges of how strong a relationship is. The two scatterplots in Figure 4.10 depict the same data, but the right-hand plot is drawn smaller in a large field. The right plot seems to show a stronger straight-line relationship. Our eyes can be fooled by changing the plotting scales or the amount of blank space around the cloud of points in a scatterplot.

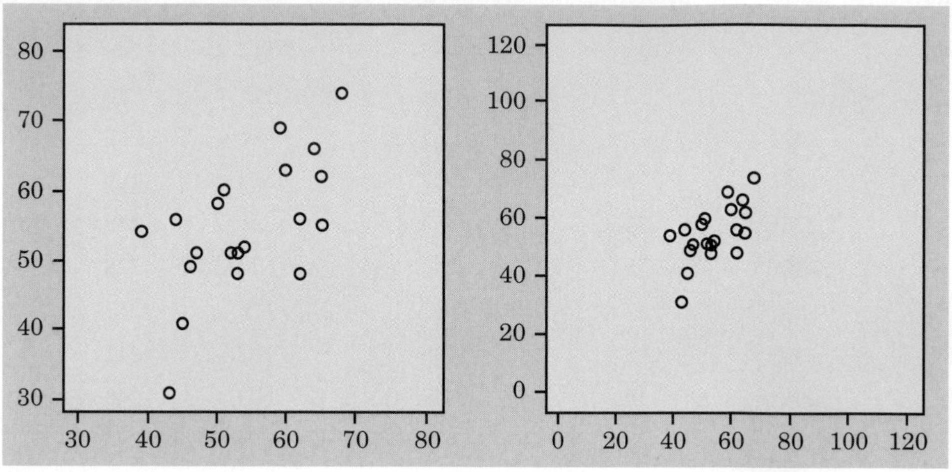

Figure 4.10 Two scatterplots of the same data. The right-hand plot suggests a stronger relationship between the variables because of the surrounding space.

We need to follow our strategy for data analysis by using a numerical measure to supplement the graph. **Correlation** is the measure we use in this setting.

Correlation

The **correlation** describes the direction and strength of a straight-line relationship between two quantitative variables. Correlation is usually written as r.

To calculate correlation, use the formula

$$r = \frac{1}{n-1}\sum\left[\left(\frac{x-\bar{x}}{s_x}\right)\left(\frac{y-\bar{y}}{s_y}\right)\right]$$

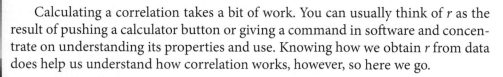

Calculating a correlation takes a bit of work. You can usually think of r as the result of pushing a calculator button or giving a command in software and concentrate on understanding its properties and use. Knowing how we obtain r from data does help us understand how correlation works, however, so here we go.

Example 4.5 Back to the backpackers

In Example 4.2 (page 146), we examined data on the relationship between students' body weights and their backpack weights prior to a hiking trip. Here are the data once again:

Body weight (lb):	120	187	109	103	131	165	158	116
Backpack weight (lb):	26	30	26	24	29	35	31	28

To calculate the correlation:

Step 1: *Find the mean and standard deviation for both x and y.* Use your calculator to check that the mean and standard deviation for each of the two variables are as follows:

Body weight:	$\bar{x} = 136.125$ pounds	$s_x = 30.296$ pounds
Pack weight:	$\bar{y} = 28.625$ pounds	$s_y = 3.462$ pounds

Step 2: *Standardize the x- and y-values.* Using the means and standard deviations from Step 1, find the standardized score (*z*-score) for each *x*-value and for each *y*-value. (Note that we used the unrounded values of z_x and z_y to compute the products in the final column.)

Value of x	Standardized score $z_x = (x - \bar{x})/s_x$	Value of y	Standardized score $z_y = (y - \bar{y})/s_y$	Product of standardized scores
120	$(120 - 136.125)/30.296 = -0.532$	26	$(26 - 28.625)/3.462 = -0.758$	$(-0.532)(-0.758) = 0.404$
187	$(187 - 136.125)/30.296 = 1.679$	30	$(30 - 28.625)/3.462 = 0.397$	$(1.679)(0.397) = 0.667$
109	$(109 - 136.125)/30.296 = -0.895$	26	$(26 - 28.625)/3.462 = -0.758$	$(-0.895)(-0.758) = 0.679$
103	$(103 - 136.125)/30.296 = -1.093$	24	$(24 - 28.625)/3.462 = -1.336$	$(-1.093)(-1.336) = 1.461$
131	$(131 - 136.125)/30.296 = -0.169$	29	$(29 - 28.625)/3.462 = 0.108$	$(-0.169)(0.108) = -0.018$
165	$(165 - 136.125)/30.296 = 0.953$	35	$(35 - 28.625)/3.462 = 1.841$	$(0.953)(1.842) = 1.755$
158	$(158 - 136.125)/30.296 = 0.722$	31	$(31 - 28.625)/3.462 = 0.686$	$(0.722)(0.686) = 0.495$
116	$(116 - 136.125)/30.296 = -0.664$	28	$(28 - 28.625)/3.462 = -0.181$	$(-0.664)(-0.181) = 0.120$
				Sum = 5.563

The first hiker's body weight is about one-half standard deviation below the mean body weight for the group of hikers. His pack weight is about three-fourths of a standard deviation below the mean backpack weight for the group. See if you can describe how the second hiker is "unusual" using the standardized values of his body weight and pack weight.

Step 3: *Multiply the standardized scores for each individual, then "average" the products.* As with the standard deviation, we "average" by dividing by $n - 1$, one fewer than the number of individuals. In this case, since $n = 8$, we divide by 7:

$$r = \frac{1}{n-1}\sum\left[\left(\frac{x-\bar{x}}{s_x}\right)\left(\frac{y-\bar{y}}{s_y}\right)\right] = \frac{1}{7}(5.563) = 0.795$$

Notice that all the products are positive except for the hiker we identify in Figure 4.11 (on the next page), who had slightly below-average weight in the group but a pack weight that was slightly above average.

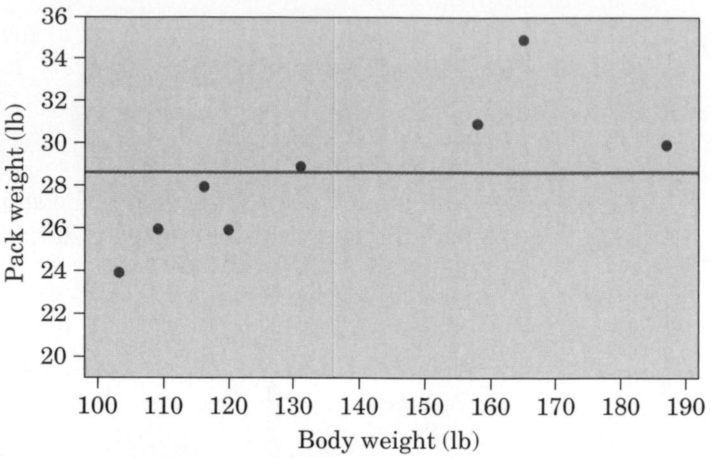

Figure 4.11 Scatterplot of backpack weight versus body weight for 8 hikers with reference lines added at the mean x-value and mean y-value.

Understanding correlation

More important than calculating r (a task better suited to a machine) is understanding how correlation measures association. Here are the facts:

- **Positive r indicates positive association between the variables, and negative r indicates negative association.** Figure 4.11 shows the original scatterplot of the backpacker data from Example 4.5 with two reference lines added: a horizontal line at the mean y-value, $\bar{y} = 28.625$, and a vertical line at the mean x-value, $\bar{x} = 136.125$. The scatterplot reveals a moderately strong positive association between body weight and backpack weight. Three hikers have above-average body weights and pack weights, so their standardized scores are positive for both x and y. Four hikers have below-average body weights and backpack weights, so both standardized scores are negative. The products of these standardized scores are all positive, however. Only one hiker contributes a small negative product of standardized scores—the one with slightly below-average body weight but slightly above-average pack weight. With so many positive products, we get a positive value for r.

"He says we've ruined his positive correlation between height and weight."

- **The correlation r always falls between -1 and 1.** Values of r near 0 indicate a very weak straight-line relationship. The strength of the relationship increases as r moves away from 0 toward either -1 or 1. Values of r close to -1 or 1 indicate that the points lie close to a straight line. The extreme values $r = -1$ and $r = 1$ occur only when the points in a scatterplot lie exactly along a straight line.

  The result $r = 0.795$ in Example 4.5 reflects the moderately strong, positive, straight-line pattern in Figure 4.11.

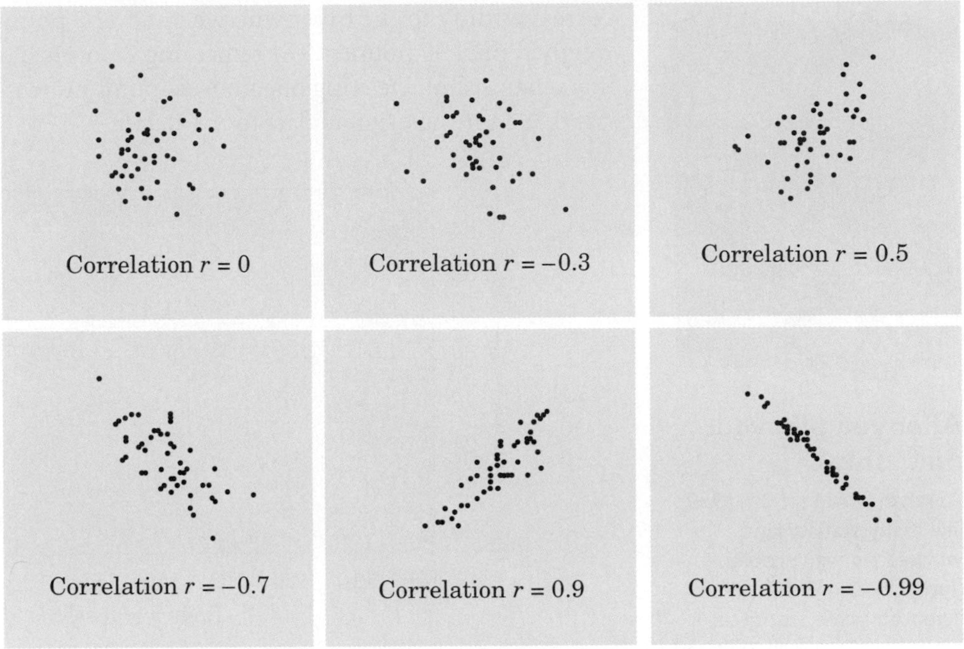

Figure 4.12 How correlation indicates the direction and strength of a straight-line relationship. Patterns closer to a straight line have correlations closer to 1 or −1.

The scatterplots in Figure 4.12 illustrate how r indicates both the direction and the strength of a straight-line relationship. Study them carefully. Note that the sign of r matches the direction of the slope in each plot, and that r approaches −1 or 1 as the pattern of the plot comes closer to a straight line.

www

How good are you at estimating the correlation by eye for a scatterplot? Try the Correlations game at **www.stat.uiuc.edu/courses/stat100/cuwu/Games.html.**

- Because r uses standardized scores for the observations, **the correlation does not change when we change the units of measurement** of x, y, or both. Measuring weight in kilograms rather than pounds in Example 4.5 would not change the correlation $r = 0.795$.

  Our descriptive measures for one variable all share the same units as the original observations. If we measure length in centimeters, the median, quartiles, mean, and standard deviation are all in centimeters. The correlation between two variables, however, has no unit of measurement; it is just a number between −1 and 1.
- **Correlation ignores the distinction between explanatory and response variables.** If we reverse our choice of which variable to call x and which to call y, the correlation does not change. Can you see why from the formula?

$$r = \frac{1}{n-1}\sum\left[\left(\frac{x-\bar{x}}{s_x}\right)\left(\frac{y-\bar{y}}{s_y}\right)\right]$$

- **Correlation measures the strength of only straight-line association between two variables.** Correlation does not describe curved relationships between variables, no matter how strong they are.
- Like the mean and standard deviation, **the correlation is strongly affected by a few outlying observations.** Use r with caution when outliers appear in the scatterplot. Look, for example, at Figure 4.13. We removed the point

corresponding to the hiker who weighed 187 pounds but whose pack weighed only 30 pounds. The remaining 7 points have a stronger straight-line relationship. Deleting one unusual point increases the correlation from $r = 0.795$ for the original data to $r = 0.946$.

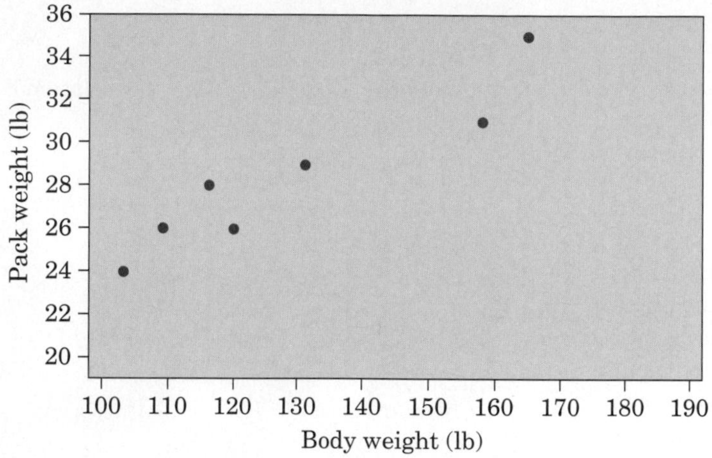

Figure 4.13 Hiker data from Figure 4.11 with the unusual point at (187, 30) deleted—note the strong, positive, linear relationship for the remaining 7 points.

There are many kinds of relationships between variables and many ways to measure them. Although correlation is very common, remember its limitations. Correlation makes sense only for quantitative variables—we can speak of the relationship between the sex of voters and the political party they prefer, but not of the correlation between these variables. Even for quantitative variables such as the length of bones, correlation measures only straight-line association.

Remember also that correlation is not a complete description of two-variable data, even when there is a straight-line relationship between the variables. You should give the means and standard deviations of both x and y along with the correlation. Because the formula for correlation uses the means and standard deviations, these measures are the proper choice to accompany a correlation.

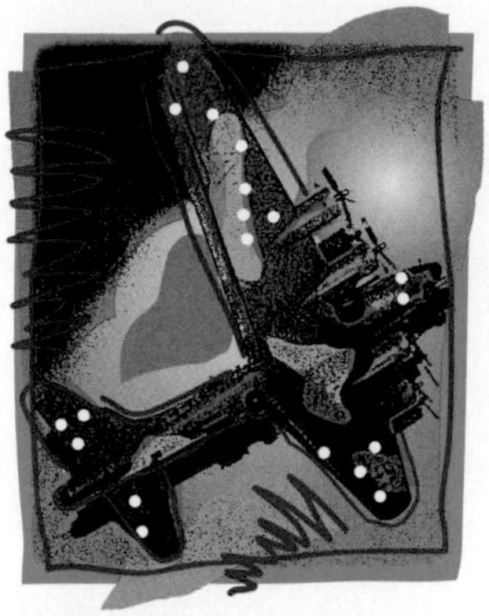

After you plot your data, think!

Abraham Wald (1902–1950), like many statisticians, worked on war problems during World War II. Wald invented some statistical methods that were military secrets until the war ended. Here is one of his simpler ideas. Asked where extra armor should be added to airplanes, Wald studied the location of enemy bullet holes in planes returning from combat. He plotted the locations on an outline of the plane. As data accumulated, most of the outline filled up. Put the armor in the few spots with no bullet holes, said Wald. That's where bullets hit the planes that didn't make it back.

Exercises

4.13 Dating and height, I A college student wonders if tall women tend to date taller men than do short women. She measures herself, her dormitory roommate, and the women in the adjoining rooms; then she measures the next man each woman dates. Here are the data (heights in inches):

Women (x):	66	64	66	65	70	65
Men (y):	72	68	70	68	71	65

(a) Make a scatterplot of these data. Based on the scatterplot, do you expect the correlation to be positive or negative? Near ± 1 or not?

(b) Find the correlation r between the heights of the men and women. Follow the method of Example 4.5.

4.14 Dating and height, II Refer to the previous exercise.

(a) How would r change if all the men were 6 inches shorter than the heights given in the table? Does the correlation tell us whether women tend to date men taller than themselves?

(b) If heights were measured in centimeters rather than inches, how would the correlation change? (There are 2.54 centimeters in an inch.)

(c) If every woman dated a man exactly 3 inches taller than herself, what would be the correlation between male and female heights?

4.15 Explaining correlation You have data on the current grade point average (GPA) of students in a basic statistics course and their scores on the first examination in the course. Say as specifically as you can what the correlation r between GPA and exam score measures.

4.16 What correlation doesn't measure

(a) Make a scatterplot of the following data:

x:	−5	−3	0	3	5
y:	0	4	5	4	0

(b) Show that the correlation is zero. Use the method of Example 4.5.

(c) The scatterplot shows a strong association between x and y. Explain how it can happen that $r = 0$ in this case.

4.17 Rank the correlations Consider each of the following relationships: the heights of fathers and the heights of their adult sons, the heights of husbands and the heights of their wives, and the heights of females at age 4 and their heights at age 18. Rank the correlations between these pairs of variables from highest to lowest. Explain your reasoning.

4.18 Matching correlations Figure 4.14 displays five scatterplots. Match each to the r below that best describes it. (Some r's will be left over.)

$r = -0.9$ $r = -0.7$ $r = -0.3$ $r = 0$ $r = 0.3$ $r = 0.7$ $r = 0.9$

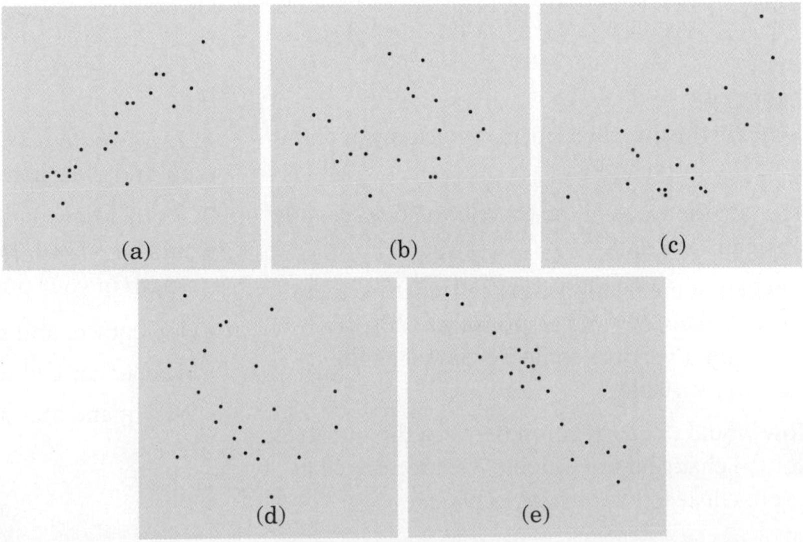

Figure 4.14 Match the correlations.

APPLICATION 4.1

Cars and correlation

The scatterplots below show some data from Environmental Protection Agency (EPA) fuel economy tests on 140 different cars that are from the same model year and that run on unleaded gas. Variables that were measured for each car include make of car, horsepower, weight (in pounds), displacement (in cubic inches), mpg:highway (highway driving miles per gallon), mpg:city (city driving miles per gallon), and carbon dioxide emitted.[7]

(a)

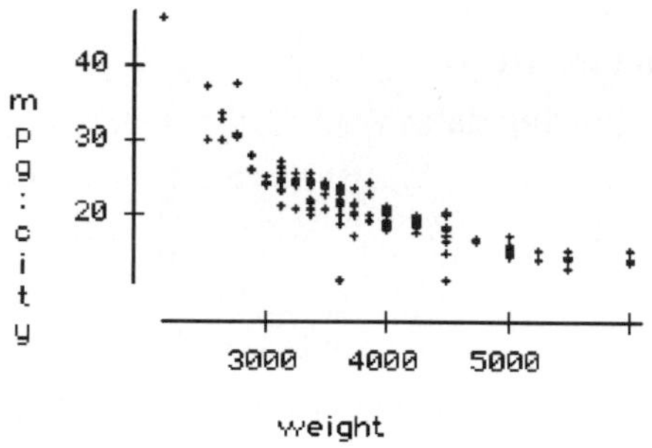

(b)

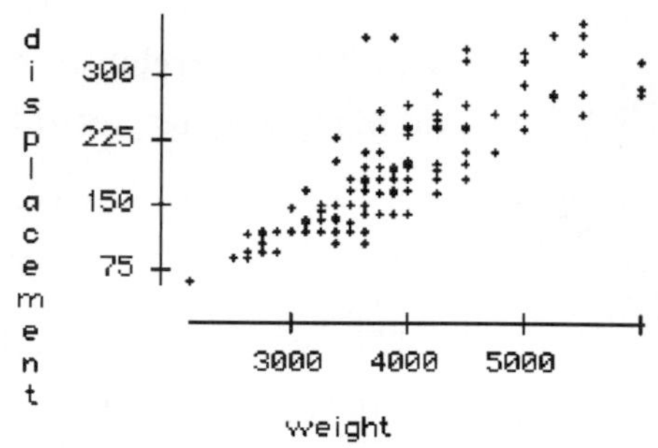

(c)

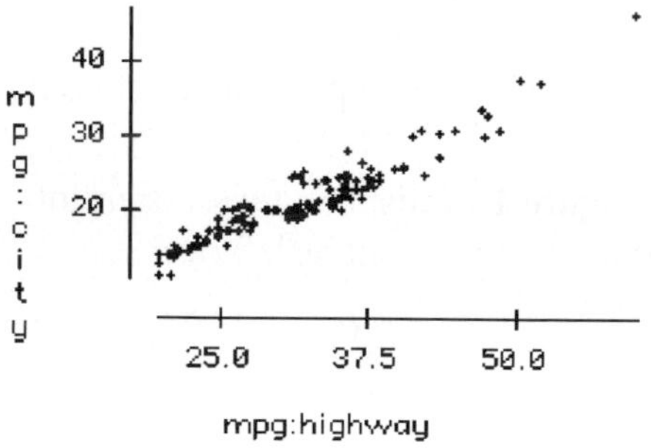

(d)

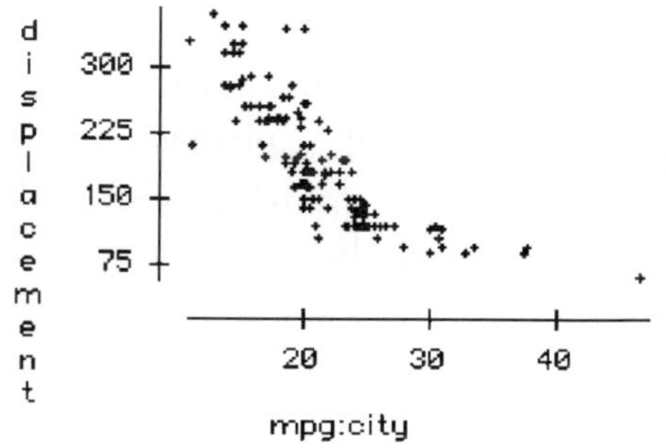

QUESTIONS

1. Describe the direction, form, and strength of the relationship in each plot.

2. Estimate the value of the correlation between the variables in each plot.

3. For each of the relationships, tell whether there are clear explanatory and response variables, or whether either variable could be viewed as the explanatory variable.

4. How would the correlation between the variables in plot (b) change if the weights were measured in kilograms instead of pounds? Explain.

5. How would the correlation in plot (c) change if the x and y variables were switched? Explain.

6. Predict how the following pairs of variables would be related. Include direction, form, and strength in your predictions.

- Horsepower and mpg:city
- Make of car and weight
- Weight and amount of carbon dioxide emitted

Section 4.1 Summary

Most statistical studies examine relationships between two or more variables. A **scatterplot** is a graph of the relationship between two quantitative variables. If you have an **explanatory** and a **response variable,** put the explanatory variable on the x (horizontal) axis of the scatterplot.

When you examine a scatterplot, look for the **direction, form, and strength** of the relationship and also for possible **outliers.** If there is a clear direction, is it a **positive association** (the scatterplot slopes upward from left to right) or a **negative association** (the plot slopes downward)? Is the form straight or curved? Are there clusters of observations? Is the relationship strong (a tight pattern in the plot) or weak (the points scatter widely)?

The **correlation r** measures the direction and strength of a straight-line relationship between two quantitative variables. Correlation is a number between -1 and 1. The sign of r shows whether the association is positive or negative. The value of r gets closer to -1 or 1 as the points cluster more tightly about a straight line. The extreme values -1 and 1 occur only when the scatterplot shows a perfectly straight line.

Section 4.1 Exercises

4.19 Born to be old? Is there a relationship between the gestational period (time from conception to birth) of an animal and its average life span? Figure 4.15 is a scatterplot of the gestational period and average life span for 43 species of animals.[8]

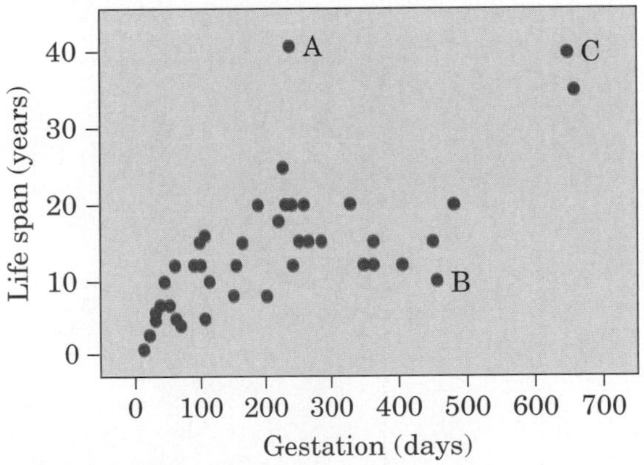

Figure 4.15 Scatterplot of the average life span and gestational period for 43 species of animals.

(a) Describe the direction, form, and strength of the scatterplot.
(b) Three "unusual" points are labeled in Figure 4.15: Point A is for the hippopotamus, Point B is for the giraffe, and Point C is for the Asian elephant. In what way is each of these animals "unusual"?
(c) If you were to add a point for humans to the plot, would it be part of the general pattern, or would it be an unusual point? Explain.

4.20 IQ and siblings The correlation between a child's score on the vocabulary portion of the Wechsler Intelligence Scale for Children (a standard IQ test) and the number of siblings a child has is $r = -0.319$.[9]

(a) Explain in words what this r says.
(b) Can you suggest an explanation for this relationship?

4.21 What number can I be?
(a) What are all the values that a correlation r can possibly take?
(b) What are all the values that a standard deviation s can possibly take?

4.22 Correlation "facts" Which of these statements are true and which are false?
(a) A correlation of 0.8 means that 80% of the points in the scatterplot lie on a line.
(b) If the correlation between two variables is zero, then there can be no relationship between them.
(c) For all books in the Library of Congress, the correlation between the thickness of the books (in inches) and their number of pages would be positive.
(d) For all of the cars registered in the state of Ohio, the correlation between their fuel efficiency (in miles per gallon) and their weight (in pounds) would be positive.
(e) If the correlation between height in inches and weight in pounds for a group of people is 0.7, then the correlation between their heights in centimeters and their weights in kilograms will still be 0.7.

4.23 Pollution from vehicle exhaust Auto manufacturers are required to test their vehicles for the amount of each of several pollutants in the exhaust. The amount of a pollutant varies even among identical vehicles, so that several vehicles must be tested. Figure 4.16 is a scatterplot of the amounts of two pollutants, carbon monoxide and nitrogen oxides, for 46 vehicles of the same model. Both variables are measured in grams of the pollutant per mile driven.[10]

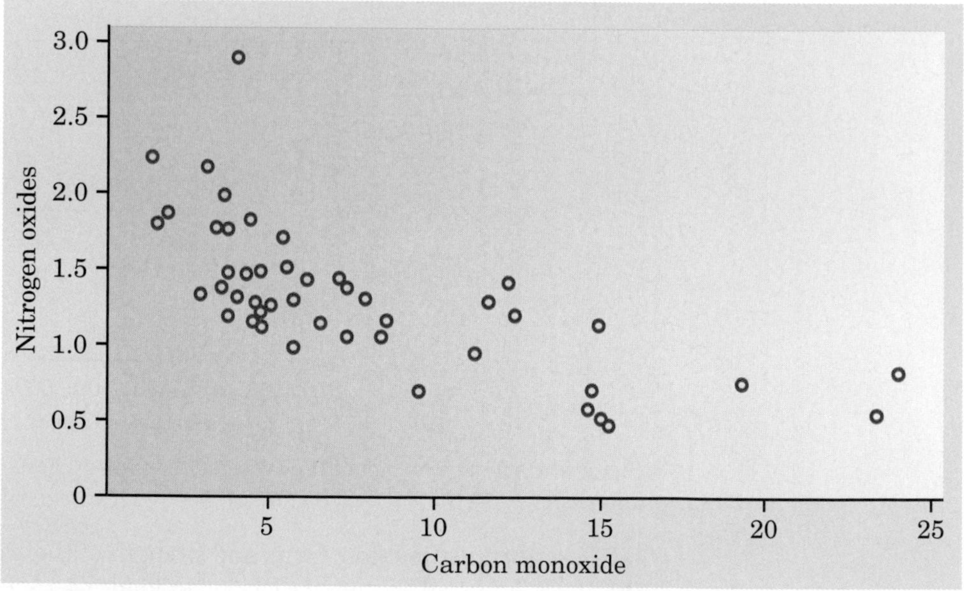

Figure 4.16 Amounts of two pollutants (measured in grams per mile driven) in the exhaust of 46 vehicles of the same model.

(a) Describe the nature of the relationship. Is the association positive or negative? Is the relation close to a straight line or clearly curved? Are there any outliers?
(b) A car magazine says, "When a car's engine is properly tuned, it emits few pollutants. If the engine is out of tune, it emits more of all pollutants. You can

find out how badly a car is polluting the air by measuring any one pollutant. If that value is acceptable, the other emissions will also be OK." Do the data in Figure 4.16 support this claim?

4.24 Guess the correlation For each of the following pairs of variables, would you expect a large negative correlation, a large positive correlation, or a small correlation? Justify your answers.

(a) The age of secondhand cars and their prices
(b) The weight of new cars and their gas mileages in miles per gallon
(c) The heights and the weights of adult men
(d) The heights and the IQ scores of adult men
(e) The heights of daughters and the heights of their mothers

4.25 Foot problems Metatarsus adductus (call it MA) is a turning in of the front part of the foot that is common in adolescents and usually corrects itself. Hallux abducto valgus (call it HAV) is a deformation of the big toe that is not common in youth and often requires surgery. Perhaps the severity of MA can help predict the severity of HAV. Table 4.3 gives data on 38 consecutive patients who came to a medical center for HAV surgery.[11] Using X-rays, doctors measured the angle of deformity for both MA and HAV. They speculated that there is a positive association: more serious MA is associated with more serious HAV.

Table 4.3 The Severity of Metatarsus Adductus and Hallux Abducto Valgus

HAV angle	MA angle	HAV angle	MA angle	HAV angle	MA angle
28	18	21	15	16	10
32	16	17	16	30	12
25	22	16	10	30	10
34	17	21	7	20	10
38	33	23	11	50	12
26	10	14	15	25	25
25	18	32	12	26	30
18	13	25	16	28	22
30	19	21	16	31	24
26	10	22	18	38	20
28	17	20	10	32	37
13	14	18	15	21	23
20	20	26	16		

(a) Use your calculator to help you make a scatterplot of the data. Which is the explanatory variable?
(b) Describe the direction, form, and strength of the relationship between MA angle and HAV angle. Are there any clear outliers in your graph?
(c) The correlation between HAV angle and MA angle is 0.30. What does this tell you?
(d) Do you think the data confirm the doctors' speculation?

4.26 Strong relationship, low correlation Make a scatterplot that shows a strong relationship between two variables but a low correlation.

4.2 Regression and Prediction

ACTIVITY 4.2A

Vitruvius and the ideal man

Materials: Ruler or measuring tape for each small group of 2 to 3 students, graph paper

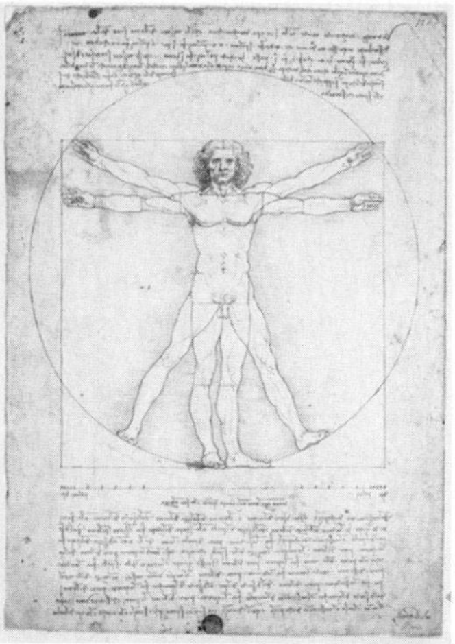

Scala/Art Resource

According to the ancient architect Vitruvius, the measurements of various parts of the human body have been set by nature to follow certain ratios. Leonardo da Vinci paid tribute to these claims in his painting entitled *Vitruvian Man.* Here is a brief excerpt from the text accompanying da Vinci's painting: "The length of a man's outspread arms is equal to his height. The greatest width of the shoulders contains in itself the fourth part of the man. From the elbow to the tip of the hand will be the fifth part of the man." In this Activity, you will make some measurements to determine whether da Vinci's statements seem to accurately reflect reality.

1. Measure the distance from the elbow to the tip of the hand (end of the middle finger) and the height for each group member to the nearest inch. Record your data.

2. Your teacher will prepare a chart for recording your data. When you have finished measuring, enter your group members' values in the chart.

3. On graph paper, make a scatterplot of the class's measurements using the distance from the elbow to the tip of the hand as the explanatory variable. Be sure to label and scale your axes.

4. Describe the direction, form, and strength of the relationship between these variables. Estimate the correlation from your scatterplot.

5. Enter the data into lists on your calculator and make a scatterplot. How does the calculator graph compare with the one you made on paper? Adjust the viewing window to make the graphs appear more similar.

6. Calculate $\bar{x}$, s_x, $\bar{y}$, s_y, and r using your calculator.

7. According to da Vinci, the distance from the elbow to the tip of the hand should be one-fifth of the height. That is, the height should be five times the distance measured on the arm. To see how close your class's data come to this prediction, graph the line $y = 5x$ on top of your scatterplot as follows:

TI-84

Press [Y=]. Define `Y1=5x` and then press [GRAPH].

TI-Nspire

On the Data & Statistics page, press (menu), choose `4:Analyze, 4:Plot Function`. Define `f1(x):=5x` and press (enter).

Describe what you see.

8. Based on your scatterplot, make a prediction for the height of a student whose distance measured on the arm is 10 inches. How does your prediction compare with those of your classmates?

Regression lines

If a scatterplot shows a straight-line relationship between two quantitative variables, we would like to summarize this overall pattern by drawing a line on the graph. A **regression line** summarizes the relationship between two variables, but only in a specific setting: when one of the variables helps explain or predict the other. That is, regression describes a relationship between an explanatory variable and a response variable.

Regression line

A **regression line** is a straight line that describes how a response variable y changes as an explanatory variable x changes. We often use a regression line to predict the value of y for a given value of x.

Example 4.6 Return of the backpackers

In Examples 4.2 and 4.5, we observed a moderately strong, positive, linear relationship between the body weights and backpack weights of 8 hikers. Figure 4.17 is another scatterplot of the data, but this time with a regression line added. The line gives a quick summary of the overall pattern.

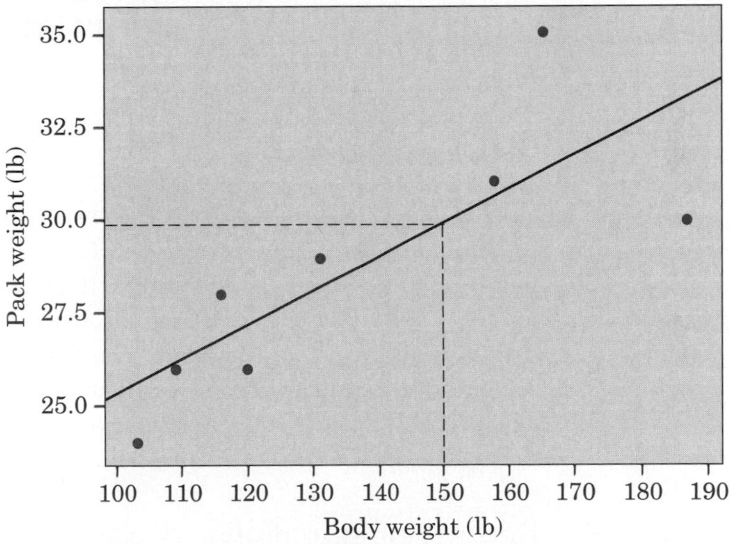

Figure 4.17 Using a straight-line pattern for prediction. The data are the body weights and backpack weights of 8 hikers.

At the last minute, another hiker is assigned to the group. This student weighs 150 pounds. Can we predict this hiker's backpack weight? The straight-line pattern connecting body weight to pack weight is strong enough that it seems reasonable to use body weight to predict backpack weight. Figure 4.17 shows how: starting at the new hiker's body weight (150 lb) on the x axis, go up to the regression line, then over to the pack weight axis. We predict a backpack weight of about 30 pounds. This is the weight his pack would have if this student's point lay exactly on the line. All the other points are relatively close to the line, so we think the missing point would also be fairly close to the line. That is, we think this prediction will be reasonably accurate.

Example 4.7 Natural-gas consumption and temperature

In Example 4.4 (page 151), we saw a very strong, negative, linear relationship between the average monthly temperature and the average amount of gas consumed per day in Joan's home. Figure 4.18 is a reproduction of the original scatterplot with a regression line added.

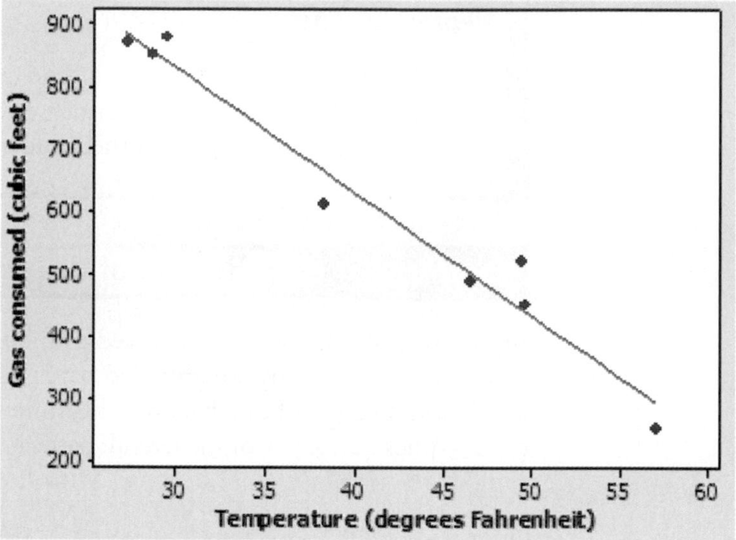

Figure 4.18 Minitab scatterplot showing a strong straight-line pattern. The data are the average monthly temperature and average amount of natural gas consumed per day in Joan's midwestern home.

We can use the regression line drawn in Figure 4.18 to predict Joan's monthly natural-gas consumption from the average monthly temperature. The points in this figure are more tightly packed around the line than are the points in the hiker scatterplot of Figure 4.17. The correlations, which measure the strengths of the straight-line relationships, are $r = 0.795$ for Figure 4.17 and $r = -0.983$ for Figure 4.18. The scatter of the points makes it clear that predictions of backpack weight will generally be less accurate than predictions of natural-gas consumption.

Regression equations

When a plot shows a straight-line relationship as strong as that in Figure 4.18, it is easy to draw a line close to the points by eye. In Figure 4.17, however, different people might draw quite different lines by eye. Because we want to predict y from x, we want a line that is close to the points in the *vertical* (y) direction. It is hard to concentrate on just the vertical distances when drawing a line by eye. What is more, drawing by eye gives us a line on the graph but not an equation for the line. We need a way to find from the data the equation of the line that comes closest to the points in the vertical direction. There are many ways to make the collection of vertical distances "as small as possible." The most common is the **least-squares method.**

Least-squares regression line

The **least-squares regression line** of y on x is the line that makes the sum of the squares of the vertical distances of the data points from the line as small as possible.

Figure 4.19 illustrates the least-squares idea. For each point in Figure 4.17, we have drawn a vertical line segment (in **bold**) from the point to the regression line. These line segments show the vertical distances of the points from the regression line. To find the least-squares line, we need to minimize the sum of the squares of these vertical distances. The squares in Figure 4.19 help you visualize these "squares of vertical distances" that we're trying to minimize. Imagine moving the line until the sum of the squares is the smallest it can be for any line. The resulting line *is* the least-squares regression line.

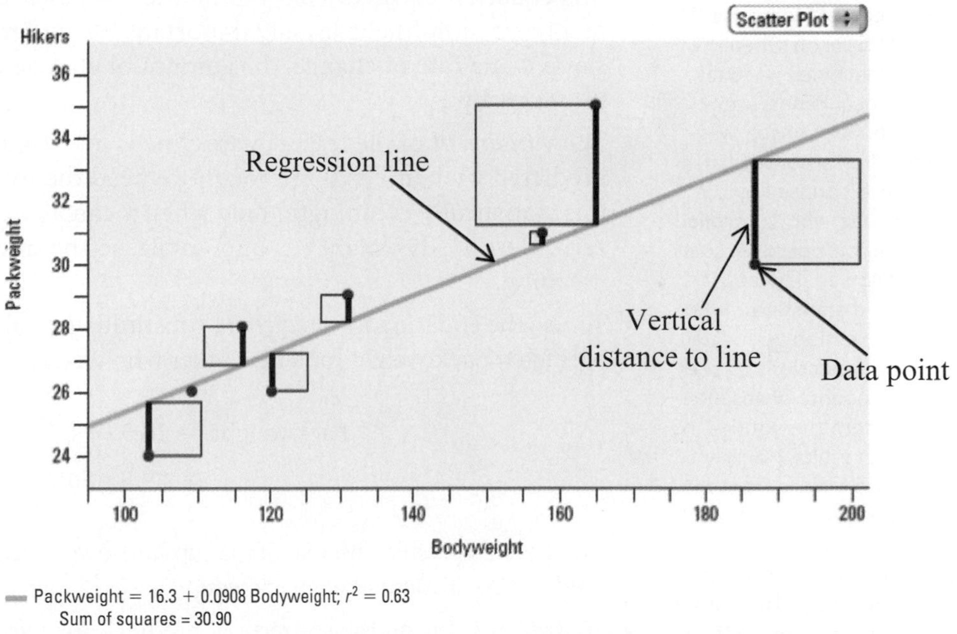

Figure 4.19 The least-squares regression line results in the smallest possible sum of squared distances of the points from the line (in this case, 30.90). Fathom software allows you to explore this fact dynamically.

The lines drawn on the scatterplots in Figures 4.17 and 4.18 are the least-squares regression lines. We won't give the formula for finding the least-squares line from data—that's a job for a calculator or computer. You should, however, be able to use the equation that technology produces.

In writing the equation of a line, x stands as usual for the explanatory variable and y for the response variable. From your previous math classes, you are probably used to seeing the equation of a line written in the form $y = mx + b$. In statistics, the equation of a line is usually written as

$$y = a + bx$$

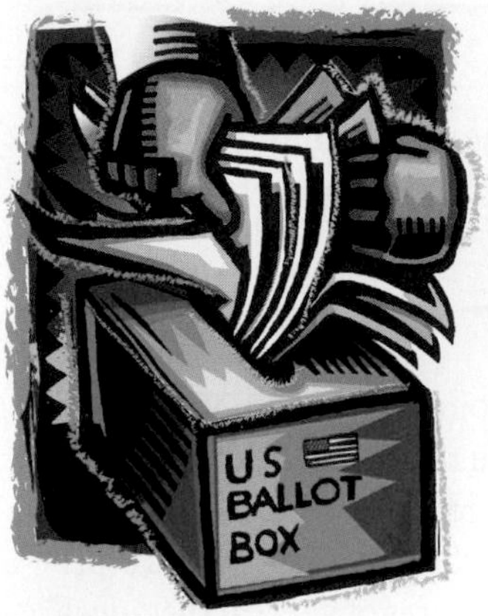

Did the vote counters cheat?

Republican Bruce Marks was ahead of Democrat William Stinson when the voting machines were tallied in their Pennsylvania election. But Stinson was ahead after absentee ballots were counted by the Democrats who controlled the election board. A court fight followed. The court called in a statistician, who used regression with data from past elections to predict the counts of absentee ballots from the voting-machine results. Marks's lead of 564 votes from the machines predicted that he would get 133 more absentee votes than Stinson. In fact, Stinson got 1025 more absentee votes than Marks. Did the vote counters cheat?

The number b is the **slope** of the line, the amount by which y changes when x increases by one unit. The number a is the **y intercept,** the value of y when $x = 0$. To use the equation for prediction, just substitute your x-value into the equation and calculate the resulting y-value.

Example 4.8 Using a regression equation

The equation of the least-squares line in Figure 4.19 is

$$\text{Pack weight} = 16.3 + 0.09(\text{Body weight})$$

- The *slope* of this line is $b = 0.09$. This means that for these hikers, pack weight is predicted to go up by 0.09 pounds when body weight goes up by 1 pound. Said another way, for every 10-pound increase in body weight, this equation predicts a 0.9-pound increase in backpack weight. The slope of a regression line is usually important for understanding the data. The slope is the rate of change, the amount of change in the predicted y when x increases by 1.
- The *y intercept* of the least-squares line is $a = 16.3$. This is the value of the predicted y when $x = 0$. Although we need the intercept to draw the line, it is statistically meaningful only when x can actually take values close to zero. Here, body weight 0 is impossible, so the intercept has no statistical meaning.
- To use the equation for *prediction,* substitute the value of x and calculate y. The predicted pack weight for a new hiker who weighs 150 pounds is

$$\begin{aligned}\text{Pack weight} &= 16.3 + 0.09(150)\\ &= 29.8 \text{ pounds}\end{aligned}$$

  Compare this result with that of the "up-and-over" method we used in Figure 4.17 to predict this student's backpack weight.
- To *draw the line* on the scatterplot, predict y for two different values of x. This gives two points. Plot them and draw the line through them.

Exercises

4.27 Predicting gas usage In Example 4.7 (page 170), we examined the relationship between the average monthly temperature and the amount of natural gas consumed in Joan's midwestern home. The equation of the least-squares line shown in Figure 4.18 is $y = 1425 - 19.87x$.

(a) Identify the slope of the line and explain what it means in this setting.

(b) Identify the y intercept of the line and explain what it means in this setting.

(c) Use the regression line to predict the amount of natural gas Joan will use in a month with an average temperature of 30°F.

Tracy A. Woodward/*Washington Post*

Exercises 4.28 to 4.30 involve the following setting. Figure 4.20 displays data on the number of slices of pizza consumed by players on a football team (the explanatory variable x) and the number of laps around the block the players could run immediately afterward (the response variable y). The line on the scatterplot is the least-squares regression line computed from these points for predicting y from x.

4.28 Pizza party regression

(a) Use the graph and some algebra to show that the equation of the least-squares line is $y = 10 - (2/3)x$.

(b) What's the slope of the line? Explain what this value means in this setting.

(c) What's the y intercept? Explain what this value means in this setting.

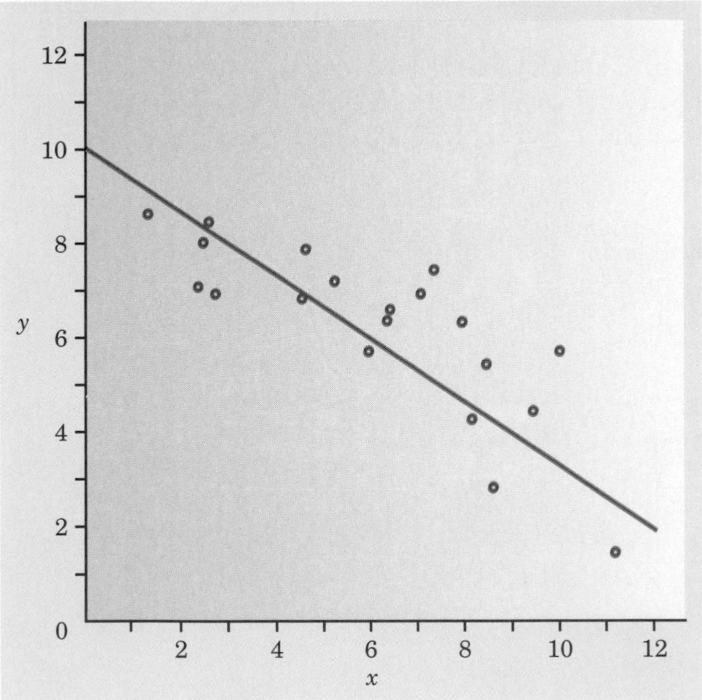

Figure 4.20 A football team's pizza regression, where $x =$ the number of slices of pizza eaten and $y =$ the number of laps the player could run afterward.

4.29 Pizza party prediction Two players, John and Ezekiel, arrived late for the party. John ate 8 slices of pizza. Ezekiel showed off by eating 16 slices.

(a) Use the regression line to predict the number of laps that John will complete.

(b) Can you use the regression line to predict how many laps Ezekiel will complete? Why or why not?

4.30 Least-squares pizza Copy Figure 4.20 onto your paper. Use the figure to help you explain the meaning of "least-squares regression line."

4.31 Least-squares idea The TI-Nspire screen shot below shows a scatterplot of five data points. Which of the following two lines "fits" the data better: $y = 1 - x$ or $y = 3 - 2x$? Justify your answer by computing the sum of the squares of the vertical distances of each point from the two lines. (*Note:* Neither of these two lines is the least-squares regression line for these data.)

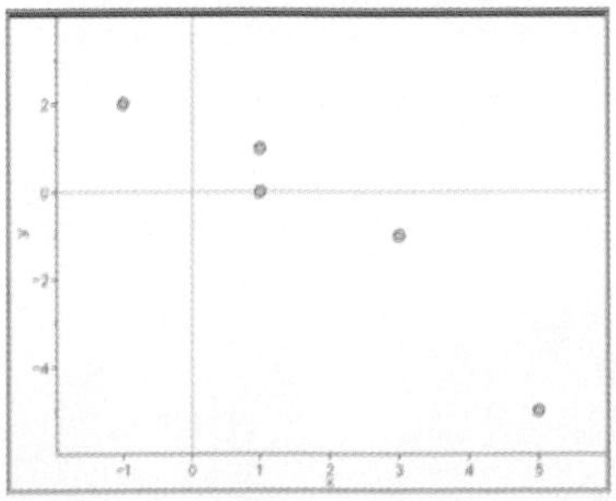

4.32 Sleep debt A researcher reported that the average teenager needs 9.3 hours of sleep per night but gets only 6.3 hours of sleep per night.[12] By the end of a 5-day school week, a teenager would accumulate about 15 hours of "sleep debt." Students in a high school statistics class were skeptical, so they gathered data on the amount of sleep debt (in hours) accumulated over time (in days) by a random sample of 25 high school students. The resulting least-squares regression equation for their data is Sleep debt = 2.23 + 3.17(Days). Do the students have reason to be skeptical of the research study's reported results? Explain.

CALCULATOR CORNER Regression and prediction

In this Calculator Corner, you will learn to use your calculator to perform least-squares regression. We will use the temperature and natural-gas data from Example 4.4 (page 151). In the previous Calculator Corner (page 152), we showed you how to make a scatterplot of the data. Follow the instructions provided there to reproduce that scatterplot now.

1. Calculate the least-squares regression line.

TI-84

Press STAT, arrow right to `CALC`, and choose `8:LinReg(a+bx)`. Then execute the command `LinReg(a+bx)` L_1, L_2, Y_1. (To find Y_1, press VARS, arrow over to `Y-VARS`, choose `1:Function` and `1:Y1`.) Compare your result with the screen shot on the right.

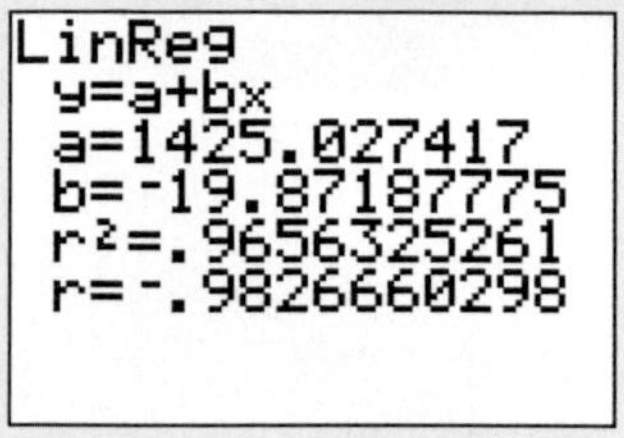

(*Note:* If the values of *r* and r^2 don't appear, you need to turn your diagnostics on. Go into the `CATALOG` by pressing 2nd 0. Scroll down until your cursor is next to `DiagnosticOn`. Then press ENTER twice. Now calculate the regression line again—*r* and r^2 should appear this time.)

TI-Nspire

On the Lists & Spreadsheet page, press (menu), and then choose `4:Statistics`, `1:Stat Calculations...`, and `4:Linear Regression (a+bx)`. In the dialog box, set `X list: temp` and `Y list: gas`, and then choose OK. Scroll down to see the value of *r*.

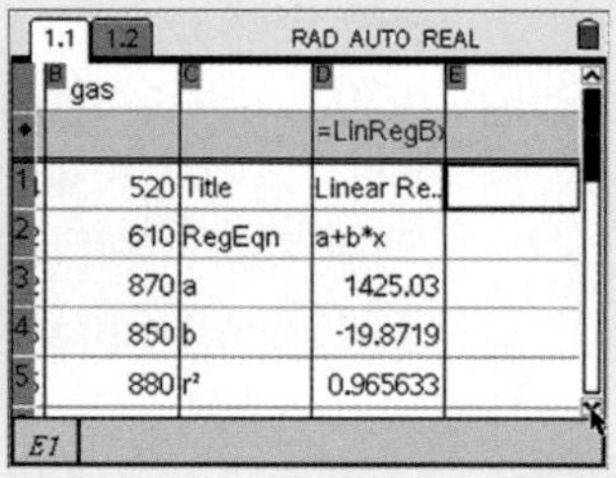

2. Graph the regression line on the scatterplot.

TI-84

Press GRAPH.

TI-Nspire

On the Data & Statistics page, press (menu), and then choose `4:Analyze`, `6:Regression`, and `2:Show Linear (a+bx)`.

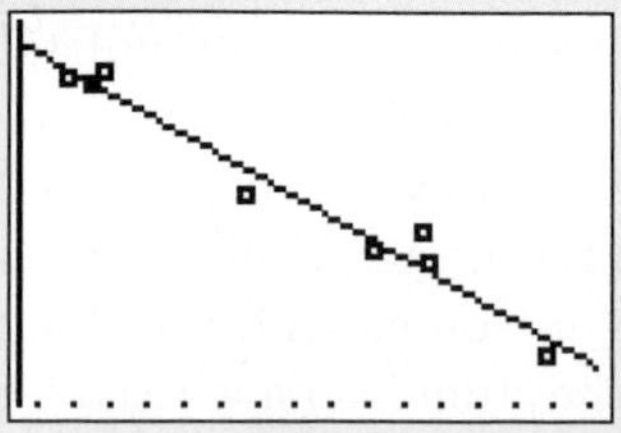

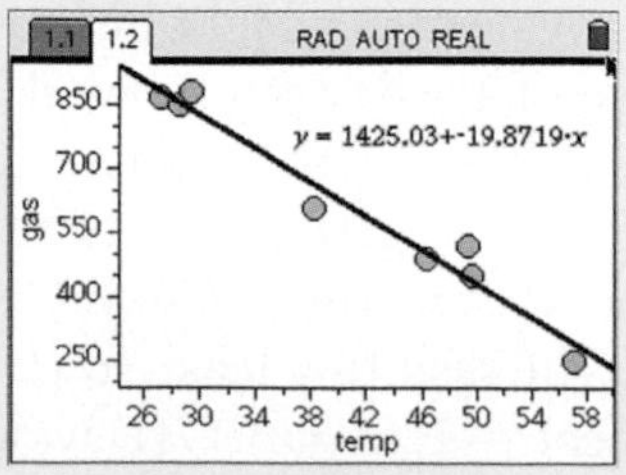

3. Use the regression line to make a prediction. Let's predict the amount of natural gas Joan will use in a month with an average temperature of 30°F.

TI-84

In the graph screen, press [TRACE] and use the down-arrow key ([▼]) to switch from the scatterplot to the regression line. You should see the equation of the least-squares line displayed at the top of the screen. Just type [3][0] [ENTER] to obtain the prediction.

TI-Nspire

On the Data & Statistics page, press (menu), and then choose `4:Analyze` and `A:Graph Trace`. Use the NavPad to move the cursor close to $x = 30$.

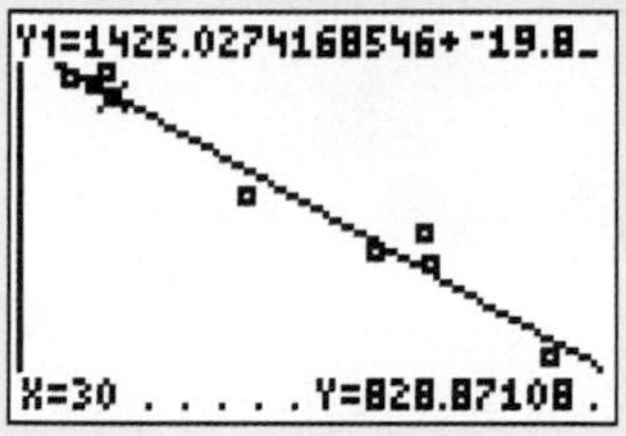

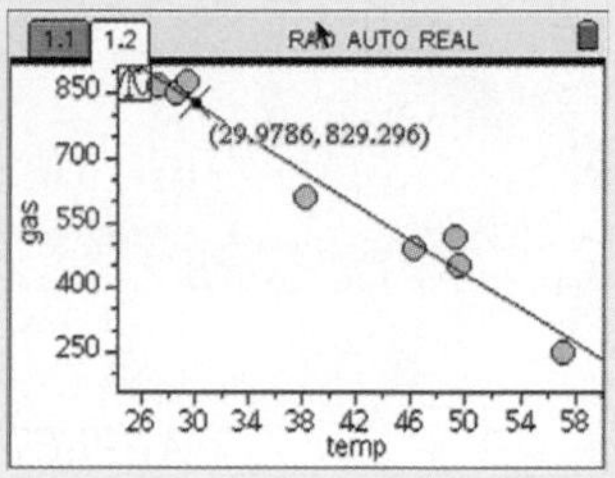

This prediction is based on fitting a regression line to the data for eight past months. Because the data points lie close to the line, we can be confident that Joan's gas consumption in such a month will be close to 829 cubic feet per day.

Correlation and regression

Correlation measures the direction and strength of a straight-line relationship. Regression draws a line to describe the relationship. Correlation and regression are closely connected, even though regression requires choosing an explanatory variable and correlation does not.

ACTIVITY 4.2B

The *Correlation and Regression* applet

APPLET

In this Activity, you will use an applet to investigate how correlation and regression are related. Go to the *Statistics Through Applications* Web site, **www.whfreeman.com/sta2e**, click on the Statistical Applets link, and launch the *Correlation and Regression* applet.

ACTIVITY 4.2B *(continued)*

ACTIVITY 4.2B *(continued)*

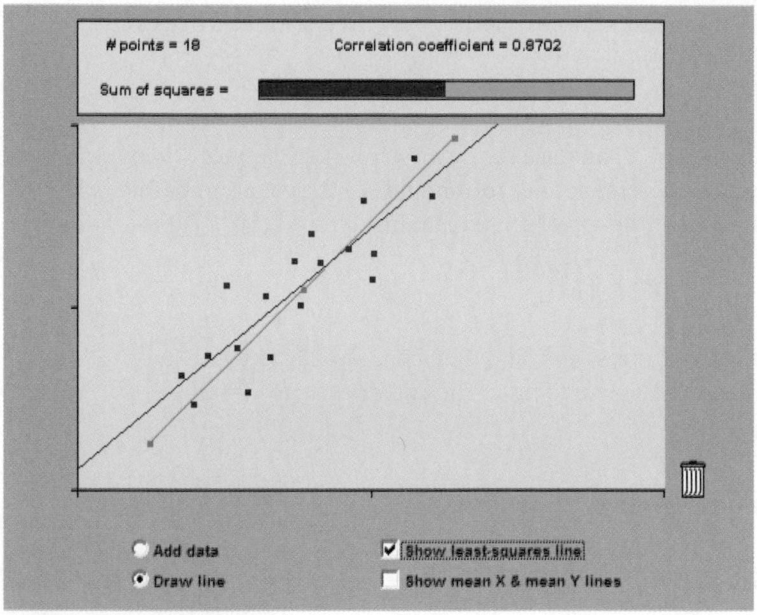

Directions

Use your mouse to click on the blank plot to make red points, creating a scatterplot. Above the plot, the applet shows the "Correlation coefficient." Below the plot are options that can be checked if you wish to draw the regression line on the plot, see the average values of x and y, or even add your own line to the plot to see how well your line competes with the regression line.

Think

1. If you plot just 2 points, what would the correlation be?

2. If you put 3 points in a tight, positively associated linear pattern in the upper-left corner of the plot, what would the correlation be? If you then add a fourth point at the lower-right corner (near the trash can), what would happen to the correlation? Where would the regression line fall?

3. If you put 10 points in the center of the plot in a U-shaped pattern, what would the correlation be? Where would the regression line fall?

Verify

4. Use the applet to test your answers to the three questions above. Did the correlation and regression lines behave as you thought they would?

APPLET TIP: You can clear the scatterplot by clicking on the trash can.

Experiment

5. Create a scatterplot of 10 points with no particular pattern at the lower-left corner of the plot. Move the points around to make the correlation as close to zero as possible. Have the applet show the least-squares regression line. What is its slope? Now add an 11th point at the upper-right corner and watch the behavior of the correlation and the regression line.

6. Clear the scatterplot and make sure the "Show least-squares line" box is unchecked. Create a cloud of about 20 points with a weak linear association (for example, a correlation of about 0.5 or -0.5). Then click the "Draw line" button. Try to position your line so it goes through the "center" of the cloud of points.

7. Now have the applet add the regression line to the plot. How does your line compare?

8. Click on the box to show the "mean X" and "mean Y" lines. Move some points in your scatterplot and observe how the "mean X" and "mean Y" lines relate to the least-squares line.

Let's summarize the important facts about correlation and regression.

- **Correlation and regression describe only linear relationships.** You can do the calculations for any data set involving two quantitative variables, but the results are useful only if the scatterplot shows a linear pattern.
- **Prediction outside the range of the available data is risky.** Suppose that you have data on a child's growth between 3 and 8 years of age. You find a strong straight-line relationship between age x and height y. If you fit a regression line to these data and use it to predict height at age 25 years, you will predict that the child will be 8 feet tall. Growth slows down and stops at maturity, so extending the straight line to adult ages is foolish.
- **Both correlation and regression are strongly affected by outliers.** Be wary if your scatterplot shows strong outliers. Figure 4.21 plots the record-high yearly precipitation in each state against that state's record-high 24-hour precipitation. Hawaii is a high outlier, with a yearly record of 704.83 inches of rain recorded at Kukui in 1982. The correlation for all 50 states in Figure 4.21 is 0.408. If we leave out Hawaii, the correlation drops to $r = 0.195$. The solid line in the figure is the least-squares line for predicting the annual record from the 24-hour record for all 50 states. If we leave out Hawaii, the least-squares line drops down to the dotted line. This line is nearly flat—there is little relation between yearly and 24-hour record precipitation once we decide to ignore Hawaii.

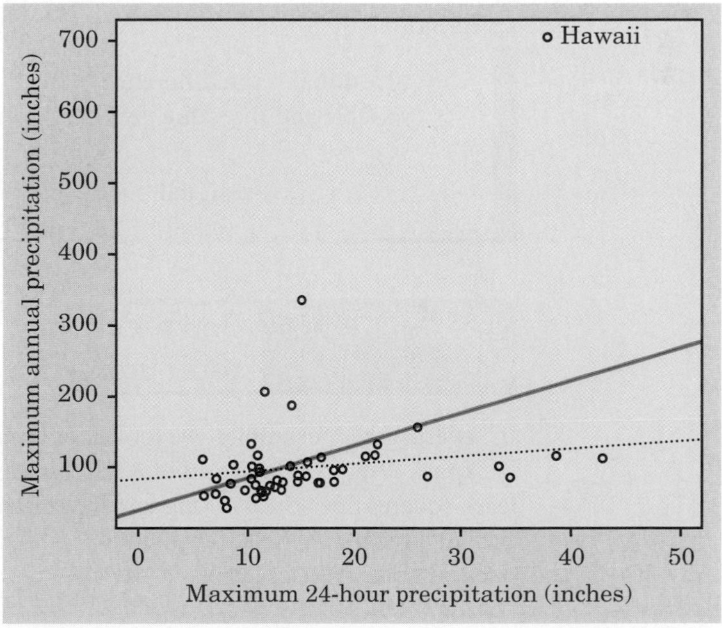

Figure 4.21 Least-squares regression lines are strongly influenced by outliers.

- **The least-squares regression line always passes through the point $(\bar{x}, \bar{y})$.** If you completed Activity 4.2B, then you observed this interesting property of the least-squares line for yourself. Of course, you already know that the least-squares line minimizes the sum of the squares of the vertical distances of the points from the line. This line has other important mathematical properties that we'll encounter later.

APPLET

- **Switching x and y has no effect on the correlation, but it does affect the least-squares line.** To see for yourself, launch the *Two Variable Statistical Calculator* applet at the book's Web site, **www.whfreeman.com/sta2e.** Choose the hiker data on the Data Sets tab. Click on the Scatterplot tab to see the graph. Check the box to "Show least-squares line." Now click on the Correlation & Regression tab. Record the correlation and the equation of the regression line. Then check the "Swap X & Y" box at the top of the page. Observe what happens to the scatterplot, the correlation, and the least-squares line.

Understanding prediction: residuals

One of the first principles of data analysis is to look for an overall pattern and also for striking deviations from the pattern. A regression line describes the overall pattern of a linear relationship between an explanatory variable and a response variable. We see deviations from this pattern by looking at the scatter of the data points about the regression line. The vertical distances from the points to the least-squares regression line are as small as possible, in the sense that they have the smallest possible sum of squares. Because they represent "leftover" variation in the response after fitting the regression line, these distances are called **residuals.**

Residual

A **residual** is the difference between an observed value of the response variable and the value predicted by the regression line. That is,

$$\text{residual} = \text{observed } y - \text{predicted } y$$

Example 4.9 Body weight and pack weight: a residual plot

In several earlier examples, we looked at the relationship between body weight and backpack weight for a group of 8 hikers. Figure 4.22 shows a scatterplot with the least-squares line added, along with a **residual plot.**

One hiker weighed 165 pounds. His pack weighed 35 pounds. The predicted pack weight for this hiker is

$$\text{Pack weight} = 16.3 + 0.09(165) = 31.15$$

The residual for this hiker is therefore

$$\begin{aligned}\text{residual} &= \text{observed } y - \text{predicted } y\\ &= 35 - 31.15 = 3.85 \text{ pounds}\end{aligned}$$

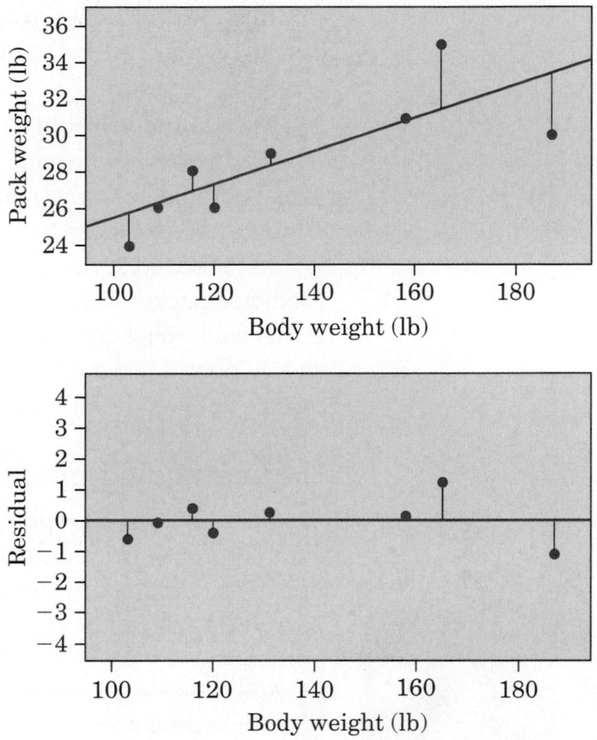

Figure 4.22 Scatterplot with least-squares line and a separate residual plot for the hiker data.

That is, the regression line underpredicts this hiker's pack weight by about 4 pounds. Since this point is above the line, its residual is positive. Points below the line have negative residuals.

Because the residuals show how far the data fall from our regression line, examining a residual plot helps us determine how well the line describes the data. The residual plot magnifies the deviations from the line to make patterns easier to see. If the regression line captures the overall pattern of the data well, two things should be true:

1. **The residual plot should show no obvious pattern.** Figure 4.23(a) is a residual plot with no pattern. It shows an unstructured scatter of points in a horizontal band centered at zero. A curved pattern in the residual plot, like the one in Figure 4.23(b), shows that the relationship is not linear.

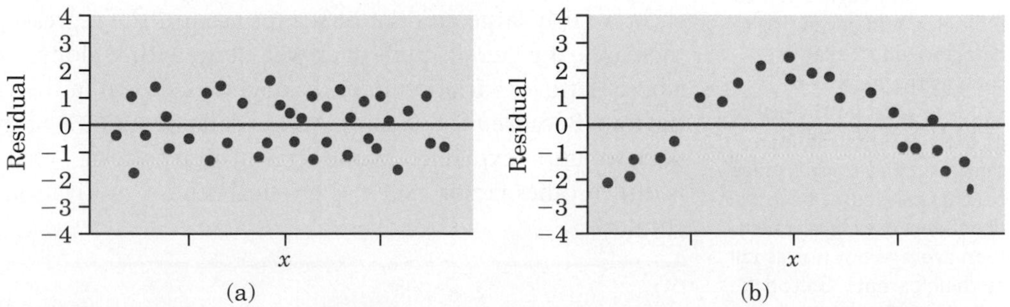

Figure 4.23 Two residual plots. In (a), the unstructured scatter of points indicates that the regression line fits the data well. In (b), the residuals have a curved pattern, so a straight line may not be the best model for the data.

In Figure 4.22, we see a slightly curved pattern like the one in Figure 4.23(b). This may suggest that the underlying relationship between body weight and pack weight isn't linear.

2. **The residuals should be relatively small in size.** A regression line in a model that fits the data well should come "close" to most of the points. That is, the residuals should be fairly small. How do we decide whether the residuals are "small enough"? We consider the size of a "typical" prediction error.

In Figure 4.22, the residuals are between -4 and 4. For these hikers, the predicted backpack weights from the least-squares line are within 4 pounds of the actual pack weights. That sounds okay. But the hikers' packs weighed between 24 and 35 pounds, so a prediction error of 4 pounds is fairly large compared with the actual pack weight for an individual.

Understanding prediction: r^2

A residual plot is a graphical tool for evaluating how well a linear model fits the data. There is also a numerical quantity that tells us how well the least-squares line predicts values of the response variable y. It turns out that the **square of the correlation, r^2,** is the right measure to use.

Regression toward the mean

To "regress" means to go backward. Why are statistical methods for predicting a response from an explanatory variable called "regression"? Sir Francis Galton (1822–1911), who was the first to apply regression to biological and psychological data, looked at examples such as the heights of children versus the heights of their parents. He found that the taller-than-average parents tended to have children who were also taller than average but not as tall as their parents. Galton called this fact "regression toward the mean," and the name came to be applied to the statistical method.

r^2 in regression

The **square of the correlation, r^2,** is the fraction of the variation in the values of y that is explained by the least-squares regression of y on x.

The idea is that when there is a linear relationship, some of the variation in y is accounted for by the fact that, as x changes, it pulls y along the regression line with it.

Example 4.10 Interpreting r^2

Look again at Figure 4.22. There is a lot of variation in the body weights of the 8 hikers, from a low of 103 pounds to a high of 187 pounds. The scatterplot shows a straight-line relationship between body weight and backpack weight, but the points show a fairly large amount of scatter around the regression line. As body weight increases, it generally pulls pack weight up with it along the line. But there is quite a bit of leftover variation in pack weight, as shown in the variation of points about the line. Because $r^2 = 0.632$ for these data, about 63% of the observed variation in pack weight is explained by the straight-line pattern. What about the other 37%? It is due to other factors, such as physical fitness, motivation, and need for "creature comforts."

In reporting a regression, it is usual to give r^2 as a measure of how successful the regression was in explaining the response. When you see a correlation, square it to get a better feel for the strength of the association. Perfect correlation ($r = -1$

or $r = 1$) means the points lie exactly on a line. Then $r^2 = 1$ and all of the variation in one variable is accounted for by the straight-line relationship with the other variable. If $r = -0.7$ or $r = 0.7$, $r^2 = 0.49$ and about half the variation is accounted for by the straight-line relationship.

CALCULATOR CORNER Making a residual plot

Both the TI-84 and the TI-Nspire automatically compute and store residuals when a regression is run. Let's pick up where we left off with the temperature and natural-gas data from the previous Calculator Corner (page 174). To make a residual plot:

TI-84

Press [2nd][Y=] to go into the `STAT PLOT` menu. In Plot1, change the `Ylist` from L_2 to `Resid`. You can choose `Resid` as follows: press [2nd][STAT] (`LIST`), arrow down to `Resid`, and press [ENTER]. To display the graph, press [ZOOM][9] (`ZoomStat`).

TI-Nspire

On the Data & Statistics page, press (menu) and then choose `4:Analyze`, `7:Residuals`, and `2:Show Residual Plot`.

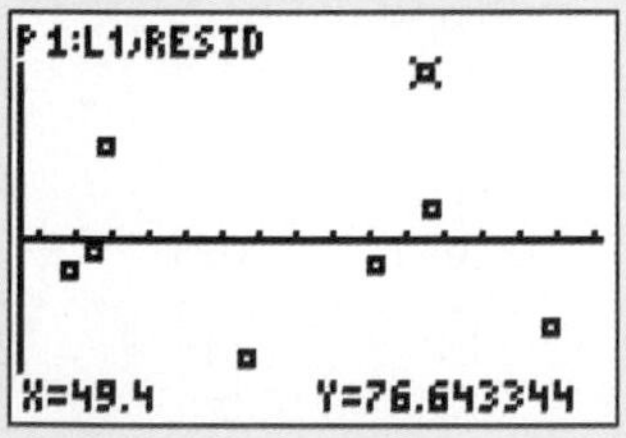

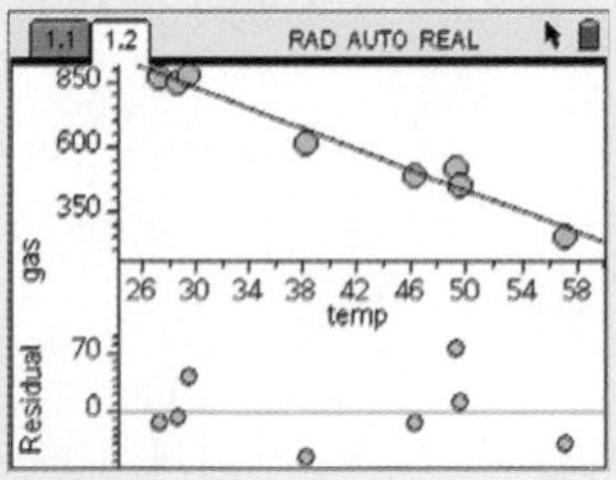

Exercises

4.33 Hiker residuals Here, one more time, are the data on body and backpack weights of the 8 hikers:

Body weight (lb):	120	187	109	103	131	165	158	116
Backpack weight (lb):	26	30	26	24	29	35	31	28

In Example 4.9, we showed you a computer-generated residual plot for these data based on the least-squares regression equation Pack weight = 16.3 + 0.09(Body weight). We also computed the residual for the hiker who weighed 165 pounds. Use the method shown there to calculate the residuals for the other 7 hikers. Then find the sum of the 8 residuals.

Result: For the least-squares line, the sum of the residuals is always 0 (up to roundoff error).

4.34 Joan's gas use residuals In the Calculator Corner, we constructed a residual plot for Joan's gas consumption data. Confirm the value of the residual shown in the TI-84 screen shot for the month with an average temperature of 49.4°F. Show your work.

4.35 r versus r^2 "When $r = 0.7$, this means that y can be predicted from x for 70% of the individuals in the sample." Is this statement true or false? Would it be true if $r^2 = 0.7$? Explain your answers.

4.36 Correlation and regression If the correlation between two variables x and y is $r = 0$, there is no straight-line relationship between the variables. It turns out that the correlation is 0 exactly when the slope of the least-squares regression line is 0. Explain why slope 0 means that there is no linear relationship between x and y. Start by drawing a line with slope 0 and explaining why in this situation x has no value for predicting y.

4.37 Beer and BAC, I How much does drinking beer increase the alcohol content of your blood? This question was addressed in an experiment at Ohio State University. Sixteen students volunteered to participate. Before the experiment, each student blew into a Breathalyzer machine to show that their blood alcohol content (BAC) was at the zero mark. The student volunteers then drank a varying number (between 1 and 9) of 12-ounce beers. How much each student drank was assigned by drawing tickets from a bowl. About 30 minutes later, an officer from the OSU Police Department measured their BAC using the Breathalyzer. Here are the data:[13]

Student:	1	2	3	4	5	6	7	8
Beers:	5	2	9	8	3	7	3	5
BAC:	0.10	0.03	0.19	0.12	0.04	0.095	0.07	0.06

Student:	9	10	11	12	13	14	15	16
Beers:	3	5	4	6	5	7	1	4
BAC:	0.02	0.05	0.07	0.10	0.085	0.09	0.01	0.05

(a) Suppose you want to estimate how a person's BAC is affected by the number of beers he or she drinks. Make a scatterplot of BAC versus beers using the data in the table above. Which variable did you choose to be the y variable and which did you choose to be the x variable? Explain.

(b) Use your calculator to find the correlation between BAC and beers. Is the correlation an appropriate measure of the strength of the association between BAC and beers? Explain briefly.

(c) Use your calculator to find the least-squares regression line for BAC on beers. Interpret the slope and y intercept of the regression line.

4.38 Beer and BAC, II Refer to the previous exercise.

(a) If a student drinks 5 beers, on average what do you predict the student's BAC will be? Show your work.

(b) Would the regression method be as accurate for predicting BAC for a person who drinks 15 beers? Explain.

(c) Use your calculator to construct a residual plot. How well does the regression line fit the data?

(d) Interpret the value of r^2 in this setting.

DATA EXPLORATION

What's Driving Car Sales?

U.S. car sales change month by month because of car prices, the economy, and other factors that affect consumers' buying habits. How do gas prices affect car sales? To answer this question, we collected data on U.S. car sales and gas prices over an 18-month period when gas prices were generally increasing.[14]

1. As gas prices go up, people may think more about gas mileage in deciding what car to buy. Figure 4.24 shows parallel boxplots of gas mileage for each size class of car—subcompact, compact, midsize, and large. Write a paragraph comparing the distributions of gas mileage for the four car classes.

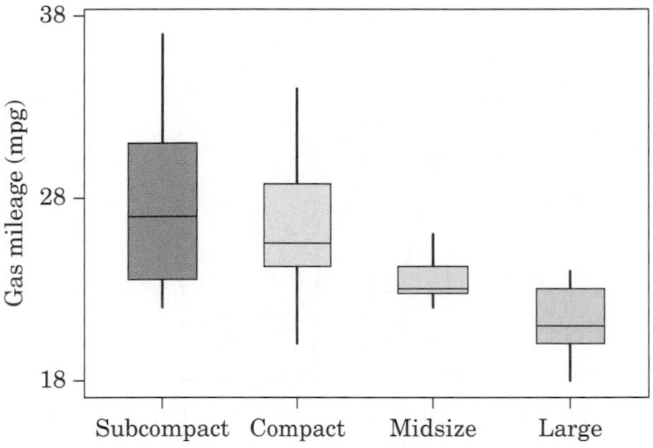

Figure 4.24 Boxplot of gas mileage for different classes of car.

A reasonable theory is that as gas prices increase, people will buy smaller cars. Figure 4.25 displays a scatterplot of the average price of regular unleaded gasoline each month and the percent of cars sold that month with gas mileage greater than 26.75 miles per gallon (mpg). The least-squares regression line has been added to the plot, and some output from a computer regression analysis appears at the top of the graph.

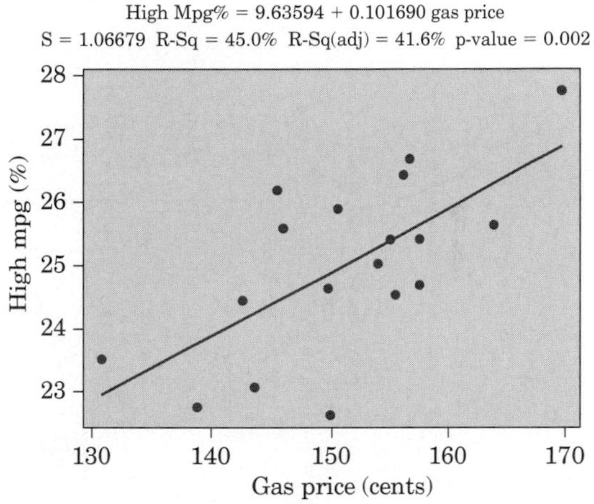

Figure 4.25 Scatterplot of monthly gas prices and percent of cars purchased that month with gas mileage greater than 26.75 mpg.

2. Interpret the slope and y intercept of the least-squares regression line in the context of the problem.

3. During one of the months in this study, the average gas price was \$1.50. Calculate the residual for that month. Show your method.

4. Explain what the value of r^2 means in this setting.

The question of causation

There is a strong relationship between cigarette smoking and death rate from lung cancer. Does smoking cigarettes *cause* lung cancer? There is a strong association between the availability of handguns in a nation and that nation's homicide rate from guns. Does easy access to handguns *cause* more murders? It says right on the pack that cigarettes cause cancer. Whether more guns cause more murders is hotly debated. Why is the evidence for cigarettes and cancer better than the evidence for guns and homicide? Let's consider a few examples before we try to answer this question.

Example 4.11 Does television extend life?

In the 1990s, researchers measured the number of television sets per person x and the life expectancy y for the world's nations. There was a high positive correlation: nations with many TV sets had higher life expectancies.

The basic meaning of causation is that by changing x we can bring about a change in y. Could we lengthen the lives of people in Botswana by shipping them TV sets? No. Rich nations have more TV sets than poor nations. Rich nations also have longer life expectancies because they offer better nutrition, clean water, and better health care. There is no cause-and-effect tie between TV sets and length of life.

"In a new attack on third-world poverty, aid organizations today began delivery of 100,000 television sets."

Correlations such as the one in Example 4.11 are sometimes called "nonsense correlations." The correlation is real. What is nonsense is the conclusion that changing one of the variables causes changes in the other. A **lurking variable**—such as national wealth in Example 4.11—that influences both x and y can create a high correlation even though there is no direct connection between x and y.

Lurking variable

A **lurking variable** is a variable that has an important effect on the relationship among the variables in a study but is not one of the explanatory variables studied.

We might call a situation like the one in Example 4.11 **common response:** both the explanatory and the response variables are responding to some lurking variable. Now consider a different example.

Example 4.12 Obesity in mothers and daughters

What causes obesity in children? Inheritance from parents, overeating, lack of physical activity, and too much television have all been named as explanatory variables.

The results of a study of Mexican American girls aged 9 to 12 years are typical. Measure body mass index (BMI), a measure of weight relative to height, for both the girls and their mothers. People with high BMIs are overweight or obese. Also measure hours of television, minutes of physical activity, and intake of several kinds of food. Result: The girls' BMIs were weakly correlated ($r = -0.18$) with physical activity and also with diet and television. The strongest correlation ($r = 0.506$) was between the BMI of daughters and the BMI of their mothers.[15]

Body type is in part determined by heredity. Daughters inherit half their genes from their mothers. There is therefore a direct causal link between the BMI of mothers and daughters. Of course, the causal link is far from perfect. The mothers' BMIs explain only 25.6% (that's r^2 again) of the variation among the daughters' BMIs. Other factors, some measured in the study and others not measured, also influence BMI. *Even when direct causation is present, it is rarely a complete explanation of an association between two variables.*

Can we use r or r^2 from Example 4.12 to say how much inheritance contributes to the daughters' BMIs? No. The reason is **confounding.** It may well be that mothers who are overweight also set an example of little exercise, poor eating habits, and lots of television. Their daughters pick up these habits to some extent, so the influence of heredity is mixed up with influences from the girls' environment. We can't say how much of the correlation between mother and daughter BMIs is due to inheritance.

Confounding

Two variables are **confounded** when their effects on a response variable cannot be distinguished from each other. The confounded variables may be either explanatory variables or lurking variables.

Figure 4.26 (on the next page) shows in outline form how a variety of underlying links between variables can explain association. The dashed line represents an observed association between the variables x and y.

Some associations are explained by a **direct cause-and-effect** link between the variables. The first diagram in Figure 4.26 shows "x causes y" by an arrow running from x to y. The second diagram illustrates *common response.* The observed association between the variables x and y is explained by a lurking variable z. Both x and y change in response to changes in z. This common response creates an association

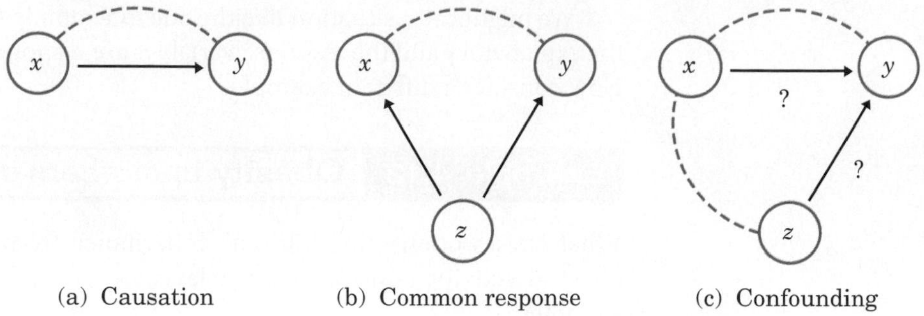

Figure 4.26 Some explanations for an observed association. A dashed line shows an association. An arrow shows a cause-and-effect link. Variable x is explanatory, y is a response variable, and z is a lurking variable.

even though there may be no direct causal link between x and y. The third diagram in Figure 4.26 illustrates *confounding*. Both the explanatory variable x and the lurking variable z may influence the response variable y. Variables x and z are themselves associated, so we cannot distinguish the influence of x from the influence of z. We cannot say how strong the direct effect of x on y is. In fact, it can be hard to say if x influences y at all.

Both common response and confounding involve the influence of a lurking variable or variables z on the response variable y. We won't overemphasize the distinction between the two kinds of relationships. Just remember that "beware the lurking variable" is good advice in thinking about relationships between variables. Here is another example of common response, in a setting where we want to do prediction.

Example 4.13 SAT scores and college grades

High scores on the SAT examinations in high school certainly do not cause high grades in college. The moderate association (r^2 is about 27%) is no doubt explained by common response to such lurking variables as academic ability, study habits, and staying sober.

The ability of SAT scores to partly predict college performance doesn't depend on causation. We need only believe that the relationship between SAT scores and college grades that we see in past years will continue to hold for this year's high school graduates. *Prediction doesn't require causation.*

Evidence for causation

There's an important fact that you'll discover in Chapter 6: *the best evidence for causation comes from well-designed experiments.* Many of the sharpest disputes in which statistics plays a role involve questions of causation that cannot be settled by experiment, however. Does taking the drug Bendectin cause birth defects? Does living near power lines cause cancer? Both of these questions concern associations between variables. And they both try to pinpoint cause and effect in a setting involving complex relationships among many interacting variables. Common response and confounding, along with the number of potential lurking variables, make observed associations misleading. However, for moral or practical reasons, it is not possible to conduct experiments to answer these questions.

Example 4.14 Do power lines increase the risk of leukemia?

Electric currents generate magnetic fields. So living with electricity exposes people to magnetic fields. Living near power lines increases exposure to these fields. Really strong fields can disturb living cells in laboratory studies. What about the weaker fields we experience if we live near power lines?

It isn't ethical to do experiments that expose children to magnetic fields. It's hard to compare cancer rates among children who happen to live in more and less exposed locations, because leukemia is rare and locations vary in many ways other than magnetic fields. We must rely on studies that compare children who have leukemia with children who don't.

A careful study of the effect of magnetic fields on children took 5 years and cost $5 million. The researchers compared 638 children who had leukemia with 620 who did not. They went into the homes and actually measured the magnetic fields in the children's bedrooms, in other rooms, and at the front door. They recorded facts about nearby power lines for the family home and also for the mother's residence when she was pregnant. *Result:* no evidence of more than a chance connection between magnetic fields and childhood leukemia.[16]

Despite the difficulties, it is sometimes possible to build a strong case for causation in the absence of experiments. The evidence that smoking causes lung cancer is about as strong as nonexperimental evidence can be.

Doctors had long observed that most lung cancer patients were smokers. Observational studies comparing smokers and "similar" nonsmokers showed a strong association between smoking and death from lung cancer. Could the association be explained by lurking variables that the studies could not measure? Might there be, for example, a genetic factor that predisposes people both to nicotine addiction and to lung cancer? Smoking and lung cancer would then be positively associated even if smoking had no direct effect on the lungs. How were these objections overcome?

Let's answer this question in general terms: what are the criteria for establishing causation when we cannot do an experiment?

- **The association is strong.** The association between smoking and lung cancer is very strong.
- **The association is consistent.** Many studies of different kinds of people in many countries link smoking to lung cancer. That reduces the chance that a lurking variable specific to one group or one study explains the association.
- **Higher doses are associated with stronger responses.** People who smoke more cigarettes per day or who smoke over a longer period get lung cancer more often. People who stop smoking reduce their risk.

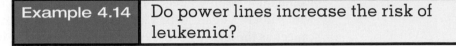

- **The alleged cause precedes the effect in time.** Lung cancer develops after years of smoking. The number of men dying of lung cancer rose as smoking became more common, with a lag of about 30 years. Lung cancer kills more men than any other form of cancer. Lung cancer was rare among women until women began to smoke. Lung cancer in women rose along with smoking, again with a lag of about 30 years, and has now passed breast cancer as the leading cause of cancer death among women.
- **The alleged cause is plausible.** Experiments with animals show that tars from cigarette smoke do cause cancer.

Medical authorities do not hesitate to say that smoking causes lung cancer. The U.S. Surgeon General has long stated that cigarette smoking is "the largest avoidable cause of death and disability in the United States." The evidence for causation is overwhelming—but it is not as strong as the evidence provided by well-designed experiments.

Exercises

4.39 Ice cream and drowning There is a positive association between ice cream sales and the number of people who drown. Does that mean eating ice cream causes people to drown? Of course not. Give a more plausible explanation that involves common response. Draw a picture like Figure 4.26 (page 186) to illustrate the relationship among the variables.

4.40 Is a little alcohol good for you? A survey of 7000 California men found little correlation between alcohol consumption and chance of dying during the $5\frac{1}{2}$ years of the study. In fact, men who did not drink at all during these years had a slightly higher death rate than did light drinkers. This lack of correlation was somewhat surprising. Explain how common response might account for the higher death rate among men who did not drink at all over a short period.

4.41 Education and income There is a strong positive correlation between years of schooling completed and lifetime earnings for American men. One possible reason for this association is causation: more education leads to higher-paying jobs. Another explanation is confounding: men who complete many years of schooling have other characteristics that would lead to better jobs even without the education. Suggest several possible confounding variables.

4.42 Snow and earthquakes A study measures the average annual snowfall (in inches) for 10 cities over the last decade along with the greatest earth movement (on the Richter scale) over this same time period. The study included 5 cities in California's San Francisco Bay Area and 5 cities from Canada's province of Ontario. The study found a very strong negative correlation between the two variables. Does this mean that a strong snowfall will prevent earthquakes? Explain your answer briefly. You may want to draw a picture like Figure 4.26 (page 186) to illustrate it.

4.43 Is math the key to success in college? Here is the opening of a newspaper account of a College Board study of 15,941 high school graduates:

> *Minority students who take high school algebra and geometry succeed in college at almost the same rate as whites, a new study says. The link between high school math and college graduation is "almost magical," says College Board*

President Donald Stewart, suggesting "math is the gatekeeper for success in college." "These findings," he says, "justify serious consideration of a national policy that all students take algebra and geometry."[17]

What lurking variables might explain the association between taking several math courses in high school and success in college? Explain why requiring algebra and geometry may have little effect on who succeeds in college.

4.44 Predicting Old Faithful eruptions Figure 4.27 shows the results from a least-squares regression on the Old Faithful eruption data from the beginning of the chapter. Length of previous eruption is the explanatory variable, and time until the next eruption is the response variable. Both variables were measured in minutes.

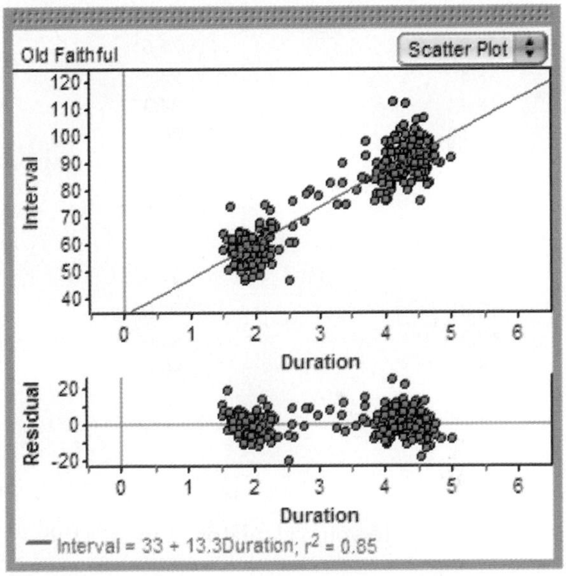

Figure 4.27 Fathom scatterplot with least-squares regression line and residual plot for the Old Faithful eruption data from the beginning of the chapter.

(a) Is the relationship between these two variables most likely a result of causation, confounding, or common response? Justify your answer.

(b) Write a brief analysis for your teacher that demonstrates your understanding of the material in this chapter. Follow the four-step statistical problem-solving process (page 19).

APPLICATION 4.2

Cricket chirps and air temperature

Can the frequency of cricket chirps be used to predict the outdoor temperature? According to one of the founding fathers of communications engineering, George Washington Pierce, the answer is yes. During his career, Pierce invented several pieces of technology that earned him patents and a lot of

APPLICATION 4.2 *(continued)*

APPLICATION 4.2 *(continued)*

money from companies like RCA and AT&T. When he retired, Pierce built a device that allowed him to record the sounds made by various insects near his New Hampshire home. In 1948, he published his research findings in a book titled *The Songs of Insects.* In this Application, you'll examine data that Pierce collected on the number of chirps per second of the striped ground cricket and the outdoor temperature in degrees Fahrenheit.[18]

Cricket chirps per second	Outdoor temperature (°F)
20.0	88.6
16.0	71.6
19.8	93.3
18.4	84.3
17.1	80.6
15.5	75.2
14.7	69.7
17.1	82.0
15.4	69.4
16.2	83.3
15.0	79.6
17.2	82.6
16.0	80.6
17.0	83.5
14.4	76.3

QUESTIONS

1. Enter the data into your calculator. Which is the explanatory variable?

2. Make a well-labeled scatterplot of the data. Describe the direction, form, and strength of the relationship. Are there any outliers?

3. Use your calculator to find the least-squares regression line for these data. Record the equation.

4. Interpret the slope and the y intercept of the least-squares line in this setting.

5. Use the equation to predict the temperature when there are 15 cricket chirps per second.

6. Construct a residual plot. Describe what you see.

7. Explain what the value of r^2 tells you in this setting.

8. How well does the regression line fit the data? Explain.

9. Is it reasonable to use the equation to predict the temperature when there are 25 cricket chirps per second? Explain.

10. Crickets make their chirping sounds by rapidly rubbing their wings together. From Pierce's data, we see that outdoor temperature increases as the number of cricket chirps increases. Can we conclude that the increased number of chirps causes the temperature to increase (maybe due to the heat generated from wings rubbing together)? Explain.

Section 4.2 Summary

Regression is the name for statistical methods that fit some model to data in order to predict a response variable from one or more explanatory variables. The simplest kind of regression fits a straight line, called a **regression line,** on a scatterplot for use in predicting y from x. The most common way to fit a line is the **least-squares method,** which finds the line that makes the sum of the squared vertical distances of the data points from the line as small as possible.

Least-squares regression is closely related to **correlation.** In particular, the **squared correlation r^2** tells us what fraction of the variation in the responses is explained by the straight-line relationship between y and x. **Residual plots** also help us assess how well a regression line fits the data. It is generally true that the success of any statistical prediction depends on the presence of strong patterns in the data. Prediction outside the range of the data is risky because the pattern may be different there.

A strong relationship between two variables is not always evidence that changes in one variable **cause** changes in the other. **Lurking variables** can create relationships through **common response** or **confounding.** If we cannot do experiments, it is often difficult to get convincing evidence for causation.

Section 4.2 Exercises

4.45 Heart attacks If you need medical care, should you go to a hospital that handles many cases like yours? Figure 4.28 presents some data for heart attacks. The figure plots mortality (the proportion of patients who died) against the number of heart attack patients treated for a large number of hospitals in a recent year. The line on the plot is the least-squares regression line for predicting mortality from number of patients.

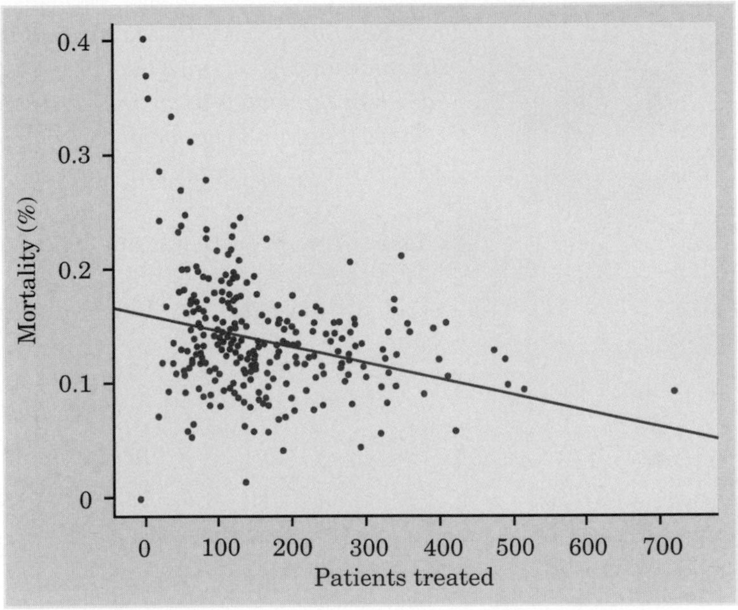

Figure 4.28 Mortality of heart attack patients and number of heart attack cases treated for a large group of hospitals.

(a) Do the plot and regression generally support the hypothesis that mortality is lower at hospitals that treat more heart attacks? Is the relationship very strong?
(b) In what way is the pattern of the plot nonlinear? Does the nonlinearity strengthen or weaken the conclusion that heart attack patients should avoid hospitals that treat few heart attacks? Why?

4.46 Beer drinking and cancer A study using data from 41 states found a positive correlation between beer consumption per person and death rates from some forms of cancer. The states with the highest death rates from these types of cancer were Rhode Island and New York. The beer consumption in those states was 80 quarts per person per year. People in South Carolina, Alabama, and Arkansas drank only 26 quarts of beer per person and had cancer rates less than one-third of those in Rhode Island and New York.

Suggest some lurking variables that may be confounded with a state's beer consumption. For a clue, look at the high- and low-consumption states given above.

4.47 Obesity in mothers and daughters A study found that the correlation between the body mass index (BMI) of young girls and their minutes of physical activity in a day was $r = -0.18$.[19] Why might we expect this correlation to be negative? What percent of the variation in BMI among the girls in the study can be explained by the straight-line relationship with minutes of activity?

4.48 Application 4.2 follow-up Refer to Application 4.2 (page 189).

(a) Suppose we want to predict the number of cricket chirps per second when the outdoor temperature is 80°F. Should we use the regression equation from Application 4.2? Why or why not?

(b) If we changed the temperature data in Application 4.2 from Fahrenheit to Celsius, how would this affect the correlation? The slope of the least-squares line? The y intercept? Justify your answers.

iStockphoto

From Rex Boggs in Australia comes an unusual data set. Before showering in the morning, he weighed the bar of soap in his shower stall. The weight goes down as the soap is used. The data appear in Table 4.4 (weights in grams). Notice that Mr. Boggs forgot to weigh the soap on some days. Exercises 4.49 to 4.51 are based on the soap data set.

Table 4.4 Weight (Grams) of a Bar of Soap Used to Shower

Day	Weight	Day	Weight	Day	Weight
1	124	8	84	16	27
2	121	9	78	18	16
5	103	10	71	19	12
6	96	12	58	20	8
7	90	13	50	21	6

Source: Rex Boggs.

4.49 Scatterplot Plot the weight of the bar of soap against day. Is the overall pattern roughly a straight line? Based on your scatterplot, is the correlation between day and weight close to 1, positive but not close to 1, close to 0, negative but not close to -1, or close to -1? Explain your answer.

4.50 Regression The equation for the least-squares regression line for the data in Table 4.4 is

$$\text{weight} = 133.2 - 6.31(\text{day})$$

(a) Explain carefully what the slope $b = -6.31$ tells us about how fast the soap lost weight.

(b) Interpret the y intercept, 133.2.

(c) Mr. Boggs did not measure the weight of the soap on Day 4. Use the regression equation to predict that weight.

(d) Draw the regression line on your scatterplot from the previous exercise.

4.51 Prediction?
(a) Use the regression equation in the previous exercise to predict the weight of the soap after 30 days. What's wrong with using the regression line to predict weight after 30 days?
(b) For these data, $r^2 = 0.996$. Explain what this value means.
(c) Make a residual plot for these data. Is a linear model appropriate in this case? Explain.

4.52 More chirping crickets In late summer 2007, Dr. Peggy LeMone decided to collect some cricket chirp and temperature data at her home in Colorado. She made several measurements of the number of cricket chirps in a 15-second period and the outdoor temperature in degrees Fahrenheit.[20] Figure 4.29 shows a scatterplot of Dr. LeMone's data with a least-squares regression line added and a residual plot.

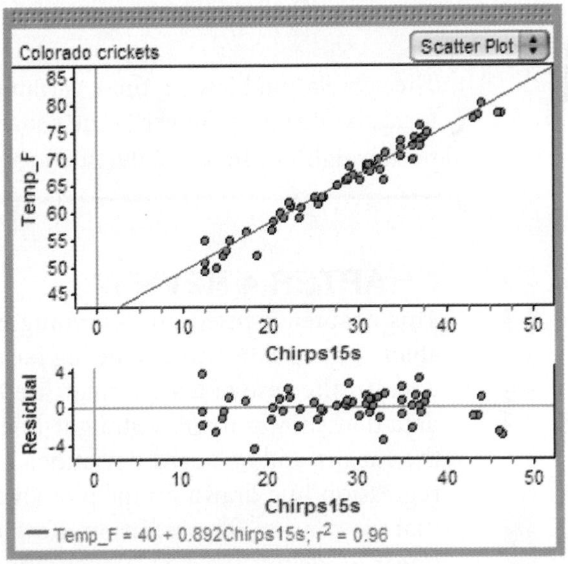

Figure 4.29 Fathom scatterplot with least-squares line and residual plot for the number of cricket chirps in 15 seconds and the outdoor temperature at Dr. LeMone's home in Colorado in 2007.

(a) How well does the regression line fit the data? Give graphical and numerical evidence to support your answer.
(b) One formula that Dr. LeMone found on the Web for predicting temperature from cricket chirps was

$$\text{temp. in degrees Fahrenheit} = \text{no. of cricket chirps in 15 seconds} + 37$$

Compare the temperature predictions for the least-squares line and this formula when there are 10, 20, 30, and 40 cricket chirps per 15 seconds.

4.53 Miscarriages among workers A study showed that women who work in the production of computer chips have abnormally high numbers of miscarriages. The union claimed that exposure to chemicals used in production caused the miscarriages. Another possible explanation is that these workers spend most of their work time standing up. Illustrate these relationships in a diagram like one of those in Figure 4.26 (page 186).

4.54 Calculating the least-squares line Do you like to know the mathematical details when you study something? Here is the formula for the least-squares regression line for predicting y from x. Start with the means and the standard deviations of the two variables and the correlation r between them. The least-squares line has equation $y = a + bx$ with

$$b = r\frac{s_y}{s_x} \qquad \text{and} \qquad a = \bar{y} - b\bar{x}$$

For the backpacker data of this chapter, we found these summary statistics:

Body weight:	$\bar{x} = 136.125$ pounds	$s_x = 30.296$ pounds
Pack weight:	$\bar{y} = 28.625$ pounds	$s_y = 3.462$ pounds

The correlation between these variables is $r = 0.795$. Use these values in the formulas just given to verify the equation of the least-squares line given earlier: pack weight $= 16.3 + 0.09$(Body weight).

CHAPTER 4 REVIEW

This chapter focused on describing relationships between two quantitative variables. Figure 4.30 retraces the big ideas. We always begin by making graphs of our data. In the case of a scatterplot, we have learned a numerical summary only for data that show a roughly straight-line pattern on the scatterplot. This summary is the means and standard deviations of the two variables and their correlation. A regression line drawn on the plot gives us a compact model of the overall pattern that we can use for prediction. Residual plots and r^2 help us assess how well the linear model fits the data. The question marks at the last two stages remind us that correlation and regression describe only straight-line relationships.

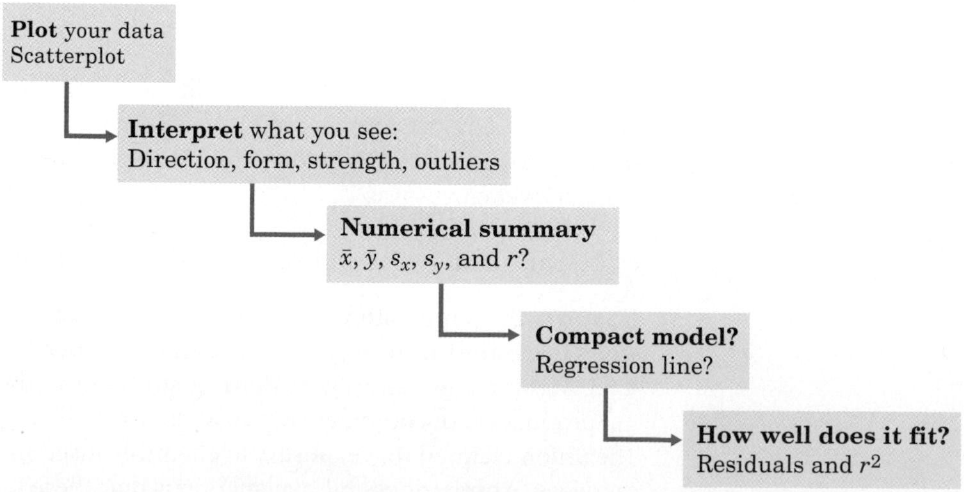

Figure 4.30 Flowchart for describing relationships.

Relationships often raise the question of causation. You will see in Chapter 6 that evidence from well-designed experiments is the "gold standard" for deciding that one variable causes changes in another variable. Section 4.2 reminded us in more detail how strong associations can appear in data even when there is no direct causation. We must always think about the possible effects of variables lurking in the background.

Here is a review list of the most important skills you should have developed from your study of this chapter.

A. SCATTERPLOTS AND CORRELATION

1. Make a scatterplot to display the relationship between two quantitative variables measured on the same subjects. Place the explanatory variable (if any) on the horizontal scale of the plot.

2. Describe the direction, form, and strength of the overall pattern of a scatterplot. In particular, recognize positive or negative associations and straight-line patterns. Recognize outliers in a scatterplot.

3. Judge whether it is appropriate to use correlation to describe the relationship between two quantitative variables. Use a calculator to find the correlation r.

4. Know the basic properties of correlation: r measures the strength and direction of only straight-line relationships; r is always a number between -1 and 1; $r = \pm 1$ only for perfect straight-line relations; r moves away from 0 toward ± 1 as the straight-line relation gets stronger.

B. REGRESSION LINES

1. Explain what the slope b and the intercept a mean in the equation $y = a + bx$ of a straight line.

2. Draw a graph of the straight line when you are given its equation.

3. Use a regression line, given on a graph or as an equation, to predict y for a given x. Recognize the danger of prediction outside the range of the available data.

4. Use a residual plot to examine how well a regression line fits the data.

5. Use r^2, the square of the correlation, to describe how much of the variation in one variable can be accounted for by a straight-line relationship with another variable.

C. STATISTICS AND CAUSATION

1. Give plausible explanations for an observed association between two variables: direct cause and effect, the influence of lurking variables, or both.

2. Assess the strength of statistical evidence for a claim of causation, especially when experiments are not possible.

CHAPTER 4 REVIEW EXERCISES

4.55 Measuring crickets For a biology project, you measure the length (centimeters) and weight (grams) of 12 crickets.
(a) Explain why you expect the correlation between length and weight to be positive.
(b) If you measured length in inches, how would the correlation change? (There are 2.54 centimeters in an inch.)

4.56 When it rains, it pours Figure 4.31 plots the highest *yearly* precipitation ever recorded in each state against the highest *daily* precipitation ever recorded in that state. The points for Alaska, Hawaii, and Texas are marked on the scatterplot.

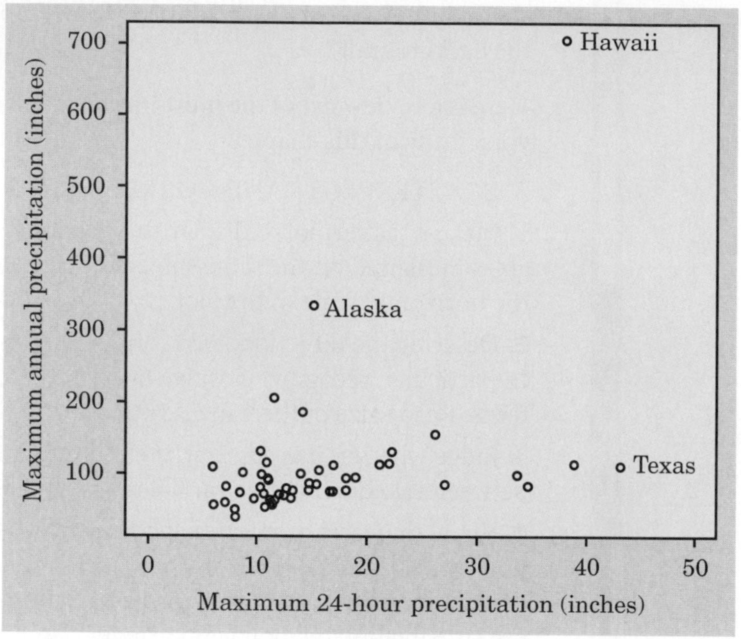

Figure 4.31 Record-high yearly precipitation at any weather station in each state plotted against record-high daily precipitation for the state.

(a) About what are the highest daily and yearly precipitation values for Alaska?
(b) Alaska and Hawaii have very high yearly maximums relative to their daily maximums. Omit these two states as outliers. Describe the nature of the relationship for the other states. Would knowing a state's highest daily precipitation be a great help in predicting that state's highest yearly precipitation?

4.57 Why so small?
(a) Make a scatterplot of the following data:

x:	1	2	3	4	10	10
y:	1	3	3	5	1	11

(b) Use the method of Example 4.5 (page 158) to show that the correlation is about 0.5.
(c) What feature of the data is responsible for reducing the correlation to this value despite a strong straight-line association between x and y in most of the observations?

4.58 SAT preparation Can intensive preparation significantly improve students' SAT scores? Here's one way to find out. Take a random sample of 100 high school students who took the most recent SAT test and ask them how much time (in hours) they spent preparing and what score they earned on the Math section. If we find a correlation of $r = 0.79$ between preparation time and SAT Math score, can we conclude that preparation causes higher scores? Why or why not?

Figure 4.32 plots the average brain weight in grams versus average body weight in kilograms for 96 species of mammals.[21] *There are many small mammals whose points at the lower left overlap. Exercises 4.59 to 4.62 are based on this scatterplot.*

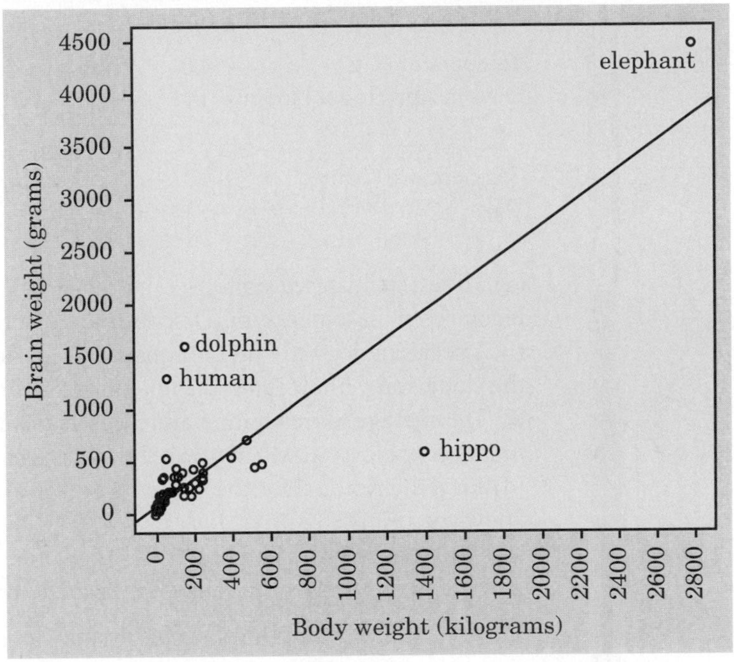

Figure 4.32 Scatterplot of the average brain weight (grams) against the average body weight (kilograms) for 96 species of mammals.

4.59 Dolphins and hippos
(a) The points for the dolphin and hippopotamus are labeled in Figure 4.32. Read from the graph the approximate body weight and brain weight for these two species.
(b) One reaction to this scatterplot is "Dolphins are smart, hippos are dumb." What feature of the plot lies behind this reaction?

4.60 Outliers The African elephant is much larger than any other mammal in the data set but lies roughly in the overall straight-line pattern. Dolphins, humans, and hippos lie outside the overall pattern. The correlation between body weight for the entire data set is $r = 0.86$.
(a) If we removed elephants, would this correlation increase, decrease, or not change much? Explain your answer.
(b) If we removed dolphins, hippos, and humans, would this correlation increase, decrease, or not change much? Explain your answer.

4.61 Brain and body The correlation between body weight and brain weight is $r = 0.86$. How well does body weight explain brain weight for mammals? Give a number to answer this question, and briefly explain what the number tells us.

4.62 Prediction The line on the scatterplot in Figure 4.32 is the least-squares regression line for predicting brain weight from body weight.
(a) Suppose that a new mammal species is discovered hidden in the rain forest with body weight 600 kilograms. Predict the brain weight for this species.
(b) The slope of this line is one of these numbers: 0.5, 1.3, or 3.2. Which number is the slope? Why?

4.63 Late bloomers? From experience, Japanese cherry trees tend to blossom early when spring weather is warm and later when spring weather is cool. Here are some data on the average March temperature (in °C) and the day in April when the first cherry blossom appeared over a 25-year period.[22]

Temperature (°C):	4.0	5.4	3.2	2.6	4.2	4.7	4.9	4.0	4.9	3.8	4.0	5.1
Days in April to 1st bloom:	14	8	11	19	14	14	14	21	9	14	13	11

Temperature (°C):	4.3	1.5	3.7	3.8	4.5	4.1	6.1	6.2	5.1	5.0	4.6	4.0
Days in April to 1st bloom:	13	28	17	19	10	17	3	3	11	6	9	11

(a) Make a well-labeled scatterplot that is suitable for predicting when the cherry trees will bloom from the temperature. Describe the direction, form, and strength of the relationship.
(b) Use technology to find the equation of the least-squares regression line. Interpret the slope and y intercept of the line in this setting.
(c) The average March temperature this year was 3.5°C. When do you predict that the first cherry blossom will appear? Show your method clearly.
(d) Find the residual for the year when the average March temperature was 4.5°C. Show your work.
(e) Use technology to construct a residual plot. Describe what you see.
(f) Find and interpret the value of r^2 in this setting.

4.64 Teacher pay and liquor sales A study found a strong positive correlation between average teacher salaries and liquor sales over a 12-month period. Does this suggest that we should not pay teachers more because they would only spend the money on more liquor? Explain your answer clearly.

PART B

PRODUCING DATA

PEANUTS reprinted by permission of United Feature Syndicate Inc.

What kind of music do you like? Rap? Rock? Alternative? Country? Some of your friends probably share your taste in music. But you and your friends are not typical. Your parents and grandparents likely have different tastes. They are not typical either. To get a true picture of the country as a whole (or even of high school students), you must recognize that the picture may not resemble you or what you see around you. You need *data*.

Data from retail sales show that the top-selling music types are rock (31.8% of all recorded music sold in 2008) and country (11.9%).[1] You might like rap (10.7%) and I might like rhythm and blues (10.2%), but that doesn't mean we have a clue about the tastes of the music-buying public as a whole. If we are in the music business, or even if we are interested in pop culture, we have to put our own tastes aside and look at the data.

You can find data in the library or on the Internet (that's where we found the music sales data). How can we know whether data can be trusted? Good data are as much a human product as shoes and MP3 players. Sloppily produced data will frustrate you as much as a sloppily made pair of shoes. You examine shoes before you buy, and you don't buy if they are not well made. Neither should you use data that are not well made. This part of this book shows how to tell if data are well made.

CHAPTER 5

Sampling and Surveys

Why can't pollsters get it right?

Follow any election closely, and you'll see numerous polls that try to predict the winner. Newscasters and political commentators frequently reference polls taken by Gallup, Zogby, and many others, and political strategists use polling data to help make decisions. Even experienced pollsters sometimes get it wrong, as we saw in the 2008 New Hampshire Democratic primary involving Hillary Clinton and Barack Obama. Heading into the primary, every poll predicted that Obama would win—some by a very large margin. In fact, Clinton won the New Hampshire primary by 3%. (Senator Obama did go on to win the presidential election, however.) Why is it so difficult to get accurate poll results?

Consider this. Suppose the pollsters choose a "representative" sample of 400 people from the population and ask them which candidate they prefer. If 60% of the sample say they will vote for candidate A, what can we say about the population percent who favor A? Is it exactly 60%? Probably not. Is it close to 60%? That depends on what we mean by "close."

If the pollsters had chosen a different sample of 400 people, they probably wouldn't get exactly 60% who favor candidate A. That's the idea of *sampling variability*. The good news is that sample results vary around the population truth. How much they vary depends mainly on the size of the sample, as you'll see later in the chapter. *Sample surveys* come with practical difficulties in addition to sampling variability, as the New Hampshire primary results remind us.

So what happened in New Hampshire? We can't be sure, but we can speculate. Maybe the pollsters selected samples that weren't representative of the population of actual voters. Some people in the sample couldn't be reached or refused to answer. What effect did this *nonresponse* have on the results? Those who did respond may not have told the truth about their voting intentions or might have changed their minds later. Sample surveys assume that we can measure people's views accurately at a moment in time. That's easy to say, but much harder to do.

5.1 Samples, Good and Bad

ACTIVITY 5.1A

How many fish are in the lake?

Materials: Large bag of Goldfish crackers and small bag of different Goldfish crackers; brown paper lunch bag; small paper cup. (Colored marbles can be used instead of Goldfish.)

In this Activity, you will learn how scientists use sampling to estimate the size of an animal population, like the number of deer in a large forest or the number of fish in a lake. You will also investigate how information obtained from a sample relates to the truth about the corresponding population.

1. Prior to the Activity, your teacher will count the number of "fish" in a large bag of Goldfish crackers and then pour all of the crackers into a brown paper lunch bag (the "lake"). Your class's goal is to determine the total number of fish in the lake.

2. Without looking in the bag, use a small paper cup to "capture" a sample of fish. Count the number of fish in your sample.

3. We would like to find a way to "tag" the fish in your sample so that you can identify them if you capture them again later. Your teacher will provide you with some different-colored Goldfish to represent these tagged fish. Count out the same number of different-colored fish as you had in your sample. Then put these tagged fish into the paper bag.

4. Thoroughly mix the Goldfish in the bag. (This imitates the fish swimming around in the lake after the tagged fish have been replaced.) Now use the paper cup to "recapture" a new sample of fish. Count the number of tagged and untagged fish in your sample.

5. With your classmates, discuss a method for calculating the total number of fish in the lake. (Here's a hint: what should be true about the overall proportion of tagged fish in the lake and the proportion of tagged fish in your sample?)

6. Repeat Step 4 four more times to obtain a total of five estimates of the number of fish in the lake. Find the average (mean) of your five estimates.

7. Your teacher will now tell you the actual number of fish in the lake. How close were your estimates to the actual size of the fish population?

8. If scientists used the *capture-recapture method* from this Activity to estimate the size of an animal population in the wild, what practical problems might they encounter?

How to sample badly

Example 5.1 The *Town Talk* takes an opinion poll

In Rapides Parish, Louisiana, only one company is allowed to provide ambulance service. The local paper, the *Town Talk*, asked readers to call in to offer their opinions on whether the company should keep its monopoly. Call-in polls are generally automated: call one telephone number to vote "Yes" and call another number to vote "No." Telephone companies often charge callers to dial these numbers.

The *Town Talk* got 3763 calls, which suggests unusual interest in ambulance service. Investigation showed that 638 calls came from the ambulance company office or from the homes of its executives. Many more no doubt came from lower-level employees. "We've got employees who are concerned about this situation, their job stability and their families and maybe called more than they should have," said a company vice president. Other sources said employees were told to "vote early and often."[1]

We will see that a sample size of 3763 is quite respectable—if the sample is properly designed. Inviting people to call (and call, and call . . .) isn't proper sample design. We are about to learn how the pros take a sample, and why it beats the *Town Talk* method.

As the *Town Talk* learned, it is easier to sample badly than to sample well. The paper relied on **voluntary response,** allowing people to call in rather than actively selecting its own sample. The result was **biased**—the sample was overweighted with people favoring the ambulance monopoly. Voluntary response samples attract people who feel strongly about the issue in question. These people, like the employees of the ambulance company, may not fairly represent the opinions of the entire population.

There are other ways to sample badly. Suppose that I sell your company several crates of oranges each week. You examine a sample of oranges from each crate to determine the quality of my oranges. It is easy to inspect a few oranges from the top of each crate, but these oranges may not be representative of the entire crate. Those on the bottom are more often damaged in shipment. If I were less than honest, I might make sure that any rotten oranges are packed on the bottom, with some good ones on top for you to inspect. If you sample from the top, your sample results are again biased—the sample oranges are systematically better than the population they are supposed to represent.

Biased sampling methods

The design of a statistical study is **biased** if it systematically favors certain outcomes.

A **voluntary response sample** chooses itself by responding to a general appeal. Write-in or call-in opinion polls are examples of voluntary response samples.

Selection of whichever individuals are easiest to reach is called **convenience sampling.**

Convenience samples and voluntary response samples are often biased.

Inspecting the oranges at the top of the crate is one example of convenience sampling. Mall interviews are another.

Example 5.2 Interviewing at the mall

Manufacturers and advertising agencies often use interviews at shopping malls to gather information about consumer habits and the effectiveness of ads. A sample of mall shoppers is fast and cheap. But people contacted at shopping malls are not representative of the entire U.S. population. They tend to be richer, for example, than the population as a whole. In addition, interviewers tend to select neat, safe-looking people to talk to. Mall samples are biased: they systematically overrepresent some parts of the population and underrepresent others. The opinions of such a convenience sample may be very different from those of the entire population.

Example 5.3 No kids for me!

Ann Landers once asked the readers of her advice column, "If you had it to do over again, would you have children?" She received nearly 10,000 responses, almost 70% saying "NO!" Can it be true that 70% of parents regret having children? Not at all. This is a voluntary response sample. People who feel strongly about an issue, particularly people with strong negative feelings, are more likely to take the trouble to respond. Ann Landers's results are strongly biased—the percent of parents who would not have children again is much higher in her sample than in the population of all parents.

"Hey, Pops, what was that letter you sent off to Ann Landers yesterday?"

Write-in and call-in opinion polls are almost sure to lead to strong bias. In fact, only about 15% of the public has ever responded to a call-in poll, and these tend to be the same people who call radio talk shows. That's not a representative sample of the population as a whole.

The Internet brings voluntary response polls to the computer nearest you. Visit **www.misterpoll.com** to become part of the sample in any of dozens of online polls. As the site says, "None of these polls are 'scientific,' but do represent the collective opinion of everyone who participates."

Exercises

5.1 Instant opinion A recent online poll posed the question: "Should female athletes be paid the same as men for the work they do?" In all, 13,147 (44%) said "Yes," another 15,182 (50%) said "No," and the remaining 1448 said "Don't know."

(a) What is the sample size for this poll?

(b) That's a much larger sample than standard sample surveys. In spite of this, we can't trust the result to give good information about any clearly defined population. Why?

5.2 An online poll In June 2008, *Parade* magazine posed the following question: "Should drivers be banned from using all cell phones?" Readers were encouraged to vote online at **parade.com**. The July 13, 2008, issue of *Parade* reported the results: 2407 (85%) said "Yes" and 410 (15%) said "No."

(a) What type of sample did the *Parade* survey obtain?

(b) Explain why this sampling method is biased. Is 85% probably higher or lower than the true percent of all adults who believe that cell phone use while driving should be banned? Why?

5.3 Sleepless nights How much sleep do high school students get on a typical school night? An interested student designed a survey to find out. To make data collection easier, the student surveyed the first 100 students to arrive at school on a particular morning. These students reported an average of 7.2 hours of sleep on the previous night.

(a) What type of sample did the student obtain?

(b) Explain why this sampling method is biased. Is 7.2 hours probably higher or lower than the true average amount of sleep last night for all students at the school? Why?

5.4 Hand-washing habits Do adults typically wash their hands after using the bathroom? In a telephone survey of 1001 U.S. adults, 92% said they always wash their hands after using a public restroom.[2] An observational study of 6076 adults in public restrooms told a slightly different story: only 77% of those observed actually washed their hands after using the restroom.[3]

(a) Why do you think the results of the two studies are so different?

(b) According to a description of the observational study, "Observers discreetly watched and recorded whether or not adults using public restrooms washed their hands. Observers were instructed to groom themselves (comb their hair, put on make-up, etc.) while observing and to rotate bathrooms every hour or so to avoid counting repeat users more than once. Observers were also instructed to wash their hands no more than 10 percent of the time." Explain why these precautions were taken by the observers.

5.5 Design your own bad sample A large high school wants to gather student opinion about parking for students on campus. It isn't practical to contact all students.

(a) Give an example of a way to choose a sample of students that is bad because it depends on voluntary response.

(b) Give another example of a bad way to choose a sample—but this time one that doesn't use voluntary response.

5.6 More capture-recapture Refer to Activity 5.1A (page 202). Mr. Washington's class takes an initial sample of 50 goldfish. After replacing the 50 tagged fish, the class obtains a second sample of fish. Of the 75 fish in this sample, 25 of the tagged fish are recaptured (along with 50 untagged fish).

(a) Compute an estimate of the total number of fish in the lake. Show your method clearly.

(b) Is the number of fish you calculated in (a) definitely, probably, probably not, or definitely not equal to the actual number of fish in the lake? Why?

(c) Explain why the capture-recapture sampling method used in this Activity is *not* biased.

Simple random samples

In a voluntary response sample, people choose whether to respond. In a convenience sample, the interviewer makes the choice. In both cases, personal choice produces bias. The statistician's remedy is to allow impersonal chance to choose the sample. A sample chosen by chance allows neither favoritism by the sampler nor self-selection by respondents. Choosing a sample by chance avoids bias by giving all individuals an equal chance to be chosen. Rich and poor, young and old, black and white, all have the same chance to be in the sample.

The simplest way to use chance to select a sample is to place names in a hat (the population) and draw out a handful (the sample). This is the idea of a **simple random sample.**

Simple random sample

A **simple random sample (SRS)** of size n consists of n individuals from the population chosen in such a way that every set of n individuals has an equal chance to be the sample actually selected.

An SRS not only gives each individual an equal chance to be chosen (thus avoiding bias in the choice) but also gives every possible sample an equal chance to be chosen. Drawing names from a hat does this. Write 100 names on identical slips of paper and mix them in a hat. This is a population. Now draw 10 slips, one after the other. This is an SRS, because any 10 slips have the same chance as any other 10.

Drawing names from a hat makes clear what it means to give each individual and each possible set of *n* individuals the same chance to be chosen. That's the idea of an SRS. Of course, we would need a very big hat to get a sample of the country's 117 million households. In practice, we use computer-generated **random digits** to choose samples. If you don't use software directly, you can use a **table of random digits** to choose small samples by hand.

> **Random digits**
>
> A **table of random digits** is a long string of the digits 0, 1, 2, 3, 4, 5, 6, 7, 8, 9 with these two properties:
>
> 1. Each entry in the table is equally likely to be any of the 10 digits 0 through 9.
> 2. The entries are independent of each other. That is, knowledge of one part of the table gives no information about any other part.

Are these random digits really random?

Not a chance. The random digits in Table B were produced by a computer program. Computer programs do exactly what you tell them to do. Give the program the same input and it will produce exactly the same "random" digits. Of course, clever people have devised computer programs that produce output that looks like random digits. These are called "pseudo-random numbers," and that's what Table B contains. Pseudo-random numbers work fine for statistical randomizing, but they have hidden nonrandom patterns that can mess up more advanced uses.

Table B at the back of the book is a table of random digits. You can think of Table B as the result of asking an assistant (or a computer) to mix the digits 0 to 9 in a hat, draw one, then replace the digit drawn, mix again, draw a second digit, and so on. The assistant's mixing and drawing save us the work of mixing and drawing when we need to randomize. Table B begins with the digits 19223950340575628713. To make the table easier to read, the digits appear in groups of five and in numbered rows. The groups and rows have no meaning—the table is just a long list of randomly chosen digits. Here's how to use the table to choose an SRS.

Example 5.4 Choosing an SRS: teens and the Internet

There has been much talk (and growing concern) about teenagers' use of the Internet. In addition to countless hours spent in chat rooms and surfing the Net, teens are viewing increasing amounts of pornography. Should stricter controls be imposed on teen Internet use?

Let's select a panel of teens to help answer this question. Ideally, we would like to have teens representing each of the 50 states on the panel. Unfortunately, budget restrictions will permit us to choose only 6 teens for the panel. So here's what we'll do. We'll ask each state to select 1 teen nominee by a process of their choosing. Then, to avoid bias, we will select an SRS of size 6 from the 50 state nominees to form the panel.

Step 1: Label. Give each state a numerical label, using as few digits as possible. Two digits are needed to label 50 states, so we use labels

01, 02, 03, . . . , 48, 49, 50

It is also correct to use labels 00 to 49 or even another choice of 50 two-digit labels. Here is a partial list of states, with labels attached:

01 Alabama	02 Alaska	03 Arizona . . .
48 Washington	49 Wisconsin	50 Wyoming

Step 2: Table. Enter Table B anywhere and read two-digit groups. Suppose we enter at line 103, which is

45467 71709 77558 00095 32863 29485 82226 90056

Each two-digit group in Table B is equally likely to be any of the 100 possible groups, 00, 01, 02, . . . , 99. So two-digit groups choose two-digit labels at random. That's just what we want.

We used only labels 01 to 50, so we ignore all other two-digit groups. The first 6 labels between 01 and 50 that we encounter in the table choose our sample. The first 10 two-digit groups in line 103 are

45	46	77	17	09	77	55	80	00	95
√	√	skip	√	√	skip	skip	skip	skip	skip

Notice that we skip six of these 10 labels because they are too high (over 50). The others are 45, 46, 17, and 09. The nominees from the states labeled 09, 17, 45, and 46 go into the sample. We need 2 more teenagers to complete the panel. The remaining 10 two-digit groups in line 103 are

32	86	32	94	85	82	22	69	00	56
√	skip	skip	skip	skip	skip	√			

The teenager from state 32 is chosen for the panel. Ignore the second 32 because that state is already in the sample. State 22's teen completes the panel.

The sample is composed of the states labeled 09, 17, 22, 32, 45, and 46. Our panel consists of the teens nominated by those states.

Golfing at random
Random drawings give everyone the same chance to be chosen, so they offer a fair way to decide who gets a rare prize—like a round of golf at the famous Old Course at St. Andrews, Scotland. A few can reserve in advance. Most must hope that chance favors them in the daily random drawing for tee times. At the height of the summer season, only 1 in 6 wins the right to pay over $100 for a round.

Using the table of random digits is much quicker than drawing names from a hat. As Example 5.4 shows, choosing an SRS has two steps.

Choose an SRS in two steps

Step 1: Label. Assign a numerical label to every individual in the population. Be sure that all labels have the same number of digits.

Step 2: Table. Use random digits to select labels at random.

You can assign labels in any convenient manner, such as alphabetical order for names of people. As long as all labels have the same number of digits, all individuals will have the same chance to be chosen. Use the shortest possible labels: one digit for a population of up to 10 members, two digits for 11 to 100 members, three digits for 101 to 1000 members, and so on. As standard practice, we recommend that you begin with label 1 (or 01 or 001, as needed). You can read digits from Table B in any order—across a row, down a column, and so on—because the table has no order. As standard practice, we recommend reading across rows.

Example 5.5 Choosing an SRS: Congress joins the teenagers

To add some clout to the teenagers' recommendations about Internet use in Example 5.4, we decide to add four members of the House of Representatives to the panel. Again, we decide to take an SRS.

Step 1: Label. Obtain an alphabetical listing of the 435 members of the House of Representatives. Assign numbers 001 to 435 as labels to the alphabetized list.

Step 2: Table. Enter Table B at line 134 (say) and read groups of three digits. The first 10 three-digit groups are

278	167	841	618	329	213	373	521	337	741
√	√	skip	skip	√	√				

Our sample consists of the representatives labeled 278, 167, 329, and 213. These individuals will join the teens on our panel.

Exercises

5.7 Apartment living You are planning a report on apartment living in a college town. You decide to select three apartment complexes at random for in-depth interviews with residents. Use Table B, starting at line 117, to select a simple random sample of three of the following apartment complexes. Explain your method clearly enough for a classmate to obtain your results.

Ashley Oaks	Country View	Mayfair Village
Bay Pointe	Country Villa	Nobb Hill
Beau Jardin	Crestview	Pemberly Courts
Bluffs	Del-Lynn	Peppermill
Brandon Place	Fairington	Pheasant Run
Briarwood	Fairway Knolls	Richfield
Brownstone	Fowler	Sagamore Ridge
Burberry	Franklin Park	Salem Courthouse
Cambridge	Georgetown	Village Manor
Chauncey Village	Greenacres	Waterford Court
Country Squire	Lahr House	Williamsburg

5.8 How do random digits behave? Which of the following statements are true of a table of random digits, and which are false? Explain your answers.

(a) There are exactly four 0s in each row of 40 digits.

(b) Each pair of digits has chance 1/100 of being 00.

(c) The digits 0000 can never appear as a group, because this pattern is not random.

5.9 An election day sample You want to choose an SRS of 25 of a city's 440 voting precincts for special voting-fraud surveillance on election day.

(a) Explain clearly how you would label the 440 precincts. How many digits make up each of your labels? What is the greatest number of precincts you could label using this number of digits?

(b) Use Table B at line 107 to choose the first 10 precincts in the SRS.

5.10 Is this an SRS? A university has 1000 male and 500 female faculty members. A survey of faculty opinion selects 100 of the 1000 men at random and then separately selects 50 of the 500 women at random. The 150 faculty members chosen make up the sample.

(a) Explain why this sampling method gives each member of the faculty an equal chance to be chosen.

(b) Nonetheless, this is not an SRS. Why not?

5.11 Drug testing Use the table of random digits (Table B) to select an SRS of 3 of the following 25 members of an athletic team for a drug test. Be sure to say where you entered the table and how you used it.

Agarwal	Fuest	Milhalko	Shen
Andrews	Fuhrmann	Moser	Smith
Baer	Garcia	Musselman	Sundheim
Berger	Healy	Pavnica	Wilson
Brockman	Hixson	Petrucelli	
Chen	Lee	Reda	
Frank	Lynch	Roberts	

D. Hurst/Alamy

5.12 Not an SRS It's sometimes not practical to take a simple random sample. Here's an example that shows why. Suppose 1000 iPhones are produced at a factory today. Management would like to ensure that the phones' display screens meet their quality control standards before shipping them to retail stores. Since it takes about 10 minutes to inspect an individual phone's display screen, managers decide to inspect a sample of 20 phones from the day's production.

(a) Explain why it would be difficult for managers to inspect an SRS of 20 iPhones that are produced today.

(b) An eager employee suggests that it would be easy to inspect the last 20 iPhones that were produced today. Why isn't this a good idea?

(c) Another employee recommends inspecting every fiftieth iPhone that is produced. Explain carefully why this sampling method is *not* an SRS.

Sampling with technology

Sample surveys use computer software to choose an SRS, but the software just automates the steps in Examples 5.4 and 5.5. The computer doesn't look in a table of random digits, because it can generate them on the spot, as Activity 5.1B illustrates.

ACTIVITY 5.1B

Simple random samples on the computer

In this Activity, you will learn how to select a simple random sample from a population of interest using a computer applet. You will also explore how results vary in repeated simple random samples from the same population.

Materials: Computer with Internet access and Java capability

APPLET

1. Go to the textbook Web site, **www.whfreeman.com/sta2e.** Click on the Statistical Applets link, and then choose the *Simple Random Sample* applet. You should see a screen like Figure 5.1.

2. Enter the number 5 next to "Population = 1 to" and click the Reset button. The "Population hopper" now contains five balls numbered 1 to 5. Enter sample size 3 and click the Sample button. Record the numbers of the balls in the "Sample bin."

3. *Think:* If you were to sample three balls from a population of size five, which is more likely to happen: getting the sample 1, 2, and 3 (in that order) or getting the sample 3, 1, and 5 (in that order)?

4. *Think:* If you were to repeatedly draw samples of size three from a population of size five, how often would the largest of the three numbers be picked first?

5. *Verify:* Test your answers to Questions 3 and 4 by making repeated use of the applet. Record the results of each sample in a table. Did the samples behave as you thought they would?

6. Draw a sample of seven balls from a population of size twenty. After the sample is drawn, what is left in the "Population hopper"? Is the sample "random"? Explain your answer.

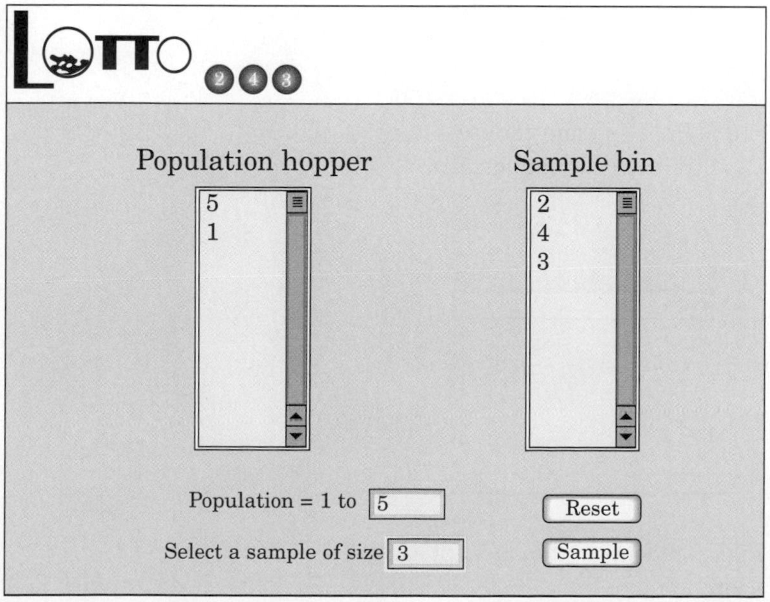

Figure 5.1 The *Simple Random Sample* applet at the book's Web site.

To see how software speeds up choosing an SRS, go to the Research Randomizer at **www.randomizer.org.** Click on "Randomize" and fill in the boxes. You can even ask the Randomizer to arrange your sample in order.

You can also use a calculator to choose an SRS.

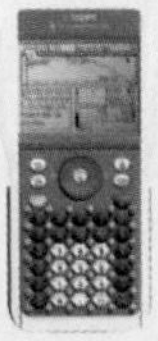

CALCULATOR CORNER How to choose an SRS

The TI-84 and TI-Nspire have built-in commands for generating "random" numbers. Just how "random" are they?

TI-84

1. Press MATH, then select PRB and 5:randInt(. Complete the command randInt(1,5) and press ENTER.

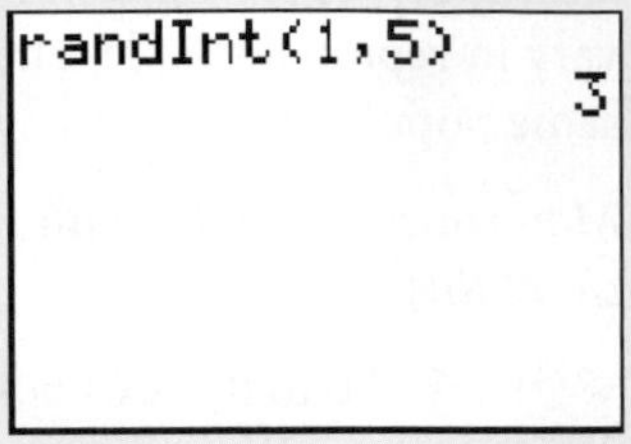

TI-Nspire

1. Open a New Document and create a calculator page. Press (home), choose 6: New Document, and 1:Add Calculator.

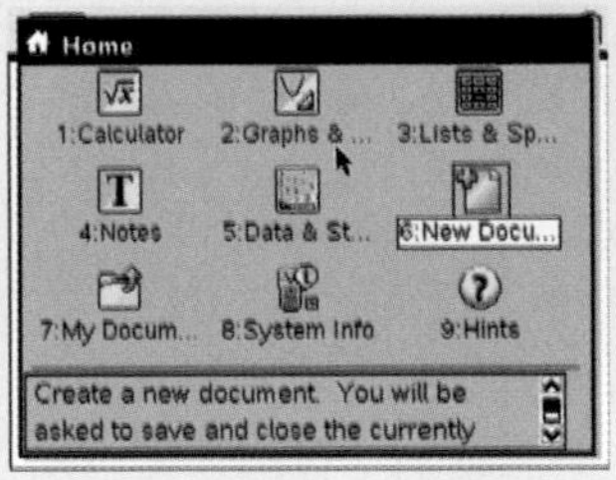

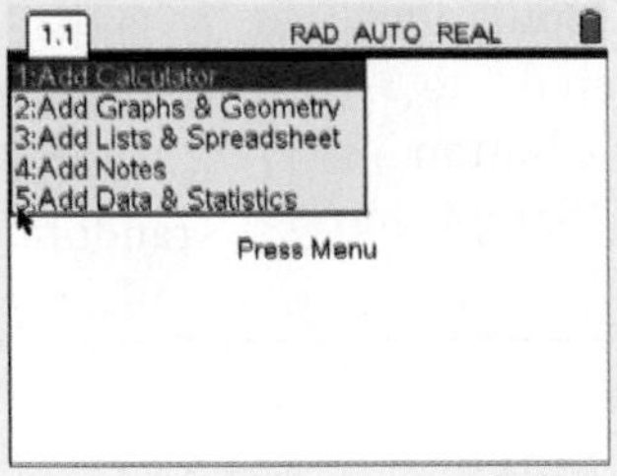

Generate a "random" integer between 1 and 5. Press (menu) and choose 4:Probability, 4:Random, and 2:Integer. Complete the command randInt(1,5) and press (enter).

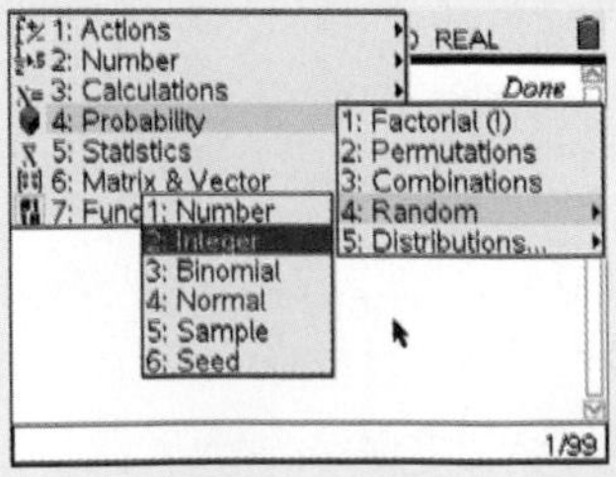

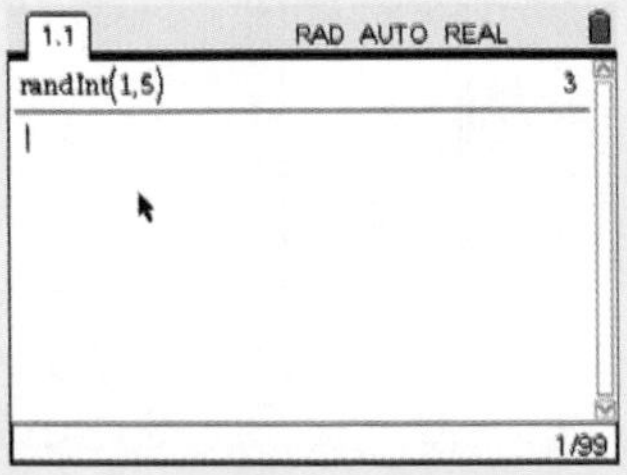

Determine by a show of hands how many students in the class got each number: 1, 2, 3, 4, and 5. Do the results seem random?

2. Press ENTER four more times. Record the sequence of five numbers you obtain. Compare your results with those of your classmates. Do the results seem random?

3. Calculators and computers use algorithms written by humans to generate "random" numbers. The TI-84 and TI-Nspire `randInt` command uses a starting value, called a "seed," to determine the next "random" integer between 1 and 5. If several students in your class obtained the same values in Steps 1 and 2, it is because their calculators were using the same seed. We can improve on this situation by having each student input a different seed to the random number generator. Here's one method: use the last four digits (####) of your home phone number as the seed (I used 0594).

TI-84

Type ####, press STO▸ MATH, choose `PRB` and `1:rand`, then press ENTER to generate the command `#### → rand`.

TI-Nspire

Press (menu), then choose `4:Probability`, `4:Random`, and `6:Seed`. Complete the command `RandSeed ####` and press (enter).

4. Now do `randInt(1,100)`. See if anyone else in the class got the same number.

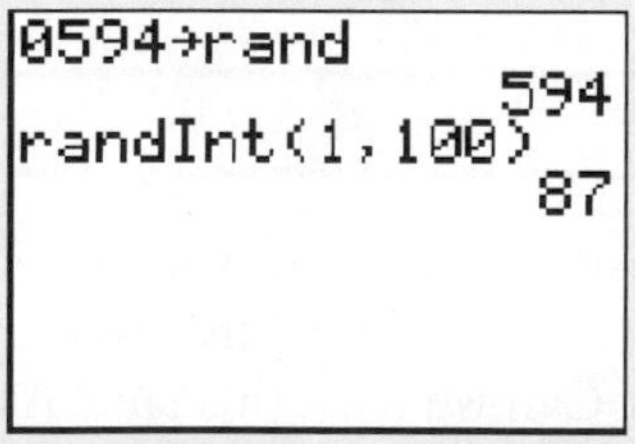

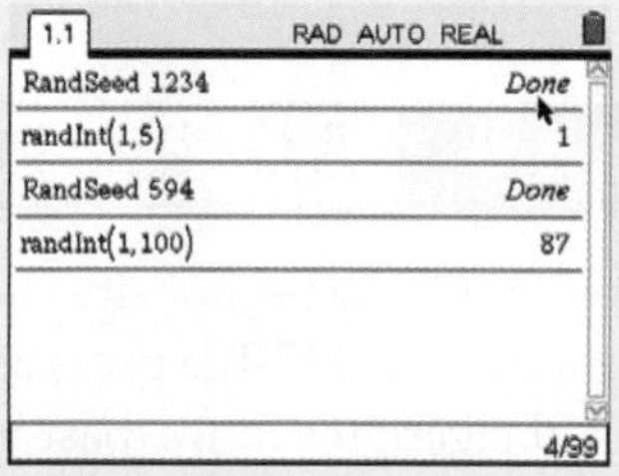

5. Now you're ready to choose an SRS of students from your class. Your teacher will provide a class roster. Number the students on the roster 1, 2, 3, . . . in some sensible way. On your calculator, execute the command `randInt(1,n)` where *n* is the number of students in your class. Note who is selected. Press ENTER four more times. Did you get five different numbers? If not, press ENTER a few more times until you have selected five people for your sample. Note who is selected.

Did you notice an important difference between the samples produced by the *Simple Random Sample* applet of Activity 5.1B and the calculator's `RandInt` command in the Calculator Corner? The applet sampled *without replacement* from the "Population hopper," but the calculator sampled *with replacement* from the specified population. As a result, the calculator sometimes selects the same number more than once in a given sample. To deal with this problem, you can generate additional random numbers as needed to replace any repeats. Alternatively, you can use a method other than `RandInt` (like `RandSamp` on the TI-Nspire) to sample without replacement. Refer to your device's reference manual or your teacher for more details.

DILBERT reprinted by permission of United Feature Syndicate, Inc.

Can you trust a sample?

The *Town Talk,* Ann Landers, and mall interviews produce samples. We can't trust results from these samples, because they are chosen in ways that invite bias. We have more confidence in results from an SRS, because it uses impersonal chance to avoid bias. The first question to ask of any sample is whether it was chosen at random. Opinion polls and other sample surveys carried out by people who know what they are doing use random sampling.

Example 5.6 A Gallup Poll

A Gallup Poll on smoking began with the question "Have you, yourself, smoked any cigarettes in the past week?" The press release reported that "just 25% of Americans say they smoked cigarettes in the past week." Can we trust this fact? Ask first how Gallup selected its sample. Later in the press release we read this: "These results are based on telephone interviews with a randomly selected national sample of 1,007 adults, aged 18 years and older, conducted July 6–9, 2006."[4]

This is a good start toward gaining our confidence. Gallup tells us what population it has in mind (people at least 18 years old living anywhere in the United States). We know that the sample from this population was of size 1007 and, most important, that it was chosen at random. There is more to say, and we will soon say it, but we have at least heard the comforting words "randomly selected." (In the 2007 and 2008 Gallup polls on smoking, only 21% of American adults said that they had smoked cigarettes in the past week.)

What should a nation do if there are not enough volunteers for military service but only a small fraction of eligible youth are needed by the military? That question last arose during the Vietnam era. Beginning in 1970, a draft lottery was used to choose draftees by random selection. Although the draft was ended in 1976, young men must still register at age 18. The draft lottery may return if the military cannot attract enough volunteers.

Actually making a random selection in the public arena isn't always easy. In principle, it is just like choosing an SRS. In practice, random digits can't be used. Because few people understand random digits, a lottery looks fairer if a respected person chooses capsules from a glass bowl in front of the TV cameras. This also prevents cheating: no one can check the table of random digits in advance to see how Cousin Joe will make out in the selection.

Physical mixing and drawing *look* random, but it can be hard to achieve a mixing that really *is* random. There is no better illustration of this than the first Vietnam era draft lottery, held in 1970.

APPLICATION 5.1

The draft lottery

Because taking an SRS of all eligible men would be hopelessly awkward, the draft lottery selected birth dates in a random order. Men born on the first date chosen would be drafted first, then those born on the second date chosen, and so on. All men aged 19 to 25 were included in the first lottery, and there were 366 birth dates. The 366 dates were placed into identical small capsules, which were put in a bowl and publicly drawn one by one.

QUESTIONS

1. Explain why this method does *not* choose an SRS from all men eligible for the draft.

The scatterplot in Figure 5.2 displays the results of the 1970 draft lottery. We have plotted the variable "day of year" on the horizontal axis, with Day 1 = January 1 and Day 366 = December 31. The variable "draft number" is plotted on the vertical axis.

2. The highlighted point has coordinates (15, 17). Explain what this point means in terms of the draft lottery.

3. Describe the relationship that you see in the scatterplot between the day of the year on which people were born and the draft lottery number from the drawing.

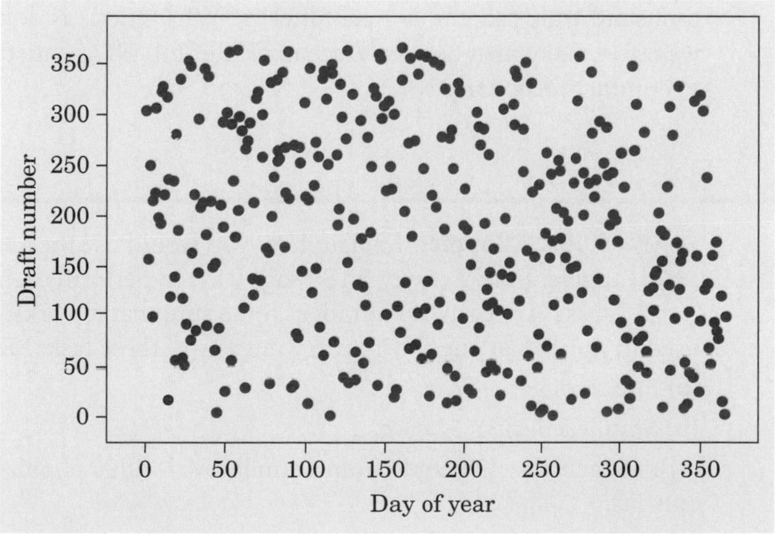

Figure 5.2 Scatterplot showing the results of the 1970 draft lottery.

APPLICATION 5.1 *(continued)*

APPLICATION 5.1 *(continued)*

Figure 5.3 shows side-by-side boxplots of the draft lottery results for each birth month.

4. Describe the relationship that you see in the boxplots between the day of the year on which people were born and the draft lottery number from the drawing.

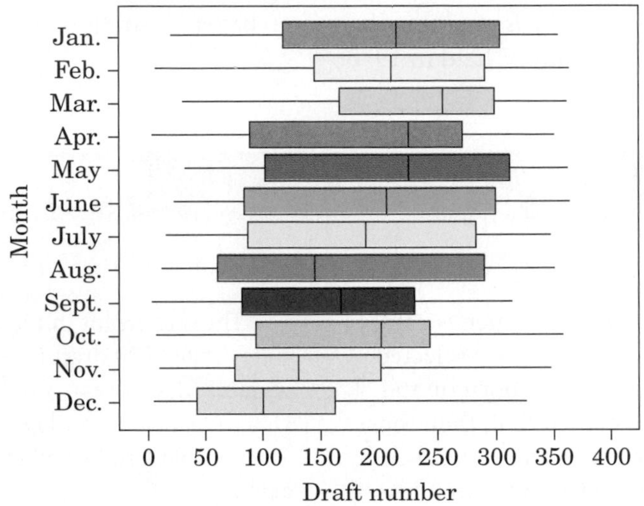

Figure 5.3 Side-by-side boxplots of the draft numbers for birth dates within each month.

As the 1970 draft lottery was taking place, news reporters quickly noticed that men born later in the year seemed to receive lower draft numbers. Statisticians soon showed that this trend was so strong that it would occur less than once in a thousand truly random annual lotteries. What's done is done, and off to Vietnam went too many men born in December. But for 1971, statisticians were brought in to conduct the lottery.

Exercises

APPLET **5.13 *SRS* applet** Explain how you would use the *Simple Random Sample* applet from Activity 5.1B (page 211) to perform each of the following tasks. Try each of your ideas to be sure that it works. Can the applet be used in more than one way to carry out any of these tasks? Explain.

(a) Flip a coin.

(b) Shuffle a deck of 52 cards.

(c) Pick a jury of 12 people from a small town with a population of 500 adult citizens.

5.14 SRS on the calculator Explain how you would use your calculator's `randInt` command to perform each of the tasks in Exercise 5.13, if possible. Try each of your ideas to be sure that it works.

5.15 Rating the police The Miami Police Department wants to know how black residents of Miami feel about police service. A researcher prepares several questions about the police. The police department chooses an SRS of 300 mailing addresses in predominantly black neighborhoods and sends a uniformed black police officer to each address to ask the questions of an adult living there.

(a) What are the population and the sample?

(b) Why are the results likely to be biased even though the sample is an SRS?

5.16 A call-in opinion poll Should the United Nations continue to have its headquarters in the United States? A television program asked its viewers to call in with their opinions on that question. There were 186,000 callers, 67% of whom said "No." A nationwide random sample of 500 adults found that 72% answered "Yes" to the same question.[5] Explain to someone who knows no statistics why the opinions of only 500 randomly chosen respondents are a better guide to what all Americans think than the opinions of 186,000 callers.

5.17 Local draft boards Prior to 1970, young men were selected for military service by local draft boards. There was a complex system of exemptions and quotas that allowed, for example, farmers' sons and married young men with children to avoid the draft. Do you think that random selection among all men of the same age is preferable to making distinctions based on marital status? Give your reasons.

5.18 Random selection? Choosing at random is a "fair" way to decide who gets some rare prize, in the sense that everyone has the same chance to win. Random choice isn't always a good idea—sometimes we don't want to treat everyone the same, because some people have a better claim. In each of the following situations, would you support choosing at random? Give your reasons in each case.

(a) The basketball arena has 4000 student seats, and 7000 students want tickets. Should we choose 4000 of the 7000 at random?

(b) The list of people waiting for liver transplants is much larger than the number of available livers. Should we let chance decide who gets a transplant?

Section 5.1 Summary

We select a **sample** in order to get information about some **population.** How can we choose a sample that represents the population fairly? **Convenience samples** and **voluntary response samples** are common but do not produce trustworthy data because these sampling methods are usually **biased.** That is, they systematically favor some parts of the population over others in choosing the sample.

The deliberate use of chance in producing data is one of the big ideas of statistics. Random samples use chance to choose a sample, thus avoiding bias due to personal choice. The basic type of random sample is the **simple random sample,** which gives all samples of the same size the same chance to be the sample we actually choose. To choose an SRS by hand, use a **table of random digits** such as Table B in the back of the book. Alternatively, you can use a computer or calculator random number generator to help choose an SRS.

Section 5.1 Exercises

5.19 How much do students earn? A university's financial aid office wants to know how much it can expect students to earn from summer employment. This information will be used in setting the level of financial aid. The population contains 3478 students who have completed at least one year of study but have not yet graduated. The university will send a questionnaire to an SRS of 100 of these students, drawn from an alphabetized list.

(a) Describe how you will label the students in order to select the sample.

(b) Use Table B, beginning at line 105, to select the first 5 students in the sample.

5.20 More randomization Most sample surveys call residential telephone numbers at random. They do not, however, always ask their questions of the person who picks up the phone. Instead, they ask about the adults who live in the residence and choose one at random to be in the sample. Why is this a good idea?

5.21 A biased sample You see a female student standing in front of the cafeteria, now and then stopping other students to ask them questions. She says that she is conducting a survey of student opinions about the quality of food in the cafeteria. Explain why this sampling method is almost certainly biased.

5.22 Sampling TVs An electronics company has 50 large flat-screen televisions ready for shipment, each labeled with one of the following serial numbers:

A1109	A2056	A2219	A2381	B0001
A1123	A2083	A2336	A2382	B0012
A1186	A2084	A2337	A2383	B0046
A1197	A2100	A2338	A2384	B1195
A1198	A2108	A2339	A2385	B1196
A2016	A2113	A2340	A2390	B1197
A2017	A2119	A2351	A2396	B1198
A2020	A2124	A2352	A2410	B1199
A2029	A2125	A2367	A2410	B1200
A2032	A2130	A2372	A2500	B1201

An SRS of 5 TVs must be chosen for inspection. Use Table B to do this, beginning at line 139. Explain your method clearly.

5.23 Gun control Since the beginning of opinion polls over 50 years ago, at least two-thirds of those surveyed said that they favored stronger controls on firearms. Specific gun control proposals have often been favored by 80% to 85% of respondents. Yet little national gun control legislation has passed, and no major national restrictions on firearms exist. Why do you think this has occurred?

5.24 A lottery for drugs? Only limited supplies of some experimental drugs are available because the drugs are very difficult to make. Sometimes it is necessary to use a lottery to decide at random which patients can receive a drug that is in short supply. Several years ago, for example, Hoffman–La Roche had enough doses of the drug Invirase for 2880 AIDS patients. Shortly after announcing this, the company received 10,000 calls from patients who wanted to enter the lottery.

Discuss this practice. Do you favor random selection? If not, how should recipients be chosen?

5.25 Read anything lately? The Denver Public Library wants to estimate the percent of Denver households with an adult who has read at least one book in the last month. The homes of 400 people who have library cards are sampled, and it turns out that 90% of these households have an adult who has read a book in the past month. Is 90% likely to be a biased estimate for the true percent of Denver households with an adult who has read at least one book in the last month? Explain why or why not.

5.26 Women count! Broadcast and print media across the country have reported on the government's release of a report on the status of working women called "Working Women Count." This report gives findings from a survey of 250,000 women. According to one news report:

> *More than 1600 businesses, unions, newspapers, magazines, and community service organizations helped distribute the survey to their members, subscribers, and patrons, which the White House announced with much fanfare. It sought women's opinions on job satisfaction, pay, benefits, and opportunities for advancement.*[6]

A second survey, asking the same questions, was conducted at the same time. However, this survey interviewed only 1200 working women chosen at random.

Which survey is likely to give a more accurate view of working women's opinions on job satisfaction, pay, benefits, and opportunities for advancement? Explain briefly.

5.2 What Do Samples Tell Us?

ACTIVITY 5.2A

Should gambling be legal?

Figure 5.4 (on the next page) shows a small population. Each circle represents an adult. The colored circles are people who disapprove of legal gambling, and the white circles are people who approve. You can check that 60 of the 100 circles are white, so in this population the proportion who approve of gambling is $60/100 = 0.6$.

1. The circles are labeled 00, 01, . . . , 99. Your teacher will assign each student in the class a different line of Table B to use for this Activity. Use your assigned line to take an SRS of size 10. Explain your method clearly enough for a classmate to repeat the process. What is the proportion of the people in your sample who approve of gambling?

2. Take 9 more SRSs of size 10 (10 samples in all). Be sure to use a different starting point in Table B for each sample. You now have 10 values of the sample proportion.

3. Because your samples have only 10 people, the only values the sample proportion can take are 0/10, 1/10, 2/10, . . . , 9/10, and 10/10. That is, the sample proportion is always one of the following values: 0, 0.1, 0.2, . . . , 0.9, or 1.

4. Taking samples of size 10 from a population of size 100 is not a setting that occurs often in the real world, but let's look at your results anyway. How many of your 10 samples estimated the population proportion of 0.6 exactly correctly? Is the true value 0.6 roughly in the center of your sample values? Explain why 0.6 would be in the center of the sample values if you took a large number of samples.

5. Calculate the average of your 10 sample proportions. How does this value compare with the proportion of adults in the population who favor legal gambling?

6. Pool results with your classmates. Make a large dotplot on the board. Describe what you see.

7. Calculate the average of the class's sample proportions. Compare with the true population proportion.

ACTIVITY 5.2A *(continued)*

ACTIVITY 5.2A *(continued)*

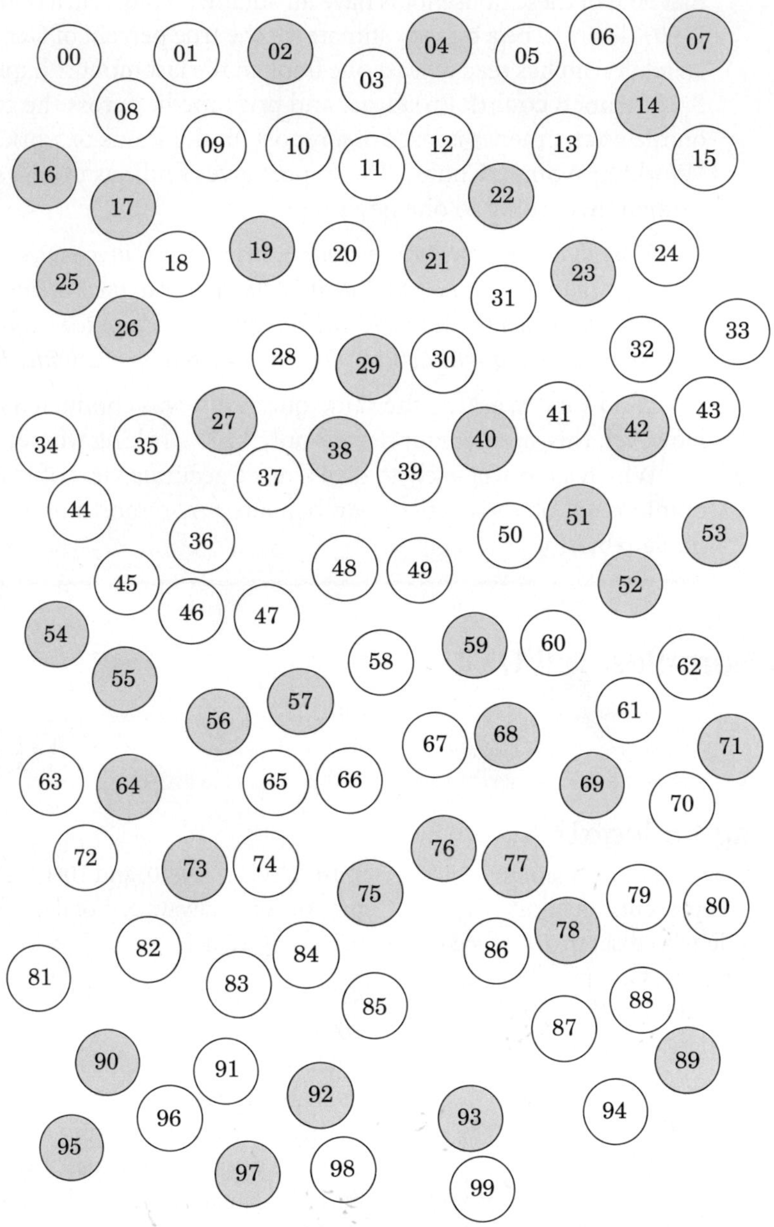

Figure 5.4 A population of 100 individuals for Activity 5.2A. Some individuals (white circles) approve of legal gambling, and others disapprove.

From sample to population

You know that lotteries are popular. How popular? Here's what the Gallup Poll says: "They can offer massive jackpots—and a ticket just costs a dollar at your neighborhood store. For many Americans, picking up a lottery ticket has become routine, despite the massive odds against striking it rich. A Gallup Poll Social Audit on gambling shows that 57% of Americans have bought a lottery ticket in the last 12 months, making lotteries by far the favorite choice of gamblers." Reading further, we find that Gallup talked with "1523 randomly selected adults" to reach these conclusions.[7]

Gallup's finding that "57% of Americans have bought a lottery ticket in the last 12 months" makes a claim about the population of 235 million adults. But Gallup doesn't know the truth about this population. The poll contacted 1523 people and found that 57% of them said they had bought a lottery ticket in the past year. Because the sample of 1523 people was chosen at random, it's reasonable to think that they represent the entire population pretty well. So Gallup turns the *fact* that 57% of the *sample* bought lottery tickets into an *estimate* that about 57% of *all adults* bought tickets. That's a basic strategy in statistics: use a fact about a sample to estimate the truth about the whole population. To think about such a strategy, we must keep straight whether a number describes a sample or a population. Here is the vocabulary we use.

Parameters and statistics

A **parameter** is a number that describes the **population.** A parameter is a fixed number, but in practice we don't know its value.

A **statistic** is a number that describes a **sample.** The value of a statistic is known when we have taken a sample, but it can change from sample to sample. We often use a statistic to estimate an unknown parameter.

So "parameter" is to "population" as "statistic" is to "sample." Want to estimate an unknown parameter? Choose a sample from the population and use a sample statistic as your estimate. That's what Gallup did.

Example 5.7 Who plays Lotto?

The proportion of all adults who bought a lottery ticket in the past 12 months is a *parameter* describing the population of 235 million adults. Call it p, for "proportion." Alas, we don't know the numerical value of p. To estimate p, Gallup took a sample of 1523 adults. The proportion of the sample who bought lottery tickets is a *statistic*. Call it $\hat{p}$ (read as "p-hat"). It happens that 868 of this sample of size 1523 bought tickets, so for this sample,

$$\hat{p} = \frac{868}{1523} = 0.57 \qquad \text{(that is, 57\%)}$$

Because all adults had the same chance to be among the chosen 1523, it seems reasonable to use the statistic $\hat{p} = 0.57$ as an estimate of the unknown parameter p. It's a fact that 57% of the sample bought lottery tickets—we know because we asked them. We don't know what percent of all adults bought tickets, but we *estimate* that about 57% did.

Exercises

5.27 Stop smoking! A random sample of 1000 people who signed a card saying they intended to quit smoking were contacted 9 months later. It turned out that 210 (21%) of the sampled individuals had not smoked over the past 6 months. Specify the population of interest, the parameter of interest, the sample, and the sample statistic in this problem.

Each boldface number in Exercises 5.28 to 5.30 is the value of either a ***parameter*** *or a* ***statistic.*** *In each case, state which it is.*

5.28 Drink Arizona On Tuesday, the bottles of Arizona iced tea filled in a plant were supposed to contain an average of **20** ounces of iced tea. Quality control inspectors sampled 50 bottles at random from the day's production. These bottles contained an average of **19.6** ounces of iced tea.

5.29 Flight safety On a New York–to–Denver flight, **8**% of the 125 passengers were selected for random security screening prior to boarding. According to the Transportation Security Administration, **10**% of airline passengers are chosen for random screening.

5.30 Sleeping ducks, I A recent report in the journal *Nature* examined whether ducks keep an eye out for predators while they sleep. The researchers, from Indiana State University, put four ducks in each of four plastic boxes, which were arranged in a row. Ducks in the two end boxes slept with one eye open **31.8**% of the time, compared with only **12.4**% of the time for the ducks in the two center boxes.

5.31 Sleeping ducks, II Is the study described in the previous exercise an example of an observational study or a comparative experiment? Explain briefly.

AP Photo/Ed Andrieski

5.32 Dead trees On the west side of Rocky Mountain National Park, many mature pine trees are dying due to infestation by pine beetles. Scientists would like to use sampling to estimate the proportion of all pine trees in the area that have been infected.

(a) Explain why it wouldn't be practical for scientists to obtain an SRS in this setting.

(b) A possible alternative would be to use every pine tree along the park's main road as a sample. Why is this sampling method biased?

(c) Suppose that a more complicated random sampling plan is carried out, and that 35% of the pine trees in the sample are infested by the pine beetle. Can scientists conclude that 35% of *all* the pine trees on the west side of the park are infested? Why or why not?

Sampling variability

If Gallup took a second random sample of 1523 adults, the new sample would have different people in it. It is almost certain that there would not be exactly 868 positive responses. That is, the value of the statistic $\hat{p}$ will *vary* from sample to sample. Could it happen that one random sample finds that 57% of adults recently bought a lottery ticket and a second random sample finds that only 37% had done so? Random samples eliminate *bias* from the act of choosing a sample, but they can estimate a population proportion badly because of the *variability* that results when we choose at random. If the variation when we take repeated samples from the same population is too great, we can't trust the results of any one sample.

We are saved by the second great advantage of random samples. The first advantage is that choosing at random eliminates favoritism. That is, random sampling reduces bias. The second advantage is that, if we took lots of random samples of the same size from the same population, the variation from sample to sample would follow a predictable pattern. This predictable pattern shows that results of bigger samples are less variable than the results of smaller samples.

Example 5.8 Lots and lots of samples

Here's another big idea of statistics: to see how trustworthy one sample is likely to be, ask what would happen if we took many samples from the same population. Let's try it and see. Suppose that in fact (unknown to Gallup) exactly 60% of all adults have bought a lottery ticket in the past 12 months. That is, the truth about the population is that $p = 0.6$. What if Gallup used the sample proportion $\hat{p}$ from an SRS of size 100 to estimate the unknown value of the population proportion p?

Figure 5.5 illustrates the process of choosing many samples and finding $\hat{p}$ for each one. In the first sample, 56 of the 100 people had bought lottery tickets, so $\hat{p} = 56/100 = 0.56$. Only 46 in the next sample had bought tickets, so for that sample $\hat{p} = 0.46$. We let technology choose 1000 samples and make a histogram of the 1000 values of $\hat{p}$. That's the graph at the right of Figure 5.5. The different values of $\hat{p}$ run along the horizontal axis. The heights of the bars show how many of our 1000 samples gave each group of values.

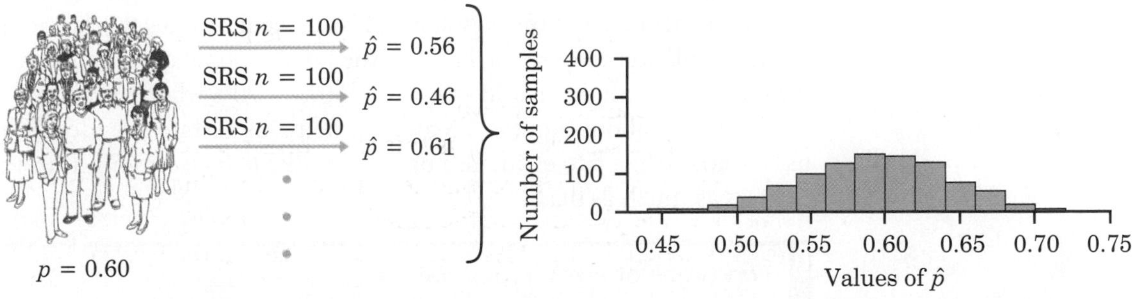

Figure 5.5 The results of many SRSs have a regular pattern. Here we draw 1000 SRSs of size 100 from the same population. The population proportion is $p = 0.6$. The sample proportions vary from sample to sample, but their values center at the truth about the population.

Of course, Gallup interviewed 1523 people, not just 100. Figure 5.6 shows the results of 1000 SRSs, each of size 1523, drawn from a population in which the true proportion is $p = 0.6$. Figures 5.5 and 5.6 are drawn on the same scale. Comparing them shows what happens when we increase the size of our samples from 100 to 1523.

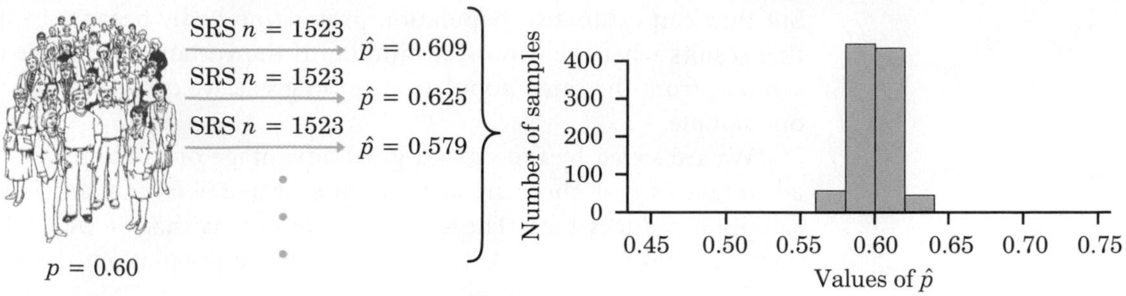

Figure 5.6 Draw 1000 SRSs of size 1523 from the same population as in Figure 5.5. The 1000 values of the sample proportion are much less spread out than was the case for the smaller samples.

Look carefully at Figures 5.5 and 5.6. We flow from the population, to many samples from the population, to the many values of $\hat{p}$ from these many samples. Gather these values together and study the histograms that display them.

- In both cases, the values of the sample proportion $\hat{p}$ vary from sample to sample, but the values are centered at 0.6. Recall that $p = 0.6$ is the true population parameter. Some samples have a $\hat{p}$ less than 0.6 and some greater, but there is no tendency to be always low or always high. That is, $\hat{p}$ has no **bias** as an estimator of p. This is true for both large and small samples.
- The values of $\hat{p}$ from samples of size 100 are much more spread out than the values from samples of size 1523. In fact, 95% of our 1000 samples of size 1523 have a $\hat{p}$ lying between 0.576 and 0.624. That's within 0.024 on either side of the population truth 0.6. Our samples of size 100, on the other hand, spread the middle 95% of their values between 0.50 and 0.69. That goes out 0.1 from the truth, about four times as far as the larger samples. So larger random samples have less **variability** than smaller samples.

We can rely on a sample of size 1523 to almost always give an estimate $\hat{p}$ that is close to the truth about the population. Figure 5.6 illustrates this fact for just one value of the population proportion, but it is true for any population. Samples of size 100, on the other hand, might give an estimate of 50% or 70% when the truth is 60%.

Thinking about Figures 5.5 and 5.6 helps us restate the idea of bias when we use a statistic like $\hat{p}$ to estimate a parameter like p. It also reminds us that variability matters as much as bias.

Two types of error in estimation

Bias is consistent, repeated deviation of the sample statistic from the population parameter in the same direction when we take many samples.

Variability describes how spread out the values of the sample statistic are when we take many samples. Large variability means that the result of sampling is not repeatable.

A good sampling method has both small bias and small variability.

We can think of the true value of the population parameter as the bull's-eye on a target, and of the sample statistic as an arrow shot at the bull's-eye. Bias and variability describe what happens when an archer shoots many arrows at the target. *Bias* means that the aim is off, and the arrows land consistently off the bull's-eye in the same direction. The sample values do not center about the population value. High *variability* means that repeated shots are widely scattered on the target. Repeated samples do not give similar results but differ widely among themselves. Figure 5.7 shows this target illustration of the two types of error.

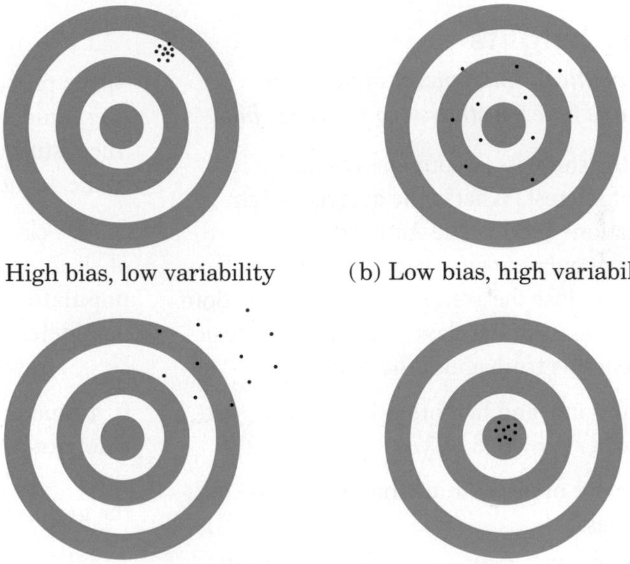

(a) High bias, low variability

(b) Low bias, high variability

(c) High bias, high variability

(d) The ideal: low bias, low variability

Figure 5.7 Bias and variability in shooting arrows at a target. Bias means the archer systematically misses in the same direction. Variability means the arrows are scattered.

Notice that low variability (repeated shots are close together) can accompany high bias (the arrows are consistently away from the bull's-eye in one direction). And low bias (the arrows center on the bull's-eye) can accompany high variability (repeated shots are widely scattered). A good sampling scheme, like a good archer, must have both low bias and low variability. Here's how we do this.

Managing bias and variability

To reduce bias, use random sampling. When we start with a list of the entire population, simple random sampling produces unbiased estimates—the values of a statistic computed from an SRS neither consistently overestimate nor consistently underestimate the value of the population parameter.

To reduce the variability of an SRS, use a larger sample. You can make the variability as small as you want by taking a large enough sample.

In practice, Gallup takes only one sample. We don't know how close to the truth an estimate from this one sample is, because we don't know what the truth about the population is. But *large random samples almost always give an estimate that is close to the truth.* Looking at the pattern of many samples shows that we can trust the result of one sample.

ACTIVITY 5.2B

Sampling heights

Materials: Identical, small pieces of paper or cardstock—one for each student in the class; hat

Think of all the students in your class as the population of interest. What is the average height of the population? Later in the Activity, you'll compute this population parameter. But for now, let's investigate how close the average height for a random sample of 4 students in the class (the sample statistic) is likely to be to the truth about the population.

1. Write your name and height, in inches, on one side of the paper provided by your teacher.

2. Put your piece of paper in the hat with those of your classmates.

3. Your teacher will mix the pieces of paper in the hat and then ask a student to draw 4 slips of paper from the hat. Notice that the 4 students chosen are a simple random sample of size 4 from the population of interest. The student should quickly record the names and heights of the 4 students in the SRS and then return all 4 slips to the hat.

4. The process in Step 3 should be repeated until every student in the class has selected an SRS of size 4. Then, students should calculate the average height of the 4 students in their samples.

5. As a class, make a large dotplot of the heights of all students in the class. We call this graph the **population distribution.** Compute the population parameter—the average height of all students in the class. Draw a vertical line through this value.

6. Immediately below the graph from Step 5, make a dotplot of the average heights from the samples of size 4. Use the same horizontal scale as you did for the graph in Step 5.

7. Is the dotplot in Step 6 centered at the population parameter? Explain why this makes sense.

8. Which distribution shows more variability around the population parameter—the population distribution or the distribution of sample averages? In general, how close are the values of the sample statistic to the population parameter?

Exercises

5.33 Bias and variability Figure 5.8 (on the facing page) shows the behavior of a sample statistic in many samples in four situations. These graphs are like those in Figures 5.5 and 5.6. That is, the heights of the bars show how often the sample statistic took various values in many samples from the same population. The true value of the population parameter is marked by an arrow on each graph. Label each of the graphs in Figure 5.8 as showing high or low bias and as showing high or low variability.

5.34 Sampling variability In thinking about Gallup's sample of size 1523, we asked, "Could it happen that one random sample finds that 57% of adults recently bought a lottery ticket and a second random sample finds that only 37% had done so?" Look at Figure 5.6 (page 224), which shows the results of 1000 samples of this size when the population truth is 60%. Would you be

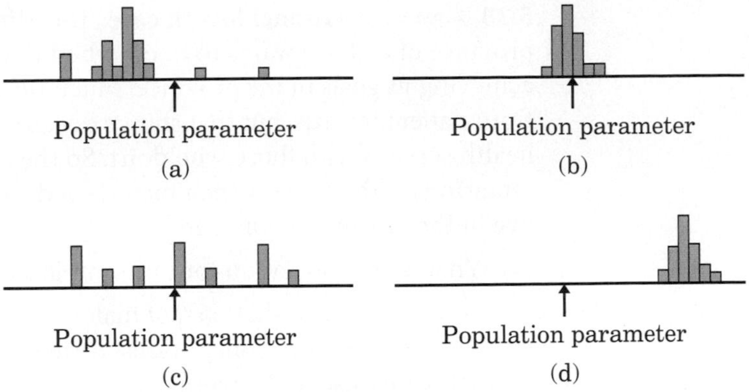

Figure 5.8 Take many samples from the same population and make a histogram of the values taken by a sample statistic. Here are the results for four different sampling methods.

surprised if a sample from this population gave 57%? Would you be surprised if a sample gave 37%?

5.35 No hands! Students from Hunter College in New York City carried out an observational study of driver behaviors at 50 different intersections. Of the 3120 drivers they observed, 23% were talking on cell phones.[8]

(a) In national surveys of driver behavior, well over half of those surveyed admit that they regularly talk on cell phones while driving. Give at least two reasons why the Hunter College study gave such different results.

(b) If the Hunter College students had observed twice as many drivers, would the percent of drivers talking on cell phones have been about the same as the results from national surveys? Why or why not?

5.36 Predict the election Just before a presidential election, a national opinion poll increases the size of its weekly sample from the usual 1500 people to 4000 people.

(a) Does the larger random sample reduce the bias of the poll result? Explain.

(b) Does it reduce the variability of the result? Explain.

5.37 A sampling simulation Let us illustrate sampling variability in a small sample from a small population. Ten of the 25 club members listed below are female. Their names are marked with asterisks in the list. The club chooses 5 members at random to receive free trips to the national convention.

Alonso	Darwin	Herrnstein	Myrdal	Vogt*
Binet*	Epstein	Jimenez*	Perez*	Went
Blumenbach	Ferri	Luo	Spencer*	Wilson
Chase*	Gonzales	Moll*	Thomson	Yerkes
Chen*	Gupta*	Morales*	Toulmin	Zimmer

(a) Draw 20 SRSs of size 5, using a different part of Table B each time. Record the number of females in each of your samples.

(b) Make a dotplot to display your results. What is the average number of females in your 20 samples?

(c) Do you think the club members should suspect discrimination if none of the 5 tickets goes to a woman? Justify your answer.

5.38 Canada's national health care The Ministry of Health in the Canadian province of Ontario wants to know whether the national health care system is achieving its goals in the province. Much information about health care comes from patient records, but that source doesn't allow us to compare people who use health services with those who don't. So the Ministry of Health conducted the Ontario Health Survey, which interviewed a random sample of 61,239 people who live in the province of Ontario.[9]

(a) What is the population for this sample survey? What is the sample?

(b) The survey found that 76% of males and 86% of females in the sample had visited a general practitioner at least once in the past year. Do you think these estimates are close to the truth about the entire population? Why?

Margin of error and all that

The **margin of error** that sample surveys announce translates sampling variability of the kind shown in Figure 5.5 (page 223) and Figure 5.6 (page 224) into a statement of how much confidence we can have in the results of a survey. Let's start with the kind of language we hear so often in the news.

What a margin of error means

"Margin of error plus or minus 3 percentage points" is shorthand for this statement:

If we took many samples using the same method we used to get this one sample, 95% of the samples would give a result within plus or minus 3 percentage points of the truth about the population.

Take this step-by-step. A sample chosen at random will usually not estimate the truth about the population exactly. We need a margin of error to tell us how close our estimate comes to the truth. But we can't be *certain* that the truth differs from the estimate by no more than the margin of error. Ninety-five percent of all samples come this close to the truth, but 5% miss by more than the margin of error. We don't know the truth about the population, so we don't know if our sample is one of the 95% that hit or one of the 5% that miss. We say we are **95% confident** that the truth lies within the margin of error.

Example 5.9 Understanding the news

Here's what the TV news announcer says: "A new Gallup Poll finds that 57% of American adults bought a lottery ticket in the last 12 months. The margin of error for the poll was 3 percentage points." Of course, 57% plus or minus 3 percent is 54% to 60%. Most people think that Gallup claims that the truth about the entire population lies in that range.

This is what Gallup actually said: "For results based on a sample of this size, one can say with 95% confidence that the error attributable to sampling and other random effects could be plus or minus 3 percentage points for adults." That is, Gallup tells us that the margin of error works for only 95% of all its samples. And "95% confidence" is shorthand for that. The news report left out the "95% confidence."

Finding the margin of error exactly is a job for statisticians. You can, however, use a simple formula to get a rough idea of the size of a sample survey's margin of error.

A quick method for the margin of error

> If you use the sample proportion $\hat{p}$ from a simple random sample of size n to estimate an unknown population proportion p, then the **margin of error** for 95% confidence is roughly equal to $1/\sqrt{n}$.

Example 5.10 What is the margin of error?

The Gallup Poll in Example 5.9 interviewed 1523 people. The margin of error for 95% confidence will be about

$$\frac{1}{\sqrt{1523}} = \frac{1}{39.03} = 0.026 \qquad \text{(that is, 2.6\%)}$$

Gallup actually announced a margin of error of 3%. Our result differs a bit from Gallup's for two reasons. First, polls usually round their announced margin of error to the nearest whole percent to keep their press releases simple. Second, our rough formula works for an SRS. We will see in the next section that most national samples are more complicated than an SRS in ways that tend to slightly increase the margin of error. Still, our quick method comes pretty close.

"A new poll shows 67% of Americans think polls are inaccurate 52% of the time, and 84% think the margin of error of 3% is inaccurate 71% of the time . . ."

Example 5.11 Estimating the margin of error

In Example 5.8 we compared the results of taking many SRSs of size $n = 100$ and many SRSs of size $n = 1523$ from the same population. We found that the spread of the middle 95% of the sample results was about four times as large for the smaller samples.

Our quick formula estimates the margin of error for SRSs of size 1523 to be about 2.6%. The margin of error for SRSs of size 100 is about

$$\frac{1}{\sqrt{100}} = \frac{1}{10} = 0.1 \qquad \text{(that is, 10\%)}$$

Because 1523 is roughly 16 times 100 and the square root of 16 is 4, the margin of error is about four times larger for samples of 100 people than for samples of 1523 people.

Our quick method also reveals an important fact about how margins of error behave. Because the sample size n appears in the denominator of the fraction, larger samples have smaller margins of error. We knew that. Because the formula uses the square root of the sample size, however, *to cut the margin of error in half, we must use a sample four times as large.*

Confidence statements

Here is Gallup's conclusion about buying lottery tickets in short form: "The poll found that 57% of adults bought a lottery ticket in the past 12 months. We are 95% confident that the truth about all adults is within plus or minus 3 percentage points of this sample result." Here is an even shorter form: "We are 95% confident that between 54% and 60% of all adults bought a lottery ticket in the last 12 months." These are **confidence statements.**

Confidence statements

A **confidence statement** has two parts: a **margin of error** and a **level of confidence.** The margin of error says how close the sample statistic lies to the population parameter. The level of confidence says what percent of all possible samples satisfy the margin of error.

A confidence statement is a fact about what happens in all possible samples; it is used to state how much we can trust the result of one sample. The phrase "95% confidence" means "We used a sampling method that gives a result this close to the truth 95% of the time." Here are some hints for interpreting confidence statements:

- *The conclusion of a confidence statement always applies to the population, not to the sample.* We know exactly how the 1523 people in the sample acted, because Gallup interviewed them. The confidence statement uses the sample result to say something about the population of all adults.

- *Our conclusion about the population is never completely certain.* Gallup's sample might be one of the 5% that miss by more than 3 percentage points.
- *A sample survey can choose to use a confidence level other than 95%.* We pay for higher confidence with a larger margin of error. For the same sample, a 99% confidence level requires a larger margin of error than 95% confidence. If you are content with 90% confidence, you get a smaller margin of error in return. Remember that our quick method gives the margin of error only for 95% confidence.
- *It is usual to report the margin of error for 95% confidence.* If a news report gives a margin of error but leaves out the confidence level, it's pretty safe to assume 95% confidence.
- *Want a smaller margin of error with the same confidence? Take a larger sample.* Remember that larger samples yield less variable sample statistics than smaller samples. You can get as small a margin of error as you want and still have high confidence by paying for a large enough sample.

Example 5.12 Do you approve of gambling?

Gallup began its poll on gambling with this question: "First, generally speaking, do you approve or disapprove of legal gambling or betting?" In addition to the 1523 adults (aged 18 and older) whose lottery habits we have explored, Gallup took a random sample of 501 teenagers (aged 13 to 17). The sample results were

Adults:	959 out of 1523 approve	$\hat{p} = 959/1523 = 0.63$
Teens:	261 out of 501 approve	$\hat{p} = 261/501 = 0.52$

After reporting these and other results, Gallup says: "For results based on a sample of this size, one can say with 95 percent confidence that the error attributable to sampling and other random effects could be plus or minus 3 percentage points for adults (18+), and plus or minus 5 percentage points for teens (13–17 year olds)."

There you have it: the sample of teens is smaller, so the margin of error for conclusions about teens is wider. We are 95% confident that between 47% (that's 52% minus 5%) and 57% (that's 52% plus 5%) of all teenagers approve of legal gambling.

Exercises

5.39 Take a bigger sample A student is planning a project on student attitudes toward part-time work while attending school. She develops a questionnaire and plans to ask 25 randomly selected students to fill it out. Her statistics teacher approves the questionnaire but suggests that the sample size be increased to at least 100 students. Why is the larger sample helpful? Back up your answer by using the quick method to estimate the margin of error for samples of size 25 and for samples of size 100.

5.40 Find the margin of error Example 5.12 tells us that Gallup asked 501 teenagers whether they approved of legal gambling; 52% said they did. Use the quick method to estimate the margin of error for conclusions about all teenagers. How does your result compare with Gallup's margin of error quoted in Example 5.12?

5.41 The Current Population Survey Though opinion polls usually make 95% confidence statements, some sample surveys use other confidence levels. The monthly unemployment rate, for example, is based on the Current Population Survey of about 50,000 households. The margin of error in the unemployment rate is announced as about two-tenths of 1 percentage point with 90% confidence. Would the margin of error for 95% confidence be smaller or larger? Why?

5.42 Polling women Many years ago, a *New York Times* Poll on women's issues interviewed 1025 women randomly selected from the United States, excluding Alaska and Hawaii. One question was "Many women have better jobs and more opportunities than they did 20 years ago. Do you think women had to give up too much in the process, or not?" Forty-eight percent of women in the sample said "Yes."[10]

(a) The poll announced a margin of error of ±3 percentage points for 95% confidence in its conclusions. Make a 95% confidence statement about the percent of all adult women who felt that women had to give up too much.

(b) Explain to someone who knows no statistics why we can't just say that 48% of all adult women felt that women had to give up too much.

(c) Explain clearly what "95% confidence" means.

5.43 Polling men and women The sample survey described in Exercise 5.42 interviewed 472 randomly selected men as well as 1025 women. The poll announced a margin of error of ±3 percentage points for 95% confidence in conclusions about women. The margin of error for results concerning men was ±5 percentage points. Why is this larger than the margin of error for women?

5.44 Is there a heaven? A news article reports that in a recent opinion poll, 81% of a sample of 1003 adults said they believe there is a heaven.

(a) Use the quick method to estimate the margin of error for a sample of this size.

(b) Make a confidence statement about the percent of all adults who believe there is a heaven.

Sampling from large populations

Gallup's sample of 1523 adults is only 1 out of every 154,000 adults in the United States. Does it matter whether 1523 is 1-in-100 individuals in the population or 1-in-154,000?

Population size doesn't matter

The variability of a statistic from a random sample does not depend on the size of the population, as long as the population is at least 10 times larger than the sample.

Why does the size of the population have little influence on the behavior of statistics from random samples? Imagine sampling harvested corn by thrusting a scoop into a large sack of corn kernels. The scoop doesn't know whether it is surrounded by a bag of corn or by an entire truckload. As long as the corn is well mixed (so that the scoop selects a random sample), the variability of the result depends only on the size of the scoop.

You can read the Gallup Organization's own explanation of why surprisingly small samples can give trustworthy results about large populations at **media.gallup.com/PDF/FAQ/HowArePolls.pdf.**

This is good news for national sample surveys like the Gallup Poll. A random sample of size 1000 or 1500 has small variability because the sample size is large. But remember that even a very large voluntary response sample or convenience sample is worthless because of bias. Taking a larger sample doesn't fix bias.

However, the fact that the variability of a sample statistic depends on the size of the sample and not on the size of the population is bad news for anyone planning a sample survey in a university or a small city. For example, it takes just as large an SRS to estimate the proportion of Ohio State University students who blog regularly as to estimate with the same margin of error the proportion of all U.S. adults who blog regularly. We can't use a smaller SRS at Ohio State just because there are 50,000 Ohio State students and 235 million adults in the United States.

APPLICATION 5.2

Doping questions cloud Americans' view of the Olympics

© Paul J. Sutton/PCN/PCN/Corbis

Positive drug tests or admissions of doping in recent years have cost Olympic athletes their records, their medals, their careers, even their freedom. Just over half of American sports fans believe that at least some Olympic athletes in track and field use performance-enhancing drugs. A *USA Today*/Gallup Poll conducted just prior to the Beijing Olympics (July 31, 2008) also shows that when a track-and-field athlete sets a world record, more than 1 in 3 sports fans are suspicious that doping helped. More than 1 in 5 fans say they are suspicious of doping when a swimmer sets a world record, and 1 in 3 believe that at least some Olympians in swimming use performance-enhancing drugs, the poll indicates.

The results reflect how scandals involving performance-enhancing drugs—notably in Major League Baseball, track and field, and professional cycling—have created a credibility gap in the public's mind.

QUESTIONS

1. The poll described in the article sampled 626 sports fans. Why do you think the pollsters asked sports fans rather than people in general?

2. Here is the actual question that the poll asked: "When you see or hear about an athlete breaking a world record in swimming or track and field, are you suspicious or not suspicious that the athlete used performance-enhancing drugs?" Does the question seem clear? Is the question slanted in a particular direction, or does it seem balanced?

3. The pie charts on the next page summarize the poll results. A bar graph might be more appropriate for comparing people's opinions for these two sports. Construct a well-labeled bar graph that could be used for easy comparison.

APPLICATION 5.2 *(continued)*

APPLICATION 5.2 *(continued)*

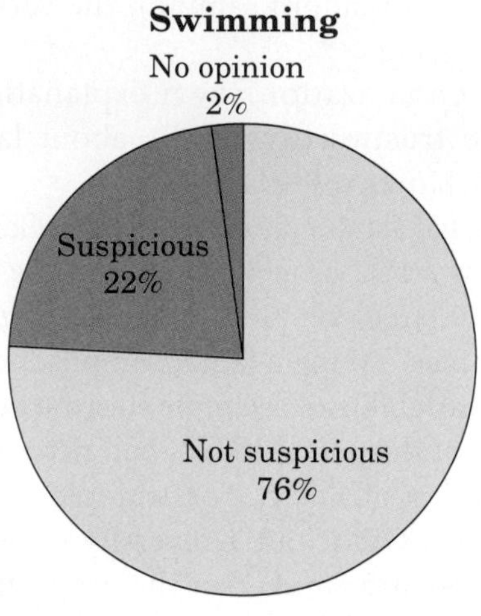

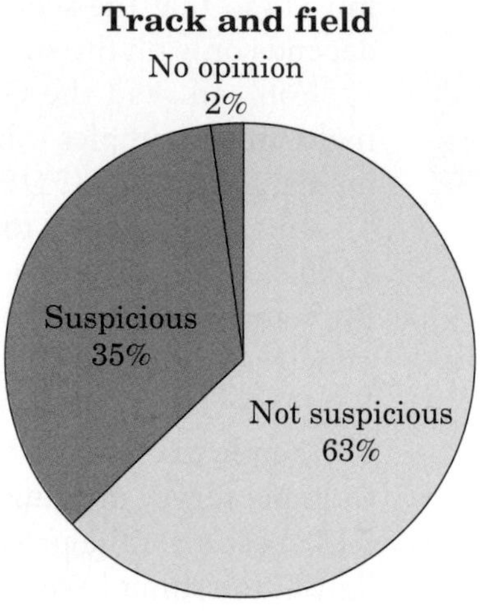

Along with the poll results, *USA Today* announced a margin of error of ± 4 percentage points.

4. Use the quick estimate for the margin of error to confirm this value.

5. Explain to someone who knows no statistics what the margin of error means in this case.

Section 5.2 Summary

The purpose of sampling is to use a sample to gain information about a population. We often use a sample **statistic** to estimate the value of a population **parameter.** This section has one big idea: to describe how trustworthy a sample is, ask "What would happen if we took a large number of samples from the same population?" If almost all samples would give a result close to the truth, we can trust our one sample even though we can't be certain that it is close to the truth.

In planning a sample survey, first aim for small **bias** by using random sampling and avoiding bad sampling methods such as voluntary response. Next, choose a large enough random sample to reduce the **variability** of the result. Using a large random sample guarantees that almost all samples will give accurate results. To say how accurate our conclusions about the population are, make a **confidence statement.** News reports often mention only the **margin of error.** Most often this margin of error is for **95% confidence.** That is, if we chose many samples, the truth about the population would be within the margin of error 95% of the time. We can estimate the margin of error for 95% confidence based on a simple random sample of size n by the formula $1/\sqrt{n}$. As this formula suggests, only the size of the sample, not the size of the population, matters. This is true as long as the population is much larger than the sample.

Section 5.2 Exercises

5.45 Strike three A survey is conducted in Chicago (population 2,800,000) using random digit dialing equipment that places calls at random to residential phones, both listed and unlisted. The purpose of the survey is to determine the percent of Chicagoans who support the "three strikes and you're out" provision of the crime bill passed by Congress. One-tenth of 1% of the adult population is interviewed, and 53%

of them favor the proposal. A second survey is taken in Dayton, Ohio (population 156,000), using the same techniques and asking the same question of one-tenth of 1% of the adults living in Dayton. Which survey is more likely to produce a sample result that is closer to the truth about the population in that city? Explain briefly.

5.46 Presidential polls California has about four times as many voters as Ohio, but both are very important states for the outcome of any presidential election. One national polling organization plans to take surveys of 1000 randomly selected voters in each of these two states on the day before the election. In which state is the survey more likely to accurately predict the state's election results? Explain.

*Each boldface number in Exercises 5.47 and 5.48 is the value of either a **parameter** or a **statistic.** In each case, state which it is.*

www

5.47 Lefties According to "Real Facts" at **Snapple.com**, **13%** of adults are left-handed. At a math teacher's conference, **16%** of those attending were left-handed.

5.48 Single-sex classes In an experiment to test the effectiveness of single-sex classrooms, girls assigned at random to a coeducational chemistry class gained an average of **12.2** points from a pretest to a posttest. Girls assigned randomly to a single-sex chemistry class taught by the same teacher gained **15.1** points.

5.49 Find the margin of error Exercise 5.38 (page 228) describes a sample survey of 61,239 adults living in Ontario. About what is the margin of error for conclusions having 95% confidence about the entire adult population of Ontario? Show your work.

5.50 Explaining confidence A student reads that we are 95% confident that the average score of young men on the quantitative part of the National Assessment of Educational Progress is 267.8 to 276.2. Asked to explain the meaning of this statement, the student says, "95% of all young men have scores between 267.8 and 276.2." Is the student right? Explain your answer.

5.51 Teens on the Net A poll of 1070 teens aged 13 to 17 finds that 742 have received personal messages online from people they don't know. The announced margin of error for this result is plus or minus 3 percentage points. The news report does not give the confidence level, but you can be quite sure that it's 95%.

(a) What is the value of the sample proportion $\hat{p}$ who have received personal messages online from people they don't know? Explain in words what the population parameter p is in this setting.

(b) Make a confidence statement about the parameter p.

5.52 Protecting teens The previous exercise describes a sample survey of 1070 teens, with margin of error $\pm 3\%$ for 95% confidence.

(a) An agency that is responsible for protecting children thinks that 95% confidence is not enough. They want to be 99% confident. How would the margin of error for 99% confidence based on the same sample compare with the margin of error for 95% confidence?

(b) A parents group is satisfied with 95% confidence but wants a smaller margin of error than $\pm 3\%$. How can we get a smaller margin of error, still with 95% confidence?

5.53 Video gaming In a 2006 poll, 40% of American adults surveyed said that they play computer or video games. Of those who played such games, 10% reported playing for 10 or more hours per week. According to the Associated Press, "The results are taken from a poll of 3,024 adults, including 1,046 who play electronic games, that was conducted April 18–27. The poll, conducted by the international polling firm Ipsos, has a margin of sampling error of plus or minus 2 percentage points, 3 points for the sample of those who play electronic games."[11]

(a) Use the quick method to show that the margins of error quoted above are approximately correct if a simple random sample was taken.

(b) The confidence level wasn't stated, but it's almost certainly 95%. Make an appropriate confidence statement about video gaming in the population of American adults.

(c) Identify two possible sources of error in this survey that the margin of error does not include.

(d) If this survey were repeated today, do you think the results would be about the same? Do some research to find whether your suspicion is correct.

5.54 Simulating summer employment Random digits can be used to simulate the results of random sampling. Suppose that you are drawing simple random samples of size 25 from a large number of college students and that 20% of the students are unemployed during the summer. To simulate this SRS, let 25 consecutive digits in Table B stand for the 25 students in your sample. The digits 0 and 1 stand for unemployed students, and other digits stand for employed students. This is an accurate imitation of the SRS because 0 and 1 make up 20% of the 10 equally likely digits.

Simulate the results of 50 samples by counting the number of 0s and 1s in the first 25 entries in each of the 50 rows of Table B. Make a histogram like the one in Figure 5.5 to display the results of your 50 samples. Is the truth about the population (20% unemployed, or 5 in a sample of 25) near the center of your graph? What are the smallest and largest counts of unemployed students you obtained in your 50 samples? What percent of your samples had either 4, 5, or 6 unemployed?

5.3 Sample Surveys in the Real World

ACTIVITY 5.3

Design your own survey

In this Activity, you will work with two or three of your classmates to design a survey. By the end of this section, you will be ready to revise your survey and to administer it at your school.

1. On your own, brainstorm a list of current topics of interest to your school community.

2. Pool ideas with your team members. Once you have agreed on a topic of interest, write three to five questions that address the topic. Try to include one question that measures a quantitative (numerical) variable.

3. Spend a few minutes editing your questions to make them as clear as possible. Have one member of the group write the revised list of questions neatly on a clean sheet of paper.

4. Exchange survey questions with another team, as instructed by your teacher. Discuss the content and clarity of the questions. Note any changes you would recommend on their question list.

5. Review any suggestions made regarding your own list. Finalize your questions.

6. Ideally, you would select a simple random sample of students from your school to complete the survey. Give two reasons why this might not be practical.

7. Make a list of other issues your team needs to address before you can administer the survey.

Keep all of your work handy for later use.

The whole truth about opinion polls

An opinion poll is administered to 1000 people chosen at random, announces its results, and announces a margin of error. Should we be happy? Maybe not. Many polls don't tell the whole truth about their samples. The Pew Research Center for the People and the Press imitated the methods of the better opinion polls and did tell the whole truth. Here it is.

Most polls are taken by telephone, dialing numbers at random to get a random sample of households. After eliminating fax and business numbers, Pew had to call 2879 residential numbers to get their sample of 1000 people. Here's the breakdown:

Never answered phone	938
Answered but refused	678
Not eligible: no person aged at least 18, or language barrier	221
Incomplete interview	42
Complete interview	1000
Total called	**2879**

Out of 2879 good residential phone numbers, 33% never answered. Of those who answered, 35% refused to talk. The overall rate of nonresponse (people who never answered, refused, or would not complete the interview) was 1658 out of 2879, or 58%. Pew called every number five times over a five-day period, at different times of day and on different days of the week. Many polls call only once, and it is usual to find that over half of those who answer refuse to talk. In the real world, a simple random sample isn't simple and may not be random. That's our issue for this section.[12]

How sample surveys go wrong

Random sampling eliminates bias in choosing a sample and allows control of variability. So once we see the magic words "randomly selected" and "margin of error," do we know we have trustworthy information before us? It certainly beats voluntary response, but not always by as much as we might hope. Sampling in the real world is more complex and less reliable than choosing an SRS from a list of names in a textbook exercise. Confidence statements do not reflect all of the sources of error that are present in practical sampling.

Errors in sampling

Sampling errors are errors caused by the act of taking a sample. They cause sample results to be different from the results of a census.

Random sampling error is the deviation between the sample statistic and the population parameter caused by chance in selecting a random sample. The margin of error in a confidence statement includes only random sampling error.

Nonsampling errors are errors not related to the act of selecting a sample from the population. They can be present even in a census.

Most sample surveys are affected by errors other than random sampling errors. These errors can introduce bias that makes a confidence statement meaningless. Good sampling technique includes the art of reducing all sources of error. Part of this art is the science of statistics, with its random samples and confidence statements. In practice, however, good statistics isn't all there is to good sampling. Let's look at sources of errors in sample surveys and at how samplers combat them.

Sampling errors

Random sampling error is one kind of sampling error. The margin of error tells us how serious random sampling error is, and we can control it by choosing the size of our random sample. Another source of sampling error is the use of *bad sampling methods,* such as voluntary response. We can avoid bad methods. Other sampling errors are not so easy to handle. Sampling begins with a list of individuals from which we will draw our sample. This list is called the **sampling frame.** Ideally, the sampling frame should list every individual in the population. Because a list of the entire population is rarely available, most samples suffer from some degree of **undercoverage.**

Undercoverage

Undercoverage occurs when some groups in the population are left out of the process of choosing the sample.

If the sampling frame leaves out certain classes of people, even random samples from that frame will be biased. Using telephone directories as the frame for a telephone survey, for example, would miss everyone with an unlisted telephone number. More than half the households in many large cities have unlisted numbers, so massive undercoverage and bias against urban areas would result. In fact, telephone surveys usually use random digit dialing equipment that dials landline telephone numbers in selected regions at random. In effect, the sampling frame contains all residential telephone numbers. Of course, this sampling frame will exclude all individuals who have only a cell phone.

Example 5.13 We undercover

Most opinion polls can't afford to even attempt full coverage of the population of all adult residents of the United States. The interviews are done by telephone, thus missing the 6% of households without phones. Only households are contacted, so that students in dormitories, prison inmates, and most members of the armed forces are left out. So are the homeless and people staying in shelters. Because calls to Alaska and Hawaii are expensive, most polls restrict their samples to the continental states. Many polls interview only in English, which leaves some immigrant households out of their samples. And most polls use sampling frames that consist only of residential phone numbers, which omits people who use cell phones exclusively.

The kinds of undercoverage found in most sample surveys are most likely to leave out people who are young or poor or who move often. Nonetheless, random digit dialing comes close to producing a random sample of households with phones outside Alaska and Hawaii. Sampling errors in careful sample surveys are usually quite small. The real problems start when someone picks up (or doesn't pick up) the phone. Now **nonsampling errors** take over.

Nonsampling errors

Nonsampling errors are those that can plague even a census. They include **processing errors,** which are mistakes in mechanical tasks such as doing arithmetic or entering responses into a computer. The spread of computer-assisted interviewing has made processing errors less common than in the past.[13]

Example 5.14 Computer-assisted interviewing

The days of the interviewer with a clipboard are past. Contemporary interviewers carry a laptop computer for face-to-face interviews or watch a computer screen as they conduct a telephone interview. Computer software manages the interview. The interviewer reads questions from the computer screen and uses the keyboard to enter the responses. The computer skips irrelevant items—once a respondent says that she has no children, further questions about her children never appear. The computer can check that answers to related questions are consistent with each other. It can even present questions in random order to avoid any bias due to always asking questions in the same order.

Computer software also manages the sampling process. It keeps records of who has responded and prepares a file of data from the responses. The tedious process of transferring responses from paper to computer, once a source of processing errors, has been eliminated. The computer even schedules the calls in telephone surveys, taking into account the respondent's time zone and honoring appointments made by people who were willing to respond but did not have time when first called.

Another type of nonsampling error is **response error,** which occurs when someone gives an incorrect response. A person may lie about her age or income or about whether she has used illegal drugs. She may remember incorrectly when asked how many packs of cigarettes she smoked last week. Someone who does not understand a question may guess at an answer rather than appear ignorant. Questions that ask people about their behavior during a fixed time period are prone to response errors due to faulty memory.[14] For example, the National Health Survey asks people how many times they have visited a doctor in the past year. Checking their responses against health records found that they failed to remember 60% of their visits to a doctor. A survey that asks about sensitive issues can also expect response errors, as the next example illustrates.

Example 5.15 The effect of race

In 1989, New York City elected its first black mayor and the state of Virginia elected its first black governor. In both cases, samples of voters interviewed as they left their polling places predicted larger margins of victory than the official vote counts. The polling organizations were certain that some voters lied when interviewed because they felt uncomfortable admitting that they had voted against the black candidate.

History may have repeated itself in 2008. Preelection polls before the New Hampshire primary projected Barack Obama as the winner over Hillary Clinton. In fact, these polls suggested that Obama would win by about 8 percentage points—an amount much larger than the margin of error for any of the polls. Clinton won the New Hampshire primary, receiving 39% of the vote to Obama's 36%.

Technology and attention to detail can minimize processing errors. Skilled interviewers greatly reduce response errors, especially in face-to-face interviews. There is no simple cure, however, for the most serious kind of nonsampling error, **nonresponse.**

Nonresponse

Nonresponse is the failure to obtain data from an individual selected for a sample. Most nonresponse happens because some people can't be contacted or because some individuals who are contacted refuse to cooperate.

"You can call, you can send email, you can stand at the door all day. The answer is still NO!"

Nonresponse is the most serious problem facing sample surveys. People are increasingly reluctant to answer questions, particularly over the phone. The rise of telemarketing, answering machines, cell phones, and caller ID drives down response to telephone surveys. Gated communities and buildings guarded by doormen prevent face-to-face interviews. Nonresponse can bias sample survey results because different groups have different rates of nonresponse. Refusals are higher in large cities and among the elderly, for example. Bias due to nonresponse can easily overwhelm the random sampling error described by a survey's margin of error.

Example 5.16 How bad is nonresponse?

The Current Population Survey has the best response rate of any poll we know: only about 6% or 7% of the households in the CPS sample don't respond. People are more likely to respond to a government survey such as the CPS, and the CPS contacts its sample in person before doing later interviews by phone.

New York, New York

New York City, they say, is bigger, richer, faster, and sometimes ruder. Maybe there's something to that. The sample survey firm Zogby International says that as a national average it takes 5 telephone calls to reach a live person. When calling to New York, it takes 12 calls. Survey firms assign their best interviewers to make calls to New York and often pay them bonuses to cope with the stress.

The General Social Survey (Example 1.8, page 11) also contacts its sample in person, and it is run by a university. Despite these advantages, its recent surveys have a 24% rate of nonresponse.

What about polls done by the media and by market research and opinion-polling firms? We don't know their rates of nonresponse, because they won't say. That's a bad sign! The Pew study we looked at suggests how bad things are. Pew got 1221 responses (of whom 1000 were in the population they targeted) and 1658 who were never at home, refused, or would not finish the interview. That's a nonresponse rate of 1658 out of 2879, or 58%. The Pew researchers were more thorough than many pollsters. Insiders say that nonresponse often reaches 75% or 80% of an opinion poll's original sample.[15]

Sample surveyors know some tricks to reduce nonresponse. Carefully trained interviewers can keep people on the line if they answer at all. Calling back after longer time periods helps. So do letters sent in advance. Letters and many callbacks slow down the survey, so opinion polls that want fast answers to satisfy the media don't use them. Even the most careful surveys find that nonresponse is a problem that no amount of expertise can fully overcome. That makes this reminder even more important.

What the margin of error doesn't say

> The announced margin of error for a sample survey covers only random sampling error. Undercoverage, nonresponse, and other practical difficulties can cause large bias that is not covered by the margin of error.

Careful sample surveys tell us this. Gallup, for example, says, "In addition to sampling error, question wording and practical difficulties in conducting surveys can introduce error or bias into the findings of public opinion polls." How true that is.

Exercises

5.55 Not in the margin of error A recent Gallup Poll found that 68% of adult Americans favor teaching creationism along with evolution in public schools. The Gallup press release says:

> *For results based on samples of this size, one can say with 95 percent confidence that the maximum error attributable to sampling and other random effects is plus or minus 3 percentage points.*[16]

Give one example of a source of error in the poll result that is not included in this margin of error.

5.56 What kind of error? Which of the following are sources of *sampling error* and which are sources of *nonsampling error?* Explain your answers.

(a) The subject lies about past drug use.

(b) A typing error is made in recording the data.

(c) Data are gathered by asking people to mail in a coupon printed in a newspaper.

5.57 Internet users then A survey of Internet users in 1995 found that males outnumbered females by nearly 2 to 1. This was a surprise, because earlier surveys had put the ratio of men to women closer to 9 to 1. Later in the article we find this information:

> *Detailed surveys were sent to more than 13,000 organizations on the Internet; 1,468 usable responses were received. According to Mr. Quarterman, the margin of error is 2.8 percent, with a confidence level of 95 percent.*[17]

(a) What was the response rate for this survey? (The response rate is the percent of the planned sample that responded.)

(b) Use the quick method (page 229) to estimate the margin of error of this survey. Is your result close to the 2.8% claimed?

(c) Is the small margin of error a good measure of the accuracy of the survey's results? Explain your answer.

5.58 Polling students A high school chooses an SRS of 100 students from the school's attendance list to interview about student life. If it selected two SRSs of 100 students at the same time, the two samples would give somewhat different results. Is this variation a source of sampling error or of nonsampling error? Will the survey's announced margin of error take this source of error into account?

5.59 Internet users now Who uses the Internet more today—males or females? Find a report of a recent survey that you believe provides accurate information on this question. Print the article if possible. Be sure to record all source information. Write a brief report following the four-step statistical problem-solving process (page 19).

5.60 Activity 5.3 follow-up Return to the survey your team constructed in Activity 5.3 (page 236). Discuss how your design protects against

(a) sampling errors.

(b) nonsampling errors.

Wording questions

A final influence on the results of a sample survey is the exact **wording of questions.** It is surprisingly difficult to word questions that are completely clear. A survey that asked about "ownership of stock" found that most Texas ranchers owned stock, though probably not the kind traded on the New York Stock Exchange.

The telemarketer's pause

People who do sample surveys hate telemarketing. We get so many unwanted sales pitches by phone that many people hang up before learning that the caller is conducting a survey rather than selling vinyl siding. Here's a tip. Both sample surveys and telemarketers dial telephone numbers at random. Telemarketers automatically dial many numbers, and their sellers come on the line only after you pick up the phone. Once you know this, the telltale "telemarketer's pause" gives you a chance to hang up before the seller arrives. Sample surveys have a live interviewer on the line when you answer.

"Do I own any stock, Ma'am? Why, I've got 10,000 head out there."

Example 5.17 A few words make a big difference

Question: How do Americans feel about government help for the poor? Only 13% think we are spending too much on "assistance to the poor," but 44% think we are spending too much on "welfare."

Question: How do the Scots feel about the movement to become independent from England? Well, 51% would vote for "independence for Scotland," but only 34% support "an independent Scotland separate from the United Kingdom."[18]

It seems that "assistance to the poor" and "independence" are nice, hopeful words. "Welfare" and "separate" are negative words. Small changes in how a question is worded can make a big difference in the response.

The wording of questions always influences the answers. If the questions are slanted to favor one response over others, we have another source of nonsampling error. A favorite trick is to ask if the subject favors some policy as a means to a desirable end: "Do you favor banning private ownership of handguns in order to reduce the rate of violent crime?" and "Do you favor imposing the death penalty in order to reduce the rate of violent crime?" are loaded questions that draw positive responses from people who are worried about crime.

Sample design in the real world

The basic idea of sampling is straightforward: take an SRS from the population and use a statistic from your sample to estimate a population parameter. Unfortunately, it's usually very difficult to actually get an SRS from the population of interest. Imagine trying to get a simple random sample of all the batteries produced in one day at a factory. Or an SRS of all U.S. high school students.

In either case, it's just not practical to choose an SRS. In the real world, most sample surveys use more complex designs.

Example 5.18 The Current Population Survey (CPS)

The CPS population consists of all households in the United States (including Alaska and Hawaii). The sample is *chosen in stages*. The Census Bureau divides the nation into 2007 geographic areas called Primary Sampling Units (PSUs). These are generally groups of neighboring counties. At the first stage, 792 PSUs are chosen. This isn't an SRS. If all PSUs had the same chance to be chosen, the sample might miss Chicago and Los Angeles. So 432 highly populated PSUs are automatically in the sample. The other 1575 are grouped into 360 **strata** by combining PSUs that are similar in various ways. One PSU is chosen at random to represent each stratum. From there, groups of nearby households in each PSU are randomly selected for the CPS.[19]

The design of the CPS illustrates several ideas that are common in real-world samples that use face-to-face interviews. Taking the sample in several stages with nearby households at the final stage saves travel time for interviewers. The most important refinement mentioned in Example 5.18 is **stratified random sampling.**

Stratified random sample

To choose a **stratified random sample:**

Step 1: Divide the sampling frame into distinct groups of individuals, called **strata.** Choose the strata because you have a special interest in these groups within the population or because the individuals in each stratum resemble each other.

Step 2: Take a separate SRS in each stratum and combine these to make up the complete sample.

We must of course choose the strata using facts about the population that are known before we take the sample. You might group a high school's students by grade level (9, 10, 11, 12) or by gender. Stratified samples have some advantages over an SRS. First, by taking a separate SRS in each stratum, we can set sample sizes to allow separate conclusions about each stratum. Second, a stratified sample usually has a smaller margin of error than an SRS of the same size. The reason is that the individuals in each stratum are more alike than the population as a whole, so working stratum by stratum eliminates some variability in the sample.

It may surprise you that stratified samples can violate one of the most appealing properties of the SRS—stratified samples need not give all individuals in the population the same chance to be chosen. Some strata may be deliberately overrepresented in the sample.

Example 5.19 Stratifying a sample of students

A large university has 30,000 students, of whom 3000 are graduate students. An SRS of 500 students gives every student the same chance to be in the sample. That chance is

$$\frac{500}{30{,}000} = \frac{1}{60}$$

We expect an SRS to contain only about 50 grad students—because grad students make up 10% of the population, we expect them to make up about 10% of an SRS. A sample of size 50, however, isn't large enough to estimate grad student opinion with reasonable accuracy. Therefore, we prefer to take a stratified random sample of 200 grad students and 300 undergraduates.

You know how to select such a stratified sample. Label the graduate students 0001 to 3000 and use Table B to select an SRS of 200. Then label the undergraduates 00001 to 27000 and use Table B a second time to select an SRS of 300 of them. These two SRSs together form the stratified sample.

In the stratified sample, each grad student has chance

$$\frac{200}{3000} = \frac{1}{15}$$

to be chosen. Each of the undergraduates has a smaller chance,

$$\frac{300}{27{,}000} = \frac{1}{90}$$

It's clear from the last few examples that designing samples is a business for experts. We won't worry about such details. The big idea is that good sample designs use chance to select individuals from the population. That is, all good samples are **probability samples.**

Probability sample

A **probability sample** is a sample chosen by chance. We must know what samples are possible and what chance, or probability, each possible sample has. Some probability samples, such as stratified samples, don't allow all possible samples from the population and may not give an equal chance to all the samples they do allow.

A stratified sample of 300 undergraduate students and 200 grad students, for example, only allows samples with exactly that makeup. An SRS would allow any 500 students. Both are probability samples. We need know only that estimates from any probability sample share the nice properties of estimates from an SRS. Confidence statements can be made without bias and have smaller margins of error as the size of the sample increases. Nonprobability samples such as voluntary response samples do not share these advantages and cannot give trustworthy information about a population. Now that we know that most nationwide samples are more complicated than an SRS, we will usually go back to acting as if good samples were SRSs. That keeps the big idea and hides the messy details.

Exercises

5.61 A sampling paradox? Example 5.19 compares two SRSs—of a university's undergraduate and graduate students. The sample of undergraduates contains a smaller fraction of the population, 1 out of 90, versus 1 out of 15 for graduate students. Yet sampling 1 out of 90 undergraduates gives a smaller margin of error than sampling 1 out of 15 graduate students. Explain to someone who knows no statistics why this happens.

5.62 Scholar-athletes To ask about their future plans, you want to interview 10 students at your high school who have received athletic scholarships to attend college. Because you believe there may be large differences among athletes in different sports, you decide to interview a stratified random sample of 7 basketball players and 3 golfers. Use Table B, beginning at line 101, to select your sample from the team rosters below. Explain your method carefully enough that a classmate could obtain your results.

BASKETBALL

Arenas	Duncan	Leslie	Robinson
Billups	Farmar	McGrady	Stoudamire
Brand	Fowles	Miller	Taurasi
Bryant	Gasol	Nowitzki	Wade
Carter	Iverson	Parker	

GOLF

Creamer	Kim	Singh
Els	Mickelson	Wie
Gulbis	Prammanasudh	Woods

5.63 Genetically modified foods An article in the journal *Science* looks at differences in attitudes toward genetically modified foods in Europe and the United States. This calls for sample surveys. The European survey chose a sample of 1000 adults in each of 17 European countries. Here's part of the description: "The Eurobarometer survey is a multistage, random-probability face-to-face sample survey."[20]

(a) What does "multistage" mean?

(b) You can see that the first stage was stratified. What were the strata?

(c) What does "random-probability sample" mean?

5.64 Wording of questions A *New York Times*/CBS News Poll asked a random sample of Americans about abortion: "Do you think there should be an amendment to the Constitution prohibiting abortions, or shouldn't there be such an amendment?" The same people were later asked, "Do you believe there should be an amendment to the Constitution protecting the life of the unborn child, or shouldn't there be such an amendment?" For one of the questions, 50% were in favor and 39%

were opposed. For the other, 29% were in favor and 62% were opposed. (The rest were uncertain.)[21] Which question do you think yielded each result? Explain why.

5.65 Closed versus open questions Two basic types of questions are closed questions and open questions. A closed question asks the subject for one or more of a fixed set of responses. An open question allows the subject to answer in his or her own words. The interviewer writes down the responses and sorts them later. An example of an open question is

How do you feel about broccoli?

An example of a closed question is

What is your opinion about broccoli? Do you

a. like it very much? *b. like it somewhat?* *c. neither like nor dislike it?*
d. dislike it somewhat? *e. dislike it very much?*

What are the advantages and disadvantages of open and closed questions?

5.66 Bad survey questions Write your own examples of bad sample survey questions.

(a) Write a biased question designed to get one answer rather than another.

(b) Write a question that is confusing, so that it is hard to answer.

5.67 Systematic random sample The final stage in a multistage sample must choose 5 of the 500 addresses in a neighborhood. You have a list of the 500 addresses in geographical order. To choose a systematic random sample, proceed as follows:

> **Step 1.** Choose 1 of the first 100 addresses on the list at random. (Label them 00, 01, . . . , 99 and use a pair of digits from Table B to make the choice.)
>
> **Step 2.** The sample consists of the address from Step 1 and the addresses 100, 200, 300, and 400 positions down the list from it.

If 71 is chosen at random in Step 1, for example, the systematic random sample consists of the addresses numbered 71, 171, 271, 371, and 471.

(a) Use Table B to choose a systematic random sample of 5 from a list of 500 addresses. Enter the table at line 130.

(b) What is the chance that any specific address will be chosen? Explain your answer.

(c) Explain why this sample is *not* an SRS.

5.68 A stratified sample A university has 2000 male and 500 female faculty members. The university president wants to poll the opinions of a random sample of faculty members. In order to give adequate attention to female faculty opinion, the president decides to choose a stratified random sample of 200 males and 200 females. The president obtains alphabetized lists of female and male faculty members.

(a) Explain how you would assign labels and use random digits to choose the desired sample. Enter Table B at line 122 and give the first 5 females and the first 5 males in your sample.

(b) What is the chance that any one of the 2000 males will be in your sample? What is the chance that any one of the 500 females will be in your sample?

(c) Each member of the sample is asked, "In your opinion, are female faculty members in general paid less than males with similar positions and qualifications?"

180 of the 200 females (90%) say "Yes."
60 of the 200 males (30%) say "Yes."

In all, 240 of the sample of 400 (60%) answered "Yes." The president therefore reports: "Based on a sample, we can conclude that 60% of the total faculty feel that female members are underpaid relative to males." Explain why this conclusion is wrong.

(d) If we took a stratified random sample of 200 male and 50 female faculty members at this university, each member of the faculty would have the same chance of being chosen. What is that chance? Explain why this sample is *not* an SRS.

DATA EXPLORATION

Do Teachers Have Tattoos?

About 1100 high school teachers attended a weeklong summer institute for teaching AP classes. After studying Example 1.10 (page 20), the teachers in the AP Statistics class wondered whether the results of the tattoo survey would be similar for teachers. They designed a survey to find out. The class opted for a sample size of 100 teachers. One of the questions on the survey was:

Do you have any tattoos on your body? (Circle one) YES NO

1. One of the first decisions the class had to make was what kind of sampling method to use. They knew that a simple random sample was the "preferred" method. With 1100 teachers in 40 different sessions, the class decided not to use an SRS. Give at least two reasons why you think they made this decision.

2. The AP Statistics class believed that there might be systematic differences in the proportions of teachers who had tattoos based on the subject areas that they taught. What sampling method would you recommend to account for this possibility?

3. In the end, the class chose to use systematic random sampling (see Exercise 5.67) to choose their sample. They used a calculator's random number generator to pick a number from 1 to 10. The result was a "7." So the class decided to give the survey to the 7th, 17th, 27th, . . . people to arrive at the cafeteria for the morning snack break. Did this sampling method give every teacher at the institute an equal chance to be selected? Why or why not? Explain why this sampling method does *not* yield an SRS of teachers at the institute.

4. Members of the class handed the survey to the randomly selected teachers. The respondents were asked to complete the survey at a table nearby, and then to fold the paper in half and place it in a box with a hole in the lid. Why do you think the class decided to administer the survey in this way rather than doing face-to-face interviews?

5. Two of the selected teachers refused to respond to the survey. How might this affect the results?

6. Some teachers did not come to the morning snack break that day, including an entire class of teachers who were on a field trip. How might this affect the survey results?

7. Of the 98 teachers who responded, 23.5% said that they had one or more tattoos. Although the sampling method used was not an SRS, the quick method still gives a reasonable estimate for the margin of error. Use the quick method to help you make a 95% confidence statement.

8. Write a brief report summarizing this survey. Follow the four-step statistical problem-solving process (page 19).

Questions to ask before you believe a poll

Opinion polls and other sample surveys can produce accurate and useful information if the pollster uses good sampling techniques and also works hard at preparing a sampling frame, wording questions, and reducing nonresponse. Many surveys, however—especially those designed to influence public opinion rather than just record it—do not produce accurate or useful information. Here are some questions to ask before you pay much attention to poll results.

- **Who carried out the survey?** Even a political party should hire a professional sample survey firm whose reputation demands that they follow good survey practices.
- **What was the population?** That is, whose opinions were being sought?
- **How was the sample selected?** Look for mention of random sampling.
- **How large was the sample?** Even better, find out both the sample size and the margin of error within which the results of 95% of all samples drawn as this one was would fall.
- **What was the response rate?** That is, what percent of the people who were selected actually provided information?
- **How were the subjects contacted?** By telephone? Mail? Face-to-face interview?
- **When was the survey conducted?** Was it just after some event that might have influenced opinion?
- **What were the exact questions asked?**

Academic survey centers and government statistical offices answer these questions when they announce the results of a sample survey. National opinion polls usually don't announce their response rate (which is often low) but do give us the other information. Editors and newscasters have the bad habit of cutting out these dull facts and reporting only the sample results. Many sample surveys by interest groups and local newspapers and TV stations don't answer these questions because their polling methods are in fact unreliable. If a politician, an advertiser, or your local TV station announces the results of a poll without complete information, be skeptical.

APPLICATION 5.3

What high school math courses help in college?

Many high school students are faced with the choice of taking a statistics course or a calculus course. Which is the better choice? The answer may be "it depends." College admissions staff often view calculus as a more "rigorous" course when reviewing applicants' transcripts. Many of those same people concede that a high school statistics course actually prepares more students for the majors they'll pursue in college and for their careers. What's a student to do?

A math department chair at a large private high school wanted to get some advice from the school's recent graduates about this issue. He contacted the school's data analysis class and asked if they would

APPLICATION 5.3 *(continued)*

APPLICATION 5.3 *(continued)*

design and carry out a survey for him. The students eagerly accepted the challenge. They began by getting a list of recent graduates (current juniors and seniors in college) with their email addresses from the school's alumni office. After eliminating duplicate names, the students were left with a list of 428 recent graduates.

Once the survey questionnaire was finalized, the class sent an email message to all 428 people on the list inviting them to take the survey online at **www.surveymonkey.com.** Some of the email messages bounced back, so students tried to track down correct contact information for those individuals. With a reminder 2 weeks later, the class eventually got 128 completed surveys.

The two questions of particular interest on the survey were:

- After Algebra 2, which mathematics courses did you take in high school?
- If you were to design your high school course schedule again, which mathematics courses would you take after Algebra 2?

The bar graph summarizes the 128 responses.

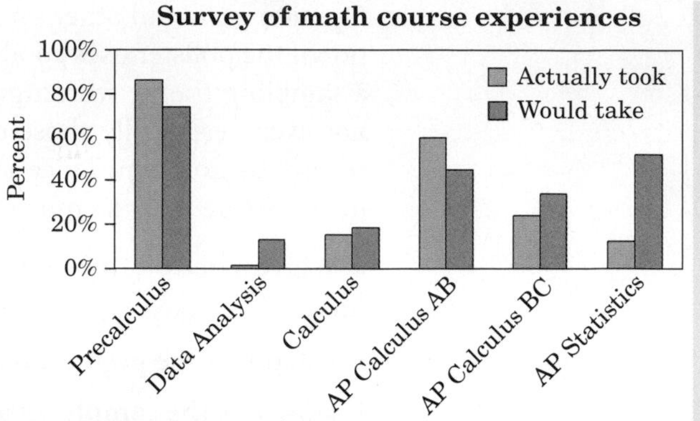

QUESTIONS

1. Why do you think the data analysis class decided not to ask the survey questions in an email and to have people respond directly to the email?

2. The class was unable to locate current email addresses for 10 graduates. What sampling issue does this raise?

3. What was the response rate for this survey? How might this nonresponse affect the results of the survey?

4. Write a few sentences describing what the bar graph tells you.

Section 5.3 Summary

Even professional sample surveys don't give exactly correct information about the population. There are many potential sources of error in sampling. The margin of error announced by a sample survey covers only **random sampling error,** the variation due to chance in choosing a random sample. Other types of error are in addition to the margin of error and can't be directly measured. **Sampling errors** come from the act of choosing a sample. Random sampling error and **undercoverage** are common types of sampling error. Undercoverage occurs when some members of the population are left out of the **sampling frame,** the list from which the sample is actually chosen.

The most serious errors in most careful surveys, however, are **nonsampling errors.** These have nothing to do with choosing a sample—they are present even in a census. The single biggest problem for sample surveys is **nonresponse:** subjects can't be contacted or refuse to answer. Mistakes in handling the data **(processing errors)** and incorrect answers by respondents **(response errors)** are other examples of nonsampling errors. Finally, the exact **wording of questions** has a big influence on the answers. People who design sample surveys use statistical techniques that help correct nonsampling errors, and they also use **probability samples** more complex than simple random samples, such as **stratified samples.**

You can assess the quality of a sample survey quite well by just looking at the basics: use of random samples, sample size and margin of error, the rate of nonresponse, and the wording of the questions.

Section 5.3 Exercises

5.69 Women engineers About 20% of the engineering students at a large university are women. The school plans to poll a sample of 200 engineering students about the quality of student life.

(a) If an SRS of size 200 is selected, about how many women do you expect to find in the sample?

(b) If the poll wants to be able to report separately the opinions of male and female students, what type of sampling design would you suggest? Why?

5.70 Critiquing a poll The *Wall Street Journal* published an article on attitudes toward the Social Security system based on a sample survey. It found, for example, that 36% of people aged 18 to 34 expected Social Security to pay nothing at all when they retire. News articles tend to be brief in describing sample surveys. Here is part of the *Wall Street Journal*'s description of this poll:

> *The* Wall Street Journal/*NBC News poll was based on nationwide telephone interviews of 2,012 adults, conducted Thursday through Sunday by the polling organizations of Peter Hart and Robert Teeter.*
>
> *The sample was drawn from 520 randomly selected geographic points in the continental U.S. Each region was represented in proportion to its population. Households were selected by a method that gave all telephone numbers, listed and unlisted, an equal chance of being included.*[22]

Page 249 lists several "questions to ask" about an opinion poll. What answers does the *Wall Street Journal* give to each of these questions?

5.71 Multistage sampling The previous exercise gives part of the description of a sample survey from the *Wall Street Journal.*

(a) It appears that the sample was taken in several stages. Why can we say this?

(b) The first stage no doubt used a stratified sample, though the *Journal* does not say this. Explain why it would be bad practice to use an SRS from a large number of "geographic points" across the country rather than a stratified sample of such points.

5.72 TV ratings The method of collecting the data can influence the accuracy of sample results. The following methods have been used to collect data on television viewing in a sample household:

(a) *Diary method.* The household keeps a diary of all programs watched and who watched them for a week, then mails in the diary at the end of the week.

(b) *Roster-recall method.* An interviewer shows the person a list of programs for the preceding week and asks which programs were watched.

(c) *Telephone-coincidental method.* The survey firm telephones the household at a specific time and asks if the television is on, which program is being watched, and who is watching it.

(d) *Automatic recorder method.* A device attached to the set records what hours the set is on and to which channel it is tuned. At the end of the week, this record is removed from the recorder.
(e) *People meter.* Each member of the household is assigned a numbered button on a hand-held remote control. Everyone is asked to push their button whenever they start or stop watching TV. The remote control signals a device attached to the set that keeps track of what channel the set is tuned to and who is watching at all times.

Discuss the advantages and disadvantages of each of these methods, especially the possible sources of error associated with each method. The Nielsen national ratings use Method (e). Local ratings (there are more than 200 local television markets) use Method (a). Do you agree with these choices? (Do not discuss choosing the sample, just collecting the data once the sample is chosen.)

5.73 Did you vote? When the Current Population Survey asked the adults in its sample of 50,000 households if they voted in the most recent presidential election, 54% said they had. In fact, only 49% of the adult population voted in that election.
(a) Use the quick method to estimate the margin of error for this survey. (Although the CPS sample is not an SRS, the quick method still gives a reasonably accurate estimate of the margin of error.)
(b) Why do you think the CPS result missed by much more than the margin of error?

5.74 Telling the truth? Many people don't give honest answers to questions about illegal or sensitive issues. One study divided a large group of adults into thirds at random. All were asked if they had ever used cocaine. The first group was interviewed by telephone: 21% said "Yes." In the group visited at home by an interviewer, 25% said "Yes." The final group was interviewed at home but answered the question on an anonymous form that they sealed in an envelope. Of this group, 28% said they had used cocaine.[23]
(a) Which result do you think is closest to the truth? Why?
(b) Give two other examples of behavior you think would be underreported in a telephone survey.

5.75 What kind of error? Each of the following is a source of error in a sample survey. Label each as *sampling error* or *nonsampling error,* and explain your answers.
(a) The telephone directory is used as a sampling frame.
(b) The person cannot be contacted in five calls.
(c) Interviewers choose people walking by on the sidewalk to interview.

5.76 Survey time! Return to Activity 5.3 (page 236). Once your teacher has approved your survey design plan, go ahead and carry it out. Write a brief report summarizing the results. Follow the four-step statistical problem-solving process (page 19).

CHAPTER 5 REVIEW

We have discussed sampling, the art of choosing a part of a population to represent the whole. We saw the potential for bias from voluntary response and convenience samples. This led us to the big idea of a simple random sample, which is summarized in the following figure.

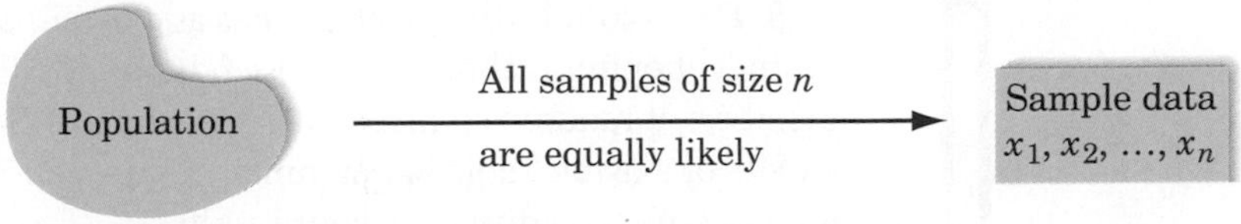

Introducing randomness into the sampling scheme means that the things we measure will vary from sample to sample. But this variation will have a predictable pattern if we repeat the procedure over and over. Increasing the size of the sample will decrease the variability of a sample statistic. But this won't help with bias.

Two types of error can affect sample surveys: sampling errors and nonsampling errors. When the sampling frame differs from the population of interest, undercoverage results. The margin of error does not account for such sampling errors. It also does not reflect nonresponse or question wording, two major causes of nonsampling error.

In many settings, an SRS would be too difficult or too costly to obtain. Many national surveys use some form of multistage sampling. If subgroups within a population differ in their opinions, a stratified random sample is appropriate.

Here is a review list of the most important skills you should have developed from your study of this chapter.

1. Identify the population and parameter of interest.

2. Recognize bias due to voluntary response samples and other inferior sampling methods.

3. Use Table B of random digits to select a simple random sample (SRS) from a population.

4. Explain how sample surveys deal with bias and variability in their conclusions.

5. Explain in simple language what the margin of error for a sample survey result tells us and what "95% confidence" means.

6. Use the quick method to get an approximate margin of error for 95% confidence. Make an appropriate 95% confidence statement about a population parameter.

7. Understand the distinction between sampling errors and nonsampling errors. Recognize the presence of undercoverage and nonresponse as sources of error in a sample survey. Recognize the effect of the wording of questions on the responses.

8. Use random digits to select a stratified random sample from a population when the strata are identified.

CHAPTER 5 REVIEW EXERCISES

5.77 **Baseball tickets** Suppose you want to know the average amount of money spent by the fans attending opening day for the Cleveland Indians baseball season. You get permission from the team's management to conduct a survey at the stadium, but they will not allow you to bother the fans in the club seating or box seat areas (the most expensive seating). Using a computer, you randomly select 500 seats from the rest of the stadium. During the game, you ask the fans in those seats how much they spent that day.

(a) Provide a reason why this survey might yield a biased result.

(b) Explain whether the reason you provided in (a) is a sampling error or a nonsampling error.

5.78 Dress code? The principal has asked your statistics class to carry out a survey of student opinion about a proposed dress code. The class decides to choose four of its members at random to meet with the principal. The class list appears below. Choose an SRS of 4 using Table B, beginning at line 145. Explain your method clearly enough for a classmate to duplicate your results.

Anderson	Fernandez	Kempthorn	Robertson
Aspin	Gupta	Liang	Rodriguez
Benitez	Gutierrez	Montoya	Siegel
Bock	Gwynn	Olds	Tompkins
Breiman	Harter	Patnaik	Vandegraff
Castillo	Henderson	Pirelli	Wang
Dixon	Hughes	Rao	
Edwards	Johnson	Rider	

5.79 We don't like one-way streets Highway planners decided to make a main street in West Lafayette, Indiana, a one-way street. The *Lafayette Journal and Courier* took a one-day poll by inviting readers to call a telephone number to record their comments. The next day, the paper reported:

> Journal and Courier *readers overwhelmingly prefer two-way traffic flow in West Lafayette's Village area to one-way streets. By nearly a 7-1 margin, callers to the newspaper's Express Yourself opinion line on Wednesday complained about the one-way streets that have been in place since May. Of the 98 comments received, all but 14 said no to one-way.*

(a) What population do you think the newspaper wants information about?
(b) Is the proportion of this population who favor one-way streets almost certainly larger or smaller than the proportion 14/98 in the sample? Why?

5.80 TV commercials The noted scientist Dr. Iconu wanted to investigate attitudes toward television advertising among American college students. He decided to use a sample of 100 students. Students in freshman psychology (PSY 001) are required to serve as subjects for experimental work. Dr. Iconu obtained a class list for PSY 001 and chose a simple random sample of 100 of the 340 students on the list. He asked each of the 100 students in the sample the following question:

> *Do you agree or disagree that having commercials on TV is a fair price to pay for being able to watch it?*

Of the 100 students in the sample, 82 marked "Agree." Dr. Iconu announced the result of his investigation by saying, "82% of American college students are in favor of TV commercials."
(a) What is the population in this example?
(b) What is the sampling frame in this example?
(c) Explain briefly why the sampling frame is or is not suitable for the question being investigated.
(d) Discuss briefly the question Dr. Iconu asked. Is it a slanted question?

(e) Discuss briefly why Dr. Iconu's announced result is misleading.
(f) Dr. Iconu defended himself against criticism by pointing out that he had carefully selected a simple random sample from his sampling frame. Is this defense relevant? Why?

5.81 Planning a survey of students The student government plans to ask a random sample of students at a large high school about their priorities for improving the school newspaper. A school counselor provides a list of the 3500 students at the school to serve as a sampling frame.
(a) How would you choose an SRS of 250 students?
(b) How would you choose a systematic sample of 250 students? (See Exercise 5.67 page 247, to learn about systematic samples.)
(c) The list shows whether students are bussed to school (2400 students) or live nearby (1100 students). How would you choose a stratified sample of 200 bussed students and 50 local students?
(d) Which of the three sampling methods would you choose? Why?

5.82 Bigger samples, please Explain in your own words the advantages of bigger random samples in a sample survey.

5.83 Should he go or stay? In December 1998, the House of Representatives impeached President Clinton, the first step in removing him from office. The Senate then conducted a trial and found the president not guilty. Here are two opinion poll questions asked after the House had acted:

> *What do you think President Clinton should do: fight the charges in the Senate, or resign from office?*
>
> *What do you think President Clinton should do: continue to serve and stand trial in the Senate, or resign from office?*

In response to the first question, 58% thought the president should resign. But only 43% of those asked the second question thought he should resign.[24] Why do you think the first wording encouraged more people to favor resignation?

5.84 The Harris Poll Here is the language used by the Harris Poll to explain the accuracy of its results: "In theory, with a sample of this size, one can say with 95 percent certainty that the results have a statistical precision of plus or minus 3 percentage points of what they would be if the entire adult population had been polled with complete accuracy."[25] What does Harris mean by "95 percent certainty"?

5.85 Life and Internet A poll of 586 adults who used the Internet in the past week were asked whether "the Internet has made your life much better, somewhat better, somewhat worse, much worse, or has it not affected your life either way." In all, 152 of the 586 said "much better."[26]
(a) What is the population for this sample survey?
(b) Use the quick method to find a margin of error. Then give a complete confidence statement for a conclusion about the population.

5.86 Steroids in baseball A 2008 *New York Times* article on public opinion about steroid use in baseball discussed the results of a sample survey. The survey found that

34% of adults think that at least half of Major League Baseball (MLB) players "use steroids to enhance their athletic performance." Another 36% thought that about a quarter of MLB players use steroids; 8% had no opinion. Here is part of the *Times*'s statement on "How the Poll Was Conducted":

> *The latest* New York Times/*CBS News Poll is based on telephone interviews conducted March 15 through March 18 with 1,067 adults throughout the United States. . . .*
>
> *The sample of telephone numbers called was randomly selected by a computer from a list of more than 42,000 active residential exchanges across the country. The exchanges were chosen to ensure that each region of the country was represented in proportion to its population.*
>
> *In each exchange, random digits were added to form a complete telephone number, thus permitting access to listed and unlisted numbers. In each household, one adult was designated by a random procedure to be the respondent for the survey.*[27]

Page 249 lists several "questions to ask" about an opinion poll. What answers does the *Times* give to each of these questions?

CHAPTER 6

Designing Experiments

A tale of three studies

An optimistic account of online learning describes a study at Nova Southeastern University in Fort Lauderdale, Florida. The authors of the study claim that students taking courses online were "equal in learning" to students taking the same courses in class. Replacing college classes with Web sites saves colleges money, so this study suggests we should all move online.[1]

Ulcers seem to accompany the stress of modern life. "Gastric freezing" is a clever treatment for stomach ulcers. The patient swallows a deflated balloon with tubes attached; then a refrigerated solution is pumped through the balloon for an hour. The idea is that cooling the stomach will reduce its production of acid and so relieve ulcers. An experiment reported in the *Journal of the American Medical Association* claimed that gastric freezing did relieve ulcer pain.[2]

Should the government provide day care for children in low-income families? If day care helps these children stay in school and hold good jobs later in life, the government would save money by paying less welfare and collecting more taxes. The Carolina Abecedarian Project (the name suggests learning the ABCs) has followed a group of children since 1972. The results show that good day care makes a big difference in later school and work.[3]

Three big issues, three studies that claim to shed light on the issues. In fact, the first of these is an observational study that can't be trusted to give good evidence about learning in online college courses. The gastric-freezing experiment now appears to be plainly misleading. Most ulcers are caused by bacteria, and cooling the stomach is not an effective treatment. The results of the Abecedarian Project, on the other hand, are about as convincing as evidence about the long-term effects of day care can be. What makes some studies—especially some experiments—convincing? Why should we ignore others? This chapter shows you what to look for.

6.1 Experiments, Good and Bad

ACTIVITY 6.1

Testing therapeutic touch

Materials: Blindfold and coin for each pair of students

For her fourth-grade science fair project, Emily Rosa designed an experiment to test whether an alternative healing method known as therapeutic touch actually works. Therapeutic touch requires the "healer" to detect and manipulate the patient's "human energy field." Emily doubted that this was possible. So she invited 21 people who had practiced therapeutic touch to participate in her experiment.

Emily's experimental design was fairly simple. She sat at one side of a table and the subject sat at the other side. A tall screen with two holes was placed in the middle of the table, so that the subject could not see Emily. The subject was asked to extend one arm, palm side up, through each of the holes. Unknown to the subject, Emily then flipped a coin to decide which of the subject's two hands she would place her hand above. She then asked the subject to identify the location of the "energy field." This process was repeated either 10 or 20 times with each of the 21 subjects.

Want to know what happened in Emily's experiment? We'll tell you after you and your classmates try a version of her experiment for yourself. Work with a partner as instructed by your teacher.

1. With your partner, decide which of you will be the "experimenter" and which will be the "subject." You will switch roles later, so don't spend too much time deciding!

2. Blindfold the subject. Have the subject sit with both arms extended, palms up.

3. The experimenter should flip a coin. If it lands "heads," place one hand about six inches above the subject's left hand. If it lands "tails," place one hand about six inches above the subject's right hand. Ask the subject to identify the presence of the energy field by saying either "left" or "right."

4. Record the outcome of the coin toss (H or T) and whether the subject picked correctly (C) or incorrectly (I). Don't tell the subject whether the identification is correct!

5. Repeat the previous two steps 19 more times, for a total of 20 trials. Organize your results in a table like this one.

Trial no.	Coin outcome (H or T)	Subject's pick (C or I)
1	T	C

6. Have the subject remove the blindfold. Share the results of the experiment.

7. Switch roles, and repeat Steps 2 through 6.

8. Pool results with your classmates. Make a dotplot that shows the percent of correct (C) identifications made by each subject.

9. Discuss what you see in the dotplot. Does any member of the class appear to be able to detect a human energy field? Why or why not?

10. Calculate the average percent of correct identifications made by all of the subjects. How does this compare with what you would expect if the subjects were just guessing?

The results of Emily Rosa's experiment in Activity 6.1 were striking: the subjects correctly identified the location of Emily's hand in only 123 of 280 trials (44%). This is pretty much what we would expect if the subjects were merely guessing. With a little help from her mother (a nurse) and a statistician, Emily published an article summarizing her research in the *Journal of the American Medical Association* on April 1, 1998.[4]

Talking about experiments

Observational studies involve passive data collection. We observe, record, or measure, but we don't interfere. **Experiments** involve active data production. Experimenters actively intervene by imposing some treatment in order to see what happens. All experiments and many observational studies are interested in the effect one variable has on another variable. In Chapter 4, we called these the **explanatory variable** and the **response variable,** respectively. Now we introduce two more terms that have common use when describing experiments—**subjects** and **treatments.**

Subjects and treatments

The individuals studied in an experiment are often called **subjects.**

A **treatment** is any specific experimental condition applied to the subjects. If an experiment has several explanatory variables, a treatment is a combination of specific values of these variables.

The following example illustrates how we use the vocabulary of experiments.

Example 6.1 The effects of day care

The Carolina Abecedarian Project described in the chapter-opening story (page 257) is an experiment. The *subjects* were 111 healthy, low-income black infants in Chapel Hill, North Carolina. All the infants received nutritional supplements and help from social workers. Half, chosen at random, were also placed in an intensive preschool program. The experiment compares these two *treatments.* The *explanatory variable* is just "preschool, yes or no." There are many *response variables,* recorded over more than 20 years, including academic test scores, college attendance, and employment. □

You will often see explanatory variables called *independent variables* and response variables called *dependent variables.* The idea is that the response variables depend on the explanatory variables. We won't use these terms, partly because "independent" has other and very different meanings in statistics.

Designing studies: what can go wrong

Do students who take a course on the Web learn as well as those who take the same course in a traditional classroom? The best way to find out is to assign some students to the classroom and others to the Web. That's an experiment. The Nova Southeastern University study described in the chapter-opening story (page 257) was not an experiment, because it imposed no treatment on the student subjects. Students chose for themselves whether to enroll in a classroom or in an online version of a course. The study simply measured their learning.

The students who chose the online course were very different from the classroom students. For example, their average score on tests on the course material given before the courses started was 40.70, against only 27.64 for the classroom students. It's hard to compare in-class versus online learning when the online students have a big head start. The effect of online versus in-class instruction is hopelessly mixed up with influences lurking in the background. In Chapter 4, we called this situation **confounding.** Figure 6.1 shows the mixed-up influences in picture form.

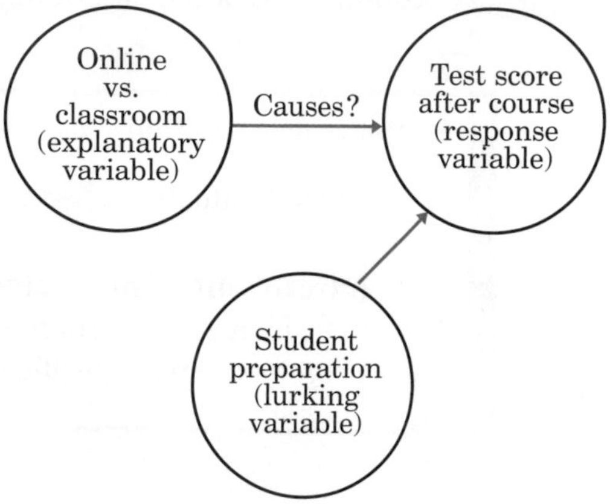

Figure 6.1 Confounding in the Nova Southeastern University study. The influence of course setting (the explanatory variable) can't be distinguished from the influence of student preparation (a lurking variable).

In the Nova Southeastern University study, student preparation (a **lurking variable**) is confounded with the explanatory variable. The study report claims that the two groups did equally well on the final test. We can't say how much of the online group's performance is due to their head start. Since the online group started with a big advantage but did no better than the more poorly prepared classroom students on the final, the study gives little evidence to support the effectiveness of Web-based instruction.

Here is another example, one in which a second experiment untangled the confounding.

Example 6.2 Gastric freezing flunks

Experiments that study the effectiveness of medical treatments on actual patients are called **clinical trials.** The clinical trial that made gastric freezing a popular treatment for stomach ulcers had this "one-track" design:

Impose treatment	→	*Measure response*
Gastric freezing	→	Reduced pain?

The patients did report reduced pain, but we can't say that gastric freezing caused the reduced pain. It might be just the **placebo effect.** A **placebo** is a dummy treatment with no active ingredients. Many patients respond favorably to *any* treatment, even a placebo. This response to a dummy treatment is the placebo effect. Perhaps the placebo effect is in our minds, based on trust in the doctor and expectations of a cure. Maybe it is just a name for the fact that many patients improve for no visible reason. The one-track design of the experiment meant that the placebo effect was *confounded* with any effect gastric freezing might have.

A second clinical trial, done several years later, divided ulcer patients into two groups. One group was treated by gastric freezing as before. The other group received a placebo treatment in which the solution in the balloon was at body temperature rather than freezing. The results: 34% of the 82 patients in the treatment group improved, but so did 38% of the 78 patients in the placebo group. This and other properly designed experiments showed that gastric freezing was no better than a placebo, and doctors stopped using it.[5]

Both observation and one-track experiments often yield useless data because of confounding with lurking variables. It is hard to avoid confounding when only observation is possible. Experiments offer better possibilities, as the second gastric-freezing experiment shows. This experiment included a group of subjects who received only a placebo. This allows us to see whether the treatment being tested does better than the placebo and so has more than the placebo effect going for it. Effective medical treatments pass the placebo test. Gastric freezing flunks the test.

Exercises

6.1 On sale A researcher wants to study the effect of price promotions on consumers' expectations. The researcher makes up two different histories of the store price of a video game for the past year. Students in an economics course view one or the other price history on a computer. Some students see a steady price, while others see regular promotions that temporarily cut the price. Then the students are asked what price they would expect to pay for the video game.

(a) Is this study an experiment? Why?

(b) Identify the explanatory and response variables, the subjects, and the treatments.

6.2 Public housing A study of the effect of living in public housing on family stability in low-income households was carried out as follows: A list of applicants accepted for public housing was obtained, together with a list of families who applied but were rejected by the housing authorities. A random sample was drawn from each list, and the two groups were observed for several years.

(a) Is this an experiment? Why?

(b) What are the explanatory and response variables?

(c) Does this study contain confounding that may prevent valid conclusions on the effects of living in public housing? Explain.

6.3 Nursing your baby An article in a women's magazine says that women who nurse their babies feel warmer and more receptive toward the infants than mothers who bottle-feed. The author concludes that nursing has desirable effects on the mother's attitude toward the child. But women choose whether to nurse or bottle-feed. Explain why this fact makes any conclusion about cause and effect untrustworthy. Use the language of lurking variables and confounding in your explanation, and draw a picture like Figure 6.1 to illustrate it.

6.4 Vitamin C and colds Last year only 10% of a group of adult men did not have a cold at some time during the winter. This year all the men in the group took 1 gram of vitamin C each day, and 20% had no colds. A writer claims that this shows that vitamin C helps prevent colds. Explain why this conclusion is shaky at best. Use the language of lurking variables and confounding in your explanation, and draw a picture like Figure 6.1 to illustrate it.

6.5 Activity 6.1 follow-up

(a) Identify the subjects, treatments, and explanatory and response variables in the therapeutic touch experiment.

(b) After Emily's results were published, therapeutic touch practitioners argued that her results were not valid. Discuss one or two arguments they might have made.

6.6 Do placebos really work? Researchers in Japan conducted an experiment on 13 individuals who were extremely allergic to poison ivy. On one arm, each subject was rubbed with a poison ivy leaf and told the leaf was harmless. On the other arm, each subject was rubbed with a harmless leaf and told it was poison ivy. All of the subjects developed a rash on the arm where the harmless leaf was rubbed. Of the 13 subjects, 11 did not have any reaction to the real poison ivy leaf.[6]

(a) What was the placebo in this experiment?

(b) Explain how the results of this study support the idea of a placebo effect.

By permission of John L. Hart FLP and Creators Syndicate, Inc.

Randomized comparative experiments

The first goal in designing an experiment is to ensure that it will show us the effect of the explanatory variables on the response variables. Confounding often prevents one-track experiments from doing this. The remedy is to compare two or more treatments. Here is an example of a new medical treatment that passes the placebo test in a direct comparison.

Example 6.3 Sickle-cell anemia

Sickle-cell anemia is an inherited disorder of the red blood cells that in the United States affects mostly blacks. It can cause severe pain and many complications. The National Institutes of Health carried out a clinical trial of the drug hydroxyurea for treatment of sickle-cell anemia. The subjects were 299 adult patients who had had at least three episodes of pain from sickle-cell anemia in the previous year.

Simply giving hydroxyurea to all 299 subjects would confound the effect of the medication with the placebo effect and other lurking variables such as the effect of knowing that you are a subject in an experiment. Instead, half of the subjects received hydroxyurea, and the other half received a placebo that looked and tasted the same. All subjects were treated exactly the same (same schedule of medical checkups, for example) except for the content of the medicine they took. Lurking variables therefore affected both groups equally and should not have caused any differences between their average responses.

The two groups of subjects must be similar in all respects before they start taking the medication. Just as in sampling, the best way to avoid bias in determining which subjects get hydroxyurea is to allow impersonal chance to make the choice. An SRS of 152 of the subjects formed the hydroxyurea group; the remaining 147 subjects made up the placebo group. Figure 6.2 outlines the experimental design.

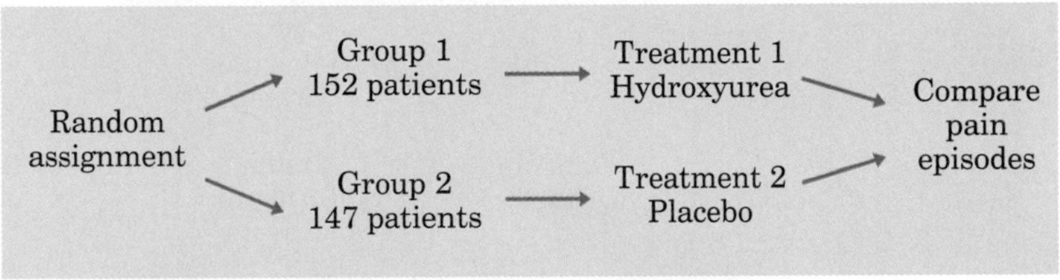

Figure 6.2 The design of a randomized comparative experiment to compare hydroxyurea with a placebo for treating sickle-cell anemia.

The experiment was stopped ahead of schedule because the hydroxyurea group had many fewer pain episodes than the placebo group. This was compelling evidence that hydroxyurea is an effective treatment for sickle-cell anemia, good news for those who suffer from this serious illness.[7]

Figure 6.2 illustrates the simplest **randomized comparative experiment,** one that compares just two treatments. The diagram outlines the essential information about the design: random assignment; one group for each treatment; the number of subjects in each group (it is generally best to keep the groups similar in size); what treatment each group gets; and the response variable we compare. You know how to carry out the random assignment of subjects to groups. Label the 299 subjects 001 to 299, then read three-digit groups from the table of random digits (Table B) until you have chosen the 152 subjects for Group 1. The remaining 147 subjects form Group 2.

The placebo group in Example 6.3 is called a **control group** because comparing the treatment and control groups allows us to control the effects of lurking variables. A control group need not receive a dummy treatment such as a placebo. Clinical trials often compare a new treatment for a medical condition, not with a placebo, but with a treatment that is already on the market. Patients who are randomly assigned to the existing treatment form the control group. To compare more than two treatments, we can randomly assign the available experimental subjects to as many groups as there are treatments. Here is an example with three groups.

Example 6.4 Stopping drunk drivers

Most people know that drunk driving is dangerous. Even so, some drivers are cited multiple times for driving while intoxicated. Is there anything that can be done to change their behavior? An experiment was designed to help answer this question.

The subjects were 300 people convicted of drunk driving three times in one year. The treatments, imposed after the third conviction, were a fine plus a suspended jail sentence plus one of (1) no treatment, (2) attend an alcoholism clinic, or (3) participate in Alcoholics Anonymous. The subjects assigned to Treatment 1 served as a control group. One response variable was whether or not the subject was arrested again in the following year. Figure 6.3 outlines the design.

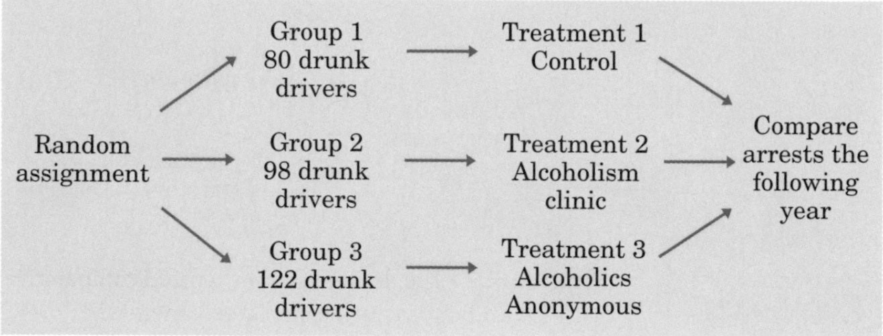

Figure 6.3 The design of a randomized comparative experiment to compare three programs to deter drunk drivers.

Ideally, 100 subjects would be assigned at random to each of the three treatments. This is probably not practical, since a treatment must be assigned immediately after the subject's third drunk-driving conviction. Instead, enter Table B at any row and begin reading off single digits. The digits 1, 2, and 3 represent treatment option 1; the digits 4, 5, and 6 indicate treatment option 2; and the digits 7, 8, and 9 signal treatment option 3. Ignore the digit 0. The first subject selected for the experiment is assigned the treatment corresponding to the first nonzero digit in the row. The second subject's treatment is determined by the second nonzero digit in the row, and so on. For example, if we use row 105 in Table B, our assignment of subjects to treatments proceeds as follows:

Subject:	1	2	3	4	5	6	7	8	9	10
Nonzero digit:	9	5	5	9	2	9	4	7	6	9
Treatment:	3	2	2	3	1	3	2	3	2	3

By the end of row 105, we have assigned 10 subjects to Treatment 1, 9 to Treatment 2, and 18 to Treatment 3. Continuing in this fashion through rows 106, 107, . . . , 113, we end up with 80 subjects receiving Treatment 1, 98 receiving Treatment 2, and 122 receiving Treatment 3.

"OK, Mr. Simms. Now it's time to test your pacemaker against a control subject."

The logic of experimental design

The randomized comparative experiment is one of the most important ideas in statistics. It is designed to allow us to draw cause-and-effect conclusions. Be sure you understand the logic:

- Random assignment produces groups of subjects that should be similar in all respects before we apply the treatments.
- A proper comparative design ensures that influences other than the experimental treatments operate equally on all groups.
- Therefore, differences in the response variable must be due to the effects of the treatments.

We use chance to choose the groups in order to eliminate any systematic bias in assigning the subjects to groups. In the sickle-cell study, for example, a doctor might subconsciously assign the most seriously ill patients to the hydroxyurea group, hoping that the untested drug will help them. That would bias the experiment against hydroxyurea. Choosing an SRS of the subjects to be Group 1 gives everyone the same chance to be in either group. We expect the two groups to be similar in all respects—age, seriousness of illness, smoker or not, and so on. Chance tends to assign equal numbers of smokers to both groups, for example, even if we don't know which subjects are smokers.

There is one important caution about randomized experiments. Like random samples, they are subject to the laws of chance. Just as an SRS of voters might by bad luck choose nearly all Republicans, a random assignment of subjects might by bad luck put nearly all the smokers in one group. We know that, if we choose large random samples, it is very likely that the sample will match the population well. In the same way, if we use many experimental subjects, it is very likely that random assignment will produce groups that match closely. More subjects means that there is less chance variation among the treatment groups and less chance variation in the outcomes of the experiment. "Use enough subjects" joins "control for lurking variables" and "randomize" as a basic principle of statistical design of experiments.

Principles of experimental design

The basic principles of statistical design of experiments are

1. **Control** the effects of lurking variables on the response. Use a comparative design and ensure that the only systematic difference between the groups is the treatment administered.
2. **Randomize**—use impersonal chance to assign subjects to treatments.
3. **Use enough subjects** in each group to reduce chance variation in the results.

Exercises

6.7 Aspirin and heart attacks, I Can aspirin help prevent heart attacks? The Physicians' Health Study, a large medical experiment involving 22,071 male physicians, attempted to answer this question. One randomly selected group of 11,037 physicians took an aspirin every second day, while the rest took a placebo. After several years the study found that subjects in the aspirin group had significantly fewer heart attacks than subjects in the placebo group.

(a) Identify the subjects, the explanatory variable and the values it can take, and the response variable.

(b) Use a diagram like the one in Figure 6.2 or Figure 6.3 to outline the design of the Physicians' Health Study. (When you outline the design of an experiment, be sure to indicate the size of the treatment groups and the response variable.)

6.8 Aspirin and heart attacks, II Refer to the previous exercise. Discuss how each of the three principles of experimental design was addressed in the Physicians' Health Study.

Note: You can get all the details of the Physicians' Health Study described in Exercises 6.7 and 6.8 at the Web site **http://phs.bwh.harvard.edu/**.

6.9 Internet phone calls It is possible to use a computer to make telephone calls over the Internet. How will the cost affect the behavior of users of this service? You will offer the service to all 200 rooms in a college dormitory. Some rooms will pay a flat rate. Others will pay higher rates at peak periods and very low rates off-peak. You are interested in the amount and time of use.

(a) Outline the design of an experiment to study the effect of rate structure. Use the diagrams in Figures 6.2 and 6.3 as models.

(b) Use Table B, starting at line 125, to assign the first 5 rooms to the flat-rate group.

6.10 Healthy turkeys Turkeys raised commercially for food are often fed the antibiotic salinomycin to prevent infections from spreading among the birds. Salinomycin can damage the birds' internal organs, especially the pancreas. A researcher believes that adding vitamin E to the diet may prevent injury. He wants to explore the effects of three levels of vitamin E added to the diet of turkeys along with the usual dose of salinomycin. There are 30 turkeys available for the study. At the end of the study, the birds will be killed and each pancreas examined under a microscope.

(a) Give a careful outline of the design of a randomized comparative experiment in this setting. Use the diagrams in Figures 6.2 and 6.3 as models.

(b) Use Table B, beginning at line 118, to carry out the random assignment required by your design.

6.11 Learning on the Web The discussion following Example 6.1 (page 259) notes that the Nova Southeastern University study doesn't tell us much about Web versus classroom learning because the students who chose the Web version were much better prepared. Describe the design of an experiment to get better information.

Joe Zellner/iStockphoto

6.12 Randomization at work To demonstrate how randomization reduces confounding, consider the following situation. A nutrition experimenter intends to compare the weight gain of newly weaned male rats fed Diet A with that of rats fed Diet B. To do this, she will feed each diet to 10 rats. She has available 10 rats of genetic Strain 1 and 10 of Strain 2. Strain 1 is more vigorous, so if the 10 rats of Strain 1 were fed Diet A, the effects of strain and diet would be confounded, and the experiment would be biased in favor of Diet A.

(a) Label the rats 00, 01, . . . , 19. Use Table B to assign 10 rats to Diet A. Do this four times, using different parts of the table, and write down the four groups assigned to Diet A.

(b) Unknown to the experimenter, the rats labeled 00, 02, 04, 06, 08, 10, 12, 14, 16, and 18 are the 10 Strain 1 rats. How many of these rats were in each of the four Diet A groups that you generated? What was the average number of Strain 1 rats assigned to Diet A?

Statistical significance

The presence of chance variation requires us to look more closely at the logic of randomized comparative experiments. We cannot say that *any* difference in the average number of pain episodes between the hydroxyurea and control groups in the sickle-cell study must be due to the effect of the drug. There will be some differences even if both treatments are the same because there will always be some differences among the individuals who are our subjects. Even though randomization eliminates systematic differences between the groups, there will still be chance differences. We should insist on a difference in the responses so large that it is unlikely to happen just because of chance variation.

Statistical significance

An observed effect so large that it would rarely occur by chance is called **statistically significant.**

The difference between the average number of pain episodes for subjects in the hydroxyurea group and the average for the control group was "highly statistically significant." That means that a difference this large would almost never happen just by chance. We do indeed have strong evidence that hydroxyurea beats a placebo in helping sickle-cell disease sufferers. You will often see the phrase "statistically significant" in published reports. It tells you that the investigators found good evidence for the effect they were seeking.

Of course, the actual results of an experiment are more important than the seal of approval given by statistical significance. The treatment group in the sickle-cell experiment had an average of 2.5 pain episodes per year, against 4.5 per year in the control group. That's a big enough difference to be important to people with the disease. A difference of 2.5 versus 2.8 would be much less interesting even if it were statistically significant.

How to live with observational studies

Does regular church attendance lengthen people's lives? Do doctors discriminate against women in treating heart disease? Does talking on a cell phone while driving increase the risk of having an accident? These are cause-and-effect questions, so we reach for our favorite tool, the randomized comparative experiment. Sorry. We can't randomly assign people to attend church or not, because going to religious services is an expression of beliefs or their absence. We can't use random digits to assign heart disease patients to be men or women. We are reluctant to require drivers to use cell phones in traffic, because talking while driving may be risky.

The best data we have about these and many other cause-and-effect questions come from observational studies. We know that observation is a weak second best to experiments, but good observational studies are far from worthless. What makes a good observational study?

First, good studies are **comparative** even when they are not experiments. We compare random samples of people who do and who don't attend religious services regularly. We compare how doctors treat men and women patients. We might compare drivers talking on cell phones with the same drivers when they are not on the phone. We can often combine comparison with **matching** in creating a control group.

Example 6.5 Painkillers and pregnancy

To see the effects of taking a painkiller during pregnancy, we compare women who did so with women who did not. From a large pool of women who did not take the drug, we select individuals who match the drug group in age, education, number of children, and other lurking variables. We now have two groups that are similar in all these ways, so that these lurking variables should not affect our comparison of the groups.

Comparison does not eliminate confounding. People who attend church or synagogue or mosque take better care of themselves than nonattenders. They are less likely to smoke, more likely to exercise, and less likely to be overweight. Matching can reduce some but not all of these differences. A direct comparison of the ages at death of attenders and nonattenders would confound any effect of religion with the effects of healthy living. So a good comparative study **measures and adjusts for confounding variables.** If we measure weight, smoking, and exercise, there are statistical techniques that reduce the effects of these variables on length of life so that (we hope) only the effect of religion itself remains.

Example 6.6 Living longer through religion

One of the better studies of the effect of regular attendance at religious services gathered data from a random sample of 3617 adults. Random sampling is a good start. The researchers then measured lots of variables, not just the explanatory variable (religious activities) and the response variable (length of life). A news article said:

> *Churchgoers were more likely to be nonsmokers, physically active, and at their right weight. But even after health behaviors were taken into account, those not attending religious services regularly still were about 25% more likely to have died.*[8]

That "taken into account" means that the final results were adjusted for differences between the two groups. Adjustment reduced the advantage of religion but still left a large benefit.

Example 6.7 Sex bias in treating heart disease?

Doctors are less likely to give aggressive treatment to women with symptoms of heart disease than to men with similar symptoms. Is this because doctors are sexist? Not necessarily. Women tend to develop heart problems much later than men, so that female heart patients are older and often have other health problems. That might explain why doctors proceed more cautiously in treating them.

This is a case for a comparative study with statistical adjustments for the effects of confounding variables. There have been several such studies, and they produce conflicting results. Some show, in the words of one doctor, "When men and women are otherwise the same and the only difference is gender, you find that treatments are very similar."[9] Other studies find that women are undertreated even after adjusting for differences between the female and male subjects.

"Statistics say that religious people live longer, so I practice a different religion every day of the week to be sure I'm covered."

As Example 6.7 suggests, statistical adjustment is tricky. Randomization creates groups that are similar in *all* variables known and unknown. Matching and adjustment, on the other hand, can't work with variables the study didn't think to measure. Even if you believe that the researchers thought of everything, you should be a bit skeptical about statistical adjustment. There's lots of room for cheating in deciding which variables to adjust for. And the "adjusted" conclusion is really something like this:

> *If female heart disease patients were younger and healthier than they really are, and if male patients were older and less healthy than they really are, then the two groups would get the same medical care.*

This may be the best we can get, and we should thank statistics for making such wisdom possible. But we end up longing for the clarity of a good experiment.

Exercises

6.13 Statistical significance, I A randomized comparative experiment examines whether a calcium supplement in the diet reduces the blood pressure of healthy men. The subjects receive either a calcium supplement or a placebo for 12 weeks. The researchers conclude that "the blood pressure of the calcium group was significantly lower than that of the placebo group." "Significant" in this conclusion means statistically significant. Explain what statistically significant means in the context of this experiment, as if you were speaking to someone who knows no statistics.

6.14 Statistical significance, II The financial aid office of a university asks a sample of students about their employment and earnings. The report says that "for academic year earnings, a significant difference was found between the sexes, with men earning more on the average. No significant difference was found between the earnings of black and white students." Explain the meaning of "a significant difference" and "no significant difference" in plain language.

6.15 Prayer and meditation You read in a magazine that "nonphysical treatments such as meditation and prayer have been shown to be effective in controlled scientific studies for such ailments as high blood pressure, insomnia, ulcers, and asthma." Explain in simple language what the article means by "controlled scientific studies" and why such studies might show that meditation and prayer are effective treatments for some medical problems.

6.16 TV harms children Observational studies suggest that children who watch many hours of television get lower grades in school and are more likely to commit crimes than those who watch less TV. Explain clearly why these studies do not show that watching TV *causes* these harmful effects. In particular, suggest some lurking variables that may be confounded with heavy TV viewing.

6.17 Family dinners and better grades? Does eating dinner with their families improve students' academic performance? According to an ABC News article, "Teenagers who eat with their families at least five times a week are more likely to get better grades in school."[10] This finding was based on an observational study conducted by researchers at Columbia University. Explain clearly why such a study cannot establish a cause-and-effect relationship. Suggest a lurking variable that may be confounded with whether families eat dinner together.

6.18 Significant doesn't mean important Suppose we're testing a new antibacterial cream, "Formulation NS." We know from previous research that, with no medication, the mean healing time (defined as the time for the scab to fall off) is 7.6 days. We make a small cut on the inner forearm of 25 volunteer college students and apply Formulation NS to the wounds. The mean healing time for these subjects is $\bar{x} = 7.1$ days. A statistical test shows that this decrease in healing time is statistically significant. Explain why this result has little practical importance. *Moral: Statistical significance is not the same as practical importance.*

APPLICATION 6.1

Good for your heart

Heart Treatment's Value Doubted

By Michael Specter
Washington Post Staff Writer

In a study with far-reaching implications for the routine treatment of heart attack patients, researchers have found that the immediate use of special balloons to force open clogged arteries is unnecessary in the vast majority of cases if the patient is treated with a clot-dissolving drug.

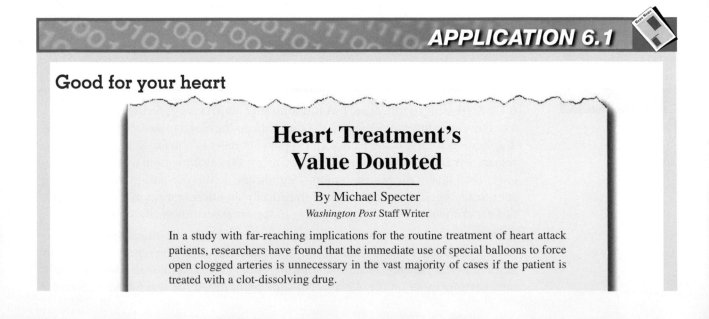

Most heart specialists have assumed that the balloon treatment, an expensive and increasingly popular procedure called balloon angioplasty, should routinely follow the use of drugs, such as TPA, that dissolve the blood clots that cause most heart attacks.

But in a study of 3,262 heart attack patients that is expected to transform the standards for treatment of the nation's leading killer, researchers at medical centers across the country found the extra measures were rarely needed. Half of the randomly selected patients were treated solely with the clot-dissolving drug TPA, while the other half were given TPA followed by angioplasty.

The results, reported in today's issue of *The New England Journal of Medicine,* showed that for most people adding angioplasty was no better than relying on the less complicated and less costly drug treatment.

Only 10 percent of the group assigned solely to drug treatment died or had another heart attack in the six weeks following treatment. By comparison, 11 percent of the group assigned to receive the combination drug and angioplasty treatment suffered the same fate. The difference could have occurred by chance. But the fact that angioplasty did not prove to be the better option was a shock even to some of the nation's most renowned heart experts.

"There is no question that it bucks the trend," said Eugene Braunwald, professor of medicine at Harvard Medical School and the study chairman for the research project. "It will spare many patients unnecessary surgery and reduce the cost of medical care greatly."[11]

QUESTIONS

1. Identify the subjects, treatments, and explanatory and response variables in this study.

2. Describe the design of the experiment in as much detail as you can.

3. Explain how the principles of control, randomization, and using enough subjects were applied in this experiment.

4. What other information would you like to have to evaluate the design of the experiment?

Comment: As Application 6.1 shows, the objective of some clinical trials is to show that there is *no* significant difference in the effects of the treatments, especially when one of the treatments is less risky or less expensive.

Section 6.1 Summary

Statistical studies often try to show that changing one variable (the **explanatory variable**) causes changes in another variable (the **response variable**). In an **experiment,** we actually set the explanatory variables ourselves rather than just observing them. Observational studies and one-track experiments that simply apply a single treatment often fail to produce useful data because **confounding** with **lurking variables** makes it impossible to say what the effect of the treatment was. The remedy is to use a **randomized comparative experiment** that compares two or more **treatments.** Use chance to decide which **subjects** get each treatment, **control** for lurking variables, and use enough subjects so that the effects of chance variations between the groups are small. Comparing two or more treatments helps avoid the **placebo effect** because it will act on all the treatment groups.

Differences among the effects of the treatments so large that they would rarely happen just by chance are called **statistically significant.** Statistically significant results from randomized comparative experiments are the best available evidence that changing the explanatory variable really **causes** changes in the response. Observational studies of cause-and-effect questions are more impressive if they **compare matched groups** and measure as many lurking variables as possible to allow **statistical adjustment.** For answering questions about causation, observational studies remain a weak second best to experiments.

Section 6.1 Exercises

6.19 Does job training work? A state institutes a job-training program for manufacturing workers who lose their jobs. After five years, the state reviews how well the program works. Critics claim that, because the state's unemployment rate for manufacturing workers was 6% when the program began and 10% five years later, the program is ineffective. Explain why higher unemployment does not necessarily mean that the training program failed. In particular, identify some lurking variables whose effect on unemployment may be confounded with the effect of the training program. Draw a picture like Figure 6.1 (page 260) to illustrate your explanation.

6.20 What a headache! Doctors identify "chronic tension–type headaches" as headaches that occur almost daily for at least six months. Can antidepressant medications or stress management training reduce the number and severity of these headaches? Are both together more effective than either alone? Investigators compared four treatments: antidepressant alone, placebo alone, antidepressant plus stress management, and placebo plus stress management.

(a) Outline the design of the experiment. Use the diagrams in Figures 6.2 (page 263) and 6.3 (page 264) as models.

(b) The headache sufferers named below have agreed to participate in the study. Use Table B at line 130 to randomly assign the subjects to the treatments.

Acosta	Duncan	Han	Liang	Padilla	Valasco
Asihiro	Durr	Howard	Maldonado	Plochman	Vaughn
Bennett	Edwards	Hruska	Marsden	Rosen	Wei
Bikalis	Farouk	Imrani	Montoya	Solomon	Wilder
Chem	Fleming	James	O'Brien	Trujillo	Willis
Clement	George	Kaplan	Ogle	Tullock	Zhang

6.21 Learning about markets, I An economics teacher wonders if playing market games online will help students understand how markets set prices. You suggest an experiment: have some students use the online games, while others talk about markets in small group discussions. The course has two sections of 30 students each, at 8:30 A.M. and 2:30 P.M. The teacher says, "Let's just have the 8:30 class do online work and the 2:30 class do discussion." Why is this a bad idea?

6.22 Learning about markets, II Refer to the previous exercise.

(a) Outline a better design for an experiment to compare the two methods of learning about economic markets. What do you suggest as a response variable? Use the diagrams in Figures 6.2 (page 263) and 6.3 (page 264) as models.

(b) Use Table B, starting at line 116, to do the randomization your design requires.

6.23 Healthy nurses? The Nurses' Health Study has surveyed a sample of over 100,000 female registered nurses every 2 years since 1976. Beginning in 1980, the study asked questions about diet, including alcohol consumption. After looking at all deaths among these nurses, the researchers concluded: "As compared with non-drinkers and heavy drinkers, light-to-moderate drinkers had a significantly lower risk of death."[12] The word "significantly" in this conclusion has a technical meaning. Explain to someone who knows no statistics what "significant" means in a statistical study.

6.24 Clumsy men? A doctor at a veterans hospital examines all of the patient records over a 9-year period and finds that twice as many men as women fell out of their hospital beds during their stay. This is put forward as evidence that men are clumsier than women.

(a) Is this an experiment or an observational study? Why?

(b) Give an example of a possible confounding variable and a clear explanation of why it is a possible confounding variable.

6.25 More statistics majors A university's statistics department wants to attract more majors. It prepares two advertising brochures. Brochure A stresses the intellectual excitement of statistics. Brochure B stresses how much more money statisticians make. Which will be more attractive to first-year students? You have a questionnaire to measure interest in majoring in statistics, and you have 50 first-year students to work with. Outline the design of an experiment to decide which brochure works better. Use the diagrams in Figures 6.2 (page 263) and 6.3 (page 264) as models.

6.26 Acupuncture and pregnancy A study sought to determine if the ancient Chinese art of acupuncture could help infertile women become pregnant.[13] One hundred sixty healthy women undergoing treatment with artificial insemination were recruited for the study. Half of the subjects were randomly assigned to receive acupuncture treatment 25 minutes before embryo transfer and again 25 minutes after the transfer. The remaining 80 subjects were instructed to lie still for 25 minutes after the embryo transfer. *Results:* In the acupuncture group, 34 women became pregnant. In the control group, 21 women became pregnant.

(a) Explain how the three principles of experimental design were addressed in this study.

(b) The difference in the percent of women who became pregnant in the two groups is statistically significant. Explain to someone who knows little statistics what this means.

(c) Why can't we conclude that acupuncture caused the difference in pregnancy rates?

6.2 Experiments in the Real World

ACTIVITY 6.2A

Mozart and mazes

Materials: *Two mazes per student (provided by your teacher), classical music CD or cassette, contemporary music CD or cassette, CD/cassette player, stopwatch, and coin for each pair of students*

In this Activity, you will take part in an experiment to determine whether listening to either classical music or contemporary music helps students complete a maze task more quickly.

ACTIVITY 6.2A *(continued)*

ACTIVITY 6.2A *(continued)*

You will work with a partner during the data collection.

1. Your teacher will distribute two copies of Maze 1 face-down to each pair of students. With your partner, decide who will work the maze first and who will time.

2. When the timer says "go," the subject should turn the paper over and connect the circles in the maze, beginning at 1 (the start) and finishing at 25 (the end). The subject should say "done" when finished. Record the time it takes for the subject to complete the maze to the nearest tenth of a second. (*Note:* The timer should *not* look at the maze!)

3. Switch roles, and repeat Step 2.

4. Next, the members of each pair must be assigned at random to the two treatments: classical music or contemporary music. Determine this with a coin flip.

5. Your teacher will play a short selection of classical music. When the music stops, the chosen member of each pair will work Maze 2 while the remaining member times, using the procedure described in Step 2.

6. Your teacher will play a short selection of contemporary music. When the music stops, the remaining member of each pair will work Maze 2 while the partner times.

7. With your partner, compare your times on Maze 1 and Maze 2. Did the music appear to help? If so, did the classical music or the contemporary music produce more improvement?

8. Pool results with your classmates. In one column, list the difference in times (Maze 2 − Maze 1) for each student who received the classical music treatment. In a second column, list the difference for each student who listened to contemporary music. Construct a graph to compare the results.

9. Compute the average difference for each of the two lists. Do you have reason to believe that classical music helped more than contemporary music? Explain.

©Paha/Dreamstime.com

In 1993, three researchers conducted an experiment to determine whether listening to Mozart's music would improve performance on spatial-reasoning tasks. Using 36 college students as subjects, Frances Rauscher and her colleagues randomly assigned 12 students to each of three treatment groups. Group 1 listened to a 10-minute selection from Mozart's Sonata for Two Pianos in D Major. Group 2 listened to a relaxation tape with mixed sounds for 10 minutes. Group 3 sat in silence for 10 minutes and served as a control group. Each subject completed a pretest two days before the treatment was given and a posttest immediately following the treatment.

The results of the experiment were surprising: students who listened to Mozart showed significant gains in their scores on spatial-reasoning tasks.[14] Rauscher, Shaw, and Ky published their findings in *Nature,* and the so-called Mozart Effect was born. Parents looking for ways to improve their child's IQ scores started buying Mozart tapes. One state provided Mozart cassettes and CDs to parents of newborn babies. Another passed legislation requiring that preschools play 30 minutes of classical music a day. Meanwhile, other researchers attempted to verify the Mozart Effect in experiments of their own. A few obtained similar results. Many, however, found no evidence of a Mozart Effect. The controversy continues, and Mozart plays on.

Equal treatment for all

Probability samples are a big idea, but sampling in practice has difficulties that just using random samples doesn't solve. Randomized comparative experiments are also a big idea, but they don't solve all the difficulties of experimenting. A sampler must know exactly what information is wanted and must compose questions that extract that information from the sample. An experimenter must know exactly what treatments and responses information is wanted about and must construct the apparatus needed to apply the treatments and measure the responses. This is what psychologists, medical researchers, and engineers mean when they talk about "designing an experiment." We are concerned with the *statistical* side of designing experiments, ideas that apply to experiments in psychology, medicine, engineering, and other areas as well. Even at this general level, you should understand the practical problems that can prevent an experiment from producing useful data.

The logic of a randomized comparative experiment assumes that all the subjects are treated alike except for the treatments that the experiment is designed to compare. Any other unequal treatment can cause bias. But treating subjects exactly alike is hard to do.

Example 6.8 Mice, rats, and rabbits

Mice, rats, and rabbits that are specially bred to be uniform in their inherited characteristics are the subjects in many experiments. Animals, like people, can be quite sensitive to how they are treated. Here are two amusing examples of how unequal treatment can create bias.

Does a new breakfast cereal provide good nutrition? To find out, compare the weight gains of young rats fed the new product and rats fed a standard diet. The rats are randomly assigned to diets and are housed in large racks of cages. It turns out that rats in upper cages grow a bit faster than rats in bottom cages. If the experimenters put rats fed the new product at the top and those fed the standard diet below, the experiment would be biased in favor of the new product.[15] Solution: Assign the rats to cages at random.

Another study looked at the effects of human affection on the cholesterol level of rabbits. All of the rabbit subjects ate the same diet. Some (chosen at random) were regularly removed from their cages to have their furry heads scratched by friendly people. The rabbits who received affection had lower cholesterol. So affection for some but not other rabbits could bias other experiments in which the rabbits' cholesterol level is a response variable.

Double-blind experiments

Placebos work. That bare fact means that medical studies must take special care to show that a new treatment is not just a placebo. Part of **equal treatment for all subjects** is to be sure that the placebo effect operates on all subjects.

Example 6.9 The powerful placebo

Want to help balding men keep their hair? Give them a placebo—one study found that 42% of balding men maintained or increased the amount of hair on their heads when they took a placebo. In another study, researchers zapped the wrists of 24 test subjects with a painful jolt of electricity. Then they rubbed a cream with no active medicine on subjects' wrists and told them the cream should help soothe the pain. When researchers shocked them again, 8 subjects said they experienced significantly less pain.[16]

When the ailment is vague and psychological, like depression, some experts think that about three-quarters of the effect of the most widely used drugs is just the placebo effect.[17] Others disagree. The strength of the placebo effect in medical treatments is hard to pin down because it depends on the exact environment. How enthusiastic the doctor is seems to matter a lot. But "placebos work" is a good place to start when you think about planning medical experiments.

The strength of the placebo effect is a strong argument for randomized comparative experiments. In the baldness study, 42% of the placebo group kept or increased their hair, but 86% of the men getting a new drug to fight baldness did so. The drug beats the placebo, so it has something besides the placebo effect going for it. Of course, the placebo effect is still part of the reason this and other treatments work.

Because the placebo effect is so strong, it would be foolish to tell subjects in a medical experiment whether they are receiving a new drug or a placebo. Knowing that they are getting "just a placebo" might weaken the placebo effect and bias the experiment in favor of the other treatments. It is also foolish to tell doctors and other medical personnel what treatment each subject received. If they know that a subject is getting "just a placebo," they may expect less than if they know the subject is receiving a promising experimental drug. Doctors' expectations change how they interact with patients and even the way they diagnose a patient's condition. Whenever possible, experiments with human subjects should be **double-blind.**

Double-blind experiments

In a **double-blind experiment,** neither the subjects nor the people who work with them know which treatment each subject is receiving.

Until the study ends and the results are in, only the study's statistician knows for sure which treatment a subject is receiving. Reports in medical journals regularly begin with words like these, from a study of a flu vaccine given as a nose spray: "This study was a randomized, double-blind, placebo-controlled trial. Participants were enrolled from 13 sites across the continental United States between mid-September and mid-November."[18] Doctors are supposed to know what this means. Now you also know.

"Dr. Burns, are you sure this is what the statisticians call a double-blind experiment?"

Refusals, nonadherers, and dropouts

Sample surveys suffer from nonresponse due to failure to contact some people selected for the sample and the refusal of others to participate. Experiments with human subjects suffer from similar problems.

Example 6.10 Minorities in clinical trials

Refusal to participate is a serious problem for medical experiments on treatments for serious diseases such as cancer. As with samples, bias can result if those who refuse are systematically different from those who cooperate.

Minorities, women, the poor, and the elderly have long been underrepresented in clinical trials. In many cases, they weren't asked. The law now requires representation of women and minorities, and data show that most clinical trials now have fair representation. But refusals remain a problem. Minorities, especially blacks, are more likely to refuse to participate. The government's Office of Minority Health says, "Though recent studies have shown that African Americans have increasingly positive attitudes toward cancer medical research, several studies corroborate that they are still cynical about clinical trials. A major impediment for lack of participation is a lack of trust in the medical establishment."[19] Some remedies for lack of trust are complete and clear information about the experiment, insurance coverage for experimental treatments, participation of black researchers, and cooperation with doctors and health organizations in black communities.

Subjects who participate but don't follow the experimental treatment, called **nonadherers,** can also cause bias. AIDS patients who participate in trials of a new drug sometimes take other treatments on their own, for example. What's more, some AIDS subjects have their medication tested privately and drop out or add other medications if they were not assigned to the new drug. This may bias the trial against the new drug.

Experiments that continue over an extended period of time also suffer **dropouts,** subjects who begin the experiment but do not complete it. If the reasons for dropping out are unrelated to the experimental treatments, no harm is done other than reducing the number of subjects. If subjects drop out because of their reaction to one of the treatments, bias can result.

Example 6.11 Dropouts in a medical study

Orlistat is a drug that may help reduce obesity by preventing absorption of fat from the foods we eat. As usual, the drug was compared with a placebo in a double-blind randomized trial. Here's what happened.

Researchers started with 1187 obese subjects. They gave a placebo to all the subjects for 4 weeks and dropped the subjects who didn't take a pill regularly. This attacked the problem of nonadherers. There were 892 subjects left. Researchers randomly assigned these subjects to orlistat or a placebo, along with a weight-loss diet. After a year devoted to losing weight, 576 subjects were still participating. On the average, the orlistat group lost 3.15 kilograms (about 7 pounds) more than the placebo group. The study kept going for another year, emphasizing maintaining the weight loss from the first year. At the end of the second year, 403 subjects were left. That's only 45% of the 892 who were randomized. Orlistat again beat the placebo, reducing the weight regained by an average of 2.25 kilograms (about 5 pounds).

Can we trust the results when so many subjects dropped out? The overall dropout rates were similar in the two groups: 54% of the subjects taking orlistat and 57% of those in the placebo group dropped out. Were dropouts related to the treatments? Placebo subjects in weight-loss experiments often drop out because they aren't losing weight. This would bias the study against orlistat because the subjects in the placebo group at the end may be those who could lose weight just by following a diet. The researchers looked carefully at the data available for subjects who dropped out. Dropouts from both groups had lost less weight than those who stayed, but careful statistical study suggested that there was little bias. Perhaps so, but the results aren't as clean as our first look at experiments promised.[20]

Exercises

6.27 Medical news When it was found that hydroxyurea reduced the symptoms of sickle-cell anemia, the National Institutes of Health released a medical bulletin. The bulletin said, "These findings are the results of data analyzed from the Multicenter Study of Hydroxyurea in Sickle Cell Anemia (MSH), which was a double-blind, placebo-controlled trial in which half of the patients received hydroxyurea and half received a placebo capsule." Explain to someone who knows no statistics what the terms "placebo-controlled" and "double-blind" mean here.

6.28 Emergency room care An article in a medical journal reports on an experiment to see if injecting an oxygen-carrying fluid in addition to performing standard emergency room procedures would help patients in shock from loss of blood. The article describes the experiment as a "randomized, controlled, single-blinded efficacy trial conducted at 18 U.S. trauma centers."[21] What do you think "single-blinded" means here? Why isn't a double-blind experiment possible?

6.29 Testing a natural remedy Although the law doesn't require it, we decide to subject Dr. Moore's Old Indiana Extract to a clinical trial. We hope to show that the extract reduces pain from arthritis. Sixty patients suffering from arthritis and needing pain relief are available. We will give a pill to each patient and ask them an hour later, "About what percent of pain relief did you experience?"

(a) Why should we not simply give the extract to all 60 patients and record the responses?

(b) Outline the design of an experiment to compare the extract's effectiveness with that of aspirin and of a placebo.

(c) Should patients be told which remedy they are receiving? How might this knowledge affect their reactions?

(d) If patients are not told which treatment they are receiving, the experiment is single-blind. Should this experiment be double-blind? Explain.

6.30 Real-world problems Explain the difference between refusals, nonadherers, and dropouts in an experimental study. Why must researchers be concerned about these three groups?

6.31 Ultrasound and birth weight In this study, researchers examined the effect of ultrasound on birth weight. Pregnant women participating in the study were randomly assigned one of two groups. The first group of women received an ultrasound; the second group did not. When the subjects' babies were born, their birth weights were recorded. The women who received the ultrasounds had heavier babies.[22]

(a) Was the experiment double-blind? Why is this important?

(b) Did the experimental design take the placebo effect into account? Why is this important?

(c) Based on your answers to (a) and (b), describe an improved design for this experiment.

6.32 Activity 6.2A follow-up Refer to the Mozart Effect experiment from Activity 6.2A (page 273).

(a) Was the experiment double-blind? Why is this important?

(b) Did the experimental design take the placebo effect into account? Why is this important?

(c) Why was the coin flip important in this experiment?

Can we generalize?

A well-designed experiment tells us that changes in the explanatory variable cause changes in the response variable. More exactly, it tells us that this happened for specific subjects in the specific environment of this specific experiment. No doubt we had bigger things in mind. We want to proclaim that our new method of teaching math does better for high school students in general or that our new drug beats a placebo for some broad class of patients. Can we **generalize** our conclusions from our little group of subjects to a wider population?

The first step is to be sure that our findings are *statistically significant,* that they are too strong to often occur just by chance. That's important, but it's a technical detail that the study's statistician can reassure us about. The serious threat is that the treatments, the subjects, or the environment of our experiment may not be realistic. Let's look at some examples.

Example 6.12 Studying frustration

A psychologist wants to study the effects of failure and frustration on the relationships among members of a work team. She forms a team of students, brings them to the psychology laboratory, and has them play a game that requires teamwork. The game is rigged so that they lose regularly. The psychologist observes the students through a one-way window and notes the changes in their behavior during an evening of game playing.

Playing a game in a laboratory for small stakes, knowing that the session will soon be over, is a long way from working for months developing a new product that never works right and is finally abandoned by your company. Does the behavior of the students in the lab tell us much about the behavior of the team whose product failed?

In Example 6.12, the subjects (students who know they are subjects in an experiment), the treatment (a rigged game), and the environment (the psychology lab) are all unrealistic if the psychologist's goal is to reach conclusions about the effects of frustration on teamwork in the workplace. Psychologists do their best to devise realistic experiments for studying human behavior, but **lack of realism** limits the usefulness of experiments in this area.

Example 6.13 Center brake lights

Cars sold in the United States since 1986 have been required to have a high center brake light in addition to the usual two brake lights at the rear of the vehicle. This safety requirement was justified by randomized comparative experiments with fleets of rental and business cars. The experiments showed that the third brake light reduced rear-end collisions by as much as 50%.

After almost a decade of actual use of center brake lights, the Insurance Institute found only a 5% reduction in rear-end collisions, helpful but much less than the experiments predicted. What happened? Most cars did not have the extra brake light when the experiments were carried out, so it caught the eye of following drivers. Now that almost all cars have the third light, it no longer captures attention. The experimental conclusions did not generalize as well as safety experts hoped because the environment changed.

Example 6.14 Are subjects treated too well?

Surely medical experiments are realistic? After all, the subjects are real patients in real hospitals really being treated for real illnesses.

Even here, there are some questions. Patients participating in medical trials get better medical care than most other patients, even if they are in the placebo group. Their doctors are specialists doing research on their specific ailment. They are watched more carefully than other patients. They are more likely to take their pills regularly because they are constantly reminded to do so. Providing "equal treatment for all" except for the experimental and control therapies translates into "provide the best possible medical care for all." The result: Ordinary patients may not do as well as the clinical trial subjects when the new therapy comes into general use. It's likely that a therapy that beats a placebo in a clinical trial will beat it in ordinary medical care, but "cure rates" or other measures of success from the trial may be optimistic.

Meta-analysis

A single study of an important issue is rarely decisive. We often find several studies in different settings, with different designs, and of different quality. Can we combine their results to get an overall conclusion? That is the idea of "meta-analysis." Of course, differences among the studies prevent us from just lumping them together. Statisticians have more sophisticated ways of combining the results. Meta-analysis has been applied to issues from the effect of secondhand smoke to whether coaching improves SAT scores.

The Carolina Abecedarian Project (Section 6.1) faces the same "too good to be realistic" question. That long and expensive experiment does show that intensive day care has substantial benefits in later life. The day care in the study was intensive indeed—lots of highly qualified staff, lots of parent participation, and detailed activities starting at a very young age, all costing about \$11,000 per year for each child. It's unlikely that society will decide to offer such care to all low-income children. The unanswered question is a big one: how good must day care be to really help children succeed in life?

When experiments are not fully realistic, statistical analysis of the experimental data cannot tell us how far the results will generalize. Experimenters generalizing from students in a lab to workers in the real world must argue based on their understanding of how people function, not based just on the data. It is even harder to generalize from rats in a lab to people in the real world. This is one reason why a single experiment is rarely completely convincing, despite the compelling logic of experimental design. The true scope of a new finding must usually be explored by a number of experiments in various settings.

A convincing case that an experiment is sufficiently realistic to produce useful information is based not on statistics but on the experimenter's knowledge of the subject matter of the experiment. The attention to detail required to avoid hidden bias also rests on subject-matter knowledge. Good experiments combine statistical principles with understanding of a specific field of study.

Experimental design in the real world

The experimental designs we have met all have the same pattern: divide the subjects at random into as many groups as there are treatments, then apply each treatment to one of the groups. These are **completely randomized designs.**

Completely randomized design

In a **completely randomized** experimental design, all the experimental subjects are allocated at random among all the treatments.

Our examples to this point have had only a single explanatory variable (drug versus placebo, classroom versus Web instruction). A completely randomized design can have any number of explanatory variables. Here is an example with two.

Example 6.15 Durable fabric

A fabrics researcher is studying the durability of a fabric under repeated washings. Because the durability may depend on the water temperature and the type of cleansing agent used, the researcher decides to investigate the effect of these two explanatory variables on durability. Variable A is water temperature and has three levels: hot (145°F), warm (100°F), and cold (50°F). Variable B is the cleansing agent and also has three levels: regular Tide, low-phosphate Tide, and Ivory Liquid. A treatment consists of washing a piece of fabric (a unit) 50 times in a home automatic washer with a specific combination of water temperature and cleansing agent. The response variable is strength after 50 washes, measured by a fabric-testing machine that forces a steel ball through the fabric and records the fabric's resistance to breaking. Figure 6.4 shows the layout of the treatments.

		Variable B Cleansing Agent		
		Regular Tide	Low-phosphate Tide	Ivory Liquid
Variable A Temperature	Hot (145°F)	Treatment 1	Treatment 2	Treatment 3
	Warm (100°F)	Treatment 4	Treatment 5	Treatment 6
	Cold (50°F)	Treatment 7	Treatment 8	Treatment 9

Figure 6.4 The treatments in the fabric durability experiment. Combinations of two explanatory variables form nine treatments.

In Example 6.15, there are nine possible treatments (combinations of a temperature and a cleansing agent). By using them all, the researcher obtains a wealth of information on how temperature alone, cleansing agent alone, and the two in combination affect the durability of the fabric. For example, water temperature may have no effect on the strength of the fabric when regular Tide is used, but after 50 washings in low-phosphate Tide, the fabric may be weaker when cold water is used instead of hot water. This kind of combination effect is called an **interaction** between cleansing agent and water temperature. Interactions can be important, as when a drug that ordinarily has no unpleasant side effects interacts with alcohol to knock out the patient who drinks a beer. Because an experiment can combine levels of several explanatory variables, interactions between the variables can be observed.

Exercises

6.33 Testing a natural remedy The National Institutes of Health is at last sponsoring proper clinical trials of some natural remedies. In one study at Duke University, 330 patients with mild depression are enrolled in a trial to compare Saint-John's-wort with a placebo and with Zoloft, a common prescription drug for depression. The Beck Depression Inventory is a common instrument that rates the severity of depression on a 0-to-3 scale.

(a) What would you use as the response variable to measure change in depression after treatment?

(b) Outline the design of a completely randomized clinical trial for this study.

(c) What other precautions would you take in this trial?

6.34 Daytime running lights Canada requires that cars be equipped with "daytime running lights," headlights that automatically come on at a low level when the car is started. Some manufacturers are now equipping cars sold in the United States with running lights. Will running lights reduce accidents by making cars more visible?

(a) Briefly discuss the design of an experiment to help answer this question. In particular, what response variables will you examine?

(b) Example 6.13 discusses center brake lights. What cautions do you draw from that example that apply to an experiment on the effects of running lights?

6.35 Diet and cancer Substances that cause cancer should not appear in our food. We don't want to experiment on people to learn what substances cause cancer, so we experiment on rats instead. The rats are specially bred to have more tumors than humans do. They are fed large doses of the test chemical for most of their natural lives, about 2 years. Briefly discuss the questions that arise in using these experiments to decide what is safe in human diets.

6.36 Dealing with cholesterol Clinical trials have shown that reducing blood cholesterol using either drugs or diet reduces heart attacks. The first researchers followed their subjects for 5 to 7 years. In order to see results as quickly as possible, the subjects were chosen from the group at greatest risk, middle-aged men with high cholesterol levels or existing heart disease. The experiments generally showed that reducing blood cholesterol does decrease the risk of a heart attack. Some doctors questioned whether these experimental results applied to many of their patients. Why?

6.37 Clean clothes Which of two brands of laundry detergent—Brand A or Brand B—cleans clothes better? Does one brand work better in both hot and cold water? Let's design an experiment to find out. A basket containing 120 pieces of dirty laundry is available for the experiment.

(a) Identify the subjects, explanatory variable(s), and response variable(s) for this experiment.

(b) Make a diagram like Figure 6.4 to describe the treatments. How many treatment combinations are there? How many pieces of laundry should be assigned to each treatment combination?

(c) Describe how you would use randomization in your experimental design.

6.38 Baking cakes A food company is preparing to market a new cake mix. It is important that the taste of the cake not be changed by small variations in baking time or temperature. In an experiment, cakes made from the mix are baked at 300°F, 320°F, or 340°F, and for 1 hour or for 1 hour and 15 minutes. Ten cakes are baked at each combination of temperature and time. A panel of tasters scores each cake for texture and taste.

(a) What are the explanatory variables and the response variables for this experiment?

(b) Make a diagram like Figure 6.4 to describe the treatments. How many treatments are there? How many cakes are needed?

(c) Explain why it is a bad idea to bake all 10 cakes for one treatment at once, then bake the 10 cakes for the second treatment, and so on. Instead, the experimenters will bake the cakes in a random order determined by the randomization in your design.

Matched pairs and block designs

Completely randomized designs are the simplest statistical designs for experiments. They illustrate clearly the principles of control and randomization. However, completely randomized designs are often inferior to more elaborate statistical designs. In particular, matching the subjects in various ways can produce more precise results than simple randomization.

One common design that combines matching with randomization is the **matched pairs design.** A matched pairs design compares just two treatments. Choose pairs of subjects that are as closely matched as possible. Randomly assign the two treatments to the subjects in each pair by tossing a coin or reading odd and even digits from Table B.

Sometimes each "pair" in a matched pairs design consists of just one subject, who gets both treatments one after the other. Each subject serves as his or her own control. The order of the treatments can influence the subject's response, so we randomize the order for each subject, again by a coin toss.

ACTIVITY 6.2B

Stepping up your heart rate

Materials: Stopwatch, two steps of height about 6 and 12 inches (you may be able to use some steps near your classroom), metronome (if available)

How is heart rate affected by physical activity? The more work your body is doing, the higher your heart rate will be. In this Activity, you will perform an experiment to examine the effect on people's heart rates of repeatedly stepping up and down on steps of two different heights.

1. One option for designing the experiment would be to randomly assign half of the students in your class to use 6-inch steps and the other half to use 12-inch steps. Explain why this is not a good idea.

2. Instead, we'll use a matched pairs design in which each subject receives both treatments. Randomly assign the students in your class to one of two treatment groups, with about equal numbers in each group. You can draw names from a hat or assign numbers and use the random

digit table or your calculator. Students assigned to Group 1 will use the 6-inch step on the first trial and the 12-inch step on the second trial. Students assigned to Group 2 will use the 12-inch step first and then the 6-inch step.

3. When your teacher instructs you to, take your pulse for 60 seconds. Record this "resting" pulse rate.

4. Your teacher will tell you when to begin the first trial. Be sure to step up and down according to the pace set by your teacher's voice (or the metronome). You will step up and down for 3 minutes.

5. At the end of the first trial, take your pulse for 60 seconds. Record the data.

6. Your teacher will give you a few minutes to "recover." Afterward, take your pulse again prior to beginning the second trial. Record this measurement.

7. For the second trial, you will switch to the other step height for 3 minutes. When you have finished, take your pulse one more time and record the result.

8. Calculate the difference in your stepping pulse rate and your resting pulse rate for each of the two trials. Did your pulse increase more when you used the higher step?

9. Make a chart on the board that shows the differences in pulse rate for each subject, like this:

Subject	Group	6″ step diff.	12″ step diff.	Diff. (12″ − 6″)

10. Construct a graph to display the differences in pulse rate for the 6- and 12-inch steps (last column in the chart). Describe what you see.

11. Calculate the average difference in pulse rate for your class. Is there evidence that heart rate increases more with greater step height? Explain.

Activity 6.2B shows you how a matched pairs experimental design can be effectively used to answer a question of interest. Here's another example of a matched pairs design that produced less clear results.

Example 6.16 Coke versus Pepsi

Pepsi wanted to demonstrate that Coke drinkers prefer Pepsi when they taste both colas blind. The subjects, all people who said they were Coke drinkers, tasted both colas from glasses without brand markings and said which they liked better. This is a matched pairs design in which each subject compares the two colas. Because responses may depend on which cola is tasted first, the order of tasting should be chosen at random for each subject.

When more than half the Coke drinkers chose Pepsi, Coke claimed that the experiment was biased. The Pepsi glasses were marked M and the Coke glasses were marked Q. Aha, said Coke, the results could just mean that people like the letter M better than the letter Q. The matched pairs design is adequate, but a more careful experiment would avoid any distinction other than Coke versus Pepsi.[23]

Matched pairs designs use the principles of comparison of treatments and randomization. However, the randomization is not complete—we do not randomly assign all the subjects at once to the two treatments. Instead, we randomize only within each matched pair. Matching reduces the effect of variation among the subjects. Matched pairs are an example of **block designs.**

Block design

A **block** is a group of experimental subjects that are known before the experiment to be similar in some way that is expected to affect the response to the treatments. In a **block design,** the random assignment of subjects to treatments is carried out separately within each block.

A block design combines the idea of creating equivalent treatment groups by matching with the principle of forming treatment groups at random. Blocks are another form of *control.* They control the effects of some outside variables by bringing those variables into the experiment to form the blocks. Here is a typical example of a block design.

Example 6.17 Men, women, and advertising

Women and men respond differently to advertising. An experiment to compare the effectiveness of 3 television commercials for the same product will want to look separately at the reactions of men and women, as well as assess the overall response to the ads.

A completely randomized design considers all subjects, both men and women, as a single pool. The randomization assigns subjects to three treatment groups without regard to their sex. This ignores the differences between men and women. A better design considers women and men separately. Randomly assign the women to three groups, one to view each commercial. Then separately assign the men at random to three groups. Figure 6.5 outlines this improved design.

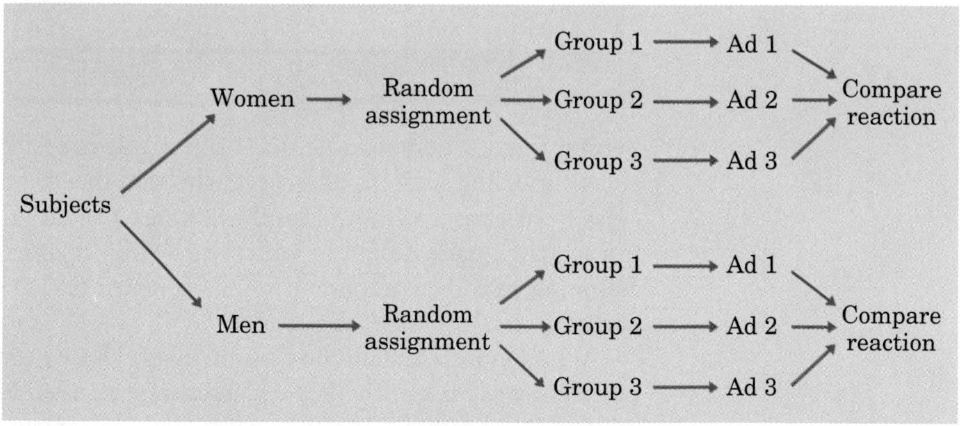

Figure 6.5 A block design to compare the effectiveness of 3 TV advertisements. Female and male subjects form 2 blocks.

A block is a group of subjects formed before an experiment starts. We reserve the word "treatment" for a condition that we impose on the subjects. We don't speak of 6 treatments in Example 6.17 even though we can compare the responses of 6 groups of subjects formed by the 2 blocks (men, women) and the 3 commercials.

Block designs are similar to stratified samples. Blocks and strata both group similar individuals together. We use two different names only because the idea developed separately for sampling and experiments. The advantages of block designs are the same as the advantages of stratified samples. Blocks allow us to draw separate conclusions about each block—for example, about men and women in the advertising study in Example 6.17. Blocking also allows more precise overall conclusions, because the systematic differences between men and women can be removed when we study the overall effects of the three commercials. The idea of blocking is an important additional principle of statistical design of experiments.

A wise experimenter will form blocks based on the most important unavoidable sources of variability among the experimental subjects. Randomization will then average out the effects of the remaining variation and allow an unbiased comparison of the treatments.

DATA EXPLORATION

Get Your Heart Beating

Are standing pulse rates generally higher than sitting pulse rates? A high school statistics class performed a set of two experiments involving 14 students to help answer this question. In this Data Exploration, you'll analyze their results.

1. In the first experiment, random assignment was used to divide the subjects into two groups of 7 students—a sitting group and a standing group. All of the students completed their assigned treatments in the same room at the same time. Then they measured their pulses for 60 seconds. A comparative dotplot and numerical summaries for these data are shown below. Write a few sentences comparing the pulse rates in the two treatment groups. What conclusion do you draw?

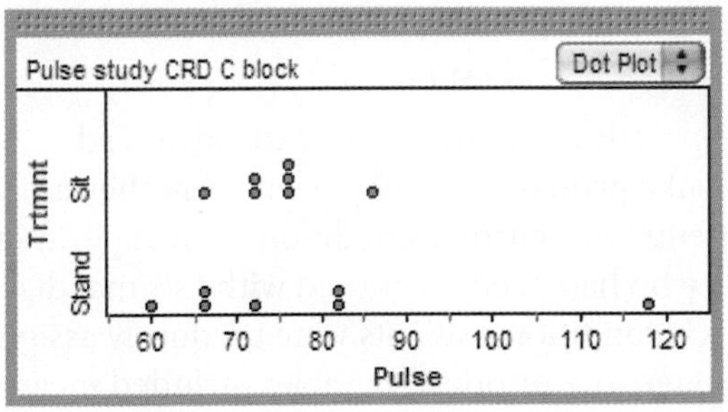

Group	n	$\bar{x}$	s	Min.	Q_1	M	Q_3	Max.
Sit	7	74.86	6.09	66	72	76	76	86
Stand	7	78	19.49	60	66	72	82	118

A second experiment was conducted using a matched pairs design with the same 14 students as subjects. This time, random assignment was used to decide which 7 students would stand first and which 7 would sit first. Once again, the students completed their assigned treatments in the same room at the same time. Then they measured their pulses for 60 seconds. Following a brief adjustment period, the subjects who sat first stood up, and those who stood first sat down. Then the subjects measured their pulse rates again for 60 seconds.

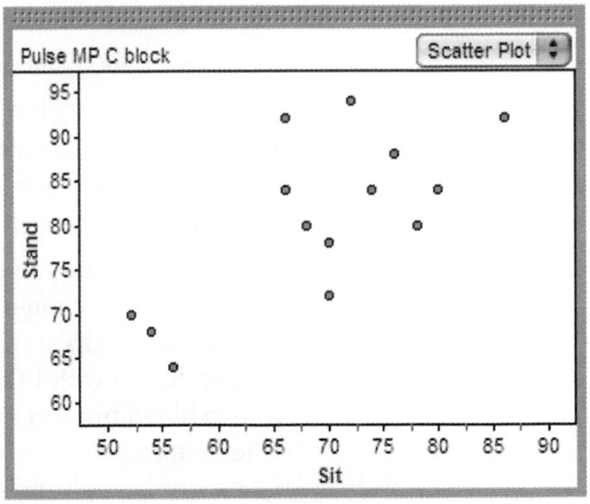

2. The scatterplot displays data from the second experiment.

(a) What does each point represent?

(b) Describe the pattern you see in the scatterplot. Explain what it means.

DATA EXPLORATION *(continued)*

DATA EXPLORATION *(continued)*

3. For each subject, the difference between standing and sitting pulse rate was calculated (difference = standing − sitting). A dotplot and numerical summaries of the difference values are shown here. Write a few sentences describing the distribution of difference in pulse rates. What conclusion do you draw now?

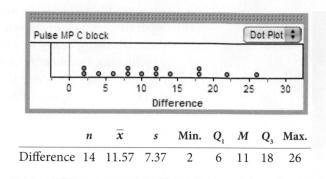

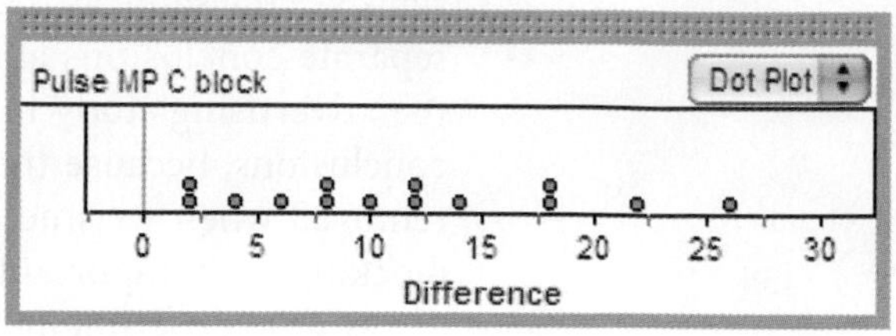

	n	$\bar{x}$	s	Min.	Q_1	M	Q_3	Max.
Difference	14	11.57	7.37	2	6	11	18	26

Exercises

6.39 In the cornfield An agriculture researcher wants to compare the yield of 5 corn varieties. The field in which the experiment will be carried out increases in fertility from north to south. The researcher therefore divides the field into 30 plots of equal size, arranged in 6 east-west rows of 5 plots each, and employs a block design with the rows of plots as the blocks.

(a) Draw a sketch of the field, divided into 30 plots. Label the rows Block 1 to Block 6.

(b) Do the randomization required by the block design. That is, randomly assign the 5 corn varieties A, B, C, D, and E to the 5 plots in each block. Mark on your sketch which variety is planted in each plot.

6.40 Comparing cancer treatments The progress of a type of cancer differs in women and men. A clinical experiment to compare three therapies for this cancer therefore treats sex as a blocking variable.

(a) You have 500 male and 300 female patients who are willing to serve as subjects. Use a diagram to outline a block design for this experiment. Use Figure 6.5 as a model.

(b) What are the advantages of a block design over a completely randomized design using these 800 subjects? What are the advantages of a block design over a completely randomized design using 800 male subjects?

6.41 Doctors and nurses Nurse-practitioners are nurses with advanced qualifications who often act much like primary-care physicians. Are they as effective as doctors at treating patients with chronic conditions? An experiment was conducted with 1316 patients who had been diagnosed with asthma, diabetes, or high blood pressure. Within each condition, patients were randomly assigned to either a doctor or a nurse-practitioner. The response variables included measures of the patients' health and of their satisfaction with their medical care after 6 months.[24]

(a) Is the diagnosis (asthma, etc.) a treatment variable or a block? Why?

(b) Is the type of care (nurse or doctor) a treatment variable or a block? Why?

6.42 More rats In Exercise 6.12 (page 266), a nutritionist had 10 rats of each of two genetic strains. The effect of genetic strain can be controlled by treating the strains as blocks and randomly assigning 5 rats of each strain to Diet A. The remaining 5 rats of each strain receive Diet B.

(a) Describe the design of this experiment.

(b) Use Table B, beginning at line 111, to do the required randomization.

6.43 Activity 6.2B follow-up Refer to the heart rate experiment from Activity 6.2B.

(a) Why did we randomize the order in which each student used the two different step heights?

(b) Examine the class data from a different angle. Do you see any evidence of a "learning effect" from the first step height to the second? Follow the four-step statistical problem-solving process (page 19) as you answer this question.

6.44 Data Exploration follow-up Refer to the Data Exploration (page 287) on standing and sitting pulse rates.

(a) What statistical advantages did the class's matched pairs design have over its completely randomized design?

(b) In the matched pairs experiment, why did the class randomize who stood first and who sat first?

APPLICATION 6.2

Design your own study

In this Application, you will design a taste-testing experiment in which subjects will be asked which of two similar items tastes better to them. However, the issue is not which item the subjects prefer, but rather *whether or not preferences are different when you know what is being tasted.* "Blinding" subjects is an important precaution in testing a new medical treatment because subjects might be influenced by the idea of being treated. We want to design a study to see if blinding is also an important issue for taste-testing experiments. For example, perhaps people will tend to state a preference for a more expensive item because they think they are supposed to like it better.

Your group will develop a complete experimental plan to determine whether blinding affects subjects' preferences. Your group's plan should describe the following:

- precisely what the two items being tested are,
- how the items will be served (for example, will corn flakes be served with nonfat, low-fat, or regular milk?),
- how you will decide which item to present first and which to present second,
- how you will determine which subjects are to be blinded and which will be told what they are tasting, and
- what other special precautions you will take to avoid confounding.

Incorporate these items into a written report describing your experiment.

Here are some examples of items you might taste-test:

- a nonfat yogurt versus a regular yogurt of the same brand,
- Kellogg's corn flakes versus a store brand of corn flakes,
- orange juice from a carton versus freshly squeezed orange juice, or
- french fries cooked in a microwave versus the same brand cooked in a conventional oven.

Remember the issue at hand: whether or not preferences are different when you know what is being tasted.

Once your teacher approves your design, carry out your experiment and analyze the data. Follow the four-step statistical problem-solving process (page 19).

Section 6.2 Summary

As with samples, experiments require a combination of good statistical design and steps to deal with practical problems. Because the **placebo effect** is strong, **clinical trials** and other experiments with human subjects should be **double-blind** whenever possible. The double-blind method helps achieve a basic requirement of comparative experiments: **equal treatment for all subjects** except for the actual treatments the experiment is comparing.

Just as samples suffer from nonresponse, experiments suffer from uncooperative subjects. Some subjects refuse to participate; others **drop out** before the experiment is complete; others, called **nonadherers,** don't follow instructions, as when some subjects in a drug trial don't take their pills. The most common weakness in experiments is that we can't **generalize** the conclusions widely. Some experiments apply unrealistic treatments, some use subjects from some special group such as college students, and all take place at some specific place and time. We need similar experiments at other places and times to confirm important findings.

Many experiments use designs that are more complex than the basic **completely randomized design** that divides all the subjects among all the treatments in one randomization. **Matched pairs designs** compare two treatments by giving one to one member of a pair of similar subjects and the other to the other member or by giving both to the same subject in random order. **Block designs** form blocks of similar subjects and assign treatments at random separately in each block. The big ideas of randomization, control, and using enough subjects remain the keys to convincing experiments.

Section 6.2 Exercises

6.45 Better air bags Car makers are interested in developing safer air bags that reduce injuries in collisions. An automobile manufacturer is trying to decide between two possible air bags for use in the company's vehicles. Six reusable crash test dummies—large male, average male, small male, large female, average female, and small female—are available for use on the company's indoor "test track." Carefully describe the design of an experiment to help the company decide which air bag is safer.

6.46 Meditation lowers anxiety An experiment that was publicized as showing that a meditation technique lowered the anxiety level of subjects was conducted as follows: The experimenter interviewed the subjects and assessed their levels of anxiety. The subjects then learned how to meditate and did so regularly for a month. The experimenter reinterviewed them at the end of the month and assessed whether their anxiety levels had decreased or not.

(a) There was no control group in this experiment. Why is this a blunder? What lurking variables may be confounded with the effect of meditation?

(b) The experimenter who diagnosed the effect of the treatment knew that the subjects had been meditating. Explain how this knowledge could bias the experimental conclusion.

(c) Briefly discuss a proper experimental design, with controls and blind diagnosis, to assess the effect of meditation on anxiety level.

6.47 Ugly french fries Few people want to eat discolored french fries. Potatoes are kept refrigerated before being cut for french fries to prevent spoiling and preserve flavor. But immediate processing of cold potatoes causes discoloring due to complex chemical reactions. The potatoes must therefore be brought to room temperature

before processing. Design an experiment in which tasters will rate the color and flavor of french fries prepared from several groups of potatoes. The potatoes will be fresh picked or stored for a month at room temperature or stored for a month refrigerated. They will then be sliced and cooked either immediately or after an hour at room temperature.
(a) What are the explanatory variables, the treatments, and the response variables?
(b) Describe and outline the design of this experiment.
(c) It is efficient to have each taster rate fries from all treatments. How will you use randomization in presenting fries to the tasters?

6.48 Drug improves coordination A drug is suspected of affecting the coordination of subjects. The drug can be administered in three ways: orally, by injection under the skin, or by injection into a vein. The potency of the drug probably depends on the method of administration as well as on the dose administered. A researcher therefore wishes to study the effects of the two explanatory variables: dose at two levels and method of administration by the three methods mentioned. The response variable is the score of the subjects on a standard test of coordination. Ninety subjects are available.
(a) List the treatments that can be formed from the two explanatory variables.
(b) Describe an appropriate completely randomized design. (Just outline the design; don't do any randomization.)
(c) The researcher could study the effect of the dose in an experiment comparing two dose levels for one method of administration. The researcher then could separately study the effect of administration by comparing the three methods for one dose level. What advantage does the experiment you designed in (a) have over these two separate experiments taken together?

6.49 Harmful auto emissions Most motor vehicles are equipped with catalytic converters to reduce harmful emissions. The ceramic used to make the converters must be baked to a certain hardness. The manufacturer must decide which of three temperatures (500°F, 750°F, and 1000°F) is best. The position of the converter in the oven (front, middle, or back) also affects the hardness. So there are two explanatory variables: temperature and placement.
(a) List the treatments in this experiment if all possible treatment combinations are used.
(b) Design a completely randomized experiment with 5 units in each group.
(c) Using Table B, beginning at line 101, do the randomization for your experiment.

6.50 Morphine and cancer Health care providers are giving more attention to relieving the pain of cancer patients. An article in the journal *Cancer* reviewed a number of studies and concluded that controlled-release morphine tablets, which release the painkiller gradually over time, are more effective than giving standard morphine when the patient needs it.[25] The "methods" section of the article begins: "Only those published studies that were controlled (i.e., randomized, double-blind, and comparative), repeated dose studies with CR morphine tablets in cancer pain patients were considered for this review." Explain the terms in parentheses to someone who knows nothing about medical trials.

6.51 Is caffeine dependence real? Many people start their day with a jolt of caffeine from coffee or a soft drink. Most experts agree that people who take in large amounts of caffeine each day may suffer from physical withdrawal symptoms if they stop ingesting their usual amounts of caffeine. Researchers recruited 11 volunteers who were caffeine dependent and who were willing to take part in a caffeine withdrawal experiment. The experiment was conducted on two 2-day periods that occurred one week apart. During one of the 2-day periods, each subject was given a capsule containing the amount of caffeine normally ingested by that subject in one day. During the other study period, the subjects were given placebos. The order in which each subject received the two types of capsules was randomized. The subjects' diets were restricted during each of the study periods. At the end of each 2-day study period, subjects were evaluated using a tapping task in which they were instructed to press a button 200 times as fast as they could.[26]

(a) Carefully describe the design of this experiment. How was blocking incorporated, and why?

(b) Why did researchers randomize the order in which subjects received the two treatments?

(c) Could this experiment have been carried out in a double-blind manner? Explain.

6.52 Bee healed! "Bee pollen is effective for combating fatigue, depression, cancer, and colon disorders." So says a Web site that offers the pollen for sale. We wonder if bee pollen really does prevent colon disorders. Here are two ways to study this question. Explain why the second design will produce more trustworthy data.

- Find 200 women who take bee pollen regularly. Match each with a woman of the same age, race, and occupation who does not take bee pollen. Follow both groups for 5 years.
- Find 400 women who do not have colon disorders. Assign 200 to take bee pollen capsules and the other 200 to take placebo capsules that are identical in appearance. Follow both groups for 5 years.

6.3 Data Ethics

ACTIVITY 6.3

Is it ethical?

Medical professionals are taught to follow the basic principle "First, do no harm." Shouldn't those who carry out statistical studies follow the same principle? Most reasonable people think so. But this may not always be as simple as it sounds. Not convinced? Discuss each of the following statistical practices with your classmates. Decide whether each is ethical or unethical.

1. A promising new drug has been developed for treating cancer in humans. Researchers want to administer the drug to animal subjects to see if there are any potentially serious side effects before giving the drug to human subjects.

2. Are companies discriminating against some individuals in the hiring process? To find out, researchers prepare several equivalent résumés for fictitious job applicants, with the only difference being the sex of the applicant. They send the fake

By permission of Dave Coverly and Creators Syndicate, Inc.

résumés to the companies advertising positions and keep track of the number of males and females who are contacted for interviews.

3. In a medical study of a new drug for migraine sufferers, volunteer subjects are randomly assigned to two groups. Members of the first group are given a placebo pill. Subjects in the second group are given the new drug. None of the subjects know whether they are taking a placebo or the active drug. Neither do any of the physicians who are interacting with the subjects know.

4. Will people try to stop someone from driving drunk? A television news program hires an actor to play a drunk driver and records the behavior of individuals who encounter the driver on hidden camera.

Are sham surgeries ethical?

"Randomized, double-blind, placebo-controlled trials are the gold standard for evaluating new interventions and are routinely used to assess new medical therapies."[27] So says an article in the *New England Journal of Medicine* that discusses the treatment of Parkinson's disease. The article isn't about the new treatment, which offers hope of reducing the tremors and lack of control brought on by the disease, but about the ethics of studying the treatment.

The law requires well-designed experiments to show that new drugs work and are safe. Not so with surgery—only about 7% of studies of surgery use randomized comparisons. Surgeons think their operations succeed—but innovators always think their innovations work. Even if the patients are helped, the placebo effect may deserve most of the credit. So we don't really know whether many common surgeries are worth the risk they carry. To find out, do a proper experiment. That includes a "sham surgery" to serve as a placebo. In the case of Parkinson's disease, the promising treatment involves surgery to implant new cells. The placebo subjects get the same surgery, but the cells are not implanted.

Is this ethical? One side says: Only a randomized comparative experiment can tell us if the new treatment works. If it doesn't work, thousands of patients will be spared operations that will do no good. If it does work, we can do those thousands of operations knowing that we offer more than a placebo. The other side says: Any surgery carries risks. Sham operations place patients at risk without the hope that the operation will help them. Doctors should not risk harm to some patients today just because other patients may benefit tomorrow.

Reasonable people disagree about sham surgery as part of a test of real surgery for Parkinson's disease. But there is wide agreement on some basic ethical principles for statistical studies. To think about the hard cases, we should first think about these principles.

First principles

The most complex issues of data ethics arise when we collect data from people. The ethical difficulties are more severe for experiments that impose some treatment on people than for sample surveys that simply gather information. Trials of new medical treatments, for example, can do harm as well as good to their subjects. Here are some basic standards of data ethics that must be obeyed by any study that gathers data from human subjects, whether sample survey or experiment.

Basic data ethics

The organization that carries out the study must have an **institutional review board** that reviews all planned studies in advance in order to protect the subjects from possible harm.

All individuals who are subjects in a study must give their **informed consent** before data are collected.

All individual data must be kept **confidential.** Only statistical summaries for groups of subjects may be made public.

The law requires that studies funded by the federal government obey these principles. But neither the law nor the consensus of experts is completely clear about the details of their application.

Institutional review boards

The purpose of an institutional review board is not to decide whether a proposed study will produce valuable information or whether it is statistically sound. The board's purpose is, in the words of one university's board, "to protect the rights and welfare of human subjects (including patients) recruited to participate in research activities." The board reviews the plan of the study and can require changes. It reviews the consent form to be sure that subjects are informed about the nature of the study and about any potential risks. Once research begins, the board monitors its progress at least once a year.

The most pressing issue concerning institutional review boards is whether their workload has become so large that their effectiveness in protecting subjects drops. When the government temporarily stopped human-subject research at Duke University Medical Center in 1999 due to inadequate protection of subjects, more than 2000 studies were going on. That's a lot of review work. There are shorter review procedures for projects that involve only minimal risks to subjects, such as most sample surveys. When a board is overloaded, there is a temptation to put more proposals in the minimal-risk category to speed the work.

Informed consent

Both words in the phrase "informed consent" are important, and both can be controversial. Subjects must be informed in advance about the nature of a study and any risk of harm it may bring. In the case of a sample survey, physical harm is not possible, but there could be emotional harm (like asking a rape victim questions about rape). The subjects should be told what kinds of questions the survey will ask and about how much of their time it will take. Experimenters must tell subjects the nature and purpose of the study and outline possible risks. Subjects must then *consent* in writing.

Example 6.18 Who can consent?

Are there some subjects who can't give informed consent? It was once common, for example, to test new vaccines on prison inmates who gave their consent in return for good-behavior credit. Now we worry that prisoners are not really free to refuse, and the law forbids medical experiments in prisons.

Children can't give fully informed consent, so the usual procedure is to ask their parents. A study of new ways to teach reading is about to start at a local elementary school, so the study team sends consent forms home to parents. Many parents don't return the forms. Can their children take part in the study because the parents did not say "No," or should we allow only children whose parents returned the form and said "Yes"?

What about research into new medical treatments for people with mental disorders? What about studies of new ways to help emergency room patients who may be unconscious or have suffered a stroke? In most cases, there is not time even to get the consent of the family. Does the principle of informed consent bar realistic trials of new treatments for unconscious patients? These are questions without clear answers. Reasonable people differ strongly on all of them. There is nothing simple about informed consent.[28] □

The difficulties of informed consent do not vanish even for capable subjects. Some researchers, especially in medical trials, regard consent as a barrier to getting patients to participate in research. They may not explain all possible risks; they may not point out that there are other therapies that might be better than those being studied; they may be too optimistic when talking with patients even when the consent form has all the right details. On the other hand, mentioning every possible risk leads to very long consent forms that really are barriers. "They are like rental car contracts," one lawyer said. Some subjects don't read forms that run five or six printed pages. Others are frightened by the large number of possible (but unlikely) disasters that might happen and so refuse to participate. Of course, unlikely disasters sometimes happen. When they do, lawsuits follow and the consent forms become even longer and more detailed.

Confidentiality

Ethical problems do not disappear once a study has been cleared by the review board, has obtained consent from its subjects, and has actually collected data about the subjects. It is important to protect the subjects' privacy by keeping all data about individuals **confidential.** The report of an opinion poll may say what percent of the 1500 respondents felt that legal immigration should be reduced. It may not report what *you* said about this or any other issue.

Confidentiality is not the same as **anonymity.** Anonymity means that subjects are anonymous—their names are not known even to the director of the study. Anonymity is rare in statistical studies. Even where anonymity is possible (mainly in surveys conducted by mail), it prevents any follow-up to improve nonresponse or inform subjects of results.

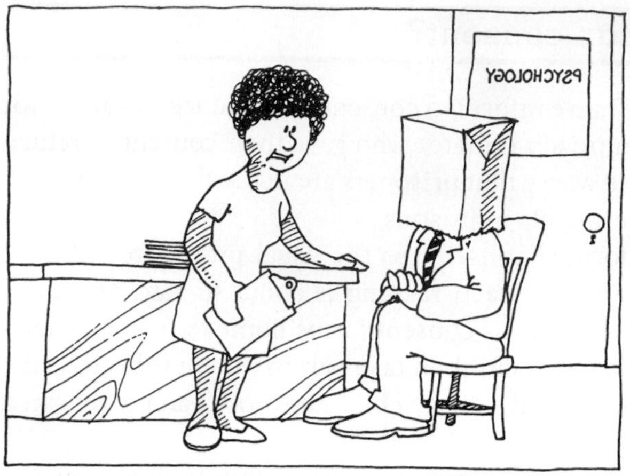

"I realize the participants in this study are to be anonymous, but you're going to have to expose your eyes."

Any breach of confidentiality is a serious violation of data ethics. The best practice is to separate the identity of the subjects from the rest of the data at once. Sample surveys, for example, use the identification only to check on who did or did not respond. In an era of advanced technology, however, it is no longer enough to be sure that each individual set of data protects people's privacy.

The government, for example, maintains a vast amount of information about citizens in many separate data bases—census responses, tax returns, Social Security information, data from surveys such as the Current Population Survey, and so on. Many of these data bases can be searched by computers for statistical studies. A clever computer search of several data bases might be able, by combining information, to identify you and learn a great deal about you even if your name and other identification have been removed from the data available for search. Privacy and confidentiality of data are hot issues among statisticians in the computer age.

Exercises

Most of the exercises in this section pose issues for discussion. There are few right or wrong answers, but there are more and less thoughtful answers.

David Gilder/FeaturePics

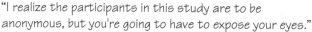

6.53 Opinion polls The presidential election campaign is in full swing, and the candidates have hired polling organizations to take regular polls to find out what the voters think about the issues. What information should the pollsters be required to give out?

(a) What does the standard of informed consent require the pollsters to tell potential respondents?

(b) The standards accepted by polling organizations also require giving respondents the name and address of the organization that carries out the poll. Why do you think this is required?

(c) The polling organization usually has a professional name such as "Samples Incorporated," so respondents don't know that the poll is being paid for by a political party or candidate. Would revealing the sponsor to respondents bias the poll? Should the sponsor always be announced whenever poll results are made public?

6.54 Minimal risk? You have been invited to serve on a college's institutional review board. You must decide whether several research proposals qualify for lighter review because they involve only minimal risk to subjects. Federal regulations say that "minimal risk" means the risks are no greater than "those ordinarily encountered in daily life or during the performance of routine physical or psychological examinations or tests." That's vague. Which of these do you think qualifies as "minimal risk"?

(a) Draw a drop of blood by pricking a finger in order to measure blood sugar.

(b) Draw blood from the arm for a full set of blood tests.

(c) Insert a tube that remains in the arm, so that blood can be drawn regularly.

6.55 Is consent needed? In which of the circumstances below would you allow collecting personal information without the subjects' consent?

(a) A government agency takes a random sample of income tax returns to obtain information on the average income of people in different occupations. Only the incomes and occupations are recorded from the returns, not the names.

(b) A social psychologist attends public meetings of a religious group to study the behavior patterns of members.

(c) The social psychologist pretends to be converted to membership in a religious group and attends private meetings to study the behavior patterns of members.

6.56 Informed consent A researcher suspects that traditional religious beliefs tend to be associated with an authoritarian personality. She prepares a questionnaire that measures authoritarian tendencies and also asks many religious questions. Write a description of the purpose of this research to be read by subjects in order to obtain their informed consent. You must balance the conflicting goals of not deceiving the subjects as to what the questionnaire will tell about them and of not biasing the sample by scaring off religious people.

6.57 Anonymous or confidential? Texas A&M, like many universities, offers free screening for HIV, the virus that causes AIDS. The announcement says, "Persons who sign up for the HIV Screening will be assigned a number so that they do not have to give their name." They can learn the results of the test by telephone, still without giving their name. Does this practice offer *anonymity* or just *confidentiality*?

6.58 Not really anonymous Some common practices may appear to offer anonymity while actually delivering only confidentiality. Market researchers often use mail surveys that do not ask the respondent's identity but contain hidden codes on the questionnaire that identify the respondent. A false claim of anonymity is clearly unethical. If only confidentiality is promised, is it also unethical to say nothing about the identifying code, perhaps causing respondents to believe their replies are anonymous?

Clinical trials

Clinical trials are experiments that study the effectiveness of medical treatments on actual patients. Medical treatments can harm as well as heal, so clinical trials spotlight the ethical problems of experiments with human subjects. Here are the starting points for a discussion:

- Randomized comparative experiments are the only way to see the true effects of new treatments. Without them, risky treatments that are no better than placebos will become common.
- Clinical trials produce great benefits, but most of these benefits go to future patients. The trials also pose risks, and these risks are borne by the subjects of the trial. So we must balance future benefits against present risks.

- Both medical ethics and international human rights standards say that "the interests of the subject must always prevail over the interests of science and society."

The quoted words are from the 1964 Helsinki Declaration of the World Medical Association, the most respected international standard. The most outrageous examples of unethical experiments are those that ignore the interests of the subjects.

Statisticians, honest and dishonest

Developed nations rely on government statisticians to produce honest data. We trust the monthly unemployment rate, for example, to guide both public and private decisions. Honesty can't be taken for granted everywhere, however. In 1998, the Russian government arrested the top statisticians in the State Committee for Statistics. They were accused of taking bribes to fudge data to help companies avoid taxes. "It means that we know nothing about the performance of Russian companies," said one newspaper editor.

Example 6.19 The Tuskegee syphilis study

In the 1930s, syphilis was common among black men in the rural South, a group that had almost no access to medical care. The Public Health Service recruited 399 poor black sharecroppers with syphilis and 201 others without the disease in order to observe how syphilis progressed when no treatment was given. Beginning in 1943, penicillin became available to treat syphilis. The study subjects were not treated. In fact, the Public Health Service prevented any treatment until word leaked out and forced an end to the study in the 1970s.

The Tuskegee study is an extreme example of investigators following their own interests and ignoring the well-being of their subjects. A 1996 review said, "It has come to symbolize racism in medicine, ethical misconduct in human research, paternalism by physicians, and government abuse of vulnerable people."[29] In 1997, President Clinton formally apologized to the surviving participants in a White House ceremony.

www For additional details of the Tuskegee syphilis study, visit **www.cdc.gov/tuskegee/**.

Because "the interests of the subject must always prevail," medical treatments can be tested in clinical trials only when there is reason to hope that they will help the patients who are subjects in the trials. Future benefits aren't enough to justify experiments with human subjects. Of course, if there is already strong evidence that a treatment works and is safe, it is unethical *not* to give it. Here are the words of Dr. Charles Hennekens of Harvard Medical School, who directed the large clinical trial that showed that aspirin reduces the risk of heart attacks:

> *There's a delicate balance between when to do or not do a randomized trial. On the one hand, there must be sufficient belief in the agent's potential to justify exposing half the subjects to it. On the other hand, there must be sufficient doubt about its efficacy to justify withholding it from the other half of subjects who might be assigned to placebos.*[30]

Why is it ethical to give a control group of patients a placebo? Well, we know that placebos often work. What is more, placebos have no harmful side effects. So in the state of balanced doubt described by Dr. Hennekens, the placebo group may be getting a better treatment than the drug group. If we *knew* which treatment was better, we would give it to everyone. When we don't know, it is ethical to try both and compare them.

Behavioral and social science experiments

When we move from medicine to the behavioral and social sciences, the direct risks to experimental subjects are less acute, but so are the possible benefits to the subjects. Consider, for example, some experiments conducted by psychologists in their study of human behavior.

Example 6.20 Group pressure

Stanley Milgram of Yale conducted a famous experiment "to see if a person will perform acts under group pressure that he would not have performed in the absence of social inducement."[31]

The subject arrives with three others who (unknown to the subject) are confederates of the experimenter. The experimenter explains that he is studying the effects of punishment on learning. The Learner (one of the confederates) is strapped into an electric chair, mentioning in passing that he has a mild heart condition. The subject and two other confederates are Teachers. They sit in front of a panel with switches with labels ranging from "Slight Shock" to "Danger: Severe Shock" and are told that they must shock the Learner whenever he fails in a memory learning task. How badly the Learner is shocked is up to the Teachers and will be the lowest level suggested by any Teacher on that trial.

All is rigged. The Learner answers incorrectly on 30 of the 40 trials. The two Teacher confederates call for a higher shock level at each failure. As the shock increases, the Learner protests, asks to be let out, shouts, complains of heart trouble, and finally screams in pain. (The "shocks" are phony and the Learner is an actor, but the subject doesn't know this.) What will the subject do? He can keep the shock at the lowest level, but the two other Teachers are pressuring him to punish the Learner more for each failure.

The subjects usually give in to the pressure. "While the experiment yields wide variation in performance, a substantial number of subjects submitted readily to pressure applied to them by the confederates." Milgram noted that, in questioning after the experiment, the subjects often admitted that they had acted against their own principles and were upset by what they had done. "The subject was then dehoaxed carefully and had a friendly reconciliation with the victim."[32]

www You can find more information about Stanley Milgram's research, including the infamous shock therapy experiment, at **www.stanleymilgram.com.**

This experiment was acceptable in the early 1960s, when Milgram did his work. Psychologists felt it gave insight into the power of group pressure to coerce individuals into acts they would not otherwise commit. Now, however, such a study would be considered clearly unethical.

Milgram's experiment illustrates the difficulties facing those who plan and review behavioral studies.

- There is no risk of physical harm to the subjects, but they would certainly risk embarrassment and lingering emotional effects. What should we protect subjects from when physical harm is unlikely? Possible emotional harm? Undignified situations? Invasion of privacy?

"I'm doing a little study on the effects of emotional stress. Now, just take the axe from my assistant."

- What about informed consent? Many behavioral experiments rely on hiding the true purpose of the study. Some, like Milgram's, actively deceive subjects. Since the subjects might change their behavior if told in advance what the investigators are looking for, they are asked to consent on the basis of vague information. They receive full information only after the experiment.

The "Ethical Principles" of the American Psychological Association require consent unless a study merely observes behavior in a public place. They allow deception only when it is necessary to the study, does not hide information that might influence a subject's willingness to participate, and is explained to subjects as soon as possible. Milgram's study (from the 1960s) does not meet current ethical standards because subjects did not give informed consent to take part in an experiment.

We see that the basic requirement for informed consent is understood differently in medicine and psychology. Here is an example of another setting with yet another interpretation of what is ethical. The subjects get no information and give no consent. They don't even know that an experiment may be sending them to jail for the night.

Example 6.21 Domestic violence

How should police respond to domestic-violence calls? In the past, the usual practice was to remove the offender and order him to stay out of the household overnight. Police were reluctant to make arrests because the victims rarely pressed charges. Women's groups argued that arresting offenders would help prevent future violence even if no charges were filed. Is there evidence that arrest will reduce future offenses? That's a question that experiments have tried to answer.

A typical domestic-violence experiment compares two treatments: arrest the suspect and hold him overnight, or warn the suspect and release him. When police officers reach the scene of a domestic-violence call, they calm the participants and investigate. Weapons or death threats require an arrest. If the facts permit an arrest but do not require it, an officer radios headquarters for instructions. The person on duty opens the next envelope in a file prepared in advance by a statistician. The envelopes contain the treatments in random order. The police either arrest the suspect or warn and release him, depending on the contents of the envelope. The researchers then monitor police records and visit the victim to see if the domestic violence reoccurs.

The first such experiment appeared to show that arresting domestic-violence suspects does reduce their future violent behavior. As a result of this evidence, arrest has become the common police response to domestic violence.

The domestic-violence experiments shed light on an important issue of public policy. Because there is no informed consent, the ethical rules that govern clinical trials and most social science studies would forbid these experiments. They were cleared by review boards because, in the words of one domestic-violence researcher, "These people became subjects by committing acts that allow the police to arrest them. You don't need consent to arrest someone."

Exercises

6.59 Sham surgery? Clinical trials like the Parkinson's disease study cited at the beginning of this section are becoming more common. One medical researcher says, "This is just the beginning. Tomorrow, if you have a new procedure, you will have to do a double-blind placebo trial."[33] Arguments for and against testing surgery just as drugs are tested are given on page 293. When would you allow sham surgery in a clinical trial of a new surgery?

6.60 AIDS clinical trials Now that effective treatments for AIDS are at last available, is it ethical to test treatments that may be less effective? Combinations of several powerful drugs reduce the level of the virus in the blood and at least delay illness and death from AIDS. But effectiveness depends on how damaged the patient's immune system is and what drugs he or she has previously taken. There are strong side effects, and patients must be able to take more than a dozen pills on time every day. Because AIDS is often fatal and the combination therapy works, we might argue that it isn't ethical to test any treatment for AIDS that doesn't give the full combination. But not doing further tests might prevent discovery of better treatments. This is a strong example of the conflict between doing the best we know for patients now and finding better treatments for other patients in the future. How can we ethically test new drugs for AIDS?[34]

6.61 Equal treatment Researchers on aging proposed to investigate the effect of supplemental health services on the quality of life of older people. Eligible patients on the rolls of a large medical clinic were to be randomly assigned to treatment and control groups. The treatment group would be offered hearing aids, dentures, transportation, and other services not available without charge to the control group. The review board felt that providing these services to some but not other persons in the same institution raised ethical questions. Do you agree?

6.62 Placebos At present there is no vaccine for a serious viral disease. A vaccine is developed and appears effective in animal trials. Only a comparative experiment with human subjects in which a control group receives a placebo can determine the true worth of the vaccine. Is it ethical to give some subjects the placebo, which cannot protect them against the disease?[35]

6.63 Tempting subjects A psychologist conducts the following experiment: she measures the attitude of subjects toward cheating, then has them play a game rigged so that winning without cheating is impossible. The computer that organizes the game also records—unknown to the subjects—whether or not they cheat. Then attitude toward cheating is retested.

Subjects who cheat tend to change their attitudes to find cheating more acceptable. Those who resist the temptation to cheat tend to condemn cheating more strongly on the second test of attitude. These results confirm the psychologist's theory.

This experiment tempts subjects to cheat. The subjects are led to believe that they can cheat secretly when in fact they are observed. Is this experiment ethically objectionable? Explain your position.

6.64 Students as subjects Students taking Psychology 001 are required to serve as experimental subjects. Students in Psychology 002 are not required to serve, but they are given extra credit if they do so. Students in Psychology 003 are required either to sign up as subjects or to write a term paper. Serving as an experimental subject may be educational, but current ethical standards frown on using "dependent subjects" such as prisoners or charity medical patients. Students are certainly somewhat dependent on their teachers. Do you object to any of these course policies? If so, which ones, and why?

APPLICATION 6.3

Hope for sale?

We have pointed to the ethical problems of experiments with human subjects, clinical trials in particular. *Not* doing proper experiments can also pose problems. Here is an example. Women with advanced breast cancer will eventually die. A promising but untried treatment appears. Should we wait for controlled clinical trials to show that it works, or should we make it available right now?

The promising treatment is "bone marrow transplant" (BMT for short). The idea of BMT is to harvest a patient's bone marrow cells, blast the cancer with very high doses of drugs, then return the harvested cells to keep the drugs from killing the patient. BMT has become popular. It is painful, expensive, and dangerous.

New anticancer drugs are first available through clinical trials, but there is no constraint on therapies such as BMT. When small, uncontrolled trials seemed to show success, BMT became widely available. But does BMT keep patients alive longer than standard treatments? We don't know, but the answer appears to be "probably not." The patients naturally want to try anything that might keep them alive. Patients would not join controlled trials that might assign them to standard treatments rather than to BMT. Results from such trials were delayed for years by the difficulty in recruiting subjects. Of the first five trials reported, four found no significant difference between BMT and standard treatments. The fifth favored BMT—but the researcher soon admitted "a serious breach of scientific honesty and integrity." The *New York Times* put it more bluntly: "he falsified data."[36]

QUESTIONS

1. Outline the design of an experiment that compares BMT to an existing treatment.

2. Describe how you would incorporate each of the three basic principles of data ethics in your design from Question 1.

3. Explain what "no significant difference between BMT and standard treatments" means.

4. Compassion seems to support making untested treatments available to dying patients. Reason responds that this approach gives patients false hope and delays development of treatments that really work. Which position do you support? Why?

Section 6.3 Summary

Ordinary honesty says you shouldn't make up data or claim to be taking a survey when you are really selling security alarms. Data ethics begin with some principles that go beyond just being honest. Studies with human subjects must be screened in advance by an **institutional review board.** All subjects must give their **informed consent** before taking part. All information about individual subjects must be kept **confidential.** These principles are a good start, but many ethical debates remain, especially in the area of experiments with human subjects. Many of the debates concern the right balance between the welfare of the subjects and the future benefits of the experiment. Remember that randomized comparative experiments can answer questions that can't be answered without them. Also remember that "the interests of the subject must always prevail over the interests of science and society."

Section 6.3 Exercises

6.65 Surveys of youth A survey asked teenagers whether they had ever consumed an alcoholic beverage. Those who said "Yes" were then asked,

How old were you when you first consumed an alcoholic beverage?

Should consent of parents be required to ask minors about alcohol, drugs, and other such issues, or is consent of the minors themselves enough? Give reasons for your opinion.

6.66 Who serves on the review board? Government regulations require that institutional review boards consist of at least five people, including at least one scientist, one nonscientist, and one person from outside the institution. Most boards are larger, but many contain just one outsider.

(a) Why should review boards contain people who are not scientists?

(b) Do you think that one outside member is enough? How would you choose that member? (For example, would you prefer a medical doctor? A member of the clergy? An activist for patients' rights?)

6.67 Animal welfare Many people are concerned about the ethics of experimentation with living animals. Some go as far as to regard any animal experiments as unethical, regardless of the benefits to human beings. Briefly explain your position on each of the following uses of animal subjects.

(a) Military doctors use goats that have been deliberately shot (while completely anesthetized) to study and teach the treatment of combat wounds. Assume that there is no equally effective way to prepare doctors to treat human wounds.

(b) Several states are considering legislation that would end the practice of using cats and dogs from pounds in medical research. Instead, the animals will be killed at the pounds.

(c) The cancer-causing potential of chemicals is assessed by exposing lab rats to high concentrations. The rats are bred for this specific purpose. (Would your opinion differ if dogs or monkeys were used?)

6.68 Informed consent The information given to potential subjects in a clinical trial before asking them to decide whether or not to participate might include any of the following. Do you feel that all of this information is ethically required? Discuss.

(a) The basic statement that an experiment is being conducted; that is, something beyond simply treating your medical problem occurs in your therapy.
(b) A statement of any potential risks from any of the experimental treatments.
(c) An explanation that random assignment will be used to decide which treatment you get.
(d) An explanation that one "treatment" is a placebo and a statement of the probability that you will receive the placebo.

6.69 AIDS trials in Africa Effective drugs for treating AIDS are very expensive, so most African nations cannot afford to give them to large numbers of people. Yet AIDS is more common in parts of Africa than anywhere else. Several clinical trials are looking at ways to prevent pregnant mothers infected with HIV from passing the infection to their unborn children, a major source of HIV infections in Africa. Some people say these trials are unethical because they do not give effective AIDS drugs to their subjects, as would be required in rich nations. Others reply that the trials are looking for treatments that can work in the real world in Africa and that they promise benefits at least to the children of their subjects. What do you think?

6.70 Anonymous or confidential? One of the most important nongovernment surveys in the United States is the General Social Survey (Example 1.7, page 9). The GSS regularly monitors public opinion on a wide variety of political and social issues. Interviews are conducted in person in the subject's home. Are a subject's responses to GSS questions anonymous, confidential, or both? Explain your answer.

6.71 Human biological materials Long ago, doctors drew a blood specimen from you as part of treating minor anemia. Unknown to you, the sample was stored. Now researchers plan to use stored samples from you and many other people to look for genetic factors that may influence anemia. It is no longer possible to ask your consent. Modern technology can read your entire genetic makeup from the blood sample.
(a) Do you think it violates the principle of informed consent to use your blood sample if your name is on it but you were not told that it might be saved and studied later?
(b) Suppose that your identity is not attached. The blood sample is known only to come from (say) "a 20-year-old white female being treated for anemia." Is it now OK to use the sample for research?
(c) Perhaps we should use biological materials such as blood samples only from patients who have agreed to allow the material to be stored for later use in research. It isn't possible to say in advance what kind of research, so this falls short of the usual standard for informed consent. Is it nonetheless acceptable, given complete confidentiality and the fact that using the sample can't physically harm the patient?

6.72 Deceiving subjects Students sign up to be subjects in a psychology experiment. When they arrive, they are told that interviews are running late and are taken to a waiting room. The experimenters then stage a theft of a valuable object left in the waiting room. Some subjects are alone with the thief, and others are in pairs—these are the treatments being compared. Will the subject report the theft?

The students had agreed to take part in an unspecified study, and the true nature of the experiment is explained to them afterward. Do you think this study is ethically OK?

CHAPTER 6 REVIEW

Sections 6.1 and 6.2 dealt with the statistical aspects of designing experiments, studies that impose some treatment in order to learn about the response. The big idea is the randomized comparative experiment. Figure 6.6 outlines the simplest design.

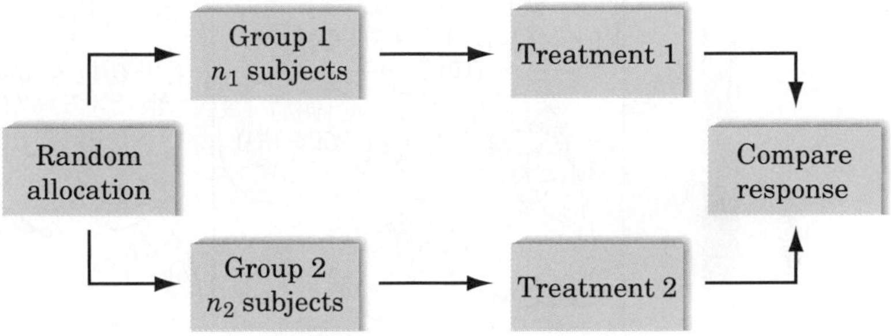

Figure 6.6 The idea of a randomized comparative experiment.

If the subjects in an experiment differ in some systematic way that may affect their responses to the treatments, blocking can be used to reduce unwanted variability. A matched pairs design, in which each subject serves as his or her own control, is a special form of blocking.

Well-designed experiments include random allocation of subjects to treatments, comparison, and use of enough subjects. To reduce the potential for bias, many designs are double-blind—neither the subjects nor those working with them know who is receiving which treatment. Confounding can blur the relationship between the explanatory and response variables and must be actively prevented.

When we collect data about people, ethical issues can be important. Section 6.3 discussed these issues and introduced three principles that apply to any study with human subjects.

Here is a review list of the most important skills you should have developed from your study of this chapter.

1. Identify the explanatory variables, treatments, response variables, and subjects in an experiment.

2. Recognize bias due to confounding of explanatory variables with lurking variables in either an observational study or an experiment.

3. Outline the design of a completely randomized experiment using a diagram like that in Figure 6.6. Such a diagram should show the sizes of the groups, the specific treatments, and the response variable.

4. Use Table B of random digits to carry out random assignment of subjects to groups in a completely randomized experiment.

5. Make use of matched pairs or other block designs when appropriate.

6. Recognize the placebo effect. Recognize when the double-blind technique should be used. Be aware of weaknesses in an experiment, especially in the ability to generalize its conclusions.

7. Explain why a randomized comparative experiment can give good evidence for cause-and-effect relationships.

8. Understand key concepts of data ethics: informed consent, confidentiality of personal data, and review of ethical considerations by an institutional review board. Discuss how these apply in specific settings.

CHAPTER 6 REVIEW EXERCISES

6.73 Testing a new drug A drug manufacturer is studying how a new drug behaves in patients. Investigators compare 2 doses: 5 milligrams (mg) and 10 mg. The drug can be administered by injection, by a skin patch, or by intravenous drip. Concentration in the blood after 30 minutes (the response variable) may depend both on the dose and on the method of administration.
(a) Make a sketch that describes the treatments formed by combining dose and method. Then use a diagram to outline a completely randomized design for this experiment.
(b) "How many subjects?" is a tough issue. What is the advantage of using larger groups of subjects?
(c) The drug may behave differently in men and women. How would you modify your experimental design from (a) to take this into account?

6.74 Should I take the class? A college allows students to choose either classroom or self-paced instruction in a basic economics course. The college wants to compare the effectiveness of self-paced and regular instruction. A professor proposes giving the same final exam to all students in both versions of the course and comparing the average score of those who took the self-paced option with the average score of students in the regular sections.
(a) Explain why confounding would make the results of this study worthless.
(b) Given 30 students who are willing to use either regular or self-paced instruction, outline an experimental design to compare the two methods of instruction. Then use Table B, starting at line 108, to carry out the randomization.

6.75 Surgery or not? To compare surgical and nonsurgical treatments of intestinal cancer, researchers examined the records of a large number of patients. Patients who received surgery survived much longer (on the average) than patients who were treated

without surgery. The study concludes that surgery is more effective than nonsurgical treatment.
(a) What are the explanatory and response variables in this study?
(b) Is this study an experiment? Why or why not?
(c) Discussion with medical experts reveals that surgery is reserved for relatively healthy patients; patients who are too ill to tolerate surgery receive nonsurgical treatment. Explain why the conclusion of the study is not supported by the data.
(d) Outline the design of a randomized comparative experiment.

6.76 Reducing health care spending Will people spend less on health care if their health insurance requires them to pay some part of the cost themselves? An experiment on this issue asked if the percent of medical costs paid by health insurance has an effect either on the amount of medical care that people use or on their health. The treatments were four insurance plans. Each plan paid all medical costs above a ceiling. Below the ceiling, the plans paid 100%, 75%, 50%, or 0% of costs incurred.[37]
(a) Outline the design of a randomized comparative experiment suitable for this study.
(b) Briefly describe the practical and ethical difficulties that might arise in such an experiment.

6.77 Do the twist The design of controls and instruments has a large effect on how easily people can use them. A student investigates this effect by asking right-handed students to turn a knob (with their right hands) that moves an indicator by screw action. There are two identical instruments, one with a right-hand thread (the knob turns clockwise) and the other with a left-hand thread (the knob must be turned counterclockwise). The response variable is the time required (in seconds) to move the indicator a fixed distance. Thirty right-handed students are available to serve as subjects. You have been asked to lay out the statistical design of this experiment. Describe your design, and carry out any randomization that is required.

Exercises 6.78 to 6.81 are based on an article in the Journal of the American Medical Association *that asks if flu vaccine works.*[38] *The article reports on a study of the effectiveness of a nasal spray vaccine called "trivalent LAIV." Here is part of the article's summary:*

Design: Randomized, double-blind, placebo-controlled trial.
Participants: A total of 4561 healthy, working adults aged 18 to 64 years recruited through health insurance plans, at work sites, and from the general population.
Intervention: Participants were randomized 2:1 to receive intranasally administered trivalent LAIV vaccine ($n = 3041$) or placebo ($n = 1520$).
Results: Vaccination also led to fewer days of work lost (17.9% reduction for severe febrile illnesses; 28.4% reduction for febrile upper-respiratory-tract illnesses) and fewer days with health-care-provider visits (24.8% reduction for severe febrile illnesses; 40.9% reduction for febrile upper-respiratory-tract illnesses).

6.78 Know these terms Explain in one sentence each what "randomized," "double-blind," and "placebo-controlled" mean in the description of the design of this study.

6.79 Experiment basics Identify the subjects, the explanatory variable, and several response variables in this study.

6.80 Design an experiment Use a diagram to outline the design of the experiment in this medical study.

6.81 Ethics What are the three "first principles" of data ethics? Explain briefly what the flu vaccine study must do to apply each of these principles.

6.82 Prison experiments The decision to ban medical experiments on federal prisoners followed the uncovering of experiments in the 1960s that exposed prisoners to serious harm. But experiments such as the vitamin C test of Example 1.9 (page 13) are also banned from federal prisons. Because of the difficulty of obtaining truly voluntary consent in a prison, is it necessary to ban even experiments in which all treatments appear harmless? What is your overall opinion of this ban on experimentation?

PART C

CHANCE

"What kind of childish nonsense are you working on now?"

Chance is all around us. You and your friend play rock-paper-scissors to determine who gets the remaining slice of pizza. A coin toss decides which team gets to receive the ball first in a football game. Many adults regularly play the lottery, hoping to win a big jackpot with a few lucky numbers. Others head to casinos or racetracks, hoping that some combination of luck and skill will pay off. People young and old play games of chance involving cards or dice or spinners. The traits that children inherit—gender, hair color, eyesight, blood type, handedness, dimples, whether they can roll their tongue—are determined by the chance involved in which genes get passed along by their parents.

A roll of dice, a simple random sample, and even the inheritance of eye color or blood type represent chance behavior that we can understand and work with. We can roll the dice again, and again, and again. The outcomes are governed by chance, but in many repetitions a pattern emerges. We use mathematics to understand the regular patterns of chance behavior when we can repeat the same chance phenomenon again and again.

The mathematics of chance is called probability. Probability is the topic of this part of the book.

CHAPTER 7

Probability: What Are the Chances?

AP Photo/Mark Humphrey

Playing the lottery

Many people play the lottery in hopes of striking it rich. They pick a few "lucky" numbers, buy a lottery ticket, and wait. The winning numbers are chosen using a chance process. Picture a bunch of numbered balls bubbling about in a container and then being randomly popped out one at a time by air pressure. Even though most people don't win, they view the lottery as "fair" because chance decides the winner. Once in a while, the lottery isn't as fair as it seems.

Several years ago, the Pennsylvania lottery was rigged by the smiling host and several stagehands. They injected paint into all balls bearing 8 of the 10 digits. This weighed them down and guaranteed that all three balls for the winning number would have the remaining 2 digits. The perpetrators then bet on all combinations of these digits. When 6-6-6 popped out, they won $1.2 million. Yes, they were caught.

To avoid similar incidents, some states switched from live drawings using numbered balls to computer-generated random numbers. The state of Tennessee did just that. They hired one company to program the random number generator and a second company to certify that the program was working properly. In 2007, a problem emerged with their Cash 3 drawing. For several consecutive weeks, the computer generated no repeated digits among the three winning numbers. Any player who had picked numbers with repeated digits was guaranteed to lose. Investigation revealed that the computer programmer entered one incorrect character that told the computer not to repeat digits. The mistake was fixed, but public trust was lost.

In 2008, the Cash 3 computer program spat out a surprising set of winning numbers over a three-day period. On Thursday, the winning number was 0-7-7. On Friday, the winning number was 7-0-7. On Saturday, the winning number was again 0-7-7. Lottery officials had the computer equipment checked again. Their conclusion: "The repeating numbers are a statistical coincidence." With a little deeper understanding of random numbers, simulation, and probability, you can decide for yourself whether the lottery was conducted fairly.

7.1 Randomness, Probability, and Simulation

ACTIVITY 7.1A

The *Probability* applet

APPLET Go to the textbook Web site **www.whfreeman.com/sta2e**, click on "Statistical Applets," and select the *Probability* applet.

Directions When you click the Toss button, the applet will toss a coin the number of times you specify. You can set the probability that the coin lands heads by entering a new value and clicking the Reset button. A sample screen shot from the applet is shown. Before you begin using the applet, try to answer the following three questions.

Think

1. If you toss the coin just once, what are the possible values for the proportion of heads?
2. If you toss the coin 400 times with probability 0.5 for heads, which is more likely to happen: (*a*) getting exactly 200 heads or (*b*) getting within a few percent of 200 heads?
3. What will the plot of the proportion of heads look like as you accumulate more and more tosses with a probability of 0.7 for heads?

APPLET TIP: You can toss the coin a maximum of 40 times at once. To make longer sequences, toss again without resetting.

Verify

Use the applet to test your answers to questions 1–3. Did the proportion of heads behave as you predicted?

Experiment

4. Set the probability of heads at 0.3 and have the applet repeatedly toss the coin until you accumulate several hundred tosses. How does the behavior of the proportion of heads at the beginning of the plot differ from that at the end of the plot?
5. Check the "Show true probability" box to have the applet draw a horizontal line on the plot at 0.3. How does the behavior of the plot of the proportion of heads relate to this line?

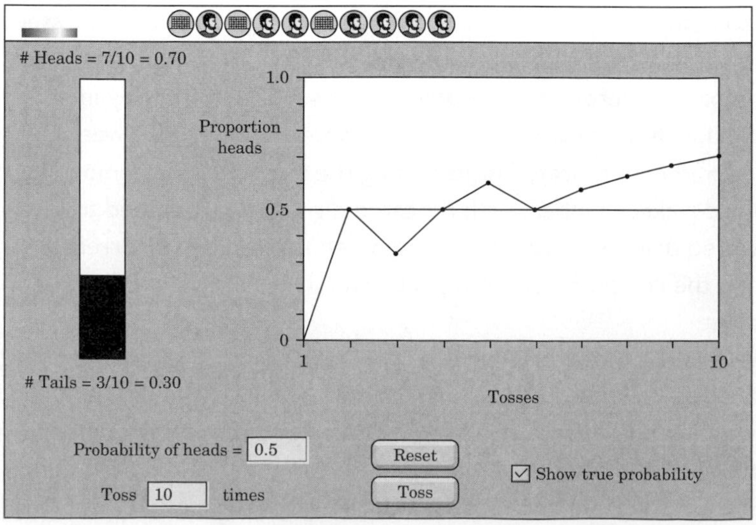

6. Change the probability of heads to 0.7 and have the applet toss the coin several hundred times again. How does the plot of the proportion of heads compare with the plot when the probability was 0.3? In what ways is it similar? In what ways does it differ?

Probability: what happens in the long run

In football, a coin toss helps determine which team gets the ball first. Why do the rules of football require a coin toss? Because tossing a coin seems a "fair" way to decide. That is, tossing a coin avoids favoritism. Favoritism is as undesirable in choosing people for a sample survey or in assigning patients to treatment and placebo groups in a medical experiment as it is in awarding first possession in a football game. That's why statisticians recommend random samples and randomized experiments. They avoid favoritism by letting chance decide who gets selected or who receives which treatment. A big fact emerges when we watch coin tosses or the results of random samples closely: *chance behavior is unpredictable in the short run but has a regular and predictable pattern in the long run.*

Toss a coin, or choose a simple random sample. The result can't be predicted in advance, because the result will vary when you toss the coin or choose the sample repeatedly. But there is still a regular pattern in the results, a pattern that emerges clearly only after many repetitions. This remarkable fact is the basis for the idea of probability.

Example 7.1 Coin tossing

When you toss a coin, there are only two possible outcomes, heads or tails. Figure 7.1 shows the results of tossing a coin 1000 times. For each number of tosses from 1 to 1000, we have plotted the proportion of those tosses that gave a head. The first toss was a head, so the proportion of heads starts at $1/1 = 1.0$. The second toss was a tail, reducing the proportion of heads to $1/2 = 0.5$ after two tosses. The next three tosses produced a tail followed by two heads, so the proportion of heads after five tosses is $3/5 = 0.6$.

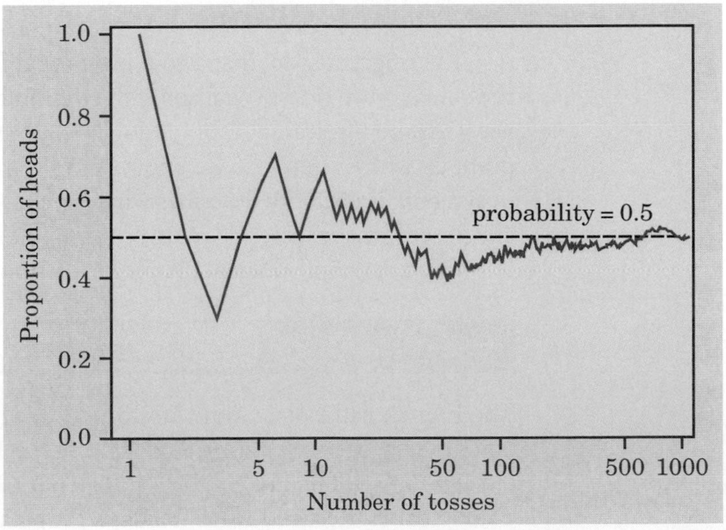

Figure 7.1 Toss a coin many times. The proportion of heads changes as we make more tosses but eventually gets very close to 0.5. This is what we mean when we say, "The probability of a head is one-half."

The proportion of tosses that produce heads is quite variable at first, but it settles down as we make more and more tosses. Eventually this proportion gets close to 0.5 and stays there. We say that 0.5 is the *probability* of a head. The probability 0.5 appears as a horizontal line on the graph.

"Random" in statistics is not a synonym for "haphazard." It is a description of a kind of order that emerges only in the long run. We encounter the unpredictable side of randomness in our everyday lives. But we rarely see enough repetitions of the same random phenomenon to observe the long-term regularity that probability describes. You can see that regularity emerging in Figure 7.1. In the very long run, the proportion of tosses that give a head is 0.5. Probability 0.5 means "occurs half the time in a very large number of trials." This is the intuitive idea of probability.

ACTIVITY 7.1B

Tossing thumbtacks

Materials: Box of thumbtacks (enough for one per student)

When you flip a fair coin, it is equally likely to land "heads" or "tails." Do thumbtacks behave in the same way? In this Activity, you will toss a thumbtack several times and observe whether it comes to rest point up (U) or point down (D). The question you are trying to answer is: what is the probability that the tossed thumbtack will land point up?

1. Before you begin the Activity, make a guess about what will happen. If you could toss your thumbtack over and over and over, what proportion of all tosses do you think would land point up (U)?

2. Toss your thumbtack 50 times onto a flat surface. Tally the number of point up (U) and point down (D) landings. Calculate the proportion of point up (U) landings you obtained.

3. What would you estimate as the probability of your thumbtack landing point up if you tossed it one more time? Explain.

4. Compare your answer to Step 3 with those of your classmates. Did everyone in the class come up with the same estimated probability? Give at least two reasons why people's probabilities might differ.

You might suspect that a coin has probability 0.5 of coming up heads just because the coin has two sides. What about the thumbtacks of Activity 7.1B? They also have two ways to land—point up or point down—but the chance that a tossed thumbtack lands point up isn't 0.5. How do we know that? From tossing a thumbtack over and over and over again. Probability describes what happens in very many trials, and we must actually observe many tosses of a coin or thumbtack to pin down a probability. In the case of tossing a coin, some diligent people have actually made thousands of tosses.

Example 7.2 Some remarkable coin tossers

The French naturalist Count Buffon (1707–1788) tossed a coin 4040 times. Result: 2048 heads, or proportion $2048/4040 = 0.5069$ for heads.

Around 1900, the English statistician Karl Pearson heroically tossed a coin 24,000 times. Result: 12,012 heads, a proportion of 0.5005.

While imprisoned by the Germans during World War II, the South African mathematician John Kerrich tossed a coin 10,000 times. Result: 5067 heads, a proportion of 0.5067.

Here's a brief summary of what we have learned so far about randomness and probability.

Randomness and probability

We call a phenomenon **random** if individual outcomes are uncertain but there is nonetheless a regular distribution of outcomes in a large number of repetitions.

The **probability** of any outcome of a random phenomenon is a number between 0 and 1 that describes the proportion of times the outcome would occur in a very long series of repetitions.

An outcome with probability 0 never occurs. An outcome with probability 1 happens on every repetition. An outcome with probability 0.5 happens half the time in a very long series of trials. Of course, we can never observe a probability exactly. We could always continue tossing the coin, for example. Mathematical probability is based on imagining what would happen in an indefinitely long series of trials.

Probability gives us a language to describe the long-term regularity of random behavior. The outcome of a coin toss and the sex of the next baby born in a local hospital are both random. So is the outcome of a random sample or a randomized experiment. The behavior of large groups of individuals is often as random as the behavior of many coin tosses or many random samples. Life insurance, for example, is based on the fact that deaths occur at random among many individuals.

Example 7.3 Probability and life insurance

We can't predict whether a particular person will die in the next year. But if we observe millions of people, deaths are random. The National Center for Health Statistics says that the proportion of men aged 20 to 24 years who die in any one year is 0.0015. This is the *probability* that a randomly selected young man will die next year. For women that age, the probability of death is about 0.0005.

If an insurance company sells many policies to people aged 20 to 24, it knows that it will have to pay off next year on about 0.15% of the policies sold to men and on about 0.05% of the policies sold to women. It will charge more to insure a man because the probability of having to pay is higher.

The history of chance

Randomness is most easily noticed in many repetitions of games of chance—rolling dice, dealing shuffled cards, spinning a roulette wheel.[1] The most common method of randomization in ancient times was "rolling the bones," which involved tossing several astragali. The astragalus (Figure 7.2) is a solid bone from the heel of animals that, when thrown, will come to rest on any of four sides. (The other two sides are rounded.) Cubical dice, made of pottery or bone, came later, but even dice existed before 2000 B.C.

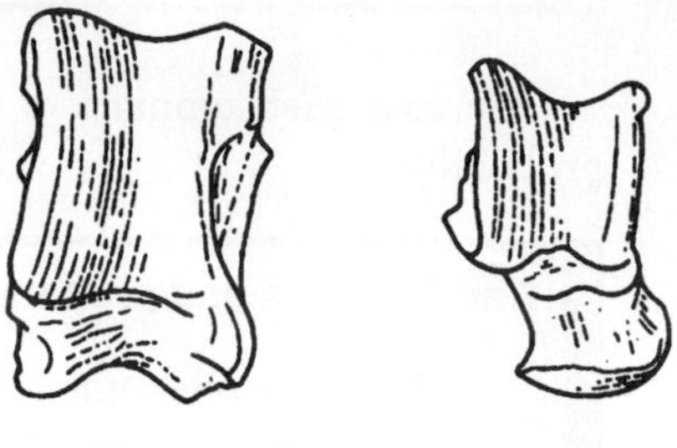

Figure 7.2 Animal heel bones (astragali), shown actual size. (From F. N. David, *Games, Gods, and Gambling,* Charles Griffin & Company, 1962. Reproduced by permission of the publishers.)

Chance devices such as astragali have been used from the beginning of recorded history. Yet none of the great ancient mathematicians studied the regular pattern of many throws of bones or dice. Perhaps this is because astragali and most ancient dice were so irregular that each had a different pattern of outcomes.

Professional gamblers, however, did notice the regular pattern of outcomes of dice or cards and tried to adjust their bets to the odds of success. "How should I bet?" is the question that launched mathematical probability. The systematic study of randomness began when seventeenth-century French gamblers asked French mathematicians for help in figuring out the "fair value" of bets on games of chance. **Probability theory,** the mathematical study of randomness, originated with Pierre de Fermat and Blaise Pascal in the seventeenth century and was well developed by the time statisticians took it over in the twentieth century.

Exercises

7.1 Spinning a quarter With your forefinger, hold a new quarter (one with a state featured on the reverse) upright, on its edge, on a hard surface. Then flick it with your other forefinger so that it spins for some time before it falls and comes to rest. Spin the coin a total of 25 times, and record the results.

(a) Based on 25 spins, estimate the probability of heads.

(b) Estimate the probability of tails.

7.2 How many tosses to get a head? When we toss a penny, experience shows that the probability (long-term proportion) of a head is close to 1/2. Suppose that we toss the penny repeatedly until we get a head. We want to know the probability that the first head comes up in an odd number of tosses (1, 3, 5, and so on). To find out, perform 20 trials, and keep a record of the number of tosses needed to get a head on each of your 20 trials.

(a) From your results, estimate the probability of a head on the first toss. What value should we expect this probability to have?

(b) Use your results to estimate the probability that the first head appears on an odd-numbered toss.

7.3 Random digits The table of random digits (Table B) was produced by a chance process that gives each digit a 1/10 chance of being a 0. Examine the first 200 digits in the table and determine what proportion are 0s. (This proportion is an estimate, based on 200 repetitions, of the true probability, which in this case is known to be 0.1.)

www If you want to find "true" random numbers, visit **www.random.org.** Check out their coin flipper, die roller, playing card shuffler, and integer generator.

7.4 From words to probabilities Probability is a measure of how likely an event is to occur. Match one of the probabilities that follow with each statement of likelihood given. (The probability is usually a more exact measure of likelihood than is the verbal statement.)

0 0.01 0.3 0.6 0.99 1

(a) This event is impossible. It can never occur.

(b) This event is certain. It will occur on every trial.

(c) This event is very unlikely, but it will occur once in a while in a long sequence of trials.

(d) This event will occur more often than not.

7.5 Rolling a die Imagine rolling a fair, six-sided die, like the kind used in many board games. What is the probability of getting a 4? Justify your answer.

7.6 Toss three coins What is the probability of getting 2 heads and 1 tail when you toss three coins? Give appropriate evidence to support your answer.

Myths about chance behavior

The idea of probability seems straightforward. It answers the question "What would happen if we did this many times?" In fact, both the behavior of random phenomena and the idea of probability are a bit subtle. We meet chance behavior constantly, and psychologists tell us that we deal with it poorly.

ACTIVITY 7.1C

Investigating randomness

1. Pretend that you are flipping a fair coin. Without actually flipping a coin, *imagine* the first toss. Write down the result you see in your mind, heads (H) or tails (T).

2. Imagine a second coin flip. Write down the result.

3. Keep doing this until you have 50 H's or T's written down. Write your results in groups of 5 to make them easier to read, like this: HTHTH TTHHT, etc.

4. A **run** is a repetition of the same result. In the example in Step 3, there is a run of two tails followed by a run of two heads in the first 10 coin flips. Read through your 50 imagined coin flips, and count the number of runs of size 2, 3, 4, etc. Record the number of runs of each size in a table like this:

Run length	2	3	4	5	6	7	8
Frequency							

5. Use your calculator to generate a similar list of 50 coin flips. Let 1 represent a head and 0 represent a tail.

- On the TI-83/84, go to the home screen and execute the command `randInt(0,1,50)→` L_1. The command `randInt` can be found under MATH/PRB/5:randInt(.
- On the TI-Nspire, open a new document and add a Lists & Spreadsheet page. Go to the

ACTIVITY 7.1C *(continued)*

ACTIVITY 7.1C *(continued)*

formula line in column A, enter the command `=randint(0,1,50)` and press ENTER.

Record the number of runs of size 2, 3, 4, and so forth in a table like this:

Run length	2	3	4	5	6	7	8
Frequency							

6. Compare the two results. Did you or your calculator have the longest run? How much longer?

The myth of short-run regularity The idea of probability is that randomness is predictable in the long run. Unfortunately, our intuition about randomness tries to tell us that random phenomena should also be predictable in the short run. When they aren't, we look for some explanation other than chance variation.[2]

Example 7.4 What looks random?

Toss a coin six times and record heads (H) or tails (T) on each toss. Which of the following outcomes is more probable?

HTHTTH TTTHHH

Almost everyone says that HTHTTH is more probable because TTTHHH doesn't "look random." But both are equally likely. That heads and tails are equally probable says that about half a long sequence of tosses will be heads. It doesn't say that heads and tails must come close to alternating. The coin has no memory. It doesn't know what past outcomes were, and it can't try to create a balanced sequence.

The outcome TTTHHH in tossing six coins looks unusual because of the runs of 3 straight tails and 3 straight heads. Runs seem "not random" to our intuition but are quite common. Here's a more striking example than tossing coins.

Example 7.5 The hot hand in basketball

Belief that runs must result from something other than "just chance" is very common. If a basketball player makes several consecutive shots, both the fans and his teammates believe that he has a "hot hand" and is more likely to make the next shot. This is wrong. Careful study has shown that runs of baskets made or missed are no more frequent in basketball than would be expected if each shot were independent of the player's previous shots. Players perform consistently, not in streaks. If a player makes half her shots in the long run, her hits and misses behave just like tosses of a coin—and that means that runs of hits and misses are more common than our intuition expects.[3]

The myth of the surprise meeting Mr. Starnes and his wife are on a two-week vacation in Hawaii. One day, at the Volcano National Park visitor center, they run into Herbie, one of Mr. Starnes's former students. "How unusual! Maybe we were destined to meet." Well, maybe not. It is certainly unlikely that the Starnes family would run into *this particular* acquaintance that day, but it is not at all unlikely that they would meet *some* acquaintance during their vacation. After all, a typical adult has about 1500 casual acquaintances. When something unusual happens, we look back and say, "Wasn't that unlikely?" We would have said the same if any of 1500 other unlikely things had happened. Here's an example where we can actually calculate the probabilities.

Example 7.6 Winning the lottery twice

Several years ago, Evelyn Marie Adams won the New Jersey state lottery for the second time, adding $1.5 million to her previous $3.9 million jackpot. The *New York Times* claimed that the odds of one person winning the big prize twice were about 1 in 17 trillion. Nonsense, said two statistics professors in a letter that appeared in the *Times* two weeks later. The chance that Evelyn Marie Adams would win twice in her lifetime is indeed tiny, but it is almost certain that *someone* among the millions of regular lottery players in the United States would win two jackpot prizes. The statisticians estimated even odds of another double winner within seven years. Sure enough, Robert Humphries won his second Pennsylvania lottery jackpot ($6.8 million total) two years later. □

The myth of the law of averages While attending a convention in Las Vegas one of the authors roamed the gambling floors, watching money disappear into the drop boxes under the tables. You can see some interesting human behavior in a casino. When the shooter in the dice game craps rolls several winners in a row, some gamblers think she has a "hot hand" and bet that she will keep on winning. Others say that "the law of averages" means that she must now lose so that wins and losses will balance out. Believers in the law of averages think that, if you toss a coin six times and get TTTTTT, the next toss must be more likely to give a head. It's true that in the long run heads will appear half the time. What is a myth is that future outcomes must make up for an imbalance like six straight tails.

Coins and dice have no memories. A coin doesn't know that the first six outcomes were tails, and it can't try to get a head on the next toss to even things out. Of course, things do even out *in the long run.* After 10,000 tosses, the results of the first six tosses don't matter. They are overwhelmed by the results of the next 9994 tosses.

"So the law of averages doesn't guarantee me a girl after seven straight boys, but can't I at least get a group discount on the delivery fee?"

Example 7.7 We want a boy

Belief in this phony "law of averages" can lead to consequences close to disastrous. A few years ago, "Dear Abby" published in her advice column a letter from a distraught mother of eight girls. It seems that she and her husband had planned to limit their family to four children. When all four were girls, they tried again—and again and again—because they wanted a boy. After seven straight girls, even her doctor had assured her that "the law of averages was in our favor 100 to 1." Unfortunately for this couple, having children is like tossing coins. Eight girls in a row is highly unlikely, but once seven girls have been born, it is not at all unlikely that the next child will be a girl—and it was.

Probability and risk

People often use "probability" in a way that includes personal judgments of how likely it is that some event will happen. They make decisions based on these judgments—some people refuse to fly because they think the probability of an accident is too high. The experts use probabilities from data to describe the risk of an unpleasant event. By looking at what happens in many repetitions, the experts conclude that flying is actually safer than driving. Individuals and society seem to ignore data. We worry about some risks that almost never occur while ignoring others that are much more likely.

The probability of rain is . . .

You work all week. Then it rains on the weekend. Can there really be a statistical truth behind our perception that the weather is against us? At least on the East Coast of the United States, the answer is "Yes." Going back to 1946, it seems that Sundays receive 22% more precipitation than Mondays. The likely explanation is that the pollution from all those workday cars and trucks forms the seeds for raindrops—with just enough delay to cause rain on the weekend.

Example 7.8 Asbestos in schools

High exposures to asbestos are dangerous. Low exposures, like those experienced by teachers and students in schools where asbestos is present in the insulation around pipes, are not very risky. The probability that a teacher who works for 30 years in a school with typical asbestos levels will get cancer from the asbestos is around 15/1,000,000. The risk of dying in a car accident during a lifetime of driving is about 15,000/1,000,000. That is, driving regularly is 1000 times more risky than teaching in a school where asbestos is present.[4]

Risk does not stop us from driving. Yet the much smaller risk from asbestos launched massive cleanup campaigns and a federal requirement that every school inspect for asbestos and make the findings public.

Why do we take asbestos so much more seriously than driving? Why do we worry about very unlikely threats such as tornadoes more than we worry about heart attacks?

- We feel safer when a risk seems under our control than when we cannot control it. We are in control (or so we think) when we are driving, but we can't control the risk from asbestos or tornadoes.
- It is hard to comprehend very small probabilities. Probabilities of 15 per million and 15,000 per million are both so small that our intuition cannot distinguish between them. Studies have shown that we generally overestimate very small risks and underestimate higher risks.
- The probabilities for risks like asbestos in the schools are not as certain as probabilities for tossing coins. They must be estimated by experts from complicated statistical studies.

Our reactions to risk depend on more than probability. We are influenced by social standards and by how we evaluate particular risks. As one writer noted, "Few of us would leave a baby sleeping alone in a house while we drove off on a 10-minute errand, even though car-crash risks are much greater than home risks."[5]

Exercises

7.7 Personal random numbers? Ask several of your friends (at least 10 people) to choose a four-digit number "at random." How many of the numbers chosen start with 1 or 2? How many start with 8 or 9? (There is strong evidence that people in general tend to choose numbers that start with low digits.)

7.8 Playing "Pick 4" The Pick 4 games in many state lotteries announce a four-digit winning number each day. The winning number is essentially a four-digit group from a table of random digits. You win if your choice matches the winning digits, in any order. The winnings are divided among all players who matched the winning digits. That suggests a way to get an edge.

(a) The winning number might be, for example, either 2873 or 9999. Explain why these two outcomes have exactly the same probability. (It is 1 in 10,000.)

(b) If you asked many people whether 2873 or 9999 is more likely to be the randomly chosen winning number, most would favor one of them. Use the information in this section to say which one and to explain why. (If you choose a number that people think is unlikely, you have the same chance to win, but you will win a larger amount because few other people will choose your number.)

7.9 Surprising? You are getting to know your new roommate, assigned to you by the college. In the course of a long conversation, you find that both of you have sisters named Deborah. Should you be surprised? Explain your answer.

7.10 Cold weather coming A TV weather man, predicting a colder-than-normal winter, said, "First, in looking at the past few winters, there has been a lack of really cold weather. Even though we are not supposed to use the law of averages, we are due." Do you think that "due by the law of averages" makes sense in talking about the weather? Why or why not?

7.11 Reacting to risks National newspapers such as *USA Today* and the *New York Times* carry many more stories about deaths from airplane crashes than about deaths from automobile crashes. Auto accidents kill about 40,000 people in the United States each year. Crashes of all scheduled air carriers, including commuter carriers, have killed between 0 and 575 people per year in recent years, including the four hijacked planes that crashed on September 11, 2001.

(a) Why do the news media give more attention to airplane crashes?

(b) How does news coverage help explain why many people consider flying more dangerous than driving?

7.12 In the long run Suppose that the first six tosses of a coin give six tails and that tosses after that are exactly half heads and half tails. What is the proportion of heads after the first six tosses? What is the proportion of heads after 100 tosses if the last 94 produce 47 heads? What is the proportion of heads after 1000 tosses if half of the last 994 produce heads? What is the proportion of heads after 10,000 tosses if half of the last 9994 produce heads? What's the point?

Simulation

The probabilities of heads and tails in tossing a coin are very close to 1/2. In principle, these probabilities come from data on many coin tosses. What about the probability that we get a run of three straight heads somewhere in 10 tosses of a coin? Do we really have to toss a coin 10 times over and over and over again? Of course not. In the next chapter, we will learn how to find this probability by *calculation from a model* that describes tossing coins. Unfortunately, the math needed to do probability calculations is often tough. Technology comes to the rescue: we can use random digits from a table, calculator, or computer to *simulate* many repetitions. This is easier than math and much faster than actually running many repetitions in the real world.

Simulation

Using random digits from a table, a calculator, or computer software to imitate chance behavior is called **simulation.**

Simulation is an effective tool for finding probabilities involving chance behavior. We can use random digits to simulate many repetitions quickly. The proportion of repetitions on which some outcome of interest occurs will eventually be close to its probability, so simulation can give good estimates of probabilities. The art of simulation is best learned by example.

Example 7.9 Doing a simulation: Shaquille O'Neal at the free-throw line

For many years, Shaquille O'Neal was one of the NBA's most dominant centers. He always had impressive statistics, and in 2000, he won all three MVP (Most Valuable Player) awards. At 300+ pounds and 7 feet 4 inches, he was a commanding presence on the court. Still, he never became a "complete" player. Despite much practice and working with special coaches, Shaq remained a 50% free-throw shooter for most of his career.

Let's assume that every time Shaq steps up to the free-throw line, the probability that he will make the shot is 0.5. Suppose we want to know how likely he is to make at least 3 free throws in a row out of 10 attempts. Three or more baskets in a row is called a "run."

Step 1: State assumptions about the chance behavior.

- Each free throw has probabilities 0.5 for a basket (a "hit") and 0.5 for a miss.
- Free throws are *independent* of each other. That is, knowing the outcome of one free throw does not change the probabilities for the outcomes of any other free throw. (Studies of consecutive free throws have shown this to be the case.)

Step 2: Assign digits to represent outcomes. Digits in Table B (in the back of the book) will stand for the outcomes, in a way that matches the probabilities from Step 1. We know that each digit in Table B has probability 0.1 of being any one of 0, 1, 2, 3,

4, 5, 6, 7, 8, or 9, and that successive digits in the table are independent. Here is one assignment of digits for free throws:

- One digit simulates one free throw.
- Digits 0 to 4 represent a made shot; digits 5 to 9 represent a missed shot.

Successive digits in the table simulate independent tosses.

Step 3: Simulate many repetitions. Ten digits simulate 10 tosses, so looking at 10 consecutive digits in Table B simulates one repetition. Read many groups of 10 digits from the table to simulate many repetitions. Be sure to keep track of whether or not the event we want (a run of 3 hits) occurs on each repetition.

Here are the first three repetitions, starting at line 101 in Table B. We have underlined all runs of 3 or more hits.

	Repetition 1	Repetition 2
Digits	1 9 2 2 3 9 5 0 3 4	0 5 7 5 6 2 8 7 1 3
Hit/miss	HM<u>HHH</u>MM<u>HHH</u>	HMMMMHMMHH
Run of 3?	Yes	No

	Repetition 3
Digits	9 6 4 0 9 1 2 5 3 1
Hit/miss	MMHHMHHMHH
Run of 3?	No

Continuing in Table B, we did 25 repetitions. In 11 of them, Shaq made at least 3 consecutive baskets. So we estimate the probability of a run by the proportion

$$\text{estimated probability} = \frac{11}{25} = 0.44$$

Of course, 25 repetitions are not enough to be confident that our estimate is accurate. But now that we understand how to do the simulation, we can tell a computer to do many thousands of repetitions. A long simulation (or some tough calculation) finds that the true probability is about 0.508.

You can use your calculator's random integer command instead of Table B to perform the simulation of Example 7.9. The command `randInt(0,9,10)` will simulate shooting one set of 10 free throws. Pressing ENTER several more times will generate additional sets of 10 free throws.

The chance behavior in Example 7.9 is typical of many probability problems because it consists of *independent trials* (the free throws) that all have the *same possible outcomes* with the *same probabilities.* Shooting 10 free throws and observing the number of heads in 10 tosses of a coin are simulated in much the same way. Independence simplifies our work because it allows us to simulate each of the 10 free throws in exactly the same way.

Independence

Two random phenomena are **independent** if knowing the outcome of one does not change the probabilities for the outcomes of the other.

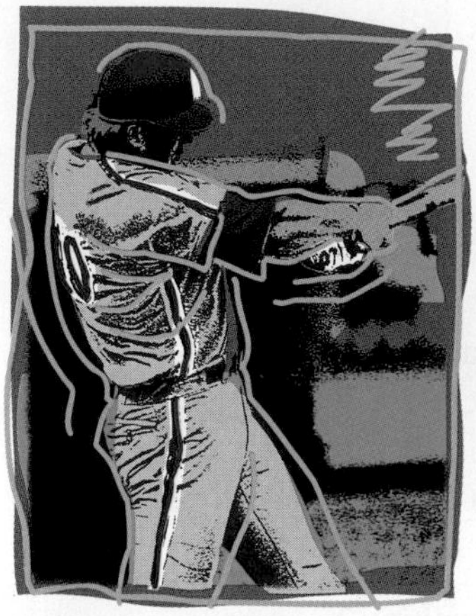

Was he good or was he lucky?

When a baseball player hits. 300, everyone applauds. A .300 hitter gets a hit in 30% of times at bat. Could a .300 year just be luck? Typical major leaguers bat about 500 times a season and hit about .260. A hitter's successive tries seem to be independent. From this model, we can calculate or simulate the probability of hitting .300. It is about 0.025. Out of 100 run-of-the-mill major league hitters, two or three each year will bat .300 because they were lucky.

Independence, like all aspects of probability, can be verified only by observing many repetitions. It is plausible that repeated tosses of a coin are independent (the coin has no memory), and observation shows that they are. It seems less plausible that successive shots by a basketball player are independent, but observation shows that they are at least very close to independent. Step 2 (assigning digits) rests on the properties of the random digit table. Here are some examples of this step.

Example 7.10 Assigning digits for simulation

(a) Choose a family at random from a group of which 80% have an Internet connection. One digit simulates one family:

$$0, 1, 2, 3, 4, 5, 6, 7 = \text{Internet connection}$$
$$8, 9 = \text{no connection}$$

(b) Choose one family at random from a group of which 83% have an Internet connection. Now two digits simulate one family:

$$00, 01, 02, \ldots, 82 = \text{Internet connection}$$
$$83, 84, 85, \ldots, 99 = \text{no connection}$$

We assigned 83 of the 100 two-digit pairs to "Internet connection" to get probability 0.83. Representing "Internet connection" by 01, 02, . . . , 83 would also be correct.

(c) Choose one family at random from a group of which 50% have a dial-up connection, 20% have no connection, and 30% have a cable or DSL connection. There are now three possible outcomes, but the principle is the same. One digit simulates one family:

$$0, 1, 2, 3, 4 = \text{dial-up}$$
$$5, 6 = \text{no connection}$$
$$7, 8, 9 = \text{cable or DSL}$$

The simulation of chance behavior is a powerful tool of contemporary science. Doing a few simulations will help increase your understanding of probability.

Exercises

7.13 Which party does it better? An opinion poll selects adult Americans at random and asks them, "Which political party, Democratic or Republican, do you think is more concerned with providing health care for the poor?" Explain carefully how you would assign digits from Table B to simulate the response of one person in each of the following situations.

(a) Of all adult Americans, 50% would choose the Democrats and 50% the Republicans.

(b) Of all adult Americans, 60% would choose the Democrats and 40% the Republicans.

(c) Of all adult Americans, 40% would choose the Democrats, 40% would choose the Republicans, and 20% would be undecided.

(d) Of all adult Americans, 53% would choose the Democrats and 47% the Republicans.

7.14 Get rid of exams! Suppose that 80% of a school's students favor abolishing exams. You ask 10 students chosen at random. What is the probability that all 10 favor abolishing exams?

(a) State assumptions for the chance behavior involved in this setting.

(b) Assign digits to represent the answers "in favor" and "against."

(c) Simulate 25 repetitions. Use your calculator or Table B. What is your estimate of the probability?

7.15 Basic simulation Use Table B to simulate the responses of 10 independently chosen adults in each of the four situations of Exercise 7.13.

(a) For situation (a), use line 110.

(b) For situation (b), use line 111.

(c) For situation (c), use line 112.

(d) For situation (d), use line 113.

7.16 Organ donors A recent opinion poll showed that about 75% of Americans would donate their organs upon death. Suppose that this is exactly true. Choosing an American at random then has probability 0.75 of getting one who would donate his or her organs. If we interview Americans separately, we can assume that their responses are independent. We want to know the probability that a simple random sample of 100 Americans will contain at least 80 who would donate their organs. Explain carefully how to do this simulation and simulate *one* repetition of the poll using your calculator or Table B. How many of the 100 would donate their organs? Explain how you would estimate the probability by simulating many repetitions.

7.17 First ace Begin with a standard deck of playing cards. Shuffle the cards thoroughly, and then draw a card. Replace the card in the deck, shuffle the deck, and draw a card again. Continue until you draw an ace, or until you draw 10 cards, whichever comes first. What is the probability of drawing an ace in 10 draws?

(a) State assumptions for the chance behavior involved in this setting.

(b) Assign digits to represent outcomes.

(c) Simulate 10 repetitions using your calculator or Table B. What is your estimate of the probability?

7.18 Rock-paper-scissors Almost everyone has played the rock-paper-scissors game when they were younger. Two players face each other and, at the count of 3, make a fist (rock), an extended hand, palm side down (paper), or a "V" with the index and middle fingers (scissors). The winner is determined by these rules: rock smashes scissors; paper covers rock; and scissors cut paper. If the two players choose different objects, then there is a winner. But both could choose the same object (for example, rock), in which case there is no winner. One of the questions that we might explore is "What is the probability that a game will result in a winner?"

(a) State assumptions for the chance behavior involved in this setting.

(b) Assign digits to represent outcomes.

(c) Simulate 20 repetitions using your calculator or Table B. What is your estimate of the probability?

APPLICATION 7.1

Luck, skill, and randomness

In 2007, actress Danica McKellar authored her first book, *Math Doesn't Suck: How to Survive Middle-School Math without Losing Your Mind or Breaking a Nail.* A *Newsweek* article about the book included a picture of McKellar standing by a chalkboard. The following string of 60 digits was handwritten on the chalkboard:

49894 84820 45868 34365 63811 77203 09179
80576 28621 35448 62270 52604

Is this a randomly generated set of digits or a human-generated set of digits? Give appropriate evidence to support your answer.

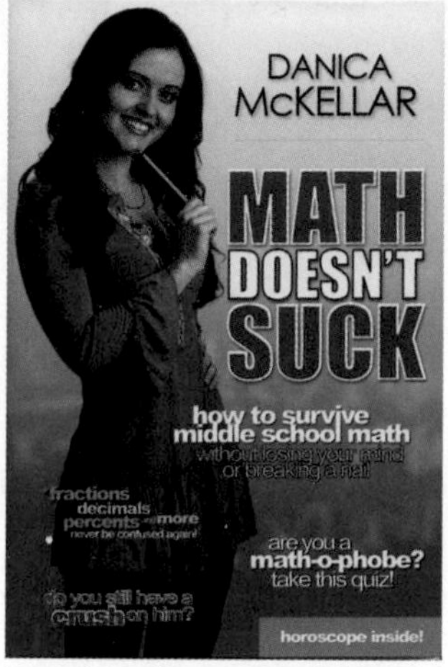

By permission of Penguin Group (USA) Inc.

Section 7.1 Summary

Some events in the world, both natural and of human design, are **random.** That is, their outcomes have a clear pattern in very many repetitions even though the outcome of any one trial is unpredictable. **Probability** describes the long-term regularity of random phenomena. The probability of an outcome is the proportion of very many repetitions on which that outcome occurs. A probability is a number between 0 (never occurs) and 1 (always occurs).

Probabilities describe only what happens in the long run. Short runs of random phenomena like tossing coins or shooting a basketball often don't look random because they do not show the regularity that in fact emerges only in very many repetitions.

We can use random digits to **simulate** chance behavior. A common situation is several **independent** trials with the same possible outcomes and probabilities on each trial. Think of several tosses of a coin or several rolls of a die. Successful simulation requires that we state our assumptions about the chance behavior, assign digits to represent outcomes, and perform many repetitions.

Section 7.1 Exercises

7.19 An unenlightened gambler
(a) A gambler knows that red and black are equally likely to occur on each spin of a roulette wheel. He observes five consecutive reds occur and bets heavily on black at the next spin. Asked why, he explains that black is "due by the law of averages." Explain to the gambler what is wrong with this reasoning.
(b) After hearing you explain why red and black are still equally likely after five reds on the roulette wheel, the gambler moves to a poker game. He is dealt five straight red

cards. He remembers what you said and assumes that the next card dealt in the same hand is equally likely to be red or black. Is the gambler right or wrong, and why?

7.20 Sports risks The probability of dying if you play high school football is about 10 per million each year you play. The risk of getting cancer from asbestos if you attend a school in which asbestos is present for 10 years is about 5 per million. If we ban asbestos from schools, should we also ban high school football? Briefly explain your position.

7.21 Rainy days The TV weatherman says, "There's a 30% chance of rain tomorrow." Explain what this statement means.

7.22 Long runs Most people are surprised at the occurrence of long runs of the same outcome in a random sequence. A science writer once said that, if you flip a fair coin 250 times, then approximately 32 runs will have at least two heads; about 16 runs will have at least three heads; 8 runs, at least four heads; 4 runs, at least five heads; 2 runs, at least six heads; and 1 run, at least 7 heads.[6] Carry out a simulation to see if the writer's claim is reasonable.

7.23 Is the Belgian euro coin "fair"? Two Polish math professors and their students spun a Belgian euro coin 250 times. It landed "heads" 140 times. One of the professors concluded that the coin was minted so that one side was heavier than the other. A representative from the Belgian mint said the result was just chance.[7] Design and carry out a simulation using random digits to help you decide who is correct.

7.24 A game of chance I have a little bet to offer you. Toss a coin 10 times. If there is no run of three or more straight heads or tails in the 10 outcomes, I'll pay you $2. If there is a run of three or more, you pay me just $1. Surely you will want to take advantage of me and play this game? Is it to your advantage to play? Design and carry out a simulation to help you answer this question.

7.25 Birth months If you choose a student from your school at random, what is the probability that he or she was born in July? Justify your answer.

7.26 First ace again In Exercise 7.17 (page 325), you performed a simulation to estimate the probability of drawing an ace in 10 draws. How would your results change if you did not replace the card in the deck after each draw?

(a) Think about this new setting. Would the probability of drawing an ace in 10 draws be higher, lower, or the same as in Exercise 7.17? Explain.

(b) Design and carry out 10 repetitions of a simulation to imitate the chance behavior in this new setting. Use Table B or your calculator. Describe the steps in your simulation clearly enough so that a classmate could repeat your process.

CALCULATOR CORNER Prob Sim APP

The Prob Sim APP for the TI-83/84 allows you to simulate tossing coins, rolling dice, spinning a spinner, drawing cards, and playing the lottery. If you don't have the APP, download it from your teacher. If you are using the TI-Nspire, you'll need to put on the TI-84 faceplate.

- To run the APP, press the [APPS] key. Then choose `Prob Sim.` You should see the introductory screen on the left at the top of the next page. Press [ENTER] to see the main menu (shown on the right at the top of the next page).

CALCULATOR CORNER *(continued)*

CALCULATOR CORNER *(continued)*

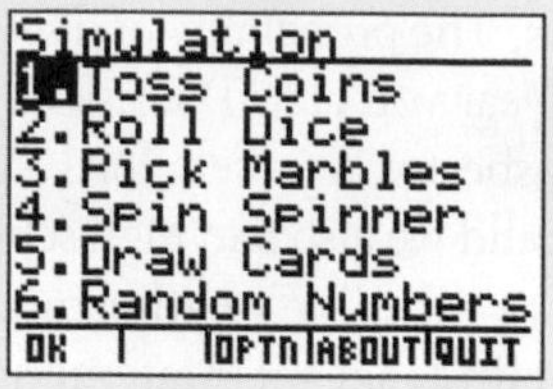

• Let's use the Prob Sim APP to carry out the simulations described in Exercises 7.17 (page 325) and 7.26, both of which involve drawing a card until you get an ace. Choose 5 : Draw Cards from the main menu. You should see the screen shown at the left below. Press WINDOW to DRAW a card. (We got the jack of spades, as shown at the right below.)

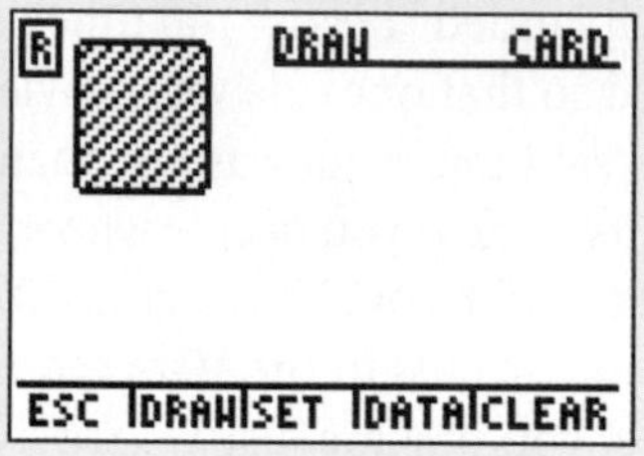

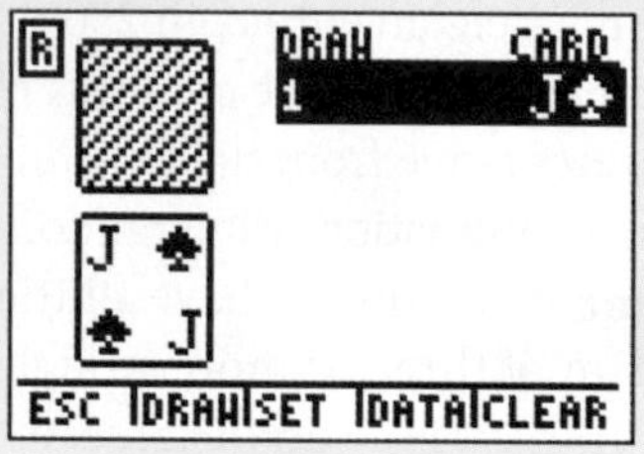

• Keep pressing WINDOW until you get an ace or until you have drawn a card 10 times.

• If you repeat this process many, many times, and keep track of the overall proportion of trials in which you get an ace within 10 draws, you'll have a good estimate of the probability for Exercise 7.17.

• In Exercise 7.26, we asked about the probability of getting an ace within 10 draws if you *don't* replace the card in the deck after each draw. To simulate this new setting, press GRAPH to CLEAR your previous trials. Press Y= to confirm the deletion.

• Now press ZOOM to adjust the APP's settings. Change the Replace option to No as shown in the screen shot below. Then press GRAPH to OK the changes.

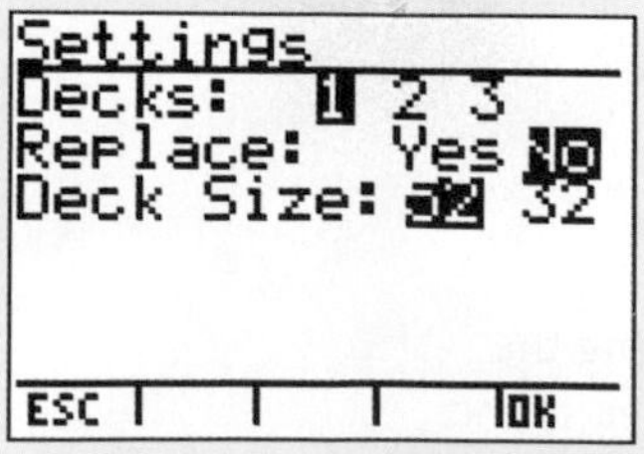

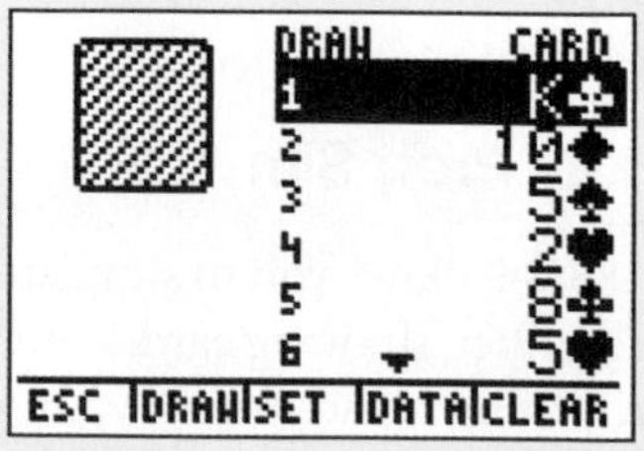

• Draw cards until you get an ace or have drawn 10 cards, as before. If you do this many times, you will get a good estimate of the probability for Exercise 7.26. The results of our first six draws are shown in the screen shot above, right.

7.2 Probability Rules

ACTIVITY 7.2

Monty's dilemma

Materials: Several decks of cards

In *Parade* magazine, a reader posed the following question to Marilyn vos Savant, author of the "Ask Marilyn" column:

> Suppose you're on a game show, and you're given the choice of three doors. Behind one door is a car, behind the others, goats. You pick a door, say #1, and the host, who knows what's behind the doors, opens another door, say #3, which has a goat. He says to you, "Do you want to pick door #2?" Is it to your advantage to switch your choice of doors?
>
> The game show in question was *Let's Make a Deal* and the host was Monty Hall.
>
> Here's the first part of Marilyn's response:
>
> Yes; you should switch. The first door has a 1/3 chance of winning, but the second door has a 2/3 chance.

Thousands of readers wrote to Marilyn to disagree with her answer. But she held her ground. In this Activity, you will use simulation to decide who was right.

1. Discuss with your classmates: should the contestant stay with the original choice or switch? Why?

2. Working with a partner, simulate the game as follows.

- Pull an ace and two 2s from the deck of cards. These represent the three "doors" with prizes (the ace represents the nice prize).
- Designate one person to serve as game show host. The host should shuffle and arrange the cards so that he or she knows where the ace is located.
- The contestant then picks a door.
- The host will then reveal one of the doors that wasn't picked and that has a goat behind it (so the host will always show a 2).
- The contestant must then decide whether to stick with the original choice or to switch to the door that hasn't been opened.
- Finally, the host reveals what is located behind the contestant's chosen door.
- Play the game 10 times, and record the results in a table like the one below. The player should be sure to stick sometimes and to switch sometimes.

Trial	Door chosen first	Stick/switch	Win/lose
1	2	Switch	Lose
2	1	Stick	Win
3	2	Stick	Lose

3. Switch roles and repeat Step 2.

4. What's the probability that you picked the door with the nice prize behind it in the first place?

5. Intuition tells us that it shouldn't make any difference whether you stick or switch. There's still a 1/3 chance that you're right. Based on the results of this Activity, would you agree or disagree?

Probability models

As Activity 7.2 illustrates, our intuition about probability is sometimes unreliable. In such cases, simulation can help clarify our thinking about chance behavior. But do we always need to repeat a chance process many times to determine the probability of a particular outcome? Fortunately, the answer is no. What we need is a **probability model** that describes the chance behavior. Our probability model should include a list of all possible outcomes (the **sample space**) and a probability for each outcome.

Probability Model

A **probability model** is a description of some chance behavior that consists of two parts: a list of all possible outcomes and the probability for each outcome.

The list of all possible outcomes of some random phenomenon is called the **sample space.**

Let's see how a probability model works in a familiar setting—tossing a coin.

Example 7.11 Probability model: tossing a coin three times

If you toss a fair coin three times, what's the probability of getting two heads and one tail? In this case, we're trying to model the chance behavior involved in tossing a coin three times. One possible outcome of this random phenomenon is heads on the first toss, tails on the second toss, and tails on the third toss, which we can represent as H-T-T. The *sample space* consists of all possible outcomes, which we can list as follows:

H-H-H H-H-T H-T-H T-H-H T-T-H T-H-T H-T-T T-T-T

Since each toss is equally likely to land heads or tails, each of these 8 possible outcomes has the same chance of occurring: 1/8. Three of these outcomes consist of two heads and one tail. So the probability of getting two heads and one tail is 3/8.

You can see from Example 7.11 that a probability model does more than just assign a probability to each outcome. It allows us to find the probability of any collection of outcomes, which we call an **event.** Events are usually designated by capital letters, like A, B, C, etc.

Event

An **event** is any collection of outcomes from some chance process.

In Example 7.11, if we let event A = getting 2 heads and 1 tail, then we can write $P(\text{A}) = 3/8$ as a shorthand for saying "the probability that event A will happen is 3/8." Consider two more events:

B = getting three heads and C = getting more heads than tails

Convince yourself that $P(\text{B}) = 1/8$. How do we find $P(\text{C})$? There are four outcomes in the sample space with more heads than tails, so $P(\text{C}) = 4/8$, or 1/2. What is the relationship between $P(\text{A})$, $P(\text{B})$, and $P(\text{C})$? Why does this make sense?

Here's one more event: D = getting at least one tail. What's $P(D)$? (Recall that "at least one" means one or more.) Confirm that $P(D) = 7/8$. What's the relationship between $P(D)$ and $P(B)$? Why does this make sense?

Basic rules of probability

Our coin-tossing example has revealed some basic rules that any probability model must obey:

- **The probability of any event is a number between 0 and 1.** The probability of an event is the long-run proportion of repetitions on which that event occurs. Any proportion is a number between 0 and 1, so any probability is also a number between 0 and 1. An event with probability 0 never occurs, and an event with probability 1 occurs on every trial. An event with probability 0.5 occurs in half the trials in the long run.
- **All possible outcomes together must have probability 1.** Because some outcome must occur on every trial, the sum of the probabilities for all possible outcomes must be exactly 1.
- **The probability that an event does *not* occur is 1 minus the probability that the event does occur.** If an event occurs in (say) 70% of all trials, it fails to occur in the other 30%. The probability that an event occurs and the probability that it does not occur always add to 100%, or 1. (This explains why $P(D) = 1 - P(B)$ in the coin-tossing example, since event D occurring is the same as event B not occurring.)
- **If two events have no outcomes in common, the probability that one or the other occurs is the sum of their individual probabilities.** If one event occurs in 40% of all trials, a different event occurs in 25% of all trials, and the two can never occur together, then one or the other occurs on 65% of all trials because 40% + 25% = 65%. (This explains why $P(A) + P(B) = P(C)$ in the coin-tossing example. Event A was getting exactly two heads; event B was getting three heads. These two events have no outcomes in common, so the probability that one or the other occurs is $P(A) + P(B)$. Note that event C—getting more heads than tails—is equivalent to "event A occurs or event B occurs.")
- **If all outcomes in the sample space are equally likely, the probability that event A occurs can be found using the formula**

$$P(A) = \frac{\text{number of outcomes corresponding to event A}}{\text{total number of outcomes in sample space}}$$

Here's an example that illustrates the use of these probability rules.

Example 7.12 Probability model: rolling two dice

Imagine rolling two fair, six-sided dice—one that's red and one that's green. Figure 7.3 (on the next page) shows the 36 possible outcomes in the sample space. How should we assign probabilities to these outcomes?

If the dice are fair, then it is reasonable to assign the same probability to each of the 36 outcomes in Figure 7.3. Because these 36 probabilities must have sum 1, each outcome must have probability 1/36.

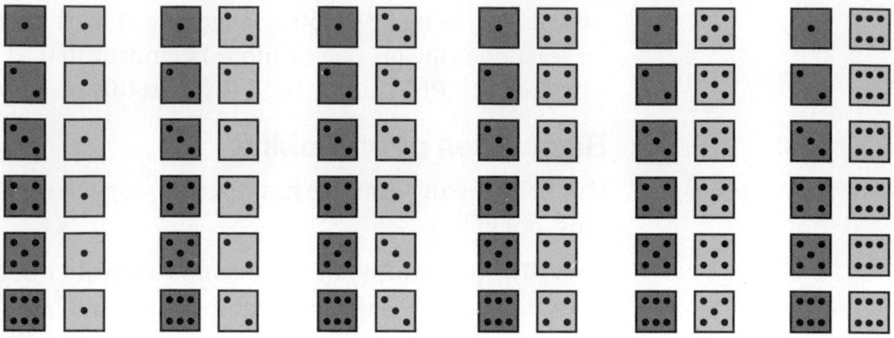

Figure 7.3 The 36 possible outcomes from rolling two dice, one red and one green.

We are interested in the sum of the spots on the up-faces of the dice. What is the probability that this sum is 5? The event A = "roll a 5" contains four outcomes, and its probability is the sum of these outcomes:

$P(A) = P(\text{roll a 5}) =$ 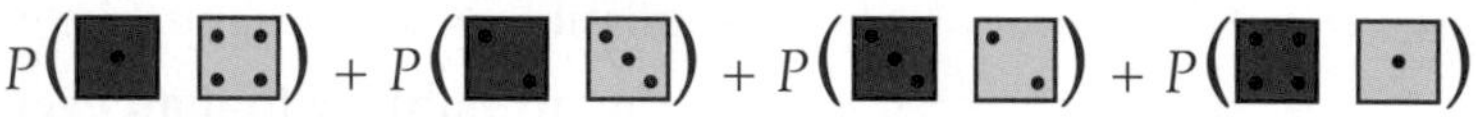

$$= \frac{1}{36} + \frac{1}{36} + \frac{1}{36} + \frac{1}{36}$$

$$= \frac{4}{36} = 0.111$$

What's the probability that the sum is *not* 5?

$$P(\text{sum is not 5}) = 1 - P(\text{sum is 5}) = 1 - \frac{4}{36} = \frac{32}{36} = 0.889$$

The rules tell us only what probability models *make sense*. They don't tell us whether the probabilities are *correct*, that is, whether they describe what actually happens in the long run. The probabilities in Example 7.12 are correct for fair dice, like those used in casinos. Inexpensive dice with hollowed-out spots are not balanced, and this probability model does not describe their behavior.

The basic probability rules apply to more than just tossing coins or rolling dice—they apply in any situation involving chance behavior, as the following example illustrates.

Example 7.13 Probability model: marital status

Choose a woman aged 25 to 29 years old at random and record her marital status. "At random" means that we give every such woman the same chance to be the one we choose. That is, we choose a random sample of size 1. The probability of any marital status is just the proportion of all women aged 25 to 29 who have that status—if we chose many women, this is the proportion we would get. Here is the set of probabilities:

Marital status:	Never married	Married	Widowed	Divorced
Probability:	0.506	0.452	0.002	0.04

This table gives a probability model for picking a young woman at random and finding out her marital status. It tells us what the possible outcomes are (there are only four) and it assigns probabilities to these outcomes. The probabilities here are the proportions of all women who are in each marital class. That makes it clear that the probability that a woman is not married is just the sum of the probabilities of the three classes of unmarried women:

$$\begin{aligned} P(\text{not married}) &= P(\text{never married}) + P(\text{widowed}) + P(\text{divorced}) \\ &= 0.506 + 0.002 + 0.04 = 0.548 \end{aligned}$$

Alternatively, we can use one of our other basic probability rules:

$$\begin{aligned} P(\text{not married}) &= 1 - P(\text{married}) \\ &= 1 - 0.452 = 0.548 \end{aligned}$$

That is, if 45.2% are married, 54.8% are not married.

Exercises

7.27 Toss 4 times Imagine tossing a fair coin 4 times.

(a) List all possible outcomes in the sample space.

(b) What probability would you assign to each outcome? Why?

(c) Event A = getting three tails and one head. Find $P(\text{A})$. Show your method.

7.28 Use your head Refer to Exercise 7.27. In two different ways, find the probability of getting at least one head. Show your work.

7.29 Pair-a-dice Imagine rolling two fair, six-sided dice—one red and one green. Use the probability model from Example 7.12 and the basic probability rules to help you find the probability of each of the following events. Show your work.

D = doubles (the same number of spots showing on both dice)

M = sum of the spots showing on the two dice is 10 or less

R = red die has a higher number of spots showing than green die

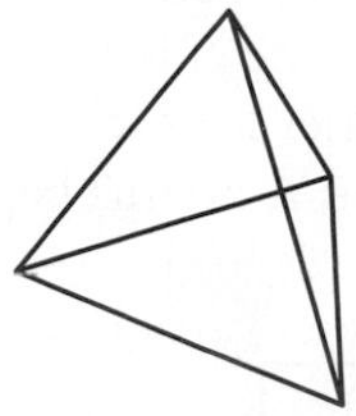

7.30 Four-sided dice A tetrahedron (see image) is a pyramid with four faces, each a triangle with all sides equal in length. Label the four faces of a tetrahedral die with 1, 2, 3, and 4 spots.

(a) Give a probability model for rolling two such dice—one blue and one yellow—and recording the number of spots on the down-face. Explain why you think your model is at least close to correct.

(b) What is the probability that the sum of the down-faces is 5? Show your work.

7.31 Causes of death Government data assign a single cause for each death that occurs in the United States. The data show that the probability is 0.45 that a randomly chosen death was due to cardiovascular (mainly heart) disease, and 0.23 that it was due to cancer.

(a) What is the probability that a death was due either to cardiovascular disease or to cancer? Show your work.

(b) What is the probability that the death was due to some other cause? Show your work.

7.32 Do husbands do their share? An opinion poll interviewed a random sample of 1025 married women. The women were asked whether their husbands did their fair share of household chores. Here are the results:

Outcome	Probability
Does more than his fair share	0.12
Does his fair share	0.61
Does less than his fair share	?

These proportions are probabilities for the random phenomenon of choosing a married woman at random and asking her opinion.

(a) What must be the probability that the woman chosen says that her husband does less than his fair share? Why?

(b) What is the probability of the event "I think my husband does at least his fair share"?

Two-way tables and probability

So far, we have learned basic tools for modeling chance behavior and for finding the probability of an event. What if we're interested in finding probabilities involving two events?

Example 7.14 Shuffle up and deal!

A standard deck of playing cards (with jokers removed) consists of 52 cards in four suits—clubs, diamonds, hearts, and spades. Each suit has 13 cards, with denominations ace, 2, 3, 4, 5, 6, 7, 8, 9, 10, jack, queen, and king. The jack, queen, and king are referred to as "face cards." Imagine that we shuffle the deck thoroughly and deal one card. Let's define two events: A = getting a face card and B = getting a heart.

We can build a two-way table that describes the possible outcomes in terms of these two events. The rows of the table relate to event A and the columns of the table relate to event B. Notice that there are only two possibilities for each event—either it happens or it doesn't. So we need two rows (A happens; A doesn't happen) and two columns (B happens; B doesn't happen).

	Heart	Not a heart
Face card		
Not a face card		

Now we need to fill in the cells of the table. Let's start with the rows, which relate to event A = getting a face card. There are a total of 12 face cards—3 in each suit times

4 suits. Since 3 of those face cards are hearts, we put a "3" in the top-left cell of the table. That leaves 9 face cards that are not hearts. This value goes in the top-right cell of the table. There are 13 hearts in all, so the bottom-left cell of the table must contain a 10. Do you see how we got the "30" in the bottom-right cell? If not, look at the marginal totals.

	Heart	Not a heart	
Face card	3	9	**12**
Not a face card	10	30	**40**
	13	**39**	**52**

What's $P(\text{A})$? Since each of the 52 cards is equally likely to be dealt, that means we have a 1/52 chance of getting any particular card. Since there are 12 face cards, $P(\text{A}) = 12/52$. Finding $P(\text{B})$ is just as easy. Since there are 13 hearts in the deck, $P(\text{B}) = 13/52$.

The **two-way table** in Example 7.14 helped organize the outcomes in the sample space in a way that allowed us to easily find $P(\text{A})$ and $P(\text{B})$. To be honest, we could have determined these probabilities without using a two-way table. All we would have to do is list the sample space in some other way and then use the same reasoning as in Example 7.14 to obtain the probabilities of these two events. But what if we wanted to find a probability involving both of these events, like the probability of getting a face card *and* a heart? In that case, the two-way table would make the task much easier.

Example 7.15 Playing cards and two-way tables

Continuing with the scenario of Example 7.14, we're interested in finding the probability of getting a face card *and* a heart. That is, we're trying to determine the probability $P(\text{A and B})$. The event "face card *and* heart" can be seen at the intersection of the "face card" row and the "heart" column in the completed two-way table of Example 7.14. There are 3 cards in the deck that are both face cards and hearts (namely, the jack, queen, and king of hearts). So the desired probability is

$$P(\text{A and B}) = P(\text{face card and heart}) = 3/52$$

What about the probability of getting a face card *or* a heart? (Note the mathematical use of the word "or" here—the card could be a face card, or a heart, or both.) There are 12 face cards in the deck, and 13 hearts. If you weren't paying close attention, you might be tempted to add these two numbers together and to give the (incorrect) answer 25/52 for the desired probability. Do you see why this is wrong? There are 3 cards in the deck that are both face cards and hearts. If we add the row total for "face card" to the column total for "heart," we are counting these three cards *twice*. How many cards are there that are face cards *or* hearts (or both)? From the two-way table, we can see that there are $10 + 3 + 9 = 22$ such cards. So the desired probability is

$$P(\text{A or B}) = P(\text{face card or heart}) = 22/52$$

There's another way to get the correct value of 22/52 for $P(\text{A or B})$ in Example 7.15. If we add $P(\text{A}) = 12/52$ and $P(\text{B}) = 13/52$, we get the incorrect answer (25/52) mentioned in the example. Why? Because we are counting three outcomes—the cards that are face cards *and* hearts—in both $P(\text{A})$ and $P(\text{B})$. If we subtract the probability $P(\text{A and B}) = 3/52$, we get

$$\begin{aligned} P(\text{A}) + P(\text{B}) - P(\text{A and B}) &= 12/52 + 13/52 - 3/52 \\ &= 22/52 \\ &= P(\text{A or B}) \end{aligned}$$

This intuitive result is known as the **general addition rule.**

General addition rule for two events

If A and B are any two events resulting from some chance process, then

$$P(\text{A or B}) = P(\text{A}) + P(\text{B}) - P(\text{A and B})$$

You might be wondering whether there is also a rule for $P(\text{A and B})$. There is, but it's not quite as intuitive. Stay tuned for that later. For now, let's see what we can do with two-way tables and the general addition rule.

iStockphoto

Example 7.16 Who has pierced ears?

Students in a college statistics class wanted to find out how common it was for young adults to have their ears pierced. They recorded data on two variables—gender and whether the student had pierced ears—for all 178 individuals in the class. The two-way table below displays the data.

	Pierced ears?		
Gender	**Yes**	**No**	**Total**
Male	19	71	**90**
Female	84	4	**88**
Total	**103**	**75**	**178**

- If we randomly select a student from the class, what's the probability that the student has pierced ears? Let event A = has pierced ears. Since each of the 178 students in the class is equally likely to be chosen, and there are 103 students with pierced ears, $P(\text{A}) = 103/178$.
- If we randomly select a student from the class, what's the probability that we choose a male with pierced ears? Let event B = is a male. We want to find $P(\text{pierced ears and male})$, that is, $P(\text{A and B})$. Looking at the intersection of the "male" row and "Yes" column, we see that there are 19 males with pierced ears. So $P(\text{A and B}) = 19/178$.

- If we randomly select a student from the class, what's the probability that we choose someone with pierced ears or a male? This time, we're interested in P(A or B). Using the general addition rule,

$$\begin{aligned} P(\text{A or B}) &= P(\text{A}) + P(\text{B}) - P(\text{A and B}) \\ &= 103/178 + 90/178 - 19/178 \\ &= 174/178 \end{aligned}$$

Can you think of another (simpler) way to find this probability?

When we found the probability of getting a male with pierced ears in Example 7.16, we could have described this as either P(A and B) or P(B and A). Why? Because "A and B" describes the same event as "B and A." Likewise, P(A or B) is the same as P(B or A). *Don't get so caught up in the notation that you lose sight of what's really happening!* If you convert the symbols to words, it's pretty obvious that P(pierced ears and male) is the same as P(male and pierced ears). Likewise, P(pierced ears or male) is clearly the same as P(male or pierced ears).

Exercises

7.33 Card tables and two-way tables Refer to Example 7.14. Let event C = getting a card that is not a face card and event D = getting a spade, club, or diamond.

(a) Find P(C) and P(D). Show your method clearly.

(b) Describe the event "C and D" in words. Then find P(C and D). Show your work.

(c) Describe the event "C or D" in words. Then find P(C or D). Show your work.

7.34 More cards Shuffle a standard deck of playing cards and deal one card, as in Example 7.14. Let event J = getting a jack and event R = getting a red card.

(a) Construct a two-way table that describes the sample space in terms of events J and R.

(b) Find P(J), P(R), and P(J and R). Show your work.

(c) Explain why $P(\text{J or R}) \neq P(\text{J}) + P(\text{R})$. Then use the general addition rule to compute P(J or R).

7.35 Sampling senators, I The two-way table below describes the members of the U.S. Senate in 2008.

	Male	Female
Democrats	40	11
Republicans	44	5

(a) Who are the individuals? What variables are being measured?

(b) If we select a U.S. senator at random, what's the probability that we choose

- a Democrat?
- a female?
- a female Democrat?
- a female or a Democrat?

7.36 Who eats breakfast? Students in an urban school were curious about how many children regularly eat breakfast. They conducted a survey, asking, "Do you eat breakfast on a regular basis?" All 595 students in the school responded to the survey. The resulting data are shown in the two-way table below.[8]

	Male	Female	Total
Eat breakfast regularly	190	110	**300**
Don't eat breakfast regularly	130	165	**295**
Total	**320**	**275**	**595**

(a) Who are the individuals? What variables are being measured?

(b) If we select a student from the school at random, what is the probability that we choose

- a female?
- someone who eats breakfast regularly?
- a female who eats breakfast regularly?
- a female or someone who eats breakfast regularly?

7.37 Playing roulette An American roulette wheel has 38 slots with numbers 1 through 36, 0, and 00 (see Figure 7.4). On the wheel, 18 of the numbered slots are red, 18 are black, and two—the 0 and 00—are green. When the wheel is spun, a metal ball is dropped onto the middle of the wheel. If the wheel is balanced, the ball is equally likely to settle in any of the numbered slots. Imagine spinning a fair wheel once. Let event B = ball lands in a black slot and event E = ball lands in an even-numbered slot. (Treat 0 and 00 as even numbers.)

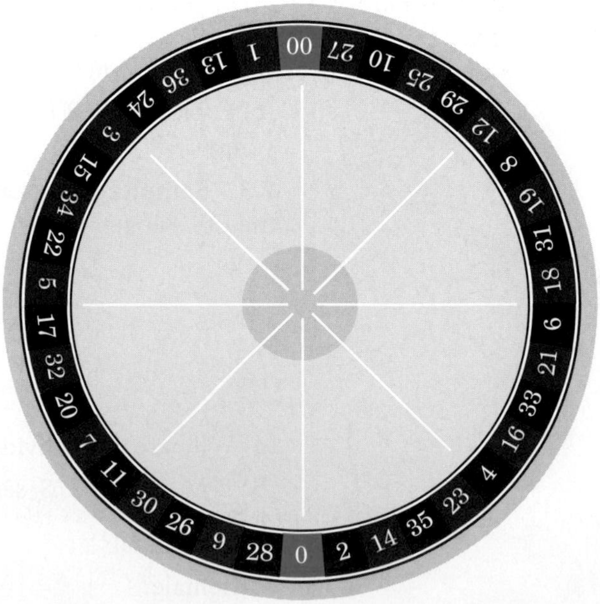

Figure 7.4 Diagram of an American roulette wheel.

(a) Make a two-way table that displays the sample space in terms of events B and E.

(b) Find *P*(B) and *P*(E). Show your method clearly.

(c) Describe the event "B and E" in words. Then find *P*(B and E). Show your work.

(d) Describe the event "B or E" in words. Then find *P*(B or E). Show your work.

7.38 More roulette Refer to Exercise 7.37. Let event G = ball lands in a green slot.

(a) Find *P*(B and G). Explain why this makes sense.

(b) Find *P*(B or G) using the general addition rule. What do you notice?

(c) Find *P*(E and G). Explain why this makes sense.

(d) Find *P*(E or G) using the general addition rule.

Venn diagrams and probability

When we're trying to find probabilities involving two events, two-way tables display the sample space in a way that makes probability calculations simple. **Venn diagrams** provide another way to illustrate the set of all possible outcomes of a chance process. A Venn diagram usually consists of two or more circles surrounded by an outer rectangle, as in Figure 7.5. Each circle represents an event. The area inside the rectangle represents the sample space of the random phenomenon (sometimes called the "universe"). An example should help illustrate how Venn diagrams work.

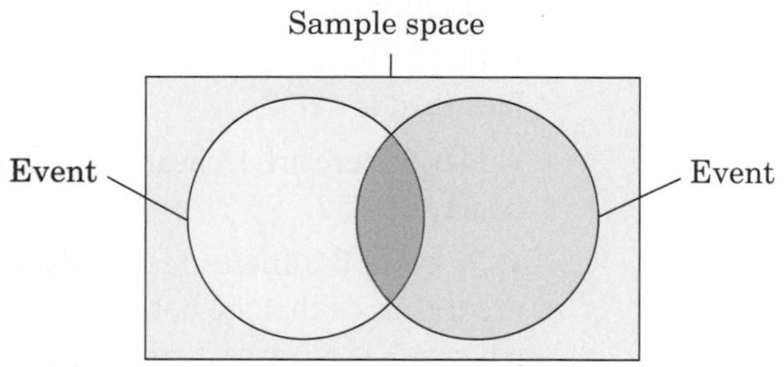

Figure 7.5 A typical Venn diagram shows the sample space and the relationship between two or more events.

Example 7.17 Venn diagrams and dealing cards

In Examples 7.14 and 7.15 (pages 334 and 335), we modeled the process of dealing a playing card from a well-shuffled deck. Our two events of interest were A = getting a face card and B = getting a heart. Earlier, we used a two-way table to show the sample space. Now we'll use a Venn diagram.

Figure 7.6 (on the next page) is a Venn diagram showing all 52 possible outcomes of this chance process. The face cards are all shown inside circle A; the hearts are all shown inside circle B. Those cards that are *both* face cards and hearts are placed in the intersection of the two circles. Which cards are outside the two circles? Those cards that are *neither* hearts nor face cards. Recall that since each card is equally likely to be the one dealt, each outcome in the sample space has a 1/52 chance of occurring.

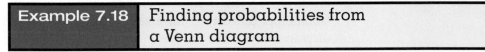

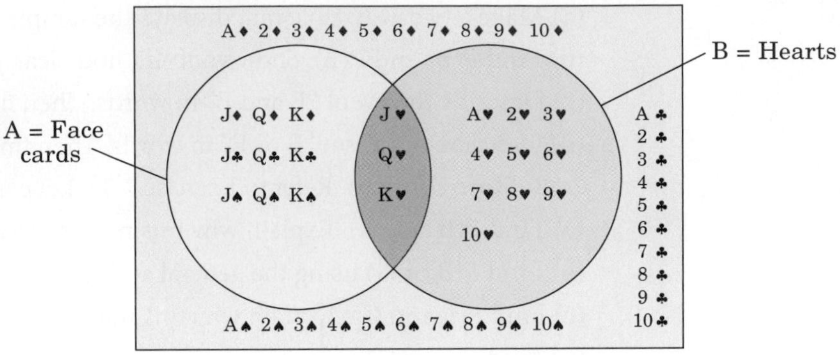

Figure 7.6 A Venn diagram showing the sample space when we deal a card from a well-shuffled deck. Our two events of interest are A = getting a face card and B = getting a heart.

How can we find probabilities from a Venn diagram? By counting the outcomes in the appropriate region(s) of the diagram, as the following example illustrates.

Example 7.18 Finding probabilities from a Venn diagram

Let's use Figure 7.6, the Venn diagram in Example 7.17, to find some probabilities.

- *P*(A): There are 12 face cards inside circle A. So the probability of getting a face card is 12/52.
- *P*(B): There are 13 hearts inside circle B. So the probability of getting a heart is 13/52.
- *P*(A and B): There are 3 cards in the intersection of circles A and B—these are the cards that are both face cards and hearts. So the probability of getting a heart that is a face card is 3/52.
- *P*(A or B): We are looking for the cards that are face cards or hearts (or both). These are the cards anywhere inside the two circles. Since there are 22 such cards, *P*(A or B) = 22/52.
- *P*(*not* A and *not* B): We want the cards that are *not* face cards and also *not* hearts. Where are they located? In the region of the Venn diagram that falls outside the two circles. Since there are 30 such cards, the desired probability is 30/52. How is this probability related to *P*(A or B)? Why does this make sense?

Venn diagrams have uses in other branches of mathematics. As a result, some common vocabulary and notation have been developed that you ought to know. The following chart shows the language that we have been using, along with standard mathematical notation and terminology, and the corresponding Venn diagrams.

Our language	Notation	Terminology	Venn diagram
not A	A^c	**complement** of A	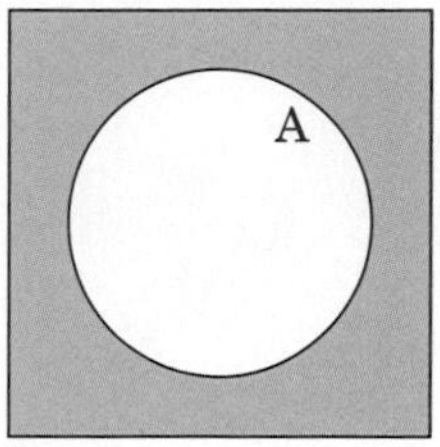
A *and* B	$A \cap B$	**intersection** of A and B	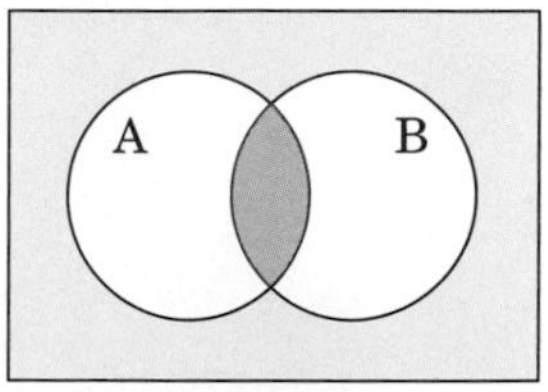
A *or* B	$A \cup B$	**union** of A and B	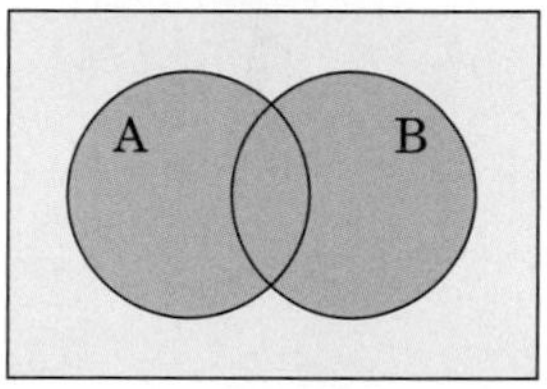

Sometimes, instead of listing the actual outcomes of a chance process in the Venn diagram, we summarize the sample space by writing the *number* of outcomes in each region of the Venn diagram. Here's an example that illustrates this approach.

Example 7.19 Another way to use Venn diagrams

In Example 7.16 (page 336), we looked at data from a survey on gender and ear piercings for a large group of college students. The chance process consisted of selecting a student in the class at random. Our events of interest were A = has pierced ears and B = is male. Here, once again, is the two-way table that summarizes the sample space:

	Pierced ears?	
Gender	**Yes**	**No**
Male	19	71
Female	84	4

How can we construct a Venn diagram that displays this same information?

There are four distinct regions in the Venn diagram shown in Figure 7.7 (on the next page). These regions correspond to the four cells in the two-way table. We can describe this correspondence in tabular form as follows:

Region in Venn diagram	In words	In symbols	Count
In the intersection of two circles	pierced ears and male	$A \cap B$	19
Inside circle A, outside circle B	pierced ears and female	$A \cap B^c$	84
Inside circle B, outside circle A	no pierced ears and male	$A^c \cap B$	71
Outside both circles	no pierced ears and female	$A^c \cap B^c$	4

We have added the appropriate counts of students to the four regions in Figure 7.7.

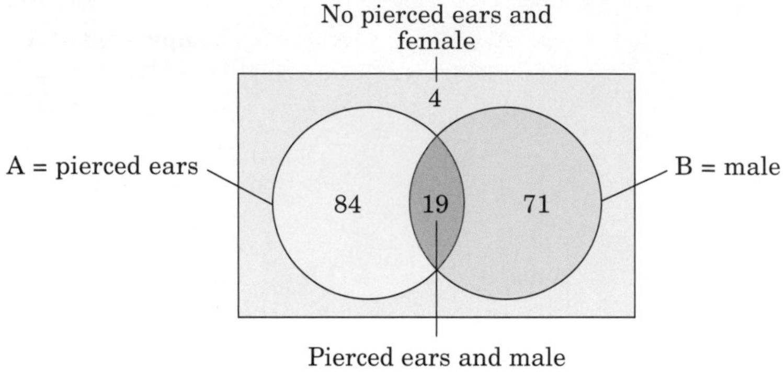

Figure 7.7 The completed Venn diagram for the large group of college students. The circles represent the two events A = has pierced ears and B = is male.

Venn diagrams can also be used to solve probability problems, as the following example illustrates.

Example 7.20 Who reads the paper?

In a large apartment complex, 35% of residents read *USA Today*. Only 20% read the *New York Times*. Five percent of residents read both papers. If we select a resident of the apartment complex at random, what's the probability that he or she doesn't read either paper?

Even though we don't know how many residents there are in the apartment complex, we can still construct a Venn diagram to model this situation. Our events of interest are A = reads *USA Today* and B = reads *New York Times*. Figure 7.8(a) shows the empty Venn diagram. In problems of this kind, it's usually a good strategy to start in the "middle." Since we know that 5% of residents read both papers, we can put this value in the intersection of the two circles, $A \cap B$. Figure 7.8(b) shows the updated Venn diagram.

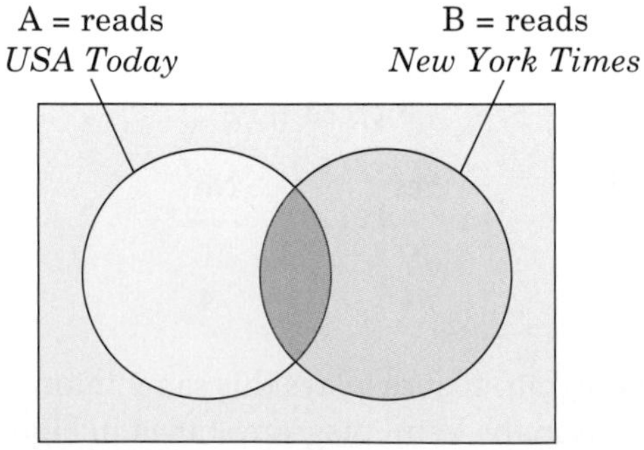

Figure 7.8(a) A Venn diagram with event A showing the residents who read *USA Today* and B showing those who read the *New York Times*.

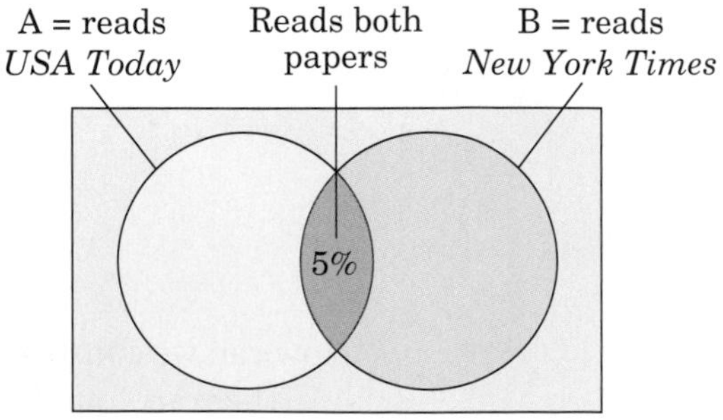

Figure 7.8(b) Updated Venn diagram showing the 5% of residents who read both papers.

Now let's work our way from the inside out. We know that 35% of residents read *USA Today*. (That is, $P(A) = 0.35$.) But we have already accounted for the 5% of people in the group that read both papers. That leaves 35% − 5% = 30% who read

USA Today but not the *New York Times.* We'll fill in that value in the corresponding region of Figure 7.8(c). Similarly, we can calculate that 20% − 5% = 15% read the *New York Times* but not *USA Today.* This value has also been added to Figure 7.8(c).

How can we figure out the percent of residents who read neither paper? Those are the people in the region of the Venn diagram that's outside both circles. Since all residents in the apartment complex fall somewhere in the diagram, we start with 100% of residents and subtract those who read either paper. That is,

$$\text{percent who read neither paper} = 100\% - 30\% - 5\% - 15\% = 50\%$$

Figure 7.8(d) shows the completed Venn diagram.

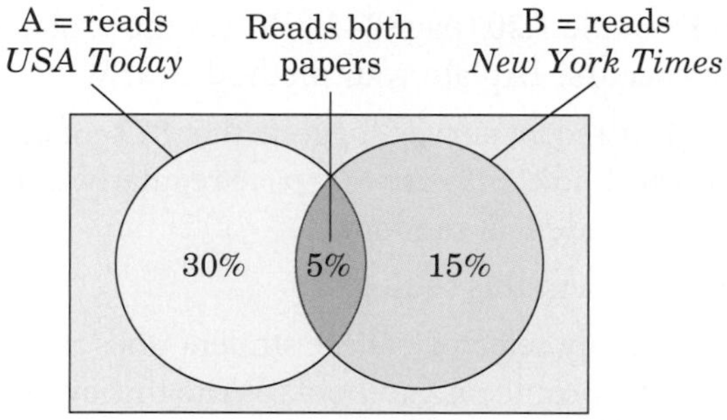

Figure 7.8(c) The updated Venn diagram showing the percents of residents who read only one of the two newspapers.

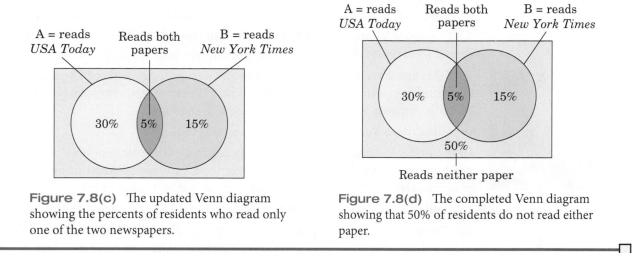

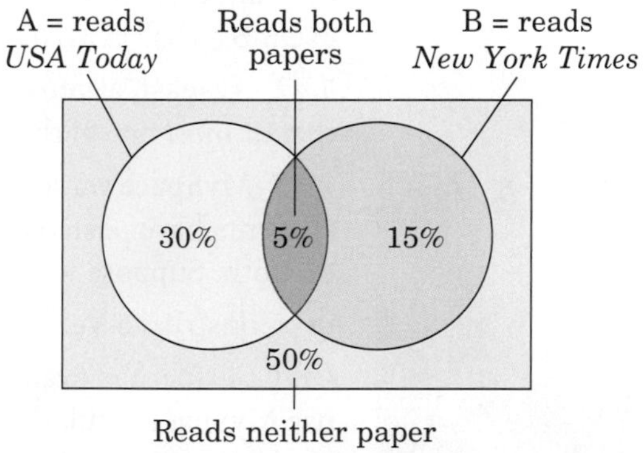

Figure 7.8(d) The completed Venn diagram showing that 50% of residents do not read either paper.

So far, we have tried to keep the wording of probability questions as consistent and straightforward as possible. There are times, though, when you will encounter more complicated language. One common example is found in questions like "What's the probability that at least one of the two events occurs?" and "What's the probability that at most one of the two events occurs?" Let's consider the first of these questions briefly in the setting of Example 7.20. We'll save the other question for an exercise.

For events A and B as described in Example 7.20, *at least one* of the two events occurs if we randomly select a resident who reads *USA Today,* the *New York Times,* or both papers. The probability that this happens is 0.50, since 50% of the residents read at least one of the two papers. Note that this is also $P(A \cup B)$, since the event $A \cup B$ includes those who read one or both of the newspapers.

Exercises

7.39 Sampling senators, II The two-way table below describes the members of the U.S. Senate in 2008. Construct a Venn diagram of these data using D = is a Democrat and F = is female as the events of interest.

	Male	Female
Democrats	40	11
Republicans	44	5

7.40 Breakfast eaters, again The two-way table below describes the 595 students who responded to a school survey about eating breakfast. Construct a Venn diagram of these data using B = eats breakfast regularly and M = is male as the events of interest.

	Male	Female	Total
Eat breakfast regularly	190	110	**300**
Don't eat breakfast regularly	130	165	**295**
Total	**320**	**275**	**595**

7.41 Playing roulette again Refer to Exercise 7.37 (page 338). Construct a Venn diagram that shows all possible outcomes of spinning the roulette wheel in terms of event B = ball lands in black slot and event E = ball lands in an even-numbered slot.

7.42 At least, at most Refer to Example 7.20 (page 342). What's the probability that *at most* one of the two events occurs? Explain your method clearly.

Chris Jackson/Getty Images

7.43 MySpace versus Facebook, I A recent survey suggests that 85% of college students have posted a profile on Facebook, 54% use MySpace regularly, and 42% do both. Suppose we select a college student at random.

(a) Construct a Venn diagram to represent this setting.

(b) Consider the event that the randomly selected college student does not use MySpace regularly or has posted a profile on Facebook. Write this event in symbolic form using the two events of interest that you chose in (a).

(c) Find the probability of the event described in (b). Explain your method.

(d) Can you make a two-way table for this setting? If so, do it. If not, explain why not.

7.44 Computers at Princeton An October 2007 census revealed that 40% of Princeton students primarily used Macintosh computers (MACs). The rest primarily used PCs. At the time of the census, 67% of Princeton students were undergraduates. The rest were graduate students. According to the census, 23% of Princeton's graduate students said that they used PCs as their primary computers. Suppose that we select a Princeton student at random.

(a) Construct a Venn diagram to represent this setting.

(b) Consider the event that the randomly selected student is a graduate student who uses a MAC. Write this event in symbolic form using the two events of interest that you chose in (a).

(c) Find the probability of the event described in (b). Explain your method.

(d) Can you make a two-way table for this setting? If so, do it. If not, explain why not.

Mutually exclusive (disjoint) events

Thus far, we have drawn our Venn diagrams with two overlapping circles. That's because in every example we've studied, there have been outcomes that fell in the intersection of the two events. In some situations, two events may not have any outcomes in common. We say that the two events are **mutually exclusive,** or **disjoint.** When that happens, a Venn diagram displaying the events should show two nonoverlapping circles. Figure 7.9 shows a Venn diagram for mutually exclusive events A and B.

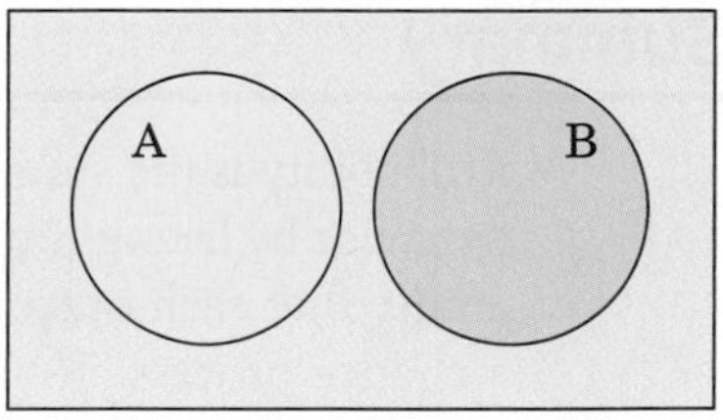

Figure 7.9 A Venn diagram showing disjoint events A and B.

Let's return to the card-dealing scenario of Examples 7.17 and 7.18. Recall that we had previously defined event B = getting a heart. If we define a new event C = getting a black card, then events B and C are mutually exclusive (disjoint). Why? Because if we deal one card from a standard deck, that card cannot be both a heart and a black card. What's $P(B \cap C)$ in this case? It's 0, of course. What is $P(B \cup C)$? By the general addition rule,

$$P(B \cup C) = P(B) + P(C) - P(B \cap C) = P(B) + P(C) \quad \text{since } P(B \cap C) = 0$$

These special results for disjoint events are worth summarizing.

Special rules for disjoint (mutually exclusive) events

If events A and B are disjoint (mutually exclusive), then

$$P(A \cap B) = 0$$
$$P(A \cup B) = P(A) + P(B)$$

APPLICATION 7.2

Genetics at work

You inherited certain physical characteristics from your parents—gender, eye and hair color, and the shape of your hairline and your earlobes. In this Application, we'll investigate the relationship between two other traits that are believed to be inherited—tongue curling and hand clasping.

QUESTIONS

1. Have everyone in your class try to curl their tongue (rolling the sides up to form a cup).

2. Now have everyone interlock the fingers of their two hands. Note which thumb (right or left) comes out on top.

Our two events of interest are T = can curl tongue and R = right thumb on top.

3. Make a Venn diagram that shows where each member of your class falls with respect to the two events of interest. Use initials to represent each student.

4. Now make a two-way table that summarizes the number of students in each of the four regions of the Venn diagram.

5. Describe a method for selecting a student from the class at random.

6. Suppose we select a student from the class at random. Find each of the following probabilities. Describe in words what your result means.
   (a) $P(T^c \cap R^c)$
   (b) $P(T^c \cup R)$

7. *Challenge:* Consider the event F = is female. See if you can construct a Venn diagram with three circles to classify the members of your class with respect to events T, R, and F.

Section 7.2 Summary

Probability is the study of chance behavior. A **probability model** describes chance behavior by listing the possible outcomes in the **sample space** and giving the probability that each outcome occurs. An **event** is a subset of the possible outcomes in a chance process. To find the probability that an event A happens, we can rely on some basic probability rules:

- $0 \leq P(A) \leq 1$
- $P(A^c) = 1 - P(A)$, where A^c is the **complement** of event A; that is, the event that A does not happen
- $P(S) = 1$, where S = the sample space
- If all outcomes in the sample space are equally likely,

$$P(A) = \frac{\text{number of outcomes corresponding to event A}}{\text{total number of outcomes in sample space}}$$

A **two-way table** or a **Venn diagram** can be used to display the sample space of some chance process when we have two events of interest, say A and B. Two-way tables and Venn diagrams can also be used to find probabilities involving events A and B, like the **union** ($A \cup B$) and **intersection** ($A \cap B$). The event $A \cup B$ ("A or B") consists of all outcomes in event A, event B, or both. The event $A \cap B$ ("A and B") consists of those outcomes in both A and B. The **general addition rule** can be used to find $P(A \cup B)$:

$$P(\text{A or B}) = P(A \cup B) = P(A) + P(B) - P(A \cap B)$$

If events A and B cannot happen together, we say that they are **mutually exclusive,** or **disjoint.** For mutually exclusive events A and B, the general addition rule simplifies to

$$P(\text{A or B}) = P(A \cup B) = P(A) + P(B)$$

Section 7.2 Exercises

7.45 Popular kids Researchers carried out a survey of fourth-, fifth-, and sixth-grade students in Michigan. Students were asked whether good grades, athletic ability, or being popular was most important to them. The two-way table below summarizes the survey data.[9]

	4th grade	5th grade	6th grade	Total
Grades	49	50	69	**168**
Athletic	24	36	38	**98**
Popular	19	22	28	**69**
Total	**92**	**108**	**135**	**335**

(a) Who are the individuals? What variables are being measured?
(b) If we select a survey respondent at random, what's the probability that we choose

- a sixth-grader?
- a student who rated good grades as most important?
- a sixth-grader who rated good grades as most important?
- a sixth-grader or a student who rated good grades as most important?

(c) How is the two-way table in this exercise different from the ones you saw in earlier examples?

7.46 TCNJ survey, I The 28 students in Mr. Starnes's introductory statistics class at The College of New Jersey (TCNJ) completed a brief survey. One of the questions asked whether each student was right- or left-handed. The two-way table below displays information on the gender and handedness of the students in the class.

	Gender		
Handedness	**F**	**M**	**Row summary**
L	3	1	4
R	18	6	24
Column summary	21	7	28

(a) Nationally, about 10% of people are left-handed. Are the students in Mr. Starnes's class similar to the nation as a whole? Justify your answer.
(b) What's the probability that a randomly selected student from the class is

- male or right-handed?
- male and right-handed?

7.47 TCNJ survey, II Refer to Exercise 7.46. Construct a Venn diagram to display the sample space for the chance process of choosing a student at random from Mr. Starnes's class. The events of interest are M = male and R = right-handed.
7.48 Mutually exclusive versus complementary For each part below, classify the statement as true or false. Justify your answer.
(a) If one event is the complement of another event, then the two events are mutually exclusive.
(b) If two events are mutually exclusive, then one event is the complement of the other.
7.49 Are you my (blood) type? Each of us has an ABO blood type, which describes whether two characteristics called A and B are present. Every human being has two blood type alleles (gene forms), one inherited from our mother and one from our father. Each of these alleles can be A, B, or O. Which two we inherit determines our blood type. Here is a table that shows what our blood type is for each combination of two alleles:

Alleles inherited	Blood type	Alleles inherited	Blood type
A and A	A	B and B	B
A and B	AB	B and O	B
A and O	A	O and O	O

Children are equally likely to inherit either allele from their mother and also equally likely to inherit either allele from their father.
(a) Mary and John both have alleles A and B. List the sample space that shows the alleles that their next child could receive.
(b) Give the possible blood types that this child could have, along with the probability for each blood type.

7.50 More on blood The table below shows the distribution of ABO blood types in the United States.[10]

Blood type:	O	A	B	AB
Relative frequency:	44%		10%	4%

(a) The entry for type A blood is missing from the table. What is this value? Explain.
(b) Suppose we select one person from the United States at random. Find the probability that the person does not have type O blood. Show your method clearly.

7.51 Even more on blood A person's blood type can be further classified based on whether a certain substance known as a Rhesus antigen is present. For example, a person with type B blood and the Rhesus antigen present is said to be type B+ (B-positive). A person with type B blood but without the Rhesus antigen present would be type B– (B-negative).
(a) People with blood type B+ can receive blood from anyone with blood type O or B. What's the probability that a randomly selected person from the United States can donate blood to someone with blood type B+?
(b) In the United States, 84% of people have the Rhesus antigen present, 8.5% of people have blood type B+ and 1.5% have blood type B–. Suppose we choose a U.S. person at random and record his or her blood type. Construct a Venn diagram that displays event B = has blood type B and event P = has the Rhesus antigen present.

7.52 Ask Marilyn again! In a *Parade* magazine column, Marilyn vos Savant was asked the following question:

> *A woman and a man (who are unrelated) each have two children. At least one of the woman's children is a boy, and the man's older child is a boy. Which is more likely: that the man has 2 boys or that the woman has 2 boys?*

What's your answer? Explain your reasoning.

7.3 Conditional Probability and Independence

ACTIVITY 7.3

Weird dice

Materials: Four special six-sided dice provided by your teacher

In this Activity, a few lucky students will get the chance to compete for a desirable reward from your teacher. After explaining the rules of the game, your teacher will ask for volunteers.

Rules

- One student at a time may challenge the teacher in the "weird dice" game.
- The student gets to pick any one of the four dice provided by the teacher to roll.
- Then the teacher must pick one of the three remaining dice to roll.
- In each round of the game, both players roll their die. The player with the highest number showing wins the round.
- The first player to win 3 rounds wins the game.

When the competition is over, your teacher will show you the number of spots on the six faces of each die from the last game that was played.

Questions

1. Suppose your teacher's chosen die and the final player's chosen die were rolled again. Use what you learned in the previous section to illustrate the sample space for this chance process.

2. Find the probability that your teacher would win the round. Show your work.

3. Working with your classmates, find the probability that your teacher would win if the two competitors played another game.

What is conditional probability?

	Pierced ears?		
Gender	**Yes**	**No**	**Total**
Male	19	71	**90**
Female	84	4	**88**
Total	**103**	**75**	**178**

Let's return to the college statistics class of Example 7.16 (page 336). Earlier, we used the two-way table at left to find probabilities involving events A = has pierced ears and B = is male for a randomly selected student. Here is a summary of our previous results.

- $P(A) = 103/178$
- $P(A \cap B) = P(A \text{ and } B) = 19/178$
- $P(A \cup B) = P(A \text{ or } B) = 174/178$

Since the probability that a randomly selected student is male is 90/178, we could add $P(B) = 90/178$ to the list. Now let's turn our attention to some other interesting probability questions.

Example 7.21 Getting acquainted with conditional probability

(a) If we know that a randomly selected student has pierced ears, what is the probability that the student is male? There are 103 students in the class with pierced ears. We can restrict our attention to this group, since we are told that the chosen student has pierced ears. Since there are 19 males among the 103 students with pierced ears, the desired probability is

$$P(\text{is male } \textit{given} \text{ has pierced ears}) = 19/103, \text{ or about } 18.4\%$$

(b) If we know that a randomly selected student is male, what's the probability that the student has pierced ears? This time, our attention is focused on the males in the class. Since only 19 of the 90 males in the class have pierced ears,

$$P(\text{has pierced ears } \textit{given} \text{ is male}) = 19/90, \text{ or about } 21.1\%$$

In this class, it's slightly more likely for a male to have pierced ears than it is for someone with pierced ears to be a male.

A probability like "the probability that a randomly selected student is male *given that* the student has pierced ears" is known as a **conditional probability.** The name comes from the fact that we are trying to find the probability that one event will occur under the condition that some other event is already known to have occurred. We often use the phrase "given that" to signal the condition. There's even a special notation to indicate a conditional probability. In the example above, we would write $P(\text{is male} \mid \text{has pierced ears})$, where the | means "given that" or "under the condition that." Since we already defined the events A = has pierced ears and B = is male, we could write the conditional probability as $P(B|A)$.

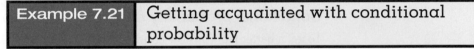

Conditional probability

The probability that one event happens under the condition that another event is already known to have happened is called a **conditional probability.** Suppose we know that event A has happened. Then the probability that event B happens *given that* event A has happened is denoted by $P(B|A)$.

With this new notation available, we can restate the results of Example 7.21 as

$$P(B|A) = 19/103 \qquad \text{and} \qquad P(A|B) = 19/90$$

Notice that the conditional probability $P(B|A) = 19/103$ is very different from the *unconditional* probability $P(B) = 90/178$. About half of the students in the class are male, but far fewer than half of the students with pierced ears are male. Also note that $P(A|B) = 19/90$ is much smaller than $P(A) = 103/178$. Over half of the students in the class have pierced ears, but only 19 out of 90 males in the class have pierced ears.

Conditional probability and independence

To get a better feel for how conditional probability works, let's return to two familiar settings involving chance behavior—tossing coins and rolling dice.

Example 7.22 Conditional probability: coins and dice

Suppose you toss a fair coin twice. Define events A = first toss is a head and B = second toss is a head. We have already established that $P(A) = 1/2$ and $P(B) = 1/2$. What's $P(B|A)$? It's the conditional probability that the second toss is a head given that the first toss was a head. The coin has no memory, so $P(B|A) = 1/2$. In this case, $P(B|A) = P(B)$. The condition that the first toss was a head does not affect the probability that the second toss is a head.

Let's contrast the coin toss scenario with the chance process of rolling two fair, six-sided dice—one red and one green. This time, we'll define events R = red die shows a 4 and S = sum of two dice is 11. Hopefully you'll agree that $P(R) = 1/6$. What's $P(S)$? From the sample space of Example 7.12 (page 331), it's 3/36. What's the conditional probability $P(S|R)$ that the sum of the two dice is 11 given that the red die shows a 4? If the red die shows a 4, then it's impossible to get a sum of 11 on the two dice. So $P(S|R) = 0$. This conditional probability is very different from the unconditional probability $P(S)$ of getting a sum of 11.

In Example 7.22, we observed that $P(B|A) = P(B)$ for events A and B in the coin toss setting. For the dice-rolling scenario, however, $P(S|R) \neq P(S)$. When knowledge that one event has happened does not change the likelihood that another event will happen, we say that the two events are **independent.**

Independent events

Two events A and B are **independent** if the occurrence of one event has no effect on the chance that the other event will occur. In other words, events A and B are independent if $P(A|B) = P(A)$ and $P(B|A) = P(B)$. If two events are not independent, we sometimes say that they are *dependent.*

There's something else interesting about events R and S in Example 7.22. See if you can figure out what it is. (Here's a hint: what's the probability $P(R \text{ and } S)$ of both events occurring?) By now, you've probably discovered that events R and S can't happen together. If event R happens and the red die shows a 4, the sum can't be 11. Likewise, if event S happens and the sum is 11, then the red die can't show a 4. That is, events R and S are mutually exclusive. **Two mutually exclusive events can *never* be independent,** because if one event happens, the other event is guaranteed not to happen.

We have one other important caution to offer. *Don't confuse independent trials in a chance process with independent events.* Both coin tosses and dice rolls result in independent trials. But in either setting, you can define events A and B that are or aren't independent.

Exercises

7.53 Sampling senators, III The two-way table at left describes the members of the U.S. Senate in 2008. Suppose we select a senator at random. Consider events D = is a Democrat and F = is female.

	Male	Female
Democrats	40	11
Republicans	44	5

(a) Find $P(D|F)$. Explain what this value means.

(b) Find $P(F|D)$. Explain what this value means.

(c) Are events D and F mutually exclusive? Justify your answer.

(d) Are events D and F independent? Justify your answer.

7.54 More breakfast eaters The two-way table below describes the 595 students who responded to a school survey about eating breakfast. Suppose we select a student at random. Consider events B = eats breakfast regularly and M = is male.

	Male	Female	Total
Eat breakfast regularly	190	110	**300**
Don't eat breakfast regularly	130	165	**295**
Total	**320**	**275**	**595**

(a) Find $P(B|M)$. Explain what this value means.

(b) Find $P(M|B)$. Explain what this value means.

(c) Are events B and M mutually exclusive? Justify your answer.

(d) Are events B and M independent? Justify your answer.

7.55 Rolling dice Suppose you roll two fair, six-sided dice—one red and one green. Define two independent events A and B related to this chance process. Show that your two events are, in fact, independent.

7.56 Tossing coins Suppose you toss a fair coin twice. Define two events A and B related to this chance process that are not independent and are not disjoint.

7.57 Teachers and advanced degrees Select an adult at random. Let A = person has earned an advanced degree (masters or PhD) and T = person's career is teaching. Rank the following probabilities from smallest to largest. Justify your answer.

$$P(A) \qquad P(T) \qquad P(A|T) \qquad P(T|A)$$

7.58 Mutually exclusive versus independent Decide whether the following statement is true or false: two events that are not mutually exclusive must be independent. Justify your answer.

Tree diagrams and conditional probability

If you toss a fair coin twice, what's the probability that you get two heads? By the reasoning of Section 7.1, it's the proportion of times you would get H-H if you tossed the coin two times over and over and over again. We could perform a simulation to estimate the true probability. Using the methods of Section 7.2, we would take a different approach to answering the question. We begin by listing the sample space for this chance process:

H-H H-T T-H T-T

Since the result of one coin toss is *independent* from the result of any other toss, the 4 outcomes in the sample space are equally likely. So $P(\text{two heads}) = P(\text{H-H}) = 1/4$. There's another way to model chance behavior that involves a sequence of outcomes, like tossing coins. It's called a **tree diagram.**

Figure 7.10 shows a tree diagram for the coin-tossing scenario. After the first toss, there are only two possible results—heads (H) or tails (T)—each with probability 1/2. These two possibilities are shown on the leftmost branches of the tree diagram. From there, we toss a second time, which results in either an H or a T, again with equal probability. This puts us on one of the four "nodes" beneath the "Second toss" label on the tree diagram after two tosses.

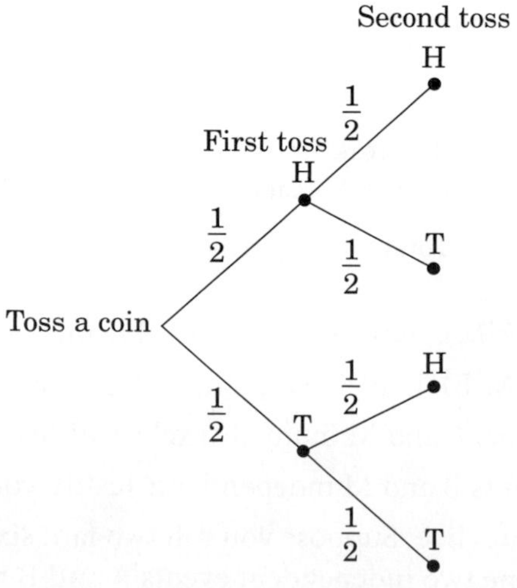

$$P(\text{two heads}) = \frac{1}{2} \cdot \frac{1}{2} = \frac{1}{4}$$

Figure 7.10 A tree diagram displaying the sample space of tossing two coins.

What's the probability of getting "heads, heads"? We already know it's 1/4. Do you see how we can obtain that value from the tree diagram? If we multiply probabilities along the top branches, we get

$$P(\text{two heads}) = \frac{1}{2} \cdot \frac{1}{2} = \frac{1}{4}$$

This same general strategy will work even when trials are *not* independent, as the following example illustrates.

Example 7.23 Looking for hearts

If we shuffle a standard deck and deal the top two cards, what's the probability that we get two hearts? We know the probability that the first card dealt is a heart is 13/52. What about the probability that the second card is a heart? That depends on whether the first card was a heart. In this case, the trials are *not* independent. This is starting to sound complicated! It's time to call on our new friend, the tree diagram.

Figure 7.11(a) shows a tree diagram for this setting with $P(\text{heart}) = 13/52$ for the first card. What's the probability of *not* getting a heart on the first card? By the complement rule, it's

$$P(\text{not heart}) = 1 - P(\text{heart}) = 1 - \frac{13}{52} = \frac{39}{52}$$

We have added this probability on the bottom branch of Figure 7.11(a).

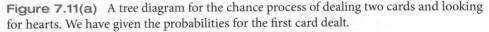

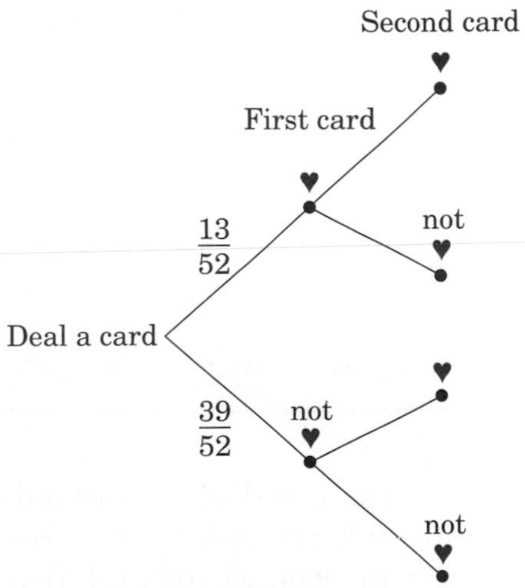

Figure 7.11(a) A tree diagram for the chance process of dealing two cards and looking for hearts. We have given the probabilities for the first card dealt.

If the first card we draw is a heart, then there are 12 hearts left among the 51 remaining cards. So the conditional probability of getting a heart on the second card *given that* we got a heart on the first card is 12/51. Using the complement rule again, the probability that the second card is *not* a heart *given that* the first card was a heart is $1 - 12/51 = 39/51$.

What if the first card we draw isn't a heart? Then there are still 13 hearts among the 51 remaining cards. That is, the probability of getting a heart on the second card *given that* we did not get a heart on the first card is 13/51. By the complement rule,

$$P(\text{no heart on second card } \textit{given} \text{ no heart on first card}) = 1 - \frac{13}{51} = \frac{38}{51}$$

Figure 7.11(b) shows the completed tree diagram. To get two hearts, we have to get a heart on the first card *and* a heart on the second card. Following along the top branches of the tree, we see that the probability is

$$P(\text{two hearts}) = P(\text{heart on first card } \textit{and} \text{ heart on second card}) = \frac{13}{52} \cdot \frac{12}{51} = 0.059$$

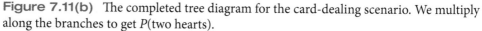

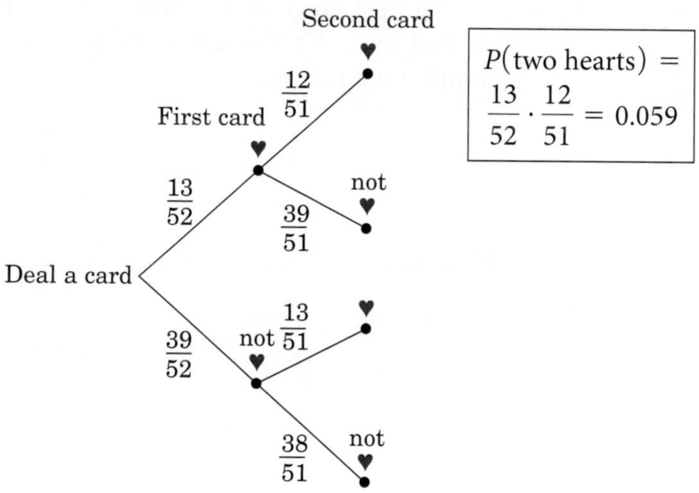

Figure 7.11(b) The completed tree diagram for the card-dealing scenario. We multiply along the branches to get *P*(two hearts).

Being dealt two hearts is a pretty unusual occurrence!

The tree diagram method can be used to find the probability *P*(A and B) that two events happen together. In Example 7.23, A = heart on first card and B = heart on second card. To calculate *P*(A and B), we multiplied *P*(heart on first card) by *P*(heart on second card *given* heart on first card). In symbols, we write

$$P(A \cap B) = P(A) \cdot P(B|A)$$

where $P(B|A)$ is the conditional probability that event B will happen given that event A has happened. This formula is known as the **general multiplication rule.**

General multiplication rule

The probability that events A and B both occur can be found using the **general multiplication rule**

$$P(A \cap B) = P(A) \cdot P(B|A)$$

where $P(B|A)$ is the conditional probability that event B occurs given that event A has already occurred.

If we rearrange the terms in the general multiplication rule, we can get a formula for the conditional probability $P(B|A)$. Start with the general multiplication rule

$$P(A \cap B) = P(A) \cdot P(B|A)$$

Divide both sides of the equation by $P(A)$. This yields

$$\frac{P(A \cap B)}{P(A)} = P(B|A)$$

or

$$P(B|A) = \frac{P(A \cap B)}{P(A)}$$

Conditional probability formula

To find the **conditional probability** $P(B|A)$, use the formula

$$P(B|A) = \frac{P(A \cap B)}{P(A)}$$

Let's illustrate the use of this formula with a familiar example.

Example 7.24 Who reads the paper, continued

In Example 7.20 (page 342), we classified the residents of a large apartment complex based on the events A = reads *USA Today* and B = reads the *New York Times*. The completed Venn diagram is reproduced in Figure 7.12 (on the next page). What's the probability that a randomly selected resident who reads *USA Today* also reads the *New York Times*?

Since we're given that the randomly chosen resident reads *USA Today,* we want to find *P*(reads *New York Times* | reads *USA Today*), or $P(B|A)$. By the conditional probability formula,

$$P(B|A) = \frac{P(A \cap B)}{P(A)}$$

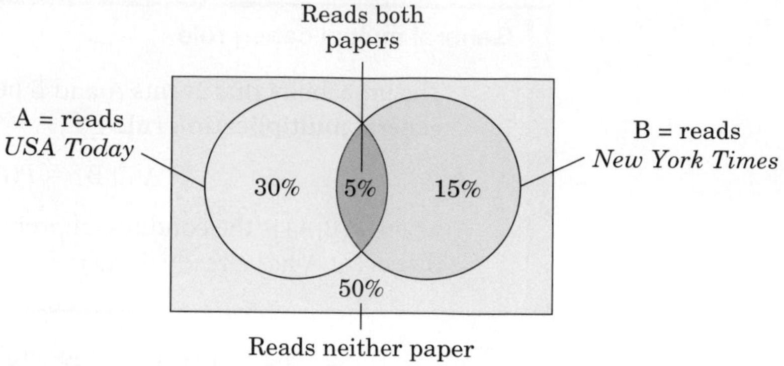

Figure 7.12 The completed Venn diagram from Example 7.20.

Since $P(A \cap B) = 0.05$ and $P(A) = 0.35$, we have

$$P(B|A) = \frac{0.05}{0.35} = 0.14$$

There's a 14% chance that a randomly selected resident who reads *USA Today* also reads the *New York Times.*

What about the special case when events A and B are independent? In the toss two coins scenario of Example 7.22, A = head on first toss and B = head on second toss. By the general multiplication rule,

$$\begin{aligned} P(A \cap B) &= P(A) \cdot P(B|A) \\ &= P(\text{head on first toss}) \cdot P(\text{head on second toss} \mid \text{head on first toss}) \\ &= \frac{1}{2} \cdot \frac{1}{2} = \frac{1}{4} \end{aligned}$$

But for the coin-toss setting, the chance of getting a head on the second toss *given that* we got a head on the first toss is just 1/2, because the coin has no memory. That is, $P(B|A) = P(B)$. In other words, A and B are independent events. We can simplify the general multiplication rule in the case of independent events A and B as follows:

$$\begin{aligned} P(A \cap B) &= P(A) \cdot P(B|A) \\ &= P(A) \cdot P(B) \text{ since } P(B|A) = P(B) \text{ for independent events} \end{aligned}$$

This result is known as the **multiplication rule for independent events.**

Multiplication rule for independent events

If A and B are independent events, then the probability that A and B both occur is

$$P(A \cap B) = P(A) \cdot P(B)$$

Note that this rule applies *only* to independent events.

Exercises

7.59 Toss four more Imagine that you toss a fair coin 4 times.

(a) Draw a tree diagram to represent this chance process.

(b) Find the probability that you get at least one head. Explain your method.

7.60 Not just hearts Refer to Example 7.23 (page 353). Find the probability that at least one of the two cards that are dealt is a heart. Show your method clearly.

7.61 Looking for hearts again Refer to Example 7.23 (page 353). Suppose we draw one card from the deck, look at it, and then return it to the deck before we draw a second card. Define events A = first card is a heart and B = second card is a heart.

(a) Draw a tree diagram to represent this chance process. Be sure to label all the branches of the tree with the appropriate probabilities.

(b) Find the probability that both cards are hearts. Show your work.

(c) Are events A and B independent? Justify your answer.

7.62 Monopoly In the game of Monopoly, a player rolls two six-sided dice on each turn. If a player is in jail, he or she must roll "doubles" (both dice show the same number) in order to get out of jail. If a player in jail does not roll doubles on three consecutive turns, he or she must pay \$50 to get out of jail.

(a) Find the probability that a player rolls doubles on a turn. Show your method.

(b) John is in jail. Find the probability that John has to pay \$50 to get out of jail because he doesn't roll doubles on three consecutive turns. Show your work.

7.63 MySpace versus Facebook, II A recent survey suggests that 85% of college students have posted a profile on Facebook, 54% use MySpace regularly, and 42% do both. Suppose we select a college student at random and learn that the student has a profile on Facebook. Find the probability that the student uses MySpace regularly. Show your work.

7.64 More computers at Princeton An October 2007 census revealed that 40% of Princeton students primarily used Macintosh computers (MACs). The rest primarily used PCs. At the time of the census, 67% of Princeton students were undergraduates. The rest were graduate students. According to the census, 23% of Princeton's graduate students said that they used PCs as their primary computers. Suppose that we select a Princeton student at random and learn that the student primarily uses a MAC. Find the probability that this person is a graduate student. Show your work.

Applications of conditional probability

King Features Syndicate 2008

Conditional probability has important applications in the areas of risk management, drug and disease testing, and quality control in manufacturing. Let's look at a couple of examples that show how an understanding of conditional probability can help people make informed decisions.

Example 7.25 Medical decisions, risks, and probability

Morris's kidneys are failing and he is awaiting a kidney transplant. His doctor gives him the following information for patients in his condition: 90% survive the transplant and 10% die. The transplant succeeds in 60% of those who survive, and the other 40% must return to kidney dialysis. The proportions who survive five years are 70% for those with a new kidney and 50% for those who return to dialysis. Morris wants to know the probability that he will survive for five years.

The tree diagram in Figure 7.13 organizes this information to give a probability model in graphical form. The tree shows the three stages and the possible outcomes and probabilities at each stage. Each path through the tree leads to either survival for five years or to death in less than five years.

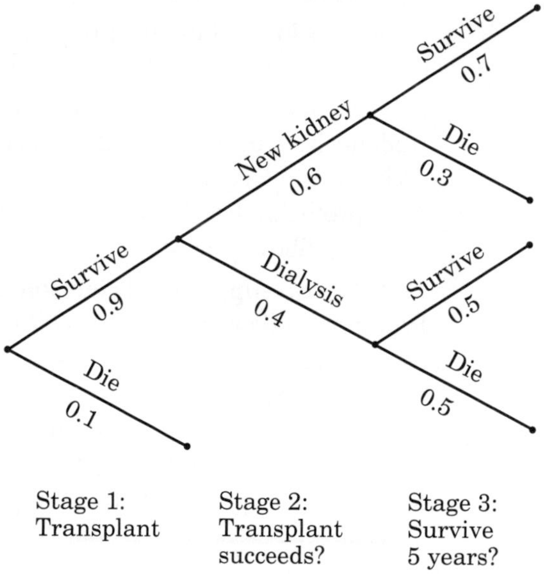

Figure 7.13 A tree diagram showing the possible outcomes and probabilities for patients in Morris's condition.

How can we calculate the probability that a patient in Morris's condition will survive for five years? In the tree diagram, there are two possible "paths" that lead to a patient surviving for five years. The first such path would require a patient to survive the transplant, have the new kidney function successfully, and then survive for five years with the new kidney. What's the probability that this sequence of events happens? There's a 90% chance that the patient survives the transplant. For those who survive the transplant, 60% have the new kidney function properly. Of those who survive the transplant and have the new kidney function properly, 70% will survive. So the probability of surviving for five years in this way is $(0.9)(0.6)(0.7) = 0.378$.

The second path would require a patient to survive the transplant, be forced to return to dialysis because the new kidney didn't function properly, and then live for five years. By multiplying along the branches of the tree, we can calculate this probability as $(0.9)(0.4)(0.5) = 0.18$.

Since either path would allow a patient to survive for five years, the desired probability is

$$0.378 + 0.18 = 0.558$$

That is, a patient in Morris's condition has about a 56% chance of surviving for five years.

Lately, we've been hearing more in the news about athletes who have tested positive for performance-enhancing drugs. From Major League Baseball, to the Tour de France, to the Olympics, we have learned about athletes who took banned substances. Sports at all levels now have programs in place to test athletes' urine or blood. There's an important question about drug testing that probability can help answer: if an athlete tests positive for a banned substance, did the athlete necessarily attempt to cheat?

Example 7.26 Athletes and drug testing

Over 10,000 athletes competed in the 2008 Olympic Games in Beijing. The International Olympic Committee wanted to ensure that the competition was as fair as possible. So they administered more than 5000 drug tests to athletes. All medal winners were tested, as well as other randomly selected competitors.

Suppose that 2% of athletes had actually taken (banned) drugs. No drug test is perfect. Sometimes the test indicates that an athlete took drugs when the athlete actually didn't. We call this a **false positive** result. Other times, the drug test indicates that an athlete is "clean" when the athlete actually took drugs. This is called a **false negative** result. Suppose that the testing procedure used at the Olympics has a false positive rate of 1% and a false negative rate of 0.5%. What's the probability that an athlete who tests positive actually took drugs?

The tree diagram in Figure 7.14 summarizes the situation. Since we assumed that 2% of all Olympic athletes took drugs, that means 98% didn't. Of those who took drugs, 0.5%, or 0.005, would "get lucky" and test negative. The remaining

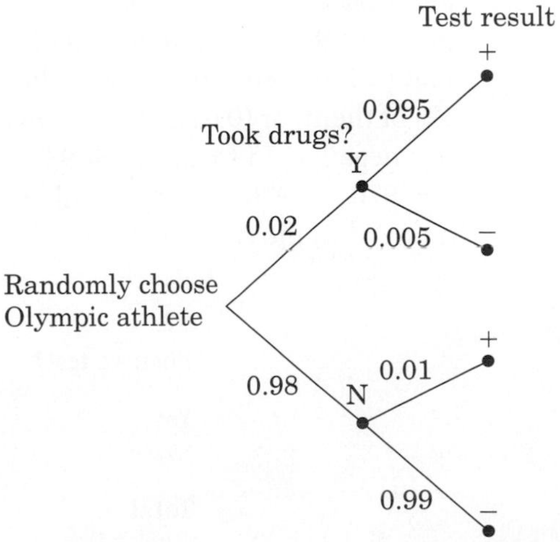

Figure 7.14 A tree diagram showing whether or not Olympic athletes took drugs and the likelihood of getting a positive or negative test result from a drug test.

99.5%, or 0.995, would test positive. Among the athletes who didn't take drugs, an "unlucky" 1%, or 0.01, would test positive. The remaining 99%, or 0.99, would (correctly) test negative.

Let's define events B = took drugs and A = positive test result. We want to find $P(B|A)$. By the conditional probability formula,

$$P(B|A) = \frac{P(A \cap B)}{P(A)}$$

To find $P(A \cap B)$, we note that this is the probability that a randomly selected athlete took drugs and got a positive test result. Multiplying along the Y and + branches of the tree,

$$P(A \cap B) = (0.02)(0.995) = 0.0199$$

To find $P(A)$, we need to calculate the probability that a randomly selected athlete gets a positive test result. There are two ways this can happen: (1) if the athlete took drugs and the test result is positive, and (2) if the athlete didn't take drugs, but the test gives a false positive. From the tree diagram, the desired probability is

$$P(A) = (0.02)(0.995) + (0.98)(0.01) = 0.0297$$

Using these two results, we can find the conditional probability:

$$P(B|A) = \frac{P(A \cap B)}{P(A)} = \frac{0.0199}{0.0297} = 0.67$$

That is, 67% of athletes who test positive actually took drugs.

Are you surprised by the final result of Example 7.26? Most people are. Sometimes a two-way table that includes counts is more convincing. To make calculations simple, we'll suppose that there were exactly 10,000 athletes at the 2008 Olympics, and that exactly 200 took banned substances. How many of those 200 would test positive? 99.5% of 200, or 199 of them. That leaves 1 who would test negative. How many of the 9800 athletes who didn't take drugs would get a false positive test result? 1% of them, or (0.01)(9800) = 98 athletes. The remaining 9800 − 98 = 9702 would test negative. In total, 199 + 98 = 297 athletes would have positive test results and 1 + 9702 = 9703 athletes would have negative test results. This information is summarized in the two-way table below.

	Took drugs?		
Positive test?	**Yes**	**No**	**Total**
Yes	199	98	**297**
No	1	9702	**9,703**
Total	**200**	**9800**	**10,000**

Given that a randomly selected athlete has tested positive, we can see from the two-way table that the conditional probability $P(\text{took drugs} \mid \text{positive test}) = 199/297 = 0.67$.

Now you can see why most drug-testing programs require an athlete to submit *two* urine or drug samples at each test. If one sample tests positive, we can't be sure that the athlete took drugs. If both samples test positive, however, we can be pretty sure that the athlete took a banned substance.

Exercises

7.65 Fill 'er up! In June 2008, 88% of automobile drivers filled their vehicles with regular gasoline, 2% purchased midgrade gas, and 10% bought premium gas.[11] Of those who bought regular gas, 28% paid with a credit card. Of customers who bought midgrade and premium gas, 34% and 42%, respectively, paid with a credit card. Suppose we select a customer at random.

(a) Construct a tree diagram to represent this situation.

(b) Find the probability that the customer paid with a credit card. Show your work.

(c) Given that the customer paid with a credit card, find the probability that she or he bought premium gas.

7.66 Desktop or laptop? A computer company makes desktop and laptop computers at factories in three states—California, Texas, and New York. The California factory produces 40% of the company's computers, the Texas factory makes 25%, and the remaining 35% are manufactured in New York. Of the computers made in California, 75% are laptops. Of those made in Texas and New York, 70% and 50%, respectively, are laptops. All computers are first shipped to a distribution center in Missouri before being sent out to stores. Suppose we select a computer at random from the distribution center.[12]

(a) Construct a tree diagram to represent this situation.

(b) Find the probability that the computer is a laptop. Show your work.

(c) Given that the computer is a laptop, find the probability that it came from the Texas factory.

7.67 Assessing risk Refer to Example 7.25 (page 358). Use the probabilities provided to make a table that shows the status of 1000 randomly selected patients with Morris's condition.

7.68 False positives and negatives Which is a more serious error in each of the following situations—a false positive result or a false negative result? Justify your answer.

(a) Testing athletes for performance-enhancing drugs.

(b) Testing people for a life-threatening disease.

7.69 Testing for AIDS, I The ELISA test can help detect whether people have the AIDS virus (HIV). As with any medical test, ELISA can give false positive and false negative results. From experience, medical researchers estimate that the ELISA test has a 0.2% false positive rate and a 0.1% false negative rate. Suppose we use the ELISA test in an area where 5% of the population has HIV.

(a) Draw a tree diagram that models this situation.

(b) Find the probability that a randomly selected person from this area would test positive. Show your work.

(c) Given that a randomly chosen person has tested positive, what's the probability that the person actually has HIV? Show your work.

7.70 Testing for AIDS, II Refer to the previous exercise. Suppose that we use the ELISA test in an area where only 0.1% of the population has HIV. Repeat parts (a), (b), and (c). Are you surprised?

APPLICATION 7.3

Baby hearing screening

According to the National Campaign for Hearing Health, deafness is the most common birth defect. Experts estimate that 3 out of every 1000 babies are born with some kind of hearing loss. Early identification of hearing loss through a screening test shortly after birth and subsequent intervention can have a significant impact on a child's cognitive development as well as financial consequences. Thus, universal newborn hearing screening (UNHS) calls for high-performing, accurate, and inexpensive testing equipment. The screening tests work by introducing a sound into a baby's ear and then measuring either the response of the ear's internal mechanisms or the electrical activity of the auditory portion of the brain. Just because a baby fails those tests, however, does not mean there is a hearing problem.

Studies conducted at the University of Utah were undertaken to assess the adequacy of a new miniature screening device called the Handtronix-OtoScreener. The accuracy of this new device was judged using as "truth" the results found using standard equipment.[13] Results are shown in the following table:

	Baby's hearing	
Test result	**Loss**	**Normal**
Loss	54	6
Normal	4	36

QUESTIONS

1. Estimate the probability that this new device will show a hearing loss for a baby who actually has hearing loss. Show your work.

2. Estimate the probability that this new device will show normal hearing for a baby who actually has normal hearing. Show your work.

3. Estimate the chance that a baby with a hearing loss will pass the hearing test (test normal) using this newer miniature screening device. Show your work.

4. Suppose we choose a baby at random from all newborns and have the baby's hearing tested using the Handtronix-OtoScreener. If the test shows that the baby has a hearing loss, what's the estimated probability that the baby really has a hearing loss?

Section 7.3 Summary

If one event has happened, the chance that another event will happen is a **conditional probability.** The notation $P(B|A)$ represents the probability that event B occurs given that event A has occurred. When the chance that event B occurs is not affected by whether event A occurs, we say that events A and B are **independent.** For independent events A and B, $P(B|A) = P(B)$ and $P(A|B) = P(A)$. If two events A and B are mutually exclusive, they cannot be independent.

When chance behavior involves a sequence of outcomes, a **tree diagram** can be used to describe the sample space. Tree diagrams can also help in finding the probability that two or more events occur together. We simply multiply along the branches that correspond to the outcomes of interest.

The **general multiplication rule** states that the probability of events A and B occurring together is

$$P(A \cap B) = P(A) \cdot P(B|A)$$

Dividing both sides of the equation by P(A), we get the **conditional probability formula**

$$P(B|A) = \frac{P(A \cap B)}{P(A)}$$

In the special case of independent events, the multiplication rule becomes

$$P(A \cap B) = P(A) \cdot P(B)$$

Section 7.3 Exercises

7.71 Tall people and basketball players Select an adult at random. Let T = person is over 6 feet tall and B = person is a professional basketball player. Rank the following probabilities from smallest to largest. Justify your answer.

$$P(T) \qquad P(B) \qquad P(T|B) \qquad P(B|T)$$

7.72 TCNJ survey, III Exercise 7.46 (page 347) described the results of a class survey. Here is the two-way table that shows the gender and handedness of the students in the class:

	Gender		
Handedness	**F**	**M**	**Row summary**
L	3	1	4
R	18	6	24
Column summary	21	7	28

(a) What percent of the males in the class are right-handed?
(b) What percent of the right-handed people in the class are male?
(c) Are the events "student is a male" and "student is right-handed" independent? Justify your answer.

7.73 The chevalier's problem In the early 1700s, French gamblers played a game in which they bet on getting at least one "1" when a fair, six-sided die was rolled four times.
(a) Draw a tree diagram that shows the sample space of this chance process.
(b) Find the probability of interest. Show your work.

7.74 The chevalier's other problem To increase interest among French gamblers, another game was devised in which players bet on getting at least 1 set of "double

1s" when rolling a pair of fair, six-sided dice 24 times. The chevalier de Mere, a frequent gambler, believed that the probability of getting the desired outcome in this game was the same as the probability in Exercise 7.73(b). He bet accordingly and lost a lot of money. Show the chevalier that it's less likely to get at least one set of "double 1s" on 24 rolls of a pair of dice than it is to get at least one "1" on four rolls of a single die.

7.75 Cats and dogs In an elementary school classroom, there are 40 students. For every student, we record whether they have a cat and whether they have a dog. Thirty students say that they have a dog, and 20 students say that they have a cat. If possible, make a two-way table in which the events D = has a dog and C = has a cat are independent if we select a student at random from the class.

7.76 Testing the test Are false positives too common in some medical tests? Researchers conducted an experiment involving 250 patients with a medical condition and 750 other patients who did not have the medical condition. The medical technicians who were reading the test results were unaware that they were subjects in an experiment.

(a) Technicians correctly identified 240 of the 250 patients with the condition. They also identified 50 of the healthy patients as having the condition. What were the false positive and false negative rates for the test?

(b) Given that a patient got a positive test result, what is the probability that the patient actually had the medical condition? Show your work.

7.77 The birthday problem If 30 unrelated people are in a room at the same time, what's the probability that at least 2 of them have the same birthday (month and day)? Make a guess before you perform any calculations.

(a) Let's start with a simpler problem. If 2 unrelated people are in a room, what's the probability that they have different birthdays? What's the probability that they have the same birthday?

(b) If 3 unrelated people are in a room, what's the probability that all 3 have different birthdays? What's the probability that at least 2 of them have the same birthday?

(c) If 30 unrelated people are in a room, what's the probability that all 30 have different birthdays? What's the probability that at least 2 of them have the same birthday?

7.78 Activity 7.3 follow-up Return to Activity 7.3 (page 348). Assume that your teacher has probability 2/3 of winning any individual round.

(a) List the sample space for the weird dice game. Find the probability for each possible outcome.

(b) Find the probability that the teacher wins the next game. Show your work.

CHAPTER 7 REVIEW

Some phenomena are random. Although their individual outcomes are unpredictable, there is a regular pattern in the long run. Using gambling devices and taking an SRS are examples of random phenomena. Probability gives us a language to describe randomness. To calculate the probability of a complicated event without using complicated math, we use random digits to simulate many repetitions. Section 7.1 discusses randomness and simulation.

When randomness is present, probability answers the question "How often in the long run?" Probability models assign probabilities to outcomes. Any such model must obey the rules of probability. Section 7.2 presents these rules, along with two useful tools for describing outcomes of chance behavior—Venn diagrams and two-way tables.

In some cases, we have partial information about the result of a chance process. This leads to the study of conditional probability in Section 7.3. Tree diagrams help organize our thinking about conditional probability in situations that involve a sequence of outcomes.

Here is a review list of the most important skills you should have developed from your study of this chapter.

A. RANDOMNESS, PROBABILITY, AND SIMULATION

1. Recognize that some phenomena are random. Probability describes the long-run regularity of random phenomena.

2. Understand the idea of the probability of an event as the proportion of times the event occurs in very many repetitions of a random phenomenon. Use the idea of probability as long-run proportion to think about probability.

3. Recognize that short runs of random phenomena do not display the regularity described by probability. Accept that randomness is unpredictable in the short run, and avoid seeking cause-and-effect explanations for random occurrences.

B. SIMULATION

1. Design a simulation using random digits to model chance behavior.

2. Estimate a probability by repeating a simulation many times.

C. PROBABILITY RULES

1. When probabilities are assigned to individual outcomes, find the probability of an event by adding the probabilities of the outcomes that make it up.

2. Use probability rules to find the probabilities of events that are formed from other events, including unions, intersections, complements, and conditional probabilities.

D. PROBABILITY MODELS

1. Use Venn diagrams, two-way tables, and tree diagrams to model chance behavior.

2. Compute probabilities using information provided in Venn diagrams, two-way tables, and tree diagrams.

CHAPTER 7 REVIEW EXERCISES

7.79 A dice game Your teacher has invented another "fair" dice game to play. Here's how it works. Your teacher will roll one fair *eight-sided* die, and you will roll a fair *six-sided* die. Each player rolls once, and the winner is the person with a higher number. In case of a tie, neither player wins.

(a) Diagram the sample space illustrating the possible outcomes on a single turn.
(b) Let A be the event "your teacher wins the first turn of the game." Find $P(A)$.
(c) Let B be the event "you get a 3 on your first roll." Find $P(A \cup B)$.
(d) Are events A and B independent? Justify your answer.

7.80 Sandblasters 2007 Eight teams of the world's best sand sculptors gathered near San Diego, California, for Sandblasters 2007: The Extreme Sand Sculpting Competition. At five points during the competition, a randomly selected team's sculpture was blown up. That team was then forced to build a new sculpture from scratch. No team could be selected for destruction more than once. Last year's winning team did not have a sculpture blown up either last year or this year. Other teams were suspicious. Should they be? Use the methods of this chapter to justify your answer. (Team Sanding Ovation won the contest with their "zipper-head" sculpture.)

Audrey Ustuzhanin/FeaturePics

7.81 Class rank Choose a college student at random and ask his or her class rank in high school. Probabilities for the outcomes are

Class rank:	Top 10%	Top quarter but not top 10%	Top half but not top quarter	Bottom half
Probability:	0.3	0.3	0.3	?

(a) What must be the probability that a randomly chosen student was in the bottom half of his or her high school class?
(b) To simulate the class standing of randomly chosen students, how would you assign digits to represent the four possible outcomes listed?

7.82 The girls Suppose that about 48% of all infants are girls. The maternity ward of a large hospital handles 1000 deliveries per year. A smaller hospital in the same city records about 50 deliveries per year. At which hospital is it more likely that between 46% and 50% of the babies born there this year will be girls? Explain briefly.

7.83 Who smokes? The question "Do you smoke?" was asked of a random sample of 100 people. Results are shown in the table below. Suppose we select one of these people at random.

	Smoke?		
Gender	**Yes**	**No**	**Total**
Male	19	41	**60**
Female	12	28	**40**
Total	**31**	**69**	**100**

(a) What's the probability that the person smokes?
(b) Given that the person is male, what's the probability that the person smokes?
(c) Are the two events "smokes" and "is male" independent? Why or why not?

7.84 More smokers Draw a Venn diagram to represent the sample space of Exercise 7.83.

7.85 *Challenger* disaster On January 28, 1986, Space Shuttle *Challenger* exploded on takeoff. All seven crew members were killed. Following the disaster, scientists and statisticians helped analyze what went wrong. They determined that the failure of o-ring joints in the shuttle's booster rockets was to blame. Under the cold conditions that day, experts estimated that the probability that an individual o-ring joint would function properly was 0.977. But there were six of these o-ring joints, and all six had to function properly for the shuttle to launch safely. Assuming that o-ring joints succeed or fail independently, find the probability that the shuttle would launch safely under similar conditions. Show your work.

7.86 Looking for metal A boy uses a homemade metal detector to look for valuable metal objects on a beach. The machine isn't perfect—it identifies only 98% of the metal objects over which it passes, and it identifies 4% of the nonmetallic objects over which it passes. Suppose that 25% of the objects that the machine passes over are metal. Find the probability that the boy has found a metal object when he receives a signal from the machine. Show your work.

7.87 Random digits Suppose you instruct your calculator to choose a random digit from 0 to 9. Define two events A and B related to this chance process that are mutually exclusive but not complements of each other.

7.88 Drug use in baseball On December 13, 2007, former senator George Mitchell released an extensive report on the use of performance-enhancing drugs in Major League Baseball. He even included a list of players who had been using such drugs. One question of interest is whether performance-enhancing drugs really help. Let's examine the performance of 19 pitchers who were reported to have used drugs. In the season when they first started using, 14 of the 19 showed improvement over the prior year. Is this convincing evidence that using drugs helped improve their performance? Design and carry out a simulation to help you answer this question. Assume that each pitcher had probability 0.5 of showing improvement without taking drugs.

CHAPTER 8

Probability Models

The house edge, and the real house edge

If you decide to gamble, you care about how often you'll win. The probability of winning tells you what proportion of a large number of bets will be winners. You care even more about how much you'll win, because winning a lot is better than winning a little. If you play a game with a 50% chance of winning $1, in the long run you'll win half the time, and average 50 cents in winnings per bet. A 10% chance of winning $100 does better—now you win only 1 time in 10 in the long run, but you average $10 in winnings per bet. Those "average winnings per bet" numbers are *expected values*. Like probabilities, expected values tell us only what will happen on the average in very many bets.

It's no secret that the expected values of games of chance are set so that the "house" wins on the average in the long run. In roulette, for example, the expected value of a $1 bet is $0.947. The house keeps the other 5.3 cents per dollar bet. Because the house plays hundreds of thousands of times each month, it will keep almost exactly 5.3 cents for every dollar bet.

What is a secret, at least to naive gamblers, is that in the real world a casino does much better than expected values suggest. In fact, casinos keep a bit over 20% of the money gamblers spend on roulette chips. That's because players who win keep on playing. Think of a player who gets back exactly 95% of each dollar bet. After one bet, he has 95 cents; after two bets, he has 95% of that, or 90.25 cents; after three bets, he has 95% of that, or 85.7 cents. The longer he keeps recycling his original dollar, the more of it the casino keeps. Real gamblers don't get a fixed percent back on each bet, but even the luckiest will lose his stake if he plays long enough. The casino keeps 5.3 cents of every dollar bet but 20 cents of every dollar that walks in the door.[1]

8.1 Probability Distributions

ACTIVITY 8.1A

Greed

Materials: One fair, six-sided die

Are you more likely to take a risk or to stay with a sure thing? In this Activity, you will get to make decisions about whether to keep playing a game of chance in hopes of getting a higher score.

Rules: The game of Greed is played with a single die, rolled by your teacher. Each game consists of 5 rounds. Your total score for the game will be the sum of your scores in the 5 rounds played. Here's how each round works.

- Everyone in the class stands up. Your teacher will throw a fair, six-sided die twice and add the numbers on the up-faces. This is your current score. If you are happy with that score, sit down and record this value as your score for the round.
- If someone is still standing, your teacher rolls the die again. Those still standing add the resulting number to their total *unless* the die shows a 5. If a 5 is rolled, anyone standing loses all their points for the round and sits down.
- Your teacher will continue to roll the die until all students have decided to sit down or until a 5 is rolled.

1. Play the game once, recording your score for each round in a table like this:

Round	1	2	3	4	5	Total
Score						

2. Your teacher will draw a stemplot on the board with stems from 0 to 9 (or higher if a student scores more than 99). After you get your total, go to the board and add your value to the stemplot.

3. Discuss the results with your classmates. Do some students' strategies sound better than the one you used? Think about how you would play differently if you played Greed again.

4. Play a second game of 5 rounds. Make a back-to-back stemplot by adding your total to the left side of the stem.

5. Compare shape, center, spread, and unusual values for the two distributions. Did the class earn higher scores in the second game than in the first?

What is a probability distribution?

A probability model describes the possible outcomes of a chance process and the likelihood that those outcomes will occur. For example, suppose we toss a fair coin 3 times. The sample space for this chance process is

H-H-H H-H-T H-T-H T-H-H H-T-T T-H-T T-T-H T-T-T

Since there are 8 equally likely outcomes, the probability is 1/8 for each possible outcome. Define the variable X = the number of heads obtained. The value of X will vary from one set of tosses to another but will always be one of the numbers 0, 1, 2, or 3. How likely is X to take each of those values? We can summarize the **probability distribution** of X as follows:

Value:	0	1	2	3
Probability:	1/8	3/8	3/8	1/8

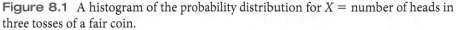

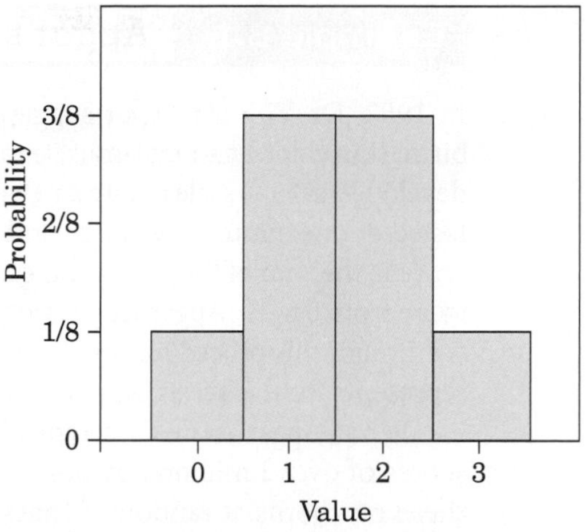

Figure 8.1 A histogram of the probability distribution for X = number of heads in three tosses of a fair coin.

Figure 8.1 shows the probability distribution of X in graphical form. Notice the symmetric shape.

We can use the probability distribution to answer questions about the variable X. What's the probability that we get at least one head in three tosses of the coin? In symbols, we want to find $P(X \geq 1)$. We could add $P(X = 1)$, $P(X = 2)$, and $P(X = 3)$ to get the answer. Or we could use the complement rule from Chapter 7:

$$\begin{aligned} P(X \geq 1) &= 1 - P(X < 1) = 1 - P(X = 0) \\ &= 1 - 1/8 = 7/8 \end{aligned}$$

A numerical variable that describes the outcomes of a chance process (like X in the coin-tossing scenario) is called a **random variable.**

Random variables and probability distributions

A **random variable** takes numerical values that describe the outcomes of some random phenomenon. The **probability distribution** of a random variable gives its possible values and associated probabilities. The probability distribution can be displayed in a table or a graph.

There is an obvious connection between probability distributions and the probability models of Chapter 7. We use the probability model for a chance process to help determine the probability distribution of a random variable. The probability model for tossing three coins tells us that $P(\text{T-T-T}) = 1/8$. Once we define the random variable X = the number of heads in three tosses, the probability distribution for X says that $P(X = 0) = 1/8$. Why? Because the chance of getting no heads (all tails) in three tosses of a coin is 1/8.

Here's an example involving something a bit more serious than tossing coins.

Example 8.1 Apgar scores: babies' health at birth

In 1952, Dr. Virginia Apgar suggested five criteria for measuring a baby's health at birth: skin color, heart rate, muscle tone, breathing, and response when stimulated. She developed a 0-1-2 scale to rate a newborn on each of the five criteria. Babies are usually tested at one minute after birth and again at five minutes after birth. A baby's Apgar score is the sum of the ratings on each of the five scales, which gives a whole number value from 0 to 10. Apgar scores are still used today to evaluate the health of newborns. (Although this procedure was later named for Dr. Apgar, the acronym APGAR also represents the five scales: Appearance, Pulse, Grimace, Activity, and Respiration.)

What Apgar scores are typical? To find out, researchers recorded the Apgar scores of over 2 million newborn babies in a single year.[2] Imagine selecting one of these newborns at random. (That's our chance process.) Define the random variable $Y =$ Apgar score of a randomly selected baby one minute after birth. The table below gives the probability distribution for Y.

Value:	0	1	2	3	4	5	6	7	8	9	10
Probability:	0.001	0.006	0.007	0.008	0.012	0.020	0.038	0.099	0.319	0.437	0.053

Figure 8.2 displays a histogram showing the probability distribution of Y. Notice the heavily left-skewed shape of the histogram.

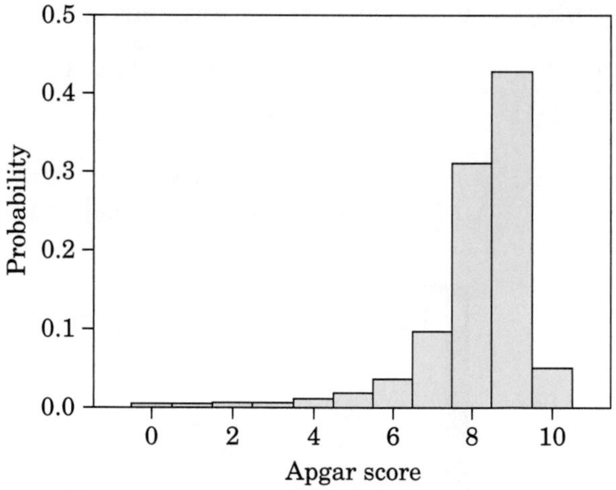

Figure 8.2 A histogram showing the probability distribution of the random variable $Y =$ Apgar score of a randomly selected newborn at one minute after birth.

Doctors decided that Apgar scores from 7 to 10 indicate a healthy baby. What's the probability that a randomly selected baby is healthy? It's $P(7 \leq Y \leq 10)$. We can calculate this probability as follows:

$$P(7 \leq Y \leq 10) = P(Y = 7) + P(Y = 8) + P(Y = 9) + P(Y = 10)$$
$$= 0.099 + 0.319 + 0.437 + 0.053 = 0.908$$

That is, we'd have about a 91% chance of randomly choosing a healthy baby.

Expected value

Gambling on chance outcomes goes back to ancient times. Both public and private lotteries were common in the early years of the United States. After disappearing for a century or so, government-run gambling reappeared in 1964, when New Hampshire introduced a lottery to raise public revenue without raising taxes. After some initial controversy, many larger states adopted the idea. Forty-two states and all Canadian provinces now sponsor lotteries.

State lotteries made gambling acceptable as entertainment. Some form of legal gambling is allowed in 48 of the 50 states. Over half of all adult Americans have gambled legally. They spend more money betting than on spectator sports, video games, theme parks, and movie tickets combined. If you are going to bet, you should understand what makes a bet good or bad. As the opening story says, we care about how much we win as well as about our probability of winning.

Example 8.2 Winning (and losing) at roulette

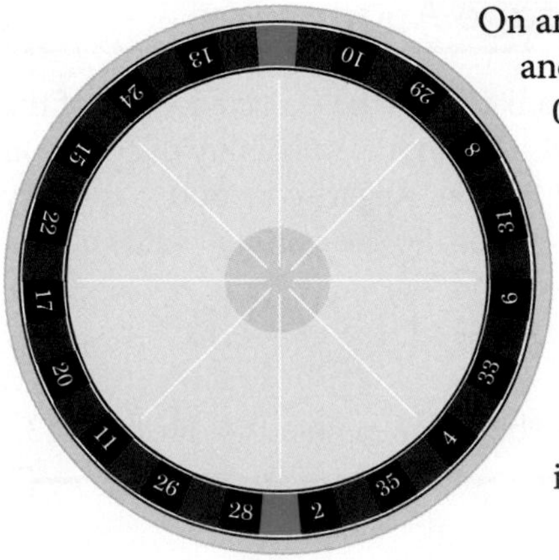

On an American roulette wheel, there are 38 slots numbered 1 through 36, plus 0 and 00. Half of the slots from 1 to 36 are red; the other half are black. Both the 0 and 00 slots are green. A player can place a simple \$1 bet on either red or black. If the ball lands in a slot of that color, the player gets the original dollar back, plus an additional dollar for winning the bet. If the ball lands in a different-colored slot, the player loses the dollar bet to the casino.

Let's define the random variable Z = amount gained on a single \$1 bet. The possible values of Z are $-\$1$ and \$1. (The player either gains a dollar or loses a dollar.) What are the corresponding probabilities? The chance that the ball lands in a slot of the correct color is 18/38. The chance that the ball lands in a different-colored slot is 20/38. Here is the probability distribution of Z:

Value:	−\$1	\$1
Probability:	20/38	18/38

What is the player's average gain? The ordinary average of the two possible outcomes −\$1 and \$1 is \$0. But \$0 isn't the average winnings because the player is less likely to win \$1 than to lose \$1. In the long run, the player gains a dollar 18 times in every 38 games played and loses a dollar on the remaining 20 of 38 bets. The player's long-run average gain for this simple bet is

$$-\$1 \cdot \frac{20}{38} + \$1 \cdot \frac{18}{38} = -\$0.05$$

You see that in the long run the player loses (and the casino gains) 5 cents per bet. □

Here is a general definition of the kind of "average outcome" we used to evaluate the bets in Example 8.2.

Expected value

The **expected value** of a random variable is found by multiplying each possible value of the variable by its probability and then summing over all possible outcomes.

In symbols, if the possible values of the variable are $x_1, x_2, \ldots, x_k$ and their probabilities are $p_1, p_2, \ldots, p_k$, the expected value is

$$\text{expected value} = x_1 \cdot p_1 + x_2 \cdot p_2 + \ldots + x_k \cdot p_k$$

An expected value is an average of the possible outcomes, but it is not an ordinary average in which all outcomes get the same weight. Instead, each outcome is weighted by its probability, so outcomes that occur more often get higher weights. The idea of expected value as an average applies to random phenomena other than games of chance.

Example 8.3 Finding the average Apgar score

What's the "average" Apgar score for a newborn baby? It's the expected value of the random variable Y that we defined in Example 8.1. From the probability distribution for Y, we see that 1 in every 1000 babies would have an Apgar score of 0, 6 in every 1000 babies would have an Apgar score of 1, and so on. So the expected value of Y is

$$0 \cdot \frac{1}{1000} + 1 \cdot \frac{6}{1000} + 2 \cdot \frac{7}{1000} + \ldots + 10 \cdot \frac{53}{1000} = 8.13$$

The mean Apgar score of a randomly selected newborn one minute after birth is 8.13.

Exercises

8.1 Random digits, I Imagine choosing a digit from 0 to 9 at random, either from Table B or with your calculator. Let $D =$ the digit selected.

(a) Explain why D is a random variable.

(b) Display the probability distribution of D in a table.

(c) Construct a histogram that shows the probability distribution of D.

8.2 Roll two dice, I Imagine that you roll two fair, six-sided dice—one red and one green. Let $T =$ the sum of the spots showing on the up-faces of the two dice.

(a) Explain why T is a random variable.

(b) Display the probability distribution of T in a table.

(c) Construct a histogram that shows the probability distribution of T.

8.3 Unhealthy babies Refer to Example 8.1. Newborns with Apgar scores of 3 and below may require medical attention. Write the probability that a randomly chosen baby has an Apgar score of 3 or below in terms of the random variable Y. Then find this probability.

8.4 Household size According to the Census Bureau, the number of people X in a randomly selected U.S. household follows the probability distribution given in the table below.

Number of people x_i:	1	2	3	4	5	6	7
Probability p_i:	0.25	0.32	0.17	0.15	0.07	0.03	0.01

(a) Sketch a histogram that displays the probability distribution of X.

(b) Calculate and interpret $P(X > 2)$.

(c) Find the expected value of X and explain what it tells you.

8.5 Random digits, II Refer to Exercise 8.1. Find the expected value of D and interpret the result.

8.6 Roll two dice, II Refer to Exercise 8.2. Find the expected value of T and interpret the result.

The law of large numbers

The definition of "expected value" says that it is an average of the possible outcomes—but an average in which outcomes with higher probability count more. We argued that the expected value is also the average outcome in another sense: it represents the long-run average we will actually see if we repeat a bet many times or choose many newborn babies at random. This is more than just intuition. Mathematicians can prove, starting from the basic rules of probability, that the expected value calculated from a probability distribution really is the "long-run average." This famous fact is called the **law of large numbers.**

The law of large numbers

If a random phenomenon with numerical outcomes is repeated many times independently, the mean of the observed outcomes approaches the expected value.

The law of large numbers is closely related to the idea of probability. In many independent repetitions, the proportion of times an outcome occurs will be close to its probability. As a result, the average outcome obtained will be close to the expected value. These facts express the long-run predictability of chance events. They are the true version of the "law of averages."

Yes!

The law of large numbers explains why gambling, which is a recreation or an addiction for individuals, is a business for a casino. The "house" in a gambling operation is not gambling at all. The average winnings of a large number of customers will be quite close to the expected value. The house has calculated the expected value ahead of time and knows what its take will be in the long run. There is no need to load the dice or stack the cards to guarantee a profit. If enough bets are placed, the law of large numbers guarantees the house a profit.

Life insurance companies operate much like casinos—they bet that the people who buy insurance will not die. Some do die, of course, but the insurance company knows the probabilities and relies on the law of large numbers to predict the average amount it will have to pay out. Then the company sets its premiums high enough to guarantee a profit.

The following Activity lets you see the law of large numbers in action.

ACTIVITY 8.1B

The *Expected Value* applet

Go to the *Statistics Through Applications* Web site, at **www.whfreeman.com/sta2e**, and launch the *Expected Value* applet.

Directions Use the "Fewer dice" and "More dice" buttons to fix the number of dice to be rolled with each trial. Next, enter the number of rolls (100 maximum) and then click the "Roll dice" button to have the applet roll the dice and plot the accumulating average of the roll totals.

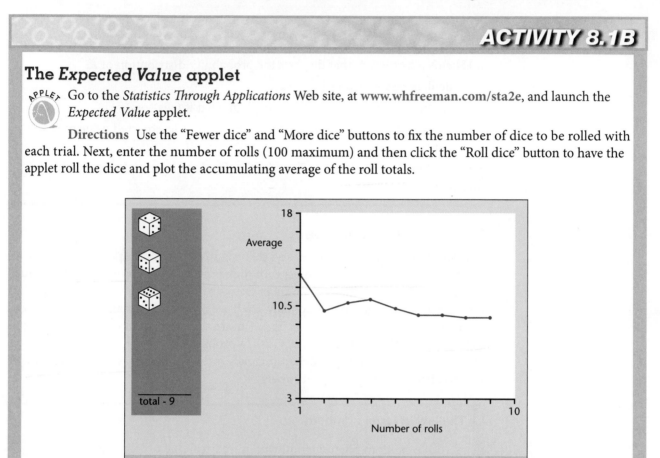

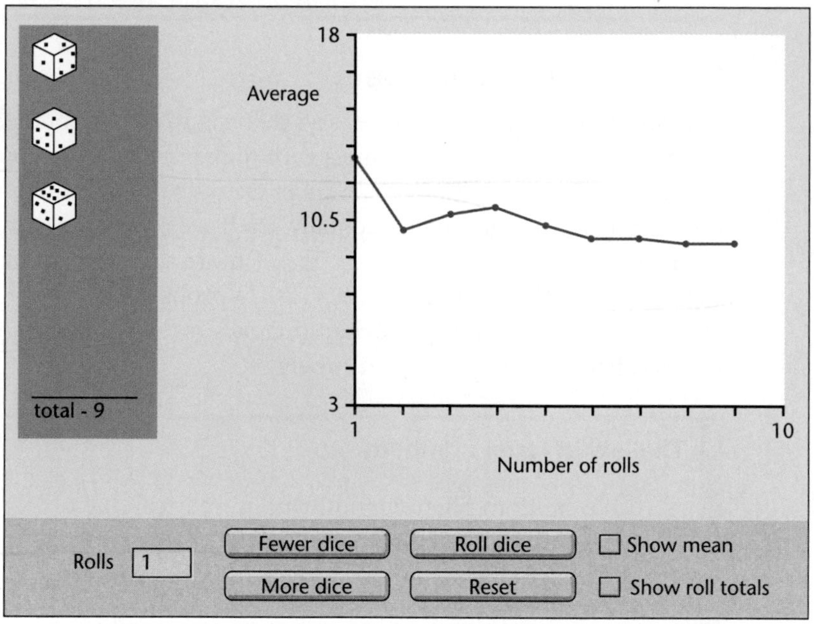

Think

1. When you roll just 1 die, what values can you get? Are they all equally likely?

2. When you roll 10 dice, what values can you get for the roll total? Are they all equally likely?

3. Which process is likely to get close to its mean value in fewer trials: (a) the accumulating averages of rolling 1 die at a time or (b) the accumulating averages of rolling 10 dice at a time?

Verify

Test your answers to the three questions above through repeated use of the applet. Did the individual roll totals and the accumulating averages behave like you thought they would?

APPLET TIP: Click the "Show roll totals" button to see the results of each individual roll (shown as blue dots).

Experiment

4. The average of the numbers from 1 to 6 is $(1 + 2 + 3 + 4 + 5 + 6)/6 = 3.5$. Set the applet to roll 1 die and click the "Show mean" box to see that the expected value for rolling a single die is 3.5. Now alternately click the "More dice" button and the "Roll dice" button while you watch the behavior of the expected value for the sum of the rolls as more dice are added. What is the pattern?

Thinking about expected values

As with probability, it is worth exploring a few fine points about expected values and the law of large numbers.

How large is a large number? The law of large numbers says that the actual average outcome of many trials gets closer to the expected value as more trials are made. It doesn't say how many trials are needed to guarantee an average outcome close to the expected value. That depends on the *variability* of the random outcomes.

The more variable the outcomes, the more trials are needed to ensure that the mean outcome is close to the expected value. Games of chance must be quite variable if they are to hold the interest of gamblers. Even a long evening in a casino has an unpredictable outcome. Gambles with extremely variable outcomes, like state lottos with their very large but very unlikely jackpots, require impossibly large numbers of trials to ensure that the average outcome is close to the expected value.

Though most forms of gambling are less variable than lotto, there's a practical answer to how the law of large numbers applies: the casino plays often enough to rely on it, but you don't. Much of the appeal of gambling is its unpredictability for the player. The business of gambling rests on the fact that the result is not unpredictable for the casino.

Is there a winning system? Serious gamblers often follow a system of betting in which the amount bet on each play depends on the outcome of previous plays. You might, for example, double your bet on each spin of the roulette wheel until you win—or until your money runs out. Such a system tries to take advantage of the fact that you have a memory even though the roulette wheel does not. Can you beat the odds with a system? No. Mathematicians have established a stronger version of the law of large numbers that says that if you do not have an infinite fortune to gamble with, your average winnings (the expected value) remain the same as long as successive trials of the game (such as spins of the roulette wheel) are independent. Sorry.

High-tech gambling

There are more than 450,000 slot machines in the United States. Once upon a time, you put in a coin and pulled the lever to spin three wheels, each with 20 symbols. No longer. Now the machines are video games with flashy graphics and outcomes produced by random number generators. Machines can accept many coins at once, can pay off on a bewildering variety of outcomes, and can be networked to allow common jackpots. Gamblers still search for systems, but in the long run the random number generator guarantees the house its 5% profit.

Finding expected values by simulation

How can we calculate expected values in practice? You know the mathematical formula, but that requires that you start with the probability of each outcome. Expected values too difficult to compute in this way can be found by simulation. The procedure is the same as before: give a probability model, use random digits to imitate it, and simulate many repetitions. By the law of large numbers, the average outcome of these repetitions will be close to the expected value.

Example 8.4 Drawing an ace

You draw cards, one after the other, from a standard deck of 52 playing cards until you draw an ace. On average, how many cards can you expect to turn over until you get an ace? That is, what is the expected value for the number of cards drawn? This is not an easy question to answer. But with simulation methods, we can obtain a fairly good estimate, although it is a bit tedious.

Clearly, the cards drawn are not independent since the first card drawn is no longer in the deck and so cannot be drawn again. If we use the random digit table, we would let the numbers 01 to 52 represent the 52 cards and arbitrarily let numbers 01, 02, 03, and 04 represent the four aces. Read pairs of digits from the table, discarding numbers in the range 53 to 99 and 00.

Because about half of the pairs of numbers would be thrown out, we will use the calculator rather than Table B. So that you can check our work, we'll show you how to seed the calculator's random digit generator to match our results. Our goal is to randomly generate integers between 1 and 52. As soon as a 1, 2, 3, or 4 is observed, we'll stop and record the number of cards drawn.

Seed the random number generator:

- **TI-84:** Type `1234`, press [STO▶][MATH], then choose `PRB` and `1:rand`. Complete the command `1234→rand` and press [ENTER].
- **TI-Nspire:** On a Calculator page, press (menu), then choose `4:Probability`, `4:Random`, and `6:Seed`. Complete the command `RandSeed 1234` and press (enter).

Generate random integers from 1 to 52 until you get an ace (1 through 4). If the calculator generates a duplicate number, ignore it (the same card can't be drawn twice), and draw again. Count the number of trials (cards) required to get an ace.

- **TI-84:** Press [MATH], then choose `PRB` and `5:randInt(`. Complete the command `randInt(1,52)` and press [ENTER]. Keep pressing [ENTER] until you get a 1, 2, 3, or 4.
- **TI-Nspire:** Press (menu), then choose `4:Probability`, `4:Random`, and `2:Integer`. Complete the command `randInt(1,52)` and press (enter). Keep pressing (enter) until you get a 1, 2, 3, or 4.

Here are the results we got in the first six repetitions of the simulation:

Repetition number	Cards drawn (* = skipped repeated number)	Number to get an ace
1	2	1
2	10, 5, 2	3
3	20, 42, *23, 52, 31, 19, 1	7
4	9, 8, 45, 46, 33, 43, 17, 47, 22, 14, 11, *5, *16, 32, 20, *31, **30, 34, *38, *51, 49, 13, *3	23
5	8, 1	2
6	41, 18, 33, 28, 44, 24, 27, *47, 14, 31, 29, 34, 32, 16, 17, 7, 42, *3	18

The average number of cards drawn for these six repetitions is 9. But there's a lot of variability. We got lucky on the first repetition and drew an ace on the first card. But it took 23 cards in the fourth repetition to draw an ace. Because of the large variability, it will take a large number of repetitions until we have some confidence in our estimate. Math or a long simulation shows that the actual expected value is 10.6. We were lucky to get a result as close to 10.6 as we did after only six repetitions.

Note: Before you continue, seed your calculator's random number generator with a unique number, such as the last four digits of your telephone number or Social Security number. For example, `1089→rand` (or `RandSeed 1089` on the TI-Nspire).

□

Exercises

8.7 16 boys! The Associated Press reported an unusual occurrence at the Canton-Potsdam Hospital in Potsdam, New York: between December 15 and December 21, all 16 newborns at the hospital were boys.

(a) Would the births of 16 consecutive boys at a different hospital in another small town in the United States be just as newsworthy? Do you think this Associated Press report is very unusual? Explain.

(b) It was also noted that on the two days before the report appeared, all 5 newborns at the Canton-Potsdam Hospital had been girls. Does this show the law of averages in action? Explain.

8.8 Spin the wheel Joey has been watching people play roulette for a few minutes. In the past 20 spins of the wheel, the ball has landed in a red slot 14 times. Joey wants to place a bet on "black" for the next spin, because he believes black is "due." He asks you for advice. What do you tell him?

8.9 Selecting a jury A county selects people for jury duty at random from a list of one million registered voters, which happens to be evenly divided in terms of gender (that is, exactly 50% of the county's registered voters are women). Which of the following is true and which is false? Explain.

(a) Out of the next 2000 names drawn, it is likely that exactly 1000 will be women.

(b) Out of the next 2000 names drawn, it is likely that the percent of women will be around 50%, to within a percent or two.

8.10 Bringing down the house? You may have heard about a group of MIT students who beat the casinos at blackjack. The 2008 movie *21* and the book *Bringing Down the House* tell their story. Do some research: how did the students use probability to get an edge on the casinos?

8.11 Duplicate birthdays There are 30 students in Austin's precalculus class, all unrelated. He wants to bet you $1 that at least 2 of the students were born on the same day of the year (same month and day but not necessarily the same year). Should you take the bet? Before you answer, you decide to conduct a simulation to see how likely that event would be. Here is the probability model:

- The birth date of a randomly chosen person is equally likely to be any of the 365 dates of the year (ignoring February 29).
- The birth dates of different people in the room are independent.

(a) Describe how you could use either Table B or your calculator to simulate 30 random birthdays from the 365 possible days.

(b) Carry out your simulation. Perform 10 repetitions. Based on your results, should you take Austin's bet?

(c) Combine your results with those from other members of the class. Are you surprised?

8.12 Birthdays revisited Refer to the previous exercise. On average, how many different people can you expect to ask before you find two with the same birthday? Use Table B or your calculator to simulate birthdays of randomly chosen people until you hit the same birthday a second time.

(a) How many people did you ask before you found two with the same birthday?

(b) Combine your results with those from other members of the class and average the results. This is an estimate of the expected value.

Probability distributions for sampling

Choosing a random sample from a population and calculating a statistic such as the sample proportion is certainly a random phenomenon. The *distribution* of the statistic tells us what values it can take and how often it takes those values. That sounds a lot like a probability distribution.

Example 8.5 A sampling distribution

Suppose you take a simple random sample (SRS) of 1523 adults. Ask each person whether they bought a lottery ticket in the last 12 months. The proportion who say "Yes,"

$$\hat{p} = \frac{\text{number who say "Yes"}}{1523}$$

is the sample proportion $\hat{p}$. Imagine repeating this sampling process 1000 times. You would collect 1000 sample proportions, one from each of the 1000 samples. The histogram in Figure 8.3 shows the distribution of 1000 sample proportions when the truth about the population is that 60% have bought lottery tickets.

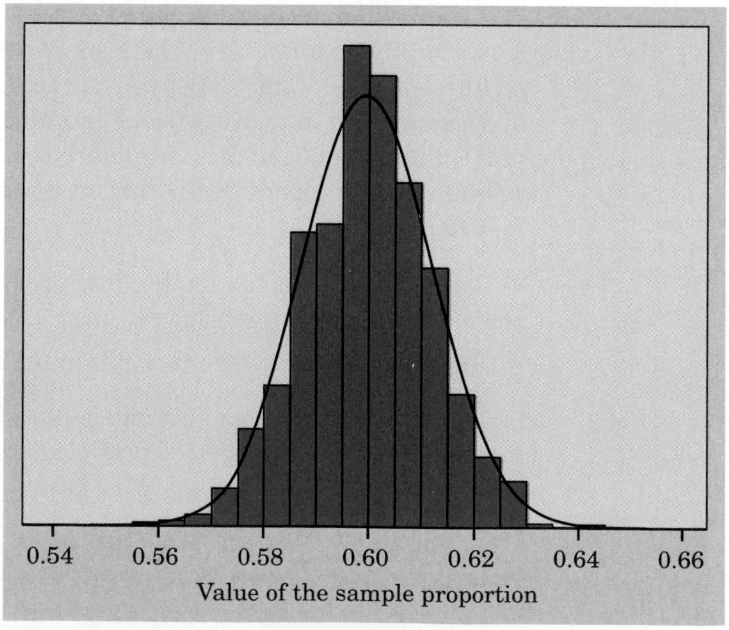

Figure 8.3 The probability distribution of a sample proportion $\hat{p}$ from SRSs of size 1523 drawn from a population in which 60% of the members would give positive answers. The histogram shows the distribution from 1000 samples. The Normal curve is the ideal pattern that describes the results of a very large number of samples.

The results of random sampling are of course random: we can't predict the outcome of one sample, but the figure shows that the outcomes of many samples have a regular pattern. The Normal curve in the figure is a good approximation to the histogram. The histogram is the result of these particular 1000 SRSs. Think of the Normal curve as the idealized pattern we would get if we kept on taking SRSs from this population forever. That's exactly the idea of probability—the pattern we would see in the very long run. *The Normal distribution assigns probabilities to the outcomes of random sampling.*

This Normal distribution has mean 0.6 and standard deviation about 0.0125. The "95" part of the 68–95–99.7 rule says that 95% of all samples will give a $\hat{p}$ falling within two standard deviations of the mean. That's within 0.025 of 0.6, or between 0.575 and 0.625. We now have more concise language for this fact: the *probability* is 0.95 that between 57.5% and 62.5% of the people in a sample will say "Yes." The word "probability" says we are talking about what would happen in the long run, in very many samples.

A statistic from a large sample has many possible values. Assigning a probability to each possible value of a random variable worked well for tossing coins or recording the Apgar score of a randomly selected baby, but it is awkward when there are thousands of possible outcomes. Example 8.5 uses a different approach: assign probabilities to intervals of outcomes by using areas under a Normal density curve. Density curves have area 1 underneath them, which lines up nicely with total probability 1. The total area under the Normal curve in Figure 8.3 is 1, and the area between 0.575 and 0.625 is 0.95, which is the probability that a sample gives a result in that interval. When a Normal curve assigns probabilities, you can calculate probabilities from the 68–95–99.7 rule or from Table A of Normal distribution areas. These probabilities satisfy the probability rules of Chapter 7.

Sampling distribution

The **sampling distribution** of a statistic tells us what values the statistic takes in repeated samples from the same population and how often it takes those values.

We think of a sampling distribution as assigning probabilities to the values the statistic can take. Because there are usually many possible values, sampling distributions are often described by a density curve such as a Normal curve.

Here's another example of sampling distributions in action.

Example 8.6 Do you approve of gambling?

An opinion poll asks an SRS of 501 teens, "Generally speaking, do you approve of legal gambling or betting?" Suppose that in fact exactly 50% of all teens would say "Yes" if asked. (This is close to what polls show to be true.) The poll's statisticians tell us that the sample proportion who say "Yes" will vary in repeated samples according to a Normal distribution with mean 0.5 and standard deviation about 0.022. This is the *sampling distribution* of the sample proportion $\hat{p}$.

What is the probability that the poll will get a sample in which 52% or more say "Yes"? Because 0.52 is not 1, 2, or 3 standard deviations away from the mean, we can't use the 68–95–99.7 rule. We will use Table A of standard Normal probabilities.

To use Table A, first turn the outcome $\hat{p} = 0.52$ into a standardized z score by subtracting the mean of the distribution and dividing by its standard deviation:

$$z = \frac{0.52 - 0.5}{0.022} = 0.91$$

Now look in Table A. The area to the left of a z-score of 0.91 is 0.8186 under a Normal curve. This means that the probability is 0.8186 that the poll gets a smaller result. By the complement rule (or just the fact that the total area under the curve is 1), this leaves probability 0.1814 for outcomes of 52% or more who say "Yes." Figure 8.4 shows the probabilities as areas under the Normal curve.

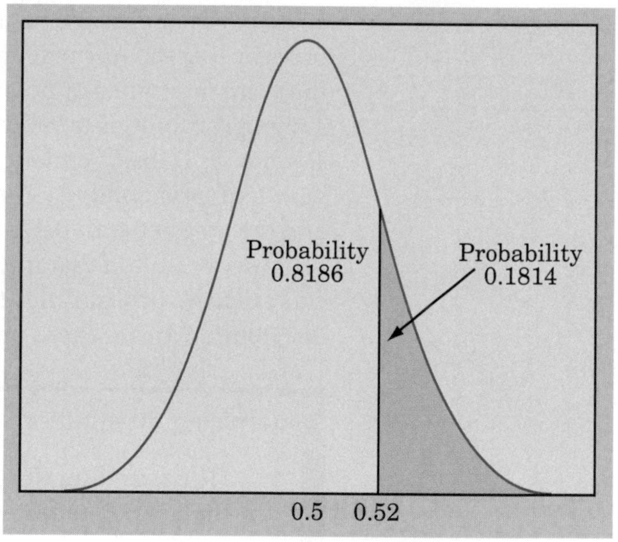

Figure 8.4 The Normal sampling distribution for the opinion poll of 501 teens. Table A tells us that the area under the curve to the left of 0.52 is 0.8186. The area to the right of 0.52 is 0.1814.

Exercises

Stephen Coburn/FeaturePics

8.13 Do you jog? An opinion poll asks an SRS of 1500 adults, "Do you happen to jog?" Suppose (as is approximately correct) that the population proportion who jog is $p = 0.15$. In a large number of samples, the proportion $\hat{p}$ who answer "Yes" will be approximately Normally distributed with mean 0.15 and standard deviation 0.009. Sketch this Normal curve and use it to answer the following questions.

(a) What percent of many samples will have a sample proportion who jog that is 0.15 or less? Explain clearly why this percent is the probability that $\hat{p}$ is 0.15 or less.

(b) What is the probability that $\hat{p}$ will take a value between 0.141 and 0.159? (Use the 68–95–99.7 rule.)

(c) What is the probability that $\hat{p}$ does not lie between 0.141 and 0.159?

(d) Would you be surprised if the sample proportion who said they jogged was $\hat{p} = 0.17$? Justify your answer.

8.14 Applying to college You ask an SRS of 1500 college students whether they applied for admission to any other college. Suppose that in fact 75% of all college students applied to colleges besides the one they are attending. (That's close to the truth.) The sampling distribution of the proportion of your sample who say "Yes" is approximately Normal with mean 0.75 and standard deviation 0.01. Sketch this Normal curve and use it to answer the following questions.

(a) Explain in simple language what the sampling distribution tells us about the results of our sample.

(b) What percent of many samples would have a $\hat{p}$ larger than 0.77? (Use the 68–95–99.7 rule.) Explain in simple language why this percent is the probability of an outcome larger than 0.77.

(c) What is the probability that your sample result will be either less than 0.73 or greater than 0.75? Show your work.

(d) Would you be surprised if the sample proportion who said they applied to another college was $\hat{p} = 0.725$? Justify your answer.

8.15 Generating a sampling distribution Let us illustrate the idea of a sampling distribution in the case of a very small sample from a very small population. The population consists of the numbers 1, 2, 4, and 9. The parameter of interest is the mean of this population. The sample is an SRS of size $n = 2$ drawn from the population.

(a) Find the mean of the four numbers in the population. This is the population mean μ.

(b) There are only six possible samples of size 2 that can be chosen. One such sample is {1, 2}. For this sample, $\overline{x} = 1.5$. This statistic, $\overline{x}$, is an estimate of the population mean μ. List the other five possible SRSs of size 2, and compute the value of $\overline{x}$ for each.

(c) Make a dotplot that shows the distribution of $\overline{x}$-values from (b). This is the sampling distribution of $\overline{x}$.

(d) Find the mean of the six $\overline{x}$-values from (b). How does this compare to the population mean μ?

Exercises 8.16 to 8.18 refer to the following setting. The Wechsler Adult Intelligence Scale (WAIS) is a common "IQ test" for adults. The distribution of WAIS scores for persons over 16 years of age is approximately Normal with mean 100 and standard deviation 15.

8.16 An IQ test, I

(a) What is the probability that a randomly chosen individual has a WAIS score of 115 or higher?

(b) In what range do the scores of the middle 95% of the adult population lie?

8.17 An IQ test, II Use Table A or your calculator to find the probability that a randomly chosen person has a WAIS score of 112 or higher. Show your work.

8.18 An IQ test, III How high must a person score on the WAIS test to be in the top 10% of all scores? Use Table A or your calculator to answer this question. Show your work.

APPLICATION 8.1

Benford's Law

1. Using the newspaper, an almanac, the *Statistical Abstract of the United States,* or similar text material, locate a compact collection of numbers such as closing stock prices or sports statistics. (Alternatively, your teacher may provide you with data on closing stock prices from the New York Stock Exchange.) Select a group of at least 100 separate numbers. Copy the table below onto your paper. Then use tally marks to record the frequency of each first digit, from 1 to 9. (*Note:* For decimals, like 0.00305, you would take the first nonzero digit.) Calculate the relative frequency for each digit.

Digit	Tally marks	Counts	Relative frequency
1			
2			
3			
4			
5			
6			
7			
8			
9			

You might think that each of the digits from 1 to 9 would occur equally often in the distribution of first digits. Were your results uniformly distributed?

Scientist Simon Newcomb (1835–1909) discovered in 1881 that the digits are not uniformly distributed. In an article that year in the *American Journal of Mathematics,* Newcomb conjectured that, with naturally occurring data, you are most likely to observe a 1 as the leading digit, and that each digit from 2 to 9 is less likely than its predecessor to occur.[3] In fact, he conjectured that many (but not all) such data have the following distribution:

Digit:	1	2	3	4	5	6	7	8	9
Probability:	0.301	0.176	0.125	0.097	0.079	0.067	0.058	0.051	0.046

Newcomb got little recognition for this remarkable result. Almost 60 years later, in 1938, Frank Benford rediscovered Newcomb's result. Benford was a physicist working for General Electric at the time. Apparently unaware of Newcomb's paper, Benford examined a large collection of data that were not produced randomly. He published his findings and his explanation, which he called Benford's Law, in another journal.[4]

Figure 8.5 is a histogram of the probabilities for each first digit. Notice that we can't locate the mean of the right-skewed distribution by eye—calculation is needed. The mean is shown in Figure 8.5 as a dotted line.

2. Calculate the average first digit (expected value) for data that obey Newcomb's probability distribution.

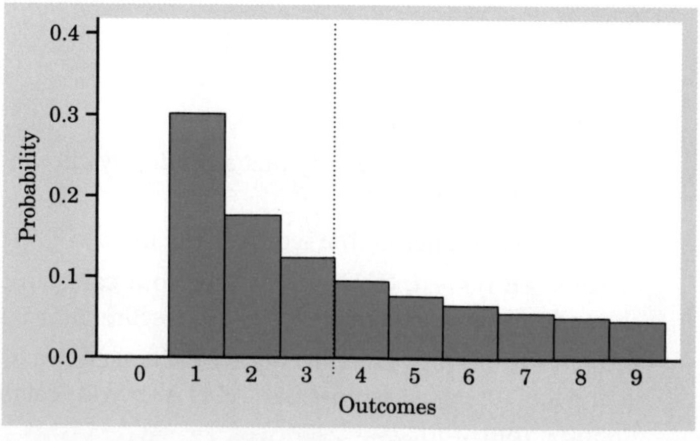

Figure 8.5 Digits between 1 and 9 chosen from records that obey Benford's Law.

3. Compare your results from step 1 with the theoretical probability for each digit. Construct a graph that shows two bars for each digit: one bar for the probability and a bar to the right of it for the relative frequency from your data. Use color or shading to distinguish the two bars. This is a good way to visually compare how well your results agree with the conjectured relative frequencies.

4. Explain clearly why the distribution of first digits in the blocks of 5 digits in Table B would *not* obey Benford's Law.

Section 8.1 Summary

A **random variable** is a numerical variable that describes the outcomes of a chance process. A **probability distribution** tells us how often (in the very long run) a random variable takes each of its possible values. We are often interested in the long-run average outcome, which we call the **expected value** of the random variable. The expected value is found as an average of all the possible outcomes, each weighted by its probability. If you don't know the outcome probabilities, you can estimate the expected value (along with the probabilities) by simulation. The **law of large numbers** says that the mean outcome in many repetitions eventually gets close to the expected value.

A second kind of probability distribution assigns probabilities as areas under a density curve, such as a Normal curve. The total probability is 1 because the total area under the curve is 1. This kind of probability model is often used to describe the **sampling distribution** of a statistic. This is the pattern of values of the statistic in many samples from the same population.

Section 8.1 Exercises

Exercises 8.19 to 8.22 refer to the following scenario. As a promotion gimmick, a drug store decides to raffle off ice cream bars in the following way. They advertise, "Ice cream bar: 0 to 43 cents. Take a chance." To get one, the customer pops a balloon and then reads the price he or she must pay on a slip of paper that was inside the balloon. Of course, the customer doesn't know what his or her chances are of getting a fairly low price for his or her ice cream bar. The manager of the drug store has marked prices on slips of paper as shown in the table. Suppose that the usual price of an ice cream bar is 40 cents.

Price:	\$0.43	\$0.42	\$0.41	\$0.40	\$0.34	\$0.20	\$0.09	\$0.02	\$0.01	\$0.00
Slips:	30	18	20	19	15	10	5	5	5	5

Mary decides to play the raffle and pops a balloon at random. Let X = the price she must pay.

8.19 Pop a balloon, I

(a) Find $P(X < \$0.10)$ and interpret your result.

(b) What is the probability that Mary will pay less than the usual cost of an ice cream bar? Write this probability in terms of the random variable X.

(c) How much money will the store take in if all of the balloons are popped? How much would the store have made on these ice cream bars without the promotion? Comment on the difference.

8.20 Pop a balloon, II Sketch a graph of the probability distribution of X.

8.21 Pop a balloon, III Compute the expected price of an ice cream bar when Mary pops a balloon. Is it to her benefit to play the raffle? Explain.

8.22 Pop a balloon, IV Make a new table of prices and numbers of slips so that the expected price a customer will pay in the pop-the-balloon game is 40 cents. Use a total of 100 slips.

8.23 Fire insurance Suppose a homeowner spends \$300 for a home insurance policy that will pay out \$200,000 if the home is destroyed by fire. Let $Y =$ the profit made by the insurance company on a single policy. From previous data, the probability that a home in this area will be destroyed by fire is 0.0002.

(a) Make a table that shows the probability distribution of Y.

(b) Compute the expected value of Y. Explain what this result means for the insurance company.

8.24 Polling women, I Suppose that 47% of all adult women think they do not get enough time for themselves. An opinion poll interviews 1025 randomly chosen women and records the sample proportion who feel they don't get enough time for themselves. This statistic will vary from sample to sample if the poll is repeated. The sampling distribution is approximately Normal with mean 0.47 and standard deviation about 0.016. Sketch this Normal curve and use it to answer the following questions.

(a) If the true population proportion is 0.47, in what range will the middle 95% of all sample results fall?

(b) What is the probability that the poll gets a sample in which fewer than 45.4% say they do not get enough time for themselves?

8.25 Polling women, II In the setting of Exercise 8.24, what is the probability of getting a sample in which more than 51% of the women think they do not get enough time for themselves? Show your work. (Use Table A or your calculator.)

8.26 Collecting cereal box prizes Several years ago, every box of Frosted Mini-Wheats contained a NASCAR driver's decal. There was a set of 6 decals, and consumers were encouraged to "collect all 6." The probability model is

- Every box of Frosted Mini-Wheats cereal contains one NASCAR driver's decal.
- A single decal is randomly selected and placed in the cereal box, one per box.
- There are 6 decals in the set.

The question we want to answer is "How many boxes of Frosted Mini-Wheats cereal can we expect to buy in order to collect the complete set of 6 NASCAR decals?"

(a) Begin with a simpler problem. Suppose there are only 2 decals. How would you use the table of random digits to simulate this problem? How would you use a single die to simulate the problem? How would you use your calculator to simulate the problem? Choose one of these methods, and carry out 10 repetitions. Then average the 10 results to estimate the expected number of boxes needed to collect a set of 2 decals.

(b) Answer the questions in (a) if you assume a set of 3 decals, and then 4 decals.

(c) Now answer the original question for a set of 6 decals. Is there a pattern? If so, describe it.

8.2 Counting and Probability

ACTIVITY 8.2

Deal or no deal?

Materials: Computer with Internet connection and projector

In the TV game show *Deal or No Deal*, a contestant begins by choosing 1 of the 26 numbered cases shown on the game board below. Each case holds a different amount of money ranging from $0.01 to $1,000,000, as shown on the game board. The amounts are assigned to the cases at random before the show. Neither the show's host, Howie Mandel, nor the banker, who offers to buy the contestant's case after each round of the game, knows what any case contains.

1. To get familiar with how the game works, go to the online version of the game at **www.nbc.com/Deal_or_No_Deal/game** and play it once as a class.

2. Before you play the game a second time, discuss the following questions with your classmates:

- What's the probability that you choose the $1,000,000 case to begin with?
- If you played the game many, many times, what is the expected value of the amount that your case contains?
- Suppose you get lucky and pick the $1,000,000 case to begin with. (Of course, you don't know what your case actually contains.) How likely do you think you would be to keep turning down the banker's offers and to actually leave the show with $1,000,000?

3. Play the game again as a class. This time, record which cases you open in each round and the banker's offer in a table like the one shown below. Decline all of the banker's offers and keep playing until you get to see what's in your case.

Round	Cases opened	Banker's offer
1	$100,000; $0.01; $300; $500; $10,000; $25	$35,100
2		$

4. Add an extra column on the right side of your table for the average amount of money in the cases that still remain on the board. Compute this value after each round. Is the banker's offer higher than, the same as, or lower than this average amount?

5. Play the game one last time as a class. This time, try to finish with as much money as you can. Did you make a good deal?

Popular TV game shows like *Deal or No Deal* and *The Price Is Right* rely on probability, expected values, and the law of large numbers to tell them how much contestants will win—not in a single show, but on average in the long run. The same is true for state governments who sponsor lotteries, where the payouts can reach tens or even hundreds of millions. Casinos invite "high rollers" to place large bets on complicated games of chance whose results are not as predictable as those of roulette or slot machines. As the games get more complex, we need more advanced methods to determine all possible outcomes and their probabilities.

In the early days of the Internet, many companies and individuals scrambled to get the domain name of their choosing. American Airlines landed **aa.com**, International Business Machines (IBM) got **ibm.com**, and Kentucky Fried Chicken snagged **kfc.com**. Not everyone was so lucky. A small business in Canton, Ohio, had to settle for a domain name that was 15 letters long. The company is The Karcher Group, which specializes in Web design. Until recently, their domain name was **thekarchergroup.com**. After waiting 10 years and paying $15,000, the company has now acquired the domain name **tkg.com**. Imagine how much easier it will be now for people to get to The Karcher Group's Web site and to send emails to company employees!

Example 8.7 Choosing a domain name

Internet domain names must have more than one character. That rules out names like **i.com**. How many such names are there? If we consider letters only, there are 26 possibilities. What if we allowed a digit or a letter? Since there are 10 digits, we'd have $26 + 10 = 36$ possible one-character domain names.

Two-character domain names are allowed. How many two-letter domain names are there? There are 26 possible choices for the first letter. Suppose we choose "a." How many choices are there for the second letter? 26, right? So there are 26 two-letter domain names that begin with "a." Similarly, there are 26 names that begin with "b," "c," "d," . . . , "z." For any choice of first letter, we get 26 possibilities for the second letter. So the total number of two-letter domain names is

$$(\text{\# of choices for first letter}) \cdot (\text{\# of choices for second letter})$$
$$= \underline{26} \cdot \underline{26} = 676$$

What if we allow letters or numbers for either character? Then there are a total of

$$(\text{\# of choices for first character}) \cdot (\text{\# of choices for second character})$$
$$= \underline{36} \cdot \underline{36} = 1296$$

possible two-character domain names. All 1296 have been claimed. □

Let's head to the ice cream parlor for some more counting.

Example 8.8 Ice cream sundaes

The 28 Choices ice cream shop offers 28 flavors of ice cream, 11 different toppings, and 3 sizes of bowl for their ice cream sundaes. In how many ways can you order a sundae with one scoop of ice cream and one topping?

Let's start with the ice cream. You have 28 flavor choices. For each choice of flavor, you can select 11 possible toppings. So there are $28 \cdot 11 = 308$ different ice cream plus topping selections. For each of those 308 concoctions, you have three possible sizes of bowl. In all, that gives you

$$(\text{\# of ice cream flavors}) \cdot (\text{\# of topping choices}) \cdot (\text{\# of size options})$$
$$= \underline{28} \cdot \underline{11} \cdot \underline{3} = 924$$

ways to order a one-scoop sundae.

Examples 8.7 and 8.8 illustrate the **multiplication counting principle.**

Multiplication counting principle

For a process involving multiple (say k) steps, if there are n_1 ways to do Step 1, n_2 ways to do Step 2, . . . , and n_k ways to do Step k, then the total number of different ways to complete the process is

$$n_1 \cdot n_2 \cdot \ \ldots \ \cdot n_k$$

This result is called the **multiplication counting principle.**

The multiplication counting principle can also help us determine how many ways there are to arrange a group of people, animals, or things. For example, suppose you have 5 framed photographs of different family members that you want to arrange in a line on top of your dresser. In how many ways can you do this? Since you are arranging the photographs in a line, let's count the options moving from left to right across the dresser. There are 5 choices for the leftmost photo. For each of these 5 possibilities, there are 4 options for the next photo to the right. By the multiplication counting principle, there are $5 \cdot 4 = 20$ different possibilities for these first two photos. For each of these 20 options, there are 3 choices for the next photo to the right, then 2 choices for the photo to the right of that one, and finally, only 1 choice left for the final photo. Again by the multiplication counting principle, that gives a total of $5 \cdot 4 \cdot 3 \cdot 2 \cdot 1 = 120$ different photo arrangements.

Expressions like $5 \cdot 4 \cdot 3 \cdot 2 \cdot 1$ occur often enough in counting problems that mathematicians invented a special name and notation for them. We write $5 \cdot 4 \cdot 3 \cdot 2 \cdot 1 = 5!$, read as "5 **factorial.**"

Factorial

For any positive integer n, we define $n!$ (read "n **factorial**") as

$$n! = n(n-1)(n-2) \ \ldots \ \cdot 3 \cdot 2 \cdot 1$$

That is, n factorial is the product of the numbers starting with n and going down to 1.

The following example shows how quickly the number of different arrangements of n individuals grows.

Example 8.9 A special offer from the teacher

At the beginning of class, your teacher makes a special offer. She will dismiss class early if you can make all possible arrangements with every student in the class in a single-file line. To try to tempt you, she lets you try with one student, then two students, and finally with three students. Should you accept the offer?

How many possible arrangements are there? If a small class has 15 students, then there are 15 choices for which student may be first in line. For each of these possibilities, there are 14 choices for who is second in line, 13 choices for who is third in line, and so on. By the multiplication counting principle, there are

$$15 \cdot 14 \cdot 13 \cdot \ ... \ \cdot 3 \cdot 2 \cdot 1 = 15!$$

possible arrangements. Instead of multiplying all 15 numbers together, we can use the factorial command on the calculator to evaluate 15!

- **TI-84:** Type `15`. Then press MATH, arrow to `PRB`, choose `4:!`, and press ENTER.
- **TI-Nspire:** On a Calculator page, type `15`, press (menu), then choose `4:Probability`, `1:Factorial(!)`, and press (enter).

There are about $1.31 \cdot 10^{12}$, or just over 1 trillion possible arrangements. (Note that the TI-Nspire shows all 13 digits of the result.)

Our advice: don't take the offer!

Exercises

8.27 More domain names Refer to Example 8.7. How many three-character Internet domain names are possible that consist of

(a) letters only?

(b) numbers or letters?

Note: All three-character domain names have been taken.

8.28 More ice cream Refer to Example 8.8. Mary stops by 28 Choices on the way home from school. She only has enough money for a sundae with one scoop of ice cream and one topping in a small bowl. How many different choices does Mary have for her sundae?

Typical 1959 NJ license plate

Typical 1999 NJ license plate

8.29 New Jersey plates, I The illustration shows what New Jersey license plates looked like in two different years—1959 and 1999. Assume that letters and numbers must be in the positions shown on each license plate: for 1959, three letters followed by three numbers; for 1999, three letters, then two numbers, and then a letter. If there are no restrictions on the letters and numbers used in any position, how many different license plates were possible in 1959? 1999? Show your work.

8.30 New Jersey plates, II Refer to the previous exercise. In 1999, New Jersey license plates were actually not allowed to have the letters D, T, or X in the first position, or the letters I, O, or Q in *any* position. With these restrictions, how many different license plates were actually possible in 1999? Show your work.

8.31 Organizing homework Suppose you have six homework assignments to complete one night.

(a) In how many different orders can you complete all of the assignments?

(b) In how many different orders can you complete all six assignments if you simply must do your statistics assignment first?

8.32 Assigned seating Mrs. Random decides to randomly assign seats for the 25 students in her statistics class.

(a) If there are 25 seats in the classroom, how many different seating assignments are possible? Show your work.

(b) Repeat part (a) for a classroom with 30 seats.

Permutations

Mr. Paradise likes to get the students in his class involved in the action. But he doesn't want to play favorites. Each day, Mr. Paradise puts the names of all 28 of his students in a hat and mixes them up. He then draws out 3 names, one at a time. The student whose name is chosen first gets to operate the display calculator for the day. The second student chosen is in charge of reading the answers to the even-numbered homework problems. The third student selected writes class notes on the Smart board. In how many different ways can Mr. Paradise randomly assign these three jobs?

There are 28 possibilities for the first name chosen, 27 possibilities for the second name, and 26 possibilities for the third name. By the multiplication counting principle, there are

$$28 \cdot 27 \cdot 26 = 19{,}656$$

different ways for Mr. Paradise to assign the three jobs. We call arrangements like this, where the order of selection matters, **permutations.** In this case, we calculated the number of permutations of 28 people taken 3 at a time. In symbols, we'll write this as ${}_{28}P_3$.

With a little clever math, we can rewrite as ${}_{28}P_3$ as follows:

$$\begin{aligned} {}_{28}P_3 &= 28 \cdot 27 \cdot 26 \\ &= \frac{28 \cdot 27 \cdot 26 \cdot 25 \cdot 24 \cdot \ldots \cdot 3 \cdot 2 \cdot 1}{25 \cdot 24 \cdot \ldots \cdot 3 \cdot 2 \cdot 1} \\ &= \frac{28!}{25!} \\ &= \frac{28!}{(28-3)!} \end{aligned}$$

In general, the number of permutations of n things taken k at a time is given by

$${}_nP_k = \frac{n!}{(n-k)!}$$

Permutation

A **permutation** is a distinct arrangement of some collection of individuals. If there are n individuals, then the notation ${}_nP_k$ represents the number of different permutations of k individuals chosen from the entire group of n. We can calculate ${}_nP_k$ using the multiplication counting principle or with the formula

$$ {}_nP_k = \frac{n!}{(n-k)!} $$

By definition, $0! = 1$.

To make the formula for ${}_nP_k$ work properly for all possible values of k, we must define $0! = 1$. Why? Consider ${}_{28}P_{28}$. This is the number of different arrangements of 28 individuals, taken 28 at a time. In other words, it's $28 \cdot 27 \cdot 26 \cdot \ldots \cdot 3 \cdot 2 \cdot 1 = 28!$ Using the permutation formula for ${}_{28}P_{28}$, we get

$$ {}_{28}P_{28} = \frac{28!}{(28-28)!} = \frac{28!}{0!} $$

If we define $0! = 1$, then the values obtained by the two methods will agree.

Combinations

With permutations, order matters. Sometimes, we're just interested in the number of ways to choose some number of individuals from a group, but we don't care about the order in which the individuals are selected. For example, suppose Mr. Paradise decides to randomly select 3 students' homework papers to grade each day. He once again puts all 28 names in a hat, mixes them up, and draws out 3 names, one at a time. In how many different ways can he choose 3 students' papers to grade?

It's tempting to say that there are $28 \cdot 27 \cdot 26 = 19{,}656$ ways for Mr. Paradise to do this. That's not correct, however. Suppose he picks Leucretia, Tim, and Kiran—in that order. That's really no different from getting Tim, then Leucretia, then Kiran. Or Kiran, then Leucretia, then Tim. How many selections consist of these same 3 students? Use the multiplication counting principle: there are 3 possibilities for the first pick, 2 possibilities for the second pick, and only 1 option for the last pick. So there are $3 \cdot 2 \cdot 1 = 3! = 6$ arrangements that consist of these same 3 students. This same argument applies for any 3 students that Mr. Paradise selects. So to get the correct number of ways for him to choose 3 students' papers to grade, we need to divide our original (wrong) answer by 6:

$$ \text{number of ways to choose 3 homework papers to grade out of 28} = \frac{28 \cdot 27 \cdot 26}{3 \cdot 2 \cdot 1} $$

With a little fancy math, we can rewrite this answer as

$$ \frac{28 \cdot 27 \cdot 26}{3 \cdot 2 \cdot 1} = \frac{{}_{28}P_3}{3!} $$

When the order in which we select individuals from a group doesn't matter, we call the number of possible selections a **combination.** In the previous paragraph, we computed the number of combinations of 28 students taken 3 at a time, which we'll write as ${}_{28}C_3$. We also saw an important connection between the number of permutations and the number of combinations in this setting:

$$ {}_{28}C_3 = \frac{{}_{28}P_3}{3!} $$

In general, the number of combinations of n things taken k at a time can be found using the formula

$$ {}_nC_k = \frac{{}_nP_k}{k!} $$

Substituting the formula we found earlier for ${}_nP_k$ yields

$$ {}_nC_k = \frac{{}_nP_k}{k!} = \frac{\left(\dfrac{n!}{(n-k)!}\right)}{k!} = \frac{n!}{k!(n-k)!} $$

Combination

A **combination** is a selection of individuals from some group where the order of selection doesn't matter. If there are n individuals, then the notation ${}_nC_k$ represents the number of different combinations of k individuals chosen from the entire group of n. We can calculate ${}_nC_k$ using the formula

$$ {}_nC_k = \frac{{}_nP_k}{k!} = \frac{n!}{k!(n-k)!} $$

Finding probabilities using counting methods

With these new counting tools—the multiplication counting principle, permutations, and combinations—to help us, we are ready to tackle some more challenging probability applications.

Example 8.10 Pick 3 lottery

Here is a simple lottery wager, the "Straight" from the Pick 3 game of the Tri-State Daily Numbers offered by New Hampshire, Maine, and Vermont. You pay \$1 and choose a three-digit number. The state chooses a three-digit winning number at random and pays you \$500 if your number is chosen. What's the probability that you win?

By the multiplication counting principle, there are $10 \cdot 10 \cdot 10 = 1000$ possible winning numbers. Your chance of matching the winning number picked randomly by the computer is 1/1000. Of course, the probability that you don't match the winning number is $1 - 1/1000 = 999/1000$. If we let X = the amount of your winnings, then the probability distribution of X is

Value:	\$0	\$500
Probability:	999/1000	1/1000

Your long-run average winnings from a ticket are

$$\$0 \cdot \frac{999}{1000} + \$500 \cdot \frac{1}{1000} = \$0.50$$

or 50 cents. You see that in the long run the state pays out half the money bet and keeps the other half.

In the popular card game Texas Hold 'Em, players are initially dealt two cards. The "best" hand a player can start with is a pair of aces (known as "pocket rockets"). What's the probability that a particular player is dealt a pair of aces? Since Texas Hold 'Em often involves several players, and thus several starting hands that must be dealt, let's begin with a simpler version.

Example 8.11 Looking for aces

Suppose we deal two cards from a well-shuffled, standard deck. What's the probability that both are aces? In Chapter 7, we used a tree diagram and the general multiplication rule to answer a similar question. Let's follow that strategy again here.

Method 1—Tree diagram and general multiplication rule Let's define the events A = ace on first card and B = ace on second card. The probability that the first card dealt is an ace is 4/52. Given that the first card was an ace, there's a 3/51 chance that the second card is an ace. Then the probability of getting two aces can be found using the general multiplication rule:

$$P(\text{A} \cap \text{B}) = P(\text{A}) \cdot P(\text{B}|\text{A})$$
$$= (4/52)(3/51) = 0.0045$$

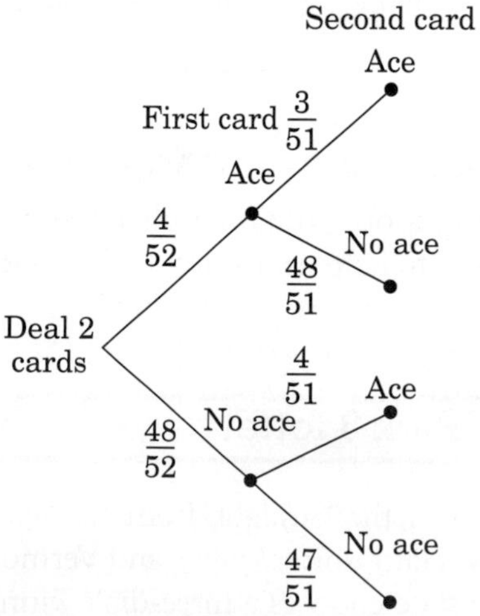

Figure 8.6 A tree diagram showing the sample space for the chance process of dealing two cards from a deck and looking for aces.

Figure 8.6 shows a tree diagram of this chance process. Getting two aces isn't very likely!

There are two completely different ways to approach this problem using the methods of Chapter 8.

Method 2—Permutations If we think of the two cards that are dealt as Card 1 and Card 2, then there are 52 possibilities for Card 1. Once Card 1 is dealt, there are only 51 possibilities for Card 2. So there are $52 \cdot 51$, or $_{52}P_2$, possible arrangements of the first two cards in the deck, each of which is equally likely after shuffling. To get two aces, we have four choices for Card 1 and then three choices remaining for Card 2. That is, there are $4 \cdot 3$, or ${}_4P_2$, arrangements in which the first two cards are aces. The probability of getting two aces is

$$P(\text{two aces}) = \frac{\text{number of arrangements with two aces}}{\text{number of arrangements of first two cards}}$$
$$= \frac{{}_4P_2}{{}_{52}P_2} = \frac{4 \cdot 3}{52 \cdot 51} = 0.0045$$

Method 3—Combinations We can also use combinations to get this result. First, we count the number of possibilities for dealing two cards from the deck. Note that the order in which the cards are dealt really doesn't matter here. All we care about is which two cards are dealt. Since we are "choosing" two cards out of 52 at random, there are ${}_{52}C_2$ different ways to do this. Each of these possible two-card sets is equally likely to be dealt. How many ways are there to get two aces? With four aces to choose from, there are ${}_4C_2$ ways to obtain two aces. So the probability of getting two aces is

$$P(\text{two aces}) = \frac{\text{number of ways to get two aces}}{\text{number of ways to deal two cards}}$$
$$= \frac{{}_4C_2}{{}_{52}C_2} = \frac{\dfrac{4!}{2!(4-2)!}}{\dfrac{52!}{2!(52-2)!}} = \frac{\dfrac{4 \cdot 3 \cdot \cancel{2!}}{2!\cancel{2!}}}{\dfrac{52 \cdot 51 \cdot \cancel{50!}}{2!\cancel{50!}}} = \frac{\dfrac{4 \cdot 3}{2 \cdot 1}}{\dfrac{52 \cdot 51}{2 \cdot 1}} = \frac{6}{1326} = 0.0045$$

In Example 8.11, we calculated the probability of getting two aces when we deal two cards from a standard deck. In other words, we found the probability that two aces will end up as the top two cards in the deck after shuffling. But in card games involving several players, like Texas Hold 'Em, the top card on the deck is dealt to Player 1, the second card to Player 2, and so on until each player has one card. Then a second card is dealt to Player 1, Player 2, and so forth until all the players have two cards. What's the probability that Player 1 gets two aces? It's the chance that there are aces in Positions 1 and $n + 1$ in the deck, where n is the number of players. Let's think about this. The probability that the top card and second card in the deck are aces is the same as the probability that the top card and the $(n + 1)$st card are aces. What is that chance? By Example 8.11, it's 0.0045. Every player at the table has the same chance (before the cards are dealt) of getting "pocket rockets."

Exercises

8.33 **Two scoops** On the way home, you decide to stop by 28 Choices ice cream parlor for a snack. With 28 flavors to choose from, how many different ways are there to order:

(a) a cone with two scoops if you care about the order of flavors?

(b) a cone with two scoops of *different* flavors if you care about the order of flavors?

(c) a cup with two scoops of *different* flavors (order doesn't matter)?

(d) a cup with two scoops of ice cream (order doesn't matter)?

8.34 Three scoops Refer to Exercise 8.33. How many different ways are there to order

(a) a cone with three scoops if you care about the order of flavors?

(b) a cone with three scoops of *different* flavors if you care about the order of flavors?

(c) a cup with three scoops of *different* flavors (order doesn't matter)?

(d) a cup with three scoops of ice cream (order doesn't matter)?

8.35 Pick 3 more Refer to Example 8.10.

(a) Find the number of ways in which the computer can pick a winning number with three different digits. Show your work.

(b) What's the probability that the computer picks a winning number with two or more matching digits? Show your work.

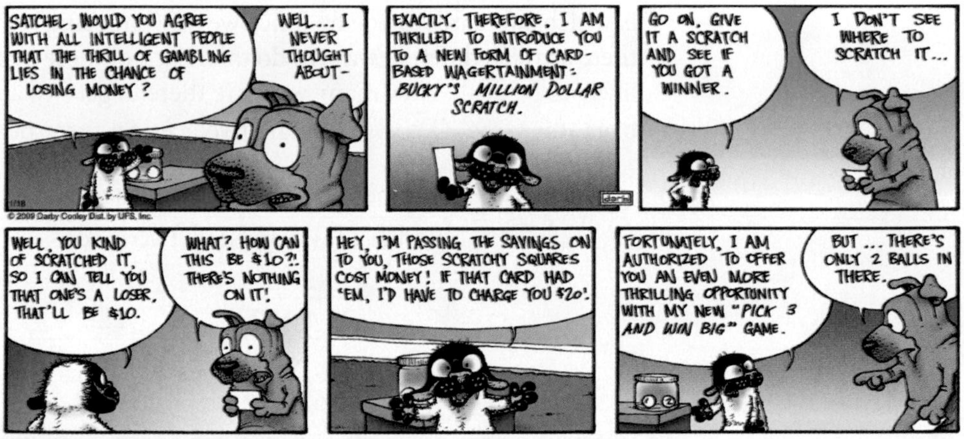

© 2009 Darby Conley, distributed by UFS, Inc.

8.36 Pick 4 Refer to Example 8.10. In the Pick 4 game of the Tri-State Daily Numbers, you pay \$1 and choose a four-digit number. The state chooses a four-digit winning number at random and pays you \$5000 if your number is chosen.

(a) What's the probability that you win? Show your method clearly.

(b) Find the expected value of your winnings from playing Pick 4.

8.37 Permutations and combinations Six friends—Aaron, Betty, Caleb, Deanne, Ernie, and Flo—are having a study session on permutations and combinations.

(a) Find the value of ${}_6P_0$. Explain why this value makes sense.

(b) Compute the value of ${}_6P_3$ and ${}_6C_3$. Create an example involving the six friends to show how these values are related.

(c) Calculate ${}_6C_2$ and ${}_6C_4$. Create an example involving the six friends to show how these values are related.

8.38 Big slick In Texas Hold 'Em, players are initially dealt two cards. One very strong starting hand is called "big slick," which consists of an ace and a king (not necessarily of the same suit).

(a) If you deal two cards from a well-shuffled deck, what's the probability of getting "big slick"? Show your method clearly.

(b) In a Texas Hold 'Em game with six players, what's the probability that Player 1 is dealt "big slick"? Justify your answer.

CALCULATOR CORNER Permutations and combinations on the calculator

The TI-84 and TI-Nspire have built-in commands for calculating the number of permutations and combinations.

TI-84

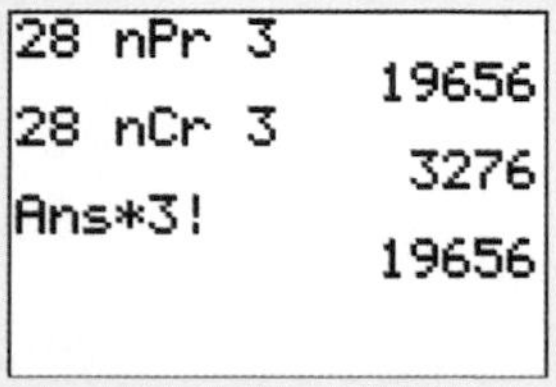

- *Permutations:* To find ${}_{28}P_3$, type `28`, then press [MATH], choose `PRB`, and `2:nPr`. Complete the command `28 nPr 3` and press [ENTER].
- *Combinations:* To find ${}_{28}C_3$, type `28`, then press [MATH], choose `PRB`, and `3:nCr`. Complete the command `28 nCr 3` and press [ENTER].

TI-Nspire

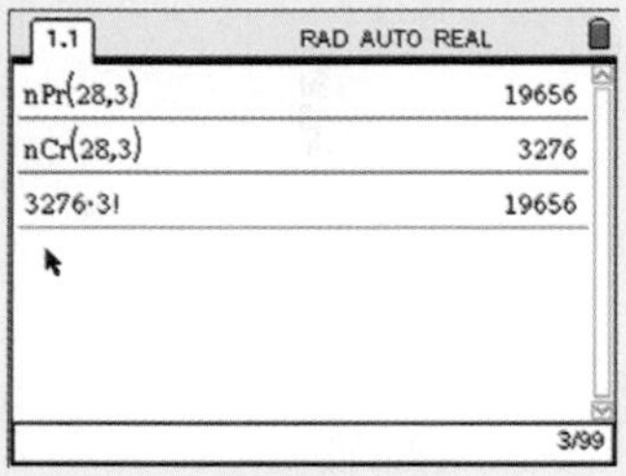

- *Permutations:* To find ${}_{28}P_3$, on a Calculator page press (menu), choose `4: Probability, 2:Permutations`. Complete the command `nPr(28,3)` and press (enter).
- *Combinations:* To find ${}_{28}C_3$, press (menu), choose `4:Probability, 3: Combinations`. Complete the command `nCr(28,3)` and press (enter).

The last line in both calculator screen shots reminds you of the relationship between permutations and combinations.

APPLICATION 8.2

iPod play list

Janine wants to set up a play list with 8 songs on her iPod. She has 75 songs to choose from, including 15 by the Jonas Brothers, 12 by Miley Cyrus, and 10 by Bow Wow.

QUESTIONS

1. How many different sets of 8 songs are possible for Janine's play list? Assume that the order of the songs doesn't matter.

2. How many different 8-song play lists can Janine create? This time, assume the order of the songs does matter.

3. How many 8-song play lists contain only songs by the Jonas Brothers? No songs by the Jonas Brothers?

Andreanna Seymore/Getty

Suppose Janine decides to let her iPod select an 8-song play list at random.

4. What's the probability that all 8 songs are by Miley Cyrus?

5. What's the probability that all 8 songs are by Miley Cyrus, the Jonas Brothers, or Bow Wow?

6. What's the probability that at least 1 of the 8 songs is by Bow Wow?

Section 8.2 Summary

Most probability problems involve counting the number of possible outcomes of a chance process. When that process involves multiple steps, the **multiplication counting principle** can be used to determine the total number of possible outcomes. The multiplication counting principle also helps us count the number of distinct arrangements of some group of individuals. Such arrangements are called **permutations.** The number of permutations of n individuals is ***n* factorial:**

$$n! = n \cdot (n-1) \cdot (n-2) \cdot \ldots \cdot 3 \cdot 2 \cdot 1$$

The number of permutations of n individuals taken k at a time is given by

$$_nP_k = \frac{n!}{(n-k)!}$$

With permutations, the order in which individuals are selected matters. When order doesn't matter, we use **combinations.** The number of combinations of n individuals taken k at a time is

$$_nC_k = \frac{_nP_k}{k!} = \frac{n!}{k!(n-k)!}$$

To solve a complicated probability problem, begin by counting all possible outcomes. Use a tree diagram, the multiplication counting principle, permutations, or combinations as needed.

Section 8.2 Exercises

8.39 Radio station call signs, I In 1912, the U.S. government began issuing licenses to radio stations. Each station was given a unique three-letter "call sign." By international agreement, the U.S. eventually received rights to all call signs beginning with the letters W, N, and K. Radio stations in the western United States were given call signs starting with K. Stations in the east were given call signs starting with W. (N was reserved for use by the U.S. Navy.)
(a) How many three-letter call signs start with the letter W? Show your method.
(b) How many three-letter call signs starting with W or K were available for U.S. radio stations? Show your work.

8.40 Radio station call signs, II Refer to Exercise 8.39. By 1922, there were more applications for radio station licenses than the number of three-letter call signs available. A radio station in New Orleans applied for and was granted the call sign WAAB.
(a) How many four-letter call signs start with the letter W? Show your method.
(b) In fact, no call sign was permitted in which the same letter appeared three times in a row. (That's why the New Orleans station got WAAB, not WAAA.) How many four-letter call signs starting with W are possible with this restriction? Show your work.
(c) How many four-letter call signs start with W or K and meet the restriction in (b)? Show your method.

8.41 Who gets picked? Mr. Paradise uses a calculator program to select students at random to answer questions during class. Each of the 28 students in his class

is equally likely to be chosen every time he runs the program. Suppose Mr. Paradise runs the program twice. The first student selected gets to answer Question 1, and the second student selected (who might be the same student) gets to answer Question 2.
(a) Use the multiplication counting principle to find the number of different ways that the program can choose who answers these two questions.
(b) Now find the number of ways in which two different students are chosen to answer the two questions.
(c) What's the probability that the program selects two different students in the class for the first two questions?

8.42 Who else gets picked? Refer to Exercise 8.41. This time, Mr. Paradise uses his program three times in a row. Find the probability that three different students in the class are chosen.

8.43 Pick Six In the New Jersey "Pick Six" lotto game, a player chooses six different numbers from 1 to 49. The six winning numbers for the lottery are chosen at random. If the player matches all six numbers, she wins the jackpot, which starts at \$2 million. Find the probability of picking all six winning numbers. Show your work.

8.44 Pick Six again Refer to Exercise 8.43. Find the probability that an unlucky player does not match any of the six winning numbers. Show your work.

8.45 Roll 5 dice Suppose you roll 5 six-sided dice at one time.
(a) Find the probability that all 5 dice show the same number of spots on the up-faces. Show your work.
(b) What's the probability that all 5 dice show a different number of spots on the up-faces? Show your work.

8.46 *Deal or No Deal?* In Activity 8.2 (page 387), you explored the TV game show *Deal or No Deal*. Now, you should be ready to analyze some of the probabilities involved in the game. Suppose you get very lucky and choose the case with \$1 million in it as your case. In the first round of the game, you must choose 6 of the remaining 25 cases to open.
(a) In how many different ways can you select 6 cases containing dollar amounts on the left side of the game board? What's the probability that this happens?
(b) If you happen to choose the 6 cases with the lowest amounts of money in them (\$0.01 to \$50), how much should the banker offer you? Justify your answer.

8.3 Binomial Distributions

ACTIVITY 8.3A

Rock-paper-scissors-art?

In 2005, Takashi Hashiyama, president of a successful Japanese electronics company, wanted to sell his company's \$20 million art collection. He decided to use one of two famous auction companies—Christie's or Sotheby's—to offer the collection for sale. Since he felt that the two companies were equally qualified, Mr. Hashiyama invited representatives from Christie's and Sotheby's to compete in a game of rock-paper-scissors.

ACTIVITY 8.3A *(continued)*

ACTIVITY 8.3A *(continued)*

© 2007 Vic Lee, distributed by King Features Syndicate, Inc.

(Using games of chance to make such decisions is common in Japanese business.) The winner would get to sell the collection. Who won, and what was their strategy?

In this Activity, you will learn more about the game of rock-paper-scissors via simulation and the rules of probability. Let's review the basic rules of the game before you begin. A "round" of the game consists of a "throw" by each of the two opposing players. There are three possible throws that a player may choose: rock, paper, or scissors. If both players make the same "throw" (for example, rock and rock), then the round is a tie. Otherwise, the winner of the round is determined as follows: rock breaks scissors, scissors cuts paper, and paper covers rock.

1. Play 10 rounds of rock-paper-scissors against your opponent. Record the result of each round in a table like the one shown.

Round	My throw	Opponent's throw	Outcome
1	Rock	Paper	Lose
2	Scissors	Scissors	Tie
3	Scissors	Paper	Win
...			

2. Count the number of rounds won by each player. Is there a winner?

3. Based on your results, does it appear that both players were selecting among the three possible throws at random in each round? Or do you think some more sophisticated "strategy" was being used by one or both players? Justify your answer.

4. Pool your results with those of your classmates to determine the highest number of rounds won by any student in the class. Do you think it's possible that a player could win this many rounds if both players were selecting throws at random in each round? Explain.

5. Work with your partner to design and carry out a simulation to help answer the question posed in Step 4. You may use your calculator or Table B of random digits. Perform 10 trials of your simulation. What conclusion do you draw? Why?

6. Now pool results with your classmates. Does your conclusion change? Why or why not?

What happened in the $20 million game of rock-paper-scissors described earlier? Christie's got a little help from the 11-year-old twin daughters of a company executive. From their experience playing the game many times at school, the girls recommended "scissors." The Christie's executive took their advice. Sotheby's chose "paper," which gave Christie's the win and the art collection.

Binomial settings and the binomial theorem

What do each of the following scenarios have in common?

- Toss a coin 5 times. Count the number of heads.
- Spin a roulette wheel 8 times. Record how many times the ball lands in a red slot.
- Play 10 games of rock-paper-scissors with a friend in which you both use a chance process (like rolling a die) to determine what you throw each time. Count how many games you win.
- Take a random sample of 100 babies born in U.S. hospitals today. Record the number of females.

In each case, we're performing repeated trials of the same chance process. The number of trials is fixed in advance. In addition, the outcome of one trial has no effect on the outcome of any other trial. That is, the trials are *independent.* We're interested in the number of times that a specific outcome (we'll call it a "success") occurs. Our chances of getting a "success" are the same on each trial. When these conditions are met, we have a **binomial setting.**

Binomial setting

A **binomial setting** arises when we perform several independent trials of the same chance process and record the number of times that a particular outcome occurs. The four conditions for a binomial setting are

- **Binary?** The possible outcomes of each trial can be classified as "success" or "failure." Note that there can be more than two possible outcomes per trial—a roulette wheel has numbered slots of *three* colors: red, black, and green. If we define "success" as having the ball land in a red slot, then "failure" occurs when the ball lands in a black or a green slot.
- **Independent?** Trials must be independent; that is, the outcome of one trial must not have any effect on the outcome of any other trial.
- **Number?** The number of trials n of the chance process must be fixed in advance.
- **Success?** On each trial, the probability p of success must be the same.

The boldface letters in the previous box give you a helpful way to remember the conditions for a binomial setting: just check the BINS!

Example 8.12 Is it a binomial setting?

Here are three scenarios involving chance behavior. Which describe binomial settings?

(a) Shuffle a deck of cards. Turn over the top card. Put the card back in the deck, and shuffle again. Repeat this process 10 times. Count the number of aces you observe.

- **Binary?** "Success" = get an ace; "Failure" = don't get an ace.
- **Independent?** Since you are replacing the card in the deck and shuffling each time, the result of one trial does not affect the outcome of any other trial.
- **Number?** You're performing $n = 10$ trials of this chance process.
- **Success?** The probability of a success is 4/52 on each trial.

So this is a binomial setting.

(b) Shuffle a deck of cards. Turn over the first 10 cards, one at a time. Count the number of aces you observe.

- **Binary?** "Success" = get an ace; "Failure" = don't get an ace.
- **Independent?** In this case, the trials are not independent. If the first card you turn over is an ace, then the next card is less likely to be an ace. That is, P(first card is an ace) = 4/52, but P(second card is an ace | first card is an ace) = 3/51.

Since the trials of this chance process are not independent, this is not a binomial setting.

(c) Shuffle a deck of cards. Turn over the top card. Put the card back in the deck, and shuffle again. Repeat this process until you get an ace. Count the number of trials required.

- As in (a), the **B**inary and **I**ndependent conditions are satisfied.
- **N**umber? The number of trials is not set in advance. You could get an ace on the first card you turn over, or it may take many cards to get an ace.

Since there is no fixed number of trials, this is not a binomial setting.

Part (c) of Example 8.12 raises another important point about binomial settings. In addition to checking the BINS, make sure that you're being asked to count the number of successes in a certain number of trials. In Part (c), for example, you're asked to count the number of *trials* until you get a success. That can't be a binomial setting.

In a binomial setting, we can define a random variable (say X) as the number of successes in n independent trials. What's the probability distribution of X? Let's see if an example can help shed some light on this question.

Example 8.13 1 in 6 wins a prize

As a special promotion for its 20-ounce bottles of soda, a soft drink company printed a message on the inside of each cap. Some of the caps said "Please try again," while others said "You're a winner!" The company advertised the promotion with the slogan "1 in 6 wins a prize." Assume that the company is telling the truth and that each bottle of soda has a 1/6 chance of having a cap that says "You're a winner!" Suppose that Alan and five of his friends—Bart, Chloe, Delores, Ed, and Frankie—each buy one 20-ounce bottle. What's the probability that none of them wins a prize?

First, let's be sure that this is a binomial setting. We'll check the BINS. Note that a trial consists of buying a soda and looking under the cap. The chance is involved in which soda bottles receive caps that say "You're a winner!"

- **B**inary? "Success" = cap says "You're a winner!"; "Failure" = cap says "Please try again."
- **I**ndependent? If the company is telling the truth, then the message printed on one bottle's cap should have no effect on which message appears on another bottle's cap.
- **N**umber? Between Alan and his friends, there are $n = 6$ trials of this chance process.
- **S**uccess? Each bottle has a $p = 1/6$ chance of having a cap that says "You're a winner!"

Since the conditions are met and we're counting the number of successes in a fixed number of trials, this is a binomial setting.

Let X = the number of prizes won by Alan and his friends. We want to find $P(X = 0)$. The complement rule tells us that the probability that a particular bottle is *not* a winner is $1 - 1/6 = 5/6$. If $X = 0$, that means that none of the six wins a prize. In other words, all six bottles must have caps that say "Please try again." Since the trials are independent, the chance that this happens is

$$(5/6)(5/6)(5/6)(5/6)(5/6)(5/6) = 0.335$$

There's about a 1/3 chance that none of the six wins a prize.

In Example 8.13, we found the probability that Alan and his five friends won nothing in the soft drink company's "1 in 6 wins a prize" contest. What's the probability that exactly one of them wins a prize? One way that can happen is if Alan's bottle of soda says "You're a winner!" while all five friends' bottles say "Please try again." The chance that this happens is

$$\underset{\text{Alan}}{(1/6)} \times \underset{\text{Bart}}{(5/6)} \times \underset{\text{Chloe}}{(5/6)} \times \underset{\text{Delores}}{(5/6)} \times \underset{\text{Ed}}{(5/6)} \times \underset{\text{Frankie}}{(5/6)}$$

since the trials are independent. Another way this can happen is if Bart wins a prize, but no one else does. The probability of this happening is

$$\underset{\text{Alan}}{(5/6)} \times \underset{\text{Bart}}{(1/6)} \times \underset{\text{Chloe}}{(5/6)} \times \underset{\text{Delores}}{(5/6)} \times \underset{\text{Ed}}{(5/6)} \times \underset{\text{Frankie}}{(5/6)}$$

Note that the probability that Bart is the only winner in the group is the same as the probability that Alan is the only winner. There are four more possibilities to consider—the ones in which Chloe, Delores, Ed, and Frankie are the only winners. Of course, the probability will be the same for each of those cases. So the chance that exactly one of the six wins a prize is

$$6 \times (1/6) \times (5/6) \times (5/6) \times (5/6) \times (5/6) \times (5/6)$$
$$= 6 \cdot \left(\frac{1}{6}\right)^1 \cdot \left(\frac{5}{6}\right)^5 = 0.402$$

Where did the 6 come from? It's the number of ways to choose which one of the six friends wins a prize. In Section 8.2, we wrote this as ${}_6C_1$, the number of combinations of 6 things taken 1 at a time. Using this notation, we can rewrite the above probability as

$$P(X = 1) = {}_6C_1 \cdot \left(\frac{1}{6}\right)^1 \cdot \left(\frac{5}{6}\right)^5 = 0.402$$

Can you see where we're headed? What's the probability $P(X = 2)$ that exactly two of the six friends win a prize? One way this can happen is for Alan and Bart to win a prize, but the other four to get a soda that says "Please try again." The chance that this occurs is

$$\underset{\text{Alan}}{(1/6)} \times \underset{\text{Bart}}{(1/6)} \times \underset{\text{Chloe}}{(5/6)} \times \underset{\text{Delores}}{(5/6)} \times \underset{\text{Ed}}{(5/6)} \times \underset{\text{Frankie}}{(5/6)}$$

Since there are ${}_6C_2$ ways to choose which two of the six friends win prizes, the probability we want is

$$P(X = 2) = {}_6C_2 \cdot \left(\frac{1}{6}\right)^2 \cdot \left(\frac{5}{6}\right)^4 = 0.201$$

Our work on this problem suggests a more general result, known as the **binomial theorem.**

Binomial theorem

In a binomial setting involving n independent trials of a chance process with probability p of success on each trial, the probability of getting exactly k successes is given by

$$P(X = k) = {}_nC_k \cdot p^k \cdot (1 - p)^{n-k}$$

Exercises

In Exercises 8.47 to 8.50, determine whether the given scenario describes a binomial setting. If so, confirm that all of the required conditions are met. If not, explain why not.

8.47 Sowing seeds Seed Depot advertises that 85% of its flower seeds will germinate (grow). Suppose that the company's claim is true. Judy buys a packet with 20 flower seeds from Seed Depot and plants them in her garden. She counts how many of the seeds germinate.

8.48 Long or short? Put the names of all the students in your class in a hat. Mix them up, and draw four names without looking. Count the number of these students whose last names have more than six letters.

8.49 Lefties, I About 10% of people are left-handed. Suppose you select students at random from your school, one at a time, until you find one who is left-handed. Record the number of students you selected.

8.50 Lefties, II About 10% of people are left-handed. Select 15 students at random from your school and count the number who are left-handed.

8.51 All 6 win Refer to Example 8.13 (page 402). Find the probability that Alan and all five of his friends win a prize. Show your work.

8.52 Three winners Refer to Example 8.13 (page 402). Find the probability that exactly three of the six friends win a prize. Show your work.

Binomial distributions

In Example 8.13, we examined the chance process of Alan and his five friends playing the "1 in 6 wins a prize" soft drink contest. We defined the random variable X = the number of prizes won by Alan and his friends. Earlier, we found $P(X = 0)$, $P(X = 1)$, and $P(X = 2)$. If you completed Exercises 8.51 and 8.52, then you found $P(X = 6)$ and $P(X = 3)$. By the binomial theorem,

$$P(X = 4) = {}_6C_4 \cdot \left(\frac{1}{6}\right)^4 \cdot \left(\frac{5}{6}\right)^2 = 0.008$$

and

$$P(X = 5) = {}_6C_5 \cdot \left(\frac{1}{6}\right)^5 \cdot \left(\frac{5}{6}\right)^1 = 0.0006$$

Here is the completed probability distribution for X:

Value:	0	1	2	3	4	5	6
Probability:	0.335	0.402	0.201	0.054	0.008	0.0006	0.00002

Figure 8.7 shows a histogram of this probability distribution, which we call a **binomial distribution.** It has a clear right-skewed shape. Why does this make sense? Each 20-ounce bottle of soda has only a 1/6 chance of being a prize winner. If Alan and his five friends buy a total of six bottles, then it's much more likely that 0, 1, or 2 of them win a prize than that 3 or more of them win prizes.

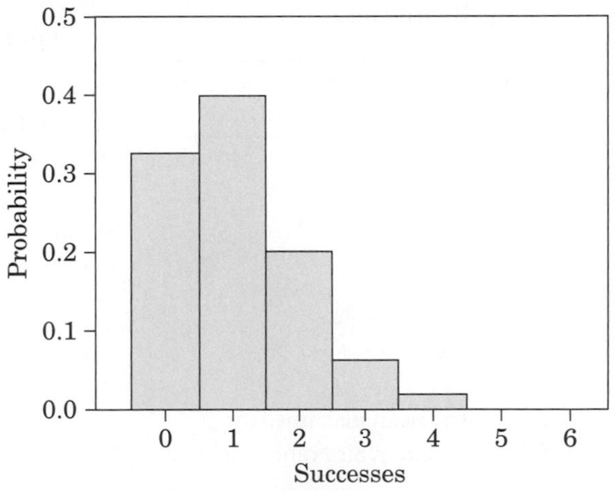

Figure 8.7 A histogram showing the probability distribution of X = number of prizes won by Alan and his friends in the "1 in 6 wins" soft drink contest.

How many prizes would we expect Alan and his friends to win? The **expected value** of X is

$$\begin{aligned} \text{expected value} &= 0 \cdot 0.335 + 1 \cdot 0.402 + 2 \cdot 0.201 + 3 \cdot 0.054 \\ &\quad + 4 \cdot 0.008 + 5 \cdot 0.0006 + 6 \cdot 0.00002 \\ &= 1.00 \end{aligned}$$

This should make sense, too! If six people play a game that gives each person a 1/6 chance of winning, then we'd expect one of them to win, on average, if the process were repeated many times.

We can generalize this result for any binomial setting.

Expected value for a binomial distribution

If a chance process involves n independent trials, with success probability p on each trial, the expected number of successes is

$$\textbf{expected value} = np$$

ACTIVITY 8.3B

Galton's Board applet

Sir Frances Galton invented the "bean machine" in the late 1880s to demonstrate what happens when a large number of beans or balls are dropped down a chute and allowed to bounce off a series of pegs. Figure 8.8 shows a small version of such a "bean machine," which we now tend to call a quincunx or **Galton's Board.** In this Activity, you will use a Java applet to see how Galton's Board works.

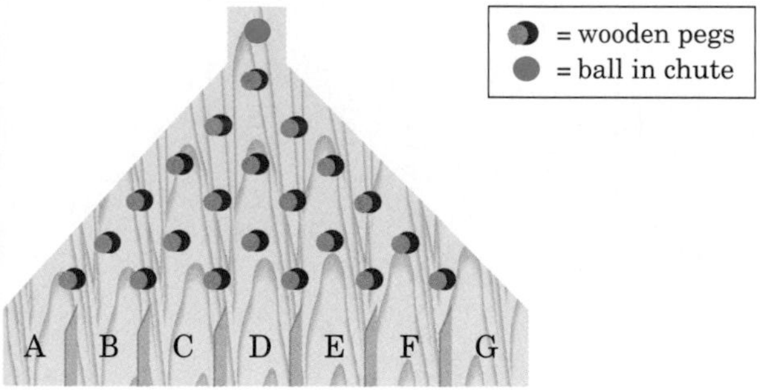

Figure 8.8 An example of a "bean machine," or Galton's Board, named for its inventor, Sir Francis Galton.

1. Go to the book's Web site, **www.whfreeman.com/sta2e**, and launch the *Galton's Board* applet.

2. Set the number of trials at 1. Click "Start" to have the applet drop one ball. Watch as the ball bounces off the pegs on its way down. What do you think is the probability that the ball falls to the left when it hits a peg? To the right?

3. Reset the applet, and set the number of trials at 100. Click "Start." After watching the 100 balls bounce down the board, what do you notice about where they have landed? Which slot(s) seem most likely? Least likely? Does that make sense to you?

Let's look more closely at Figure 8.8. When a ball is released, it hits the first peg, and then has an equal chance of falling to the left or to the right. The ball then proceeds down to the next row of pegs, hits one of them, and falls left or right with equal probability. This continues until the ball settles in one of the "bins" at the bottom of the board.

What is the probability that a ball lands in Slot A? For this to happen, the ball must fall to the left all 6 times it hits a peg. The probability that this will happen is

$$\left(\frac{1}{2}\right)^6 = 0.015625$$

Can you figure out the probability that a ball lands in Slot G?

What's the probability that a ball lands in Slot B? For this to happen, the ball must fall to the right once and to the left five times, in some order. If we designate L to mean "falls to the left" and R to mean "falls to the right," then there are six possible "routes" for the ball to travel to land in Slot B:

RLLLLL LRLLLL LLRLLL LLLRLL LLLLRL LLLLLR

The probability that the ball follows the route RLLLLL is

$$\left(\frac{1}{2}\right)^1 \cdot \left(\frac{1}{2}\right)^5$$

Likewise, the probability that the ball follows the route LRLLLL is

$$\left(\frac{1}{2}\right)^1 \cdot \left(\frac{1}{2}\right)^5$$

In fact, this is the probability that it follows any of the six specific routes. So the chance that the ball will land in Slot B is given by

$$6 \cdot \left(\frac{1}{2}\right)^1 \cdot \left(\frac{1}{2}\right)^5 = 0.09375$$

Does this calculation look familiar? It uses the binomial theorem with $n = 6$, $p = 1/2$, and $k = 1$. But why is this a binomial setting? Consider the chance process of a ball bouncing its way down Galton's Board. At each row of the board, the ball encounters a peg and falls either left or right. If we think of each peg as a "trial," then we can define a "success" as falling to the right. There are a fixed number of trials: $n = 6$. The trials are independent, and the probability of success on each trial is 1/2. If we define the variable X = the number of times the ball falls to the right on its way down the board, then X has a binomial distribution. Let's find the probability distribution of X.

Example 8.14 Galton's Board: finding probabilities

Earlier, we found the probability that the ball falls to the left all six times: 0.015625. That's $P(X = 0)$. We also determined that $P(X = 1) = 0.09375$. Now let's find the probability that a ball lands in Slot C. For this to happen, the ball must fall to the right at two pegs and to the left at four pegs. So we want $P(X = 2)$. By the binomial theorem,

$$P(X = 2) = {}_6C_2 \cdot \left(\frac{1}{2}\right)^2 \cdot \left(\frac{1}{2}\right)^4 = 0.234375$$

Similarly, the probability that a ball lands in Slot D is

$$P(X = 3) = {}_6C_3 \cdot \left(\frac{1}{2}\right)^3 \cdot \left(\frac{1}{2}\right)^3 = 0.3125$$

What's the probability that the ball lands in Slot E? It's

$$P(X = 4) = {}_6C_4 \cdot \left(\frac{1}{2}\right)^4 \cdot \left(\frac{1}{2}\right)^2 = 0.234375$$

That's the same as $P(X = 2)$. Does it make sense to you that a ball is equally likely to land in Slot C or Slot E? By similar logic, the chance that a ball lands in Slot F is

$$P(X = 5) = {}_6C_5 \cdot \left(\frac{1}{2}\right)^5 \cdot \left(\frac{1}{2}\right)^1 = 0.09375 = P(X = 1)$$

and the chance that a ball lands in Slot G is

$$P(X = 6) = {}_6C_6 \cdot \left(\frac{1}{2}\right)^6 \cdot \left(\frac{1}{2}\right)^0 = 0.015625 = P(X = 0)$$

Putting it all together, the probability distribution of X is

Value x_i:	0	1	2	3	4	5	6
Probability p_i:	0.015625	0.09375	0.234375	0.3125	0.234375	0.09375	0.015625

Figure 8.9 displays this probability distribution. Notice the perfectly symmetric shape!

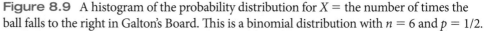

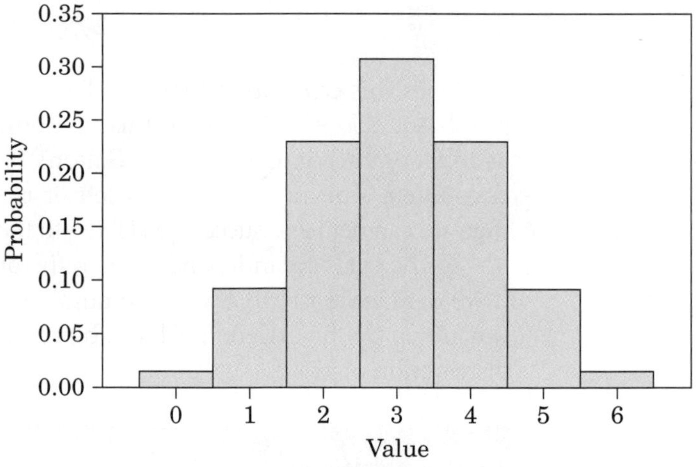

Figure 8.9 A histogram of the probability distribution for X = the number of times the ball falls to the right in Galton's Board. This is a binomial distribution with $n = 6$ and $p = 1/2$.

Exercises

8.53 Blood types, I In the United States, 44% of adults have type O blood. Suppose we choose seven U.S. adults at random.

(a) What's the probability that all seven people have type O blood? Show your method.

(b) What's the probability that exactly four of those selected have type O blood? Show your method.

8.54 Blood types, II Refer to Exercise 8.53. Let X = the number of people chosen with type O blood.

(a) Find the probability distribution of X. Display it in a table.

(b) Graph the probability distribution of X. Describe what you see.

8.55 Sowing seeds again Seed Depot advertises that 85% of its flower seeds will germinate (grow). Suppose that the company's claim is true. Judy buys a packet with 20 flower seeds from Seed Depot and plants them in her garden.

(a) How many seeds should Judy expect to germinate? Explain.

(b) Find the probability that exactly the expected number of seeds from (a) actually grow. Show your work.

(c) Find the probability that fewer than the expected number of seeds from (a) actually grow. Show your work.

8.56 Sowing more seeds Refer to Exercise 8.55.

(a) If only 15 seeds actually germinate, should Judy be suspicious that the company's claim is not true? Compute an appropriate probability to support your answer.

(b) If only 12 seeds actually germinate, should Judy be suspicious that the company's claim is not true? Compute an appropriate probability to support your answer.

8.57 Coin tosses Imagine that you toss a fair coin six times.

(a) Find the probability that you get three heads and three tails. Show your work.

(b) Find the probability that you get four heads and two tails. Show your work.

8.58 Bigger bean machine Draw a Galton's Board with two more rows of pegs than the one in Figure 8.8. Label the slots at the bottom of the board A through I from left to right.

(a) Find the probability that a ball lands in Slot C. Show your work.

(b) Find the probability that a ball lands in Slot F. Show your work.

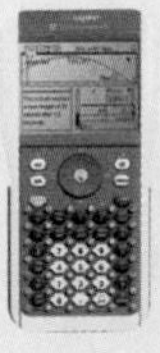

CALCULATOR CORNER Binomial distributions

Both the TI-84 and the TI-Nspire have built-in functions that can help you find probabilities involving binomial distributions. In a binomial setting involving n independent trials with success probability p on each trial, we can find the probability of getting exactly k successes using the binomial theorem:

$$P(X = k) = {}_nC_k \cdot p^k \cdot (1 - p)^{n-k}$$

The equivalent command on the calculator is `binompdf(n,p,k)`. There is also a command (`binomcdf`) for finding the probability of getting k or fewer successes in n trials.

Let's use the "1 in 6 wins a prize" binomial setting from Example 8.13 (page 402) to illustrate how these commands can be used.

CALCULATOR CORNER *(continued)*

CALCULATOR CORNER *(continued)*

1. To find the probability that Alan and his friends win a total of two prizes, $P(X = 2)$:

TI-84

- From the home screen, press 2nd VARS (DISTR) and choose `binompdf(`.
- Complete the command `binompdf(6,1/6,2)` and press ENTER.

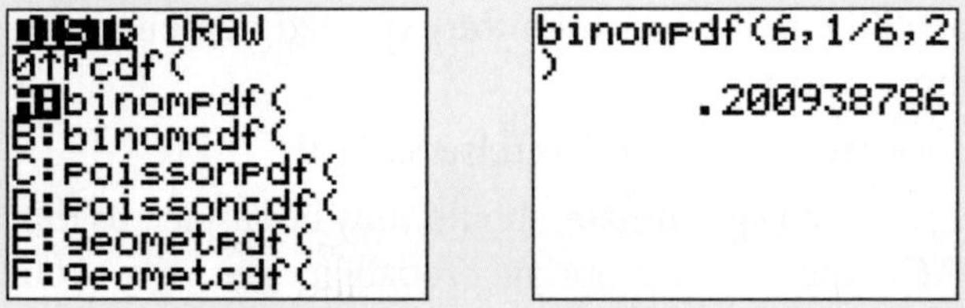

TI-Nspire

- On a Calculator page, press menu, choose `4:Probability`, `5:Distributions...`, and `D:Binomial Pdf`.
- Enter the number of trials (6), success probability (1/6), and X Value (2) in the dialog box as shown. Then choose OK.

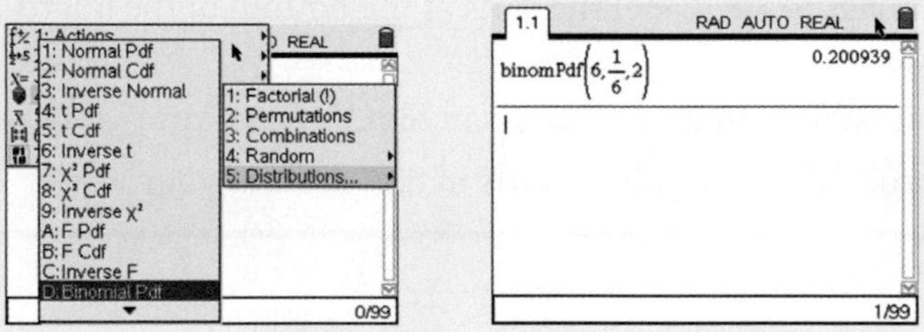

2. To find the probability that Alan and his friends win two prizes or fewer, $P(X \leq 2)$:

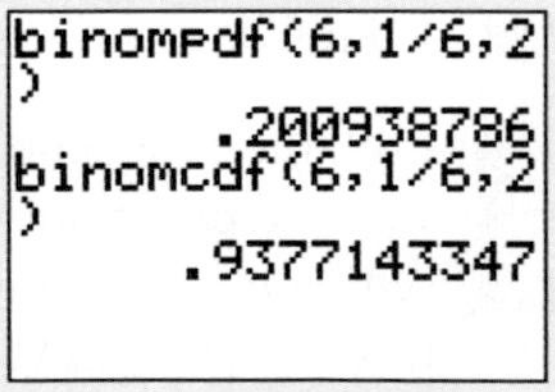

TI-84

- Press 2nd VARS (DISTR) and choose `binomcdf(`.
- Complete the command `binomcdf(6,1/6,2)` and press ENTER.

TI-Nspire

- Press menu, choose `4:Probability, 5:Distributions...`, and `E: Binomial Cdf`.
- Enter the number of trials (6), success probability (1/6), Lower Bound (0), and Upper Bound (2) in the dialog box as shown.

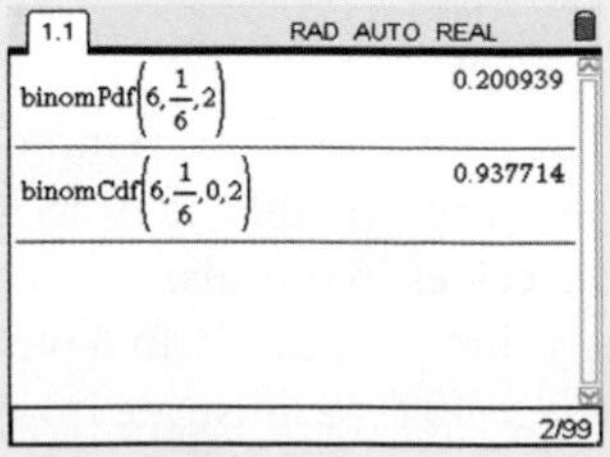

3. Can you figure out how to use the `binomcdf` command to find the probability that Alan and his friends win two or more prizes, $P(X \geq 2)$?

APPLICATION 8.3

The lady tasting tea

© Pietus/Dreamstime.com

One summer afternoon in the late 1920s, a tea party was held in Cambridge, England. A lady at the party claimed that she could tell the difference between tea to which milk had been added and milk to which tea had been added. Many of those at the party were doubtful. A few of the more open-minded gentlemen thought about designing an experiment to test the lady's claim. One of those gentlemen was Ronald A. Fisher, widely considered to be the father of experimental design. With his guidance, the tea-tasting experiment was carried out successfully using 10 cups of tea.[5] Want to know what happened? Not so fast! You'll appreciate the ending of the story better if you learn about the experiment first.

QUESTIONS

1. If you had been at the party that day, how would you have designed an experiment using 10 cups of tea to test the lady's claim? Explain the reasoning behind your design.

Assume for a moment that the lady can't actually tell the difference between the two types of beverages—tea with milk added and milk with tea added. Then for each of the 10 cups presented in the experiment, she would just be guessing. Suppose the experiment is conducted in such a way that each cup presented to the lady is equally likely to contain either beverage. Let X be the number of cups of tea that the lady identifies correctly.

2. Show that this is a binomial setting. (Check the BINS.)

3. Find and interpret $P(X = 5)$.

4. Find and interpret $P(X > 5)$.

5. How many correct identifications would you have required before concluding that the lady was not just guessing? Justify your answer using probability.

6. In the actual experiment, the lady identified all 10 cups of tea correctly. Fisher and his friends concluded that the lady could actually tell how the tea had been prepared. Do you agree? Why or why not?

Section 8.3 Summary

A process that involves counting the number of successful outcomes in a set number of independent trials of the same chance phenomenon is a **binomial setting.** To confirm that a chance process describes a binomial setting, check the four required conditions—**B**inary, **I**ndependent, **N**umber, and **S**uccess. Binary means that the outcomes can be classified as either "success" or "failure." The trials of the chance process must be independent, and the number of trials must be fixed in advance. On each trial, the probability of success p must be the same.

In a binomial setting, we often define a numerical variable (say X) to represent the number of successes. The **binomial theorem** tells us how to calculate the probability of getting exactly k successes in n independent trials, each with probability p of success:

$$P(X = k) = {}_nC_k \cdot p^k \cdot (1 - p)^{n-k}$$

The probability distribution of X is called a **binomial distribution.**

Galton's Board can be used to show the possible routes a ball can take as it bounces down through several rows of pegs. To find the probability that a ball lands in a particular slot on Galton's Board, use the binomial theorem.

Section 8.3 Exercises

8.59 Going to the dogs, I *Ladies Home Journal* magazine reported that 66% of all dog owners greet their dog before greeting their spouse or children when they return home at the end of the workday. Suppose that 12 dog owners are selected at random.
(a) Show that the four requirements for a binomial setting are met.
(b) Find the probability that exactly 7 of the 12 dog owners greet their dog first when they arrive home. Show your work.
(c) Find the probability that at least 5 of the 12 dog owners greet their dog first when they arrive home. Show your work.

8.60 Going to the dogs, II Refer to the previous exercise.
(a) What is the expected number of dog owners who greet their dog first when they arrive home? Justify your answer.
(b) Make a table that shows the probability distribution of the variable X = the number of owners who greet their dog first.
(c) Construct a histogram that shows the probability distribution for X. Describe what you see.

8.61 More lefties Suppose that exactly 10% of the students at your school are left-handed. Imagine selecting an SRS of 15 students from the school population.
(a) Find the probability that exactly 3 students in the sample are left-handed. Show your work.
(b) Find the probability that 3 or fewer students in the sample are left-handed. Show your work.
(c) Would you be surprised if the SRS contained 4 left-handed students? Justify your answer using probability.

8.62 Righties and lefties Refer to the previous exercise. Let X = the number of right-handed students in your sample.
(a) Find and interpret $P(X = 12)$. How does this compare with your answer to 8.61(a)?
(b) Write a probability expression involving X that is equivalent to the question in 8.61(b).
(c) Find the probability that the number of right-handed students in your sample is more than the expected value of X. Show your work.

8.63 Orange M&M's, I According to the Mars candy company, 20% of its plain M&M's candies are orange. Assume the company's claim is true. Suppose you reach into a large bag of plain M&M's (without looking) and pull out eight candies.
(a) Would you be surprised if none of the candies were orange? Compute an appropriate probability to support your answer.
(b) Would you be surprised if exactly two of the candies were orange? Compute an appropriate probability to support your answer.

(c) Would you be surprised if four of the candies were orange? Compute an appropriate probability to support your answer.

8.64 Orange M&M's, II Refer to Exercise 8.63. Let X = the number of orange M&M's chosen.

(a) Find the probability distribution of X. Display it in a table.

(b) Graph the probability distribution of X. Describe what you see.

8.65 Galton's Board, I Refer to Example 8.14 (page 407). A clever person has redesigned the "bean machine" of Figure 8.8. The pegs have been adjusted so that a ball is more likely to fall to the right (0.6) than to the left (0.4).

(a) Find the probability that a ball lands in Slot A. Show your work.

(b) Find the probability that a ball lands in Slot C. Show your work.

(c) In which slot is the ball most likely to land? Justify your answer with probability.

8.66 Galton's Board, II Refer to the previous exercise. Let X = the number of times that the ball falls to the right.

(a) Make a histogram that shows the probability distribution of X. How is this graph different from the one in Figure 8.9 (page 408)?

(b) Suppose 100 balls are dropped down the chute, one at a time, onto this modified Galton's Board. How many of the balls would you expect to land in Slot D? Explain.

CHAPTER 8 REVIEW

Sometimes it is helpful to define a random variable that summarizes the outcomes of a chance process. A probability distribution specifies the values that the random variable can take and how often it takes those values. We can use the probability distribution to determine the expected value of the random variable, a number that describes the long-run average outcome. If the random variable can take all values in an interval, we find probabilities using areas under a density curve (like the Normal curve).

To solve a probability problem, we often need to determine the number of possible outcomes of a chance process. We can use the multiplication counting principle to do this when the chance behavior occurs in several stages. Both permutations and combinations involve choosing individuals from an existing population. With permutations, we're interested in the order of selection; with combinations, we're not.

A binomial setting occurs when several independent trials of the same chance process are performed. If we define a variable to represent the number of trials on which a successful outcome occurs, that variable follows a binomial distribution. We can use the binomial theorem to find probabilities associated with a binomial distribution.

Here is a review list of the most important skills you should have developed from your study of this chapter.

A. EXPECTED VALUES AND PROBABILITY DISTRIBUTIONS

1. Understand the idea of expected value as the average of numerical outcomes in a great many repetitions of a random phenomenon.
2. Find the expected value of a random variable from a probability distribution.

3. Estimate an expected value by repeating a simulation many times.

4. When probabilities are assigned by a Normal curve, find the probability of an event by finding an area under the curve.

B. COUNTING TECHNIQUES AND PROBABILITY

1. Use the multiplication counting principle to determine the number of possible outcomes of a chance process involving multiple steps.

2. Count the number of distinct arrangements of a group of individuals using permutations.

3. Count the number of distinct selections of a group of individuals using combinations.

4. Distinguish counting situations in which order matters from those where it doesn't. Explain the relationship between ${}_nP_k$ and ${}_nC_k$.

5. Solve probability problems using the multiplication counting principle, permutations, and combinations.

C. BINOMIAL DISTRIBUTIONS

1. Determine whether a given chance process describes a binomial setting.

2. Use the binomial theorem to calculate probabilities involving a binomial distribution.

3. Compute the expected value of a binomial distribution.

CHAPTER 8 REVIEW EXERCISES

8.67 Pastabilities A popular Italian restaurant chain claimed that you could get "42 different pasta-sauce combinations." Explain what you think they meant by this claim.

8.68 What are the chances? Open your local telephone directory to any page in the residential listing. Look at the last four digits of each telephone number, the digits that specify an individual number within an exchange given by the first three digits. Note the first of these four digits in each of the first 100 telephone numbers on the page.
(a) How many of the digits were 1, 2, or 3? What is the approximate probability that the first of the four "individual digits" in a telephone number is 1, 2, or 3?
(b) If all 10 possible digits had the same probability, what would be the probability of getting a 1, 2, or 3? Based on your work in (a), do you think the first of the four "individual digits" in telephone numbers is equally likely to be any of the 10 possible digits? Justify your answer.

8.69 Phone numbers, I In North America, phone numbers have the form XXX-XXX-XXXX. The first three digits give the area code, and the second three digits indicate the exchange.
(a) Nick lives in western North Carolina, where the area code is 828. If there were no restrictions on the remaining seven digits, how many phone numbers would be possible in the 828 area code? Show your method.
(b) In fact, exchanges cannot begin with 0 or 1. How many possible numbers are there in the 828 area code, subject to this restriction? Show your method.

(c) How many of the possible phone numbers in (b) include at least one 9? Show your work.

8.70 Phone numbers, II Refer to the previous exercise. Suppose that a computer assigns the seven digits in Nick's phone number at random, subject to the restriction in Exercise 8.69(b). Find the probability that Nick's phone number contains no repeated digits. Show your work.

8.71 Free throws A basketball player claims to make 80% of her free throws. Suppose this probability is the same for each free throw she attempts. In one game, she shoots 7 free throws. Let $X =$ the number of free throws on which she makes a basket.
(a) Show that this is a binomial setting.
(b) Find the probability that she makes 4 or fewer baskets on 7 free throws. Show your work.
(c) If you were watching this game and saw the player get a basket on 4 out of 7 free throws, would you be suspicious about her claim of being an 80% free-throw shooter? Justify your answer.

8.72 Getting a flush If you deal five cards from a well-shuffled deck, what's the probability that they're all the same suit? (Card players call this kind of hand a "flush.")
(a) Start by calculating the probability that all five cards are diamonds. Show your work.
(b) Now find the probability of getting a flush. Show your work.

8.73 We like opinion polls Are Americans interested in opinion polls about the major issues of the day? Suppose that 40% of all adults are very interested in such polls. (According to sample surveys that ask this question, 40% is about right.) A polling firm chooses an SRS of 1015 people. If they do this many times, the percent of the sample who say they are very interested will vary from sample to sample following a Normal distribution with mean 40% and standard deviation 1.5%.
(a) What is the probability that one such sample gives a result within $\pm 3\%$ of the truth about the population? Show your method.
(b) Find the probability that one sample misses the truth about the population by 4% or more. (This is the probability that the sample result is either less than 36% or greater than 44%.) Show your work.

8.74 It pays to eat cereal Some time ago, Cheerios cereal boxes displayed a dollar bill on the front of the box and a cartoon character who said, "Free $1 bill in every 20th box." Mae Lee wants to know how many boxes of Cheerios she can expect to buy in order to get one of the "free" dollar bills.
(a) Describe how you would establish a correspondence between boxes of cereal that do and do not contain a dollar bill and numbers from Table B.
(b) Describe how you could use your calculator to simulate buying boxes of Cheerios and checking for dollar bills.
(c) Carry out a simulation to answer Mae Lee's question. Perform enough repetitions so that when you pool your results with your classmates, you will have at least 25 repetitions (don't forget that the more repetitions, the better the results).

8.75 Keno, I The game of Keno is played with 80 balls numbered 1 through 80. During each play of the game, the casino selects 20 of the balls at random. In a simple version of the game, a player pays $1 and chooses one number between

1 and 80. If the player's number matches one of the 20 randomly selected balls, the player wins \$3 (and gets the original \$1 back). If not, the player loses the original \$1 bet.

(a) Find the probability that the player's chosen number is a winner. Show your work.

(b) Let X = the amount gained by the player in a single game of Keno. Give the probability distribution of X in tabular form.

(c) Find the expected value of X.

8.76 Keno, II Refer to the previous exercise. Suppose a player decides to play 25 games of this simple version of Keno. Let Y = the number of games won by the player.

(a) Show that this is a binomial setting.

(b) Find and interpret $P(Y = 5)$. Show your work.

(c) Find and interpret $P(Y > 5)$. Show your work.

(d) How much money should the player expect to win or lose in 25 games? Explain.

PART D

INFERENCE

"Howard's been like this ever since he heard that having a dog lowers your blood pressure."

To *infer* is to draw a conclusion from evidence. Statistical *inference* draws a conclusion about a population from evidence provided by a sample. Drawing conclusions in mathematics is a matter of starting from a hypothesis and using logical argument to prove without doubt that the conclusion follows. Statistics isn't like that.

Statistical conclusions are uncertain, because the sample isn't the entire population. So statistical inference has to state conclusions and also say how uncertain they are. We use the language of probability to express uncertainty. Because inference must both give conclusions and say how uncertain they are, it is the most technical part of statistics.

Texts and courses intended to *train people to do* statistics spend most of their time on inference. Our aim in this book is to *help you understand* statistics, which takes less technique but often more thought. We will look only at a few basic techniques of inference. The techniques are simple, but the ideas are subtle, so prepare to think. To start, think about what you already know and don't be too impressed by elaborate statistical techniques: even the fanciest inference cannot remedy fundamental flaws such as voluntary response samples or uncontrolled experiments.

CHAPTER 9

Introduction to Inference

Don't get mad

Know someone who is prone to anger? Nature has a way to calm such people: they get heart disease more often. Several observational studies have discovered a link between anger and heart disease. The best study looked at 12,986 people, both black and white, chosen at random from four communities.[1] When first examined, all subjects were between the ages of 45 and 64 and were free of heart disease. Let's focus on the 8474 people in this sample who had normal blood pressure.

A short psychological test, the Spielberger Trait Anger Scale, measured how easily each person became angry. There were 633 people in the high range of the anger scale, 4731 in the middle, and 3110 in the low range. Now follow these people forward in time for almost six years and compare the rate of heart disease in the high and low groups. There are some lurking variables: people in the high-anger group are somewhat more likely to be male, to have less than a high school education, and to be smokers and drinkers. After adjusting for these differences, high-anger people were 2.2 times as likely to get heart disease and 2.7 times as likely to have an acute heart attack as low-anger people.

We know that the numbers 2.2 and 2.7 won't be exactly right for the population of all people aged 45 to 64 years with normal blood pressure. News reports of the study cited those numbers, but the full report in the medical journal *Circulation* gave confidence intervals. With 95% confidence, high-anger people are between 1.36 and 3.55 times as likely to get heart disease as low-anger people. They are between 1.48 and 4.90 times as likely to have an acute heart attack. The intervals remind us that any statement we make about the population is uncertain because we have data only from a sample. For the sample, we can say "exactly 2.2

times as likely." For the whole population, the sample data allow us to say only "between 1.36 and 3.55 times as likely," and only with 95% confidence. To go behind the news to the real thing, in medicine and in other areas, we must speak the language of inference.

9.1 What Is a Confidence Interval?

ACTIVITY 9.1A

The candy machine

www

Imagine a very large candy machine filled with orange, brown, and yellow candies. Suppose that 45% of the candies in the machine are orange. When you insert money, the machine dispenses a sample of 25 candies. Would you be surprised if your sample contained 10 orange candies (that's 40% orange)? How about 5 orange candies (20% orange)? In this Activity, you will use a Java applet to investigate the sample-to-sample variability in the proportion of orange candies dispensed by the machine.

1. Launch the Reese's Pieces applet at **www.rossmanchance.com/applets/Reeses/ReesesPieces.html**. Click on the "Draw Samples" button. An animated simple random sample of 25 candies should be dispensed. Figure 9.1 shows the results of one such sample. Was your sample proportion of orange candies (look at the value of $\hat{p}$ in the applet window) close to the actual population proportion, $p = 0.45$?

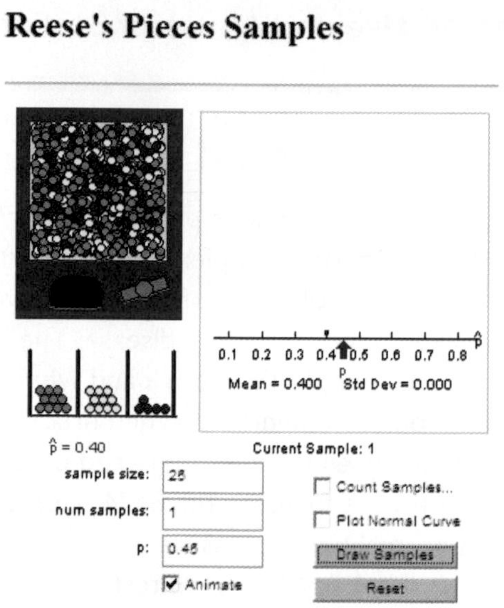

Figure 9.1 The result of taking one simple random sample of 25 candies from a large candy machine in which 45% of the candies are orange.

2. Click "Draw Samples" 9 more times, so that you have a total of 10 sample results. Look at the dotplot of your $\hat{p}$-values. What is the mean of your 10 sample proportions? What is their standard deviation?

3. To take many more samples quickly, enter 190 in the "num samples" box. Click on the "Animate" box to turn the animation off. Then click "Draw Samples." You have now taken a total of 200 samples of 25 candies from the machine. Describe the shape, center, and spread of the distribution of sample proportions ($\hat{p}$) shown in the dotplot.

4. Would you be surprised if a sample of 25 candies from the machine contained 10 orange candies (that's 40% orange)? How about 5 orange candies (20% orange)? Explain.

5. How would the distribution of the sample proportion $\hat{p}$ change if the machine dispensed 50 candies each time instead of 25? "Reset" the applet. Take 200 samples of 50 candies. Describe the shape, center, and spread of the distribution of sample proportions ($\hat{p}$).

6. Would you be surprised if a sample of 50 candies from the machine contained 20 orange candies (that's 40% orange)? How about 10 orange candies (20% orange)? Explain.

Estimating

Statistical inference draws conclusions about a population based on data from a sample. One kind of conclusion answers questions like "What percent of teens smoke?" or "What is the mean survival time for patients with a particular type of cancer?" These questions ask about a number (a percent, a mean) that describes a population. Numbers that describe a population are **parameters.** To estimate a population parameter, choose a sample from the population and use a **statistic,** a number calculated from the sample, as your estimate. Here's an example.

Example 9.1 Teen smoking

How common is behavior that puts people at risk of lung cancer? The 2007 Youth Risk Behavior Survey questioned a nationally representative sample of 14,041 students in grades 9 to 12. Of these, 2808 said they had smoked cigarettes at least one day in the past month. That's 20.0% of the sample.[2] This result may be biased by reluctance to tell the truth about any kind of risky behavior. For now, assume that the people in the sample told the truth. Based on these data, what can we say about the percent of all high schoolers who smoked?

Our population is high school students in the United States. The *parameter* is the proportion that have smoked cigarettes in the past month. Call this unknown parameter p, for "proportion." The *statistic* that estimates the parameter p is the **sample proportion**

$$\hat{p} = \frac{\text{count in the sample}}{\text{sample size}}$$
$$= \frac{2808}{14{,}041} = 0.200$$

A basic strategy in statistical inference is to use a sample statistic to estimate a population parameter. Once we have the sample data, we estimate that the proportion of all high school students who have smoked cigarettes in the past month p is "about 20.0%" because the proportion in the sample was exactly 0.200. We can only estimate that the truth about the population is "about" 20.0% because we know that the sample result is unlikely to be exactly the same as the true population proportion. A **95% confidence interval** makes that "about" precise.

> **95% confidence interval**
>
> A **95% confidence interval** is an interval calculated from sample data that is guaranteed to capture the true population parameter in 95% of all samples.

Let's look at how to calculate a confidence interval for a population proportion. Then we'll reflect on what we have done and generalize a bit.

Estimating with confidence

We want to estimate the proportion p of the individuals in a population who have some characteristic—they are orange (candies), or they smoke (teens), for example. Let's call the characteristic we are looking for a "success." We use the proportion $\hat{p}$ (read as "p-hat") of successes in a simple random sample (SRS) to estimate the proportion p of successes in the population. How good is the statistic $\hat{p}$ as an estimate of the parameter p? To find out, we ask, "What would happen if we took many samples?" Well, we know that $\hat{p}$ would vary from sample to sample. If you did Activity 9.1A, you saw that this sampling variability isn't haphazard. It has a clear pattern in the long run, a pattern that is pretty well described by a Normal curve. Here are the facts.

> **Sampling distribution of a sample proportion**
>
> The **sampling distribution** of a statistic is the distribution of values taken by the statistic in all possible samples of the same size from the same population.
>
> Take an SRS of size n from a large population that contains proportion p of successes. Let $\hat{p}$ be the **sample proportion** of successes,
>
> $$\hat{p} = \frac{\text{count of successes in sample}}{n}$$
>
> Then, if the sample is large enough:
>
> - The sampling distribution of $\hat{p}$ is **approximately Normal.**
> - The **mean** of the sampling distribution is p.
> - The **standard deviation** of the sampling distribution is
>
> $$\sqrt{\frac{p(1-p)}{n}}$$

These facts can be proved by mathematics, so they are a solid starting point. Figure 9.2 summarizes them in a form that also reminds us that a sampling distribution describes the results of lots of samples from the same population.

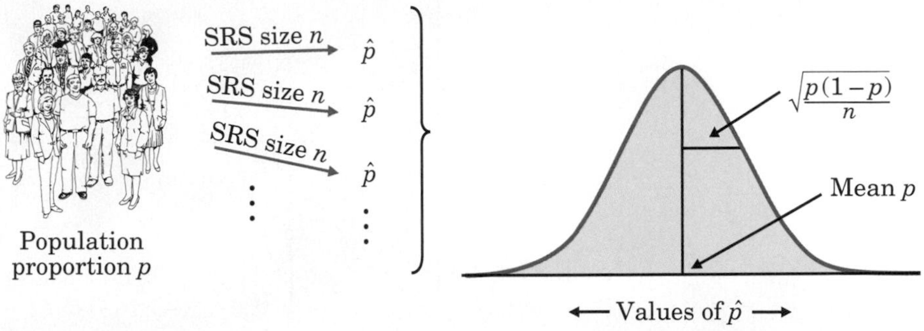

Figure 9.2 Repeat many times the process of selecting an SRS of size n from a population in which the proportion p are successes. The values of the sample proportion of successes $\hat{p}$ have this Normal sampling distribution.

Example 9.2 Risky behavior

Suppose the truth is that 21% of high school students smoke. Then in the setting of Example 9.1, $p = 0.21$. The Youth Risk Behavioral Survey sample of size $n = 14{,}041$ would, if repeated many times, produce sample proportions that closely follow the Normal distribution with

$$\text{mean} = p = 0.21$$

and

$$\begin{aligned}\text{standard deviation} &= \sqrt{\frac{p(1-p)}{n}} \\ &= \sqrt{\frac{0.21(0.79)}{14{,}041}} = 0.0034\end{aligned}$$

The center of this Normal distribution is at the truth about the population. That's the absence of bias in random sampling once again. The standard deviation is small because the sample is quite large. So almost all samples will produce a statistic that is close to the true p. In fact, the 95 part of the 68–95–99.7 rule says that 95% of all sample outcomes will fall between

$$\text{mean} - 2 \text{ standard deviations} = 0.21 - 2(0.0034) = 0.21 - 0.0068 = 0.2032$$

and

$$\text{mean} + 2 \text{ standard deviations} = 0.21 + 2(0.0034) = 0.21 + 0.0068 = 0.2168$$

Figure 9.3 (on the next page) displays these facts.

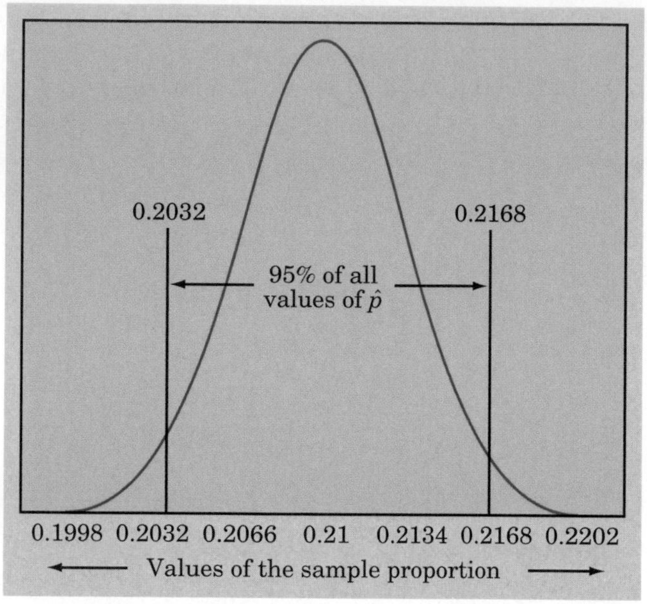

Figure 9.3 Repeat many times the process of selecting an SRS of size 14,041 from a population in which the proportion $p = 0.21$ are successes. The middle 95% of the values of the sample proportion will lie between 0.2032 and 0.2168.

So far, we have just put numbers on what we already knew. We can trust the results of large random samples because almost all such samples give results that are close to the truth about the population. The numbers say that in 95% of all samples of size 14,041, the statistic $\hat{p}$ and the parameter p are within 0.0068 of each other. We can put this another way: 95% of all samples give an outcome $\hat{p}$ such that the population truth p is captured by the interval from $\hat{p} - 0.0068$ to $\hat{p} + 0.0068$. The 0.0068 came from substituting $p = 0.21$ into the formula for the standard deviation of $\hat{p}$. For any value of p, the general fact is: **When the population proportion has the value p, about 95% of all samples catch p in the interval extending 2 standard deviations on either side of $\hat{p}$.** That's the interval

$$\hat{p} \pm 2\sqrt{\frac{p(1-p)}{n}}$$

Is this the 95% confidence interval we want? Not quite. The interval can't be found just from the data because the standard deviation involves the population proportion p, and in practice we don't know p. In Example 9.2, we applied the formula using $p = 0.21$, but this may not be the true p.

What should we do? Well, the standard deviation of the statistic does depend on the parameter p, but it doesn't change a lot when p changes. Let's go back to Example 9.2 and redo the calculation for other values of p. Here's the result:

Value of p:	0.19	0.20	0.21	0.22	0.23
Standard deviation:	0.0033	0.00338	0.00344	0.0035	0.00355

We see that if we guess a value of p reasonably close to the true value, the standard deviation found from the guessed value will be about right. We know that, when we take a large random sample, the statistic $\hat{p}$ is almost always close to the parameter p. So we will use $\hat{p}$ as the guessed value of the unknown p. Now we have an interval that we can calculate from the sample data.

95% confidence interval for a proportion

Choose an SRS of size n from a large population that contains an unknown proportion p of successes. Call the proportion of successes in this sample $\hat{p}$. An approximate 95% confidence interval for the parameter p is

$$\hat{p} \pm 2\sqrt{\frac{\hat{p}(1-\hat{p})}{n}}$$

Example 9.3 A confidence interval for youthful smokers

The Youth Risk Behavioral Survey random sample of 14,041 high school students found that 2808 had smoked cigarettes at least one day in the past month, a sample proportion $\hat{p} = 0.200$. The 95% confidence interval for the proportion of all high school students who had smoked in the past month is

$$\begin{aligned}\hat{p} \pm 2\sqrt{\frac{\hat{p}(1-\hat{p})}{n}} &= 0.200 \pm 2\sqrt{\frac{0.200(0.800)}{14{,}041}} \\ &= 0.200 \pm 2(0.0034) \\ &= 0.200 \pm 0.0068\end{aligned}$$

or the interval

$$(0.1932, 0.2068)$$

Interpret this result as follows: we got this interval by using a method that catches the true unknown population proportion in 95% of all samples. The shorthand is: we are **95% confident** that the true proportion of high school students who smoked in the past month lies between 19.32% and 20.68%.

Exercises

9.1 A student survey Tonya wants to estimate what proportion of the seniors in her school plan to attend the prom. She interviews an SRS of 50 of the 750 seniors in her school. She finds that 36 plan to attend the prom.

(a) What population does Tonya want to draw conclusions about?

(b) What does the population proportion p represent in this setting?

(c) What is the numerical value of the sample proportion $\hat{p}$ from Tonya's sample?

9.2 Our dirty little secret More than 90% of people surveyed say they always wash their hands after using public restrooms. But observers posted in public restrooms found that only 4679 of 6076 observed individuals actually washed their hands.[3]

(a) What population does inference concern here?

(b) Explain clearly what the population proportion p represents in this setting.

(c) What is the numerical value of the sample proportion $\hat{p}$?

9.3 Marijuana use A 2007–2008 survey of teens conducted by the Partnership for a Drug-Free America found that 33% of 6511 U.S. teenagers surveyed reported use of marijuana. The report states, "With 95% confidence, we can say that between 31.4% and 34.6% of all American teenagers have used marijuana."[4] Explain to someone who knows no statistics what the phrase "95% confidence" means in this report.

9.4 Do you drink the cereal milk? A *USA Today* poll asked a random sample of 1012 U.S. adults what they do with the milk in the bowl after they have eaten the cereal. Suppose that 70% of all U.S. adults actually drink the cereal milk.

(a) Find the mean and standard deviation of the proportion $\hat{p}$ of the sample who say they drink the cereal milk.

(b) Draw a Normal curve like the one in Figure 9.3 that shows the sampling distribution of $\hat{p}$ in this case.

(c) Of the 1012 adults surveyed, 67% said they drink the cereal milk. Plot this sample result on the graph from (b).

(d) Using your Normal curve, find the probability of obtaining a sample in which 67% or fewer say they drink the cereal milk. Do you think it's plausible that the true population proportion is $p = 0.70$? Explain.

9.5 Polling women A *New York Times* Poll on women's issues interviewed 1025 women randomly selected from the United States, excluding Alaska and Hawaii. Of the women in the sample, 482 said they do not get enough time for themselves. Although the samples in national polls are not SRSs, they are close enough that our method gives approximately correct confidence intervals.

(a) Explain in words what the parameter p is in this setting.

(b) Use the poll results to give a 95% confidence interval for p.

(c) Write a short explanation of your findings in (b) for someone who knows no statistics.

9.6 Gun violence The Harris Poll asked a sample of 1009 adults which causes of death they thought would become more common in the future. Gun violence topped the list: 706 members of the sample thought deaths from guns would increase. Although the samples in national polls are not SRSs, they are close enough that our method gives approximately correct confidence intervals.

(a) Say in words what the population proportion p is for this poll.

(b) Find a 95% confidence interval for p.

(c) Harris announced a margin of error of plus or minus 3 percentage points for this poll result. How well does your work in (b) agree with this margin of error?

Understanding confidence intervals

Our 95% confidence interval for a population proportion has the familiar form

estimate ± margin of error

We know that news reports of sample surveys, for example, usually give the estimate and the margin of error separately: "A new Gallup Poll shows that 66% of women favor new laws restricting guns. The margin of error is plus or minus four percentage points." This result suggests that between 62% and 70% of the population of women favor these laws. We also know that news reports usually leave out the level of confidence.

Not all confidence intervals have this form. Here's a complete description of a **confidence interval.**

> **Confidence interval**
>
> A **level *C* confidence interval** for a parameter has two parts:
>
> - An **interval** calculated from the data.
> - A **confidence level** *C*, which gives the probability that the interval will capture the true parameter value in repeated samples.

ACTIVITY 9.1B

The *Confidence Intervals* applet

APPLET Go to the *Statistics Through Applications* Web site **www.whfreeman.com/sta2e** and launch the *Confidence Intervals* applet. Although this applet involves constructing confidence intervals for the population mean, μ, the ideas of this Activity apply to both means and proportions.

Directions Click the "Sample" button to have the applet take a simple random sample and construct a confidence interval. Click the "Sample 50" button to have the applet take 50 different simple random samples and make 50 different confidence intervals from the resulting data.

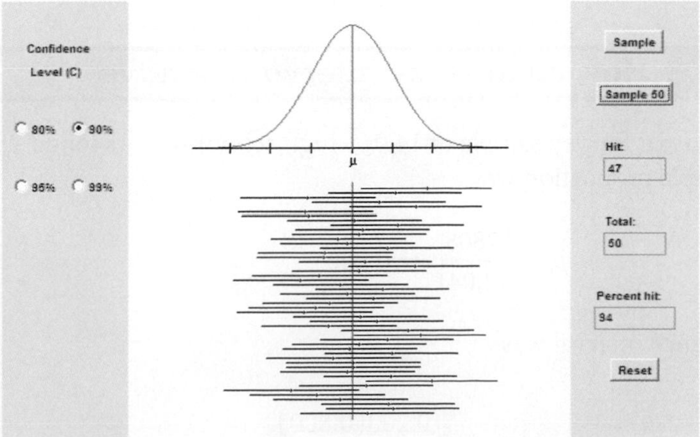

Think

1. If you make just one 90% confidence interval, will it necessarily capture the parameter value?

2. If you make 10 different 90% confidence intervals, how many of them do you think will capture the parameter?

3. If you make a thousand 90% confidence intervals, how many of them do you think will capture the parameter?

ACTIVITY 9.1B *(continued)*

ACTIVITY 9.1B *(continued)*

APPLET TIP: The cumulative totals are cleared when you click the "Reset" button. When you change the confidence level, the totals pertain to the displayed intervals.

Test

4. Test your answers to the preceding three questions through repeated use of the applet. How far did the results vary from what you expected in Questions 2 and 3? How far did the results vary from what you expected on a percentage basis?

Explore

5. Set the confidence level to 80%, and sample 50 confidence intervals. How many of these intervals fail to capture the parameter (they are drawn in red on the computer screen)?

6. Increase the confidence level in succession to 90%, then to 95%, and then to 99% while watching the changing behavior of the 50 intervals. What happens to the lengths of the intervals? What happens to the number of intervals that fail to capture the parameter?

7. Have the applet make 50 new confidence intervals. The dots in the center of the intervals mark the sample mean. Are the intervals produced by the applet symmetric around the sample mean?

The tick marks on the horizontal axis below the Normal curve are each separated by one standard deviation of the sample mean. According to the 68–95–99.7 rule, how many of the 50 sample means would you expect to fall within one standard deviation of the parameter? How many of the 50 sample means did fall within one standard deviation?

8. How is the margin of error related to the length of the interval? Explain.

9. Examine the 50 intervals at the 95% confidence level. What is the typical margin of error for these intervals measured in units of the standard deviation of the mean? Explain how this is related to the Normal curve.

There are many formulas for confidence intervals that apply in different situations. Be sure you understand how to interpret a confidence interval. The interpretation is the same for any formula, and you can't use a calculator or a computer to do the interpretation for you. As Activity 9.1B illustrates, **confidence intervals use the central idea of probability: ask what would happen if we repeated the sampling many times.** The 95% in a 95% confidence interval is a probability, the probability that the method produces an interval that does capture the true parameter.

Example 9.4 How confidence intervals behave

The Youth Risk Behavior Survey sample of 14,041 high school seniors found 2808 smokers, so the sample proportion was

$$\hat{p} = \frac{2808}{14{,}041} = 0.200$$

and the 95% confidence interval was

$$\hat{p} \pm 2\sqrt{\frac{\hat{p}(1-\hat{p})}{n}} = 0.200 \pm 2\sqrt{\frac{0.200(0.800)}{14{,}041}} = 0.200 \pm 0.0068$$

Draw a second sample from the same population. It finds 2877 of its 14,041 respondents who smoke. For this sample,

$$\hat{p} = \frac{2877}{14{,}041} = 0.205$$

$$\hat{p} \pm 2\sqrt{\frac{\hat{p}(1-\hat{p})}{n}} = 0.205 \pm 2\sqrt{\frac{0.205(0.795)}{14{,}041}} = 0.205 \pm 0.0068$$

Draw another sample. Now the count is 2719 and the sample proportion and confidence interval are

$$\hat{p} = \frac{2719}{14{,}041} = 0.194$$

$$\hat{p} \pm 2\sqrt{\frac{\hat{p}(1-\hat{p})}{n}} = 0.194 \pm 2\sqrt{\frac{0.194(0.806)}{14{,}041}} = 0.194 \pm 0.0067$$

Keep sampling. Each sample yields a new estimate and a new confidence interval. *If we sample forever, 95% of these intervals will capture the true parameter.* This is true no matter what the true value is. Figure 9.4 summarizes the behavior of the confidence interval in graphical form.

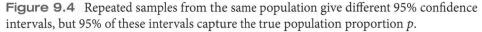

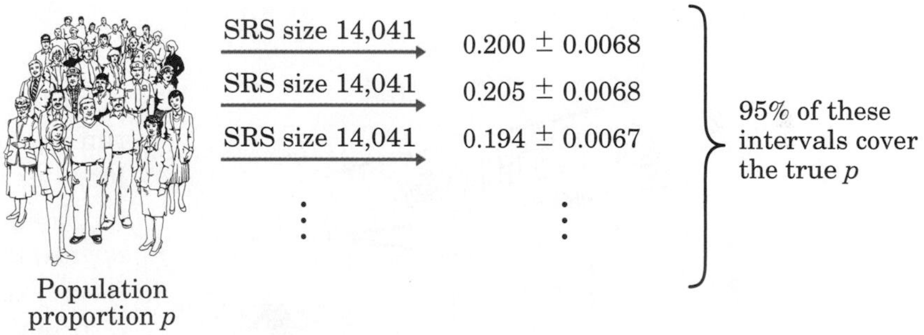

Figure 9.4 Repeated samples from the same population give different 95% confidence intervals, but 95% of these intervals capture the true population proportion p.

Figures 9.4 and 9.5 (on the next page) show how confidence intervals behave. Example 9.4 and Figure 9.4 remind us that repeated samples give different results and that we are guaranteed only that 95% of the samples give a correct result.

Figure 9.5 goes behind the scenes. The vertical line is the true value of the population proportion p. The Normal curve at the top of the figure is the sampling distribution of the sample statistic, which is centered at the true p. We are behind the scenes because in real-world statistics we don't know p. The 95% confidence intervals from 25 SRSs appear one after the other. The central dots are the values of $\hat{p}$, the centers of the intervals. The arrows on either side span the confidence interval. In the long run, 95% of the intervals will cover the true p and 5% will miss. Of the 25 intervals in Figure 9.5, 24 hit and 1 misses. (Remember that probability

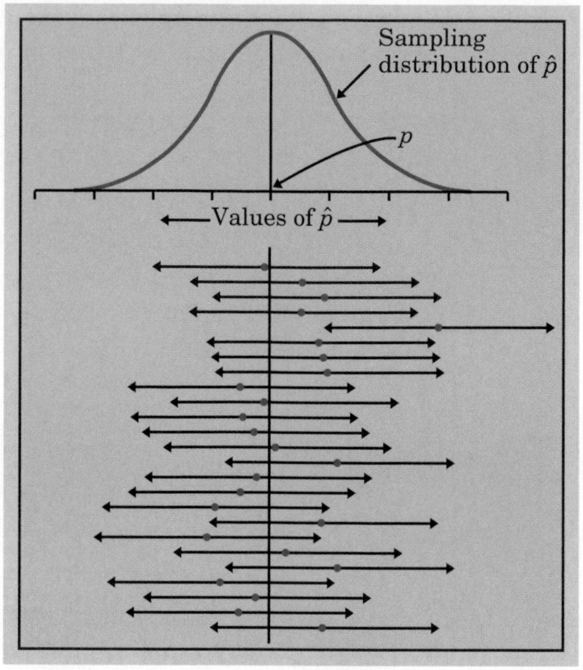

Figure 9.5 Twenty-five samples from the same population give these 95% confidence intervals. In the long run, 95% of all such intervals cover the true population proportion, marked by the vertical line.

describes only what happens in the long run—we don't expect exactly 95% of 25 intervals to capture the true parameter.)

Don't forget that our interval is only *approximately* a 95% confidence interval. It isn't exact for two reasons. The sampling distribution of the sample proportion $\hat{p}$ isn't exactly Normal. And we don't get the standard deviation of $\hat{p}$ exactly right, because we used $\hat{p}$ in place of the unknown p. Both of these difficulties become less serious as the sample size n gets larger. So our formula is good only for large samples.

What's more, *our method assumes that the population is really big—at least 10 times the size of the sample.* Professional statisticians use more elaborate methods that take the size of the population into account and work even for small samples.[5] But our method works well enough for many practical uses. More important, it shows how we get a confidence interval from the sampling distribution of a statistic. That's the reasoning behind any confidence interval.

Confidence intervals for a population proportion

We used the 95 part of the 68–95–99.7 rule to get a 95% confidence interval for the population proportion. Perhaps you think that a method that works 95% of the time isn't good enough. You want to be 99% confident. For that, we need to mark off the central 99% of a Normal distribution. For any probability C between 0 and 1, there is a number z^* such that any Normal distribution has probability C within z^* standard deviations of the mean. Figure 9.6 shows how the probability C and the number z^* are related.

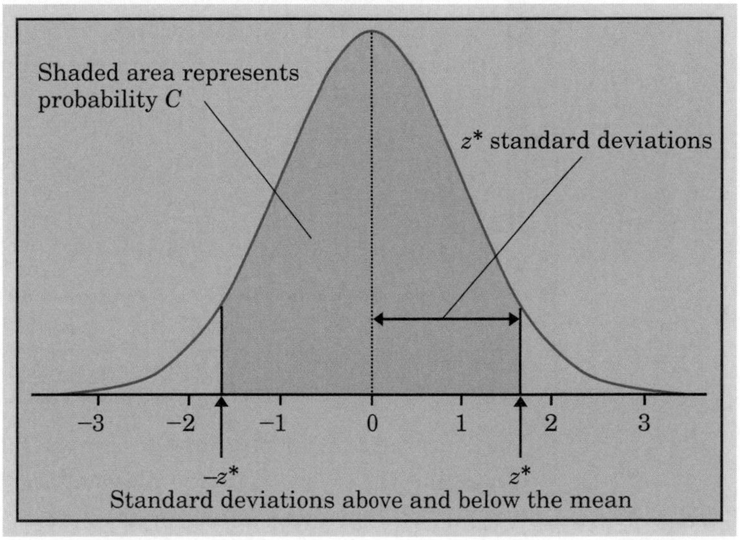

Figure 9.6 Critical values z^* of the Normal distributions. In any Normal distribution, there is area (probability) C under the curve between $-z^*$ and z^* standard deviations away from the mean.

Table 9.1 gives the numbers z^* for various choices of C. For convenience, the table gives C as a confidence level in percent. The numbers z^* are called **critical values** of the Normal distributions. Table 9.1 shows that any Normal distribution has probability 99% within ±2.58 standard deviations of its mean. The table also shows that any Normal distribution has probability 95% within ±1.96 standard deviations of its mean. The 68–95–99.7 rule uses 2 in place of the critical value $z^* = 1.96$. That is good enough for practical purposes, but the table gives the more exact value.

Table 9.1 Critical Values of the Normal Distribution

Confidence level C	Critical value z^*	Confidence level C	Critical value z^*
50%	0.67	90%	1.64
60%	0.84	95%	1.96
70%	1.04	99%	2.58
80%	1.28	99.9%	3.29

From Figure 9.6 we see that, with probability C, the sample proportion takes a value within z^* standard deviations of p. That is just to say that, with probability C, the interval extending z^* standard deviations on either side of the observed $\hat{p}$ captures the unknown p. Using the estimated standard deviation of $\hat{p}$ produces the following result.

Confidence interval for a population proportion

Choose an SRS of size n from a population of individuals of which proportion p are successes. The proportion of successes in the sample is $\hat{p}$. When n is large, an **approximate level C confidence interval** for p is

$$\hat{p} \pm z^* \sqrt{\frac{\hat{p}(1 - \hat{p})}{n}}$$

where z^* is the **critical value** for probability C from Table 9.1.

Example 9.5 A 99% confidence interval

The Youth Risk Behavior Survey random sample of 14,041 high school students found that 2808 were smokers. We want a 99% confidence interval for the proportion p of all high school students who have smoked cigarettes in the past month. Table 9.1 says that for 99% confidence, we must go out $z^* = 2.58$ standard deviations. Here are our calculations:

$$\hat{p} = \frac{2808}{14{,}041} = 0.200$$

$$\begin{aligned} \hat{p} \pm z^* \sqrt{\frac{\hat{p}(1 - \hat{p})}{n}} &= 0.200 \pm 2.58 \sqrt{\frac{0.200(0.800)}{14{,}041}} \\ &= 0.200 \pm 2.58(0.00338) \\ &= 0.200 \pm 0.0087 \end{aligned}$$

So our 99% confidence interval is (0.1913, 0.2087).

We are 99% confident that the true population proportion is between 19.13% and 20.87%. That is, we got this range of percents by using a method that gives a correct answer 99% of the time.

Who is a smoker?

When estimating a proportion p, be sure you know what counts as a "success." The news says that 20% of adolescents smoke. Shocking. It turns out that this is the percent who smoked at least once in the past month. If we say that a smoker is someone who smoked on at least 20 of the past 30 days, only 8% of adolescents qualify.

Compare Example 9.5 with the calculation of the 95% confidence interval in Example 9.3. The only difference is the use of the critical value 2.58 for 99% confidence in place of 2 for 95% confidence. That makes the margin of error for 99% confidence larger and the confidence interval wider. Higher confidence isn't free—we pay for it with a wider interval. Figure 9.6 reminds us why this is true. To cover a higher percent of the area under a Normal curve, we must go farther out from the center.

Exercises

9.7 Interpreting poll results Here's a quote from a news report about a sample survey: "A new Gallup Poll shows that 66% of women favor new laws restricting guns. The margin of error is plus or minus four percentage points." The news report omitted the confidence level, which was 95%. Explain what this poll result means to someone who knows little statistics. Be sure to discuss the meaning of both the interval and the confidence level.

9.8 Confidence level visual Figure 9.5 (page 430) shows the meaning of 95% confidence. Trace the sampling distribution in Figure 9.5. Then draw 10 intervals that illustrate the meaning of 80% confidence. Be sure to consider the lengths of your intervals.

9.9 98% confidence Table 9.1 (page 431) gives critical values z^* for several confidence levels. Find z^* for a 98% confidence interval using Table A. Show your method clearly.

9.10 93% confidence Table 9.1 (page 431) gives critical values z^* for several confidence levels. Find z^* for a 93% confidence interval using Table A. Show your method clearly.

Kevin Russ/iStockphoto

9.11 Going to the prom The prom committee of Exercise 9.1 (page 425) wants to know how many seniors they can expect to attend the prom so that they can develop a budget. An SRS of 50 of the 750 seniors in the school yielded 36 who said they plan to attend the prom.

(a) Construct and interpret a 95% confidence interval for the true proportion of seniors who will attend the prom.

(b) The committee asks if you can provide a "more precise" interval. Should you use a 90% confidence level or a 99% confidence level? Explain.

(c) Construct a new interval using the confidence level you chose in (b). Interpret the resulting interval for the prom committee in simple language.

9.12 Random digits We know that the proportion of 0s among a large set of random digits is $p = 0.1$ because all 10 possible digits are equally probable. The entries in a table of random digits are a sample from the population of all random digits. To get an SRS of 200 random digits, look at the first digit in each of the 200 five-digit groups in lines 101 to 125 of Table B in the back of the book.

(a) How many of these 200 digits are 0s? Give a 95% confidence interval for the proportion of 0s in the population from which these digits are a sample.

(b) Does your interval include the true parameter value, $p = 0.1$? What does this tell you?

CALCULATOR CORNER Confidence interval for a population proportion

You can use your TI-84/TI-Nspire to construct a confidence interval for a population proportion. We will use the teen smoking data from Examples 9.1 to 9.5 to illustrate.

CALCULATOR CORNER *(continued)*

CALCULATOR CORNER *(continued)*

TI-84

- Press STAT, choose TESTS, and then choose 1-PropZInt.
- Enter the number of teen smokers, 2808, as the count x and the number of teens in the sample, 14041, as the sample size n. Specify the confidence level as .95. Then highlight "Calculate" and press ENTER.

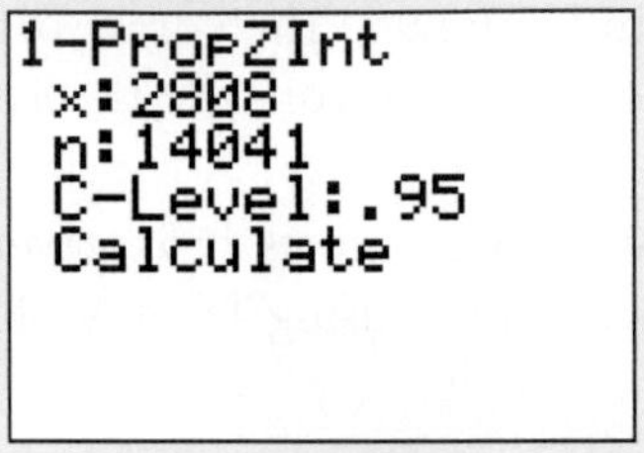

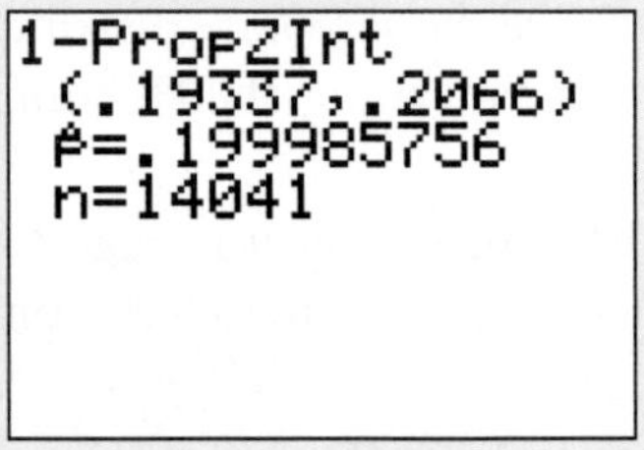

TI-Nspire

You can calculate confidence intervals in a Calculator page or a Lists & Spreadsheet page. We'll show the steps for a Calculator page here.

- On a Calculator page, press menu, and choose 5:Statistics, 6:Confidence Intervals..., and 5:1-Prop z Interval.
- In the dialog box, enter the number of teen smokers, 2808, as the number of "Successes" x and the number of teens in the sample, 14,041, as the sample size n. Specify the confidence level as 0.95. Then click OK.

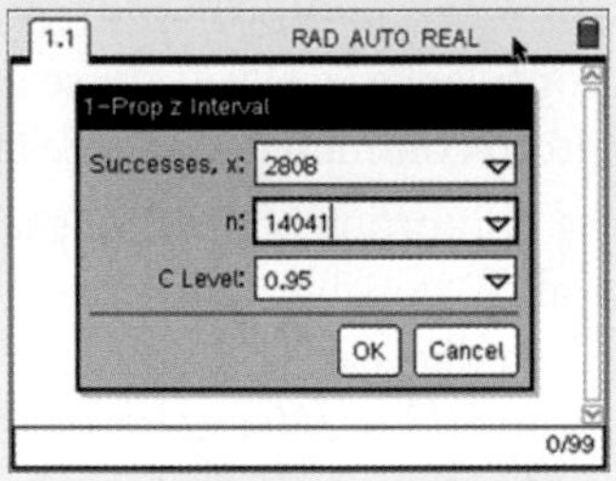

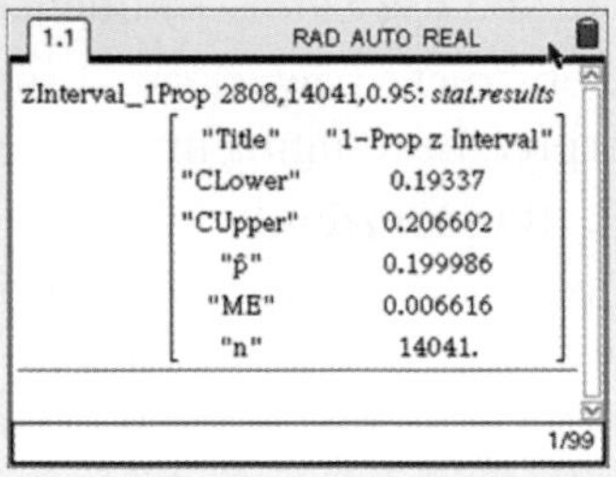

DATA EXPLORATION

No Child Left Behind

In 2002, the Bush administration sponsored legislation known as the "No Child Left Behind" act, which attempted to improve public education by mandating a testing program and minimum standards. The act left it up to each state to develop a statewide testing program in reading and math and a minimum proficiency level. Schools that didn't meet the minimum performance level would be classified as failing schools. Students at those schools would be able to transfer to other, supposedly better, schools. Many states established 40% as the minimum passing rate.

Principals and school system administrators don't want their schools to be declared failing schools, so many states have adopted a controversial procedure to buy some "wiggle room" on test results.

Here is their position. Instead of considering test results for one year to be the entire population of interest, they argue that test results should be viewed as a *sample* of school performance from a population of multiple years. Each year's results come from a sample of students gathered at one point in time, they say. Consequently, it makes sense to construct confidence intervals for the population passing rate. At this writing, 35 states use the confidence interval approach or a variation of it to interpret test scores.

Critics argue that using the confidence interval approach to determine which schools pass distorts the truth. For example, a small high school in Oregon met that state's requirement of a 40% pass rate on the state reading test even though only 28% of the school's students passed the test that year. And in Maryland, an elementary school met that state's standards even though only 31% of its pupils passed a state math test when a 41.5% pass rate was required.[6]

1. Suppose the Oregon school tested 50 students this year, and that 28% of this sample passed. Calculate the resulting 95% confidence interval. Explain why the school could be considered to have met Oregon's requirement.

2. What if the Oregon school had tested 75 students this year instead of 50? Calculate a new 95% confidence interval, still assuming a 28% pass rate in the sample. Explain why the school would not meet the state requirement now.

3. Here's another idea: the school could use a 99% confidence level instead of 95%. Recalculate the confidence interval from Question 2, this time using 99% confidence. What would you conclude?

In fact, if the Oregon school had up to 92 students take the test and 28% passed, then the state's 40% requirement would be within the 99% confidence interval and the school would be considered passing. Life is good. Thirteen states use a 99% confidence interval.

4. Show how it is possible for the Maryland elementary school to meet the state's 41.5% requirement even though only 31% of the school's students passed the state math test that year.

5. Do you think a confidence interval approach to interpreting test scores is appropriate? What do you think is the most compelling reason for or against this practice?

Determining sample size

National survey organizations like the Gallup Poll typically sample between 1000 and 1500 American adults who are interviewed by telephone. Why do they choose such sample sizes? The following example shows you why.

Example 9.6 Sample size for a desired margin of error

National polls like to use the 95% confidence level because it has become a standard and people are used to it. Similarly, pollsters like to report results with a margin of error of ±3 percentage points. Media reports of polls frequently have information about how the poll was conducted and how to interpret the results. How many people do the pollsters need to interview in order to give a ±3% margin of error at the 95% confidence level? The details require some algebra.

The form for the confidence interval is

$$\hat{p} \pm z^* \sqrt{\frac{\hat{p}(1-\hat{p})}{n}}$$

where $\hat{p}$ is the proportion of people who respond in a certain way and the term following the ± sign is the margin of error. For the 95% confidence interval, $z^* = 1.960$, correct to three decimal places.

Since we want the margin of error to be 3 percentage points, we have to solve the equation

$$z^*\sqrt{\frac{\hat{p}(1-\hat{p})}{n}} = 0.03$$

Notice that we use the decimal form for 3%. Substituting, we have

$$1.96\sqrt{\frac{\hat{p}(1-\hat{p})}{n}} = 0.03$$

Multiplying both sides by $\sqrt{n}$ and then dividing both sides by 0.03 yields

$$\frac{1.96}{0.03}\sqrt{\hat{p}(1-\hat{p})} = \sqrt{n}$$

Squaring both sides of the equation gives

$$\left(\frac{1.96}{0.03}\right)^2 \hat{p}(1-\hat{p}) = n$$

The largest value of n will be required if $\hat{p} = 0.5$, since that will make the product $\hat{p}(1-\hat{p})$ as large as possible. With $\hat{p} = 0.5$, we have

$$\left(\frac{1.96}{0.03}\right)^2 (0.5)(1-0.5) = n$$

$$1067.111 = n$$

Since you can't interview 0.111 of a person, you have to round up to the next whole number of people: 1068. Why not round to the nearest whole number—in this case, 1067? Because a smaller sample size will result in a larger margin of error, possibly more than the desired 3% for the poll.

APPLICATION 9.1

The beads

Before class, your teacher will prepare a large population of different-colored beads and put them into a container that you cannot see inside. Your goal is to estimate the actual proportion of beads in the population that have a particular color (specified by your teacher).

QUESTIONS

1. Your teacher will mix the beads in the container thoroughly. Without looking, a member of your class should use a small cup to scoop out a sample of beads. Explain why this sampling method should result in a simple random sample from the population.

2. Count the total number of beads selected and the number having the specified color. What proportion of the beads in the sample have the specified color?

3. Calculate and interpret a 95% confidence interval for the proportion of beads in the population that have the specified color.

4. Is it possible that your confidence interval from Question 3 does not capture the actual population proportion? Explain your answer.

Section 9.1 Summary

Statistical inference draws conclusions about a population on the basis of data from a sample. Because we don't have data for the entire population, our conclusions are uncertain. A **confidence interval** estimates an unknown parameter in a way that tells us how uncertain the estimate is. The interval itself says how closely we can pin down the unknown parameter. The **confidence level** is a probability that says how often in many samples the method produces an interval that does contain the parameter. Confidence intervals are based on the **sampling distribution** of a statistic, which shows how the statistic varies in repeated sampling.

In this section we found one specific confidence interval, for the proportion p of "successes" in a population, based on an SRS from the population. To calculate such an interval, we use **critical values** z^* from a Normal distribution. You will find more advice on interpreting confidence intervals in Section 9.3.

Section 9.1 Exercises

9.13 Determining sample size, I Suppose a polling organization wants to report their results using a 99% confidence level with a margin of error of ±3 percentage points. Use the method of Example 9.6 to find the number of people they would need to interview.

9.14 Determining sample size, II Suppose a polling organization wants to report their results at the 95% confidence level with a margin of error of ±4 percentage points. Use the method of Example 9.6 to find the number of people they would need to interview.

9.15 The good life Roper ASW, a market research and consulting firm, conducted in-person interviews with 2004 adults. When asked if they felt they had achieved the "good life," only 180 of the respondents said "Yes." This was despite the fact that a majority of respondents had the things they said constituted the good life: a house, good health, a car, and children. What is the margin of error for this poll? (Assume a 95% confidence level.) Show your work.

9.16 Losing weight A Gallup Poll in November 2008 found that 59% of the people in its sample said "Yes" when asked, "Would you like to lose weight?" Gallup announced: "For results based on the total sample of national adults, one can say with 95% confidence that the margin of (sampling) error is ±3 percentage points."[7]

(a) What is the 95% confidence interval for the percent of all adults who want to lose weight?

(b) Explain clearly to someone who knows no statistics why we can't just say that 59% of all adults want to lose weight.

(c) Then explain clearly what "95% confidence" means.

(d) As Gallup indicates, the 3% margin of error for this poll includes only sampling variability (what they call "sampling error"). What other potential sources of error (Gallup calls these "nonsampling errors") could affect the accuracy of the 59% estimate?

A baseball poll

Polls can sometimes yield surprising and entertaining results. A *Sports Illustrated* poll conducted in summer 2003 asked 550 Major League Baseball players to name the greatest living player. Barry Bonds was named most often (39.9%). Next was Alex Rodriguez (12.8%), followed by Willie Mays (12.1%), Nolan Ryan (7.1%), Hank Aaron (6.7%), and Pete Rose (3.6%). Babe Ruth was named by 8 of the players (1.5%). Babe Ruth died in 1948.

9.17 Harley motorcycles Harley-Davidson motorcycles make up 14% of all the motorcycles registered in the United States. You plan to interview an SRS of 600 motorcycle owners.
(a) Describe the shape, center, and spread of the sampling distribution of the sample proportion $\hat{p}$ who own Harleys.
(b) How likely is your sample to contain 18.2% or more who own Harleys? Show your work.
(c) How likely is your sample to contain less than 12% Harley owners? Show your work.

9.18 The quick method The quick method of Section 5.2 uses $1/\sqrt{n}$ as a rough estimate for the margin of error for a 95% confidence interval for a population proportion. The margin of error from the quick method is actually a bit larger than needed. It differs most from the more accurate method of this chapter when $\hat{p}$ is close to 0 or 1. An SRS of 600 motorcycle registrations finds that 68 of the motorcycles are Harley-Davidsons. Calculate a 95% confidence interval for the proportion of all motorcycles that are Harleys by the quick method and then by the method of this chapter. How much larger is the quick-method margin of error?

9.19 Do college students pray? Social scientists asked 127 undergraduate students "from courses in psychology and communications" about prayer and found that 107 prayed at least a few times a year.[8]
(a) To what population is it reasonable to generalize the results of this survey? Justify your answer.
(b) Calculate and interpret a 99% confidence interval for the population proportion p.

9.20 Wildlife management Wildlife biologists know that males of most big-game species can be "harvested" without affecting the total population because they tend to mate with several females a season. It is the female that requires the most protection. In a recent year, Virginia hunters killed 928 black bears during the fall hunting season. Of these, 328 were females.[9] Consider the 928 bears a random sample of the population of all bears killed over many seasons.
(a) What is the sample proportion of females harvested?
(b) Find a 95% confidence interval for the true proportion of female bears killed in Virginia over many seasons.
(c) What is the margin of error for your confidence interval in (b)? What two factors determine the margin of error?

9.2 What Is a Significance Test?

ACTIVITY 9.2A

Is this deck fair?

Materials: Three decks of playing cards, provided by your teacher

Your teacher has what appear to be three standard decks of playing cards. In this Activity, your task is to try to determine whether each deck is "fair" by collecting data about the distribution of black and red cards in the deck.

1. If a deck is fair, then the actual proportion of red cards in the deck is $p =$ _______. For each of the three decks that your teacher will present, start out assuming that the deck is fair.

2. *The first deck:* Your teacher will have one student draw a card, note the color, and replace the card in the

deck. Write RED and BLACK on the board and use a tally mark to record the result of the first draw. After the deck has been shuffled, a second student should draw a card and tally the result. Continue until your class reaches a consensus on the proportion p of red cards in the deck. Do you think this deck is fair?

3. *The second deck:* Your teacher will introduce a second deck. Repeat the process described in Step 2 until the class decides whether the deck is fair. How many draws did it take?

4. *The third deck:* Your teacher will introduce a third deck. Repeat the process described in Step 2 until the class decides whether the deck is fair. How many draws did it take? Once the class has made a decision about all three decks, your teacher will reveal how you did.

The reasoning of statistical tests

Confidence intervals are one of the two most common types of statistical inference. Use a confidence interval when your goal is to estimate a population parameter. The second common type of inference, called **significance tests,** has a different goal: to assess the evidence provided by data about some claim concerning a population. Here's a simple example.

Example 9.7 Shooting free throws

The local hot-shot playground basketball player claims to make 80% of his free throws. "Show me," you say. He shoots 20 free throws and makes 8 of them. "Aha," you conclude, "if he makes 80%, he would almost never make as few as 8 of 20. So I don't believe his claim."

Your reasoning is based on asking what would happen if the shooter's claim were true and we repeated the sample of 20 free throws many times—he would almost never make as few as 8. This outcome is so unlikely that it gives strong evidence that his claim is not true.

You can say how strong the evidence against the shooter's claim is by giving the probability that he would make as few as 8 out of 20 free throws if he really makes 80% in the long run. This probability is 0.0001. The shooter would make as few as 8 of 20 only once in 10,000 tries in the long run if his claim to make 80% is true. The small probability convinces you that his claim is false.

Example 9.7 shows you the reasoning of statistical tests at the playground level: *an outcome that is very unlikely if a claim is true is good evidence that the claim is not true.*

Statistical inference uses data from a sample to draw conclusions about a population. So once we leave the playground, statistical tests deal with claims about a population. Tests ask if sample data give good evidence *against* a claim. A test says, "If we took many samples and the claim were true, we would rarely get a result like this." To get a numerical measure of how strong the sample evidence is, replace the vague term "rarely" by a probability. Here is an example of this reasoning at work.

Example 9.8 Is the coffee fresh?

People of taste are supposed to prefer fresh-brewed coffee to the instant variety. On the other hand, perhaps many coffee drinkers just want their caffeine fix. A skeptic claims that only half of all coffee drinkers prefer fresh-brewed coffee. Let's do an experiment to test this claim.

Each of 50 subjects tastes two unmarked cups of coffee and says which he or she prefers. One cup in each pair contains instant coffee; the other, fresh-brewed coffee. The statistic that records the result of our experiment is the proportion of the sample who say they like the fresh-brewed coffee better. We find that 36 of our 50 subjects choose the fresh coffee. That is,

$$\hat{p} = \frac{36}{50} = 0.72 = 72\%$$

To make a point, let's compare our outcome $\hat{p} = 0.72$ with another possible result. If only 28 of the 50 subjects like the fresh coffee better than the instant coffee, the sample proportion is

$$\hat{p} = \frac{28}{50} = 0.56 = 56\%$$

Surely 72% is stronger evidence against the skeptic's claim than 56%. But how much stronger? Is even 72% in favor in a sample convincing evidence that a majority of the *population* prefer fresh coffee? Statistical tests answer these questions.

Here's the answer in outline form:

- **The claim.** The skeptic claims that only half of all coffee drinkers prefer fresh-brewed coffee. That is, he claims that the population proportion p is only 0.5. *Suppose for the sake of argument that this claim is true.*
- **The sampling distribution (page 422).** If the claim $p = 0.5$ were true and we tested many random samples of 50 coffee drinkers, the sample proportion would vary from sample to sample according to (approximately) the Normal distribution with

$$\text{mean} = p = 0.5$$

and

$$\begin{aligned}\text{standard deviation} &= \sqrt{\frac{p(1-p)}{n}}\\ &= \sqrt{\frac{0.5(0.5)}{50}}\\ &= 0.0707\end{aligned}$$

Figure 9.7 displays this Normal curve.

- **The data.** Place the sample proportion on the sampling distribution. You see in Figure 9.7 that $\hat{p} = 0.56$ isn't an unusual value but that $\hat{p} = 0.72$ is

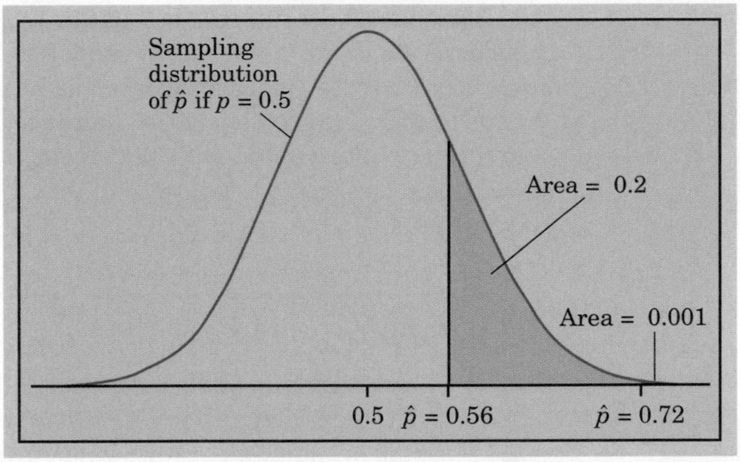

Figure 9.7 The sampling distribution of the proportion of 50 coffee drinkers who prefer fresh-brewed coffee. This distribution would hold if the truth about all coffee drinkers is that 50% prefer fresh coffee. The shaded area is the probability that the sample proportion is 56% or greater.

unusual. We would rarely get 72% of a sample of 50 coffee drinkers preferring fresh-brewed coffee if only 50% of all coffee drinkers felt that way. So the sample data do give evidence against the claim.

- **The probability.** We can measure the strength of the evidence against the claim by a probability. What is the probability that a sample gives a $\hat{p}$ this large or larger if the truth about the population is that $p = 0.5$? If $\hat{p} = 0.56$, this probability is the shaded area under the Normal curve in Figure 9.7. This area is 0.20. But our sample actually gave $\hat{p} = 0.72$. The probability of getting a sample outcome this large is only 0.001, an area too small to see in Figure 9.7. An outcome that would occur just by chance in 20% of all samples is *not* strong evidence against the claim. But an outcome that would happen only 1 in 1000 times *is* good evidence.

Be sure you understand why this evidence is convincing. There are two possible explanations of the fact that 72% of our subjects prefer fresh to instant coffee:

1. The skeptic is correct ($p = 0.5$) and by bad luck a very unlikely outcome occurred.
2. The population proportion favoring fresh coffee is greater than 0.5, so the sample outcome is about what would be expected.

We cannot be certain that Explanation 1 is untrue. Our taste test results *could* be due to chance alone. But the probability that such a result would occur by chance is so small (0.001) that we are quite confident that Explanation 2 is right.

Stating hypotheses

A significance test starts with a careful statement of the claims we want to compare. In most studies, we hope to show that some definite effect is present in the population.

In Example 9.8, we suspect that a majority of coffee drinkers prefer fresh-brewed coffee. A statistical test begins by supposing that the effect we seek is *not* present. We then look for evidence against this possibility and in favor of the effect we hope to find. The first step in a significance test is to state a claim that we will try to find evidence *against.* We call this claim our **null hypothesis.** The statement we hope or suspect is true instead of the null hypothesis is called the **alternative hypothesis.**

Null and alternative hypotheses

The claim being tested in a statistical test is called the **null hypothesis.** The test is designed to assess the strength of the evidence against the null hypothesis. Usually the null hypothesis is a statement of "no effect" or "no difference."

The claim about the population that we are trying to find evidence *for* is the **alternative hypothesis.**

We abbreviate the null hypothesis as H_0 (read as "H-naught") and the alternative hypothesis as H_a. Both of these hypotheses are statements about the population and so must be stated in terms of a population parameter.

In Example 9.8, the parameter is the proportion p of all coffee drinkers who prefer fresh to instant coffee. The null hypothesis is

$$H_0 : p = 0.5$$

The alternative hypothesis is that a majority of the population favor fresh coffee. In terms of the population parameter, this is

$$H_a : p > 0.5$$

ACTIVITY 9.2B

The *Significance Test* applet

APPLET The *Significance Test* applet on the book's Web site (**www.whfreeman.com/sta2e**) animates Example 9.7. That example asks if a basketball player's actual performance gives evidence against the claim that he makes 80% of his free throws. The parameter in question is the proportion p of free throws that the player will make if he shoots free throws forever. The population is all free throws the player will ever shoot.

The "null hypothesis" is always the same, that the player makes 80% of shots taken: we write this as $H_0 : p = 0.80$. The applet does not do a formal statistical test. Instead, it allows you to ask the player to shoot until you are reasonably confident that the true percent of shots made is or is not very close to 80%. The shooter claims that he makes 80% of his free throws. To test his claim, we have him shoot 20 free throws.

1. Set the applet to take 20 shots. Check "Show null hypothesis" so that the shooter's claim is visible in the graph.

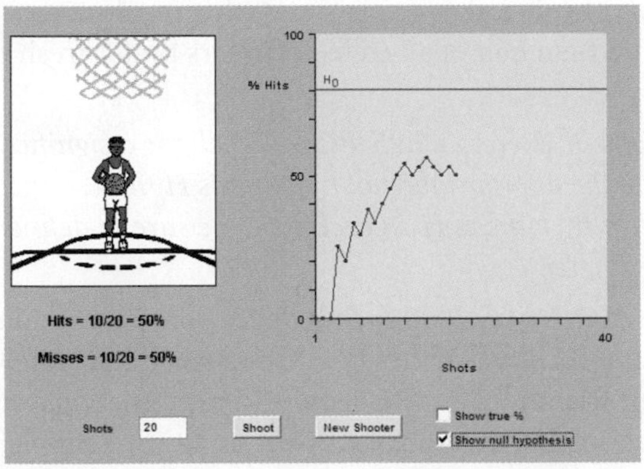

2. Click "Shoot." How many of the 20 shots did he make? Are you convinced that the shooter really makes less than 80%?

3. If you are not convinced, click "Shoot" again for 20 more shots. Keep going until *either* you are convinced that he doesn't make 80% of his shots *or* it appears that his true percent made is pretty close to 80%. How many shots did you watch him shoot? How many did he make? What did you conclude?

4. Click "Show true %" to reveal the truth. Was your conclusion correct?

Comment: You see why statistical tests say how strong the evidence is *against* some claim. If the shooter makes only 10 of 40 shots, you are pretty sure he can't make 80% in the long run. But even if he makes exactly 80 of 100, his true long-term percent might be 78% or 81% instead of 80%. It's hard to be convinced that he makes exactly 80%.

Exercises

9.21 A fair deck? Activity 9.2A (page 438) asks you to determine whether a deck of cards is fair by looking at the proportion of red cards you obtain in a simple random sample. Suppose that the deck of cards is fair and that you take an SRS of 25 cards.

(a) What are the mean and standard deviation of the sampling distribution of $\hat{p}$, the sample proportion of red cards?

(b) If your sample contains 16 red cards, what is the value of $\hat{p}$?

(c) Sketch a graph like Figure 9.7 (page 441) that shows the sampling distribution. Mark the location of the sample proportion $\hat{p}$ from (b) on the horizontal axis of your sketch.

(d) How likely is it that you would get a value of $\hat{p}$ as unusual as the one you did if the deck is fair? What do you conclude about whether the deck is fair?

9.22 Free throws Activity 9.2B shows you how to use the *Significance Test* applet at the book's Web site, **www.whfreeman.com/sta2e**, to test a basketball player's claim that he makes 80% of his free throws. Suppose that the player's claim is true and that you use the applet to shoot 20 shots.

(a) What are the mean and standard deviation of the sampling distribution of $\hat{p}$, the sample proportion of made free throws?

(b) If the player actually makes 8 shots, what is the value of $\hat{p}$?

(c) Sketch a graph like Figure 9.7 (page 441) that shows the sampling distribution. Mark the location of the sample proportion $\hat{p}$ from (b) on the horizontal axis of your sketch.

(d) How likely is it that you would get a value of $\hat{p}$ as unusual as the one you did if the player's claim is true? What do you conclude about whether the player is an 80% free-throw shooter?

Ukapala/Dreamstime.com

9.23 Fresh coffee, I Show how to calculate the two probabilities given in Example 9.8.

9.24 Fresh coffee, II In Example 9.8, 72% of the subjects in the experiment preferred fresh coffee. Explain why this result does not *prove* that more than half of all coffee drinkers prefer fresh to instant coffee.

The situations in Exercises 9.25 and 9.26 call for a significance test. State the appropriate null hypothesis H_0 and alternative hypothesis H_a in each case. Be sure to define your parameter.

9.25 Do you argue? A May 2005 Gallup Poll report on a national survey of 1028 teenagers revealed that 72% of teens said they rarely or never argue with their friends.[10] You wonder whether this national result would be true in your school. You conduct your own survey of a simple random sample of students at your school.

9.26 One potato, two potato A potato chip producer and a supplier of potatoes agree that each shipment of potatoes must meet certain quality standards. If the producer is convinced that more than 8% of the potatoes in the shipment have blemishes, the truck will be sent away to get another load of potatoes from the supplier. Otherwise, the entire truckload will be accepted.

P-values

A significance test looks for evidence against the null hypothesis and in favor of the alternative hypothesis. The evidence is strong if the outcome we observe would rarely happen if the null hypothesis is true but is more probable if the alternative hypothesis is true. For example, it would be surprising to find 36 of 50 subjects favoring fresh coffee if in fact only half of the population feels this way. How surprising? A significance test answers this question by giving a probability: the probability of getting an outcome at least as far as the actually observed outcome is from what we would expect when H_0 is true. We call this probability the ***P*-value.** What counts as "far from what we would expect" depends on H_a as well as H_0. In the taste test, the probability we want is the probability that 36 or more of 50 subjects favor fresh coffee. If the null hypothesis $p = 0.5$ is true, this probability is very small (0.001). That's good evidence that the null hypothesis is not true.

P-value

The probability, computed assuming that H_0 is true, that the sample outcome would be as extreme or more extreme than the actually observed outcome is called the ***P*-value** of the test. The smaller the *P*-value is, the stronger is the evidence against H_0 provided by the data.

In practice, most statistical tests are carried out by computer software that calculates the *P*-value for us. It is usual to report the *P*-value in describing the results of studies in many fields. You should therefore understand what P-values say even if you don't do statistical tests yourself, just as you should understand what "95% confidence" means even if you don't calculate your own confidence intervals.

Example 9.9 Count Buffon's coin

The French naturalist Count Buffon (1707–1788) tossed a coin 4040 times. He got 2048 heads. The sample proportion of heads is

$$\hat{p} = \frac{2048}{4040} = 0.507$$

That's a bit more than one-half. Is this evidence that Buffon's coin was not balanced? This is a job for a significance test.

The truth about Count Buffon's coin-tossing experiment.

The hypotheses. The null hypothesis says that the coin is balanced ($p = 0.5$). We did not suspect a bias in a specific direction before we saw the data, so the alternative hypothesis is just "the coin is not balanced." The two hypotheses are

$$H_0: p = 0.5$$
$$H_a: p \neq 0.5$$

The sampling distribution. If the null hypothesis is true, the sample proportion of heads has approximately the Normal distribution with

$$\text{mean} = p = 0.5$$

$$\text{standard deviation} = \sqrt{\frac{p(1-p)}{n}} = \sqrt{\frac{0.5(0.5)}{4040}} = 0.00787$$

Figure 9.8 shows this sampling distribution with Buffon's sample outcome $\hat{p} = 0.507$ marked. The picture already suggests that this is not an unlikely outcome that would give strong evidence against the claim that $p = 0.5$.

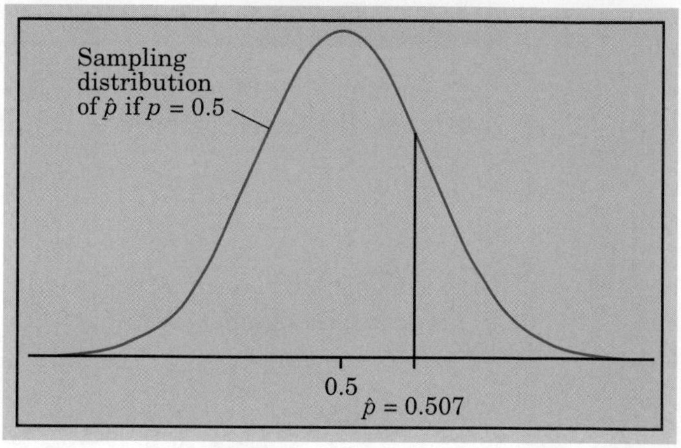

Figure 9.8 The sampling distribution of the proportion of heads in 4040 tosses of a balanced coin. Count Buffon's result, proportion 0.507 heads, is marked.

The *P*-value. How unlikely is an outcome as far from 0.5 as Buffon's $\hat{p} = 0.507$? Because the alternative hypothesis allows p to lie on either side of 0.5, values of $\hat{p}$ far from 0.5 in either direction provide evidence against H_0 and in favor of H_a. The *P*-value is therefore the probability that the observed $\hat{p}$ lies as far from 0.5 *in either direction* as the observed $\hat{p} = 0.507$. Figure 9.9 shows this probability as an area under the Normal curve. It is $P = 0.37$.

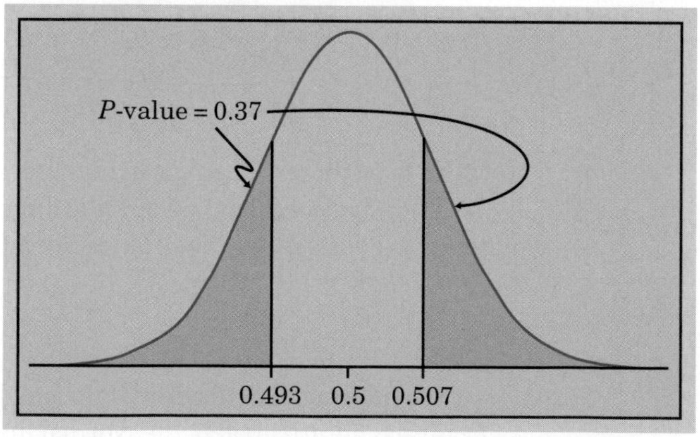

Figure 9.9 The *P*-value for testing whether Count Buffon's coin was balanced. This is the probability, calculated assuming a balanced coin, of a sample proportion as far or farther from 0.5 as Buffon's result of 0.507.

Conclusion. A truly balanced coin would give a result this far or farther from 0.5 in 37% of all repetitions of Buffon's trial. His result gives no reason to think that his coin was not balanced.

The alternative $H_a: p > 0.5$ in Example 9.8 is a **one-sided alternative** because the effect we seek evidence for says that the population proportion is greater than one-half. The alternative $H_a: p \neq 0.5$ in Example 9.9 is a **two-sided alternative** because we ask only whether or not the coin is balanced. Whether the alternative is one-sided or two-sided determines whether sample results, extreme in one or in both directions, count as evidence against H_0 and in favor of H_a.

www

Try the Java applet at **www.math.usu.edu/~schneit/CTIS/PValue/**. It allows you to test whether a coin is fair by simulating repeated tosses.

Statistical significance

We can decide in advance how much evidence against H_0 we will insist on. The way to do this is to say how small a *P*-value we require. The decisive value of *P* is called the **significance level.** It is usual to write it as α, the Greek letter alpha. If we choose $\alpha = 0.05$, we are requiring that the data give evidence against H_0 so strong that it would happen no more than 5% of the time (1 time in 20) when H_0 is true. If we choose $\alpha = 0.01$, we are insisting on stronger evidence against H_0, evidence so strong that it would appear only 1% of the time (1 time in 100) if H_0 is in fact true.

> **Statistical significance**
>
> If the *P*-value is smaller than α, we say that the data are **statistically significant at level** α.

"Significant" in the statistical sense does not mean "important." It means simply "not likely to happen just by chance." We used these words in Section 6.1 (page 267). Now we have attached a number to statistical significance to say what "not likely" means. You will often see significance at level 0.01 expressed by the statement "The results were significant ($P < 0.01$)." Here *P* stands for the *P*-value.

We don't need to make use of traditional levels of significance such as 5% and 1%. The *P*-value is more informative because it allows us to assess significance at any level we choose. For example, a result with $P = 0.03$ is significant at the $\alpha = 0.05$ level but not significant at the $\alpha = 0.01$ level. Nonetheless, the traditional significance levels are widely accepted guidelines for "how much evidence is enough." We might say that $P < 0.10$ indicates "some evidence" against the null hypothesis, $P < 0.05$ is "moderate evidence," and $P < 0.01$ is "strong evidence." Don't take these guidelines too literally, however. We will say more about interpreting tests in the next section.

Calculating *P*-values

Finding the *P*-values we gave in Examples 9.8 and 9.9 requires doing Normal distribution calculations using Table A. That was addressed in Section 3.1. In practice, software or a calculator does the calculation for us, but here is an example that shows how to use Table A.

Example 9.10 Tasting coffee

The hypotheses. In Example 9.8, we want to test the hypotheses

$$H_0 : p = 0.5$$
$$H_a : p > 0.5$$

Here p is the proportion of the population of all coffee drinkers who prefer fresh coffee to instant coffee.

The sampling distribution. If the null hypothesis is true, then $p = 0.5$. For samples of size 50, $\hat{p}$ follows a Normal distribution with mean 0.5 and standard deviation 0.0707.

The data. A sample of 50 people found that 36 preferred fresh coffee. The sample proportion is $\hat{p} = 0.72$.

The *P*-value. The alternative hypothesis is one-sided on the high side. So the *P*-value is the probability of getting an outcome at least as large as 0.72. Figure 9.7 (page 441) displays this probability as an area under the Normal sampling distribution curve. To find any Normal curve probability, move to the standard scale. The standard score for the outcome $\hat{p} = 0.72$ is

$$\text{standardized score} = \frac{\text{observation} - \text{mean}}{\text{standard deviation}}$$

$$z = \frac{0.72 - 0.5}{0.0707} = 3.11$$

Table A says that the area under a Normal curve to the left of 3.11 (in the standard scale) is 0.9991. The area to the right is therefore $1 - 0.9991 = 0.0009$, and that is our *P*-value.

Conclusion. The small P-value means that the data provide very strong evidence that a majority of the population prefer fresh coffee.

Exercises

9.27 Baggage check! Thousands of travelers pass through Guadalajara airport each day. Before leaving the airport, each passenger must pass through Customs. Customs officials want to be sure that passengers do not bring illegal items into the country. But they do not have time to search every traveler's luggage. Instead, they require each person to press a button. Either a red or a green bulb lights up. If the red light shows, the passenger will be searched by Customs agents. A green light means "go ahead." Customs officers claim that the probability that the light turns red on any press of the button is 0.30. A weary traveler waiting for her luggage watches the next 100 people press the button. The light turns red 20 times.

(a) State null and alternative hypotheses for testing the claim made by Customs officials.

(b) Describe the sampling distribution assuming the null hypothesis from (a) is true.

(c) How extreme is the sample outcome? Find the *P*-value.

(d) Explain your conclusions in nontechnical language.

9.28 Affirmative action The *Chronicle of Higher Education* commissioned a comprehensive survey of American attitudes toward colleges and universities. Of the 1000 American adults aged 25 to 64 surveyed, 640 said that colleges should not admit minority students who have lower grades than other qualified candidates.[11] Is this sufficient evidence to conclude that a majority of adults have this view of affirmative action?

(a) State null and alternative hypotheses.

(b) Describe the sampling distribution.

(c) How extreme is the sample outcome? Find the *P*-value.

(d) Explain your conclusions in nontechnical language.

© Godfer/Dreamstime.com

9.29 Bullies in middle school If one study is any indication, many middle school students are not shy about admitting aggressive behavior in school. A University of Illinois study on aggressive behavior surveyed 558 students in a midwestern middle school. When asked to describe their behavior in the last 30 days, 445 students said their behavior included physical aggression, social ridicule, teasing, name-calling, and issuing threats. This behavior was not defined as bullying in the questionnaire.[12]

(a) Is this evidence that more than three-quarters of the students at that middle school engage in bullying behavior? Carry out a significance test to answer this question. State hypotheses, describe the sampling distribution, and find the *P*-value. State your conclusions.

(b) Does this study provide strong evidence of frequent bullying in America's middle schools? Explain.

9.30 Count Buffon's coin again Refer to Example 9.9 (page 445).

(a) Construct and interpret a 95% confidence interval for the overall proportion of times that Buffon's coin would land heads up.

(b) Is $p = 0.5$ a plausible value for the population proportion? Justify your answer.

9.31 *P*-values and significance, I A test of the null hypothesis $H_0 : p = 0.3$ yields $z = -1.77$.

(a) If the alternative hypothesis is $H_a : p < 0.3$, find the *P*-value for this test. Is this result significant at the $\alpha = 0.05$ level? At the $\alpha = 0.01$ level? Explain.

(b) If the alternative hypothesis is $H_a : p \neq 0.3$, find the *P*-value for this test. Is this result significant at the $\alpha = 0.05$ level? At the $\alpha = 0.01$ level? Explain.

9.32 *P*-values and significance, II A test of the null hypothesis $H_0 : p = 0.68$ yields $z = 2.35$.

(a) If the alternative hypothesis is $H_a : p > 0.68$, find the *P*-value for this test. Is this result significant at the $\alpha = 0.05$ level? At the $\alpha = 0.01$ level? Explain.

(b) If the alternative hypothesis is $H_a : p \neq 0.68$, find the *P*-value for this test. Is this result significant at the $\alpha = 0.05$ level? At the $\alpha = 0.01$ level? Explain.

CALCULATOR CORNER Significance test for a population proportion

The TI-84/TI-Nspire can be used to test a claim about a population proportion. Consider Example 9.9 about Count Buffon's coin-tossing. In $n = 4040$ tosses, Buffon observed $X = 2048$ heads. Recall that our hypotheses were

$$H_0 : p = 0.5$$
$$H_a : p \neq 0.5$$

To perform a significance test:

TI-84

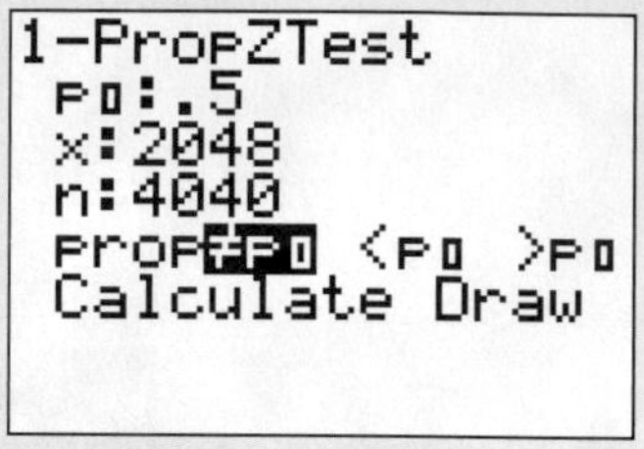

- Press STAT, choose TESTS, and then choose 1-PropZTest.
- On the 1-PropZTest screen, enter these values: $p_0 = 0.5$, $x = 2048$, and $n = 4040$. Specify the alternative hypothesis as prop≠p_0.
- If you select the "Calculate" option and press ENTER, you will see that the z statistic is 0.88 and the P-value is 0.3783.
- If you select the "Draw" option, you will see the screen shown at the right, below.

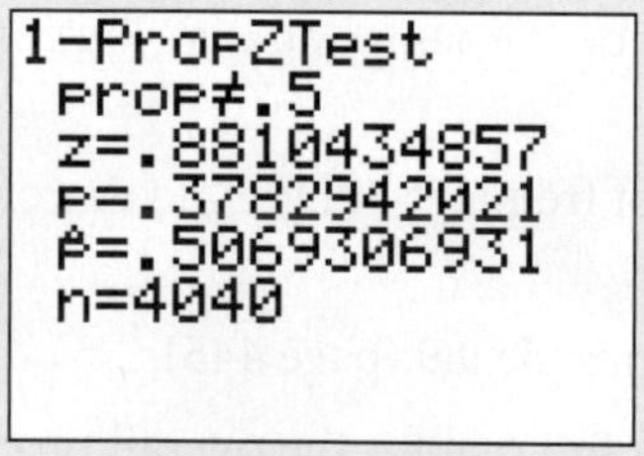

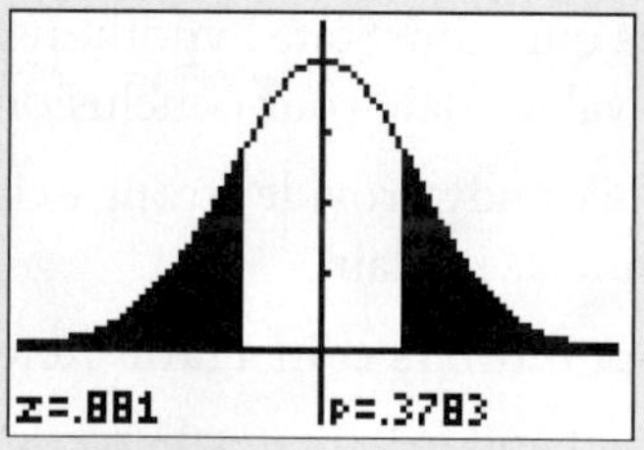

TI-Nspire

You can perform significance tests in a Calculator page or a Lists & Spreadsheet page. There are some slight differences, which we'll illustrate now.

- On a Calculator page, press (menu), and choose 5:Statistics, 7:Stat Tests..., and 5:1-Prop z Test.
- In the dialog box, enter these values: $p_0 = 0.5$, $x = 2048$, and $n = 4040$. Specify the alternative hypothesis as H_a : prop $\neq p_0$. Then click OK.

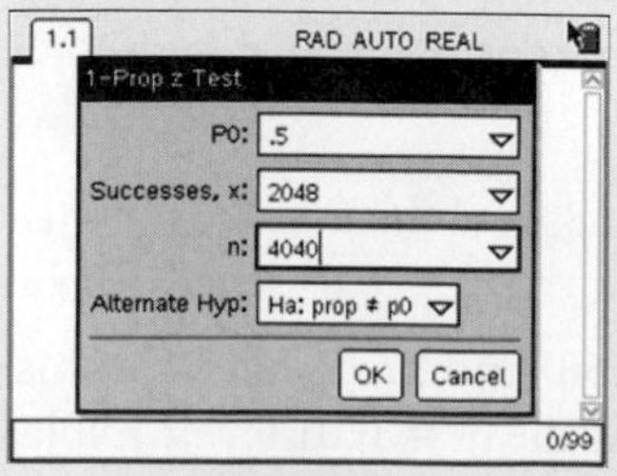

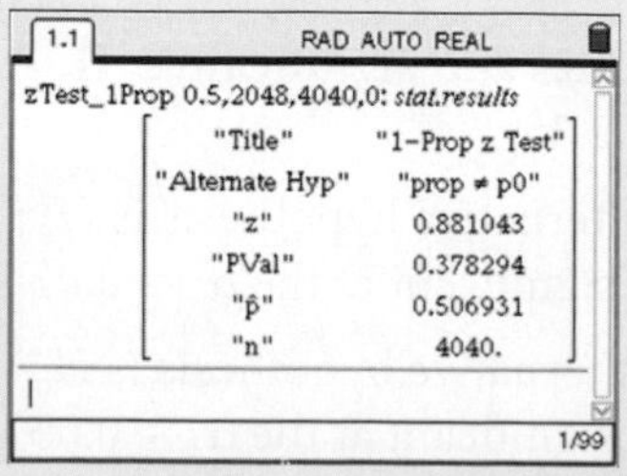

- You can also carry out the hypothesis test in a Lists & Spreadsheet page. Press (menu), and choose `4:Statistics`, `4:Stat Tests...`, and `5:1-Prop z Test`.
- In the dialog box, enter these values: $p_0 = 0.5$, $x = 2048$, and $n = 4040$. Specify the alternative hypothesis as H_a: prop $\neq p_0$. Choose any empty column for the "1st Result Column." If you click the box next to "Draw," you will get a picture of the standard Normal curve with appropriate shading, along with the values of z and P.

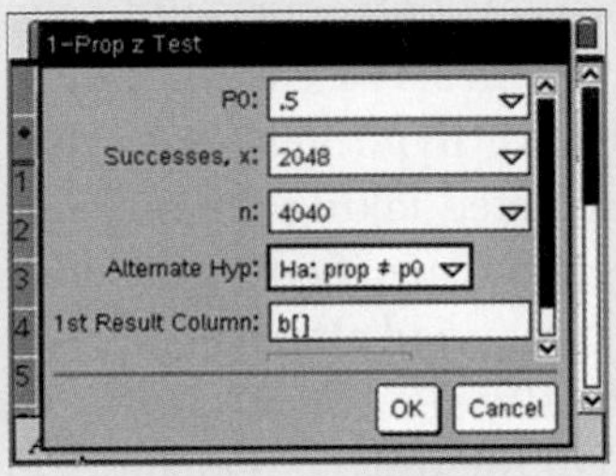

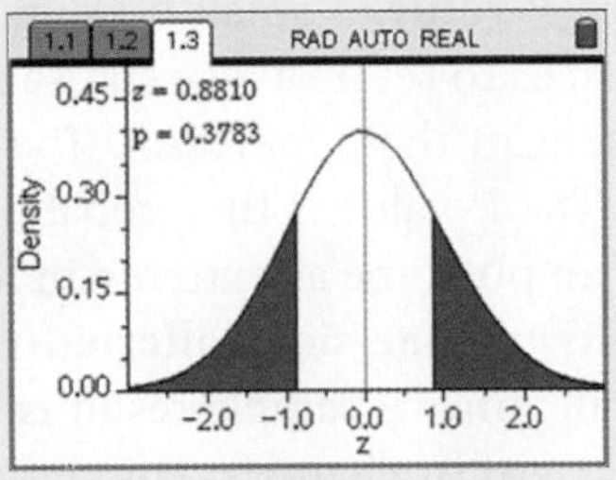

APPLICATION 9.2

Pair-a-dice!

Materials: Pair of dice (one red, one white) provided by your teacher

Students pair off, Player 1 with the red die and Player 2 with the white die. Each player rolls the die, and the higher number wins that round. A tie is ignored and the dice are rolled again. The purpose is to look at the proportion of wins by the red die over the long term.

- Roll the dice 10 times and use tally marks to record the winner.

Red wins:
White wins:

To make sure that no one has an unfair advantage, exchange the dice and roll another 10 times. Exchange the dice again and roll 10 more times, and then exchange again and roll 10 more times. Continue to record your results. You should have a total of 40 rolls.

- Combine your results with the other teams to obtain a grand total of red wins and white wins.

QUESTIONS

1. Our assumption is that each die should win about half the time in the long run. State null and alternative hypotheses in terms of the proportion of times the red die should win.
2. Draw a graph of the sampling distribution of $\hat{p}$ assuming that H_0 is true. Mark the location of the class's $\hat{p}$-value on the sketch.
3. Calculate the P-value. Is there sufficient evidence to reject the null hypothesis?
4. State your conclusion clearly.
5. Construct and interpret a 95% confidence interval for the parameter p. What additional information does this confidence interval provide?
6. Carefully inspect the red die. Can you explain the results that you observed?

Section 9.2 Summary

A confidence interval estimates an unknown parameter. A **significance test** assesses the evidence for some claim about the value of an unknown parameter. In practice, the purpose of a statistical test is to answer the question "Could the effect we see in the sample just be an accident due to chance, or is it good evidence that the effect is really there in the population?"

Significance tests answer this question by giving the probability that a sample effect as large as the one we see in this sample would arise just by chance. This probability is the ***P*-value.** A small *P*-value says that our outcome is unlikely to happen just by chance. To set up a test, state a **null hypothesis** that says the effect you seek is not present in the population. The **alternative hypothesis** says that the effect is present. The *P*-value is the probability, calculated taking the null hypothesis to be true, of an outcome as extreme in the direction specified by the alternative hypothesis (either a **one-sided alternative** or a **two-sided alternative**) as the actually observed outcome. A sample result is **statistically significant at the 5% level** if it would occur just by chance no more than 5% of the time in repeated samples.

This section concerns the basic reasoning of tests and the details of tests for hypotheses about a population proportion. There is more discussion of the practical interpretation of statistical tests in the next section.

Section 9.2 Exercises

The situations in Exercises 9.33 and 9.34 call for a significance test. State the appropriate null hypothesis H_0 and alternative hypothesis H_a in each case. Be sure to define your parameter.

9.33 No homework?! Mr. Starnes believes that fewer than 75% of the students at his school completed their math homework last night. The math teachers inspect the homework assignments from a random sample of students at the school to help Mr. Starnes test his claim.

9.34 Lefties Sally reads a newspaper report claiming that 12% of all adults in the United States are left-handed. She wonders whether 12% of the students at her large public high school are left-handed.

9.35 Where's the blunder? Write a sentence or two to explain the blunder in each of the following statements.

(a) Austin says that a result that is significant at the $\alpha = 0.05$ level must also be significant at the $\alpha = 0.01$ level.

(b) Asked to explain the meaning of "statistically significant at the $\alpha = 0.05$ level," Carla says: "This means that the probability that the null hypothesis is true is less than 0.05."

9.36 Is the Belgian euro coin fair? Exercise 7.23 (page 327) asked you to use simulation to determine whether a Belgian euro coin that landed "heads" 140 times out of 250 spins was minted fairly. One of the Polish math professors who spun the coin concluded that the coin was not fair. A representative from the Belgian mint argued that the result was due to chance.

(a) Carry out a significance test to determine whether the coin was fair. Follow the steps in Example 9.10 (page 448). What do you conclude?

(b) Construct and interpret a 95% confidence interval. What additional information does the confidence interval provide?

9.37 *P*-values and significance, III A test of the null hypothesis $H_0: p = 0.81$ versus the one-sided alternative $H_a: p > 0.81$ yields $z = 2.36$.

(a) Find the *P*-value for this test.

(b) Is this result significant at the $\alpha = 0.05$ level? At the $\alpha = 0.01$ level? Explain.

9.38 *P*-values and significance, IV A test of the null hypothesis $H_0: p = 0.81$ versus the two-sided alternative $H_a: p \neq 0.81$ yields $z = 2.36$.

(a) Find the *P*-value for this test.

(b) Is this result significant at the $\alpha = 0.05$ level? At the $\alpha = 0.01$ level? Explain.

9.39 Are cell phones distracting? The American Automobile Association (AAA) commissioned a study by researchers at the University of North Carolina on sources of distraction for drivers. The study tracked 70 drivers aged 18 to 80 for a week. Miniature cameras were placed in the drivers' cars, with their knowledge, and their driving behavior was randomly viewed for a week. The first three hours of each tape were discarded in the hope that drivers would act more naturally later in the week. The researchers considered a wide range of activities to be distracting, such as leaning over to reach for something, fiddling with radio controls, attending to babies, talking to passengers, applying makeup, and opening and reading their mail. The study found that 21 of the subjects used cell phones while their vehicles were moving.[13]

(a) Is there significant evidence that fewer than half of all drivers use cell phones while their vehicles are moving? Carry out appropriate inference, and report your results and conclusions.

(b) Do you believe that this study is subject to bias? In what regard?

9.40 Hack-a-Shaq Any NBA fan has heard about the free-throw shooting woes of the NBA's most dominant center, Shaquille O'Neal. Over his NBA career, Shaq has made 53.3% of his free throws. One off-season, Shaq worked with an assistant coach on his free-throw technique. During the first two games of the next season, Shaq made 26 out of 39 free throws. Do these results provide evidence that Shaq has significantly improved his free-throw shooting? Carry out a significance test to help answer this question.

9.3 Use and Abuse of Statistical Inference

ACTIVITY 9.3

Did you get enough blues?

Materials: One 1.69-ounce bag of milk chocolate M&M's per student

The M&M/Mars Company, headquartered in Hackettstown, New Jersey, makes milk chocolate candies. In 1995, they decided to replace the tan-colored M&M's with a new color. After conducting an extensive national survey, they chose blue as the new color. The company mixes the colors for each type of M&M's candy (plain, peanut, etc.) according to a specific percent distribution. At the time we're writing, the company claims that the distribution of colors in bags of plain M&M's candies is 24% blue, 20% orange, 16% green, 14% yellow, 13% red, and 13% brown. You

ACTIVITY 9.3 *(continued)*

ACTIVITY 9.3 *(continued)*

may want to check this information at the M&M's Web site, **www.mms.com**, before you continue. In this Activity, we'll investigate the company's claim about the proportion of M&M's that are blue.

1. Your teacher will supply each student with a 1.69-ounce bag of M&M's candies. Open your bag and carefully count the number of M&M's of each color as well as the total number of M&M's in the bag. Record your counts in a table like this:

Color:	Blue	Orange	Green	Yellow	Red	Brown
Total Count:						

2. Carry out a significance test of the company's claim about the actual proportion p of blue M&M's based on the data in your sample (bag). Use a 5% significance level ($\alpha = 0.05$). What conclusion do you draw?

3. Pool results with your classmates. What percent of the class rejected the company's claim? Are you surprised?

4. Construct and interpret a 95% confidence interval for the population proportion of blue M&M's based on your sample. Is your result consistent with your significance test in Step 2? What additional information does your confidence interval give?

5. Some students may reject the company's claim based on their sample data even though the company's claim is actually true. Discuss with your classmates how this might happen.

6. Your teacher will draw a horizontal axis at the top of the chalkboard with a scale that goes from 0 to 0.6. Each student should draw his or her confidence interval below the axis. What proportion of the class's confidence intervals contained 0.24? Are you surprised?

7. Keep your data from this Activity handy. You will need it again in the next chapter.

Using inference wisely

We have met the two major types of statistical inference: confidence intervals and significance tests. We have, however, seen only one inference method of each type, designed for inference about a population proportion p. There are libraries of both books and software filled with methods for inference about various parameters in various settings. The reasoning of confidence intervals and significance tests remains the same, but the details can seem overwhelming. The first step in using inference wisely is to understand your data and the questions you want to answer and to fit the method to its setting. Here are some tips on inference, adapted to the one setting we are familiar with.

The design of the data production matters "Where do the data come from?" remains the first question to ask in any statistical study. Any inference method is intended for use in a specific setting. For our confidence interval and test for a proportion p:

- The data must be a simple random sample (SRS) from the population of interest. When you use these methods, you are acting as if the data are an SRS. In practice, it is often not possible to actually choose an SRS from the population. Your conclusions may then be open to challenge. In Activity 9.3, for example, each individual bag of plain M&M's candies may *not* be equivalent to an SRS. As a result, several students in your class may reject the null hypothesis $H_0: p = 0.24$ even though the company is telling the truth about the population proportion of blue M&M's. Likewise, your class's 95%

confidence intervals for p may not have a 95% capture rate if each student's data do not come from an SRS.

- These methods are not correct for sample designs more complex than an SRS, such as stratified samples. There are other methods that fit these settings.
- There is no correct method for inference from data haphazardly collected with bias of unknown size. Fancy formulas cannot rescue badly produced data.
- Other sources of error, such as dropouts and nonresponse, are important. Remember that confidence intervals and tests use the data you give them and ignore these practical difficulties.

Dropping out

An experiment found that weight loss is significantly more effective than exercise for reducing high cholesterol and high blood pressure. The 170 subjects were randomly assigned to a weight-loss program, an exercise program, or a control group. Only 111 of the 170 subjects completed their assigned treatment, and the analysis used data from these 111. Did the dropouts create bias? Always ask about details of the data before trusting inference.

Know how confidence intervals behave A confidence interval estimates the unknown value of a parameter and also tells us how uncertain the estimate is. All confidence intervals share these behaviors:

- The confidence level says how often the *method* catches the true parameter in very many uses. We never know whether a specific data set gives us an interval that contains the true parameter. All we can say is that "we got this result from a method that works 95% of the time." Our data set might be one of the 5% that produce an interval that misses the parameter. If that risk bothers you, use a 99% confidence interval.
- High confidence is not free. A 99% confidence interval will be wider than a 95% confidence interval based on the same data. There is a trade-off between how closely we can pin down the parameter and how confident we are that we have caught the parameter.
- Larger samples give shorter intervals. If we want high confidence and a short interval, we must take a larger sample. The length of our confidence interval for p goes down in proportion to the square root of the sample size. To cut the interval in half, we must take four times as many observations. This is typical of many types of confidence intervals.

Know what statistical significance says Many statistical studies hope to show that some claim is true. A clinical trial compares a new drug with a standard drug because the doctors hope that patients given the new drug will do better. A psychologist studying gender differences suspects that women will do better than men (on the average) on a test that measures social-networking skills. The purpose of significance tests is to weigh the evidence that the data give in favor of such claims. That is, a test helps us know if we have found what we were looking for.

To do this, we ask what would happen if the claim were not true. That's the null hypothesis—no difference between the two drugs, no difference between women and men. A significance test answers only one question: "How strong is the evidence that the null hypothesis is not true?" A test answers this question by giving a P-value. The P-value tells us how unlikely our data would be if the null hypothesis were true. Data that are very unlikely are good evidence that the null hypothesis is not true. We never know whether the hypothesis is true for this specific population. All we can say is, "Data like these would occur only 5% of the time if the hypothesis were true."

This kind of indirect evidence against the null hypothesis (and for the effect we hope to find) is less straightforward than a confidence interval.

Know what your methods require Our test and confidence interval for a proportion p require that the population be much larger than the sample. They also require that the sample itself be reasonably large so that the sampling distribution of the sample proportion is close to Normal. We have said little about the specifics of these requirements because the reasoning of inference is more important. Just as there are inference methods that fit stratified samples, there are methods that fit small samples and small populations. If you plan to use statistical inference in practice, you will need help from a statistician (or need to learn lots more statistics) to manage the details.

Most of us read about statistical studies more often than we actually work with data ourselves. Concentrate on the big issues, not on the details of whether the authors used exactly the right inference methods. Does the study ask the right questions? Where did the data come from? Do the results make sense? Does the study report confidence intervals so you can see both the estimated values of important parameters and how uncertain the estimates are? Does it report P-values to help convince you that findings are not just good luck?

Exercises

mediablitzimages Limited/Alamy

9.41 Activity 9.3 follow-up Refer to Activity 9.3 (page 453). Statistics classes like yours have examined the M&M's color distribution stated by the M&M/Mars Company for many years. In response to consumer questions about the accuracy of their claims, the company wrote:

While we mix the colors as thoroughly as possible, the above ratios may vary somewhat, especially in the smaller bags. This is because we combine the various colors in large quantities for the last production stage (printing). The bags are then filled on high-speed packaging machines by weight, not by count.

Which of the four big ideas of this section is the company addressing?

9.42 A television poll A television news program conducts a call-in poll about a proposed city ban on handgun ownership. Of the 2372 calls, 1921 oppose the ban. The station, following recommended practice, makes a confidence statement: "81% of the Channel 13 Pulse Poll sample opposed the ban. We can be 95% confident that the true proportion of citizens opposing a handgun ban is within 1.6% of the sample result."

(a) Show that the station's confidence interval calculation is correct.

(b) Nonetheless, the station's conclusion is not justified. Why?

9.43 Only-child presidents Joe is writing a report on the backgrounds of American presidents. He looks up information on how many of the 44 presidents were only children. Because Joe took a statistics course, he uses these results to get a 95% confidence interval for the true proportion p of these 44 U.S. presidents who were only children. This makes no sense. Why not?

9.44 Who will win? A poll taken shortly before an election finds that 52% of the voters favor candidate Shrub over candidate Snort. The poll has a margin of sampling error of plus or minus 3 percentage points at 95% confidence. The poll press release says the election is too close to call. Why?

9.45 Dropping out The marginal nugget on page 455 describes an experiment comparing the effectiveness of weight loss and exercise in reducing cholesterol level and blood pressure. Explain how the 59 people who dropped out may have created bias in the results of the study.

9.46 Inference conditions Explain why each of the two conditions for performing inference about a population proportion p—population much larger than sample size and sample size large enough—is important.

The woes of significance tests

The purpose of a significance test is usually to give evidence for the presence of some effect in the population. The effect might be a probability of heads different from one-half for a coin or a longer mean survival time for patients given a new cancer treatment. If the effect is large, it will show up in most samples—the proportion of heads among our tosses will be far from one-half, or the patients who get the new treatment will live much longer than those in the control group. Small effects, such as a probability of heads only slightly different from one-half, will often be hidden behind the chance variation in a sample. This is as it should be: big effects are easier to detect. That is, the *P*-value will usually be small when the population truth is far from the null hypothesis.

The "woes" of testing start with the fact that a test just measures the strength of evidence against the null hypothesis. It says nothing about how big or how important the effect we seek in the population really is. For example, our hypothesis might be "This coin is balanced." We express this hypothesis in terms of the probability p of getting a head as $H_0: p = 0.5$. No real coin is exactly balanced, so we know that this hypothesis is not exactly true. If this coin has probability $p = 0.502$ of a head, we might say that for practical purposes it is balanced. A statistical test doesn't think about "practical purposes." It just asks if there is evidence that p is not exactly equal to 0.5. The focus of tests on the strength of the evidence against an exact null hypothesis is the source of much confusion in using tests.

Pay particular attention to the size of the sample when you read the result of a significance test. Here's why:

- Larger samples make tests of significance more sensitive. If we toss a coin hundreds of thousands of times, a test of $H_0: p = 0.5$ will often give a very low *P*-value when the truth for this coin is $p = 0.502$. The test is right—it found good evidence that p really is not exactly equal to 0.5—but it has picked up a difference so small that it is of no practical interest. **A finding can be statistically significant without being practically important.**
- On the other hand, tests of significance based on small samples are often not sensitive. If you toss a coin only 10 times, a test of $H_0 : p = 0.5$ will often give a large *P*-value even if the truth for this coin is $p = 0.7$. Again the test is right—10 tosses are not enough to give good evidence against the null hypothesis. **Lack of significance does not mean that there is no effect,**

only that we do not have good evidence for an effect. Small samples often miss effects that are really present in the population.

Whatever the truth about the population—whether $p = 0.7$ or $p = 0.502$—more observations allow us to pin down p more closely. If p is not 0.5, more observations will give more evidence of this, that is, a smaller P-value. Because significance depends strongly on the sample size as well as on the truth about the population, statistical significance tells us nothing about how large or how practically important an effect is. Large effects (like $p = 0.7$ when the null hypothesis is $p = 0.5$) often give data that are insignificant if we take only a small sample. Small effects (like $p = 0.502$) often give data that are highly significant if we take a large sample. Let's return to a favorite example to see how significance changes with sample size.

Example 9.11 Count Buffon's coin again

Count Buffon tossed a coin 4040 times and got 2048 heads. His sample proportion of heads was

$$\hat{p} = \frac{2048}{4040} = 0.507$$

Is Buffon's coin balanced? The hypotheses are

$$H_0 : p = 0.5$$
$$H_a : p \neq 0.5$$

The test of significance works by locating the sample outcome $\hat{p} = 0.507$ on the sampling distribution that describes how $\hat{p}$ would vary if the null hypothesis were true. Figure 9.10 repeats Figure 9.8 (page 446). It shows that the observed $\hat{p} = 0.507$ is not surprisingly far from 0.5 and therefore is not good evidence against the hypothesis that the true p is 0.5. The P-value, which is 0.37, just makes this precise.

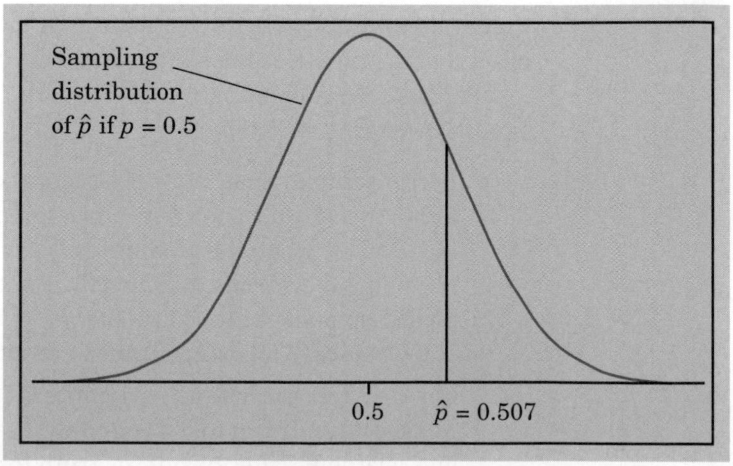

Figure 9.10 The sampling distribution of the proportion of heads in 4040 tosses of a coin if in fact the coin is balanced. Sample proportion 0.507 is not an unusual outcome.

Suppose that Count Buffon got the same result, $\hat{p} = 0.507$, from tossing a coin 1000 times and also from tossing a coin 100,000 times. The sampling distribution of $\hat{p}$ when the null hypothesis is true always has mean 0.5, but its standard deviation gets smaller as the sample size n gets larger. Figure 9.11 displays the three sampling distributions, for $n = 1000$, $n = 4040$, and $n = 100{,}000$. The middle curve in this figure is the same Normal curve as in Figure 9.10, but drawn on a scale that allows us to show the very tall and narrow curve for $n = 100{,}000$. Locating the sample outcome $\hat{p} = 0.507$ on the three curves, you see that the same outcome is more or less surprising depending on the size of the sample.

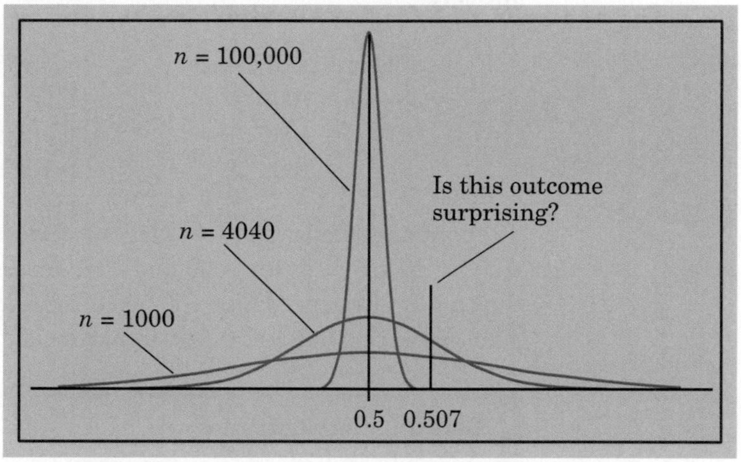

Figure 9.11 The three sampling distributions of the proportion of heads in 1000, 4040, and 100,000 tosses of a balanced coin. Sample proportion 0.507 is not unusual in 1000 or 4040 tosses but is very unusual in 100,000 tosses.

The P-values are $P = 0.66$ for $n = 1000$, $P = 0.37$ for $n = 4040$, and $P = 0.000009$ for $n = 100{,}000$. Imagine tossing a balanced coin 1000 times repeatedly. You will get a proportion of heads at least as far from one-half as Buffon's 0.507 in about two-thirds of your repetitions. If you toss a balanced coin 100,000 times, however, you will almost never (9 times in a million repeats) get an outcome this unbalanced.

The outcome $\hat{p} = 0.507$ is not evidence against the hypothesis that the coin is balanced if it comes up in 1000 tosses or in 4040 tosses. It is completely convincing evidence if it comes up in 100,000 tosses.

The moral of Example 9.11 is clear: sample size matters!

Beware the naked *P*-value

The *P*-value of a significance test depends strongly on the size of the sample, as well as on the truth about the population.

It is bad practice to report a *P*-value without also giving the sample size and a statistic or statistics that describe the sample outcome.

The advantages of confidence intervals

Example 9.11 suggests that we not rely on significance alone in understanding a statistical study. Just knowing that the sample proportion was $\hat{p} = 0.507$ helps a lot. You can decide whether this deviation from one-half is large enough to interest you. Of course, $\hat{p} = 0.507$ isn't the exact truth about the coin, just the chance result of Count Buffon's tosses. So a confidence interval, whose width shows how closely we can pin down the truth about the coin, is even more helpful. Here are the 95% confidence intervals for the true probability of a head p, based on the three sample sizes in Example 9.11. You can check that the method of Section 9.1 gives these answers.

Number of tosses	95% confidence interval
$n = 1000$	0.507 ± 0.031, or 0.476 to 0.538
$n = 4040$	0.507 ± 0.015, or 0.492 to 0.522
$n = 100{,}000$	0.507 ± 0.003, or 0.504 to 0.510

The confidence intervals make clear what we know (with 95% confidence) about the true p. The intervals for 1000 and 4040 tosses include 0.5, so we are not confident the coin is unbalanced. For 100,000 tosses, however, we are confident that the true p lies between 0.504 and 0.510. In particular, we are confident that it is not 0.5.

Give a confidence interval

Confidence intervals are more informative than tests because they actually estimate a population parameter. They are also easier to interpret. It is good practice to give confidence intervals whenever possible.

Significance at the 5% level isn't magical

The purpose of a significance test is to describe the degree of evidence provided by the sample against the null hypothesis. The P-value does this. But how small a P-value is convincing evidence against the null hypothesis? This depends mainly on two circumstances:

- *How plausible is H_0?* If H_0 represents an assumption that the people you must convince have believed for years, strong evidence (small P) will be needed to persuade them.
- *What are the consequences of rejecting H_0?* If rejecting H_0 in favor of H_a means making an expensive changeover from one type of product packaging to another, you need strong evidence that the new packaging will boost sales.

These criteria are a bit subjective. Different people will often insist on different levels of significance. Giving the P-value allows each of us to decide individually if the evidence is sufficiently strong.

Users of statistics have often emphasized standard levels of significance such as 10%, 5%, and 1%. For example, courts have tended to accept 5% as a standard in discrimination cases. This emphasis reflects the time when tables of critical values rather than computer software dominated statistical practice. The 5% level

($\alpha = 0.05$) is particularly common. **There is no sharp border between "significant" and "insignificant," only increasingly strong evidence as the *P*-value decreases.** There is no practical distinction between the *P*-values 0.049 and 0.051. It makes no sense to treat $P < 0.05$ as a universal rule for what is significant.

Exercises

Exercises 9.47 to 9.49 refer to the following brief newspaper article.[14]

> ### Depression Pill Seems to Help Smokers Quit
>
> BOSTON—Taking an antidepression medicine appears to double smokers' chances of kicking the habit, a study found. The Food and Drug Administration approved the marketing of this medicine, called Zyban or bupropion, to help smokers in May. The results of several studies with the drug, including one published in today's issue of the *New England Journal of Medicine,* were made public then.
>
> The newly published study was conducted on 615 volunteers who wanted to give up smoking and were not outwardly depressed. They took either Zyban or dummy pills for 6 weeks. A year later, 23 percent of those getting Zyban were still off cigarettes, compared with 12 percent in the comparison group.

9.47 Significant, but important? The results of this experiment were significant at the $\alpha = 0.05$ level. In your opinion, are the results practically important? Justify your position.

9.48 Generalizability To what population can the results of this study be generalized? Explain.

9.49 Causation? Can we conclude that taking Zyban *causes* people to quit smoking? Justify your answer.

9.50 *P*-value or significance level? In performing a significance test, the researcher can choose between adopting a fixed significance level or calculating a *P*-value. Explain how a *P*-value gives more information.

9.51 Searching for significance You perform 1000 significance tests using $\alpha = 0.05$. Assuming that all null hypotheses are true, about how many of the test results would you expect to be statistically significant? Explain how you obtained your answer.

9.52 Plagiarizing An online poll posed the following question:

> *It is now possible for school students to log on to Internet sites and download homework. Everything from book reports to doctoral dissertations can be downloaded free or for a fee. Do you believe giving a student who is caught plagiarizing an "F" for their assignment is the right punishment?*

Of the 20,125 people who responded, 14,793 clicked "Yes." That's 73.5% of the sample. Based on this sample, a 95% confidence interval for the percent of the population who would say "Yes" is 73.5% ± 0.61%. Why is this confidence interval worthless?

APPLICATION 9.3

Return of the free-throw shooter

APPLET In Activity 9.2B (page 442), we introduced the *Significance Test* applet at the book's Web site, **www.whfreeman.com/sta2e.** Now that you have studied some of the cautions for performing inference about a population proportion, let's return to the applet once again.

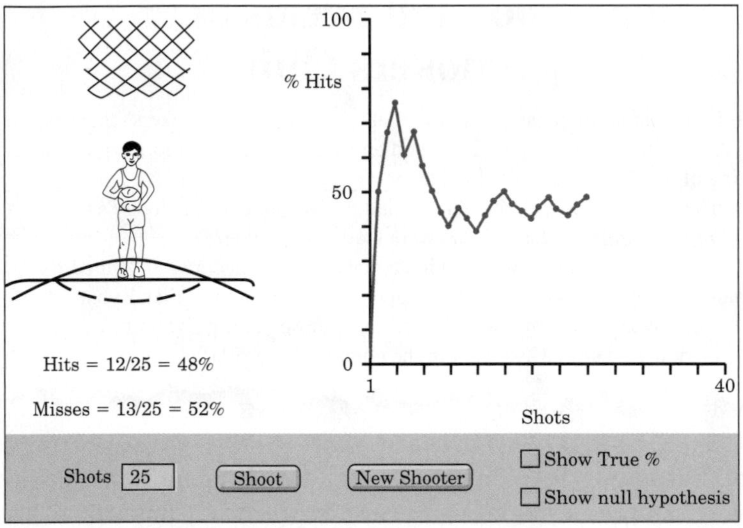

APPLET TIP: The null hypothesis is always p = 0.8, and the alternative hypothesis is p < 0.8 since the applet never fixes p at a value higher than 0.8.

QUESTIONS

1. Use the applet to have the shooter make five shots. Based on these data, can you decide whether the shooter can hit 80% of his shots overall? Explain why or why not.

2. Add another five shots to the total. Can you decide at this point?

3. Continue to add five shots at a time until you are confident that you know whether the null hypothesis is true or false. How many shots did it take? Perform a significance test using the sample of free throws that the player has shot. Do the results of the significance test confirm your belief?

4. Check the "Show True %" box to see the true value of p. Record this value.

5. Remove the true value of p from the plot and pick a new shooter. Repeat the process of using the applet to add five shots at a time until you feel confident that you know whether the null hypothesis is true or false. Did it take more shots or fewer shots this time? View and record the true value of p.

6. Repeat Step 5 several more times. Did it take longer for you to determine whether the null hypothesis was true or false when p was closer to 0.8 or when it was farther from 0.8? Explain why this makes sense.

Section 9.3 Summary

Statistical inference is less widely applicable than exploratory analysis of data. Any inference method requires the right setting, in particular the right design for a random sample or randomized experiment. Understanding the meaning of confidence levels and statistical significance helps avoid improper conclusions. Increasing the number of observations has a straightforward effect on confidence intervals: the interval gets shorter for the same level of confidence. More observations usually drive down the P-value of a test when the truth about the population stays the same, making tests harder to interpret than confidence intervals. A finding with a small P-value may not be practically interesting if the sample is large, and an important truth about the population may fail to be significant if the sample is small. Avoid depending on fixed significance levels such as 5% to make decisions.

Section 9.3 Exercises

9.53 Cholesterol and breast cancer The *Journal of Women's Health* reported on a study of 7528 women aged 65 and older. The women were all participants in a study of fractures due to osteoporosis, a disease that weakens the bones. At one clinic visit during the study, researchers asked the women whether they took any cholesterol-lowering drugs (such as the popular statins). None of the women had breast cancer at the time the researchers collected information about their use of cholesterol-lowering medication. The scientists then followed the women, for nearly seven years on average, to see who developed breast cancer. After accounting for age and weight (two factors related to breast cancer risk), the researchers found that statin users were about 75% less likely to develop breast cancer than women who weren't taking any cholesterol-lowering medication.[15] A news report said the study suggested that cholesterol-lowering medications might reduce women's risk of breast cancer. Can we be confident in this conclusion? Explain.

9.54 What is significance good for? Which of the following questions does a significance test answer?

(a) Is the sample or experiment properly designed?

(b) Is the observed effect due to chance?

(c) Is the observed effect important?

9.55 Searching for ESP A researcher looking for evidence of extrasensory perception (ESP) tests 500 subjects. Four of these subjects do significantly better ($P < 0.01$) than random guessing.

(a) Is it proper to conclude that these four people have ESP? Explain your answer.

(b) What should the researcher now do to see whether any of these four subjects have ESP?

9.56 Is this convincing? You are planning to test a new vaccine for a virus that now has no vaccine. Since the disease is usually not serious, you will expose 100 volunteers to the virus. After some time, you will record whether or not each volunteer has been infected.

(a) Explain how you would use these 100 volunteers in a designed experiment to test the vaccine. Include all important details of designing the experiment (but don't actually do any random allocation).

(b) You hope to show that the vaccine is more effective than a placebo. The experiment gave a *P*-value of 0.25. Explain carefully what this means.

(c) Your fellow researchers do not consider this evidence strong enough to recommend regular use of the vaccine. Do you agree?

9.57 Why are larger samples better? Statisticians prefer large samples. Describe briefly the effect of increasing the size of a sample (or the number of subjects in an experiment) on each of the following.

(a) The margin of error of a 95% confidence interval.

(b) The *P*-value of a test when H_0 is false and all facts about the population remain unchanged as n increases.

9.58 Low pressure The Automobile Association of America (AAA) reports that 85% of all cars on the road are driving on underinflated tires. Two researchers, Geraldine and Patti, collect data on tire pressure from a random sample of cars that were brought in for oil changes.
(a) Geraldine wants to perform a significance test of AAA's claim. Patti would rather calculate a confidence interval. Which would you recommend? Why?
(b) Will Geraldine and Patti be able to generalize their results to the population of all cars on the road? Why or why not?

9.59 Why we seek significance Asked why statistical significance appears so often in research reports, a student says, "Because saying that results are significant tells us that they cannot easily be explained by chance variation alone." Do you think that this statement is essentially correct? Explain your answer.

9.60 Online directory While surfing the Internet one day, Raul finds this claim about the site **www.whitepages.com**: "We cover 80% of U.S. adults." Raul thinks the site is exaggerating.
(a) Describe the details of a study that Raul could realistically perform to investigate this claim.
(b) If Raul performs a significance test using the data from the study in (a), is it possible that the conclusion he draws about the claim is wrong? Explain.
(c) If Raul calculates a confidence interval using the data from the study in (a), is it possible that the conclusion he draws about the claim is wrong? Explain.

CHAPTER 9 REVIEW

Statistical inference draws conclusions about a population on the basis of sample data and uses probability to indicate how reliable the conclusions are. A confidence interval estimates an unknown parameter. A significance test shows how strong the evidence is for some claim about a parameter. Sections 9.1 and 9.2 present the reasoning of confidence intervals and tests and give details for inference about a population proportion p.

The probabilities in both confidence intervals and tests tell us what would happen if we used the formula for the interval or test very many times. A confidence level is the probability that the formula for a confidence interval actually produces an interval that contains the unknown parameter. A 95% confidence interval gives a correct result 95% of the time when we use it repeatedly. Figure 9.12 illustrates the reasoning using the approximate 95% confidence interval for a population proportion p.

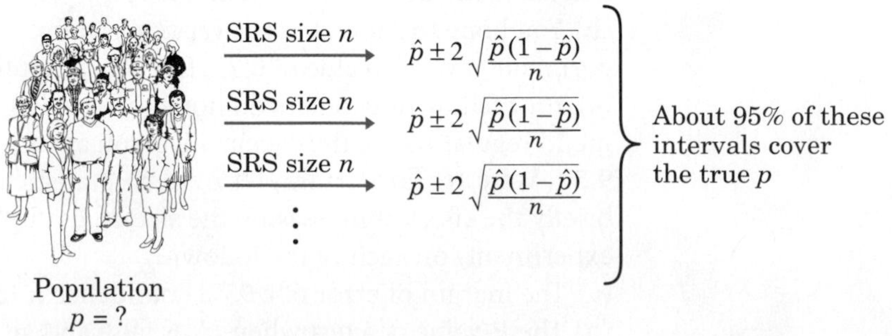

Figure 9.12 The idea of a confidence interval.

A P-value is the probability that the test would produce a result at least as extreme as the observed result if the null hypothesis really were true. Figure 9.13 illustrates the reasoning, placing the sample proportion $\hat{p}$ from our one sample on the Normal curve that shows how $\hat{p}$ would vary in all possible samples if the null hypothesis were true. A P-value tells us how surprising the observed outcome is. Very surprising outcomes (small P-values) are good evidence that the null hypothesis is not true.

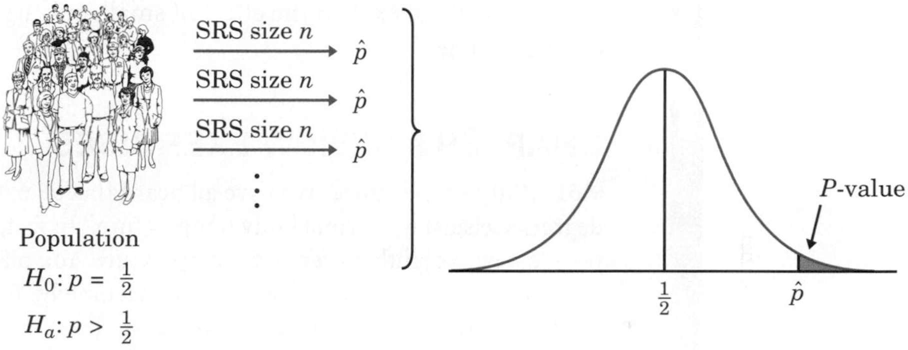

Figure 9.13 The idea of a significance test.

To detect wise and unwise uses of inference, you must know the basic reasoning and also be aware of some potential pitfalls. Section 9.3 will help.

Here is a review list of the most important skills you should have developed from your study of this chapter.

A. SAMPLING DISTRIBUTIONS

1. Explain the idea of a sampling distribution. See Figure 9.2 (page 423).

2. Use the Normal sampling distribution of a sample proportion and the 68–95–99.7 rule to find probabilities involving $\hat{p}$.

B. CONFIDENCE INTERVALS

1. Explain the idea of a confidence interval. See Figure 9.12.

2. Explain in nontechnical language what is meant by "95% confidence" and other statements of confidence in statistical reports.

3. Use the basic formula $\hat{p} \pm 2\sqrt{\hat{p}(1-\hat{p})/n}$ to obtain an approximate 95% confidence interval for a population proportion p.

4. Understand how the margin of error of a confidence interval changes with the sample size and the level of confidence.

5. Detect major mistakes in applying inference, such as improper data production, selecting the best of many outcomes, ignoring high nonresponse and outliers.

C. SIGNIFICANCE TESTS

1. Explain the idea of a significance test. See Figure 9.13.

2. State the null and alternative hypotheses in a testing situation when the parameter in question is a population proportion p.

3. Explain in nontechnical language the meaning of the *P*-value when you are given the numerical value of *P* for a test.

4. Explain the meaning of "statistically significant at the 5% level" and other statements of significance. Explain why significance at a specific level such as 5% is less informative than a *P*-value.

5. Recognize that significance testing does not measure the size or importance of an effect.

6. Recognize and explain the effect of small and large samples on the significance of an outcome.

CHAPTER 9 REVIEW EXERCISES

9.61 Body temperature We have all heard that 98.6 degrees Fahrenheit (or 37 degrees Celsius) is "normal body temperature." In fact, there is evidence that most people have a slightly lower body temperature. You plan to measure the body temperature of a random sample of people very accurately. You hope to show that a majority have temperatures lower than 98.6 degrees.
(a) Say clearly what the population proportion p stands for in this setting.
(b) In terms of p, what are your null and alternative hypotheses?

9.62 Significant versus important Explain how a result can be statistically significant but have little practical importance.

9.63 We love football! A recent Gallup Poll conducted telephone interviews with a random sample of adults aged 18 and older. Data were obtained for 1000 people. Of these, 37% said that football is their favorite sport to watch on television.
(a) Explain to someone who knows no statistics why we can't just say that 37% of adults would say that football is their favorite sport to watch on television.
(b) Construct and interpret a 95% confidence interval for the population proportion p.
(c) The poll announced a margin of error of ± 3 percentage points. Do you agree?
(d) This poll was taken in December, an exciting part of the football season. Do you think a similar poll conducted in June might produce different results? Explain why or why not.

9.64 Roulette A roulette wheel has 18 red slots among its 38 slots. You observe many spins and record the number of times that red occurs. Now you want to use these data to test whether the probability p of a red has the value that is correct for a fair roulette wheel.
(a) State the hypotheses H_0 and H_a that you will test.
(b) In 50 spins, the ball landed in a red slot 31 times. Carry out a significance test and report your conclusion.
(c) The casino manager uses your data to produce a 99% confidence interval for p and gets (0.44, 0.80). He says that this interval provides convincing evidence that the wheel is fair. How do you respond?

9.65 Senior citizens Table 2.4 (page 44) records the percent of residents aged 65 or older in each of the 50 states. You can check that this percent is 14% or higher in 10 of the states. So the sample proportion of states with at least 14% of elderly residents is $\hat{p} = 10/50 = 0.20$. Explain why it does not make sense to calculate a 95% confidence interval for the population proportion p.

9.66 Comparing package designs A company compares two package designs for a laundry detergent by placing bottles with both designs on the shelves of several markets. Checkout scanner data on more than 5000 bottles bought show that more shoppers bought Design A than Design B. The difference is statistically significant ($P = 0.02$). Can we conclude that consumers strongly prefer Design A? Explain your answer.

9.67 Online gaming, I A random sample of 1100 teenagers (aged 12 to 17) were asked whether they played games online; 775 said that they did.
(a) Construct and interpret a 99% confidence interval for the population proportion p.
(b) Suppose that the results of the survey were used to construct separate 99% confidence intervals for boys and girls. Would the margins of error for those two confidence intervals be the same as, larger than, or smaller than the margin of error for the interval you constructed in (a)? Justify your answer.
(c) In the context of this exercise, describe a real-world issue that isn't included in the margin of error.

9.68 Online gaming, II Refer to the previous exercise. How large a sample would you need to take to estimate p within 2% at a 99% confidence level? Show your work.

9.69 Abstinence? The Gallup Youth Survey asked a random sample of 439 U.S. teens aged 13 to 17 whether they thought that young people should wait to have sex until marriage.[16] The number who said "Yes" was 246.
(a) Do these results give convincing evidence that a majority of U.S. teens would say that young people should wait to have sex until they're married? Carry out a significance test to help answer this question.
(b) Construct and interpret a 95% confidence interval for the proportion of all teens who would say "Yes" if asked this question.
(c) Do the inference methods in (a) and (b) tell us whether the teens who responded to the survey were telling the truth? Explain.

9.70 Alternative medicine A nationwide random survey of 1500 adults asked about attitudes toward "alternative medicine" such as acupuncture, massage therapy, and herbal therapy. Among the respondents, 660 said they would use alternative medicine if traditional medicine was not producing the results they wanted. Do these data provide good evidence that more than 1/3 of all adults would use alternative medicine if traditional medicine did not produce the results they wanted?
(a) State the hypotheses to be tested.
(b) If your null hypothesis is true, what is the sampling distribution of the sample proportion $\hat{p}$? Sketch this distribution.
(c) Mark the actual value of $\hat{p}$ on the curve. Does it appear surprising enough to give good evidence against the null hypothesis?
(d) Carry out the significance test called for in detail. Show the steps of the inference procedure (hypotheses, sampling distribution, data, P-value, conclusion) clearly.

9.68 Comparing package designs. [illegible] package designs [illegible] shelves of [illegible] stores. Checkout [illegible] more than 5000 homes bought [illegible] shoppers bought Design A and Design B [illegible] the difference is statistically significant [illegible] Design A? Explain [illegible]

9.67 Online gaming. [illegible] random sample of 1100 teens ages [illegible] asked whether they played [illegible] said that they did.

(a) Construct and interpret a [illegible] confidence interval for the population proportion [illegible]

[illegible] pose that the results of this [illegible] used to construct separate [illegible] confidence intervals [illegible] Would these be the same as [illegible] answer? Explain.

(c) [illegible] context of this exercise, describe a [illegible] would [illegible] not included in the margin of error.

9.68 Online gaming. [illegible] How large a sample would [illegible] to estimate [illegible] with a 95% confidence [illegible]

9.69 [illegible] The [illegible] survey asked a random sample of [illegible] aged 13 to 17 [illegible] that young people [illegible] The number who said "Yes" was 1[illegible]

(a) [illegible] give convincing evidence that a majority of [illegible] teens would [illegible] test to [illegible]

(b) Construct and interpret a 95% confidence interval for the proportion of all teens who [illegible] asked this question.

(c) [illegible] methods in (a) and (b) [illegible] the teens who [illegible]

9.70 [illegible] random survey of 1500 adults asked about [illegible]

[illegible] did not provide [illegible]

(c) [illegible] what is the [illegible] distribution of [illegible] Sketch this distribution.

(d) Mark the actual value [illegible] Does [illegible] give good evidence [illegible]

(e) Carry out the significance test [illegible] the steps in the [illegible]

CHAPTER 10

Inference in Practice

Do dogs resemble their owners?

Some people look a lot like their pets. Maybe they deliberately choose animals that match their appearance. Or maybe we're perceiving similarities that aren't really there. Researchers at the University of California, San Diego, decided to investigate. They designed an experiment to test whether dogs and their owners resembled one another. The researchers believed that the resemblance might differ for owners of purebred and nonpurebred dogs.

A total of 45 dogs and their owners were photographed separately at three dog parks. Then researchers "constructed triads of pictures, each consisting of one owner, that owner's dog, and one other dog photographed at the same park." The subjects in the experiment were 28 undergraduate psychology students. Each subject was presented with the individual sets of photographs and asked to identify which dog belonged to the pictured owner. A dog was classified as resembling its owner if more than half of the 28 undergraduate students matched dog to owner.[1]

Here are some results. Of the 45 dogs photographed, 23 were classified as resembling their owners. That's just over half of the dogs in the study. If we assume that these dogs are representative of the larger canine population, we could use the methods of Chapter 9 to construct a confidence interval for the proportion of all dogs that resemble their owners.

What about the researchers' belief that the resemblance between dog and owner might differ for purebred and nonpurebred dogs? Of the 45 dogs in the study, 25 were purebreds and 20 were not. For the purebred dogs, 16 resembled their owners. As a proportion, that's $16/25 = 0.64$. For the nonpurebred dogs, 7 resembled their owners. That's $7/25 = 0.28$ as a proportion. Does such a large difference in proportions provide convincing evidence in favor of the researchers' belief?

To answer this question, we need a new kind of significance test—one that allows us to compare proportions for two groups or populations. This new test, called a *chi-square test,* is the subject of Section 10.1. In Section 10.2, we'll introduce confidence intervals and significance tests for a population mean.

10.1 Chi-Square Tests: Goodness of Fit and Two-Way Tables

ACTIVITY 10.1

Return of the M&M's!

Materials: One 1.69-ounce bag of milk chocolate M&M's per student or data from Activity 9.3

The M&M/Mars Company says that the color distribution for its plain M&M's candies is 24% blue, 20% orange, 16% green, 14% yellow, 13% red, and 13% brown. If you completed Activity 9.3 (page 453), then you have already tested the company's claim about the proportion of blue candies, $p = 0.24$. The purpose of this Activity is to compare the entire color distribution of M&M's in your individual bag with the advertised distribution. We want to see if there is sufficient evidence to dispute the company's claim.

1. Make a table on your paper like the one shown below. Fill in the counts, by color, and the total number of M&M's in *your* bag in the "Observed" column. We have entered values from a bag that we purchased for demonstration purposes. Leave the other columns blank for now.

Color	Observed (O)	Expected (E)	$O - E$		
Blue	9	$(0.24)(61) = 14.64$	$9 - 14.64 = -5.64$		
Orange	10	$(0.20)(61) = 12.20$	$10 - 12.20 = -2.20$		
Green	15	$(0.16)(61) = 9.76$	$15 - 9.76 = 5.24$		
Yellow	12	$(0.14)(61) = 8.54$	$12 - 8.54 = 3.46$		
Red	8	$(0.13)(61) = 7.93$	$8 - 7.93 = 0.07$		
Brown	7	$(0.13)(61) = 7.93$	$7 - 7.93 = -0.93$		
Total	**61**	**61**	**0**		

2. Assuming that the company's claim is true, how many of each color would you expect in *your* bag? We had a total of 61 M&M's in *our* bag. How many blue M&M's should we expect in a bag with 61 candies? If 24% of all plain M&M's are blue, then we'd expect $(0.24)(61) = 14.64$ blue M&M's, on average, in a bag with 61 M&M's. In a similar way, we can calculate the expected values for the other five colors. Follow our model, and calculate these expected values (to two decimal places) for your bag. Record your results in the "Expected" column of your table.

3. How close are your observed counts to the expected values? To answer this question, calculate the difference for each color: Observed − Expected. Record these differences in a column labeled $O - E$. Find the total of your six $O - E$ values. What do you notice?

4. In Step 3, the sum of the difference values $O - E$ should be 0 (up to roundoff error). Does that mean your individual bag matches the color distribution claimed by M&M/Mars perfectly? Of course not. You get a total difference of 0 because the positive

and negative values of $O - E$ cancel each other. How can we fix this problem? We could use absolute value, but instead, we choose to square the difference values (does that remind you of anything earlier in the text?). Compute the values of $(O - E)^2$ and fill in the next column of your table. What's your total?

Color	Observed (O)	Expected (E)	$O - E$	$(O - E)^2$	$(O - E)^2/E$
Blue	9	(0.24)(61) = 14.64	9 − 14.64 = −5.64	$(-5.64)^2$ = 31.81	31.81/14.64 = 2.173
Orange	10	(0.20)(61) = 12.20	10 − 12.20 = −2.20	$(-2.20)^2$ = 4.84	4.84/12.20 = 0.397
Green	15	(0.16)(61) = 9.76	15 − 9.76 = 5.24	$(5.24)^2$ = 27.46	27.46/9.76 = 2.81
Yellow	12	(0.14)(61) = 8.54	12 − 8.54 = 3.46	$(3.46)^2$ = 11.97	11.97/8.54 = 1.402
Red	8	(0.13)(61) = 7.93	8 − 7.93 = 0.07	$(0.07)^2$ = 0.005	0.005/7.93 = 0.0006
Brown	7	(0.13)(61) = 7.93	7 − 7.93 = −0.93	$(-0.93)^2$ = 0.86	0.86/7.93 = 0.108
Total	**61**	**61**	**0**	**76.945**	**6.8906**

5. In the last column of your table, divide the values in the $(O - E)^2$ column by the corresponding expected values (E). Then find the total. This final result is called the *chi-square statistic* and is denoted by χ^2. Keep this number handy—you will use it later in the chapter.

6. If your sample reflects the distribution advertised by the M&M/Mars Company, there should be very little difference between the observed counts and the expected values. Therefore, the calculated values making up the sum χ^2 should be very small. Are the entries in the last column all about the same, or do any values stand out because they are much larger than the others? Did you get many more or many fewer of a particular color than expected?

7. Compare your value of χ^2 with the results obtained by your classmates. How do we know whether a particular χ^2 statistic is large enough to provide convincing evidence against the company's claim? We need to learn more about chi-square before we can answer this question.

Goodness of fit and the chi-square statistic

Suppose you open a 1.69-ounce bag of plain M&M's and discover that of the 61 M&M's in the bag, only 9 are blue. In your sample of size 61, the proportion of blue M&M's is $\hat{p} = 9/61 = 0.148$. Since the company claims that 24% of all plain M&M's are blue, you might feel that you didn't get your fair share of blues. You could use the z test described in Chapter 9 to test the hypotheses

$$H_0 : p = 0.24$$
$$H_a : p \neq 0.24$$

where p is the population proportion of blue M&M's. Here are the results: $z = -1.69$; P-value $= 0.09$. Remember: the P-value tells us the probability of getting a sample result as "surprising" as the one we did if the null hypothesis is true. In this case, $P = 0.09$ says that roughly 1 in every 11 bags of 61 M&M's would have a proportion of blue candies that differed from $p = 0.24$ by at least as much as our bag did. The P-value gives marginal evidence against the company's claim about the overall proportion of blue M&M's.

You could then perform additional significance tests for each of the remaining colors. But this would be inefficient. More important, performing six individual tests wouldn't tell you how likely it is that *six* sample proportions would differ from the values stated by M&M/Mars as much as your sample does. There is a single test that can be applied to see if the observed distribution of sample data is significantly different in some way from the stated population distribution. It is called the **chi-square** (χ^2) **test for goodness of fit.** Like any other test, we must begin by stating hypotheses.

Example 10.1 Hypotheses for a goodness of fit test

Let's continue with the M&M's scenario. The null hypothesis in a chi-square goodness of fit test should state a claim about the distribution of a single categorical variable in the population of interest. In the case of M&M's, the categorical variable we're measuring is color. The appropriate null hypothesis is

H_0: The company's stated color distribution for plain M&M's candies is correct; that is,

$$p_{\text{blue}} = 0.24, p_{\text{orange}} = 0.20, p_{\text{green}} = 0.16, p_{\text{yellow}} = 0.14, p_{\text{red}} = 0.13, p_{\text{brown}} = 0.13.$$

The corresponding alternative hypothesis is

H_a: The company's stated color distribution for plain M&M's candies is not correct; that is, at least two of the stated proportions are incorrect.

The idea of the goodness of fit test is this: we compare the **observed counts** for our sample with the counts that would be expected if the color distribution claimed by M&M/Mars is correct. The more the observed counts differ from the expected counts, the more evidence we have against the company's claim. In general, the **expected counts** for a categorical variable can be obtained by multiplying the proportion of the distribution for each category by the sample size.

Example 10.2 Computing expected counts for a goodness of fit test

If the company's claim is true, 24% of all plain M&M's produced are blue. Does that mean that every 1.69-ounce bag of plain M&M's will contain exactly 24% blue candies? Of course not. Some bags will have more than 24% blue candies; other bags will have less than 24%. On average, the bags will contain 24% blue M&M's. For bags with 61 M&M's, the average number of blue M&M's should be $(0.24)(61) = 14.64$. This is our *expected count* of blue M&M's.

Using this same method, we can find the expected counts for the other color categories:

Orange: $(0.20)(61) = 12.20$ Green: $(0.16)(61) = 9.76$ Yellow: $(0.14)(61) = 8.54$

Red: $(0.13)(61) = 7.93$ Brown: $(0.13)(61) = 7.93$

Did you notice that the expected count in Example 10.2 sounds a lot like the expected value of a random variable from Chapter 8? That's no coincidence. The number of M&M's of a specific color in a bag of 61 candies is a random variable. As a result, we can find its expected value.

To see if the data give evidence against the null hypothesis, we need to compare the observed counts from our sample with the expected counts. If the observed counts are far from the expected counts, that's the evidence we were seeking. The chi-square test of goodness of fit uses a statistic (called the **chi-square statistic**) that measures how far apart the observed and expected counts are.

Chi-square statistic

The **chi-square statistic** χ^2 is a measure of how far the observed counts for a categorical variable are from the expected counts. The formula for the statistic is

$$\chi^2 = \sum \frac{(\text{observed count} - \text{expected count})^2}{\text{expected count}}$$

The symbol Σ means "sum over all the categories."

The chi-square statistic is a sum of terms, one for each value of the categorical variable. In the M&M's example, 9 of the candies in our bag were blue. The expected count of blues is 14.64. So the term in the chi-square statistic from this category is

$$\frac{(\text{observed count} - \text{expected count})^2}{\text{expected count}} = \frac{(9 - 14.64)^2}{14.64}$$
$$= \frac{31.81}{14.64} = 2.173$$

Example 10.3 Calculating the chi-square statistic

Here are the observed and expected counts for the M&M's data of Activity 10.1 side by side:

Color	Observed	Expected
Blue	9	14.64
Orange	10	12.20
Green	15	9.76
Yellow	12	8.54
Red	8	7.93
Brown	7	7.93

We can now find the chi-square statistic, adding six terms for the six color categories:

$$\begin{aligned}\chi^2 &= \frac{(9-14.64)^2}{14.64} + \frac{(10-12.20)^2}{12.20} + \frac{(15-9.76)^2}{9.76} + \frac{(12-8.54)^2}{8.54} \\ &\quad + \frac{(8-7.93)^2}{7.93} + \frac{(7-7.93)^2}{7.93} \\ &= 2.173 + 0.397 + 2.813 + 1.402 + 0.0006 + 0.108 \\ &= 6.89\end{aligned}$$

Because χ^2 measures how far the observed counts are from what would be expected if H_0 were true, large values are evidence against H_0. Is $\chi^2 = 6.89$ a large value? You know the drill: compare the observed value 6.89 against the sampling distribution that shows how χ^2 would vary if the null hypothesis were true. This sampling distribution is *not* a Normal distribution. It is a right-skewed distribution that allows only positive values because χ^2 can never be negative. Moreover, the sampling distribution differs depending on the number of possible values for the categorical variable (that is, on the number of categories). Here are the facts.

The chi-square distributions

The sampling distribution of the chi-square statistic χ^2 when the null hypothesis is true is called a **chi-square distribution.**

The chi-square distributions are a family of distributions that take only positive values and are skewed to the right. A specific chi-square distribution is specified by giving its **degrees of freedom.**

The chi-square test for goodness of fit uses critical values from the chi-square distribution with degrees of freedom = the number of categories − 1.

Figure 10.1 shows the density curves for three members of the chi-square family of distributions. As the degrees of freedom (df) increase, the density curves become less skewed and larger values become more probable. We can't find *P*-values as areas under a chi-square curve by hand, though a computer or calculator can do it for us.

Table 10.1 is a shortcut to avoid finding the *P*-value. It shows how large the chi-square statistic χ^2 must be in order to be significant at various levels. This isn't as good as an actual *P*-value, but it is often good enough. Each number of degrees of freedom has a separate row in the table. We see, for example, that a chi-square statistic with 3 degrees of freedom is significant at the 5% level if it is greater than 7.81 and is significant at the 1% level if it is greater than 11.34.

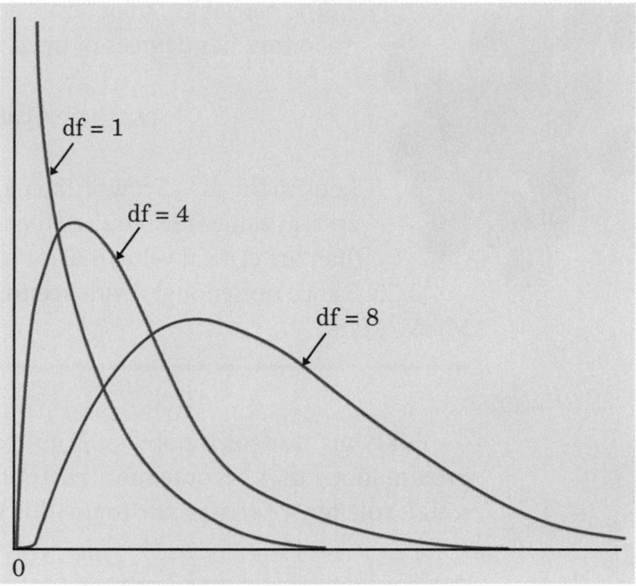

Figure 10.1 The density curves for three members of the chi-square family of distributions. The sampling distributions of chi-square statistics belong to this family.

Table 10.1 To Be Significant at Level α, a Chi-Square Statistic Must Be Larger Than the Table Entry for α

	Significance level α						
df	**0.25**	**0.20**	**0.15**	**0.10**	**0.05**	**0.01**	**0.001**
1	1.32	1.64	2.07	2.71	3.84	6.63	10.83
2	2.77	3.22	3.79	4.61	5.99	9.21	13.82
3	4.11	4.64	5.32	6.25	7.81	11.34	16.27
4	5.39	5.99	6.74	7.78	9.49	13.28	18.47
5	6.63	7.29	8.12	9.24	11.07	15.09	20.51
6	7.84	8.56	9.45	10.64	12.59	16.81	22.46
7	9.04	9.80	10.75	12.02	14.07	18.48	24.32
8	10.22	11.03	12.03	13.36	15.51	20.09	26.12
9	11.39	12.24	13.29	14.68	16.92	21.67	27.88

We have reproduced Table 10.1 in the back of the book (Table E) for easier reference.

Example 10.4 The M&M's study: conclusion

Our bag of 61 plain M&M's had fewer blues and more greens than expected if the company's claimed color distribution is correct. Comparing observed and expected counts gave the chi-square statistic $\chi^2 = 6.89$. The last step is to assess significance.

Marta Rostek/stock.xchng

The variable "color" has 6 possible values. The chi-square statistic therefore has degrees of freedom

$$\text{number of categories} - 1 = 6 - 1 = 5$$

Look in the df = 5 row of Table 10.1. We see that $\chi^2 = 6.89$ is larger than the critical value 6.63 required for significance at the $\alpha = 0.25$ level but smaller than the critical value 7.29 for $\alpha = 0.20$. So the *P*-value is between 0.20 and 0.25. That's not enough evidence to refute the color distribution proposed by M&M/Mars.

Like our test for a population proportion, the chi-square test uses some approximations that become more accurate as we take more observations. Here is a rough rule for when it is safe to use this test.[2]

Expected counts required for the chi-square test

You can safely use the chi-square test when no more than 20% of the expected counts are less than 5 and all individual expected counts are 1 or greater.

The M&M's study easily passes this test: the smallest expected count is 7.93.

Exercises

Exercises 10.1 to 10.5 refer to the following setting. The M&M/Mars Company reports that their M&M Peanut Chocolate Candies have the following distribution: 23% each of blue and orange, 15% each of green and yellow, and 12% each of red and brown. Joey bought a bag of peanut M&M's and counted out the following distribution of colors: 12 blue, 7 orange, 13 green, 4 yellow, 8 red, and 2 brown.

10.1 Hypotheses State an appropriate pair of hypotheses for testing the company's claim about the color distribution of peanut M&M's.

10.2 Expected counts Joey's bag contained 46 peanut M&M's. Calculate the expected count for each color assuming that the company's claim is true. Show your work.

10.3 Test statistic Calculate the chi-square statistic for Joey's sample. Show your work.

10.4 *P*-value Which chi-square distribution should we use? (Specify the degrees of freedom.) What's the *P*-value for this test? Refer to Table E (in the back of the book).

10.5 Conclusion What conclusion do you draw based on the results of this test? Justify your answer.

10.6 Activity 10.1 follow-up Use your M&M's data from Activity 10.1 (page 470) to perform a chi-square goodness of fit test for the company's claimed color distribution. What do you conclude?

Using the chi-square test for goodness of fit

One of the most common applications of the chi-square goodness of fit test is in the field of genetics. Scientists want to investigate the genetic characteristics of offspring that result from mating (also called "crossing") parents with known genetic makeups. Scientists use rules about dominant and recessive genes to predict the proportion of offspring that will fall in each possible genetic category. Then, the researchers mate the parents and classify the resulting offspring. The chi-square goodness of fit test helps the scientists assess the validity of their hypothesized distribution of proportions.

Example 10.5 Genetics and the chi-square goodness of fit test

Biologists wish to cross two tobacco plants having genetic makeup Gg, indicating that each has one dominant gene (G) and one recessive gene (g) for color. Each offspring plant will receive one gene for color from each parent. The following table, often called a *Punnett square*, shows the possible combinations of genes received by the offspring:

	Parent 2 passes on:	
Parent 1 passes on:	**G**	**g**
G	GG	Gg
g	Gg	gg

The Punnett square suggests that the expected ratio of green (GG) to yellow-green (Gg) to albino (gg) tobacco plants should be 1:2:1. In other words, the biologists predict that 25% of the offspring will be green, 50% will be yellow-green, and 25% will be albino. To test their hypothesis about the distribution of offspring, the biologists mate 84 pairs of yellow-green parent plants. Of 84 offspring, 23 plants are green, 50 are yellow-green, and 11 are albino. Do these data differ significantly from what the biologists have predicted?

We can now follow the steps for a significance test, familiar from Section 9.2.

The hypotheses. The chi-square goodness of fit test starts with these hypotheses:

H_0: The biologists' predicted color distribution for tobacco plant offspring is correct; that is, $p_{\text{green}} = 0.25, p_{\text{yellow-green}} = 0.5, p_{\text{albino}} = 0.25$.

H_a: The biologists' predicted color distribution is not correct; that is, at least two of the proposed proportions are incorrect.

The sampling distribution. There are three possible values for the categorical variable "offspring color": green, yellow-green, and albino. We will use critical values from the chi-square distribution with degrees of freedom

$$\text{df} = \text{number of categories} - 1 = 3 - 1 = 2$$

The data. First find the expected cell counts. For example, the expected count of green offspring is (0.25)(84) = 21. Here is the complete table of observed and expected counts side by side:

Offspring color	Observed	Expected
Green	23	(0.25)(84) = 21
Yellow-green	50	(0.50)(84) = 42
Albino	11	(0.25)(84) = 21

Looking at these counts, we see that there were many more yellow-green tobacco offspring than expected and many fewer albino plants than expected.

The *P*-value. The chi-square statistic is

$$\chi^2 = \frac{(23-21)^2}{21} + \frac{(50-42)^2}{42} + \frac{(11-21)^2}{21}$$
$$= 0.190 + 1.524 + 4.762 = 6.476$$

To find the *P*-value, look at the df = 2 line of Table E (in the back of the book). The observed chi-square $\chi^2 = 6.476$ is larger than the critical value 5.99 for $\alpha = 0.05$ but smaller than the critical value 9.21 for $\alpha = 0.01$. So the *P*-value is between 0.01 and 0.05. (Technology can give the actual *P*-value. It is $P = 0.0392$.)

Conclusion. We have significant evidence ($P < 0.05$) that the biologists' hypothesized distribution of color for tobacco plant offspring is incorrect.

CALCULATOR CORNER Goodness of fit tests

You can use the TI-84/TI-Nspire to perform the calculations for a chi-square goodness of fit test. We'll use the M&M's data from Activity 10.1 (page 470) to illustrate the steps.

1. Enter the observed counts and expected counts in two separate lists.

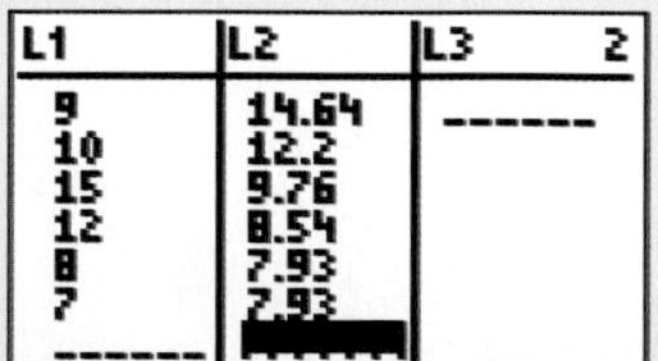

TI-84

- Clear lists L1 and L2.
- Enter the observed counts in list L1. Calculate the expected counts separately and enter them in list L2.

TI-Nspire

- On a new Lists & Spreadsheet page, name column A "obsd" and column B "expd."
- Enter the observed counts in the "obsd" list. Calculate the expected counts separately and enter them in list "expd."

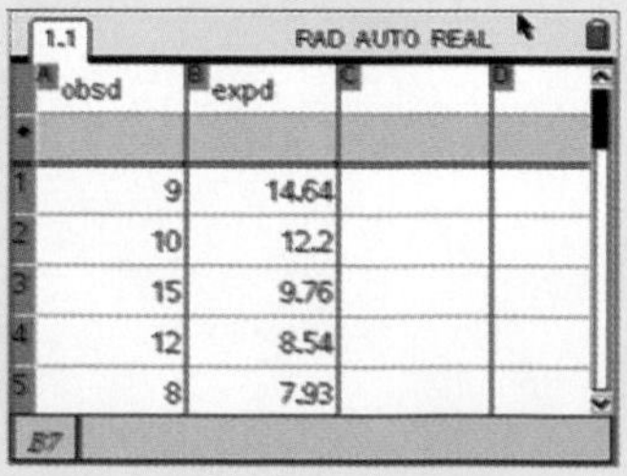

2. Perform a chi-square goodness of fit test.

TI-84

Press STAT, arrow over to `TESTS`, and choose `D:`χ^2`GOF-Test`. Enter the inputs shown. If you choose `Calculate`, you'll get a screen with the test statistic,

P-value, and df. If you choose the `Draw` option, you'll get a picture of the appropriate chi-square distribution with the test statistic marked and with the area corresponding to the *P*-value shaded.

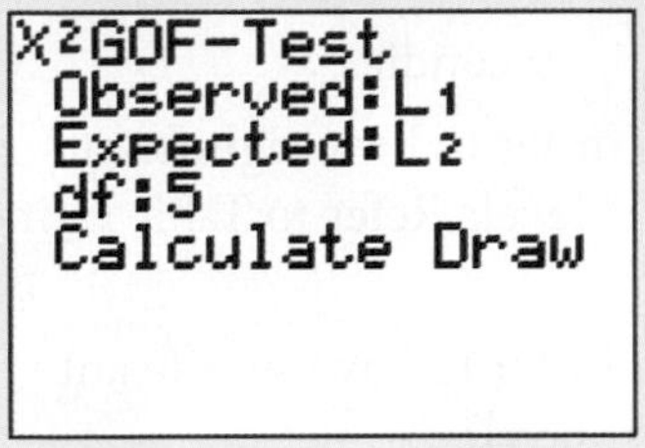

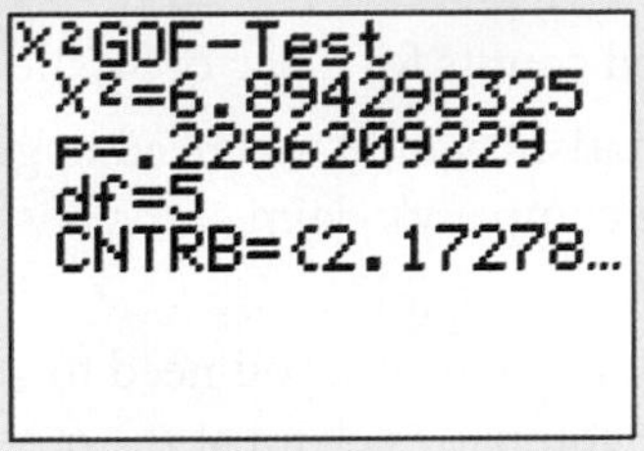

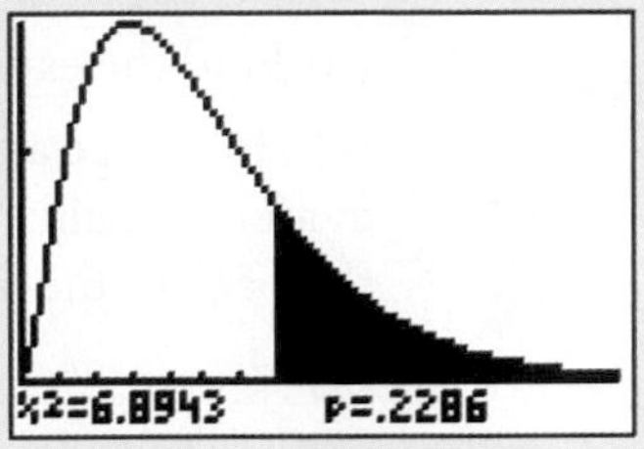

TI-Nspire

Press (menu), then choose `4:Statistics`, `4:Stat Tests...`, and `7:`χ^2`GOF`. Enter the inputs shown. If you choose `OK`, you'll get the results screen shown. If you click the "Plot data" button, you'll also get a picture of the appropriate chi-square distribution with the test statistic marked and with the area corresponding to the *P*-value shaded.

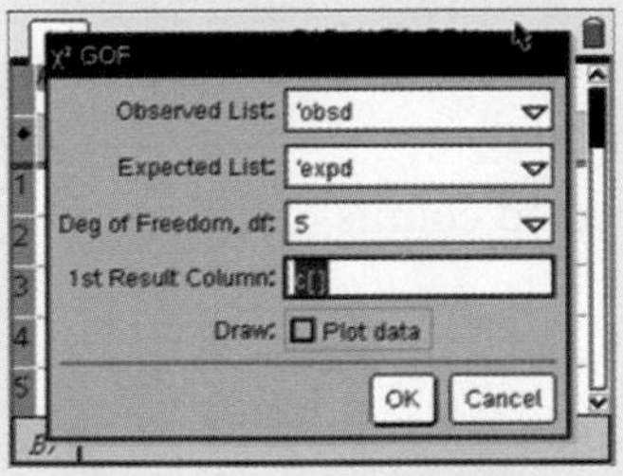

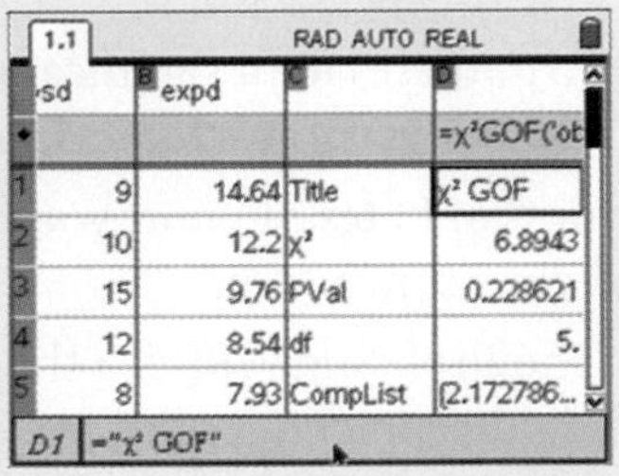

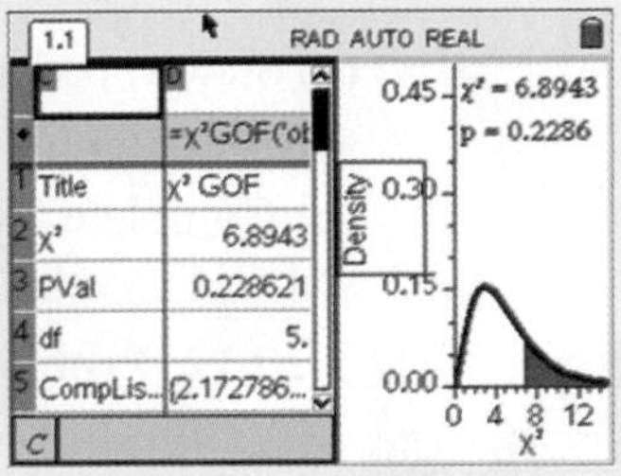

Exercises

Use Table E (in the back of the book) to help you with Exercises 10.7 to 10.12.

10.7 Mendel and the peas Gregor Mendel (1822–1884), an Austrian monk, is considered the father of genetics. Mendel studied the inheritance of various traits in pea plants. One such trait is whether the pea is smooth or wrinkled. Mendel predicted a ratio of 3 smooth peas for every 1 wrinkled pea. In one experiment, he observed 423 smooth and 133 wrinkled peas. Carry out a chi-square goodness of fit test for Mendel's prediction. What do you conclude?

10.8 Seagulls by the seashore Do seagulls show a preference for where they stand? To answer this question, biologists examined a small piece of shore whose area was made up of 56% sand, 29% mud, and 15% rocks. They recorded where each of 200 seagulls was standing on a particular afternoon. In all, 128 seagulls stood on the sand, 61 stood in the mud, and 11 stood on the rocks. Carry out a chi-square goodness of fit test. What do you conclude?

10.9 Skittles Statistics teacher Jason Molesky contacted the M&M/Mars Company to ask about the color distribution for Skittles. Here is an excerpt of the response he received:

> *The original flavor blend for the SKITTLES BITE SIZE CANDIES is lemon, lime, orange, strawberry and grape. They were chosen as a result of consumer preference tests we conducted. The flavor blend is 20 percent of each flavor.*

(a) State appropriate hypotheses for a significance test of the company's claim.

(b) Find the expected counts for a bag of Skittles with 60 candies.

(c) How large a χ^2 statistic would you need to get in order to have significant evidence against the company's claim at the $\alpha = 0.05$ level? Refer to Table E (in the back of the book).

(d) How large a χ^2 statistic would you need to get in order to have significant evidence against the company's claim at the $\alpha = 0.01$ level?

10.10 More Skittles Refer to the previous exercise.

(a) Create a set of observed counts for a bag with 60 candies that results in a *P*-value between 0.01 and 0.05. Show the calculation of your chi-square statistic.

(b) What conclusion would you draw in this case? Explain.

10.11 Is your random number generator working? Use your calculator's `randInt` function to generate 200 digits from 0 to 9 and store them in a list.

(a) State appropriate hypotheses for a significance test to determine whether your calculator's random number generator is working properly.

(b) Complete a goodness of fit test. Report your expected counts, chi-square statistic, *P*-value, and your conclusion.

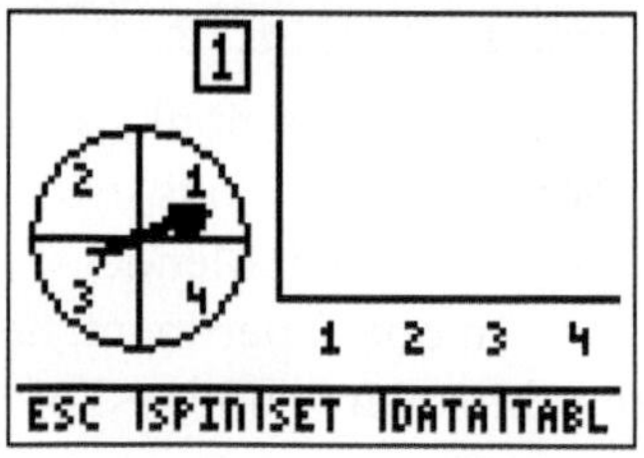

10.12 Prob Sim The Prob Sim APP for the TI-84 allows you to simulate tossing coins, rolling dice, picking marbles, spinning a spinner, drawing cards, and playing the lottery. (If you have a TI-Nspire, just switch to the TI-84 faceplate.)

- To run the APP, press the [APPS] key, and then choose `Prob Sim`. You should see the introductory screen shown in the margin. Press [ENTER] to see the main menu.
- Choose `4. Spin Spinner`. Spin the spinner a total of 200 times with 4 sets of 50 spins. Record the number of times that the spinner lands in each of the four numbered sections.
- Perform a significance test of the hypothesis that this program yields an equal proportion of 1s, 2s, 3s, and 4s. (If you do not have the Prob Sim APP, use the following sample data: 51 1s, 39 2s, 53 3s, 57 4s.)

Relationships between categorical variables: two-way tables

In Chapter 7, we used **two-way tables** to display the sample space of a random phenomenon involving two events. Two-way tables can also be used to describe the relationship between two categorical variables, as the following example illustrates.

Example 10.6 Do angry people have more heart disease?

People who get angry easily tend to have more heart disease. That's the conclusion of a study that followed a random sample of 12,986 people from three locations for about four years.[3] All subjects were free of heart disease at the beginning of the study. The subjects took the Spielberger Trait Anger Scale, which measures how prone a person is to sudden anger. Here is a two-way table that summarizes the data for the 8474 people in the sample who had normal blood pressure. CHD stands for "coronary heart disease." This includes people who had heart attacks and those who needed medical treatment for heart disease.

	Low anger	**Moderate anger**	**High anger**	**Total**
CHD	53	110	27	190
No CHD	3057	4621	606	8284
Total	3110	4731	633	8474

Anger rating on the Spielberger scale is a categorical variable that takes three possible values: low, medium, and high. Whether someone gets heart disease is also a categorical variable. The two-way table of Example 10.6 shows the relationship between anger rating and heart disease for a random sample of 8474 people. How can we make sense of the information contained in this table?

First, *look at the distribution of each variable separately.* The distribution of a categorical variable says how often each outcome occurred. The "Total" column at the right of the table contains the totals for each of the rows. These row totals give the distribution of heart disease for all people in the study, all anger levels combined. The "Total" row at the bottom of the table gives the distribution of anger rating for all 8474 people, CHD and non-CHD combined. It is often clearer to present these distributions using percents.

We might report the distribution of heart disease as

$$\text{percent CHD} = \frac{190}{8474} = 0.0224 = 2.24\%$$

$$\text{percent no CHD} = \frac{8284}{8474} = 0.9776 = 97.76\%$$

Likewise, we could give the distribution of Spielberger anger rating as

$$\text{percent Low} = \frac{3110}{8474} = 0.3670 = 36.70\%$$

$$\text{percent Moderate} = \frac{4731}{8474} = 0.5583 = 55.83\%$$

$$\text{percent High} = \frac{633}{8474} = 0.0747 = 7.47\%$$

The two-way table contains more information than the two distributions of heart disease and anger rating alone. The nature of the relationship between these two variables cannot be deduced from the separate distributions but requires the full table. **To describe relationships among categorical variables, calculate appropriate percents from the counts given.**

Example 10.7 Anger and heart disease

Since we're interested in whether angrier people tend to get heart disease more often, we can compare the percents of people who got heart disease (CHD) in each of the three anger categories:

Low anger	**Moderate anger**	**High anger**
$\frac{53}{3110} = 0.0170 = 1.70\%$	$\frac{110}{4731} = 0.0233 = 2.33\%$	$\frac{27}{633} = 0.0427 = 4.27\%$

There is a clear trend: as the anger score increases, so does the percent who suffer heart disease. The bar graph in Figure 10.2 displays these results visually. Is this relationship between anger and heart disease statistically significant?

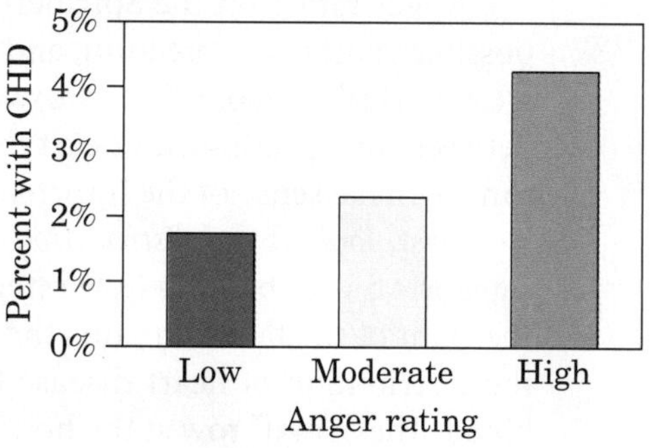

Figure 10.2 Bar graph comparing the percent of people in each anger category who got coronary heart disease (CHD).

In working with two-way tables, you must calculate lots of percents. Here's a tip to help decide what fraction gives the percent you want. Ask, "What group represents the total that I want a percent of?" The count for that group is the denominator of the fraction that leads to the percent. In Example 10.7, we wanted the percent *of each anger level* who got CHD, so the total counts in the three anger categories are the denominators.

Inference for a two-way table: hypotheses and expected counts

We often gather data and arrange them in a two-way table to see if two categorical variables are related to each other. The sample data are easy to investigate: turn them into percents and look for an association between the row and column variables. Is the association in the sample evidence of an association between these variables in the entire population? Or could the sample association easily arise just from the luck of random sampling? This is a question for a significance test.

Our null hypothesis is that there is no association between the two categorical variables. The alternative hypothesis is that there *is* an association between the variables. For the observational study of Examples 10.6 and 10.7, we want to test the hypotheses

H_0: There is no association between anger and CHD.
H_a: There is an association between anger and CHD.

To test H_0, we compare the observed counts in a two-way table with the *expected counts*, the counts we would expect—except for random variation—if H_0 were true.

Example 10.8 Anger and heart disease: finding expected counts

Here again is the two-way table that summarizes the data from the heart disease (CHD) and anger study.

	Low anger	**Moderate anger**	**High anger**	**Total**
CHD	53	110	27	190
No CHD	3057	4621	606	8284
Total	3110	4731	633	8474

Earlier, we determined that 190/8474 = 0.0224, or 2.24%, of the individuals in this observational study got CHD. If the null hypothesis is true, then there is no association between CHD and anger level. In that case, we would expect 2.24% of the individuals in each anger category to get CHD. For the 3110 people classified as "Low anger," we would expect 0.0224(3110) = 69.73 to get CHD.

Another way to write this calculation is

$$\frac{190}{8474} \cdot 3110 = 69.73$$

This is equivalent to

$$\frac{190 \cdot 3110}{8474} = 69.73$$

Do you see where the numbers in this calculation are coming from in the two-way table? The numerator is the *row total* for CHD times the *column total* for "Low anger." The denominator is the *table total*, 8474. We can find the expected counts for the other cells in the table in a similar way:

CHD, Low	**CHD, Moderate**	**CHD, High**
$\frac{190 \cdot 3110}{8474} = 69.73$	$\frac{190 \cdot 4731}{8474} = 106.08$	$\frac{190 \cdot 633}{8474} = 14.19$
no CHD, Low	**no CHD, Moderate**	**no CHD, High**
$\frac{8284 \cdot 3110}{8474} = 3040.27$	$\frac{8284 \cdot 4731}{8474} = 4624.92$	$\frac{8284 \cdot 633}{8474} = 618.81$

As we saw in Example 10.8, there is a rule that makes it easy to find expected counts for a two-way table. Here it is.

Expected counts

The **expected count** in any cell of a two-way table when H_0 is true is

$$\text{expected count} = \frac{\text{row total} \times \text{column total}}{\text{table total}}$$

Exercises

mpgphoto/FeaturePics

10.13 Extracurricular activities and grades, I North Carolina State University studied student performance in a course required by its chemical engineering major. One question of interest was the relationship between time spent in extracurricular activities and whether a student earned a C or better in the course. Here are the data for the 119 students who answered a question about extracurricular activities:[4]

	Extracurricular activities (hours per week)		
	<2	**2–12**	**>12**
C or better	11	68	3
D or F	9	23	5

(a) Calculate percents that describe the nature of the relationship between time spent on extracurricular activities and performance in the course.

(b) Draw a bar graph that shows these percents graphically. Be sure to label your graph appropriately. Give a brief summary in words.

10.14 Extracurricular activities and grades, II Refer to Exercise 10.13.

(a) Write the null and alternative hypotheses for the question of interest.

(b) Calculate the expected number of C's or better for those who spend 2 to 12 hours per week in extracurricular activities.

10.15 Trying to quit, I An observational study of 177 people who were trying to quit smoking looked at whether or not alcohol consumption was a factor in relapse (resuming smoking).[5] Here are the results:

Alcohol consumption	Smoked	Did not smoke
Yes	20	13
No	48	96

The entry 20 means that 20 of the study participants consumed alcohol and smoked during the cessation period.

(a) Find the row and column totals.

(b) Among the drinkers, what percent smoked? Among the teetotalers, what percent smoked? What is your preliminary conclusion?

(c) Draw a bar graph that shows these percents graphically. Be sure to label your graph appropriately.

10.16 Trying to quit, II Refer to Exercise 10.15.

(a) Write the null and alternative hypotheses for the question of interest.

(b) Find the expected counts.

10.17 Python eggs, I How is the hatching of water python eggs influenced by the temperature of the snake's nest? Researchers assigned newly laid eggs to one of three water temperatures: hot, neutral, or cold. Hot duplicates the extra warmth provided by the mother python, and cold duplicates the absence of the mother. Here are the data on the number of eggs and the number that hatched:[6]

	Eggs	Hatched
Cold	27	16
Neutral	56	38
Hot	104	75

(a) Make a two-way table of temperature by outcome (hatched or not).

(b) Calculate the percent of eggs in each group that hatched. The researchers anticipated that eggs would not hatch in cold water. Do the data support that anticipation?

10.18 Python eggs, II Refer to Exercise 10.17.

(a) Write the null and alternative hypotheses for the question of interest.

(b) What is the expected count of hatched pythons in the hot group?

Chi-square test for two-way tables

To see if the data give evidence against the null hypothesis of "no relationship," compare the counts in the two-way table with the counts we would expect if there really were no relationship. To measure how far apart the observed and expected counts are, use the chi-square statistic.

Chi-square test for two-way tables

To test the null hypothesis

H_0: There is no association between two categorical variables.

against the alternative hypothesis

H_a: There is an association between two categorical variables.

compute the chi-square statistic

$$\chi^2 = \sum \frac{(\text{observed count} - \text{expected count})^2}{\text{expected count}}$$

This time, the symbol Σ means "sum over all the cells in the table."

The **chi-square test for a two-way table** with r rows and c columns uses critical values from the chi-square distribution with $(r - 1)(c - 1)$ degrees of freedom.

Example 10.9 Anger and heart disease: chi-square test

Here are the observed and expected counts for the CHD and anger study side by side:

	Observed			Expected		
	Low	Moderate	High	Low	Moderate	High
CHD	53	110	27	69.73	106.08	14.19
No CHD	3057	4621	606	3040.27	4624.92	618.81

We can now follow the steps for a significance test, familiar from Section 9.2.

The hypotheses. The chi-square method tests these hypotheses:

H_0 : There is no association between anger and heart disease.
H_a : There is an association between anger and heart disease.

The sampling distribution. All the expected cell counts are larger than 5, so we can safely apply the chi-square test. The two-way table of anger versus CHD has 2 rows and 3 columns. We will use critical values from the chi-square distribution with degrees of freedom df $= (2 - 1)(3 - 1) = 2$.

The data. Looking at the observed and expected counts, we see that the high-anger group has more CHD than expected and the low-anger group has less CHD than expected. This is consistent with what the percents in Example 10.7 (page 482) show.

The *P*-value. The chi-square statistic is

$$\chi^2 = \frac{(53 - 69.73)^2}{69.73} + \frac{(110 - 106.08)^2}{106.08} + \frac{(27 - 14.19)^2}{14.19} + \frac{(3057 - 3040.27)^2}{3040.27} + \frac{(4621 - 4624.92)^2}{4624.92} + \frac{(606 - 618.81)^2}{618.81}$$
$$= 4.014 + 0.145 + 11.564 + 0.092 + 0.003 + 0.265 = 16.08$$

Look at the six terms that we sum to get χ^2. Most of the total comes from just one cell: high-anger people have more CHD than expected.

Look at the df $= 2$ line of Table E (in the back of the book). The observed chi-square $\chi^2 = 16.08$ is larger than the critical value 13.82 for $\alpha = 0.001$. So the *P*-value is less than 0.001. Statistical software can give the actual *P*-value. It is $P = 0.0003$.

Conclusion. We have highly significant evidence ($P < 0.001$) that anger and CHD are associated.

Can we conclude that proneness to anger *causes* heart disease? The anger and heart disease study is an observational study, not an experiment. It isn't surprising that some lurking variables are confounded with anger. For example, people prone to anger are more likely than others to be men who drink and smoke. The study report used advanced statistics to adjust for many differences among the three anger groups. The adjustments raised the P-value from $P = 0.0003$ to $P = 0.002$ because the lurking variables explain some of the heart disease. But this is still good evidence for a relationship. Because the study started with a random sample of people who had no heart disease and followed them forward in time, and because many lurking variables were measured and accounted for, it does give some evidence for causation. The next step might be an experiment that involves showing anger-prone people how to change. Will this reduce their risk of heart disease?

CALCULATOR CORNER Chi-square tests for two-way tables

We will perform a chi-square test of the anger and heart disease data in Examples 10.6 through 10.9 on the TI-84/TI-Nspire.

1. Enter the observed counts in a matrix.

TI-84

- Press [2nd][x^{-1}] (MATRIX), cursor over to EDIT, and choose 1: [A].
- Specify a 2 × 3 matrix. Enter the observed counts from the two-way table in the appropriate cells of the matrix.

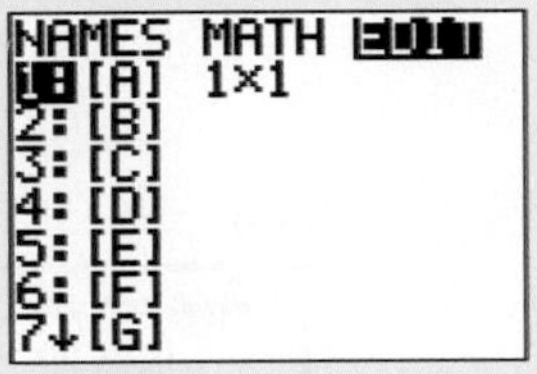

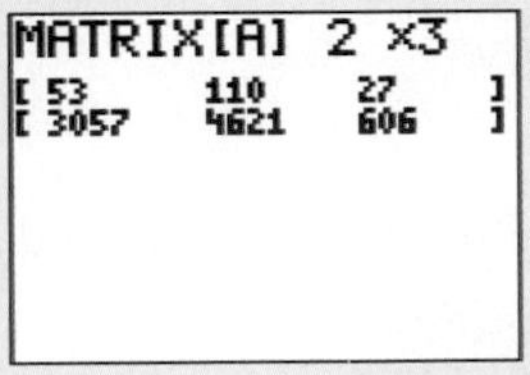

TI-Nspire

- Start with a blank Calculator page.
- Go into the Catalog (press [catalog]) and use the NavPad to move over to the templates tab. Use the NavPad to move to the m-by-n matrix... template and press [click] or [enter] to select it.
- Enter the number of rows (2) and the number of columns (3) and choose OK.
- Enter the observed counts from the two-way table in the appropriate cells of the matrix. Use the [tab] key to move to the next cell.
- Store the matrix as a variable: press [ctrl][sto→ var] (STO) [A][enter].

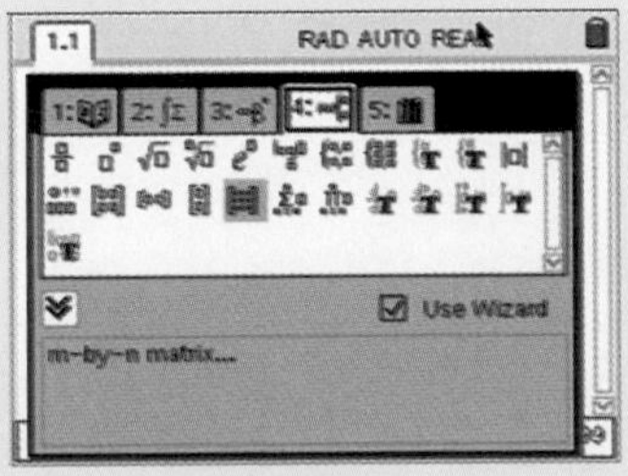

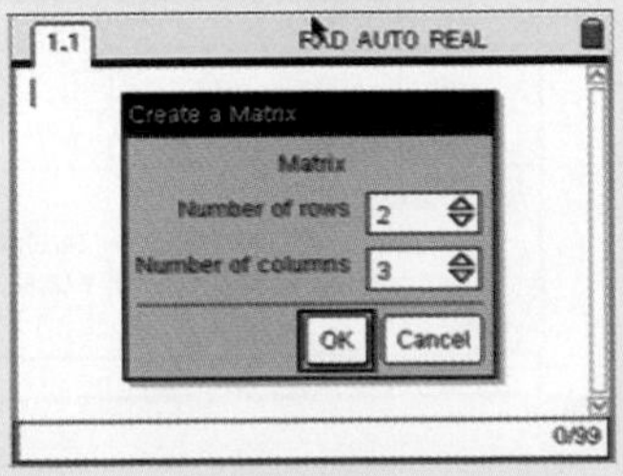

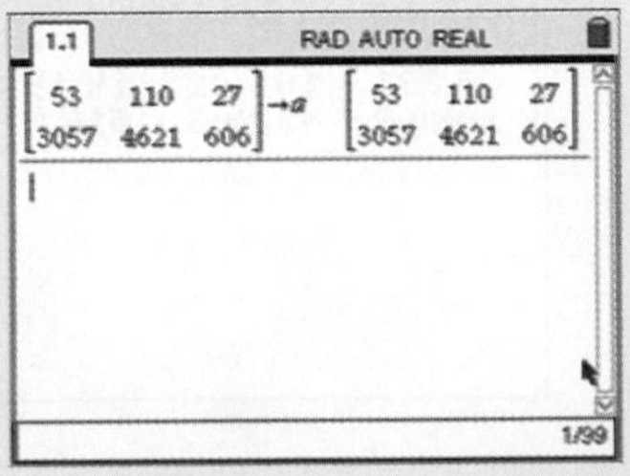

CALCULATOR CORNER *(continued)*

CALCULATOR CORNER *(continued)*

2. Perform the chi-square test.

TI-84

• Specify the chi-square test, the matrix where the observed counts are found (default is [A]), and the matrix where the expected counts will be stored (default is [B]).

• Press STAT, arrow over to `TESTS`, and choose `C:`χ^2`-Test`.

• Choose "Calculate" or "Draw" to carry out the test. If you choose "Calculate," you should get the results shown below. If you choose "Draw," the chi-square curve with 2 degrees of freedom will be drawn, the area in the tail will be shaded, and the *P*-value will be displayed.

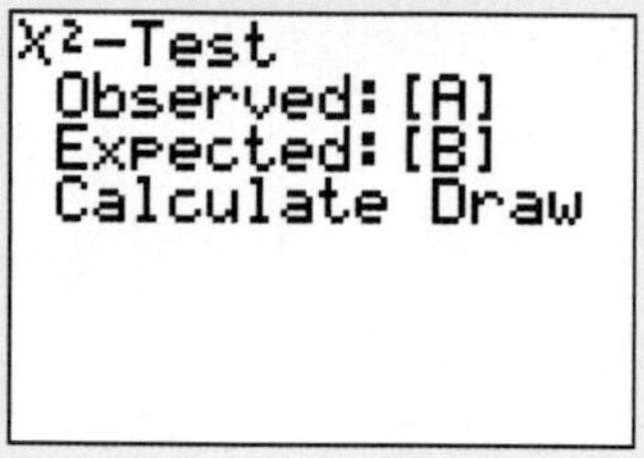

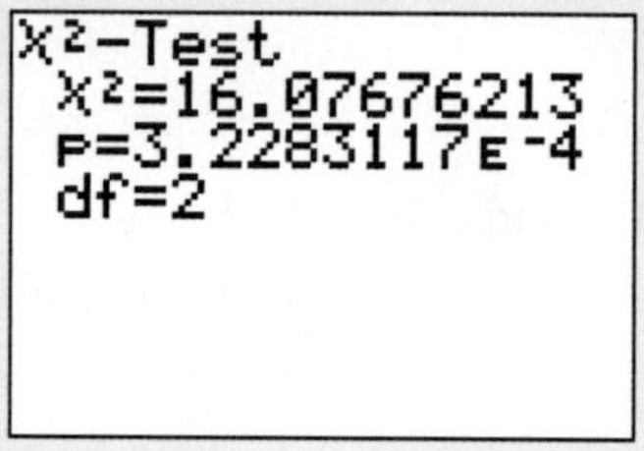

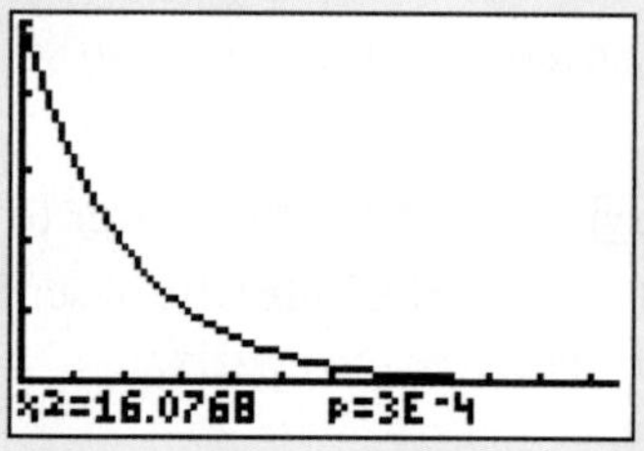

TI-Nspire

• Press menu, choose `5:Statistics`, `7:Stat Tests...`, and `8:`χ^2 `2-way Test`.

• In the dialog box, enter `a` as the "Observed Matrix" and choose `OK`.

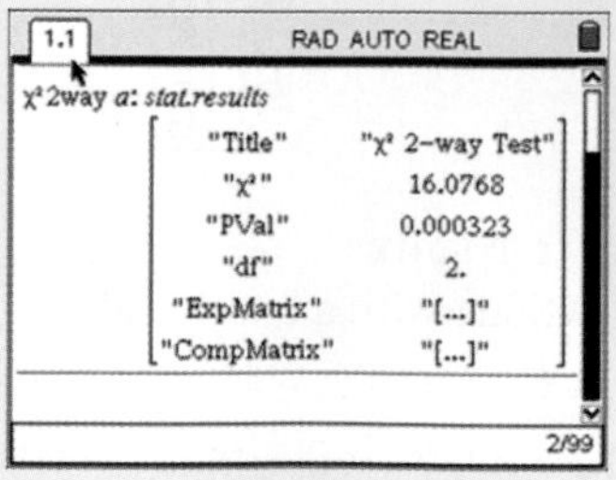

3. To see the expected counts, simply ask for a display of the matrix [B].

TI-84

• Press 2nd x^{-1} (`MATRIX`), cursor over to `EDIT`, and choose `2:[B]`.

• Verify that these calculator results agree with the results in Example 10.9.

TI-Nspire

• Press sto→ var, choose *`stat.ExpMatrix`*, and press enter.

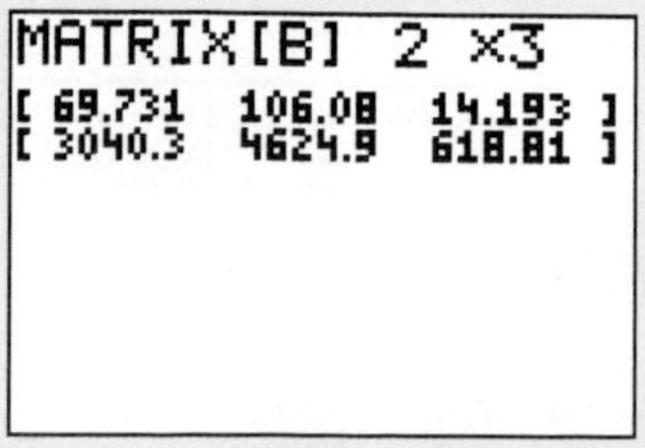

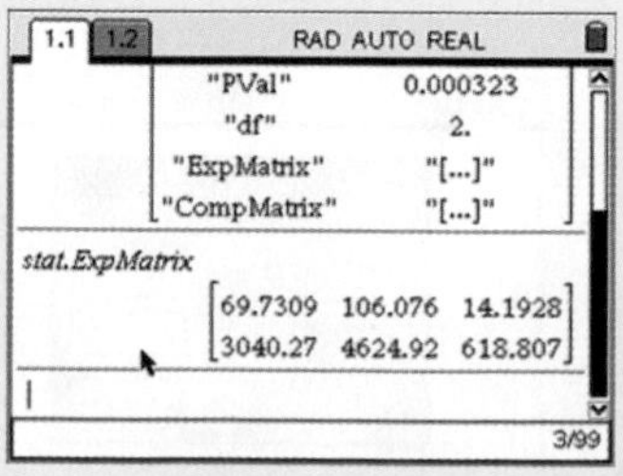

Exercises

10.19 Extracurricular activities and grades, III In Exercises 10.13 and 10.14 (page 484), you described the relationship between extracurricular activities and success in a required course. Here once again is the two-way table describing the 119 students who responded to a survey:

	Extracurricular activities (hours per week)		
	<2	2–12	>12
C or better	11	68	3
D or F	9	23	5

Is the observed association between these variables statistically significant? To find out, proceed as follows.

(a) Find the chi-square statistic. Which cells contribute most to this statistic?

(b) What are the degrees of freedom? Use Table E (in the back of the book) to say how significant the chi-square test is.

(c) Write a brief conclusion for your study.

10.20 Trying to quit, III In Exercise 10.15 (page 484), you saw an apparent association between alcohol consumption and smoking. In Exercise 10.16 (page 485) you wrote hypotheses and calculated expected counts. Here again is the two-way table that summarizes the data from the observational study:

Alcohol consumption	Smoked	Did not smoke
Yes	29	13
No	48	96

(a) Calculate a χ^2 statistic and determine the P-value.

(b) Write your conclusion from this study in plain language.

10.21 Python eggs, III Exercise 10.17 (page 485) presented data on the hatching of python eggs in water at three different temperatures. Here again are the data from that experiment. (*Note:* This is *not* a two-way table!)

	Eggs	Hatched
Cold	27	16
Neutral	56	38
Hot	104	75

Does temperature have a significant effect on hatching? Write a clear summary of your work and your conclusion.

10.22 Smoking by students and their parents How are the smoking habits of students related to their parents' smoking? Here is a two-way table from a survey of students in eight Arizona high schools:[7]

	Student smokes	Student does not smoke
Both parents smoke	400	1380
One parent smokes	416	1823
Neither parent smokes	188	1168

(a) Write the null and alternative hypotheses for the question of interest.

(b) Find the expected cell counts. Write a sentence that explains in simple language what "expected counts" are.

(c) Find the chi-square statistic, its degrees of freedom, and the *P*-value.

(d) What is your conclusion about significance?

10.23 Cocaine addiction is hard to break, I Cocaine addicts need the drug to feel pleasure. Perhaps giving them a medication that fights depression will help them stay off cocaine. A three-year study compared an antidepressant called desipramine with lithium (a standard treatment for cocaine addiction) and a placebo. The subjects were 72 chronic users of cocaine who wanted to break their drug habit. Twenty-four of the subjects were randomly assigned to each treatment. Here are the counts and percents of the subjects who succeeded in staying off cocaine during the study:

Group	Treatment	Subjects	Successes	Percent
1	Desipramine	24	14	58.3
2	Lithium	24	6	25.0
3	Placebo	24	4	16.7

(a) Compare the effectiveness of the three treatments. Use percents and draw a bar graph.

(b) Construct a two-way table that shows the relationship between treatment and success in staying off cocaine.

10.24 Cocaine addiction is hard to break, II Refer to the previous exercise. Is the relationship between treatment and success statistically significant? Carry out a significance test. What do you conclude?

APPLICATION 10.1

Do dogs resemble their owners?

In the chapter-opening story (page 469), we discussed the details of an experiment to test whether dogs and their owners resemble one another. Here is a two-way table that summarizes the results:

	Resembles owner	Doesn't resemble owner
Purebred dogs	16	9
Nonpurebred dogs	7	13

QUESTIONS

1. Compute appropriate percents and draw a well-labeled bar graph to compare the results. What do the data suggest?

2. State an appropriate pair of hypotheses for testing the research question.

3. Calculate the expected counts. Are we safe performing a chi-square test?

4. Compute the chi-square statistic. Show your work.

5. Determine the P-value using Table E (in the back of the book). Be sure to state the degrees of freedom.

6. What conclusion do you draw?

7. Follow the steps in the Calculator Corner (page 487) to perform a chi-square test. The results should confirm your answers to Questions 3, 4, and 5.

Section 10.1 Summary

Categorical variables group individuals into classes. The **chi-square test for goodness of fit** tests the null hypothesis that a single categorical variable has a specified distribution. The **expected count** for any category is obtained by multiplying the proportion of the distribution in each category times the sample size.

To display the relationship between two categorical variables, make a **two-way table** of counts in the classes. We describe the nature of an association between categorical variables by comparing selected percents. The **chi-square test for two-way tables** tells us whether an observed association in a two-way table is statistically significant. The expected count in any cell of a two-way table when H_0 is true is

$$\text{expected count} = \frac{\text{row total} \times \text{column total}}{\text{table total}}$$

To compare the observed and expected counts, calculate the **chi-square statistic:**

$$\chi^2 = \sum \frac{(\text{observed count} - \text{expected count})^2}{\text{expected count}}$$

The sampling distribution of the χ^2 statistic is not Normal. It is a new, right-skewed distribution called a **chi-square distribution.** There is a whole family of chi-square distributions having different **degrees of freedom (df).** For a goodness of fit test, df = the number of categories minus 1. For a two-way table with r rows and c columns, df = $(r - 1)(c - 1)$.

The P-value for a chi-square test is the area under the density curve to the right of χ^2. Large values of χ^2 are evidence against H_0.

Section 10.1 Exercises

10.25 Finding *P*-values Use Table E (in the back of the book) to find the P-value for a chi-square test in each of the following situations.

(a) Goodness of fit test for a categorical variable with 10 possible values; $\chi^2 = 19.62$

(b) Two-way table with 3 rows and 4 columns; $\chi^2 = 7.04$

(c) Two-way table with 2 rows and 2 columns; $\chi^2 = 1.41$

10.26 Munching Froot Loops Kellogg's Froot Loops cereal comes in six fruit flavors: orange, lemon, cherry, raspberry, blueberry, and lime. Charise poured out her morning bowl of cereal and methodically counted the number of cereal pieces of each flavor. Here are her data:

Flavor:	Orange	Lemon	Cherry	Raspberry	Blueberry	Lime
Count:	28	21	16	25	14	16

Test the null hypothesis that the population of Froot Loops produced by Kellogg's contains an equal proportion of each flavor.

10.27 Majors for men and women in business, I A study of the career plans of young women and men sent questionnaires to all 722 members of the senior class in the College of Business Administration at the University of Illinois. One question asked which major within the business program the student had chosen. Here are the data from the students who responded:[8]

	Female	Male
Accounting	68	56
Administration	91	40
Economics	5	6
Finance	61	59

Describe the differences between the distributions of majors for women and men with percents, with a graph, and in words.

10.28 Majors for men and women in business, II Exercise 10.27 gives the responses to questionnaires sent to all 722 members of a business school graduating class.

(a) Two of the observed cell counts are small. Do these data satisfy our guidelines for safe use of the chi-square test?

(b) Is there a statistically significant relationship between the gender and major of business students? Carry out a test and state your conclusion.

(c) What percent of the students did not respond to the questionnaire? How does this nonresponse weaken the conclusions that can be drawn from these data?

Alain Turgeon/FeaturePics

10.29 Birds in the trees Researchers studied the behavior of birds that were searching for seeds and insects in an Oregon forest. In this forest, 54% of the trees were Douglas firs, 40% were ponderosa pines, and 6% were other types of trees. The researchers made 156 observations of red-breasted nuthatches: 70 were seen in Douglas firs, 79 in ponderosa pines, and 7 in other types of trees.[9] Do these data suggest that nuthatches prefer particular types of trees when they're searching for seeds and insects? Carry out a chi-square goodness of fit test to help answer this question.

10.30 Preventing domestic violence A study conducted in Charlotte, North Carolina, tested the effectiveness of three police responses to spousal abuse: (1) advise and possibly separate the couple, (2) issue a citation to the offender, and (3) arrest the offender. Police officers were trained to recognize eligible cases. When presented with an eligible case, a police officer called the dispatcher, who would randomly select one of the three available treatments to be administered. There were 650 cases in the study. Each case was classified according to whether the abuser was subsequently arrested within 6 months of the original incident.[10]

	No subsequent arrest	Subsequent arrest
Advise and separate	187	25
Citation	181	43
Arrest	175	39

(a) Why was randomization important in this study? Be specific.
(b) Is there sufficient evidence from this study to conclude that the treatment imposed on an abuser is related to the likelihood of the abuser's subsequent arrest? Give appropriate evidence to support your position.

10.31 Is this coin fair? A statistics student suspected that his penny was not a fair coin, so he held it upright on a tabletop with a finger of one hand and spun the penny repeatedly by flicking it with the index finger of the other hand. In 200 spins of the coin, it landed with tails side up 122 times.
(a) Perform a goodness of fit test to see if there is sufficient evidence to conclude that spinning the coin does not produce an equal proportion of heads and tails.
(b) Use a one-proportion inference procedure to determine whether spinning the coin is equally likely to result in heads or tails.
(c) Compare your results for parts (a) and (b).

10.32 Totals aren't enough Here are the row and column totals for a two-way table with two rows and two columns:

a	b	50
c	d	50
60	40	100

Find *two different* sets of counts a, b, c, and d for the body of the table that give these same totals. This shows that the relationship between two variables cannot be obtained from the two individual distributions of the variables.

10.2 Inference about a Population Mean

ACTIVITY 10.2

Exploring sampling distributions

Professor David Lane of Rice University has developed a wonderful applet that probably ranks number 1 among high school and college students studying statistics. It provides a dynamic way for students to understand a fundamental concept in statistics: sampling distributions. And it's just plain fun to play with. Let's try it.

1. Go to the Web site **www.onlinestatbook.com/stat_sim/sampling_dist/index.html/**. When the BEGIN button appears on the left side of the screen, click on it. You will then see a yellow page entitled "Sampling Distributions."

2. There are choices for the population distribution: Normal, uniform, skewed, and custom. The default is Normal. Click the "Animated" button. What happens? Click the button several more times. What do the black boxes represent? What is the blue square that drops down onto the plot below? What does the red horizontal band under the population histogram tell us?

Notice the left panel. Important numbers are displayed there. Did you notice that the colors of the numbers match up with the objects to the right? As you make things happen, the numbers change accordingly, like an automatic scorekeeper.

ACTIVITY 10.2 *(continued)*

ACTIVITY 10.2 *(continued)*

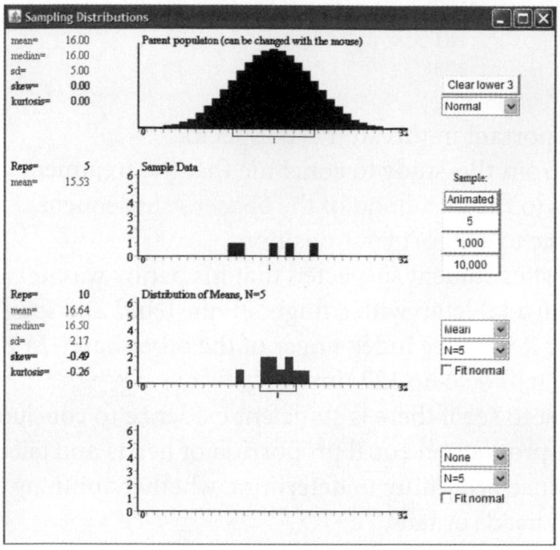

3. Click on "Clear lower 3" to start clean. Then click on the "5" button under "Sample:" repeatedly until you have simulated 100 repetitions (look for "Reps = 100" on the left panel in black letters). Does the sampling distribution (blue bars) have a recognizable shape? Click the box next to "Fit normal." Compare the center (mean) of the population distribution with the center of the sampling distribution.

4. Click "Clear lower 3." Click on 1000 samples. Click four more times until you have 5000 samples. What is the range of the data? How well does the Normal curve fit your histogram now?

5. Click "Clear lower 3." Select sample size $n = 16$. Repeat Steps 3 and 4, but with the bigger sample size. Then "Clear lower 3" and do the same for sample size $n = 25$. Complete the statements: *The center of the sampling distribution is* ________ [how related to] *the center of the population distribution. As the sample size increases, the spread of the sampling distribution* ________.

There is a very explicit relationship between the sample size and the spread of the sampling distribution of the sample mean. You will learn what that relationship is later in this section.

6. Clear the page, and select "Skewed" population. Repeat the preceding steps for the skewed distribution as you increase the sample size.

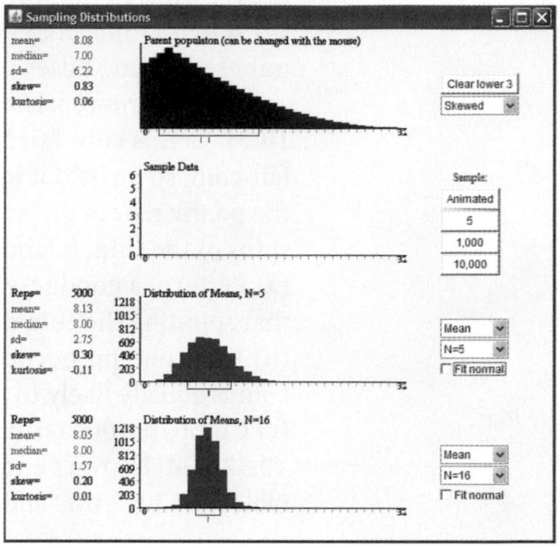

7. Clear the page, and select "Custom" distribution. Click on a point on the population histogram to insert a bar of that height. Or click on a point on the horizontal axis, and drag up to define a bar. Make a distribution as strange as you can. (*Note:* You can shorten a bar or get rid of it completely by clicking on the top of the bar and dragging down to the axis.) Then repeat the preceding steps for your custom distribution as you increase the sample size. Cool, huh?

8. What did you discover about the shape and spread of the sampling distribution? Does something special happen as you gradually increase the sample size? Fill in the blank to describe this phenomenon: *As the sample size increases, the sampling distribution of the sample mean gets closer and closer to* ________.

There is a name for this important result: the *central limit theorem*. We will learn more about the central limit theorem shortly.

Although the reasoning of confidence intervals and significance tests is always the same, specific methods vary greatly. The form of an inference procedure depends first on the parameter you want information about—a population proportion or mean or median or whatever. The second influence is the design of the sample or experiment. Estimating a population proportion from a stratified sample requires a different method than if the data come from an SRS. This section presents a confidence interval and a significance test for inference about a population mean when the data are an SRS from the population.

The sampling distribution of a sample mean

What is the mean number of hours your school's seniors study each week? What is their mean grade point average? We often want to estimate the mean of a population. To distinguish the population mean (a parameter) from the sample mean $\bar{x}$, we write the population mean as μ (the Greek letter mu). We use the mean $\bar{x}$ of an SRS to estimate the unknown mean μ of the population.

Like the sample proportion $\hat{p}$, the sample mean from a large SRS has a sampling distribution that is close to Normal. The **sampling distribution of $\bar{x}$** has μ as its mean. The standard deviation of $\bar{x}$ depends on the standard deviation of the population, which is usually written as σ (the Greek letter sigma). Using mathematics we can discover the following facts.

Sampling distribution of a sample mean

Choose an SRS of size n from a population in which individuals have mean μ and standard deviation σ. Let $\bar{x}$ be the mean of the sample. Then:

- The **mean** of the sampling distribution of $\bar{x}$ is equal to μ. In symbols, $\mu_{\bar{x}} = \mu$.
- The **standard deviation** of the sampling distribution is $\sigma/\sqrt{n}$. In symbols, $\sigma_{\bar{x}} = \sigma/\sqrt{n}$.

It isn't surprising that the values that $\bar{x}$ takes in many samples are centered at the true mean μ of the population. That's the lack of bias in random sampling once again. The other fact about the sampling distribution makes precise a very important property of the sample mean: *the mean of several observations is less variable than individual observations.* Figure 10.3 (on the next page) illustrates this property. It compares the distribution of a single observation with the sampling distribution of the mean $\bar{x}$ of 10 observations. Both have the same center, but the distribution of the mean of 10 observations is less spread out.

In Figure 10.3, the distribution of individual observations is Normal. If that is true, then the sampling distribution of $\bar{x}$ is exactly Normal for any size sample, not just approximately Normal for large samples. What if the population distribution is not Normal? As Activity 10.2 illustrates, if we take large enough samples, the sampling distribution of the sample mean will be approximately Normal. This remarkable statistical fact is known as the **central limit theorem.** The central limit theorem lies behind the use of Normal sampling distributions for sample means.

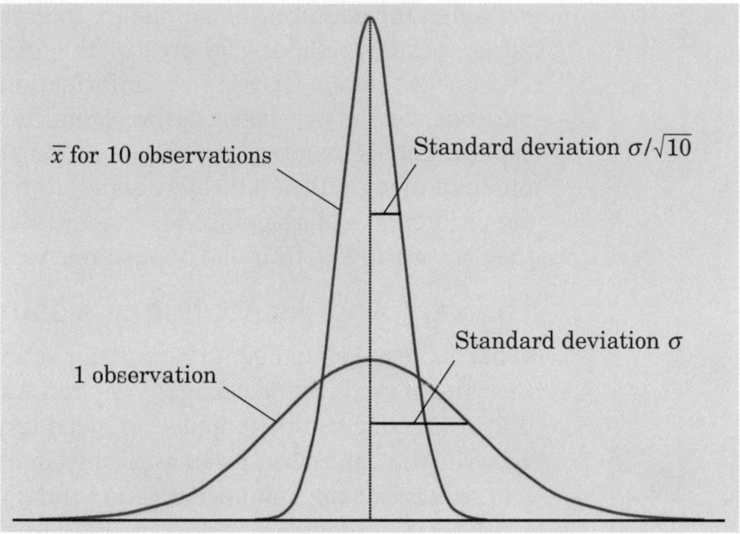

Figure 10.3 The sampling distribution of the sample mean $\bar{x}$ of 10 observations compared with the distribution of a single observation.

Example 10.10 The central limit theorem in action

Figure 10.4 shows the central limit theorem in action. The top-left density curve describes individual observations from a population. It is strongly right-skewed. Distributions like this describe the time it takes to repair a household appliance, for example. Most repairs are quickly done, but some are lengthy.

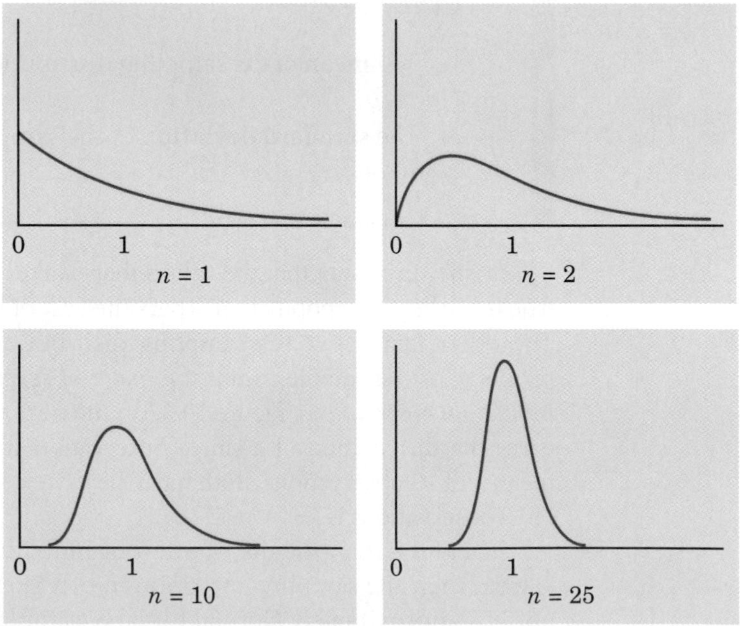

Figure 10.4 The distribution of sample means becomes more Normal as the size of the sample increases. The distribution of individual observations ($n = 1$) is far from Normal. The distributions of means for samples of size $n = 2$, $n = 10$, and $n = 25$ observations move closer to the Normal shape.

The other three density curves in Figure 10.4 show the sampling distributions of the sample mean for samples of size $n = 2$, $n = 10$, and $n = 25$ observations from this population. As the sample size n increases, the shape becomes more Normal. The mean remains fixed and the standard deviation decreases, following the pattern $\sigma/\sqrt{n}$. The distribution for $n = 10$ observations is still somewhat skewed to the right but is beginning to resemble a Normal curve. The density curve for $n = 25$ is yet more Normal. The contrast between the shapes of the population distribution and of the sampling distribution of the mean for $n = 10$ or $n = 25$ observations is striking.

Exercises

10.33 The idea of a sampling distribution Figure 9.2 (page 423) shows the idea of the sampling distribution of a sample proportion in picture form. Draw a similar picture that shows the idea of the sampling distribution of a sample mean $\bar{x}$.

10.34 Averages versus individuals Scores on the American College Testing (ACT) college admissions examination vary Normally with mean $\mu = 18$ and standard deviation $\sigma = 6$. The range of reported scores is 1 to 36.

(a) What range contains the middle 95% of all individual scores? Explain.

(b) If the ACT scores of 25 randomly selected students are averaged, what range contains the middle 95% of the averages $\bar{x}$? Justify your answer.

10.35 A sampling distribution, I We used our calculator's `rand` function to generate samples of 100 random numbers between 0 and 1 repeatedly. Here are the sample means for 50 samples of size 100:

0.532	0.450	0.481	0.508	0.510	0.530	0.499	0.461	0.543	0.490
0.497	0.552	0.473	0.425	0.449	0.507	0.472	0.438	0.527	0.536
0.492	0.484	0.498	0.536	0.492	0.483	0.529	0.490	0.548	0.439
0.473	0.516	0.534	0.540	0.525	0.540	0.464	0.507	0.483	0.436
0.497	0.493	0.458	0.527	0.458	0.510	0.498	0.480	0.479	0.499

The sampling distribution of $\bar{x}$ is the distribution of the means from all possible samples. We actually have the means from 50 samples.

(a) Make a histogram of these 50 observations. Does the distribution appear to be roughly Normal, as the central limit theorem says will happen for large enough samples?

(b) Sketch a graph of the density curve for the population distribution. What is the value of the population mean?

10.36 A sampling distribution, II Exercise 10.35 presents 50 sample means $\bar{x}$ from 50 random samples of size 100. Find the mean and standard deviation of these 50 values. Then answer these questions.

(a) The mean of the population from which the 50 samples were drawn is $\mu = 0.5$ if the random number generator is working properly. What do you expect the mean of the distribution of $\bar{x}$'s from all possible samples to be? Is the mean of these 50 samples close to this value?

(b) The standard deviation of the distribution of $\bar{x}$ from samples of size $n = 100$ is supposed to be $\sigma/\sqrt{100} = \sigma/10$, where σ is the standard deviation of individuals in the population. Use this fact with the standard deviation you calculated for the 50 $\overline{x}$'s to estimate σ. (In fact, $\sigma = 0.289$.)

10.37 Student attitudes The Survey of Study Habits and Attitudes (SSHA) is a psychological test that measures students' study habits and attitude toward school. Scores range from 0 to 200. The mean score for U.S. college students is about 115, and the standard deviation is about 30. A teacher suspects that older students have better attitudes toward school. She gives the SSHA to 25 students who are at least 30 years old. Assume that scores in the population of older students are Normally distributed with standard deviation $\sigma = 6$. The teacher wants to test the hypotheses

$$H_0: \mu = 115$$
$$H_a: \mu > 115$$

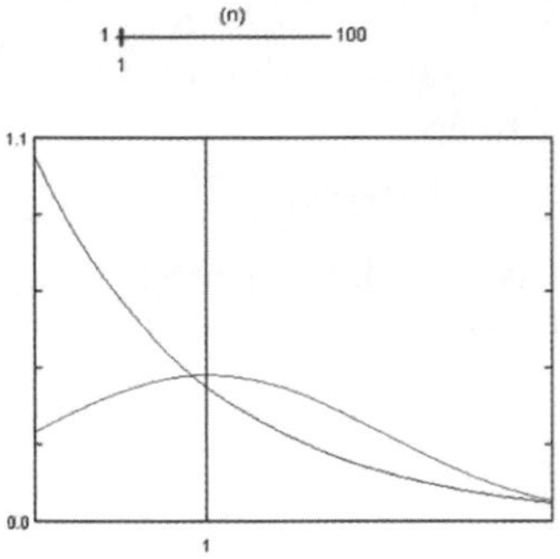

(a) What is the sampling distribution of the mean score $\overline{x}$ of a sample of 25 older students if the null hypothesis is true? Sketch the density curve of this distribution. (*Hint:* Sketch a Normal curve first, then mark the axis using what you know about locating μ and σ on a Normal curve.)

(b) Suppose that the sample data give $\overline{x} = 118.6$. Mark this point on the axis of your sketch. In fact, the outcome was $\overline{x} = 125.7$. Mark this point on your sketch. Using your sketch, explain in simple language why one outcome is good evidence that the mean score of all older students is greater than 115 and why the other outcome is not.

(c) Shade the area under the curve that is the *P*-value for the sample result $\overline{x} = 118.6$.

APPLET

10.38 The CLT applet Go to the textbook Web site, **www.whfreeman.com/sta2e**, and click on "Statistical Applets." Launch the *Central Limit Theorem* applet. You should see a screen like the one shown in the margin. Click and drag the slider to change the sample size, and watch how the idealized curve for the sampling distribution changes with it. Write a few sentences describing what is happening.

Confidence intervals for a population mean

If we want to estimate an unknown population mean μ, we start by taking a simple random sample from the population of interest and then calculate the sample mean $\overline{x}$. We use $\overline{x}$ as our estimate for μ because the values of $\overline{x}$ in repeated samples center on μ. That is, the sampling distribution of $\overline{x}$ has mean μ. The standard deviation of the sampling distribution is $\sigma/\sqrt{n}$. If the population distribution is Normal or the sample size n is sufficiently large, the sampling distribution will have a Normal shape.

Now we can find **confidence intervals for** μ following the same reasoning that led us to confidence intervals for a proportion p in Section 9.1. The big idea is that, to cover the central area C under a Normal curve, we must go out a distance z^* on either side of the mean. Figure 10.5 reminds you how C and z^* are related.

Our confidence interval has the same basic form as before:

$$\text{estimate} \pm \text{margin of error}$$

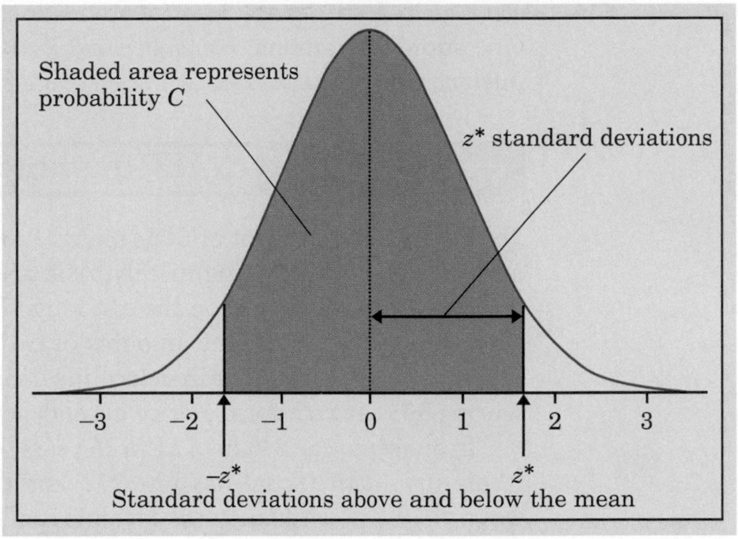

Figure 10.5 Critical values z^* of the Normal distributions. In any Normal distribution, there is area (probability) C under the curve between $-z^*$ and z^* standard deviations away from the mean.

However, the details are a little different now that we are working with means instead of proportions. The correct interval is

$$\bar{x} + z^* \frac{\sigma}{\sqrt{n}}$$

There's just one problem. The standard deviation of $\bar{x}$ depends on both the sample size n and the standard deviation σ of individuals in the population. In practice, we know n but not σ. When n is large, the sample standard deviation s is close to σ and can be used to estimate it, just as we use the sample mean $\bar{x}$ to estimate the population mean μ. The estimated standard deviation of $\bar{x}$ is therefore $s/\sqrt{n}$. For large sample sizes (say $n \geq 50$), an *approximate* level C confidence interval for an unknown population mean μ is therefore

$$\bar{x} \pm z^* \frac{s}{\sqrt{n}}$$

Confidence interval for a population mean

Choose an SRS of size n from a large population of individuals having mean μ. The mean of the sample observations is $\bar{x}$. When n is large, an **approximate level C confidence interval for μ** is

$$\bar{x} \pm z^* \frac{s}{\sqrt{n}}$$

where z^* is the critical value for confidence level C from Table D (in the back of the book).

The cautions we noted in estimating p apply here as well. The formula is valid only when an SRS is drawn and the sample size n is reasonably large. The margin of error again decreases only at a rate proportional to $\sqrt{n}$ as the sample size n increases. One additional caution: *remember that $\bar{x}$ and s are strongly influenced by outliers. Inference using $\bar{x}$ and s is suspect when outliers are present.* Always look at your data.

Example 10.11 NAEP quantitative scores

The National Assessment of Educational Progress (NAEP) includes a short test of quantitative skills, covering mainly basic arithmetic and the ability to apply it to realistic problems. Scores on the test range from 0 to 500. For example, a person who scores 233 can add the amounts of two checks appearing on a bank deposit slip; someone scoring 325 can determine the price of a meal from a menu; a person scoring 375 can transform a price in cents per ounce into dollars per pound.

In a recent year, 840 men 21 to 25 years of age were in the NAEP sample. Their mean quantitative score was $\bar{x} = 272$, and the standard deviation of their scores was $s = 59$. These 840 men are a simple random sample from the population of all young men. On the basis of this sample, what can we say about the mean score in the population of all 9.5 million young men of these ages?

The 95% confidence interval for μ uses the critical value $z^* = 1.96$ from Table D (in the back of the book). The interval is

$$\begin{aligned} \bar{x} \pm z^* \frac{s}{\sqrt{n}} &= 272 \pm 1.96 \frac{59}{\sqrt{840}} \\ &= 272 \pm (1.96)(2.036) = 272 \pm 4.0 \end{aligned}$$

We are 95% confident that the mean score for all young men lies between 268 and 276.[11]

Tests for a population mean

As with confidence intervals, the reasoning that leads to significance tests for hypotheses about a population mean μ follows the reasoning that led to tests about a population proportion p. The big idea is to use the sampling distribution that the sample mean $\bar{x}$ would have if the null hypothesis were true. Locate the $\bar{x}$ from your data on this distribution and see if it is unlikely. A value of $\bar{x}$ that would rarely appear if H_0 were true is evidence that H_0 is not true. The steps in performing a significance test about μ are the same as those in tests for a proportion.

Example 10.12 Executives' blood pressures

The National Center for Health Statistics reports that the mean systolic blood pressure for males 35 to 44 years of age is 128. The medical director of a large company looks at the medical records of a sample of 72 executives in this age group and finds that the mean systolic blood pressure is $\bar{x} = 126.1$ and that the standard deviation is $s = 15.2$. Is this evidence that the company's executives have a different mean blood pressure from the general population?

The hypotheses. The null hypothesis is "no difference" from the national mean. The alternative is two-sided, because the medical director did not have a particular

Catching cheaters

Lots of students take a long multiple-choice exam. Can the computer that scores the exam also identify papers that are suspiciously similar? Clever people have created a measure that takes into account not just identical answers but the popularity of those answers and the total score on the similar papers. The measure has close to a Normal distribution, and the computer flags pairs of papers with a measure outside ±4 standard deviations as significant.

direction in mind before examining the data. So the hypotheses about the unknown mean μ of the executive population are

$$H_0: \mu = 128$$
$$H_a: \mu \neq 128$$

The sampling distribution. *If the null hypothesis is true,* the sample mean $\bar{x}$ has approximately the Normal distribution with mean μ and standard deviation

$$\frac{s}{\sqrt{n}} = \frac{15.2}{\sqrt{72}} = 1.79$$

The data. The sample mean is $\bar{x} = 126.1$. The standardized score for this outcome is

$$\text{standardized score} = \frac{\text{observation} - \text{mean}}{\text{standard deviation}}$$
$$z = \frac{126.1 - 128}{1.79} = -1.06$$

We know that an outcome a little more than 1 standard deviation away from the mean of a Normal distribution is not very surprising. The last step is to make this formal.

The *P*-value. Figure 10.6 locates the sample outcome -1.06 (in the standard scale) on the Normal curve that represents the sampling distribution if H_0 is true. The two-sided *P*-value is the probability of an outcome at least this far out in either direction. This is the shaded area under the curve. According to Table A, the area to

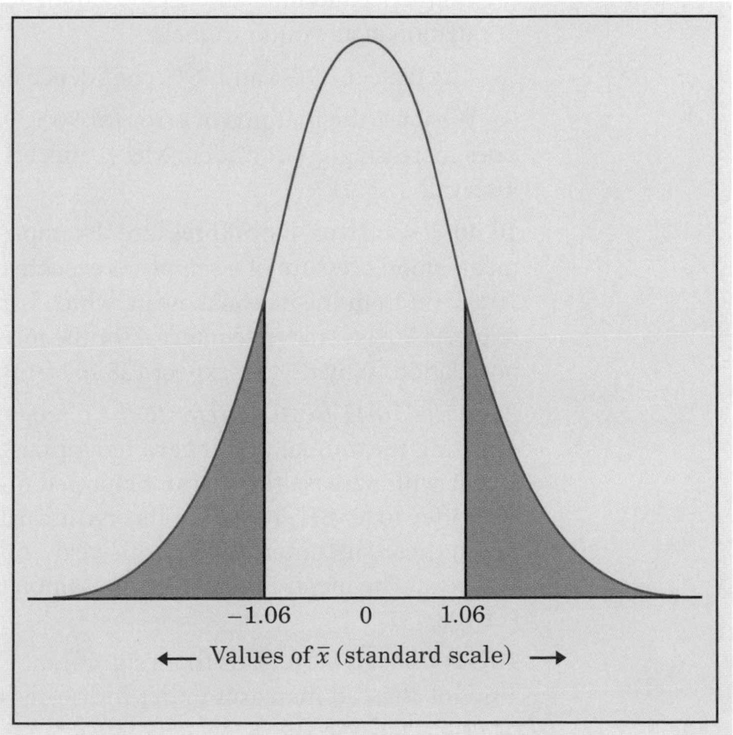

Figure 10.6 The *P*-value for a two-sided test when the standardized score for the sample mean is -1.06.

the left of $z = -1.06$ is 0.1446. The area to the left of -1.06 and to the right of 1.06 is double this, or 0.2892. This is our *P*-value.

Conclusion. The large *P*-value gives us no reason to think that the mean blood pressure of the company's executive population differs from the national average.

The test in Example 10.12 assumes that the 72 executives in the sample are an SRS from the population of all middle-aged male executives in the company. We should check this assumption by asking how the data were produced. If medical records are available only for executives with recent medical problems, for example, the data are of little value for our purpose. It turns out that all executives are given a free annual medical exam, and that the medical director selected 72 exam results at random.

The data in Example 10.12 do *not* establish that the mean blood pressure μ for this company's executives *is* 128. We sought evidence that μ differed from 128 and failed to find convincing evidence. That is all we can say. No doubt the mean blood pressure of the entire executive population of this company is not exactly equal to 128. A large enough sample would give evidence of the difference, even if it were very small.

Exercises

10.39 Confidence level and margin of error The NAEP test (Example 10.11, page 500) was also given to a sample of 1077 women aged 21 to 25 years. Their mean quantitative score was 275, and the standard deviation was 58.

(a) Construct and interpret a 95% confidence interval for the mean score μ in the population of all young women.

(b) Calculate the 90% and 99% confidence intervals for μ.

(c) What are the margins of error for 90%, 95%, and 99% confidence? How does increasing the confidence level affect the margin of error of a confidence interval?

10.40 Executives' blood pressure Example 10.12 (page 500) found that the mean blood pressure of a sample of executives did not differ significantly at the 10% level from the national mean, which is 128. Use the data in that example to give a 90% confidence interval for the mean of the company's executive population. Why do you expect 128 to be inside this interval?

Exercises 10.41 to 10.43 refer to the following setting. A bank wonders whether omitting the annual credit card fee for customers who charge at least \$2400 in a year will increase the amount charged on its credit cards. The bank makes this offer to an SRS of 200 of its credit card customers. It then compares how much these customers charge this year with the amount that they charged last year. The mean increase in the sample is \$332, and the standard deviation is \$108.

10.41 Charge more, I Is there significant evidence at the 1% level that the mean amount charged increases under the no-fee offer? State H_0 and H_a and carry out a test.

10.42 Charge more, II Calculate and interpret a 99% confidence interval for the mean amount charges would have increased if this benefit had been extended to all such customers.

10.43 Charge more, III In Exercises 10.41 and 10.42, you carried out the calculations for a test and confidence interval based on a bank's experiment in changing the rules for its credit cards. You ought to ask some questions about this study.

(a) The distribution of the amount charged is skewed to the right, but outliers are prevented by the credit limit that the bank enforces on each card. Why can we use a test and confidence interval based on a Normal sampling distribution for the sample mean $\bar{x}$?

(b) The bank's experiment was not comparative. The increase in amount charged over last year may be explained by lurking variables rather than by the rule change. What are some plausible reasons why charges might go up? Outline the design of a comparative randomized experiment to answer the bank's question.

10.44 Normal body temperature In Application 1.2 (page 25), we described an observational study that investigated whether 98.6°F is really "normal" body temperature. Allen Shoemaker, from Calvin College, produced a data set with many of the same characteristics as the original temperature readings. His data set consists of one oral temperature reading for each of 130 individuals. The mean and standard deviation of the 130 temperature readings are $\bar{x} = 98.25$ and $s = 0.73$.

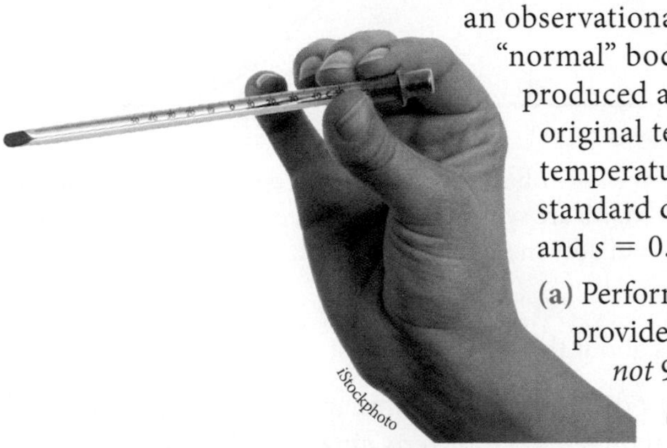
iStockphoto

(a) Perform a significance test to determine whether these data provide convincing evidence that "normal" body temperature is *not* 98.6°F. Show each step of your test clearly.

(b) Calculate a 95% confidence interval for the mean body temperature μ in the population of healthy adults. What additional information does the confidence interval provide?

Inference in practice: the *t* distribution

What would cause the head brewer of the famous Guinness brewery in Dublin, Ireland, not only to use statistics but also to invent new statistical methods? The search for a better beer, of course.

William S. Gosset (1876–1937) joined Guinness as a brewer in 1899. He soon became involved in experiments and in statistics to understand the data from those experiments. What are the best varieties of barley and hops for brewing? How should they be grown, dried, and stored? The results of the field experiments, as you can guess, varied. Gosset faced the problem we noted earlier: he didn't know the population standard deviation σ. He also observed that when the sample size was small, replacing σ by s and using a Normal distribution to estimate or test a claim about a population mean μ wasn't accurate enough.

After much work, Gosset developed what we now call the ***t* distributions.** Unlike the standard Normal distribution, there is a different t distribution for each sample size n. We specify a particular t distribution by giving its **degrees of freedom (df).** When we perform inference about μ using a t distribution, the appropriate degrees of freedom is df $= n - 1$.

Figure 10.7 compares the density curves of the standard Normal distribution and the t distributions with 2 and 9 degrees of freedom.

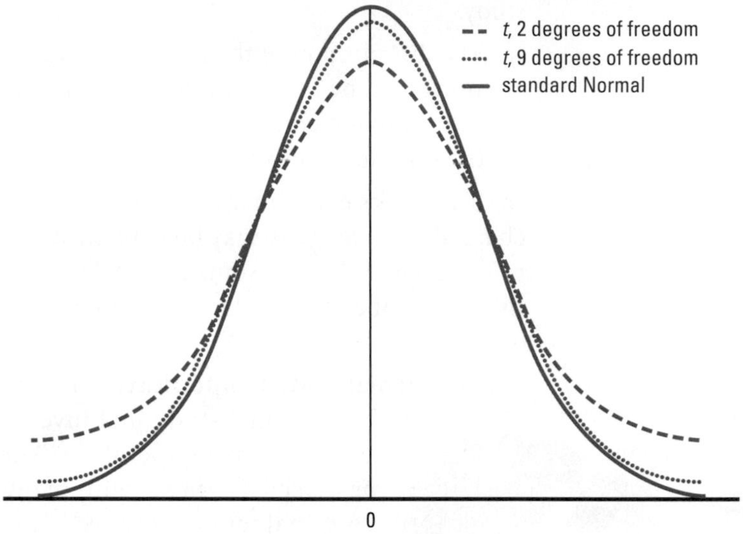

Figure 10.7 Density curves for the t distributions with 2 and 9 degrees of freedom and for the standard Normal distribution. All are symmetric with center 0. The t distributions have more probability in the tails than does the standard Normal.

The figure illustrates these facts about the t distributions:

- The density curves of the t distributions are similar in shape to the standard Normal curve. They are symmetric about zero, single-peaked, and bell-shaped.
- The spread of the t distributions is a bit greater than that of the standard Normal distribution. The t distributions in Figure 10.7 have more area (probability) in the tails and less in the center than does the standard Normal.
- As the degrees of freedom increase, the t density curves approach the standard Normal distribution. This happens because s estimates σ more accurately as the sample size increases. So using s in place of σ causes little extra variation when the sample is large. (That explains why it was acceptable to use z instead of t to perform inference about μ in our earlier examples—the sample sizes were all very large.)

Table C in the back of the book gives critical values t^* for the t distributions. Each row in the table contains critical values for one of the t distributions; the degrees of freedom appear at the left of the row. For confidence intervals, several of the more common confidence levels C (in percent) are given at the bottom of the table. By looking down any column, you can check that the t critical values approach the Normal values as the degrees of freedom increase.

Example 10.13 Finding t^* using Table C

	Upper-tail probability p			
df	.05	.025	.02	.01
10	1.812	2.228	2.359	2.764
11	1.796	2.201	2.328	2.718
12	1.782	2.179	2.303	2.681
z^*	1.645	1.960	2.054	2.326
	90%	95%	96%	98%
	Confidence level C			

Suppose you want to construct a 95% confidence interval for the mean μ of a Normal population based on an SRS of size $n = 12$. What critical value t^* should you use?

In Table C, an excerpt of which is shown, we consult the row corresponding to df $= n - 1 = 11$. We move across that row to the entry that is directly above 95% confidence level at the bottom of the chart. The desired critical value is $t^* = 2.201$. Notice that the corresponding critical value from the z distribution is $z^* = 1.96$.

Confidence intervals in practice: the *t* interval

Earlier, we used the formula

$$\bar{x} \pm z^* \frac{s}{\sqrt{n}}$$

to construct a confidence interval for μ when the sample size n is large. What if n is small? In that case, we have to use critical values from the t distribution with $n - 1$ degrees of freedom in place of the z critical values. Our formula becomes

$$\bar{x} \pm t^* \frac{s}{\sqrt{n}}$$

In fact, even for large sample sizes, this ***t* interval** "works better" than the z interval we showed you earlier. That is, 95% confidence intervals constructed using t critical values will tend to capture μ closer to 95% of the time than intervals based on z critical values. *Our advice: use t for inference about means and z for inference about proportions.*

The *t* interval

Draw an SRS of size n from a population having unknown mean μ. A **level C confidence interval for μ** is

$$\bar{x} \pm t^* \frac{s}{\sqrt{n}}$$

where t^* is the critical value for the t distribution with $n - 1$ degrees of freedom. This interval is exactly correct when the population distribution is Normal. It is approximately correct as long as no outliers or strong skewness are present.

The t interval is similar in both reasoning and computational detail to the z interval we used earlier. So we will now pay more attention to questions about using these methods in practice.

Example 10.14 Measuring stream health

The level of dissolved oxygen in a river is an important indicator of the water's ability to support aquatic life. A researcher collects water samples at 15 randomly chosen locations along a stream and measures the dissolved oxygen. Here are the results in milligrams per liter:

4.53 5.04 3.29 5.23 4.13 5.50 4.83 4.40 5.42 6.38 4.01 4.66 2.87 5.73 5.55

We will construct and interpret a 95% confidence interval for the mean dissolved oxygen level in this stream.

What is μ in this setting? It's the mean dissolved oxygen level in the stream. In order to use a t interval to estimate μ, we need to know that the population distribution is Normal or that there are no outliers or strong skewness in the data.

Figure 10.8 shows a dotplot and boxplot of the dissolved oxygen (DO) measurements. Both graphs are slightly skewed to the left. Since there are no outliers or strong skewness present, we should be safe calculating a t interval.

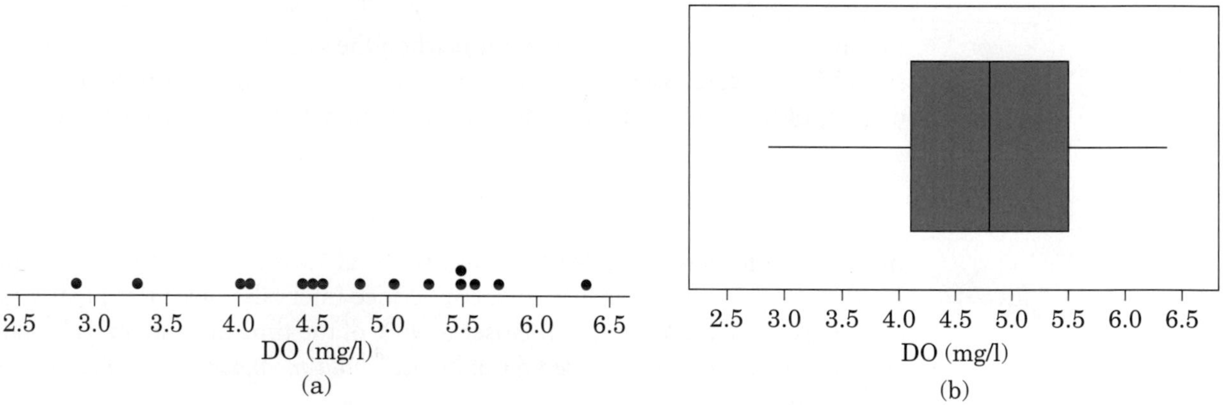

Figure 10.8 (a) A dotplot and (b) a boxplot of the dissolved oxygen level at 15 randomly selected locations along a stream.

For these data, $\bar{x} = 4.77$ mg/l and $s = 0.94$ mg/l. To find the critical value t^* for a 95% confidence interval, we look at the df $= 15 - 1 = 14$ row in Table C. We find that $t^* = 2.145$. So our desired confidence interval is

	Upper-tail probability p		
df	.05	.025	.02
14	1.761	2.145	2.264
15	1.753	2.131	2.249
16	1.746	2.120	2.235
	90%	95%	96%
	Confidence level C		

$$\bar{x} \pm t^* \frac{s}{\sqrt{n}} = 4.77 \pm 2.145 \frac{0.94}{\sqrt{15}}$$
$$= 4.77 \pm 0.52$$

We are 95% confident that the mean DO level in the stream is between 4.25 and 5.29 mg/l.

Why did the researcher in Example 10.14 collect water samples at randomly chosen locations along the stream? Because random sampling allows us to generalize from the sample results to the population of interest. If the researcher had measured the DO level at 15 locations downstream from a farm, for example, the sample mean DO level $\bar{x}$ might not be a very good estimate for the true mean DO level μ in the stream. *Except in the case of very small samples, the assumption that the data are an SRS from the population of interest is more important than the assumption that the population distribution is Normal.*

Significance tests in practice: the *t* test

In tests as in confidence intervals, we allow for unknown σ by using the standard error and replacing z by t. Now we can do a realistic analysis of data produced to test a claim about an unknown population mean.

Example 10.15 Is caffeine dependence real?

An experiment was conducted with 11 people diagnosed as being dependent on caffeine. Each subject was barred from coffee, colas, and other substances containing caffeine. Instead, they took capsules containing their normal caffeine intake. During a different time period, they took placebo capsules. The order in which subjects took caffeine and the placebo was randomized. Table 10.2 contains data on one of several tests given to the subjects.[12] "Depression" is the score on the Beck Depression Inventory. Higher scores show more symptoms of depression. Do these data provide convincing evidence that depriving caffeine-dependent people of caffeine causes them to become more depressed?

Table 10.2 Results of a Caffeine-Deprivation Study

Subject	Depression score (caffeine)	Depression score (placebo)	Difference (placebo − caffeine)
1	5	16	11
2	5	23	18
3	4	5	1
4	3	7	4
5	8	14	6
6	5	24	19
7	0	6	6
8	0	3	3
9	2	15	13
10	11	12	1
11	1	0	−1

The hypotheses. The population of interest is all people who are dependent on caffeine. Our null hypothesis is that caffeine deprivation doesn't affect depression for these individuals. The alternative is one-sided, because researchers believed that depriving people of caffeine would increase their level of depression.

Our parameter μ is the mean *difference* (placebo − caffeine) in depression scores that would be reported if all individuals in the population took both the caffeine capsule and the placebo. So the hypotheses about the unknown mean μ are

$$H_0: \mu = 0$$
$$H_a: \mu > 0$$

The sampling distribution. Since σ is unknown and the sample size is small ($n = 11$), we have to use the t distribution with $11 - 1 = 10$ degrees of freedom in place of the standard Normal curve to carry out the test. Is the population distribution of differences Normal? We don't know this, so we examine the sample data. Figure 10.9 shows a dotplot and boxplot of the differences (placebo − caffeine) in depression scores for our 11 subjects. There are no obvious outliers or strong skewness in the sample data. We should be safe using t procedures.

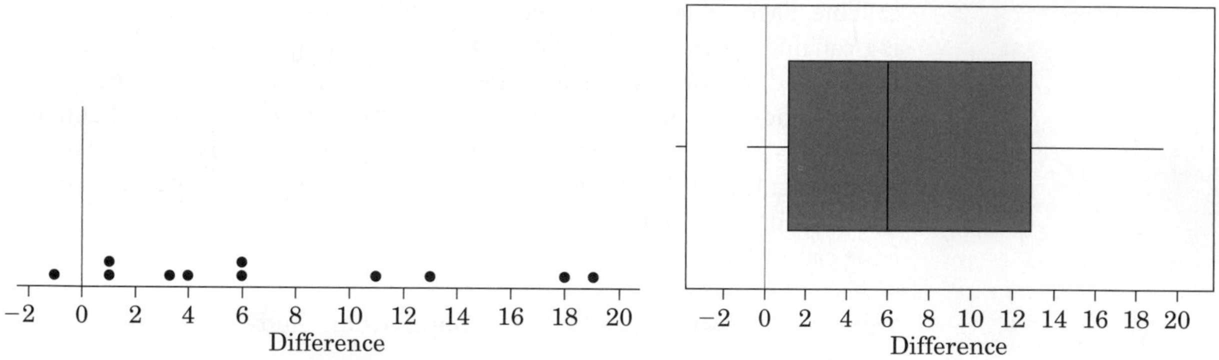

Figure 10.9 Dotplot and boxplot of the differences in Beck Depression Inventory scores for the subjects in a caffeine-dependence experiment.

The data. For our 11 subjects, $\bar{x} = 7.364$ and $s = 6.918$. The corresponding standard error is

$$\frac{s}{\sqrt{n}} = \frac{6.918}{\sqrt{11}} = 2.086$$

If the null hypothesis is true, $\mu = 0$ and our standardized score is

$$\text{standardized score} = \frac{\text{observation} - \text{mean}}{\text{standard deviation}}$$

$$t = \frac{7.364 - 0}{2.086} = 3.53$$

The *P*-value. Figure 10.10 locates the standardized value $t = 3.53$ on the t distribution with 10 degrees of freedom. The *P*-value for our one-sided test is the shaded area to the right of 3.53. Table C says that this area is between 0.005 and 0.0025. (Technology gives a *P*-value of 0.0027.)

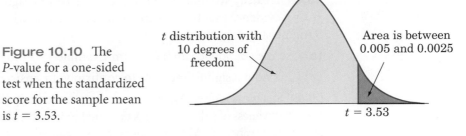

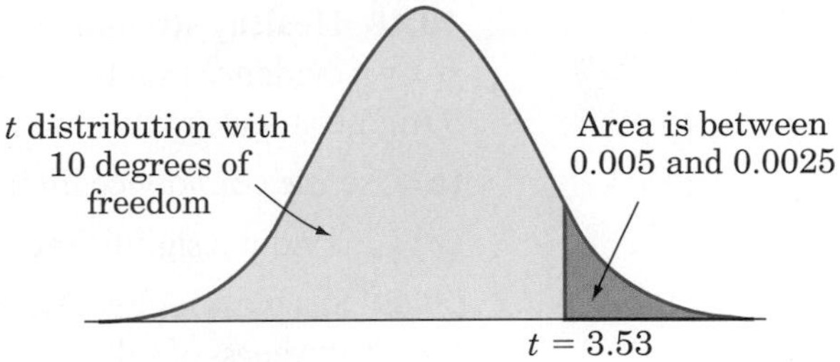

Figure 10.10 The *P*-value for a one-sided test when the standardized score for the sample mean is $t = 3.53$.

Upper-tail probability p			
df	.005	.0025	.001
9	3.250	3.690	4.297
10	3.169	3.581	4.144
11	3.106	3.497	4.025

Conclusion. A *P*-value this low (between 0.005 and 0.0025) gives quite strong evidence against the null hypothesis. We reject H_0 and conclude that withholding caffeine from caffeine-dependent individuals may lead to depression.

Many experiments involve individuals who are *not* chosen at random from the population of interest, like the caffeine-dependent subjects in Example 10.15. In such cases, we may not be able to generalize our findings to the population of interest. By randomly assigning the treatments, however, the researchers could say that the large mean difference in depression scores was due to the different treatments.

Did you notice that the data in Example 10.15 came from a matched pairs experiment? Comparative studies like this one are quite common. When we are dealing with paired data, such as the two depression scores for each subject in this experiment, we can use a *t* interval or *t* test on the observed differences. More advanced methods known as *two-sample t procedures* can be used to perform inference about the means of two groups or populations.

Exercises

10.45 Finding critical values What critical value t^* from Table C should be used for a confidence interval for the population mean μ in each of the following situations?

(a) A 99% confidence interval based on $n = 12$ observations

(b) A 90% confidence interval from an SRS of 30 observations

(c) A 95% confidence interval from a sample of size 18

10.46 Blood pressure A randomized comparative experiment studied the effect of diet on blood pressure. Researchers divided 54 healthy white males at random into two groups. One group received a calcium supplement; the other, a placebo. At the beginning of the study, the researchers measured many variables on the subjects. The paper reporting the study gives $\bar{x} = 114.9$ and $s = 9.3$ for the seated systolic blood pressure of the 27 members of the placebo group.

(a) Calculate and interpret a 95% confidence interval for the mean blood pressure of the population from which the subjects were recruited.

(b) The formula you used in part (a) requires an important assumption about the 27 men who provided the data. What is this assumption?

10.47 Caffeine and depression Refer to Example 10.15 (page 507). The results of our significance test suggest that caffeine deprivation leads to increased depression for caffeine-dependent people. Calculate a 95% confidence interval for the population mean μ. What additional information does the confidence interval provide?

10.48 Healthy streams Refer to Example 10.14 (page 506). Do the data provide strong evidence that the stream has a mean dissolved oxygen content of less than 5 mg per liter?

(a) Use the confidence interval from Example 10.14 to answer this question.

(b) Carry out a significance test to confirm your answer to (a).

10.49 Sleepless nights An experiment was carried out with 10 patients to investigate the effectiveness of a drug that was designed to increase sleep time. The data below show the number of additional hours of sleep gained by each subject after taking the drug.[13] (A negative value indicates that the subject got less sleep after taking the drug.)

1.9	0.8	1.1	0.1	−0.1	4.4	5.5	1.6	4.6	3.4

Is there convincing evidence that the drug is effective? Carry out a significance test, and state your conclusion clearly.

10.50 Darwin's plants Charles Darwin, author of *The Origin of Species* (1859), designed an experiment to compare the effects of cross-fertilization and self-fertilization on the size of plants. He planted pairs of very similar seedling plants, one self-fertilized and one cross-fertilized, in the same pot at the same time. After a period of time, Darwin measured the heights (in inches) of all the plants. Here are his data:[14]

Pair	Cross	Self	Pair	Cross	Self
1	23.5	17.4	9	18.3	16.5
2	12.0	20.4	10	21.6	18.0
3	21.0	20.0	11	23.3	16.3
4	22.0	20.0	12	21.0	18.0
5	19.1	18.4	13	22.1	12.8
6	21.5	18.6	14	23.0	15.5
7	22.1	18.6	15	12.0	18.0
8	20.4	15.3			

Perform a complete analysis of Darwin's data that includes both a significance test and a confidence interval. Follow the four-step statistical problem-solving process (page 19).

CALCULATOR CORNER Inference for means

You can perform the calculations for confidence intervals and significance tests on the TI-84/TI-Nspire. Here is a brief summary of the process using the data from the caffeine study of Example 10.15 (page 507).

1. Type the 11 difference values (placebo − caffeine) into a list.

TI-84
Put the data in list L1.

TI-Nspire
Open a Lists & Spreadsheet page. Enter the data in a list called "diff."

2. Plot your data. Inspect the shape, center, and spread and look for possible outliers.

TI-84

Define Plot 1 to be a boxplot of the values in list L1. Use `ZoomStat` to display the graph.

TI-Nspire

Press (menu), then choose `3:Data`, `5:Quick Graph`. You should see a dotplot of the data. To change it to a boxplot, press (menu), then choose `1:Plot Type`, `2:Box Plot`.

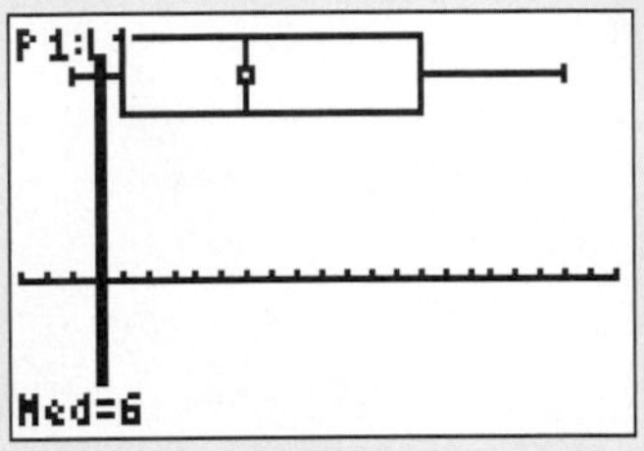

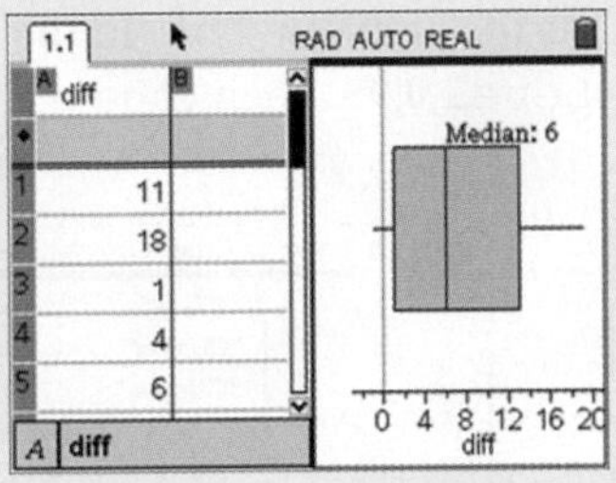

3. Proceed to inference. There are no outliers or strong skewness, so we can use t procedures for inference. For a *significance test*, our hypotheses are

$$H_0: \mu = 0$$
$$H_a: \mu > 0$$

TI-84

Press STAT, choose `TESTS`, and then choose `2:T-Test`. Adjust your settings as shown and choose `Calculate`.

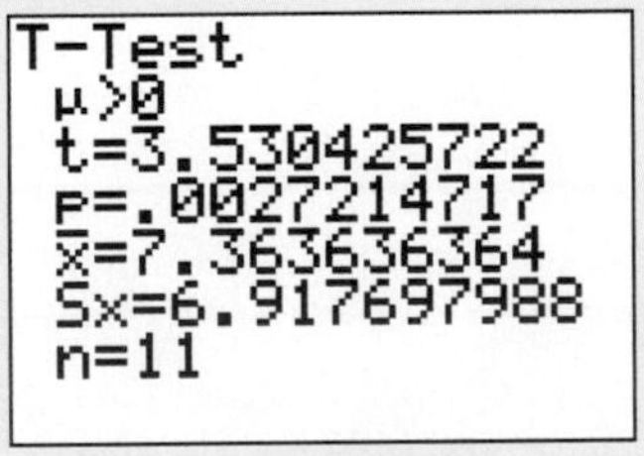

TI-Nspire

You can perform significance tests in a Calculator page or a Lists & Spreadsheet page. We'll give instructions for a Calculator page.

- Press (menu), and choose `5:Statistics`, `7:Stat Tests ...`, and `2:t Test`. Choose "Data" as the input method.
- Enter the values shown in the dialog box.

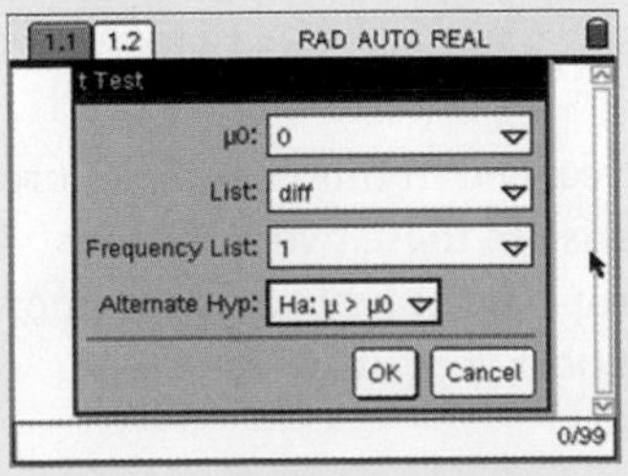

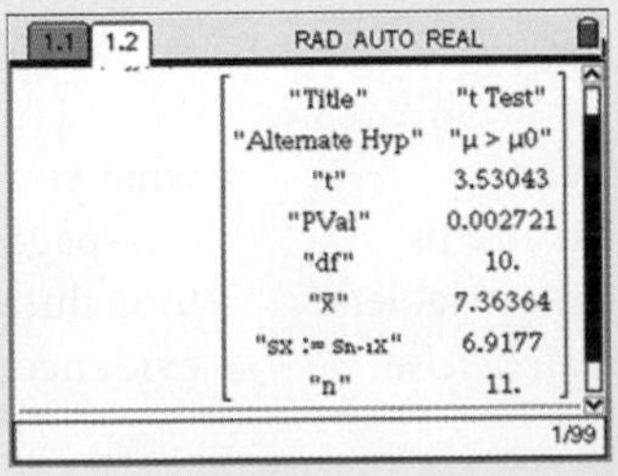

CALCULATOR CORNER *(continued)*

CALCULATOR CORNER *(continued)*

To construct a *confidence interval:*

TI-84

Press STAT, choose TESTS, and then choose 8:TInterval. Enter 0.95 for the confidence level ("C-Level") and choose Calculate.

TI-Nspire

Press menu, and choose 5:Statistics, 6:Confidence Intervals..., and 2:t Interval. Choose "Data" as the input method. In the dialog box, select "diff" as the List and enter C Level 0.95. Then choose OK.

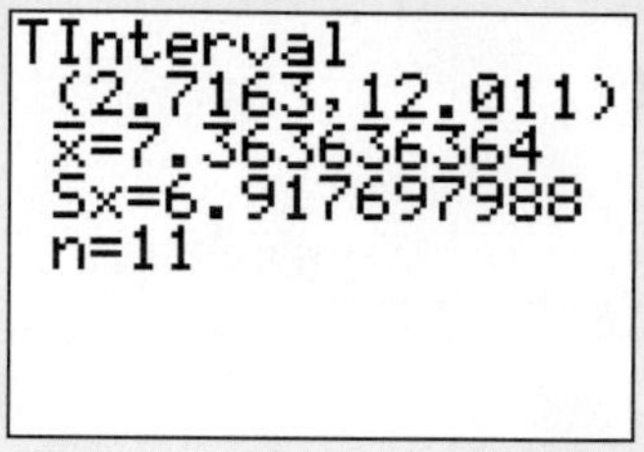

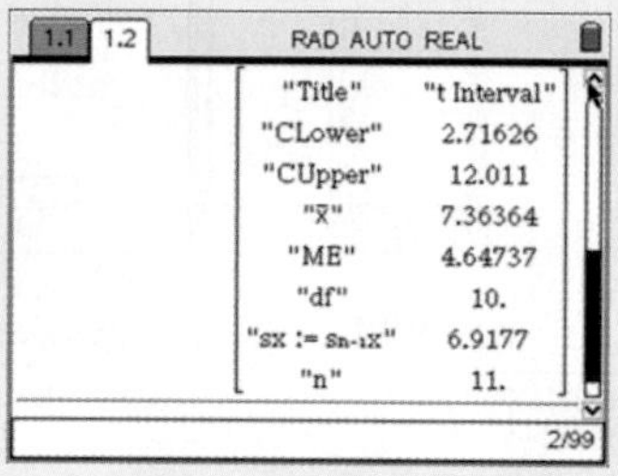

As you prepare to conclude your study of statistics, here's a final Data Exploration that allows you to put your understanding of inference to the test (pardon the pun).

DATA EXPLORATION

I'm Getting a Headache!

The makers of Aspro brand aspirin want to be sure that their tablets contain the right amount of active ingredient (acetylsalicylic acid). So they inspect a sample of 36 tablets from a batch of production. When the production process is working properly, Aspro tablets have an average of $\mu = 320$ milligrams (mg) of active ingredient. Here are the amounts (in mg) of active ingredient in the 36 selected tablets:

319	328	321	324	322	320	324	321	320	324	322	317
321	320	322	318	326	316	316	326	325	320	316	319
319	321	322	319	326	320	324	320	318	321	322	318

What do these data tell us about the mean acetylsalicylic acid content of the tablets in this batch? Should the company distribute these tablets to drugstores or dispose of the entire batch? Use what you have learned in this chapter to prepare a one-page response to these two questions. Be sure to include appropriate graphical and numerical evidence to support your answers.

APPLICATION 10.2

Nitrogen in tires—a lot of hot air?

www In Chapter 1 (page 21), we described an experiment performed by Consumers Union to test whether filling car tires with nitrogen instead of compressed air would reduce pressure loss. Let's put together what you have learned about inference with the statistical problem-solving process.

I. Ask a question of interest. Does filling automobile tires with nitrogen instead of compressed air reduce pressure loss?

1. The Hypotheses. State an appropriate pair of hypotheses for a significance test.

II. Produce data. Consumers Union designed a study to test whether nitrogen-filled tires would maintain pressure better than air-filled tires. They obtained two tires from each of several brands and then filled one tire in each pair with air and one with nitrogen. All tires were inflated to a pressure of 30 pounds per square inch and then placed outside for a year. At the end of the year, Consumers Union measured the pressure in each tire.

III. Analyze data. The amount of pressure lost (in pounds per square inch) during the year for the air-filled and nitrogen-filled tires of each brand is shown in the table below.

2. The sampling distribution. What inference procedure is appropriate in this setting? Make a graph to check that conditions for using that procedure are satisfied.

3. The data. Calculate the test statistic assuming that H_0 is true.

4. The *P*-value. State the number of degrees of freedom. Then sketch and shade an appropriate density curve, and find the *P*-value.

IV. Interpret results.

5. Conclusion. What conclusion do you draw from your significance test? Calculate and interpret a confidence interval. What additional information do you get from the confidence interval?

Brand	Air	Nitrogen
BF Goodrich Traction T/A HR	7.6	7.2
Bridgestone HP50 (Sears)	3.8	2.5
Bridgestone Potenza EL400	2.1	1.0
Bridgestone Potenza G009	3.7	1.6
Bridgestone Potenza RE950	4.7	1.5
Continental Premier Contact H	4.9	3.1
Cooper Lifeliner Touring SLE	5.2	3.5
Dayton Daytona HR	3.4	3.2
Falken Ziex ZE-512	4.1	3.3
Fuzion Hrl	2.7	2.2
General Exclaim	3.1	3.4
Goodyear Assurance Tripletred	3.8	3.2
Hankook Optimo H418	3.0	0.9
Kumho Solus KH16	6.2	3.4
Michelin Energy MXV4 Plus	2.0	1.8
Michelin Pilot XGT H4	1.1	0.7
Pirelli P6 Four Seasons	4.4	4.2
Sumitomo HTR H4	1.4	2.1
Yokohama Avid H4S	4.3	3.0
BF Goodrich Traction T/A V	5.5	3.4
Bridgestone Potenza RE950	4.1	2.8
Continental ContiExtreme Contact	5.0	3.4
Continental ContiProContact	4.8	3.3
Cooper Lifeliner Touring SLE	3.2	2.5
General Exclaim UHP	6.8	2.7
Hankook Ventus V4 H105	3.1	1.4
Michelin Energy MXV4 Plus	2.5	1.5
Michelin Pilot Exalto A/S	6.6	2.2
Michelin Pilot HX MXM4	2.2	2.0
Pirelli P6 Four Seasons	2.5	2.7
Sumitomo HTR+	4.4	3.7

This data set is available at the book's Web site, **www.whfreeman.com/sta2e.**

Section 10.2 Summary

We estimate a population mean μ using the sample mean $\bar{x}$ of an SRS from the population. Confidence intervals and significance tests for μ are based on the **sampling distribution of $\bar{x}$.** This distribution has mean μ and standard deviation $\sigma/\sqrt{n}$. When the sample size n is large, the **central limit theorem** says that the sampling distribution of $\bar{x}$ will be approximately Normal.

To perform inference about a population mean μ, we can use z critical values if the sample size is large. If the sample size is small, or if we want more accurate results, we use a ***t* distribution** with $n - 1$ degrees of freedom for confidence intervals and tests. Our results will be quite accurate when the population is Normally distributed and will be accurate enough when no outliers or strong skewness are present.

Section 10.2 Exercises

10.51 Explaining confidence A student reads that a 95% confidence interval for the mean NAEP quantitative score for men aged 21 to 25 is 267.8 to 276.2. Asked to explain the meaning of this interval, the student says, "95% of all young men have scores between 267.8 and 276.2." Is the student right? Justify your answer.

10.52 What's the average height? One hundred and fifty students attend the lecture of a college statistics course on the first day of class. The course instructor wants to estimate the average height of the students in the room and asks the nine students in the front row to write down their heights. The average height of these nine students turns out to be 64 inches with a standard deviation of 3 inches.

(a) What is the parameter in this problem? What is the sample statistic?

(b) Would it be appropriate to use the Normal distribution and the formula $\bar{x} \pm z^* \sigma/\sqrt{n}$ to make a confidence statement about the average height of the students in the class, based on the data given in the problem? If yes, construct the appropriate confidence interval. If no, explain what is wrong.

10.53 Airline passengers get heavier In response to the increasing weight of airline passengers, the Federal Aviation Administration (FAA) told airlines to assume that passengers average 190 pounds in the summer, including clothes and carry-on baggage. But passengers vary, and the FAA did not specify a standard deviation. A reasonable standard deviation is $\sigma = 35$ pounds. Weights are not Normally distributed, especially when the population includes both men and women, but they are not very non-Normal. A commuter plane carries 30 passengers.

(a) Explain why you cannot calculate the probability that a randomly selected passenger on this flight weighs more than 200 pounds.

(b) You can calculate the probability that the *mean* weight of the passengers on the flight exceeds 200 pounds. Explain why. Then do it.

10.54 One tail or two? What null and alternative hypotheses should you test in each of the following situations?

(a) Experiments on learning in animals sometimes measure how long it takes mice to find their way through a maze. The mean time is 18 seconds for one

By permission of Dave Coverly and Creators Syndicate, Inc.

particular maze. A researcher thinks that a loud noise will cause the mice to complete the maze faster. She measures how long each of several mice takes with a noise as stimulus.

(b) Last year, your company's service technicians took an average of 2.6 hours to respond to trouble calls from business customers who had purchased service contracts. Do this year's data show a significantly different average response time?

10.55 Pleasant smells, I Do pleasant odors help work go faster? Twenty-one subjects worked a paper-and-pencil maze wearing a mask that was either unscented or carried the smell of flowers. Each subject worked the maze three times with each mask, in random order. (This is a matched pairs design.) Here are the differences in their average times (in seconds), unscented minus scented. If the floral smell speeds work, the difference will be positive because the time with the scent will be lower.[15]

−7.37	−3.14	4.10	−4.40	19.47	−10.80	−0.87
8.70	2.94	−17.24	14.30	−24.57	16.17	−7.84
8.60	−10.77	24.97	−4.47	11.90	−6.26	6.67

(a) We hope to show that work is faster on the average with the scented mask. State null and alternative hypotheses in terms of the mean difference in times μ for the population of all adults.

(b) Using a calculator, find the mean and standard deviation of the 21 observations. Did the subjects work faster with the scented mask? Is the mean improvement big enough to be important?

(c) Make a stemplot of the data (round to the nearest whole second). Are there outliers or other problems that might hinder inference?

(d) Test the hypotheses you stated in (a). Is the improvement statistically significant?

10.56 Pleasant smells, II Return to the data in Exercise 10.55. Calculate a 95% confidence interval for the mean improvement in time to solve a maze when wearing a mask with a floral scent. Are you confident that the scent does improve mean working time?

10.57 Study more! A student group claims that first-year students at a university study 2.5 hours per night during the school week. A skeptic suspects that they study less than that on the average. A survey of a random sample of 269 first-year students finds that $\bar{x} = 137$ minutes and $s = 65$ minutes. What conclusion do you draw? Give appropriate evidence to support your conclusion.

10.58 IQ at BCU The admissions director at Big City University has a new idea. He wants to use the IQ scores of current students as a marketing tool. The director gives an IQ test to an SRS of 50 of the university's 5000 freshmen. The mean IQ score for the sample is $\bar{x} = 112$ and the standard deviation is $s = 15$. What can the director say about the mean score μ of the population of all 5000 freshmen? Show your work.

CHAPTER 10 REVIEW

This chapter discusses a few more specific inference procedures. Section 10.1 introduces the chi-square distributions and two types of chi-square tests. The goodness of fit test is used to determine whether a specified distribution for a single categorical variable is correct. Relationships between two categorical variables can be summarized in two-way tables. The chi-square test for two-way tables tests the null hypothesis that there is no association between two categorical variables.

Section 10.2 deals with inference for a population mean μ. Confidence intervals and significance tests rely on the sampling distribution of $\overline{x}$, which has a Normal shape if the population distribution is Normal or the sample size is large. When no outliers or strong skewness are present, the t distributions are used to calculate intervals and perform tests.

Here is a review list of the most important skills you should have developed from your study of this chapter.

A. CHI-SQUARE GOODNESS OF FIT

1. Explain what null hypothesis the chi-square goodness of fit statistic tests in a specific setting.

2. Calculate expected cell counts, the chi-square statistic, and its degrees of freedom from the data.

3. Use Table 10.1 for chi-square distributions to assess significance. Interpret the test result in context.

B. TWO-WAY TABLES

1. Use percents and bar graphs to describe the relationship between any two categorical variables starting from the counts in a two-way table.

2. Explain what null hypothesis the chi-square statistic tests for a specific two-way table.

3. Calculate expected cell counts, the chi-square statistic, and its degrees of freedom from a two-way table.

4. Use Table 10.1 for chi-square distributions to assess significance. Interpret the test result in the setting of a specific two-way table.

C. SAMPLING DISTRIBUTIONS

1. Use the Normal sampling distribution of a sample mean $\overline{x}$ to find probabilities involving $\overline{x}$.

D. CONFIDENCE INTERVALS

1. Use the formula $\overline{x} \pm z^*s/\sqrt{n}$ to obtain an approximate confidence interval for a population mean μ when the sample size is large. Use the formula $\overline{x} \pm t^*s/\sqrt{n}$ to obtain a more accurate confidence interval for a population mean μ when there are no outliers or strong skewness in the data.

E. SIGNIFICANCE TESTS

1. Carry out a one-sided or a two-sided test about a population mean μ. Use the z distribution if the sample size is large to get an approximately correct P-value. Use

the t distribution with $n - 1$ degrees of freedom to obtain a more accurate P-value if there are no outliers or strong skewness.

CHAPTER 10 REVIEW EXERCISES

10.59 Longer-lasting batteries A company that produces AA batteries makes the following statistical announcement: "Our AA batteries last 7.5 hours plus or minus 20 minutes, and our confidence in that interval is 95%."[16] They tested a random sample of 40 batteries using a special device designed to imitate real-world use and recorded data on battery lifetime.
(a) Determine the sample mean and standard deviation.
(b) A reporter translates the statistical announcement into "plain English" as follows: "If you buy one of this company's AA batteries, there is a 95% chance that it will last between 430 and 470 minutes." Comment on this interpretation.
(c) Your friend, who has just started studying statistics, claims that, if you select 40 more AA batteries at random from those manufactured by this company, there is a 95% probability that the mean lifetime will fall between 430 and 470 minutes. Do you agree?
(d) Give a statistically correct interpretation of the confidence interval that could be published in a newspaper report.

10.60 Representative sample? For a class project, a statistics student is required to take an SRS of students from his large high school to take part in a survey. The student's sample consists of 54 freshmen, 66 sophomores, 56 juniors, and 30 seniors. The school roster shows that 29% of the students enrolled at the school are freshmen, 27% are sophomores, 25% are juniors, and 19% are seniors. Use a goodness of fit test to examine how well this student's sample reflects the population at his high school.

10.61 Call the paramedics! Vehicle accidents can result in serious injuries to drivers and passengers. Police, firefighters, and paramedics respond to these emergencies as quickly as possible. Slow response times can have serious consequences for accident victims. Several cities have begun to monitor paramedic response times. In one such city, the mean response time to all accidents involving life-threatening injuries last year was $\mu = 6.7$ minutes. The city manager shares this information with emergency personnel and encourages them to "do better" next year. At the end of the following year, the city manager selects a simple random sample of 400 calls involving life-threatening injuries and examines the response times. For this sample, the mean response time was $\bar{x} = 6.48$ minutes with a standard deviation of $s = 2$ minutes. Do these data provide good evidence that response times have decreased since last year? Carry out a significance test to find out.

10.62 Tall girls Based on information from the National Center for Health Statistics, the heights of 10-year-old girls closely follow a Normal distribution with mean 54.5 inches and standard deviation 2.7 inches.
(a) Suppose we select one 10-year-old girl at random. Find the probability that she is over 56 inches tall. Show your work.
(b) Suppose we select nine 10-year-old girls at random. Find the probability that the mean height $\bar{x}$ of this sample exceeds 56 inches. Show your work.

10.63 Popular kids, I Who were the popular kids at your elementary school? Did they get good grades or have good looks? Were they good at sports? A study was performed to examine the factors that determine social status for children in grades 4,

5, and 6. Researchers administered a questionnaire to a sample of 478 elementary school students. One of the questions they asked was "What would you most like to do at school: make good grades, be good at sports, or be popular?" The two-way table below summarizes the students' responses.[17]

	Goal		
Gender	**Grades**	**Popular**	**Sports**
Female	130	91	30
Male	117	50	60

(a) Calculate appropriate percents for comparing male and female students' goals.
(b) Make a well-labeled bar graph to compare male and female responses. Write a few sentences describing the relationship between gender and goals.

10.64 Popular kids, II Refer to the previous exercise. Is there convincing evidence of an association between gender and goals for elementary school students? Carry out a test and report your conclusion.

By permission of xkcd.com

10.65 Good news for chocolate lovers? A German study concluded that dark chocolate might reduce blood pressure.[18] The subjects in the study were patients with untreated mild hypertension. Six subjects ate a three-ounce white-chocolate bar each day for two weeks, while seven subjects ate a three-ounce dark-chocolate bar. Dark chocolate contains plant substances called polyphenols, which scientists think are responsible for the heart-healthy attributes of red wine. Polyphenols also have been shown to lower blood pressure in animals. Prior to the study, the participants had average systolic and diastolic blood pressure readings of about 153 over 84. The systolic blood pressure readings for the dark-chocolate group dropped an average of 5 points to 148. Assume a standard deviation of 6 points for the dark-chocolate group.
(a) Is there sufficient evidence to conclude that eating a three-ounce dark chocolate bar every day will lower one's systolic blood pressure? Begin to answer this question by writing null and alternative hypotheses for this experiment.
(b) Calculate the test statistic and the *P*-value. Is there sufficient evidence to reject your null hypothesis?
(c) Write your conclusion in plain language.
(d) Do you have any concerns about the design of this experiment? If you were replicating this experiment, would you do anything differently? Explain.

10.66 Sleepy students? A sample of 28 college students responded to the survey question "How much sleep did you get last night?" Here are the data (in hours):

9	6	8	6	8	8	6	6.5	6	7	9	4	3	4
5	6	11	6	3	6	6	10	7	8	4.5	9	7	7

(a) Calculate a 95% confidence interval for the population mean μ.
(b) What do we need to know about these students in order to interpret your result in (a)?

10.67 Acupuncture and pregnancy A study was performed to determine if the ancient Chinese art of acupuncture could help infertile women become pregnant.[19] One hundred and sixty healthy women undergoing artificial insemination were recruited for the study. Half of the subjects were randomly assigned to an acupuncture treatment group, and the other half were assigned to a control group. The subjects in the treatment group received acupuncture 25 minutes before artificial insemination and again 25 minutes afterward. Subjects in the control group were instructed to lie still for 25 minutes after artificial insemination. Results are shown in the two-way table below.

	Acupuncture group	**Control group**
Pregnant	34	21
Not pregnant	46	59
Total	**80**	**80**

(a) Was this an observational study or an experiment? Justify your answer.
(b) Is there convincing evidence of a relationship between pregnancy status and whether acupuncture was administered? Carry out a significance test and report your conclusion.
(c) Why can't we conclude that acupuncture caused an increase in pregnancy?

10.68 Mercury in tuna As we discussed in Chapter 2 (page 56), some of the tuna that people eat may contain high levels of mercury. The Food and Drug Administration will take action (such as removing the product from store shelves) if the mercury concentration in a six-ounce can of tuna is 1.00 parts per million (ppm) or above. Defenders of Wildlife collected a sample of 164 cans of tuna from stores across the United States and sent them to a laboratory for testing. The Fathom histogram displays the mercury concentration in the sampled cans. For these data, $\bar{x} = 0.285$ ppm and $s = 0.3$ ppm.

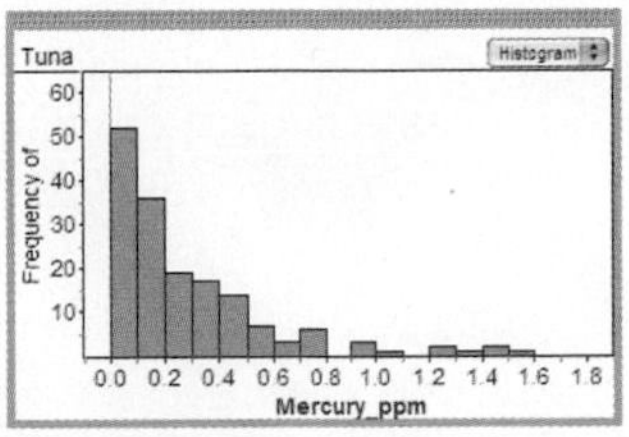

(a) Explain why it would be reasonable to construct a confidence interval based on z critical values here, in spite of the histogram's shape.
(b) What is the typical mercury concentration in cans of tuna sold in stores? Construct and interpret a 95% confidence interval.

10.[illegible] Sleepy students? A sample of 28 college students [illegible] responded to the survey question "How much sleep did you get last night?" Here are the data (in hours):

[illegible]

(a) Calculate a 9[illegible]% confidence interval for the population mean [illegible].
(b) [illegible] need to know about these students in order to [illegible] your results?

[illegible] pregnancy? A study was [illegible] to determine if the an[illegible] could help women become pregnant. One hundred and [illegible] were randomly assigned to treatment or control. Half of the subjects were [illegible] assigned to the acupuncture treatment group, and the other half were assigned to a control group. The subjects in the [illegible] group received [illegible] 25 minutes before [illegible] and [illegible] 25 minutes after [illegible]. Subjects in the control group were instructed to lie still for 25 minutes [illegible] Results are shown in the two-way table below.

	Acupuncture group	Control group
Pregnant	[illegible]	[illegible]
Not pregnant	[illegible]	[illegible]
Total	[illegible]	[illegible]

(a) [illegible] national study [illegible] expect [illegible]
(b) [illegible] convincing evidence of a relationship between pregnancy [illegible] whether acupuncture was administered? Carry out a significance test and report your conclusion.
(c) Why can we say that acupuncture caused an increase in pregnancy [illegible]?

10.[illegible] [illegible] (page [illegible]) [illegible] that [illegible] each of [illegible] levels of [illegible] and Drug [illegible] will [illegible] [illegible] [illegible] [illegible] concentration [illegible]

(a) Explain why it would be reasonable to construct a confidence interval [illegible] critical value [illegible] of [illegible] shape.
(b) What is the [illegible] treatment? [illegible] and interpret a 9[illegible]% confidence interval.

NOTES AND DATA SOURCES

PREFACE

1. From the National Council of Teachers of Mathematics (NCTM) Standards Web site, http://standards.nctm.org/.
2. The quotations are from a summary of the committee's report that was unanimously endorsed by the Board of Directors of the American Statistical Association. The full report is George Cobb, "Teaching statistics," in L. A. Steen (ed.), *Heeding the Call for Change: Suggestions for Curricular Action,* Mathematical Association of America, 1990, pp. 3–43.

CHAPTER 1

1. "Computer games 'burn up calories,'" BBC News, http://news.bbc.co.uk, February 19, 2007.
2. "Traffic safety facts 2006: young drivers," published by the National Highway Traffic Safety Administration on their Web site, www.nhtsa.dot.gov.
3. Nanci Hellmich, "Extra weight shaves years off lives," *USA Today,* January 7, 2003.
4. Joachim Schüz et al., "Cellular telephone use and cancer risk: update of a nationwide Danish cohort," *Journal of the National Cancer Institute,* 98 (2006), pp. 1707–1713.
5. Example 1.2 is suggested by Maxine Pfannkuch and Chris J. Wild, "Statistical thinking and statistical practice: themes gleaned from professional statisticians," unpublished manuscript, 1998.
6. The study is by J. David Cassidy et al., "Effect of eliminating compensation for pain and suffering on the outcome of insurance claims for whiplash injury," *New England Journal of Medicine,* 342 (2000), pp. 1179–1186.
7. The estimates of the census undercount come from Howard Hogan, "The 1990 post-enumeration survey: operations and results," *Journal of the American Statistical Association,* 88 (1993), pp. 1047–1060.
8. Read "Is vitamin C really good for colds?" in *Consumer Reports,* February 1976, pp. 68–70, for a discussion of this conclusion.The article also reviews the Toronto study and the need for controlled experiments.
9. J. E. Muscat et al., "Handheld cellular telephone use and risk of brain cancer," *Journal of the American Medical Association,* 284 (2000), pp. 3001–3007.
10. S. G. Stolberg, "Link found between behavioral problems and time in child care," *New York Times,* April 19, 2001. The study is the National Institute of Child Health and Human Development Study of Early Child Care.
11. Activity 1.2 is based on a suggested science fair project entitled "Do the eyes have it?" at the Science Buddies Web site, www.sciencebuddies.org.
12. Our steps in the statistical problem-solving process were influenced by *Guidelines for Assessment and Instruction in Statistics Education (GAISE) Report: A Pre-K–12 Curriculum Framework,* American Statistical Association, 2007.
13. Tattoo survey data from *The Harris Poll®*, No. 15, www.harrisinteractive.com, February 12, 2008.
14. We obtained the tire pressure loss data from the *Consumer Reports* Web site: http://blogs.consumerreports.org/cars/2007/10/tires-nitrogen-.html.
15. The article describing this study is Nikhil Swaminathan, "Gender jabber: do women talk more than men?" *Scientific American,* July 6, 2007.
16. Application 1.2 is based on P. A. Mackowiak, S. S. Wasserman, and M. M. Levine, "A critical appraisal of 98.6 degrees F, the upper limit of the normal body temperature, and other legacies of Carl Reinhold August Wunderlich," *Journal of the American Medical Association,* 268, No. 12 (September 23–30, 1992), pp. 1578–1580.
17. We found Allen Shoemaker's data in "What's normal? temperature, gender, and heart rate," *Journal of Statistics Education,* 4, No. 2 (July 1996), www.amstat.org/publications/jse/.
18. Barbara A. Dennison, Tara A. Erb, and Paul L. Jenkins, "Among low-income preschool children television viewing and television in bedroom associated with overweight risk," *Pediatrics* 109 (2002), pp. 1028–1035.
19. Samantha's paper towel experiment is described at www.selah.k12.wa.us/soar/sciproj2001/SamanthaP.html.
20. The cereal data came from the Data and Story Library, http://lib.stat.cmu.edu/DASL/.
21. T. M. Amabile, "Children's artistic creativity: detrimental effects of competition in a field setting." *Personality and Social Psychology Bulletin* 8 (1982), pp. 573–578.
22. Data are from the NBA Web site, www.nba.com.
23. The original paper is T. M. Amabile, "Motivation and creativity: effects of motivational orientation on creative writers," *Journal of Personality and Social Psychology* 48, No. 2 (February 1985), pp. 393–399. The data for Exercise 1.42 came from Fred Ramsey and Dan Schafer, *The Statistical Sleuth,* 2nd edition, Duxbury, 2002.

CHAPTER 2

1. Data on income and education from the March 2008 CPS supplement, downloaded from the Census Bureau Web site using the government's FERRET system. These are raw survey data, which include many people with no income or negative income. To simplify comparisons, we report only four education classes.
2. The Harris Poll online survey is described in "Despite understanding risks many U.S. adults still use cell phones while driving," published June 6, 2006, at www.harrisinteractive.com.
3. Data from the 2009 *Statistical Abstract of the United States.*

4. Data provided by the College Board.
5. Data provided by Professor Dennis Pearl, Ohio State University, Columbus.
6. The 2008–2009 academic year tuition and fees are from the College Board's *Annual Survey of Colleges,* 2008–2009.
7. Data from the government's Survey of Earned Doctorates, in the online supplement to Jeffery Mervis, "Wanted: a better way to boost numbers of minority PhDs," *Science,* 281 (1998), pp. 1268–1270.
8. Figure 2.11 is based on an episode in the Annenberg/Corporation for Public Broadcasting telecourse *Against All Odds: Inside Statistics.*
9. Data from the report "Is our tuna family-safe?" prepared by Defenders of Wildlife, 2006.
10. Percent low-birth-weight data from the 2002 *Statistical Abstract of the United States.*
11. Thanks to Sanderson Smith from Santa Barbara City College for the idea for Activity 2.2A.
12. Data from *Guidelines for Assessment and Instruction in Statistics Education: Pre-K–12 Report,* published by the American Statistical Association, 2007.
13. Data from the Higher Education Research Institute's report "The American freshman: national norms for fall 2007," published January 2008.
14. Data from Jessica Utts and Robert Heckard, *Mind on Statistics,* 3rd edition, Cengage Learning, 2007.
15. Advanced Placement exam results from AP Central Web site, http://apcentral.collegeboard.com.
16. We got the idea for Activity 2.3 from David Lane's case study "Who is buying iMacs?" which we found online at www.ruf.rice.edu/~lane/case_studies/imac/index.html.
17. Data from the Department of Energy Web site, www.eia.doe.gov/oil_gas/petroleum/data_publications/wrgp/mogas_history.html.
18. The information in Exercise 2.53 comes from the report "Cell phones key to teens' social lives, 47% can text with eyes closed," published online at www.marketingcharts.com. The results come from a Harris Interactive online survey conducted in July 2008.
19. R. J. Newan, "Snow job on the slopes," *US News and World Report,* December 17, 1994, pp. 62–65.
20. "Colleges inflate SATs and graduation rates in popular guidebooks," *Wall Street Journal,* April 5, 1995.
21. Letter by J. L. Hoffman, *Science,* 255 (1992), p. 665. The original article appeared on December 6, 1991.
22. The graphs in Application 2.3 came from the article "SEO lie factor," posted online by Marios Alexandrou at www.allthingssem.com/seo-lie-factor/.
23. *Science,* 189 (1975), p. 373.
24. *Lafayette* (Ind.) *Journal and Courier,* October 23, 1988.
25. *Condé Nast Traveler* magazine, June 1992.
26. *Purdue* (West Lafayette, Ind.) *Exponent,* September 21, 1977.
27. *Fine Gardening,* September/October 1989, p. 76.
28. *New York Times,* April 21, 1986.
29. College Board Online, www.collegeboard.com.
30. Data from W. S. Gosset, "The probable error of a mean," *Biometrika,* 6 (1908), pp. 1–25.

CHAPTER 3

1. Will Shortz, "A few words about sudoku, which has none," *New York Times,* August 28, 2005.
2. Information on bone density in the reference populations was found on the Web at http://courses.washington.edu/bonephys/opbmd.html.
3. Data from Gary Community School Corporation, courtesy of Celeste Foster, Purdue University School of Education.
4. The IQ scores in Figure 3.15 were collected by Darlene Gordon, Purdue University School of Education.
5. Ulric Neisser, "Rising scores on intelligence tests," *American Scientist,* September–October 1997, online edition at www.americanscientist.org/.
6. We found the information on birth weights of Norwegian children on the National Institute of Environmental Health Sciences Web site: http://eb.niehs.nih.gov/bwt/subcfreq.htm.

CHAPTER 4

1. From the Electronic Encyclopedia of Statistical Examples and Exercises (EESEE) story "Is Old Faithful Faithful?"
2. Christer G. Wiklund, "Food as a mechanism of density-dependent regulation of breeding numbers in the merlin *Falco columbarius,*" *Ecology,* 82 (2001), pp. 860–867.
3. Information about the sources used to obtain the data for Figure 4.6 can be found under "Documentation" at the Gapminder Web site, www.gapminder.org.
4. From the EESEE story "Is It Tough to Crawl in March?"
5. Figure 4.9 is based on data from the Web site of Professor Kenneth French of Dartmouth, mba.tuck.dartmouth.edu/pages/faculty/ken.french/data_library.html.
6. Data from Daren Starnes's Advanced Placement Statistics class, 1998–1999.
7. We got the 1999 fuel economy data from Professor Dennis Pearl, The Ohio State University. He downloaded the data from the EPA Web site at www.epa.gov.
8. *The World Almanac and Book of Facts* (2009).
9. From a study of 4511 white, non-Hispanic children reported by Judith Blake, "Number of siblings and educational attainment," *Science,* 245 (1989), pp. 32–36.
10. T. J. Lorenzen, "Determining statistical characteristics of a vehicle emissions audit procedure," *Technometrics,* 22 (1980), pp. 483–493.
11. Alan S. Banks et al., "Juvenile hallux abducto valgus association with metatarsus adductus," *Journal of the American Podiatric Medical Association,* 84 (1994), pp. 219–224.
12. These figures are cited by Dr. William Dement at "The Sleep Well" Web site, www.standford.edu/~dement/.
13. From the EESEE story "Blood Alcohol Content."
14. From the EESEE story "What's Driving Car Sales?"
15. Laura L. Calderon et al., "Risk factors for obesity in Mexican-American girls: dietary factors, anthropometric factors, physical activity, and hours of television viewing," *Journal of the American Dietetic Association,* 96 (1996), pp. 1177–1179.
16. M. S. Linet et al., "Residential exposure to magnetic fields and acute lymphoblastic leukemia in children," *New England Journal of Medicine,* 337 (1997), pp. 1–7.

17. Quotation from a Gannett News Service article appearing in the *Lafayette* (Ind.) *Journal and Courier,* April 23, 1994.
18. We obtained the cricket chirp data for Application 4.2 from Rex Boggs's Exploring Data Web site at http://exploringdata.cqu.edu.au/datasets.htm.
19. See Note 15.
20. The cricket chirp data come from the GLOBE Chief Scientist's Blog at www.globe.gov/fsl/scientistsblog/2007/10/.
21. The data plotted in Figure 4.32 come from G. A. Sacher and E. F. Staffelt, "Relation of gestation time to brain weight for placental mammals: implications for the theory of vertebrate growth," *American Naturalist,* 108 (1974), pp. 593–613. We found them in Fred L. Ramsey and Daniel W. Schafer, *The Statistical Sleuth: A Course in Methods of Data Analysis,* Duxbury, 1997, p. 228.
22. We found the data on cherry blossoms in the paper "Linear equations and data analysis," which was posted on the North Carolina School of Science and Mathematics Web site, www.ncssm.edu.

PART B OPENER

1. We found these year-end sales data at the Recording Industry Association of America's Web site, www.riaa.com.

CHAPTER 5

1. "Acadian ambulance officials, workers flood call-in poll," *Baton Rouge Advocate,* January 22, 1999.
2. The hand-washing survey described in Exercise 5.4 was conducted by Harris Interactive on behalf of the American Society for Microbiology and the Soap and Detergent Association. Harris Interactive contacted 1001 U.S. adults via telephone during August 2007. We found the results of the survey at www.washup.org.
3. Harris Interactive conducted the observational study described in Exercise 5.4 on behalf of the American Society for Microbiology and the Soap and Detergent Association. Harris Interactive observed 6076 adults in public restrooms at major U.S. public attractions (a train station, a baseball stadium, an aquarium, and a farmers' market) during August 2007.
4. "Smoking habits stable; most would like to quit," Gallup Poll press release by Jeffrey M. Jones, July 18, 2006. Read online at www.gallup.com/poll/.
5. D. Horvitz in his contribution to "Pseudo-opinion polls: SLOP or useful data?" *Chance,* 8, No. 2 (1995), pp. 16–25.
6. From "Working women count," a report released by the U.S. Department of Labor on October 15, 1997. For more information, visit their Web site, www.dol.gov.
7. Gallup Organization, "Gambling in America–1999: a comparison of adults and teenagers," Topline and Trends detailed report; read online at www.gallup.com.
8. Corky Siemaszko, "Fools of the road," *Daily News,* November 14, 2007.
9. Warren McIsaac and Vivek Goel, "Is access to physician services in Ontario equitable?" Institute for Clinical Evaluative Sciences in Ontario, October 18, 1993.
10. *New York Times,* August 21, 1989.
11. The results of the survey were announced in the article "Glance at breakdown of video-game poll results," at www.foxnews.com, on May 9, 2006.
12. Gregory Flemming and Kimberly Parker, "Race and reluctant respondents: possible consequences of non-response for pre-election surveys," Pew Research Center for the People and the Press, 1997, found at www.people-press.org.
13. For more detail on nonsampling errors, along with references, see P. E. Converse and M. W. Traugott, "Assessing the accuracy of polls and surveys," *Science,* 234 (1986), pp. 1094–1098.
14. For more detail on the limits of memory in surveys, see N. M. Bradburn, L. J. Rips, and S. K. Shevell, "Answering autobiographical questions: the impact of memory and inference on surveys," *Science,* 236 (1987), pp. 157–161.
15. The nonresponse rate for the CPS comes from "Technical notes to household survey data published in Employment and Earnings," found on the Bureau of Labor Statistics Web site at www.bls.gov/CPS/. The General Social Survey reports its response rate on its Web site, www.norc.org/homepage.htm. The claim that "pollsters say response rates have fallen as low as 20 percent in some recent polls" is made in Don Van Natta, Jr., "Polling's 'dirty little secret': no response," *New York Times,* November 11, 1999.
16. The quotation is typical of Gallup polls, from the Gallup Web site, www.gallup.com.
17. P. H. Lewis, "Technology" column, *New York Times,* May 29, 1995.
18. The responses on welfare are from a *New York Times*/CBS News Poll reported in the *New York Times,* July 5, 1992. Those for Scotland are from "All set for independence?" *Economist,* September 12, 1998.
19. The most recent account of the design of the CPS is *Design and Methodology,* Current Population Survey Technical Paper 66, October 2006. Available in print or online at www.census.gov/apsd/techdoc/cps/cps-main.html. The account in Example 5.18 omits many complications, such as the need to separately sample "group quarters" like college dormitories.
20. From the online "Supplementary Material" for G. Gaskell et al., "Worlds apart? The reception of genetically modified foods in Europe and the U.S.," *Science,* 285 (1999), pp. 384–387.
21. From the *New York Times,* August 18, 1980.
22. John Simons, "For risk takers, system is no longer sacred," *Wall Street Journal,* March 11, 1999.
23. D. Goleman, "Pollsters enlist psychologists in quest for unbiased results," *New York Times,* September 7, 1993.
24. Richard Morin, "It depends on what your definition of 'do' is," *Washington Post,* December 21, 1998.
25. See www.harrisinteractive.com.
26. Gallup Poll described on the Gallup Web site, www.gallup.com.
27. Jay Schreiber and Megan Thee, "Fans concerned about steroid use and believe it's widespread, poll shows," *New York Times,* March 31, 2008.

CHAPTER 6

1. Allan H. Schulman and Randi L. Sims, "Learning in an online format versus an in-class format: an experimental study," *T.H.E. Journal,* June 1999, pp. 54–56.
2. L. L. Miao, "Gastric freezing: an example of the evaluation of medical therapy by randomized clinical trials," in J. P. Bunker, B. A. Barnes, and F. Mosteller (eds.), *Costs, Risks and Benefits of Surgery,* Oxford University Press, 1977, pp. 198–211.
3. Details of the Carolina Abecedarian Project, including references to published work, can be found online at www.fpg.unc.edu/~abc/.
4. Linda Rosa et al., "A close look at therapeutic touch," *Journal of the American Medical Association,* 279 (1998), pp. 1005–1010.
5. See Note 2.
6. From the Electronic Encyclopedia of Statistical Examples and Exercises (EESEE) story "Anecdotes of Placebos."
7. Samuel Charache et al., "Effects of hydroxyurea on the frequency of painful crises in sickle cell anemia," *New England Journal of Medicine,* 332 (1995), pp. 1317–1322.
8. Marilyn Ellis, "Attending church found factor in longer life," *USA Today,* August 9, 1999.
9. Dr. Daniel B. Mark, in Associated Press, "Age, not bias, may explain differences in treatment," *New York Times,* April 26, 1994. Dr. Mark was commenting on Daniel B. Mark et al., "Absence of sex bias in the referral of patients for cardiac catheterization," *New England Journal of Medicine,* 330 (1994), pp. 1101–1106. See the correspondence from D. Douglas Miller and Leslee Shaw, "Sex bias in the care of patients with cardiovascular disease," *New England Journal of Medicine,* 331 (1994), p. 883, for comments on an opposed study.
10. "Family dinner linked to better grades for teens: survey finds regular meal time yields additional benefits," written by John Mackenzie for ABC News's *World News Tonight,* September 13, 2005.
11. *Washington Post,* March 9, 1989.
12. C. S. Fuchs et al., "Alcohol consumption and mortality among women," *New England Journal of Medicine,* 332 (1995), pp. 1245–1250.
13. W. E. Paulus et al., "Influence of acupuncture on the pregnancy rate in patients who undergo assisted reproductive therapy," *Fertility and Sterility,* 77, No. 4 (2002), pp. 721–724.
14. The original study on the Mozart Effect is described in F. H. Rauscher, G. L. Shaw, and K. N. Ky, "Music and spatial task performance," *Nature,* 365 (1993), p. 611. For an opposing perspective, see Kenneth Steele, "The 'Mozart Effect': an example of the scientific method in operation," *Psychology Teacher Network,* November–December 2001.
15. The fact about rats comes from E. Street and M. B. Carroll, "Preliminary evaluation of a new food product," in J. M. Tanur et al. (eds.), *Statistics: A Guide to the Unknown,* 3rd edition, Wadsworth, 1989, pp. 161–169.
16. The placebo effect examples are from Sandra Blakeslee, "Placebos prove so powerful even experts are surprised," *New York Times,* October 13, 1998.
17. The "three-quarters" estimate is cited by Martin Enserink, "Can the placebo be the cure?" *Science,* 284 (1999), pp. 238–240. An extended treatment is Anne Harrington (ed.), *The Placebo Effect: An Interdisciplinary Exploration,* Harvard University Press, 1997.
18. Kristin L. Nichol et al., "Effectiveness of live, attenuated intranasal influenza virus vaccine in healthy, working adults," *Journal of the American Medical Association,* 282 (1999), pp. 137–144.
19. "Cancer clinical trials: barriers to African American participation," *Closing the Gap,* newsletter of the Office of Minority Health, December 1997–January 1998.
20. Michael H. Davidson et al., "Weight control and risk factor reduction in obese subjects treated for 2 years with orlistat: a randomized controlled trial," *Journal of the American Medical Association,* 281 (1999), pp. 235–242.
21. Edward P. Sloan et al., "Diaspirin cross-linked hemoglobin (DCLHb) in the treatment of severe traumatic hemorrhagic shock," *Journal of the American Medical Association,* 282 (1999), 1857–1864.
22. See Note 6.
23. "Advertising: the cola war," *Newsweek,* August 30, 1976, p. 67.
24. Mary O. Mundinger et al., "Primary care outcomes in patients treated by nurse practitioners or physicians," *Journal of the American Medical Association,* 238 (2000), pp. 59–68.
25. Carol A. Warfield, "Controlled-release morphine tablets in patients with chronic cancer pain," *Cancer,* 82 (1998), pp. 2299–2306.
26. From the EESEE case study "Is Caffeine Dependence Real?"
27. Thomas B. Freeman et al., "Use of placebo surgery in controlled trials of a cellular-based therapy for Parkinson's disease," *New England Journal of Medicine,* 341 (1999), pp. 988–992. Freeman supports the Parkinson's disease trial. The opposition is represented by Ruth Macklin, "The ethical problems with sham surgery in clinical research," *New England Journal of Medicine,* 341 (1999), pp. 992–996.
28. The difficulties of interpreting guidelines for informed consent and for the work of institutional review boards in medical research are a main theme of Beverly Woodward, "Challenges to human subject protections in U.S. medical research," *Journal of the American Medical Association,* 282 (1999), pp. 1947–1952. The references in this paper point to other discussions.
29. *Report of the Tuskegee Syphilis Study Legacy Committee,* May 20, 1996. A detailed history is James H. Jones, *Bad Blood: The Tuskegee Syphilis Experiment,* Free Press, 1993.
30. Dr. Hennekens's words are from an interview in the Annenberg/Corporation for Public Broadcasting video series *Against All Odds: Inside Statistics.*
31. Stanley Milgram, "Group pressure and action against a person," *Journal of Abnormal and Social Psychology,* 69 (1964), pp. 137–143.
32. Stanley Milgram, "Liberating effects of group pressure," *Journal of Personality and Social Psychology,* 1 (1965), pp. 127–134.

33. Dr. C. Warren Olanow, chief of neurology, Mt. Sinai School of Medicine, quoted in Margaret Talbot, "The placebo prescription," *New York Times Magazine,* January 8, 2000, pp. 34–39, 44, 58–60.
34. For extensive background, see Jon Cohen, "AIDS trials ethics questioned," *Science,* 276 (1997), pp. 520–523. Search the archives at www.sciencemag.org for recent episodes in the continuing controversies of Exercises 6.60 and 6.69.
35. See Paul Meier, "The biggest public health experiment ever," in Judith M. Tanur et al. (eds.), *Statistics: A Guide to the Unknown,* Holden-Day, 1972, pp. 2–13, for a discussion of the experiment that established the effectiveness of the Salk polio vaccine, a case that fits this exercise.
36. Gina Kolata and Kurt Eichenwald, "Business thrives on unproven care leaving science behind," *New York Times,* October 3, 1999. Background and details about the first clinical trials appear in a National Cancer Institute press release "Questions and answers: high-dose chemotherapy with bone marrow or stem cell transplants for breast cancer," April 15, 1999. That one of the studies reported there involved falsified data is reported by Denise Grady, "Breast cancer researcher admits falsifying data," *New York Times,* February 5, 2000.
37. Exercise 6.76 is based on Christopher Anderson, "Measuring what works in health care," *Science,* 263 (1994), pp. 1080–1082.
38. See Note 18.

CHAPTER 7

1. More historical data can be found in the opening chapters of F. N. David, *Games, Gods and Gambling,* Charles Griffin and Co., 1962. The historical information given here comes from this excellent and entertaining book.
2. For a discussion and amusing examples, see A. E. Watkins, "The law of averages," *Chance,* 8, No. 2 (1995), pp. 28–32.
3. R. Vallone and A. Tversky, "The hot hand in basketball: on the misperception of random sequences," *Cognitive Psychology,* 17 (1985), pp. 295–314.
4. Estimated probabilities from R. D'Agostino, Jr., and R. Wilson, "Asbestos: the hazard, the risk, and public policy," in K. R. Foster, D. E. Bernstein, and P. W. Huber (eds.), *Phantom Risk: Scientific Inference and the Law,* MIT Press, 1994, pp. 183–210. See also the similar conclusions in B. T. Mossman et al., "Asbestos: scientific developments and implications for public policy," *Science,* 247 (1990), pp. 294–301.
5. The quotation is from R. J. Zeckhauser and W. K. Viscusi, "Risk within reason," *Science,* 248 (1990), pp. 559–564.
6. Ivars Peterson, "Random home runs," *Science News,* 159, No. 26 (June 30, 2001). Available at www.sciencenews.org.
7. "Euro coin accused of unfair flipping," *New Scientist,* January 4, 2002.
8. We got the data from a talk entitled "Two-Way Tables: Introducing Probability Using Real Data," presented by Gail Burrill at the Mathematics Education into the 21st Century Project, Czech Republic, September 2003. She cited her source as H. Kranendonk, P. Hopfensperger, and R. Scheaffer, *Exploring Probability,* Dale Seymour Publications, 1999.
9. From the EESEE story "What Makes a Pre-teen Popular?"
10. Data from the Stanford School of Medicine Blood Center Web site at http://bloodcenter.stanford.edu/about_blood/blood_types.html.
11. We got these data from the Energy Information Administration on their Web site at http://tonto.eia.doe.gov/dnav/pet/pet_sum_mkt_dcu_SCT_m.htm.
12. Thanks to Michael Legacy for suggesting this exercise.
13. From the EESEE story "Baby Hearing Screening."

CHAPTER 8

1. Andrew Pollack, "In the gaming industry, the house can have bad luck, too," *New York Times,* July 25, 1999, reports that Mirage Resorts issued a quarterly profit warning due in part to its bad luck at baccarat.
2. The Apgar score data come from the National Center for Health Statistics *Monthly Vital Statistics Report,* 30, No. 1, Supplement, May 6, 1981.
3. S. Newcomb, "Note on the frequency of the use of digits in natural numbers," *American Journal of Mathematics,* 4 (1881), p. 39.
4. F. Benford, "The law of anomalous numbers," *Proceedings of the American Philosophical Society,* 78 (1938), pp. 551–572.
5. The idea for Application 8.3 came from David Salsburg, *The Lady Tasting Tea,* W. H. Freeman and Co., 2001.

CHAPTER 9

1. Janice E. Williams et al., "Anger proneness predicts coronary heart disease risk," *Circulation,* 101 (2000), pp. 2034–2039.
2. The 2007 national school-based Youth Risk Behavioral Survey was carried out by the Centers for Disease Control to monitor the prevalence of youth behaviors that most influence health.
3. Report of the American Society for Microbiology: "2007 hand washing fact sheet." We found the report online at www.asm.org.
4. Data from the Web site www.drugfreeamerica.org.
5. For the state of the art in confidence intervals for p, see Alan Agresti and Brent Coull, "Approximate is better than 'exact' for interval estimation of binomial proportions," *American Statistician,* 52 (1998), pp. 119–126. The authors note that the accuracy of our confidence interval for p can be greatly improved by simply "adding 2 successes and 2 failures." That is, $\hat{p}$ = (count of successes + 2) / (n + 4). Texts on sample surveys give confidence intervals that take into account the fact that the population has finite size and also give intervals for sample designs more complex than an SRS.
6. Diane Rado and Darnelle Little, "Schools toying with test results," *Chicago Tribune,* September 29, 2003, available online at www.chicagotribune.com.
7. This and similar results of Gallup polls are from the Gallup Organization Web site, www.gallup.com.
8. John Paul McKinney and Kathleen G. McKinney, "Prayer in the lives of late adolescents," *Journal of Adolescence,* 22 (1999), pp. 279–290.

9. Hunting data provided by the Virginia Department of Game and Inland Fisheries, Richmond, Virginia.
10. Julie Ray, "Few teens clash with friends," May 3, 2005, on the Gallup Poll Web site, www.gallup.com.
11. Jeffrey Selingo, "What Americans think about higher education," *Chronicle of Higher Education,* 49, No. 34 (May 2, 2003), pp. A10–A15.
12. Dorothy Espelage et al., "Factors associated with bullying behavior in middle school students," *Journal of Early Adolescence,* 19, No. 3 (August 1999), pp. 341–362.
13. J. Stutts et al., "Distractions in everyday driving," University of North Carolina Highway Safety Research Center, June 2003.
14. The newspaper report summarizes findings from "A comparison of sustained-release bupropion and placebo for smoking cessation," *New England Journal of Medicine,* 337, No. 17 (October 23, 1997), pp. 1195–1202.
15. Jane A. Cauley et al., "Lipid-lowering drug use and breast cancer in older women: a prospective study," *Journal of Women's Health,* 12, No. 8 (October 2003), pp. 749–756.
16. Linda Lyons, "Teens: sex can wait," December 14, 2004, from the Gallup Organization Web site, www.gallup.com.

CHAPTER 10

1. M. M. Roy and N. J. S. Christenfeld, "Do dogs resemble their owners?" *Psychological Science,* 15, No. 5, (May 2004), pp. 361–363.
2. There are many computer studies of the accuracy of chi-square critical values. For a brief discussion and some references, see Section 3.2.5 of David S. Moore, "Tests of chi-squared type," in Ralph B. D'Agostino and Michael A. Stephens (eds.), *Goodness-of-Fit Techniques,* Marcel Dekker, 1986, pp. 63–95.
3. Janice E. Williams et al., "Anger proneness predicts coronary heart disease risk," *Circulation,* 101 (2000), pp. 2034–2039.
4. Richard M. Felder et al., "Who gets it and who doesn't: a study of student performance in an introductory chemical engineering course," *1992 ASEE Annual Conference Proceedings,* American Society for Engineering Education, 1992, pp. 1516–1519.
5. We found these data at Patrick Meirmans's Web site, where he attributes the study to a paper by H. R. Schiffman in a 1982 issue of the *Journal of Counseling and Psychology.*
6. R. Shine, T. R. L. Madsen, M. J. Elphick, and P. S. Harlow, "The influence of nest temperatures and maternal brooding on hatchling phenotypes in water pythons," *Ecology,* 78 (1997), pp. 1713–1721.
7. S. V. Zagona (ed.), *Studies and Issues in Smoking Behavior,* University of Arizona Press, 1967, pp. 157–180.
8. Francine D. Blau and Marianne A. Ferber, "Career plans and expectations of young women and men," *Journal of Human Resources,* 26 (1991), pp. 581–607.
9. R. W. Mannan and E. C. Meslow, "Bird populations and vegetation characteristics in managed and old-growth forests, northeastern Oregon," *Journal of Wildlife Management,* 48 (1984), pp. 1219–1238.
10. Exercise 10.30 comes from the Electronic Encyclopedia of Statistical Examples and Exercises (EESEE) story "Domestic Violence."
11. Francisco L. Rivera-Batiz, "Quantitative literacy and the likelihood of employment among young adults," *Journal of Human Resources,* 27 (1992), pp. 313–328.
12. E. C. Strain et al., "Caffeine dependence syndrome: evidence from case histories and experimental evaluation," *Journal of the American Medical Association,* 272 (1994), pp. 1604–1607.
13. W. S. Gosset, "The probable error of a mean," *Biometrika,* 6 (1908), pp. 1–25. We obtained the sleep data from the Data and Story Library (DASL) Web site, http://lib.stat.cmu.edu/DASL/. They cite as a reference R. A. Fisher, *The Design of Experiments,* 3rd edition, Oliver and Boyd, 1942, p. 27.
14. D. J. Hand et al., *A Handbook of Small Data Sets,* Chapman & Hall, 2006.
15. A. R. Hirsch and L. H. Johnston, "Odors and learning," *Journal of Neurological and Orthopedic Medicine and Surgery,* 17 (1996), pp. 119–126. We found the data in the EESEE case study "Floral Scents and Learning."
16. From program 19, "Confidence Intervals," in the *Against All Odds* video series.
17. Exercise 10.63 is based on the EESEE story "What Makes Pre-teens Popular?"
18. Dirk Taubert, Reinhard Berkels, Renate Roesen, and Wolfgang Klaus, "Chocolate and blood pressure in elderly individuals with isolated systolic hypertension," *Journal of the American Medical Association,* 290 (2003), pp. 1029–1030.
19. W. E. Paulus et al., "Influence of acupuncture on the pregnancy rate in patients who undergo assisted reproductive therapy," *Fertility and Sterility,* 77, No. 4 (2002), pp. 721–724.

SOLUTIONS TO ODD-NUMBERED EXERCISES

Chapter 1

1.1 (a) Students in a statistics class. (b) Categorical: gender, homeroom teacher, grade level, calculator number. Quantitative: score on Test 1.

1.3 Answers will vary; for example, whether a household uses its recycling bin.

1.5 (a) Likely a survey of pizza shops was taken in which data showing average daily pizza sales were collected and then an estimate of the area involved in that calculation was made. (b) Assuming 300,000,000 people in the United States, this is about 1 square foot of pizza per person per year (43,560 sq. ft./acre $\times$ 18 acres $\times$ 365 days/300,000,000 $\approx$ 1), which seems plausible.

1.7 (a) Population: all wood being considered for purchase. Sample: the 5 pieces of wood from each batch. (b) Population: all claims filed with the insurance company in a given month. Sample: the sample of claims selected.

1.9 Population: all apples in the truckload. Sample: three large buckets of apples selected. Individuals: apples. Variable(s): most likely weight, color, number of blemishes.

1.11 (a) Television shows. (b) Viewers (in millions), household rating (in points; 1 point = 1,145,000 households), network, rank. (c) 25,000 "Peoplemeter" homes.

1.13 An experiment, because a treatment (tasting muffins) is imposed on the subjects.

1.15 Answers will vary, examples are given: (a) What make of car is most popular at your school? (b) Does listening to relaxing music for ten minutes immediately prior to a test improve students' scores?

1.17 (a) Answers will vary. (b) An observational study using police records.

1.19 An observational study; no treatment was imposed.

1.21 (a) An observational study; two groups were measured on the same variable; no treatment was imposed. (b) Population: people in leadership positions. Variables: fitness level and leadership ability. (c) Volunteers are likely to have a higher fitness level.

1.23 (a) An experiment, since each student is assigned to wear either red or white clothes. (b) Clothing color and number of bee stings. (c) Since a chance process was used to determine the students' clothing color, the two groups should be pretty balanced in terms of all other variables that might affect the number of bee stings (size, smell, hair color, etc.). If the average number of bee stings is much larger for the students wearing red clothes, then the experiment gives strong evidence that the color red causes bees to sting more often. (d) There are ethical concerns associated with conducting this experiment.

1.25 The *y* scale does not start at 0, and the values 13, 14, and 15 occur twice, distorting the difference due to gender.

1.27 Variability due to fill type is present since each brand listed actually refers to two tires, one for each fill type; variability due to brand is present since many different brands are being compared; variability due to tire is included since several of the brands are repeated and there are at least two tires of each listed brand.

1.29 Take a random sample of teens across the area of interest. Make sure their responses are anonymous.

1.31 (a) Answers will vary. One possibility: "Do a majority of Americans eat meat on a daily basis?" (b) Observational study, since no treatments were applied. (c) Individuals: surveyed Americans. Variable measured: how often the individual eats meat per week; categorical.

1.33 (a) Paper towel brand (four brands were compared); towels of the same brand (four towels of each brand are tested). (b) **I. Question of interest:** Which paper towel brand absorbs best? **II. Produce data:** Four towels of each brand are placed in a container with 200 ml of water, and the amount absorbed is squeezed out and recorded for each towel. **III. Analyze data:** Scott had the highest average absorbency but also the greatest variability.

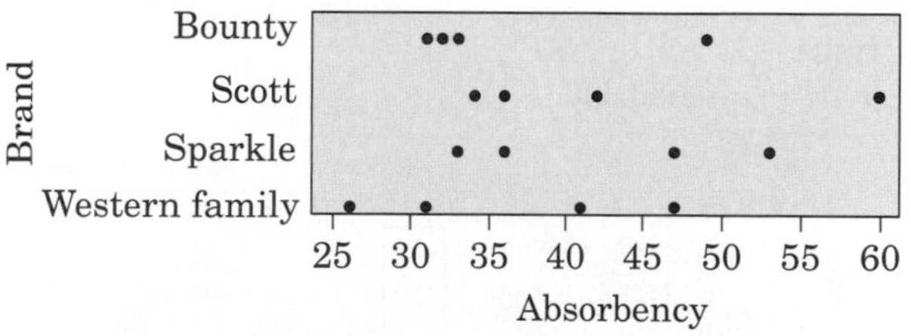

IV. Interpret results: Absorbency varies among the four brands; Scott had the greatest absorbency.

1.35 (a) An experiment, since the children were randomly assigned to the two treatments (judged by experts and prizes awarded or shared at an art party). (b) To avoid judgment based on perceived effort due to the assigned group. (c) The two groups and the judges. (d) Not necessarily. We do not know how the students were assigned to groups. If a chance process was not used, it is possible that the two groups differed systematically from each other before the treatments were administered, in which case the large difference in ratings could be due to this initial difference rather than to the treatments themselves.

1.37 (**a**) Population: passengers who take a trip that involves more than one airline. Sample: 12% of the tickets sold. (**b**) They will probably not result in exactly the correct amount of money for each airline. With a good sample, however, the sample results should be close to the truth about the population.

1.39 (**a**) Answers will vary. One possibility: "Does the pill reduce the amount of methane gas produced by cows?" (**b**) Randomly select cows on a farm; randomly assign half to receive the pill; the other half receive nothing. Provide the cows with the same feed and record the number of burps from each cow during a fixed time period. Record the amount of methane gas released in the burps if possible.

1.41 (**a**) Answers will vary. One possibility: "How much allowance does a child aged 5–12 receive in a week?" (**b**) Observational study, since no treatment was applied. (**c**) Individuals: children aged 5–12 who participated in the study. Variable: child's allowance. (**d**) Most likely due to rounding.

1.43 **I. Question of interest:** Are poems about laughter more creative if written for internal, rather than external, reasons? **II. Produce data:** Students were randomly divided into an external-reasons group and an internal-reasons group and were asked to write a poem about laughter. Their poems were rated by 12 poets using a creativity scale. **III. Analyze data:** The creativity scores were categorized by group and plotted on a dotplot. **IV. Interpret results:** Based on the dotplot, the average ratings were higher for the internal-reasons group.

Chapter 2

2.1 (**a**) 49,680,000.
(**b**)

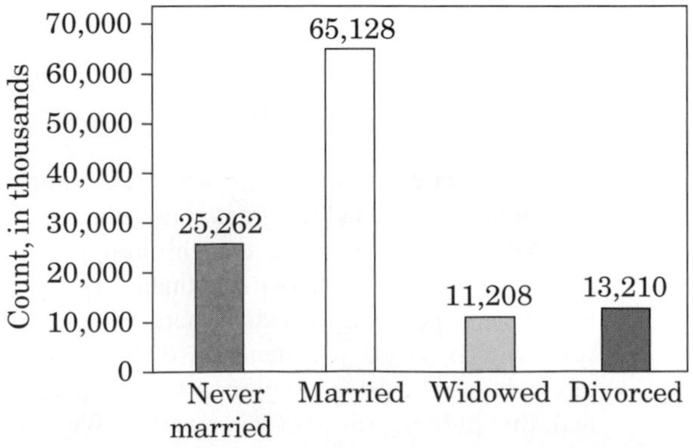

(**c**) Pie chart would be correct since categories represent distinct parts making up a whole.

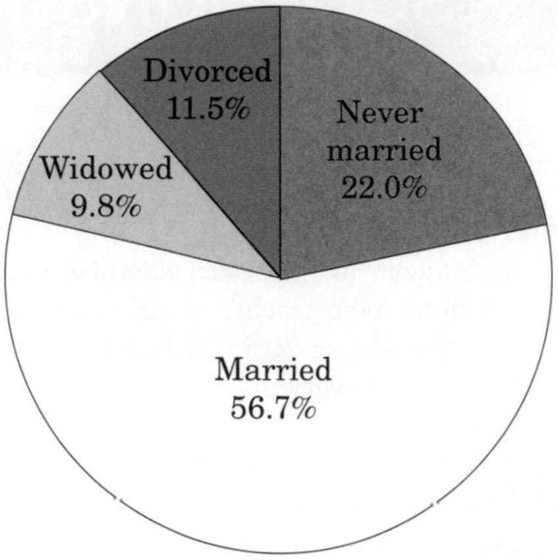

2.3 (**a**) 8.7%.
(**b**)

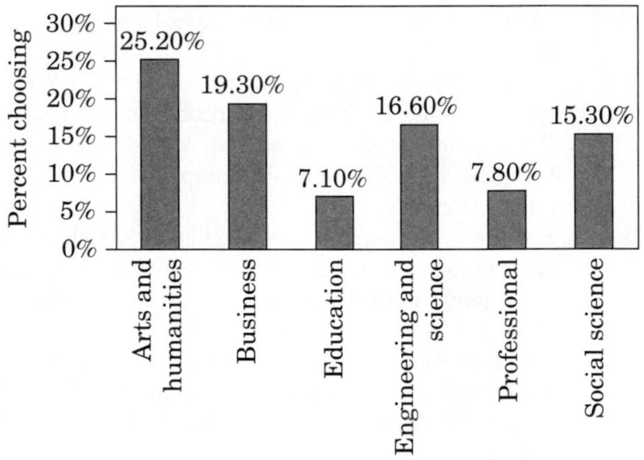

2.5 (**a**) No, the data compare four separate quantities, not four parts of a whole. (**b**) The percentage of drivers who use cell phones while driving decreases for the older generations.

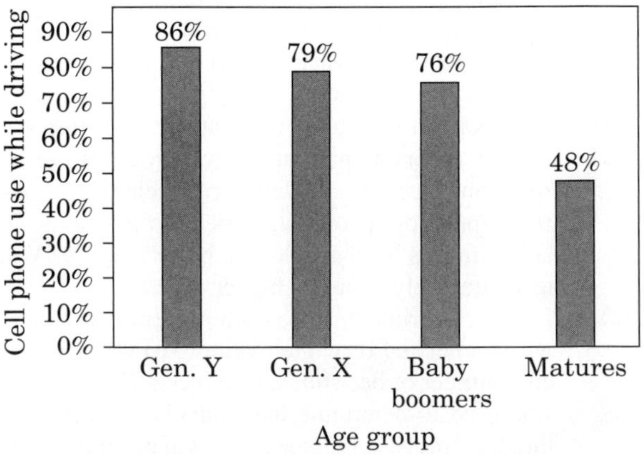

2.7 **(a)**

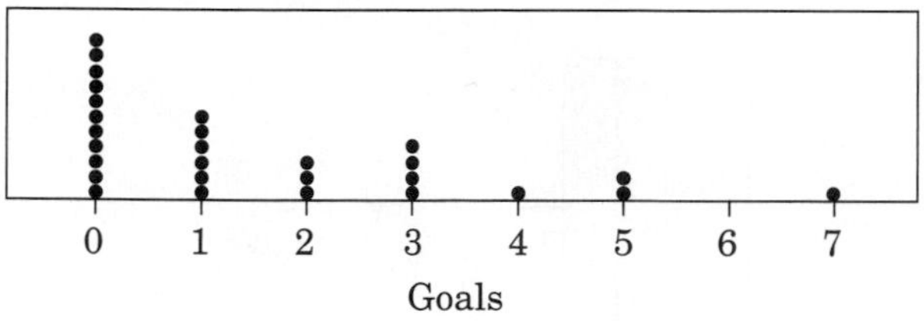

(b) Skewed to the right with a center of 1 goal and a spread of 7. 7 may be an outlier.

2.9 Centered around 19 mpg. Without the outlying values of 10 and 11, the data are somewhat skewed right.

2.11 **(a)** 16%, possibly high due to the Mormon church. **(b)** Roughly symmetric around 13% with a spread of about 3.5%. **(c)** Less, since the range is only 4.6%, whereas it is 10% for the older adults.

2.13 Skewed to the right with a center between 5 and 10 and a spread of about 55.

2.15 In a histogram the class widths should be the same. In Joe's histogram the class widths vary.

2.17 **(a)**

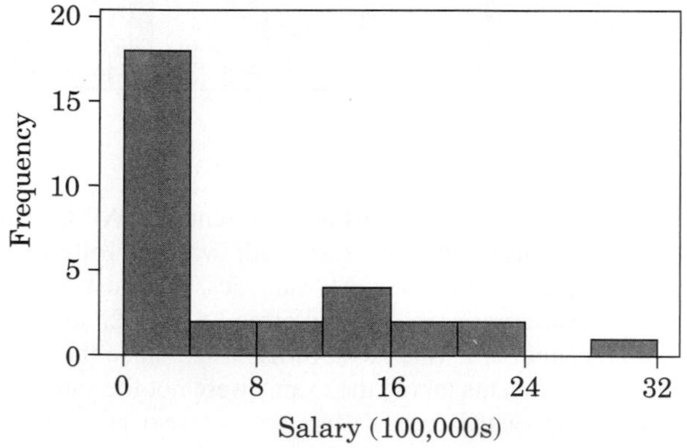

(b) Skewed to the right since there are a few very high salaries. **(c)** $27,610,000.

2.19 **(a)**

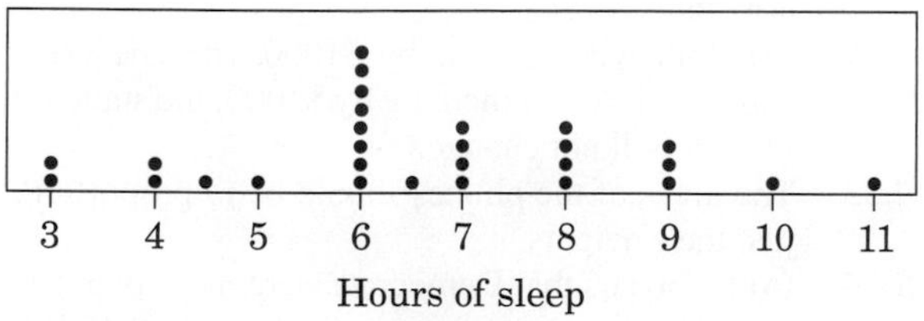

(b) Roughly symmetric with a center of 6 hours and a spread of 8 hours. No outliers.

2.21 **(a)** Skewed to the right with a center between 7 and 8 and a spread of about 5% **(b)** When making summary statements about the data, it is usually more informative to use percents rather than counts. **(c)** Possibly the poverty and teen pregnancy rates.

2.23 **(a)** Yes, since the categories represent distinct parts making up a whole.

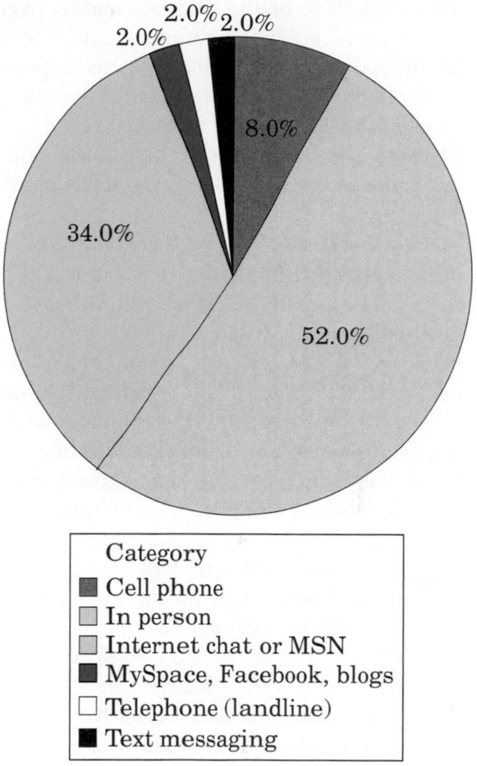

(b) Majority of the students prefer communicating with their friends in person.

2.25 **(a)**

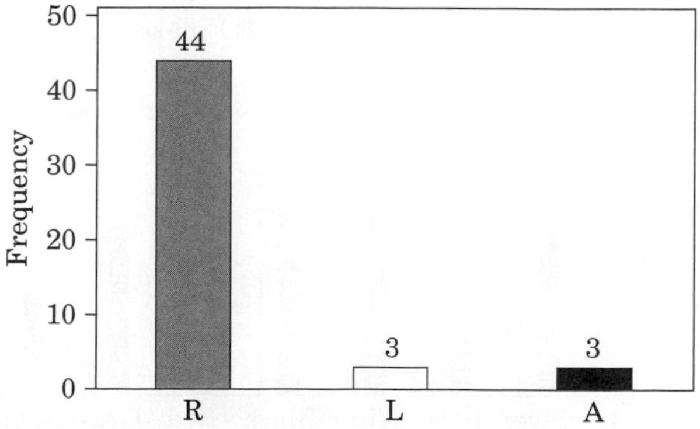

(b) 6% is our best estimate based on the sample percent who are left-handed.

2.27 $50,233 is the income such that half of the incomes of U.S. households in 2007 were less and half were more than this value.

2.29 **(a)** 7. About half of the students made/received fewer than 7 phone calls, and about half made/received more than 7. **(b)** $Q_1 = 3.5$, $Q_3 = 11.5$, $IQR = 8$. The middle 50% of the data have a range of 8. **(c)** 35 is an outlier.

2.31 **(a)** Answers may vary slightly: Minimum = −8, $Q_1 = 0$, Median = 2.5, $Q_3 = 22$, Maximum = 102. **(b)** About 25% of the students sent/received more calls than texts. The difference, texts − calls, was less than 2.5 for half of the students and more than 2.5 for half of the students. About 25% of the students sent/received 22 or more texts than calls.

2.33 Mean = 1600. The average daily metabolic rate of the 7 men taking part in the study of dieting was 1600 calories.

2.35 The mean is higher because there is a relatively small percent of households that make much higher incomes. The mean is affected by these values, but not the median.

2.37 **(a)** 2. A left-skewed distribution implies mean < median; a small spread implies a relatively small SD. **(b)** 3. A symmetric distribution implies that the mean is close to the median; a large spread implies a relatively large SD. **(c)** 1. A right-skewed distribution implies mean > median; a large spread implies a relatively large SD.

2.39 Since the distribution is skewed to the right, the median and *IQR* will give better summaries of the center and spread of the distribution.

2.41 **(a)** Mean = 87.188, Median = 87.5. **(b)** Standard deviation = 3.2, *IQR* = 3.25.

2.43 Distributions are similar except silver was the most popular for full-size/intermediate cars and not as popular for SUVs/trucks. White was the most popular option for SUVs/trucks and not as popular for full-size/intermediate cars. Beige/brown was the least popular for both car types.

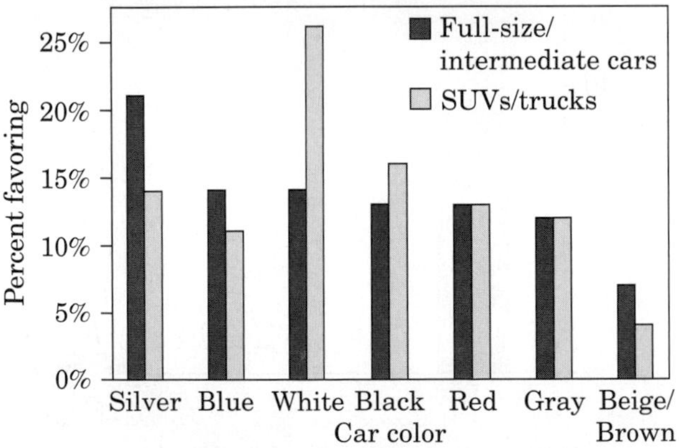

2.45 **(a)** Minimum = 3, $Q_1 = 4$, Median = 8, $Q_3 = 12$, Maximum = 55. **(b)** 27, 31, 34, and 55 are outliers.

(c)

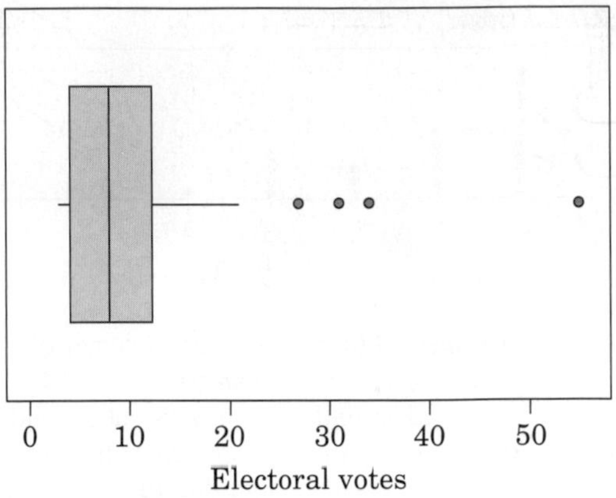

2.47 **(a)**

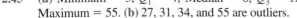

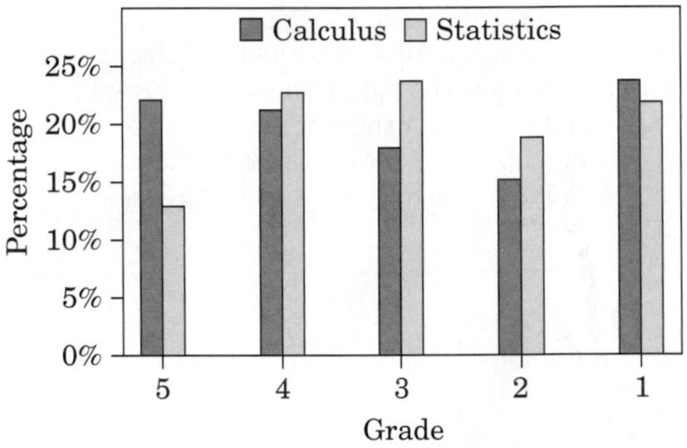

(b) The distributions are different. For AP Calculus the most common exam grade was a 1, followed by 5, 4, 3, and 2. For AP Statistics, the most common exam grade was a 3, followed by 4, 1, 2, and 5. We cannot tell which exam was easier, since the groups of students taking the exams were not the same.

2.49 The distribution differs for the two regions. The median poverty rate for the northern states was approximately 3% less than for the southern states. Q_3 for the North was less than Q_1 for the South. The spread of poverty rates was also less for the northern states.

2.51 **(a)** Both will increase by \$1000. **(b)** The extremes and quartiles will increase by \$1000; the standard deviation will not change.

2.53 The areas of the phones should be in proportion, not just their heights.

2.55 **(a)** It appears that Democrats were more than 5 times as likely to support the court's decision than Republicans or Independents.

(b)

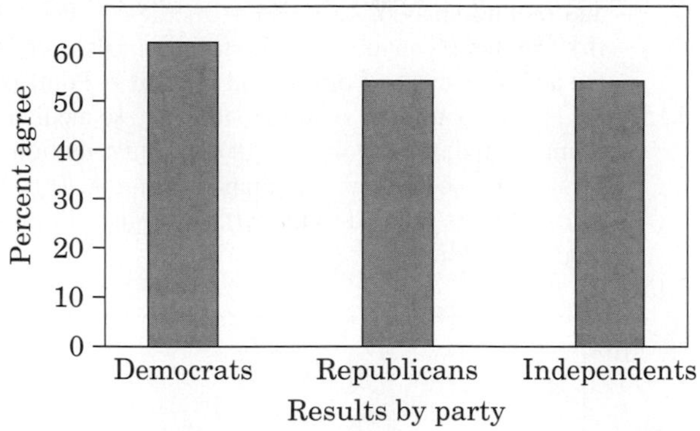

2.57 We now see that the trend has been much steadier over time.

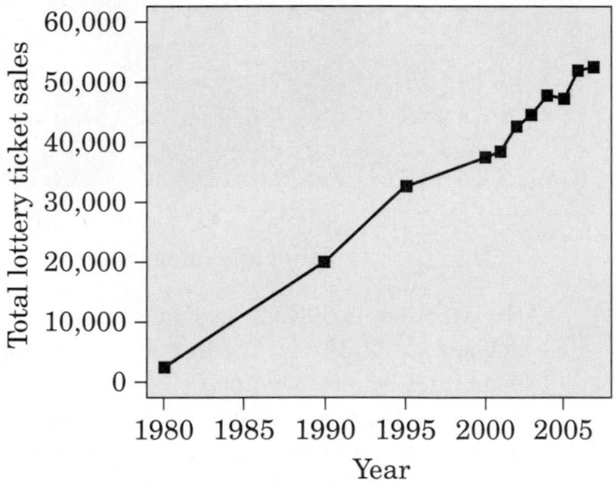

2.59 The size of the graphics is not in proportion to the percents. The last picture is below the dotted line, which gives the incorrect impression of a decrease for this point.

2.61 **(a)** The vertical scale does not begin at 0, overemphasizing the difference in cholesterol.

(b)

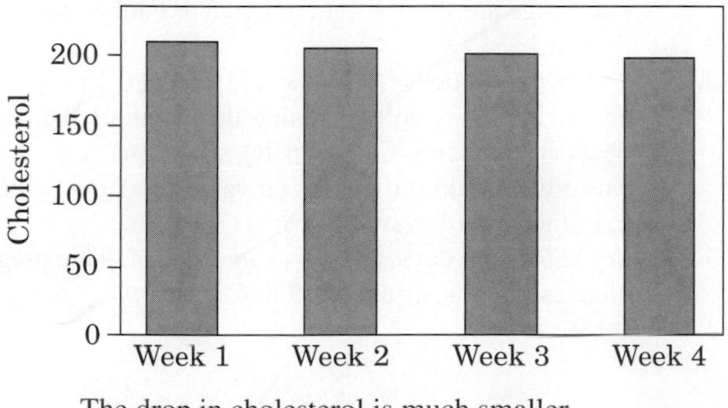

The drop in cholesterol is much smaller.

2.63 If every single adult had failed the test, 100% would have failed. It is not possible to have more than 100% failing the test.

2.65 A 14.8% increase. Poverty did not necessarily become more common during these years, because the population also increased during this time.

2.67 **(a)** In either case, the stemplot shows a symmetric distribution centered at the upper 60s.

```
4 | 69                4 | 69
5 | 36678             5 | 3
6 | 003344567778      5 | 6678
7 | 0112347889        6 | 003344
8 | 01358             6 | 567778
9 | 00                7 | 011234
                      7 | 7889
                      8 | 013
                      8 | 58
                      9 | 00
```

(b) The shape is symmetric without outliers. Mean = 70.39 and SD = 12.23. The spread is about 50 years.

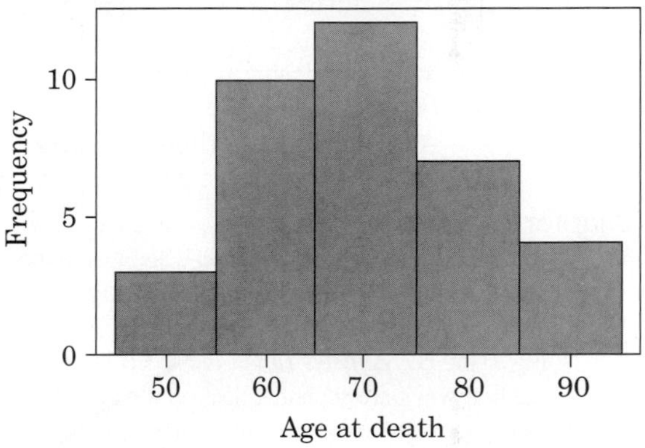

(c) Answers will vary.

2.69 The incorrect value is for coronary heart disease. It should be 27.5%, not 2.75%.

2.71 No. The cones make the differences appear larger than they really are. The *area* of the pictures should be proportional, not their heights.

2.73

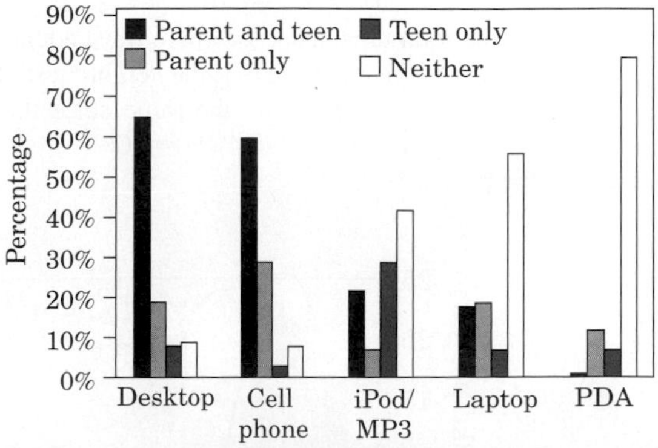

2.75 **I. Question of interest:** Does the drug increase sleep time? **II. Produce data:** An experiment was carried out with 10 patients. The number of additional hours of sleep gained by each subject after taking the drug was recorded. **III. Analyze data:** Minimum = −0.1,

$Q_1 = 0.8$, Median $= 1.75$, $Q_3 = 4.4$, Maximum $= 5.5$, Mean $= 2.33$, Standard Deviation $= 2.00$. **IV. Interpret results:** The data show that the drug was effective. On average, the participants gained at least some additional sleep. However, there should have been a control group taking a placebo to ensure that the results can be attributed to the treatment.

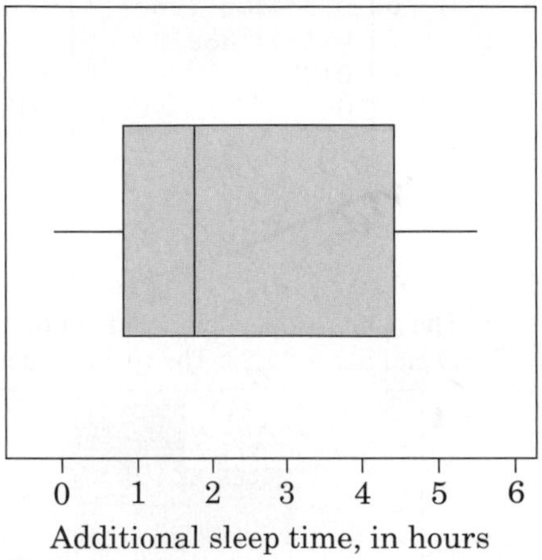

Chapter 3

3.1 **(a)** 87% of the girls her daughter's age weigh the same or less than her daughter does and 67% of girls her daughter's age are her daughter's height or shorter. **(b)** According to the *Los Angeles Times,* the speed limits on California highways are such that 85% of the vehicle speeds on those stretches of road are no greater than the speed limit.

3.3 **(a)** Francine's bone density is 1.45 standard deviations below the mean hip bone density of 956 g/cm^2 for 25-year-old women. **(b)** 5.517 g/cm^2.

3.5 **(a)** 79.3 percentile. Roughly 79% of the team had salaries less than or equal to Lidge's salary. **(b)** $z = 0.79$. Lidge's salary was 0.79 standard deviations above the mean salary of $3,388,617.

3.7 **(a)** The total area under the curve must equal 1. Since the base of the rectangle is 1, the height must also be 1. **(b)** 0.5. **(c)** 0.5. They are the same since the density curve is symmetric. **(d)** (0.4)(1) = 40%.

3.9 One possibility:

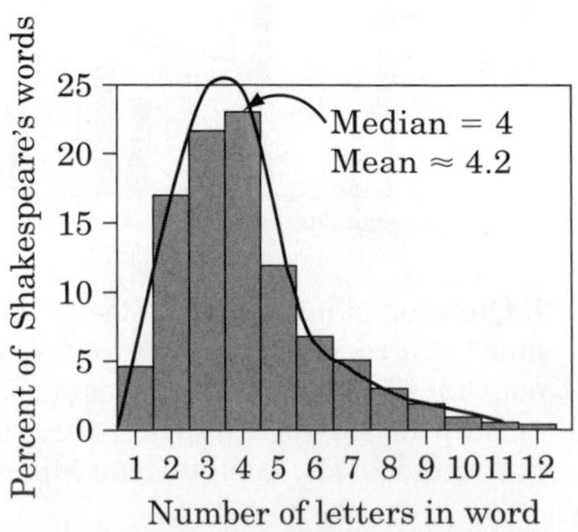

3.11 **(a)** The overall shape of the distribution is symmetric and mound-shaped, so mean = median = Point A. **(b)** The overall shape of the distribution is skewed to the left, so mean = Point A and median = Point B.

3.13 **(a)** The distribution is skewed to the right, so median = Point B, mean = Point C. **(b)** The distribution is symmetric, so mean = median = Point A. **(c)** The distribution is skewed to the left, so mean = Point A, median = Point B.

3.15 **(a)**

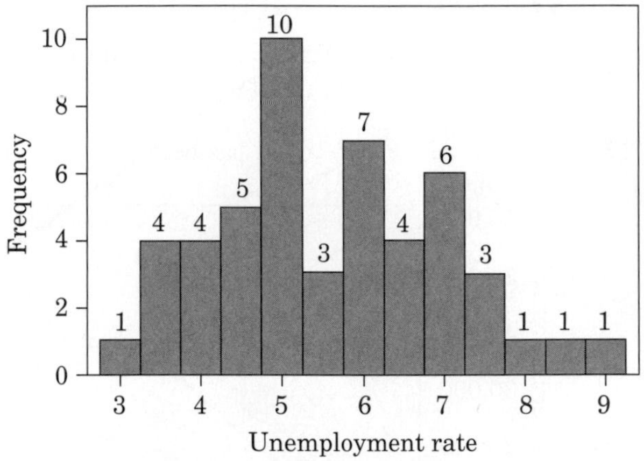

(b) Mean = 5.598%; median = 5.45%; standard deviation = 1.391%. The distribution of unemployment rates is fairly symmetric, with a center of about 5.5% and a spread of 5.6%. **(c)** 80th percentile. 80% of the unemployment rates are less than or equal to Illinois's unemployment rate. Illinois has a pretty high unemployment rate compared with the remaining 49 states. **(d)** Texas; $z = -0.36$.

3.17 **(a)** Mean = 170 cm; standard deviation = 7.5 cm. **(b)** 188.75 cm.

3.19 **(a)** $z = 0.67$. Paul is slightly taller than average for his age. His height is 0.67 standard deviations above the average male height for his age. **(b)** 75% of boys Paul's age are shorter than or equal to Paul in height.

3.21 89 and 133.

3.23 144 has a z-score of 3. We would expect 0.15% to be above 144. It is not surprising that none of the students in the sample had an IQ score this high, because we would only expect a value this large about 1.5 times out of every thousand.

3.25 **(a)** 327 to 345 days. **(b)** $z = 1 \Rightarrow 16\%$ of horse pregnancies last longer than 339 days.

3.27 **(a)** 0.3557.

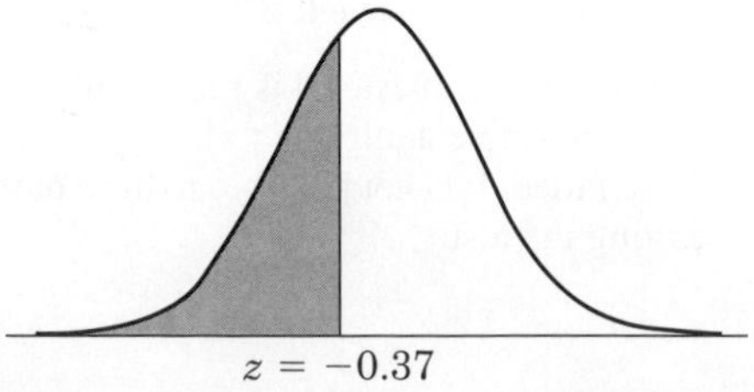

(b) 0.6443.

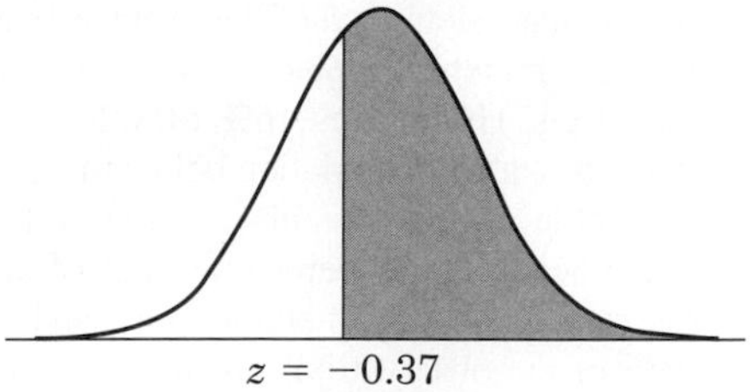

(c) 0.9842.

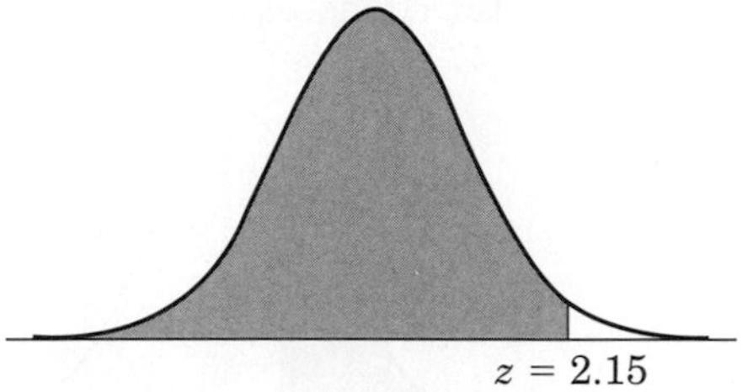

(d) 0.0158.

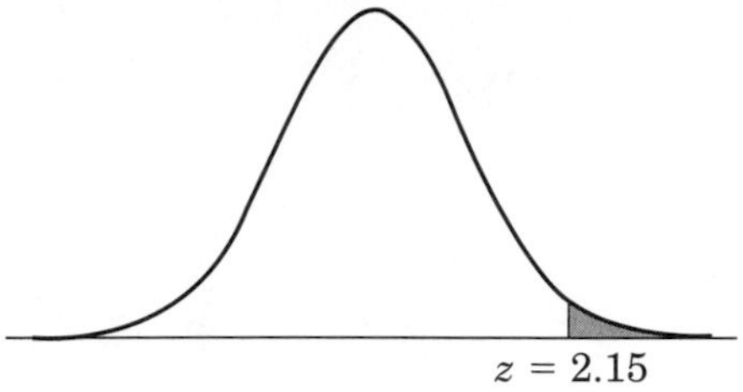

3.29 **(a)** 0.8587.

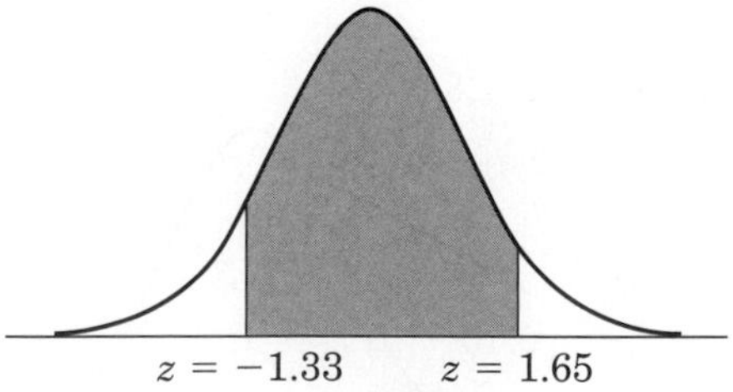

(b) 0.2718.

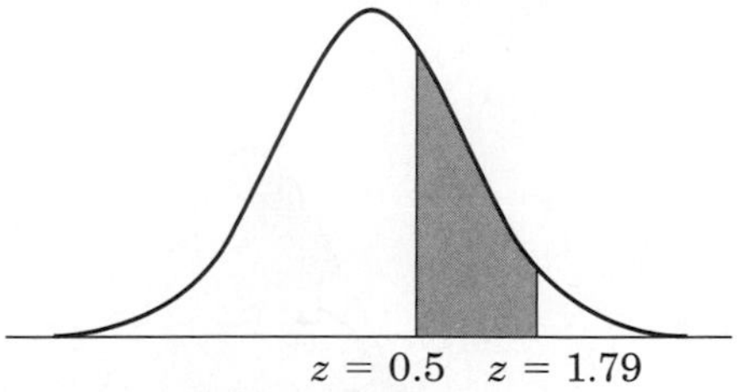

3.31 **(a)** $z = -0.84$.

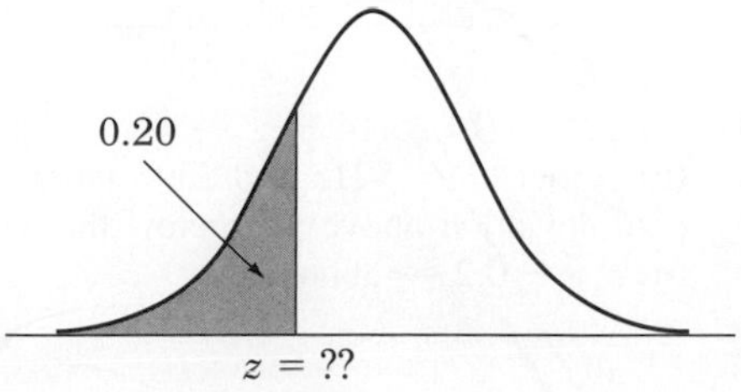

(b) $z = 0.13$.

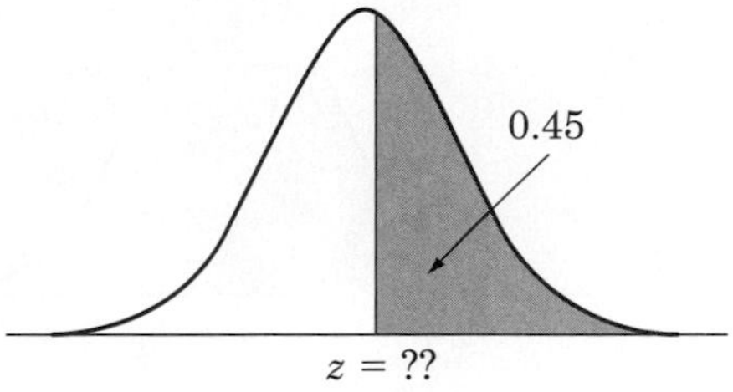

3.33 **Step 1:** *State the problem.* Let x = the distance that Tiger's ball travels. x has a Normal distribution with $\mu = 304$ and $\sigma = 8$. Find the probability that $x \leq 317$. **Step 2:** *Standardize and draw a picture.* For $x = 317$, $z = 1.63$.

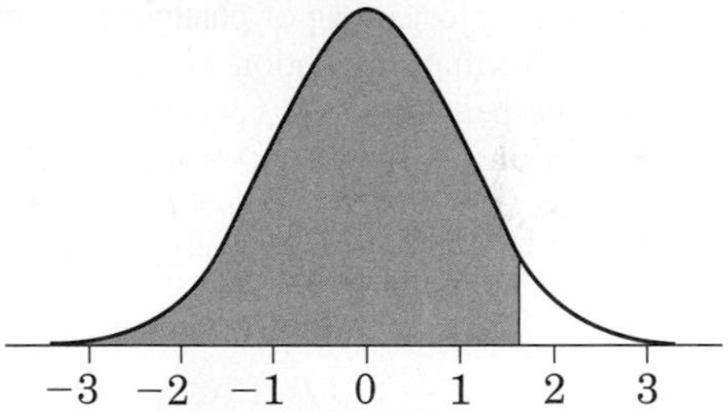

Step 3: *Use the table.* From Table A, the probability that $z \leq 1.63$ is 0.9484. The probability that Tiger is safe hitting his driver is about 0.95. **Step 4:** *Conclusion.* Since the probability that Tiger's drive misses the creek is about 0.95, we can conclude that Tiger is safe hitting his driver.

3.35 **Step 1:** *State the problem.* Let x = a randomly selected female's score on the SAT Math test. x has a Normal distribution with $\mu = 500$ and $\sigma = 111$. Find the probability that $x > 533$. **Step 2:** *Standardize and draw a picture.* For $x = 533$, $z = 0.30$.

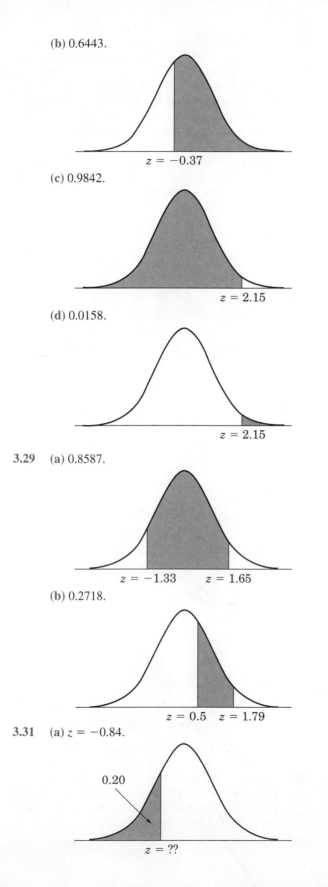

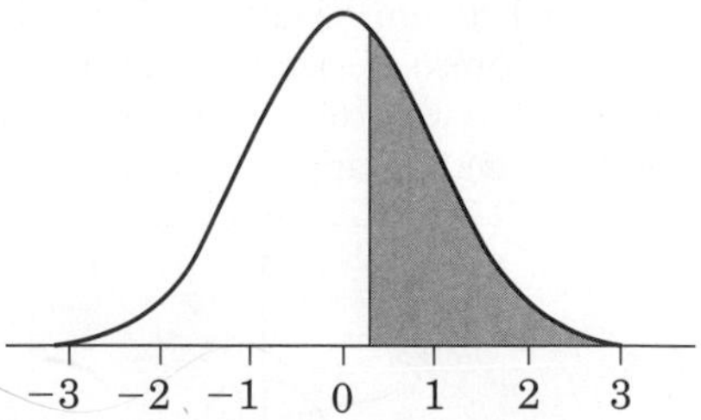

Step 3: *Use the table.* From Table A, the probability that $z < 0.30$ is 0.6179. The proportion of observations greater than 0.30 is then 0.3821. **Step 4:** *Conclusion.* Approximately 38% of females scored higher than the male mean.

3.37 **(a) Step 1:** *State the problem.* Let x = the weight of a randomly selected bag of potatoes from the shipment. x has a Normal distribution with $\mu = 10$ and $\sigma = 0.5$. Find the percent of bags weighing less than 10.25 pounds. **Step 2:** *Standardize and draw a picture.* For $x = 10.25$, $z = 0.5$.

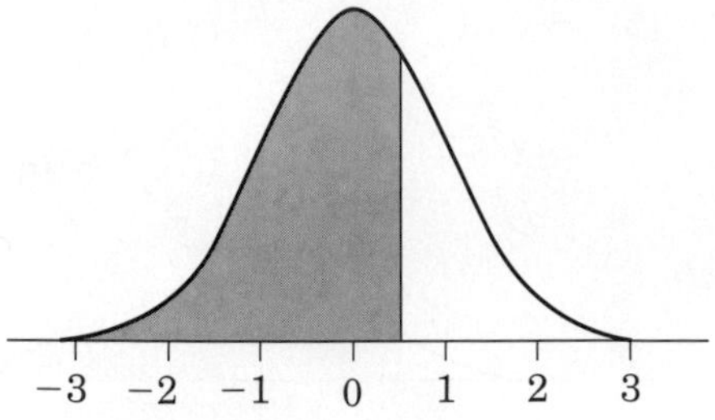

Step 3: *Use the table.* From Table A, the percent of observations less than 0.5 is 0.6915. **Step 4:** *Conclusion.* Approximately 69% of the bags in the shipment weighed less than 10.25 pounds.

(b) Step 1: *State the problem.* Let x = the weight of a randomly selected bag of potatoes from the shipment. x has a Normal distribution with $\mu = 10$ and $\sigma = 0.5$. Find the percent of bags weighing between 9.5 and 10.25 pounds. **Step 2:** *Standardize and draw a picture.* From part **(a)**, $P(x < 10.25) = P(z < 0.5)$. For $x = 9.5$, $z = -1$. The area we want to find is depicted below.

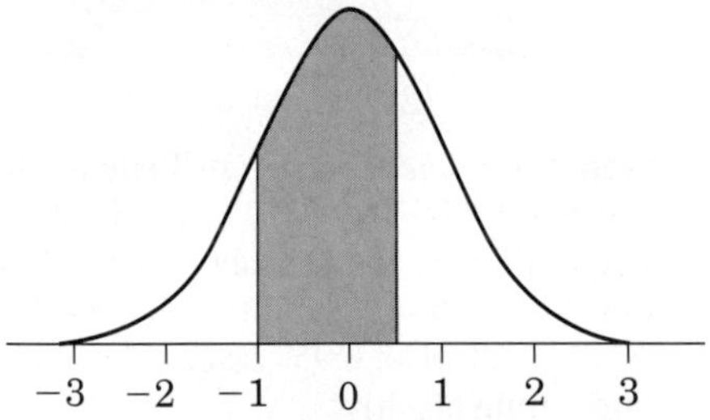

Step 3: *Use the table.* From part **(a)**, $P(z < 0.5) = 0.6915$. From Table A, the percent of observations less than -1 is 0.1587. Thus, the percent of bags weighing between 9.5 and 10.25 pounds is 0.5328. **Step 4:** *Conclusion.* Approximately 53% of the bags in the shipment weighed between 9.5 and 10.25 pounds.

3.39 68% of women are between 62.5 and 67.5 inches tall, 95% are between 60 and 70 inches tall, and 99.7% are between 57.5 and 72.5 inches tall.

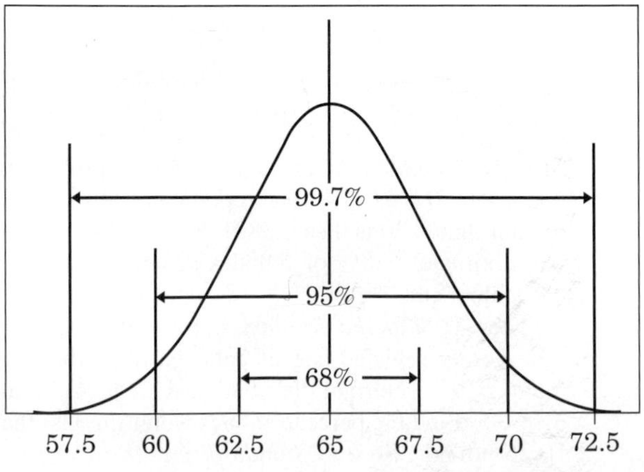

3.41 In a Normal distribution, 68% of the values are within one standard deviation of the mean. Since the distribution is symmetric, 34% of these values must be below the mean. That leaves 16% of values that fall more than one standard deviation below the mean. Since the percentile rank of an observation is the percent of terms below it, the percentile rank of an observation one standard deviation below the mean is 16. Similarly, 95% of the observations are within two standard deviations of the mean, leaving 97.5% below a value that is two standard deviations above the mean.

3.43 **(a)** $z = -0.47$.

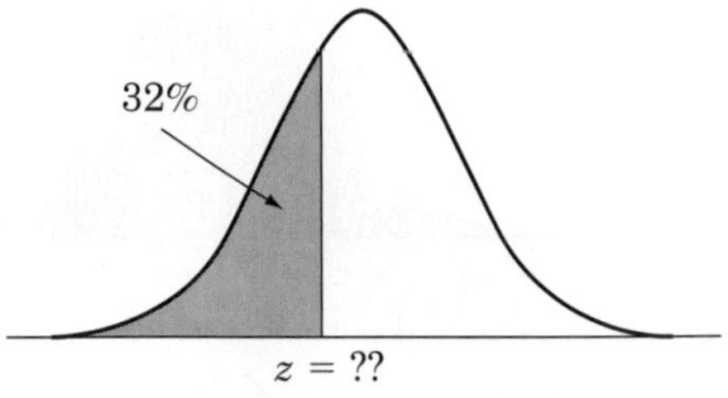

(b) $z = 0.25$.

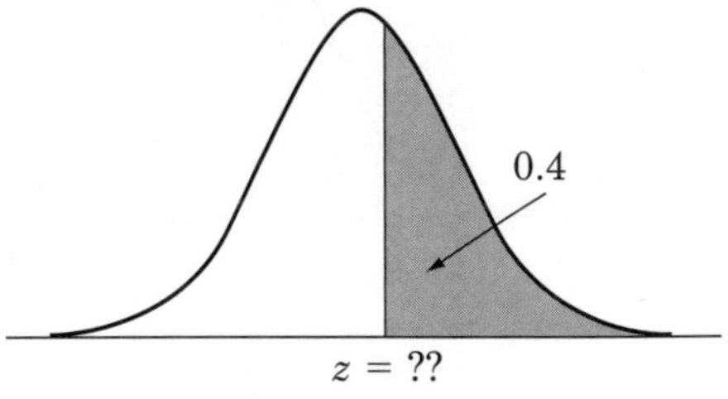

(c) $z = -1.28$.

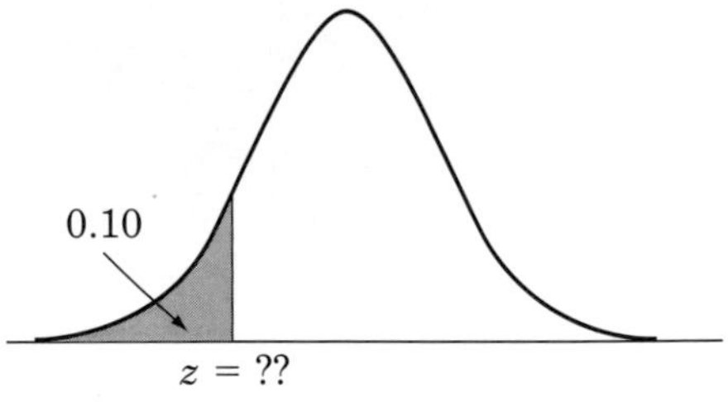

(d) $z = 0.95$.

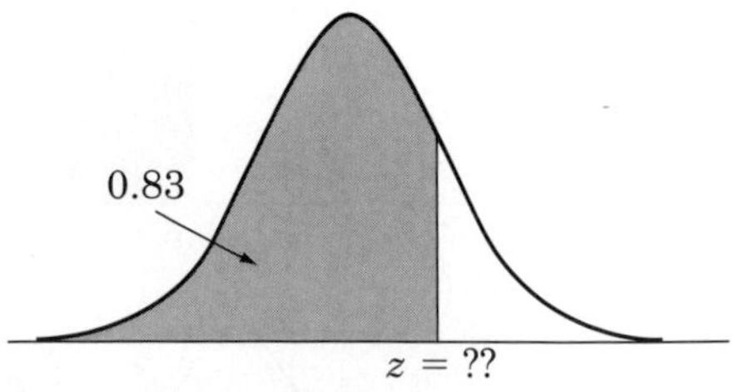

3.45 **(a)** 2.28%. **(b)** 25.14%.

3.47 **(a)** About 68%. \$215 and \$395 represent one standard deviation above and below the mean of \$305. **(b)** $z = -0.2 \Rightarrow$ about 42%.

3.49 **(a)** 0.9878.

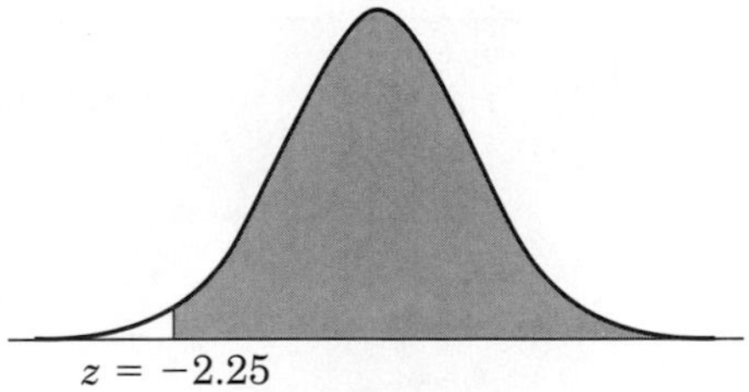

(b) 0.9494.

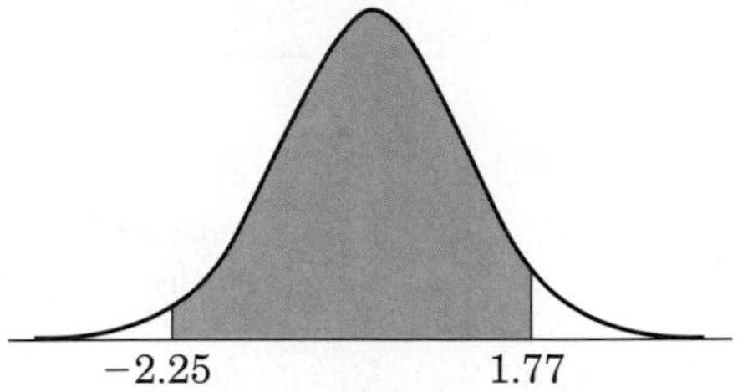

(c) 0.9981.

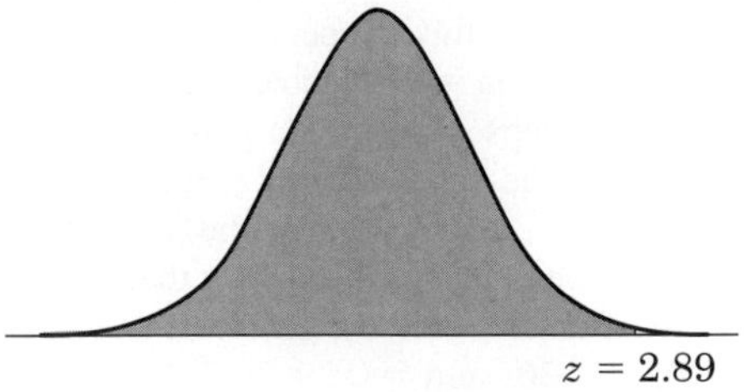

(d) 0.0730.

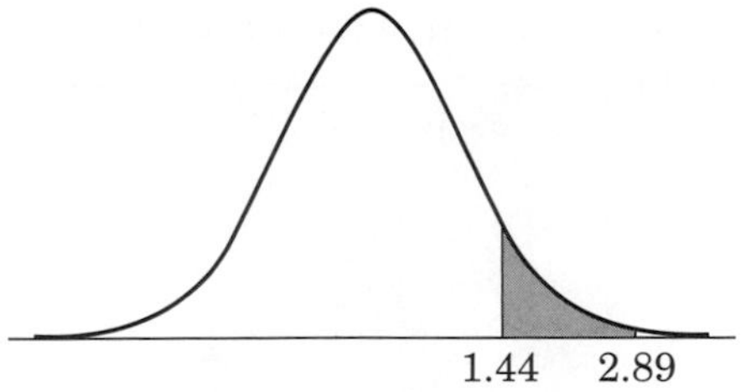

3.51 **(a)** Since $z = -2.29$, about 1% of babies will be identified as low birth weight. **(b)** Q_1: $-0.67 = (x - 3668)/511 \Rightarrow x = (-0.67 \times 511) + 3668 = 3325.63$; median = mean = 3668. Q_3: $0.67 = (x - 3668)/511 \Rightarrow x = (0.67 \times 511) + 3668 = 4010.37$.

3.53 **(a)** Bimodal. Mean = median = B. **(b)** Skewed to the right. Median = A and mean = B.

3.55 **(a)** Within 2 standard deviations of the mean, that is, between −21% and 45%. **(b)** $z = -0.73$. The area is 0.2327. The market is down for about 23% of the years.

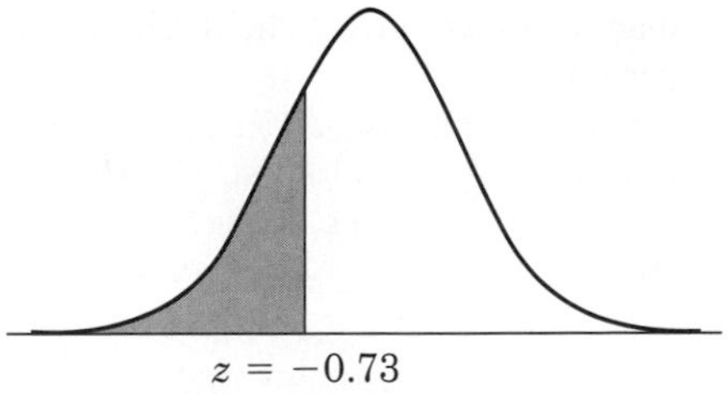

(c) $z = 0.18$ and $z = 0.79$. The area between $z = 0.18$ and $z = 0.79$ is 0.2138.

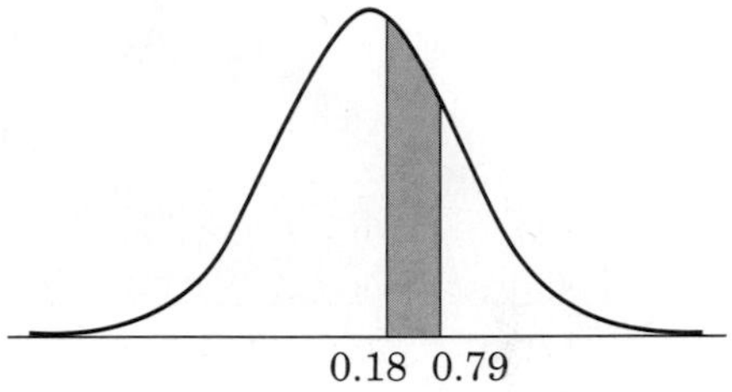

Chapter 4

4.1 **(a)** Response: grade. Explanatory: time of study. **(b)** Relationship. **(c)** Response: yield. Explanatory: rain. **(d)** Relationship.

4.3 **(a)** There is a moderate, positive, linear relationship between IQ scores and GPA. **(b)** About 103 and 0.5. **(c)** A, B: moderate IQ but low GPA. C: low IQ but high GPA.

4.5 **(a)** The number of males returning each year may vary dramatically, so it would be more appropriate to use percents for comparing various years.
(b)

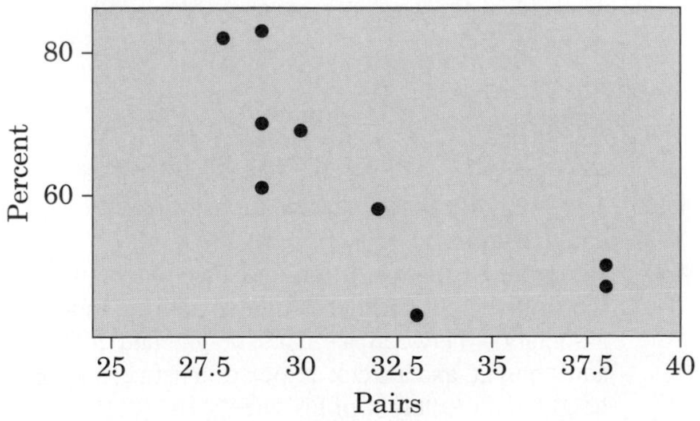

(c) There is a definite downward trend, but it is more curved than linear. Yes.

4.7 **(a)** States with higher (lower) median household income will also have higher (lower) mean personal income. **(b)** Except for the District of Columbia, the relationship is moderately strong, positive, and linear. **(c)** UT: given the mean income per person, the expected median household income is higher than expected. D.C. and CT: the median household income is lower than expected given the high mean income per person.

4.9 (a) Stocks; highest, 50%; lowest, −30%. Treasury bills: highest, 15%; lowest, 1%. (b) The relationship is weak at best. No evidence that high interest rates are bad for stocks.

4.11 (a)

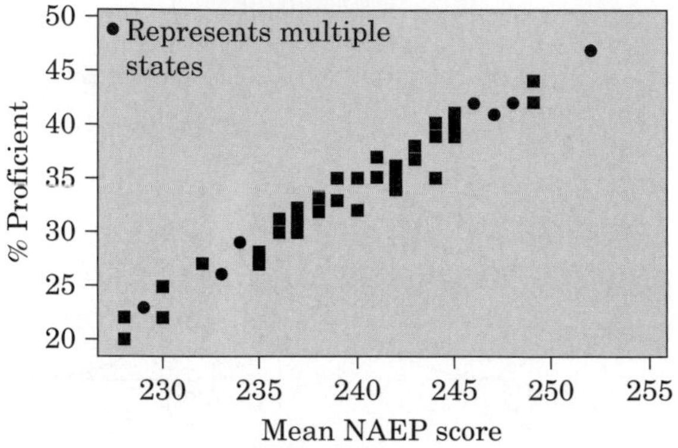

(b) Linear, positive, fairly strong. (c) Answers will vary.

4.13 (a) Positive but not close to 1.

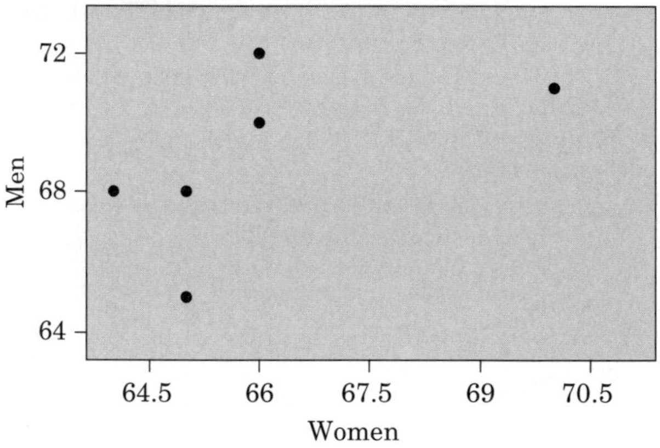

(b) $r = 0.57$.

4.15 The strength of the linear relationship between GPA and first-exam score.

4.17 We would expect a strong association between a woman's height and her height as a girl, a moderate association between a man's height and his adult son's height, and a weak association between a man's height and the height of his wife.

4.19 (a) The direction is positive, but the form appears curved, not linear. The strength is moderate. (b) Hippopotamus, Asian elephant: life spans are longer than expected given their gestation periods. Giraffe: tends to follow the curvilinear shape, possibly a little shorter life span than expected. (c) The point would be at about (280, 70) and would not be part of the general pattern.

4.21 (a) −1 to 1. (b) Any nonnegative number.

4.23 (a) Negative; slightly curved; yes, at about (4, 2.9). (b) No.

4.25 (a) MA is explanatory.

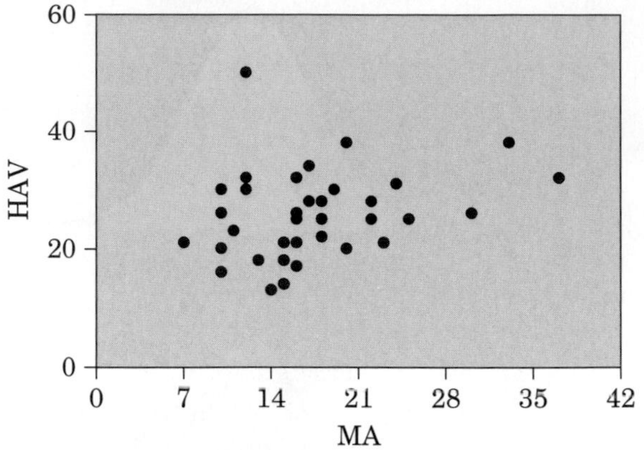

(b) Weak, positive, linear relationship; (12, 50) is an outlier. (c) There is a positive association, but it is not very strong. (d) Not really.

4.27 (a) −19.87. We predict the amount of gas consumed to decrease by 19.87 cubic feet for every degree increase in the average monthly temperature. (b) 1425. When the average monthly temperature is 0°F, the predicted gas consumption is 1425 cubic feet. (c) 828.9 cubic feet.

4.29 (a) 4.7 laps. (b) No. The number of slices eaten is outside the observed data range.

4.31 $y = 1 - x$. The sum of the squares of the vertical distances is 3, but it is 18 for the other line.

4.33 Residuals: −1.1, −3.13, −0.11, −1.57, 0.91, 3.85, 0.48, 1.26; sum = 0.59.

4.35 False. No. 70% of the variation in y is explained by the linear regression model.

4.37 (a) x = number of beers consumed. y = BAC.

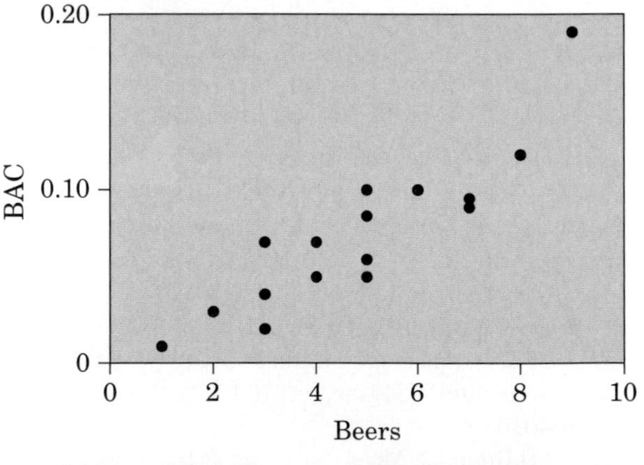

(b) $r = 0.894$. Yes, since the relationship appears very linear. (c) Predicted BAC = −0.013 + 0.018 (number of beers). Slope: for every additional beer consumed, a student's BAC increases by 0.018, according to our model. Since 0 is outside the range of the data, it should not be interpreted.

4.39 Temperature (z) is likely to be a lurking variable: in warmer weather, more people go swimming and thus there are more drowning deaths (y). Ice cream sales (x) are likely to be higher during this time as well.

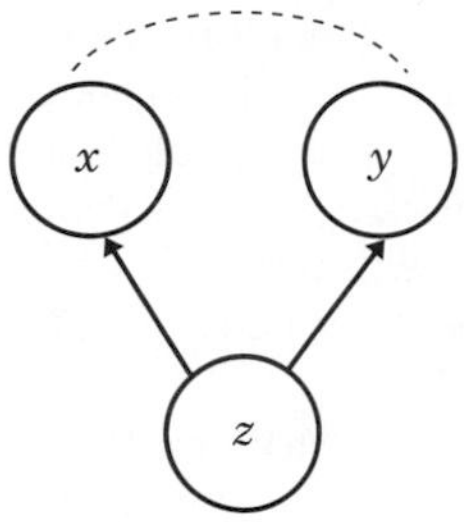

4.41 For example, motivation level, intelligence, socioeconomic status.

4.43 Stronger students are more likely to choose such math courses; weaker students may avoid them.

4.45 **(a)** The line shows a general decrease but not a strong relationship. **(b)** The general pattern is concave upward (bowl-shaped). This pattern strengthens the conclusion to avoid hospitals that treat few heart attacks, but the hospitals that treat the most patients are not necessarily better than those that treat a moderate number of patients.

4.47 Inactive girls are more likely to be obese. 3.2%.

4.49 Close to -1.

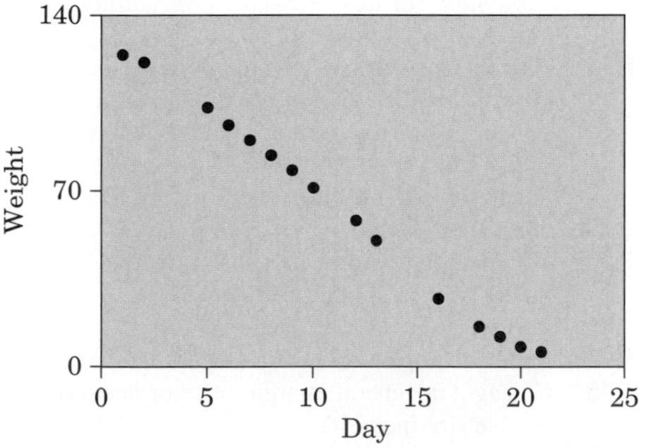

4.51 **(a)** -56.1 grams; prediction outside the range of the available data is risky. **(b)** 99.6% of the variation in the soap weights is explained by the least-squares regression of weight on day. **(c)** Yes.

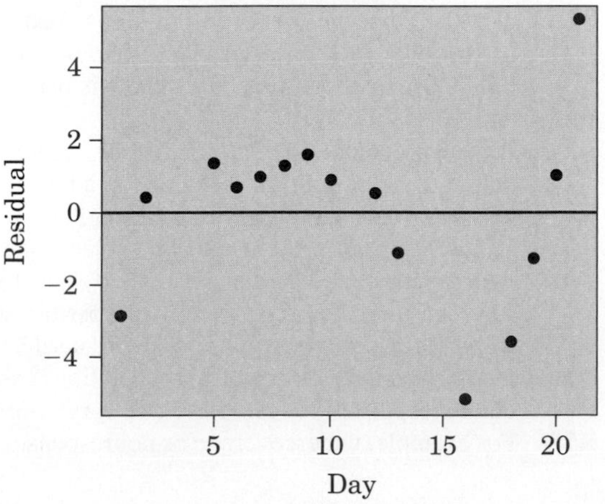

4.53 Time spent standing is a confounding variable.

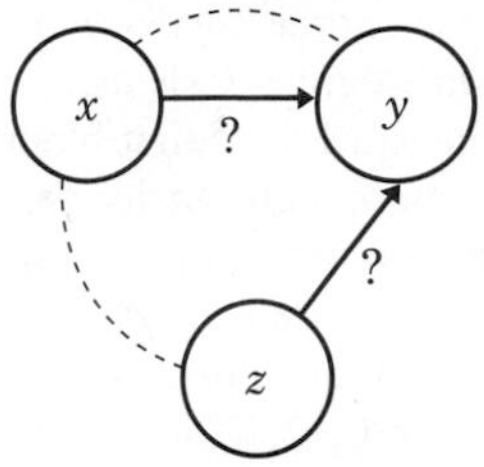

(x and z are confounded)

4.55 **(a)** Longer crickets should weigh more than shorter crickets. **(b)** It would not change.

4.57 **(a)**

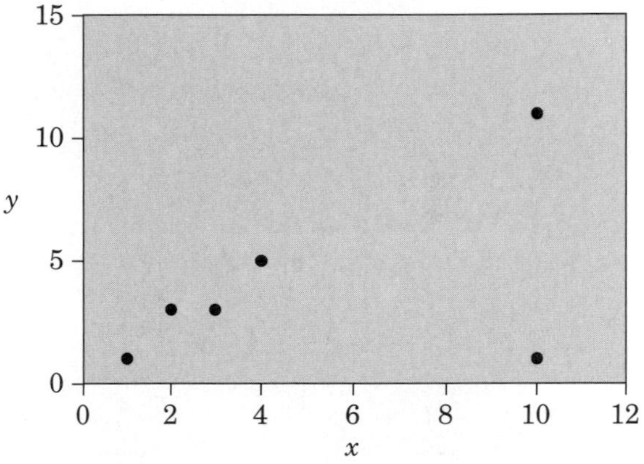

(b) $r = 0.48$. **(c)** The outlier at (10, 1).

4.59 **(a)** Dolphins: body weight, 190 kg; brain weight, 1600 g. Hippos: body weight, 1400 kg; brain weight, 600 g. **(b)** Dolphins and hippos are outliers in the plot. Dolphins have unusually high brain weights for their body weights. Hippos are heavy but have unusually low brain weights.

4.61 74% of the variation in brain weight is explained by the least-squares regression of brain weight on body weight.

4.63 **(a)** Fairly strong, negative, linear association.

Predicted number of days = 33.12–4.69° temperature

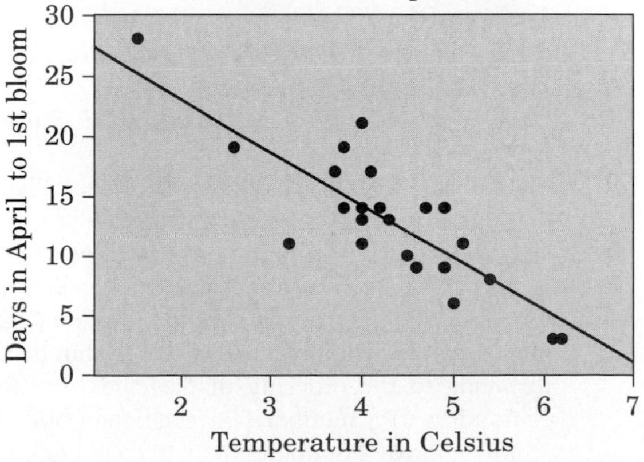

(b) Predicted number of days = 33.12 − 4.69 (temperature). For every 1 degree increase in average March temperature, in degrees Celsius, we predict the number of days in April until first bloom to decrease by 4.69. The y intercept is outside the range of data and therefore has no meaningful interpretation. (c) 16.7. (d) −2.015. (e) There is no discernible pattern in the residuals. They are clustered about 0 in a random fashion. (f) $r^2 = 0.72$. 72% of the variation in the number of days in April until the first cherry blossom appears is explained by the least-squares regression of the number of days in April until first bloom on the average temperature in March.

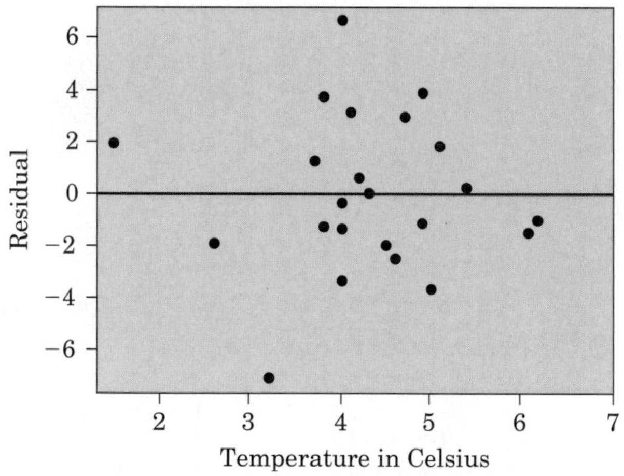

Chapter 5

5.1 (a) 29,777. (b) Voluntary response sample.

5.3 (a) Convenience sample. (b) It's unlikely that the first 100 students to arrive are representative of the student population in general. Probably higher, since students who arrive first are likely to be those who got a good night's sleep the night before.

5.5 Possible answers: (a) A call-in poll. (b) Interviewing students as they enter the student center.

5.7 Apartment complexes 16, 32, 18: Fairington, Waterford Court, and Fowler.

5.9 (a) Use labels 001 to 440. (b) 395, 020, 118, 167, 300, 360, 241, 065, 188, 365.

5.11 Using line 101: 19, 22, 05 (Petrucelli, Shen, and Brockman).

5.13 (a) Assign 1 to heads and 2 to tails. Enter "2" next to "Population =1 to" and click Reset. Enter a sample size of 1 and click Sample. Number returned identifies heads or tails. (b) Enter "52" next to "Population = 1 to" and click Reset. Enter a sample size of 52 and click Sample. The ordering of the 52 numbers corresponds to the ordering of the cards. (c) Assign each adult to a number between 1 and 500. Enter "500" next to "Population = 1 to" and click Reset. Enter a sample size of 12 and click Sample. Numbers returned correspond to jurors.

5.15 (a) All black adult residents of Miami; SRS of 300 adults from predominately black neighborhoods. (b) Out of fear, people may not respond honestly; also a black police officer is asking the questions.

5.17 Answers will vary.

5.19 (a) Label from 0001 to 3478. (b) 2940, 0769, 1481, 2975, 1315.

5.21 It's a convenience sample.

5.23 Supporters of an individual's right to bear arms have tremendous political influence. Opinion polls on such subjects do not always result in legislative action.

5.25 Yes, since the library surveyed card holders.

5.27 Population: all smokers who signed a card saying they intended to quit smoking. Parameter: proportion signing the card who had not smoked in the past six months. Sample: random sample of 1000 people who signed the card. Statistic: 21%.

5.29 Statistic, parameter.

5.31 This is a comparative experiment because the researchers assigned the ducks to the groups being compared (inside boxes versus outside boxes).

5.33 (a) High bias, high variability. (b) Low bias, low variability. (c) Low bias, high variability. (d) High bias, low variability.

5.35 Possible answers: (a) Hunter College survey was an observational study, not a national survey; intersections in New York are not likely to be representative of the nation as a whole. (b) No. The discrepancy is due to sampling differences, not sample size.

5.37 Answers will vary.

5.39 The margin of error is half as big with the larger sample (0.1 instead of 0.2).

5.41 Larger.

5.43 Smaller sample size means larger margin of error.

5.45 Chicago; in general, margin of error decreases as the sample size increases.

5.47 Parameter, statistic.

5.49 0.004.

5.51 (a) 0.69. The proportion of teens aged 13 to 17 nationwide who have received personal messages online from people they don't know. (b) We are 95% confident that the proportion of teens aged 13 to 17 nationwide who have received personal messages online from people they don't know is between 66% and 72%.

5.53 (a) Quick method: 0.02, 0.03. (b) We are 95% confident that the proportion of American adults who play computer or video games is between 38% and 42%. Of those American adults who play electronic games, we are 95% confident that between 7% and 13% play for 10 or more hours per week. (c) Possible answers: sampling method (how were the adults selected, and are they representative of the population?) and inaccurate responses. (d) Answers will vary.

5.55 For example, undercoverage or nonresponse.

5.57 **(a)** 11.3%. **(b)** 2.6%; yes. **(c)** No, since a voluntary response sample was used.

5.59 Answers will vary.

5.61 The margin of error depends on the sample size, not the population size.

5.63 **(a)** The sample was chosen in stages; it is not an SRS of all Europeans. **(b)** The 17 countries. **(c)** There is random probability sampling in at least one stage of the sampling process.

5.65 Closed questions allow for easier tabulation of results but may omit response choices that individuals would prefer to select. Open questions give individuals more freedom to respond, but the tabulation of results could then be more difficult.

5.67 **(a)** 69, 169, 269, 369, and 469. **(b)** 1/100. **(c)** Not every sample of five addresses is equally likely to be chosen.

5.69 **(a)** 40. **(b)** Stratified random samples, to allow for separate conclusions about males and females.

5.71 **(a)** We are told that each region was proportionately represented in the sample. **(b)** An SRS could result in heavy representation from some areas and light (or no) representation from other areas. This could lead to a biased estimate.

5.73 **(a)** 0.004. **(b)** Some people probably said that they voted when they actually didn't.

5.75 **(a)** Sampling error because of undercoverage. **(b)** Nonsampling error due to nonresponse. **(c)** Sampling error due to convenience sample.

5.77 **(a)** You are sampling only from the lower-priced ticket holders. **(b)** Undercoverage is a sampling error.

5.79 **(a)** Population of West Lafayette, Indiana. **(b)** Larger, since this voluntary response poll probably overrepresents those who are opposed to the one-way street.

5.81 **(a)** Assign labels 0001 through 3500 and use random digits. **(b)** Randomly select 1 of the first 14 students, then take every 14th student from that point on. **(c)** Choose SRSs of size 200 and 50 from each group. **(d)** Possible answer: choose a stratified sample to ensure that attitudes of bussed and local students are represented fairly.

5.83 The first question pictures a pugnacious president "fighting" to save his job; the second portrays a more reasonable alternative: allow a process to occur that will determine the outcome.

5.85 **(a)** Adults who used the Internet in the week prior to the survey. **(b)** 0.041. We are 95% confident that the true proportion of all adults who used the Internet in the week prior to the survey and who believe that the Internet has made their life "much better" is in the interval 0.22 to 0.30.

Chapter 6

6.1 **(a)** An experiment because the researcher imposes a treatment: one of two price histories. **(b)** The explanatory variable is whether the student sees a steady price or temporary price cuts. The response variable is the price the student would expect to pay. The subjects are the students in the course. The treatments are the two price histories.

6.3 A mother's prior attitude toward her baby, which probably influenced her choice of feeding method, is the lurking variable; the prior attitude is confounded with the effect of the nursing.

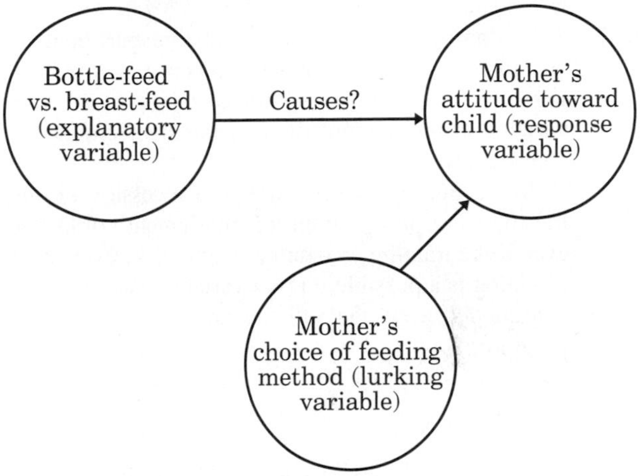

6.5 **(a)** Therapeutic touch practitioners; hand position (left or right); choice of hand (correct or incorrect). **(b)** The setting lacked realism; conditions were not similar to those typically faced by therapeutic touch practitioners.

6.7 **(a)** The subjects were 22,071 male physicians; the explanatory variable is medication (aspirin or placebo); the response variable is whether or not the subject had a heart attack.
(b)

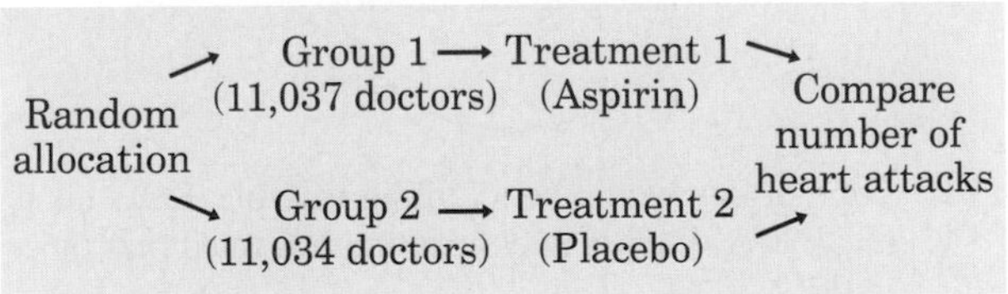

6.9 **(a)** Randomly allocate the 200 rooms to two groups of 100 rooms. First group gets the flat rate; second group gets the varied rates. Compare Internet telephone usage.

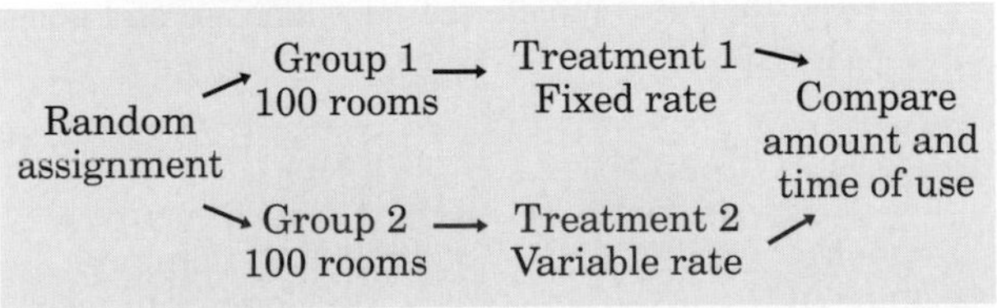

(b) Rooms 119, 033, 199, 192, and 148 are the first five selected.

6.11 Possible answer: students registering for a course should be randomly assigned to a classroom or online version of the course. Scores on a standardized test can then be compared.

6.13 The difference in blood pressures was so great that it was unlikely to have occurred by chance (if calcium is not effective).

6.15 The effects of factors other than the nonphysical treatment have been eliminated or accounted for, so the differences in improvement observed between the subjects can be attributed to the differences in treatments.

6.17 This is an observational study, not an experiment, so we are unable to measure the effects of possible lurking and/or confounding variables. A possible lurking variable is the amount of parental involvement the teens receive.

6.19 If the economy worsened due to a recession during the five-year period, then unemployment could rise even if the training program was effective. Consumer spending is a possible lurking variable that would be confounded with the effectiveness of the training program.

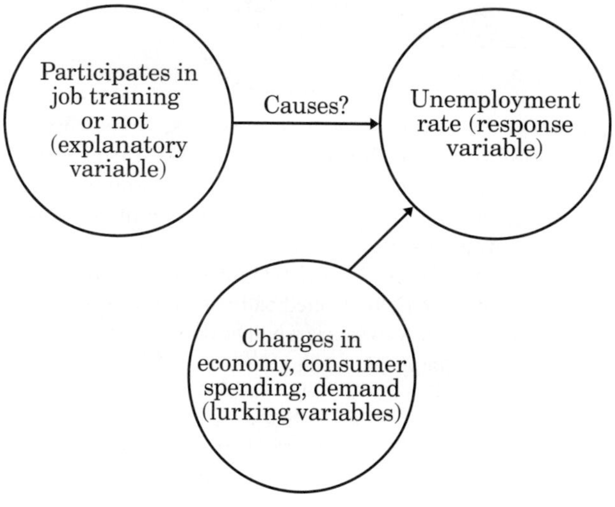

6.21 Time of day would be a lurking variable. Students in the 8:30 A.M. class may be more motivated that those in the 2:30 P.M. class.

6.23 "Significant" means that a difference as large as the one we observed is unlikely to occur by chance.

6.25 Randomly assign 25 students to Group 1 (Brochure A) and 25 students to Group 2 (Brochure B). Have every student complete the questionnaire and compare responses.

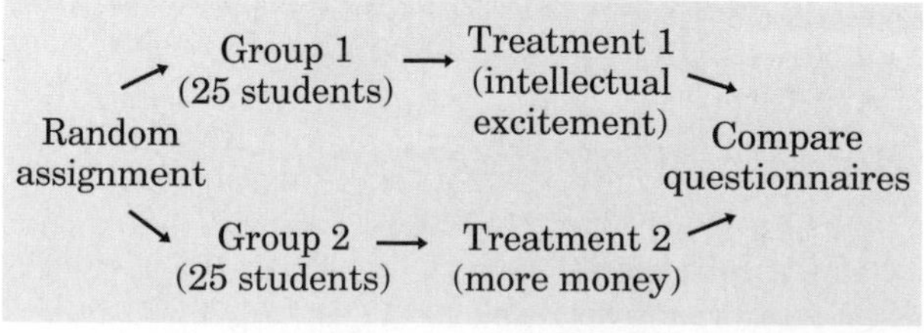

6.27 "Placebo-controlled" means that the control group received a placebo. "Double-blind" means that neither the patients nor those who interacted with them during the experiment knew who received hydroxyurea and who received the placebo.

6.29 **(a)** The placebo effect. **(b)** Use a three-treatment, completely randomized design. **(c)** No. **(d)** Since the patient assesses the effectiveness of the treatment, the experiment does not need to be double-blind.

6.31 **(a)** No. It's possible that women who see an ultrasound picture are more likely to take better care of themselves during pregnancy. **(b)** No. Without a placebo, it's unknown whether the results are due to the treatment or the idea of the treatment. **(c)** A blinded experiment using a fake ultrasound as the placebo.

6.33 **(a)** Difference in score (after minus before) on the Beck Depression Inventory. **(b)** Give the Beck Depression Inventory to all subjects before and after. Randomly assign subjects into three groups of size 110. Group 1 gets Saint-John's-wort, Group 2 gets placebo, and Group 3 gets Zoloft. **(c)** Use a double-blind design.

6.35 For example, do rat tumors arising from high exposures for a relatively short period indicate that humans would get tumors from low doses for a longer period?

6.37 **(a)** Subjects are the pieces of dirty laundry; the explanatory variables are the brands of laundry detergent and the water temperature; the response is a cleanliness score. **(b)** Four treatments. Thirty pieces.

	Detergent Brand	
Temp.	**A**	**B**
Hot	Treatment 1	Treatment 2
Cold	Treatment 3	Treatment 4

(c) Randomly assign 30 pieces of dirty laundry to each of the four treatment combinations.

6.39 **(a)** Answers will vary on the random assignment. **(b)** Label the five varieties A, B, C, D, and E. Let the digits 1 to 5 correspond to the letters. Go to any line of Table B, select the first four unique digits between 1 and 5 and assign the corresponding letters to the rows, west to east. The remaining variety will be assigned to the easternmost plot of the row. Use a new line in Table B for each row.

6.41 **(a)** A block, because the diagnosis is an existing difference between subjects. **(b)** A treatment that is being randomly assigned to the subjects.

6.43 **(a)** In order to account for an effect due to whether a particular step height was used first or second. **(b)** Answers will vary.

6.45 Use a matched pairs (block) design with each dummy as a block. Randomize the order of use of the two air bags. Repeat several times if possible with each dummy.

6.47 **(a)** The explanatory variables are storage and cooking. Storage levels are fresh, a month at room temperature, or a month refrigerated. Cooking levels are cooked immediately or cooked after an hour at room temperature. The response variables are the ratings of the color and flavor. (b) Prepare a random selection of potatoes for each of the ways described in (a). Rate and compare the treatment groups in terms of flavor and color. (c) The different combinations of treatments will be presented in a random order.

6.49 (a) Nine treatments: 500°-front, 500°-middle, . . . , 1000°-back. (b) Randomly assign five converters to each of the nine groups. (c) Converters 19, 22, 39, 34, and 05 get 500°-front.

6.51 (a) This is a matched pairs design. Blocking accounts for any possible effects of the treatment order on the response. (b) To avoid bias that may result from giving all subjects the same treatment first. (c) Yes.

6.53 (a) The subjects should be told what kinds of questions the survey will ask and about how much time it will take. (b) For example, respondents may wish to contact the organization if they feel they have been treated unfairly by the interviewer. (c) Respondents should not know the poll's sponsor (responses might be affected), but sponsors may be revealed when results are published without affecting the results.

6.55 Answers will vary.

6.57 This offers anonymity, since names are never revealed.

6.59 Answers will vary.

6.61 Answers will vary.

6.63 Answers will vary.

6.65 Answers will vary.

6.67 Answers will vary.

6.69 Answers will vary.

6.71 Answers will vary.

6.73 (a) Use a completely randomized design with equal numbers of patients assigned to each of the six possible treatments.

Method of administration	Dosage 5	Dosage 10
Injection	A	B
Skin patch	C	D
Intravenous drip	E	F

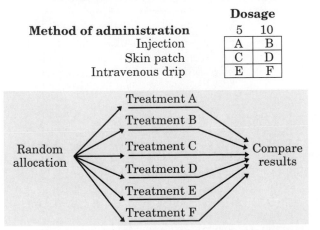

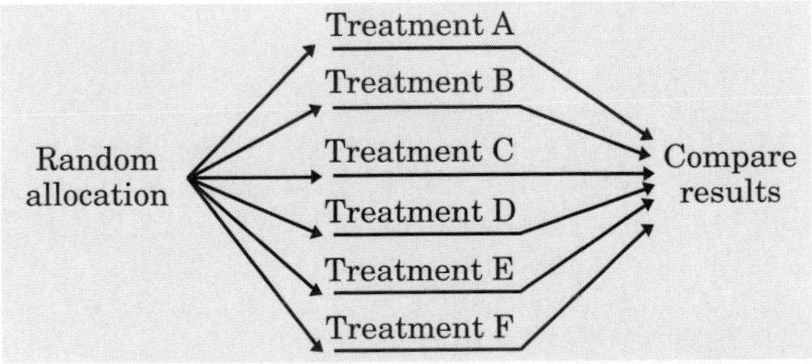

(b) With more subjects there is less chance variation in the results and a greater likelihood of detecting actual differences between treatments. (c) Block by gender, then randomly assign about equal numbers to the six treatments within each block.

6.75 (a) Type of treatment; survival time. (b) No, this is an observational study; we are examining available data. (c) The effect of the surgery is confounded with the initial health of the patient. (d) Randomly assign half the patients to surgery and the other half to nonsurgical treatment.

6.77 Matched pairs design. Randomize the order of left- and right-hand thread for each subject.

6.79 4561 healthy, working adults; flu vaccine administered; number of days of lost work, number of days with health care provider visits.

6.81 Institutional review board, informed consent, and confidentiality.

Chapter 7

7.1 Answers will vary. One example: in 25 spins we obtained 16 tails and 9 heads. Thus: (a) $P(\text{heads}) = 0.36$. (b) $P(\text{tails}) = 0.64$.

7.3 $21/200 = 0.105$.

7.5 1/6.

7.7 Answers will vary.

7.9 No. It's simply one of several thousand "surprising" random phenomena that happen to occur.

7.11 (a) Answers will vary. For example, almost all passengers in an airplane crash are killed, but many people survive automobile accidents. (b) Answers will vary. For example, news coverage gets public attention. The news fixation with airplane crashes can lead to the belief that airplane flying is more dangerous.

7.13 (a) Assign Democrats to 0 to 4 and Republicans to 5 to 9. (b) Assign Democrats to 0 to 5 and Republicans to 6 to 9. (c) Assign Democrats to 0 to 3, Republicans to 4 to 7, and undecideds to 8 and 9. (d) Assign 00 to 52 to Democrats and 53 to 99 to Republicans.

7.15 (a) Four Democrats and six Republicans. (b) Three Democrats and seven Republicans. (c) Two Democrats, four Republicans, and four undecided. (d) There are six Democrats and four Republicans.

7.17 (a) The face of the card obtained on each draw is independent of the other draws. Each card in the deck has the same probability of being drawn. (b) Assign the digits 00 to 51 to represent the 52 cards in a deck, ignore the digits 52 to 99. Let 00, 01, 02, and 03 represent the four aces; the assignment of the remaining 48 cards is immaterial. (c) Answers will vary. Beginning on line 101 and moving to the next new line for each repetition, we find that an ace is obtained in 10 draws in 6 of the 10 repetitions, giving an estimate of 0.6.

7.19 (a) The roulette wheel has no memory. The probability of getting a red on the next trial is the same as on each of the prior trials. (b) The gambler is wrong. The probability of a red or black card coming up next depends on what cards have already been drawn. That is, the deck does not have a memory.

7.21 When the weather conditions are like those seen today, it has rained the following day about 30% of the time.

7.23 Answers will vary. However, 140 heads in 250 spins is quite unusual for a fair coin.

7.25 $31/365 \approx 0.085$, assuming that it is equally likely for a person to be born on any day of the year.

7.27 **(a)** HHHH HHHT HHTH HTHH THHH HHTT HTHT HTTH THHT THTH TTHH HTTT THTT TTHT TTTH TTTT. **(b)** 1/16. **(c)** $P(A) = 4/16 = 0.25$.

7.29 $P(D) = 6/36 = 1/6$; $P(M) = 1 - P(\text{not M}) = 1 - 3/36 = 33/36$; $P(R) = 15/36$.

7.31 **(a)** Since the events are deemed to be mutually exclusive, the probability is $0.45 + 0.23 = 0.68$. **(b)** $1 - 0.68 = 0.32$.

7.33 **(a)** $P(C) = 40/52$. $P(D) = 39/52$. **(b)** Drawing a spade, club, or diamond that is not a face card. $P(C \text{ and } D) = 30/52$. **(c)** Drawing a spade, club, or diamond OR drawing a heart that is not a face card. $P(C \text{ or } D) = (39 + 10)/52 = 49/52$ or $P(C \text{ or } D) = 40/52 + 39/52 - 30/52 = 49/52$.

7.35 **(a)** The 100 U.S. senators. The senators' gender and political affiliation. **(b)** $P(\text{Democrat}) = (40 + 11)/100 = 51/100$. $P(\text{female}) = (11 + 5)/100 = 16/100$. $P(\text{female and Democrat}) = 11/100$. $P(\text{female or Democrat}) = 51/100 + 16/100 - 11/100 = 56/100$.

7.37 **(a)**

	Black	Not black	Total
Even	10	10	20
Not even	8	10	18
Total	18	20	38

(b) $P(B) = 18/38$. $P(E) = 20/38$. **(c)** The event that the ball lands in a black, even spot. $P(B \text{ and } E) = 10/38$. **(d)** The event that the ball lands in either a black spot or an even spot. $P(B \text{ or } E) = 18/38 + 20/38 - 10/38 = 28/38$.

7.39

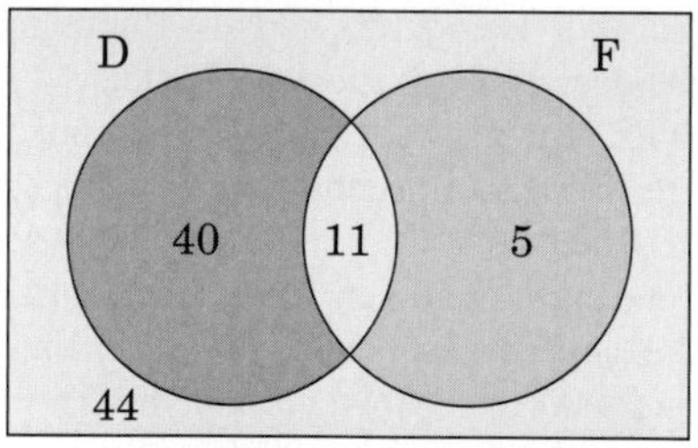

7.41

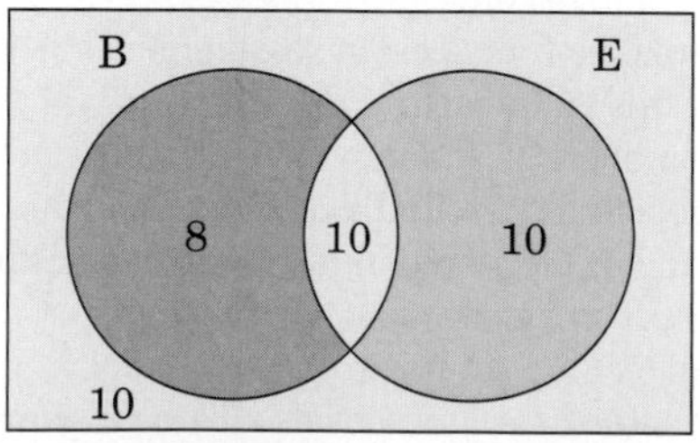

7.43 **(a)**

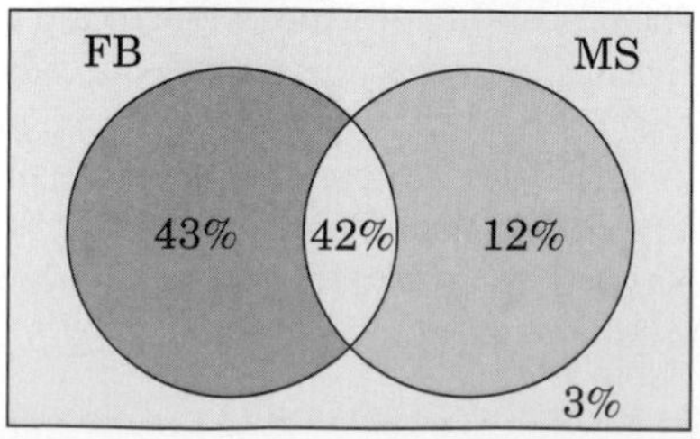

(b) $MS^c \cup FB$. **(c)** $P(MS^c \cup FB) = (1 - 0.54) + 0.85 - 0.43 = 0.88$.

(d)

	MySpace	
Facebook	Yes	No
Yes	0.42	0.43
No	0.12	0.03

7.45 **(a)** The surveyed students in fourth, fifth, and sixth grade. Which grade the student is in and which is most important to the student: good grades, athletic ability, or being popular. **(b)** $P(\text{sixth grader}) = 135/335 \approx 0.40$. $P(\text{good grades most important}) = 168/335 \approx 0.50$. $P(\text{sixth-grader and good grades most important}) = 69/335 \approx 0.21$. $P(\text{sixth-grader or good grades most important}) = 135/335 + 168/335 - 69/335 \approx 0.70$. **(c)** There are more than two responses for each variable measured.

7.47

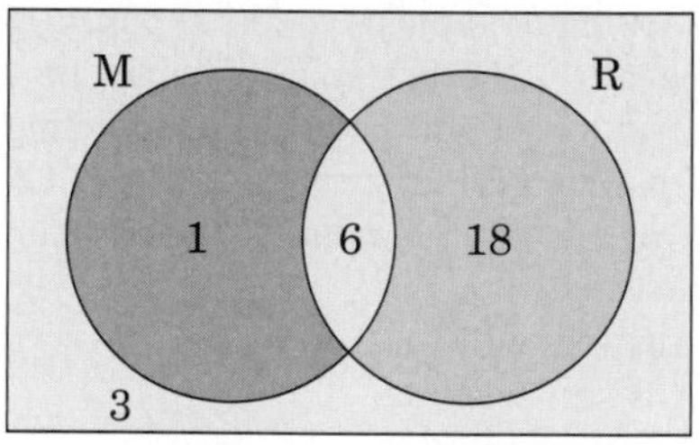

7.49 **(a)** AA AB BA BB.
(b)

Blood type	Probability
A	$1/4 = 0.25$
AB	$2/4 = 0.5$
B	$1/4 = 0.25$

7.51 **(a)** $P(B \text{ or } O) = P(B) + P(O) = 0.54$.
(b)

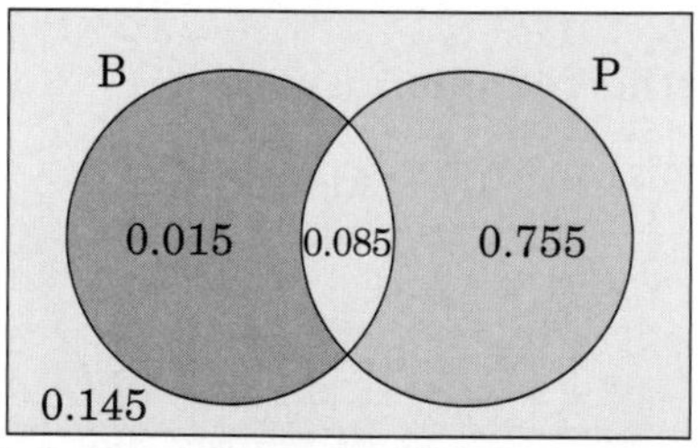

7.53 **(a)** $P(D|F) = 11/16$. $P(D|F)$ is the probability that a randomly selected senator is a Democrat, given that the senator is female. **(b)** $P(F|D) = 11/51$. $P(F|D)$ is the probability that a randomly selected senator is female, given that the senator is a Democrat. **(c)** No, $P(D \cap F) = 11/100 = 0.11 \neq 0$. **(d)** No, $P(D|F) = 11/16 \neq P(D) = 51/100$.

7.55 Answers will vary. For example, define A to be the event that the red die is even and B to be the event that the green die is even. Then $P(A \cap B) = 9/36 = 1/4$ is equal to $P(A) \cdot P(B) = 1/2 \cdot 1/2 = 1/4$.

7.57 Answers will vary.

7.59 **(a)**

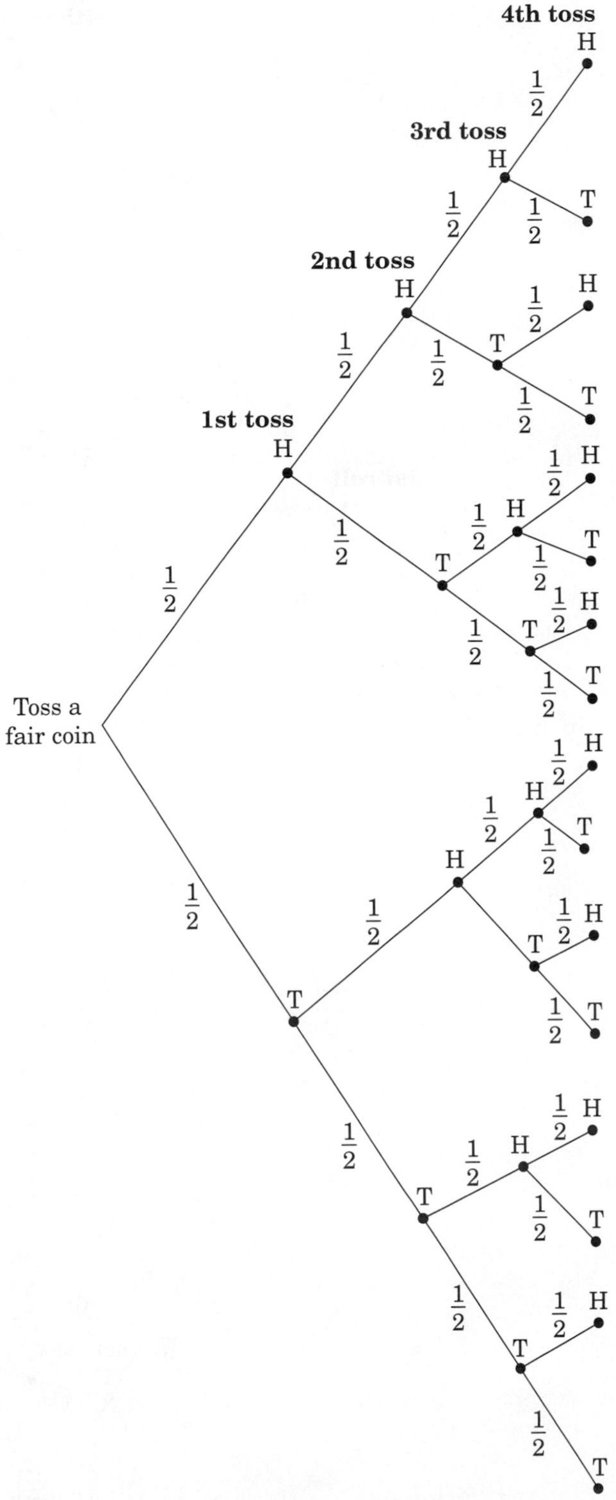

(b) 15/16. Only 1 out of 16 outcomes has no heads.

7.61 **(a)**

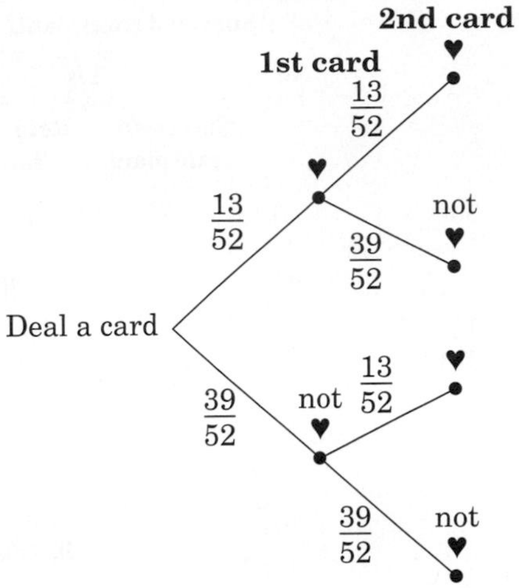

(b) $P(\text{A} \cap \text{B}) = (13/52)(13/52) = 0.0625$. **(c)** Yes, since we are drawing with replacement. $P(\text{B}|\text{A}) = 13/52 = P(\text{B})$.

7.63 $P(\text{MS}|\text{FB}) = P(\text{MS} \cap \text{FB})/P(\text{FB}) = 0.42/0.85 = 0.49$.

7.65 **(a)**

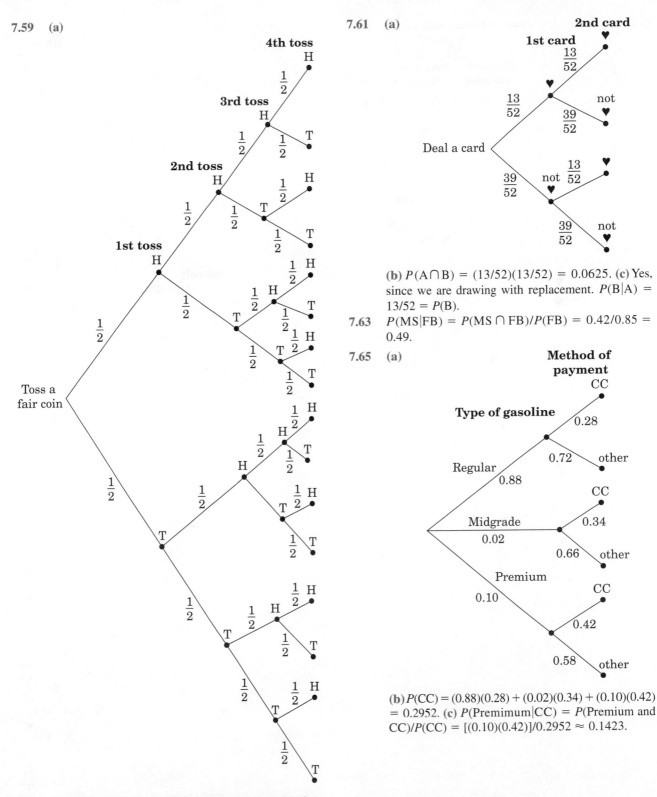

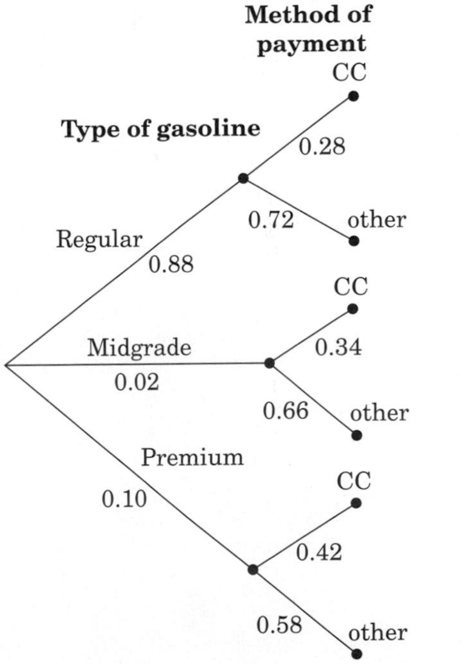

(b) $P(\text{CC}) = (0.88)(0.28) + (0.02)(0.34) + (0.10)(0.42) = 0.2952$. **(c)** $P(\text{Premimum}|\text{CC}) = P(\text{Premium and CC})/P(\text{CC}) = [(0.10)(0.42)]/0.2952 \approx 0.1423$.

7.67

	Survived transplant?			
	No	Yes		
		Successful transplant	Return to dialysis	Total
Survive 5 years	0	378	180	558
Die within 5 years	100	162	180	442
Total	100	540	360	1000

7.69 **(a)**

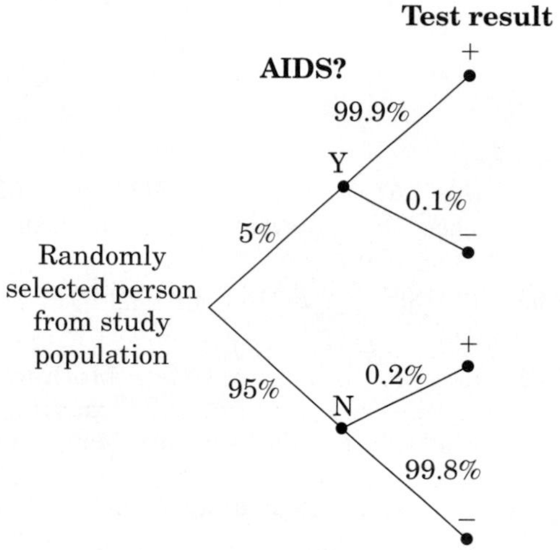

(b) $P(+) = (0.05)(0.999) + (0.95)(0.002) = 0.05185$.
(c) $P(\text{HIV}|+) = P(\text{HIV and } +)/P(+) = [(0.05)(0.999)]/0.05185 \approx 0.963$.

7.71 $P(\text{B}|\text{T}) < P(\text{B}) < P(\text{T}) < P(\text{T}|\text{B})$.

7.73 **(a)**

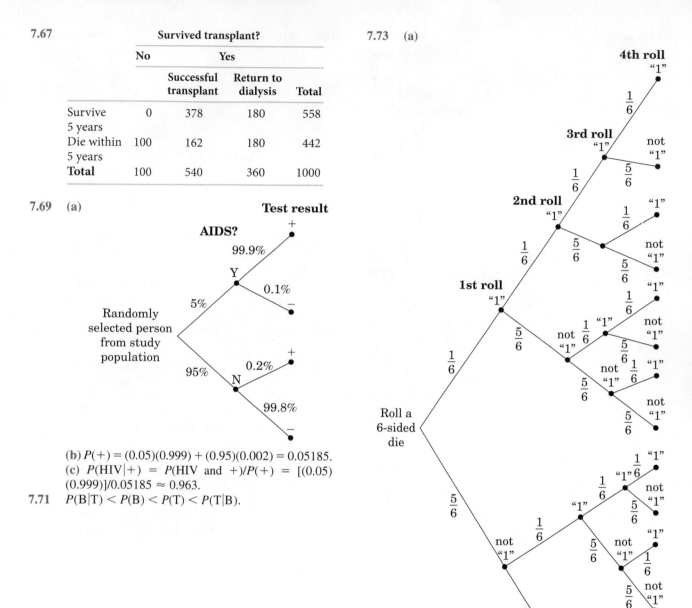

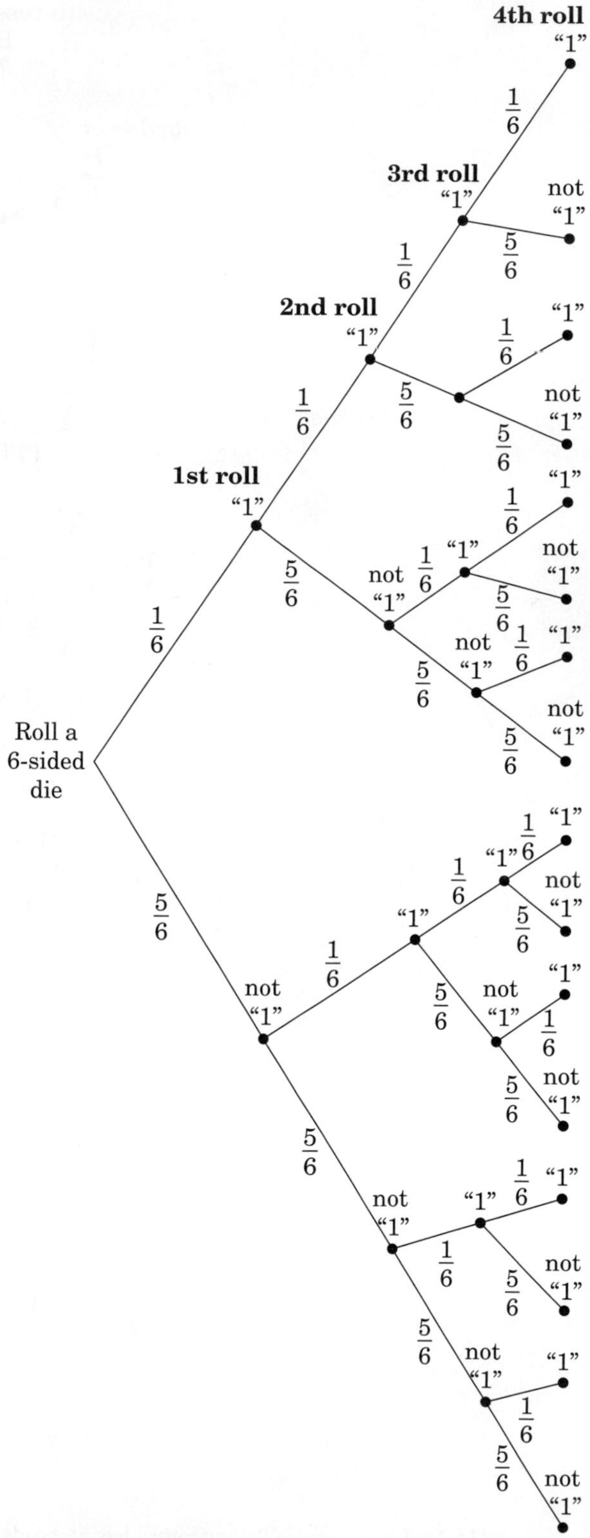

(b) $P(\text{at least one "1"}) = 1 - P(\text{no "1"s}) = 1 - (5/6)(5/6)(5/6)(5/6) \approx 0.52$.

7.75

	Cat?		
Dog?	**Yes**	**No**	**Total**
Yes	15	15	30
No	5	5	10
Total	20	20	40

7.77 **(a)** P(different birthdays) = P(2nd individual was not born on the 1st individual's birthday) = $364/365 \approx 0.9973$. P(same birthdays) = P(2nd individual was born on the 1st individual's birthday) = $1/365 \approx 0.0027$. **(b)** P(3 different birthdays) = $(364/365)(363/365) \approx 0.9918$. P(at least two have the same birthday) = 1 − P(3 different birthdays) ≈ 0.0082. **(c)** P(30 different birthdays) $= \dfrac{364 \cdot 363 \cdot \ldots \cdot 336}{365^{29}} \approx 0.2937$. P(at least 2 of 30 have the same birthday) = 1 − P(30 different birthdays) ≈ 0.7063.

7.79 **(a)** The winning die is boldfaced for each outcome.

Your 6-sided die	**Teacher's 8-sided die** 1	2	3	4	5	6	7	8
1	(1, 1)	(1, **2**)	(1, **3**)	(1, **4**)	(1, **5**)	(1, **6**)	(1, **7**)	(1, **8**)
2	(**2**, 1)	(2, 2)	(2, **3**)	(2, **4**)	(2, **5**)	(2, **6**)	(2, **7**)	(2, **8**)
3	(**3**, 1)	(**3**, 2)	(3, 3)	(3, **4**)	(3, **5**)	(3, **6**)	(3, **7**)	(3, **8**)
4	(**4**, 1)	(**4**, 2)	(**4**, 3)	(4, 4)	(4, **5**)	(4, **6**)	(4, **7**)	(4, **8**)
5	(**5**, 1)	(**5**, 2)	(**5**, 3)	(**5**, 4)	(5, 5)	(5, **6**)	(5, **7**)	(5, **8**)
6	(**6**, 1)	(**6**, 2)	(**6**, 3)	(**6**, 4)	(**6**, 5)	(6, 6)	(6, **7**)	(6, **8**)

(b) P(A) = 27/48 = 0.5625. **(c)** $P(A \cup B)$ = 27/48 + 8/48 − 5/48 = 30/48 = 0.625. **(d)** No. $P(A|B)$ = 5/8 = 0.625 ≠ P(A) = 0.5625.

7.81 **(a)** 1 − 0.3 − 0.3 − 0.3 = 0.1. **(b)** Top 10%: 0, 1, 2. Top quarter but not top 10%: 3, 4, 5. Top half but not top quarter: 6, 7, 8. Bottom half: 9.

7.83 **(a)** P(smokes) = 31/100 = 0.31. **(b)** P(smokes|male) = 19/60 ≈ 0.32. **(c)** P(smokes) ≠ P(smokes|male) but they are very close. It's possible that these events are independent but that the sample does not give an exact equality for these probabilities.

7.85 P(safe launch) = P(all six o-rings function properly) = $(0.977)^6 \approx 0.87$.

7.87 Answers will vary. One example: Define A to be the event that an even digit less than 8 is selected (0, 2, 4, 6) and B to be the event that an odd digit greater than 1 is selected (3, 5, 7, 9). These events cannot occur simultaneously and are therefore mutually exclusive, but together they do not contain the entire sample space and thus cannot be complements.

Chapter 8

8.1 **(a)** D is a random variable because it takes numerical values (0 to 9) that describe the outcomes of randomly selecting digits from Table B or using your calculator.

(b)

Value d_i	0	1	2	3	4	5	6	7	8	9
Probability p_i	1/10	1/10	1/10	1/10	1/10	1/10	1/10	1/10	1/10	1/10

(c)

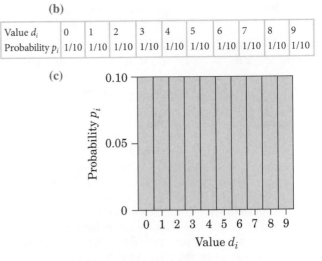

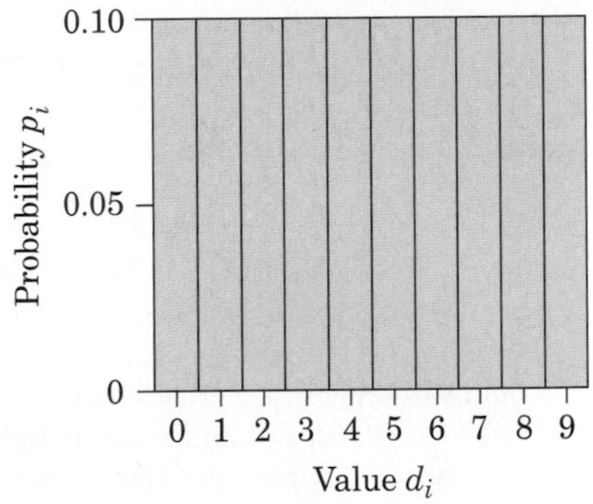

8.3 $P(Y \le 3) = 0.022$.

8.5 The expected value of D is 4.5. In the long run, the average value of the randomly chosen digits will be 4.5.

8.7 **(a)** If boys and girls are equally likely to be born, a string of 16 boys is quite unusual (1/65,536), so such a run at any small hospital should be newsworthy. However, given the large number of births across the nation, it is likely that *some* hospital is going to have a run of 16 boys in a row, so the fact that it happened somewhere isn't all that surprising. **(b)** The law of averages says nothing about short-run streaks. The overall proportion of boys and of girls over time will be about 50%, but streaks of a particular gender, even of length 16 and certainly of length 5, are to be expected.

8.9 It is *possible* that exactly 1000 will be women but it won't always be true, so statement (a) is false. With a sample of size 2000, we would expect the count of women to be quite close to the theoretical value of 1000, so (b) is true.

8.11 **(a)** Table B: Assign each day of the year a three-digit number, 001 to 365. Ignoring 000 and 366 to 999, proceed through the table noting three-digit numbers in the range until you have obtained 30 birth dates. Check for duplicates. Calculator: `randInt (1,365,30)` → `L1`. Check for duplications. **(b)** Answers will vary. For example, in one set of 20 repetitions of the calculator simulation, we obtained 15 samples that contained a duplication. You probably shouldn't take the bet.

8.13 **(a)** 50% of samples will have a sample proportion of 0.15 or less since half of the area under the curve is to the left of 0.15. **(b)** About 68%. **(c)** 0.32. **(d)** Yes. The probability of getting at least 17% is 0.0132.

8.15 **(a)** 4.

(b)

Sample	{1,4}	{1,9}	{2,4}	{2,9}	{4,9}
$\bar{x}$	2.5	5	3	5.5	6.5

(c)

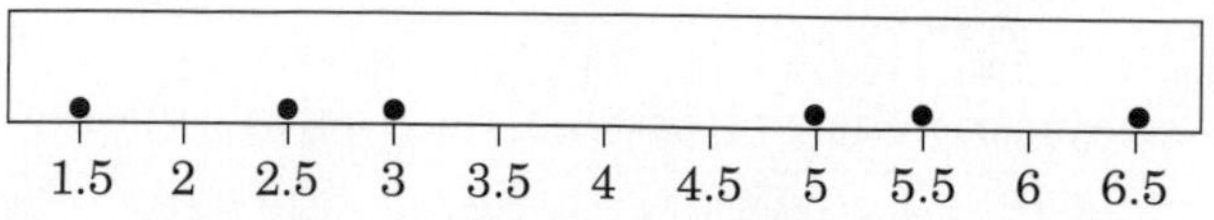

(d) 4. They are the same.

8.17 $P(X > 112) = P(z > (112 - 100)/15 = 0.8) = 0.2119$.

8.19 **(a)** 20/132. The probability of getting an ice cream bar for less than \$0.10 is approximately 0.15. **(b)** $P(X <$ \$0.40) = 45/132. **(c)** If all of the 132 balloons are popped, the store makes \$43.96. Without the promotion, the store makes \$52.80. The store would make more money without the promotion, assuming they would still sell 132 ice cream bars.

8.21 The *expected price* of an ice cream bar is \$0.33. It is to her benefit to play the raffle since her expected price is less than the usual price of \$0.40.

8.23 **(a)**

y	$P(Y = y)$
\$300	0.9998
−\$199,700	0.0002

(b) The expected value of Y = \$260. On average, the insurance company expects to earn \$260 for each home insurance policy it sells.

8.25 $z = (0.51 - 0.47)/0.016 = 2.5$. $P(Z > 2.5) = 0.0062$.

8.27 **(a)** Letters only: $26 \cdot 26 \cdot 26 = 17{,}576$ possibilities. **(b)** Numbers or letters: $36 \cdot 36 \cdot 36 = 46{,}656$ possibilities.

8.29 1959: $26 \cdot 26 \cdot 26 \cdot 10 \cdot 10 \cdot 10 = 17{,}576{,}000$ different license plates. 1999: $26 \cdot 26 \cdot 26 \cdot 10 \cdot 10 \cdot 26 = 45{,}697{,}600$ different license plates.

8.31 **(a)** $6! = 720$. **(b)** $5! = 120$.

8.33 **(a)** $28 \cdot 28 = 784$. **(b)** ${}_{28}P_2 = 756$. **(c)** ${}_{28}C_2 = 378$. **(d)** ${}_{28}C_2 + 28 = 378 + 28 = 406$.

8.35 **(a)** ${}_{10}P_3 = 720$. **(b)** $1 - 0.72 = 0.28$ (this is 1 minus the probability that the winning number has 3 different digits).

8.37 **(a)** ${}_6P_0 = 1$. There is one way to arrange none of the 6 items. **(b)** ${}_6P_3 = 120$ and ${}_6C_3 = 20$. Answers will vary. Note that once the 3 friends are chosen (20 possibilities) there are 6 $(3 \cdot 2 \cdot 1)$ ways to order the three positions so that there are $20 \cdot 6 = 120$ permutations of 3 out of the 6. **(c)** ${}_6C_2 = 15$ and ${}_6C_4 = 15$. These values are the same because once you pick 2 friends out of the 6, the remaining 4 are automatically grouped.

8.39 **(a)** $26 \cdot 26 = 676$. **(b)** $2 \cdot 26 \cdot 26 = 1352$.

8.41 **(a)** $28 \cdot 28 = 784$. **(b)** ${}_{28}P_2 = 756$. **(c)** $756/784 \approx 0.96$.

8.43 $1/{}_{49}C_6 \approx 0.00000007$.

8.45 **(a)** $6/(6 \cdot 6 \cdot 6 \cdot 6 \cdot 6) = 6/7776 \approx 0.00077$. **(b)** There are $6 \cdot 5 \cdot 4 \cdot 3 \cdot 2 = 720$ rolls such that all five dice show a different number of spots on the up-faces. The probability is therefore $(6 \cdot 5 \cdot 4 \cdot 3 \cdot 2)/(6 \cdot 6 \cdot 6 \cdot 6 \cdot 6) = 720/7776 \approx 0.09$.

8.47
- Binary? Success = seed germinates. Failure = seed does not germinate.
- Independent? Possibly, although it's possible that if one seed does not germinate, it's more likely that others in the packet will not grow either.
- Number? 20 seeds
- Success? The probability that each seed germinates is 85%, assuming the advertised percent is true.

Assuming that independence does hold, this is a binomial setting.

8.49
- Binary? Success = person is left-handed. Failure = person is right-handed.
- Independent? Since students are selected randomly, their handedness is independent.
- Number? There is not a fixed number of trials for this chance process since you continue to sample until you find a left-handed student.

Since the number of trials is not fixed, this is not a binomial setting.

8.51 $P(X = 6) = {}_6C_6(1/6)^6(5/6)^{6-6} \approx 0.00002$.

8.53 **(a)** $P(X = 7) = {}_7C_7(0.44)^7(0.56)^{7-7} \approx 0.0032$. **(b)** $P(X = 4) = {}_7C_4(0.44)^4(0.56)^{7-4} \approx 0.2304$.

8.55 **(a)** $n \cdot p = (20)(0.85) = 17$ seeds. **(b)** $P(X = 17) = {}_{20}C_{17}(0.85)^{17}(0.15)^{20-17} \approx 0.2428$. **(c)** $P(X < 17) = 1 - P(X \geq 17) = 1 - 0.6477 = 0.3523$.

8.57 Let X represent the number of heads observed in the six tosses. **(a)** $P(X = 3) = {}_6C_3(0.5)^3(0.5)^{6-3} = 0.3125$. **(b)** $P(X = 4) = {}_6C_4(0.5)^4(0.5)^{6-4} = 0.2344$.

8.59 **(a)**
- Binary? Success = owner greets the dog first. Failure = owner does not greet the dog first.
- Independent? One owner's greeting habits do not affect the greeting habits of other owners.
- Number? 12 dog owners are randomly selected.
- Success? The probability of greeting the dog first remains constant from one owner to the next, 0.66.

This is a binomial setting.

(b) $P(X = 7) = {}_{12}C_7(0.66)^7(0.34)^{12-7} = 0.1963$. **(c)** $P(X \geq 5) = 1 - P(X \leq 4) = 1 - 0.0213 = 0.9787$.

8.61 Let X = number of left-handed students in the SRS of 15 students. **(a)** $P(X = 3) = {}_{15}C_3(0.1)^3(0.9)^{15-3} = 0.1285$. **(b)** $P(X \leq 3) = 0.1285 + 0.2669 + 0.3432 + 0.2059 = 0.9445$. **(c)** $P(X \geq 4) = 1 - P(X \leq 3) = 1 - 0.9445 = 0.0555$. Since the probability of this event is quite small, it would be surprising for the SRS to contain 4 or more left-handed students. Note that the probability of obtaining exactly 4 left-handed students is 0.0428.

8.63 Let X = number of orange M&M's in a sample of size 8. **(a)** $P(X = 0) = {}_8C_0(0.2)^0(0.8)^{8-0} = 0.1678$, not surprising. **(b)** $P(X = 2) = {}_8C_2(0.2)^2(0.8)^{8-2} = 0.2936$, not surprising. **(c)** $P(X = 4) = {}_8C_4(0.2)^4(0.8)^{8-4} = 0.0459$, surprising.

8.65 Note that X = the number of times that the ball falls to the right is a binomial random variable with $n = 6$

and $p = 0.6$. (a) $P(X = 0) = {}_6C_0(0.6)^0(0.4)^{6-0} = 0.0041$. (b) $P(X = 2) = {}_6C_2(0.6)^2(0.4)^{6-2} = 0.1382$. (c) Slot E.

Slot	Value x	Probability p_i
A	0	0.0041
B	1	0.0369
C	2	0.1382
D	3	0.2765
E	4	0.3110
F	5	0.1866
G	6	0.0467

8.67 It's likely that they mean that there are n types of pasta and m types of sauce where $nm = 42$.

8.69 (a) $10^7 = 10{,}000{,}000$. (b) $8 \cdot 10^6 = 8{,}000{,}000$. (c) There are $7 \cdot 9^6 = 3{,}720{,}087$ phone numbers with no 9s. It follows that there are $8{,}000{,}000 - 3{,}720{,}087 = 4{,}279{,}913$ phone numbers with at least one 9.

8.71 (a)

- Binary? Success = makes the free-throw basket. Failure = does not make the free-throw basket.
- Independent? Since the probability of making a basket is the same for each free throw, the events are independent.
- Number? 7 free-throw attempts are made.
- Success? The probability of making a free-throw basket remains the same for each attempt, 0.8.

This is a binomial setting.
(b) $P(X \le 4) = 1 - P(X > 4) = 1 - (0.2753 + 0.3670 + 0.2097) = 1 - 0.852 = 0.148$. (c) Assuming that she is an 80% free-throw shooter, the probability that she makes 4 or fewer baskets is 0.148 (part (b)), which is not small enough to raise suspicion.

8.73 (a) By the 68–95–99.7 rule, the probability is 0.95. (b) For 36%, $z = (0.36 - 0.4)/0.015 = -2.67$; and for 44%, $z = (0.44 - 0.4)/0.015 = 2.67$. From Table A, the probability is $2 \cdot (0.0038) = 0.0076$.

8.75 (a) $20/80 = 0.25$.
(b)

Amount gained x_i	\$3	−\$1
Probability p_i	0.25	0.75

(c) The expected value of X is $\$3(0.25) - \$1(0.75) = \$0$.

Chapter 9

9.1 (a) The seniors in her school. (b) It is the true proportion of seniors in Tanya's school who actually plan to attend the senior prom. (c) $\hat{p} = 36/50 = 0.72$.

9.3 It means that the process we used to generate this interval will successfully capture the true population proportion 95% of the time.

9.5 (a) It is the true proportion of women living in the United States (excluding Alaska and Hawaii) who feel that they don't get enough time to themselves.
(b)
$$\hat{p} = \frac{482}{1025} = 0.47$$
$$0.47 \pm 2\sqrt{\frac{(0.47)(0.53)}{1025}} = (0.439, 0.501).$$
(c) We are 95% confident that the true proportion of American women who feel that they are not getting enough time for themselves is in the interval (0.439, 0.501).

9.7 We are 95% confident that the true proportion of women in favor of new laws restricting guns is in the interval (0.62, 0.70). The 95% confidence level means that the process used to generate this interval will successfully capture the true population proportion 95% of the time.

9.9 Since $(1 - 0.98)/2 = 0.01$, z^* for a 98% confidence interval can be found by looking for an area of $1 - 0.01 = 0.99$. The closest area is 0.9901, corresponding to a critical value of 2.33.

9.11 (a) $\hat{p} = 36/50 = 0.72$. A 95% confidence interval is given by
$$0.72 \pm 1.96\sqrt{\frac{(0.72)(0.28)}{50}} = (0.596, 0.844)$$
(b) A 90% confidence interval will be more precise (narrower) but also have lower confidence than either a 95% or a 99% confidence interval. (c) $z^* = 1.645$. The 90% confidence interval is given by
$$0.72 \pm 1.645\sqrt{\frac{(0.72)(0.28)}{50}} = (0.616, 0.824)$$
We are 90% confident that the true proportion of seniors who will attend the prom lies in the interval (0.616, 0.824).

9.13 $z^* = 2.58$. Using $\hat{p} = 0.5$ as our estimate of the population proportion, $n = (2.58/0.03)^2 \cdot \hat{p}(1 - \hat{p}) = (2.58/0.03)^2 \cdot 0.5(0.5) = 1849$.

9.15 $\hat{p} = 180/2004 = 0.0898$. Thus,
$$\text{the margin of error} = 1.96\sqrt{\frac{(0.0898)(0.9102)}{2004}} = 0.0125.$$

9.17 (a) The sampling distribution of $\hat{p}$ is approximately Normal with mean 0.14 and
$$\text{standard deviation} = \sqrt{\frac{(0.14)(0.86)}{600}} = 0.0142$$
(b) $z = (0.182 - 0.14)/0.0142 = 2.96$. $P(Z \ge 2.96) = 0.0015$. It would be very surprising. (c) $z = (0.12 - 0.14)/0.0142 = -1.41$. The probability of obtaining a sample with a sample proportion of 0.12 or less is 0.0793. It would be somewhat unlikely, but not too surprising.

9.19 (a) Undergraduate students taking courses in psychology and communications since this is the population from which the sample was drawn. **(b)** $z^* = 2.58$. A 99% confidence interval is then

$$\hat{p} \pm z^* \sqrt{\frac{\hat{p}(1-\hat{p})}{n}}$$

$$= \frac{107}{127} \pm 2.58\sqrt{\frac{(107/127)[1-(107/127)]}{127}}$$

$$= (0.7591, 0.9259)$$

We are 99% confident that the proportion of undergraduate students from courses in psychology and communications who pray at least a few times a year is between 0.76 and 0.93.

9.21 **(a)** Mean $= p = 1/2$ and

$$\text{standard deviation} = \sqrt{\frac{p(1-p)}{n}} = \sqrt{\frac{0.5(0.5)}{25}} = 0.1$$

(b) $\hat{p} = 16/25 = 0.64$.

(c)

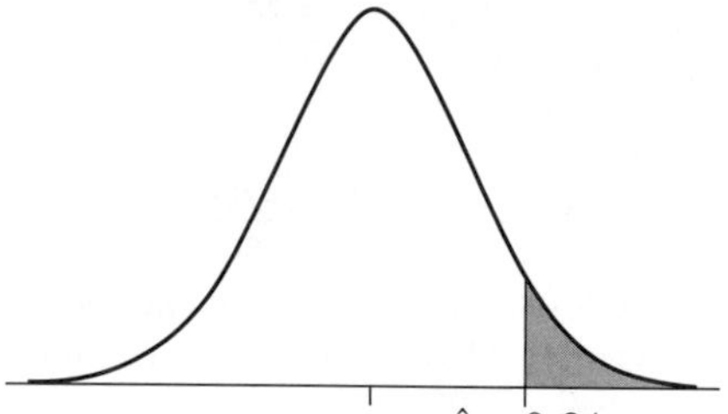

(d) $P(\hat{p} > 0.64) = P\left(z > \frac{0.64 - 0.5}{0.1}\right) = P(z > 1.4)$
$= 1 - 0.9192 = 0.0808$.

It would be somewhat unlikely, but not so small to conclude that the deck isn't fair.

9.23 $P(\hat{p} > 0.56) = P\left(z > \frac{0.56 - 0.5}{0.0707}\right) = P(z > 0.85)$
$= 1 - 0.8023 \approx 0.20$ and

$P(\hat{p} > 0.72) = P\left(z > \frac{0.72 - 0.5}{0.0707}\right) = P(z > 3.11)$
$= 1 - 0.9991 \approx 0.001$

9.25 The parameter is the proportion p of all teenagers who say they rarely or never argue with their friends. $H_0: p = 0.72$. $H_a: p \neq 0.72$.

9.27 **(a)** The parameter is the proportion p of presses of the button that result in a red light. $H_0: p = 0.30$. $H_a: p \neq 0.30$. **(b)** Assuming the null hypothesis is true, the sampling distribution of $\hat{p}$ has mean $= p = 0.30$ and

$$\text{standard deviation} = \sqrt{\frac{p(1-p)}{n}} = \sqrt{\frac{0.3(0.7)}{100}} = 0.0458$$

(c) $\hat{p} = 20/100 = 0.2$. Assuming $p = 0.3$, the probability of obtaining a sample proportion this extreme is

$$2 \cdot P(\hat{p} < 0.2) = 2 \cdot P\left(z < \frac{0.2 - 0.3}{0.0458}\right)$$
$$= 2 \cdot P(z < -2.18) = 0.0292$$

(d) If the proportion of presses of the button that result in a red light is truly 0.3, the probability of observing 20 or fewer red lights in 100 presses is about 1.5%. This is unlikely enough to cause one to suspect that the custom officers' claim is false.

9.29 **(a)** Let p = the proportion of middle school students who engage in bullying behavior. Then $H_0: p = 0.75$ and $H_a: p > 0.75$.

$$\hat{p} = \frac{445}{558} = 0.7975$$

$$z = (0.7975 - 0.75)/\sqrt{\frac{(0.75)(0.25)}{558}} = \frac{0.0475}{0.0183}$$
$$= 2.59 \Rightarrow P\text{-value} = 0.0048$$

Since the P-value is so small, the alternative hypothesis that more than 75% of middle school students engage in bullying behavior is strongly supported. **(b)** Not necessarily. The questionnaire did not define bullying; the students simply responded to questions that supposedly describe bullying. There is also some question as to how well middle school students would understand questions about their behavior over the last 30 days. Additionally, the data come from only one school in the Midwest and may not be representative of all middle schoolers in the United States.

9.31 **(a)** P-value $= P(z < -1.77) = 0.0384$. Since $0.01 <$ P-value < 0.05, we reject the null hypothesis and conclude that the population proportion is less than 0.3 if our level of significance is 5%, but not if it is 1%. **(b)** P-value $= 2 \cdot P(z < -1.77) = 0.0768$. There is insufficient evidence to reject the null hypothesis in favor of the alternative hypothesis at either the 0.01 or a 0.05 significance level.

9.33 The parameter is the proportion p of students at the school who completed their math homework last night. $H_0: p = 0.75$ and $H_a: p < 0.75$.

9.35 **(a)** Significant at the 5% level means that our observed value would occur by chance less than 5% of the time if the null hypothesis were true. This does not mean it will occur by chance less than 1% of the time, which is what it means to be significant at the 1% level. **(b)** This means that the probability of obtaining a sample result as extreme as that observed, assuming the null hypothesis is true, is less than 5%. The null hypothesis is either true or it's not true—there's no probability associated with the actual status of the null hypothesis.

9.37 **(a)** The P-value is one-sided on the high side, P-value $= P(z > 2.36) = 1 - 0.9909 = 0.0091$. **(b)** Since the P-value is less than both 5% and 1%, we reject the null hypothesis and conclude $p > 0.81$ for either a 5% or a 1% level of significance.

9.39 **(a)** **The hypotheses:** Let p be the true proportion of drivers who use cell phones while driving. Then $H_0: p = 0.5$ and $H_a: p < 0.5$. **The sampling distribution:** If the null hypothesis is true, $\hat{p}$ follows a Normal distribution with mean = 0.5 and standard deviation = $\sqrt{[(0.5)(0.5)]/70} = 0.0598$. **The data:** $\hat{p} = 21/70 = 0.30$. **The *P*-value:** $z = (0.3 - 0.5)/0.0598 = -3.34 \Rightarrow P\text{-value} = 0.0004$. **Conclusion:** This is a very small *P*-value and provides strong evidence against the null hypothesis. Based on these results, we conclude that fewer than half of all drivers use cell phones while driving. **(b)** The drivers were aware that they were being taped and might well have allowed themselves fewer distractions than if they were not being taped. Discarding the first three hours probably helped, but it is still likely that driver behavior was influenced by the presence of the cameras.

9.41 The idea of this section that the company is addressing is "the design of the data production matters." This is because the small bags are not truly SRSs from the larger population of M&M's.

9.43 You use sample data to construct confidence intervals for unknown population values. In this case, we know the exact number of presidents who were only children; inference is not necessary.

9.45 The 59 dropouts could have created bias in the results if there was a systematic reason for their dropping out. If the unmotivated participants dropped out, those remaining could have more extreme results than are typical.

9.47 The difference in the results was 11% (23% − 12%), which would be practically significant to a smoker who is trying to quit.

9.49 This study gives some evidence that Zyban is effective at helping people quit smoking. Before drawing a cause-and-effect conclusion, however, we would need to check that the difference in the percents who quit smoking is too large to be explained by chance variation.

9.51 We would expect to incorrectly reject the null hypothesis in about 5%, or 1000(0.05) = 50, of the tests.

9.53 No. This is an observational study, and we have no way of knowing whether taking statin drugs or some other factor reduced the likelihood of developing breast cancer.

9.55 **(a)** No. We would expect about 1% of the sample, five people, to have results that deviate enough by chance to have a *P*-value of less than 0.01. **(b)** The four should be tested again. The chances of their being able to get significant results twice in a row if they do not have ESP are much smaller than 0.01.

9.57 **(a)** The margin of error must get smaller as n gets larger because n is in the denominator. **(b)** As n increases, z will be farther out in the tail of the Normal distribution, and the *P*-value will therefore decrease.

9.59 Yes. Significant results have a low probability of occurring if the null hypothesis is true. While they *could* have occurred by chance, they are *unlikely* to have done so.

9.61 **(a)** It stands for the true proportion of people who have body temperatures lower than 98.6°F. **(b)** $H_0: p = 0.5$. $H_a: p > 0.5$.

9.63 **(a)** 37% is a sample proportion and is therefore subject to sampling error.

(b) $$\hat{p} \pm z^*\sqrt{\frac{\hat{p}(1-\hat{p})}{n}} = 0.37 \pm 1.96\sqrt{\frac{0.37(0.63)}{1000}} = 0.37 \pm 0.030$$

We are 95% confident that the proportion of adults aged 18 and older who claim that football is their favorite sport to watch on television is between 0.34 and 0.40. **(c)** Yes. **(d)** Yes. People may be swept into the excitement during the playoff season and claim football is their favorite sport to watch on television yet have a different opinion when a different sport is "in season."

9.65 You do not need a confidence interval when you *know* the population proportion. You construct a confidence interval only when you are using sample data to make inferences about the population.

9.67 **(a)** $$\hat{p} \pm z^*\sqrt{\frac{\hat{p}(1-\hat{p})}{n}} = \frac{775}{1100} \pm 2.58\sqrt{\frac{\frac{775}{1100}\left(1-\frac{775}{1100}\right)}{1100}} = 0.705 \pm 0.035$$

We are 99% confident that the population proportion of teenagers who play games online is between 66.5% and 73.5%. **(b)** They would be larger due to the smaller sample sizes for each of the proportions. **(c)** If asked the question directly, teenagers may not respond truthfully. They may be worried about getting in trouble with their parents, being embarrassed in front of their peers, etc.

9.69 **(a)** **The hypotheses:** $H_0: p = 0.5$ versus $H_a: p > 0.5$, where p = the proportion of all U.S. teens who would say that young people should wait to have sex until marriage. **The sampling distribution:** If the null hypothesis is true, $\hat{p}$ follows a Normal distribution with mean = 0.5 and standard deviation = $\sqrt{(0.5)(0.5)/439} = 0.0239$. **The data:** $\hat{p} = 246/439 = 0.5604$. **The *P*-value:** $z = (0.5604 - 0.5)/0.0239 = 2.53 \Rightarrow P\text{-value} = P(z > 2.53) = 0.0057$. **Conclusion:** This is a very small *P*-value and provides strong evidence against the null hypothesis. The data give convincing evidence that a majority of U.S. teens would say that young people should wait to have sex until they're married.

(b) $$\hat{p} \pm z^*\sqrt{\frac{\hat{p}(1-\hat{p})}{n}} = 0.5604 \pm 1.96\sqrt{\frac{0.5604(1-0.5604)}{439}} = 0.5604 \pm 0.0464$$

We are 95% confident that the population proportion of teenagers who would say that young people should wait to have sex until they're married is between 51.4% and 60.7%. **(c)** No. The inference methods assume that the responses are truthful; this may or may not be the case. Assuring the teens that the survey is anonymous is one way to achieve more truthful responses.

Chapter 10

10.1 H_0: The company's stated color distribution for their peanut M&M's is correct; that is, $p_{\text{blue}} = 0.23$, $p_{\text{orange}} = 0.23$, $p_{\text{green}} = 0.15$, $p_{\text{yellow}} = 0.15$, $p_{\text{red}} = 0.12$, $p_{\text{brown}} = 0.12$. H_a: The company's stated color distribution for their peanut M&M's is not correct.

10.3 $X^2 = 11.3724$.

10.5 Using a significance level of 0.05, there is enough evidence to reject the null hypothesis and conclude that the distribution of colors does not follow the company's stated color distribution.

10.7 **The hypotheses:** H_0: Mendel's hypothesis is correct; that is, $p_{\text{wrinkled}} = 0.25$, $p_{\text{smooth}} = 0.75$. H_a: The proportions of smooth and wrinkled peas differ from Mendel's hypothesized values. **The sampling distribution:** Degrees of freedom = 1. **The data:** The expected counts are smooth = 417 and wrinkled = 139. **The *P*-value:** $X^2 = 0.3453$. The *P*-value is greater than 0.25. **Conclusion:** There is not enough evidence to reject Mendel's hypothesized distribution of pea type.

10.9 **(a)** H_0: Skittles candies are equally distributed among the five flavors: lemon, lime, orange, strawberry, and grape; that is, $p_{\text{lemon}} = p_{\text{lime}} = p_{\text{orange}} = p_{\text{strawberry}} = p_{\text{grape}} = 0.20$. H_a: Skittles candy flavors are not equally distributed. **(b)** 12 candies of each of the five flavors. **(c)** Greater than 9.49. **(d)** Greater than 13.28.

10.11 **(a)** H_0: The digits are equally likely to occur; that is, $p_0 = p_1 = \ldots = p_9 = 0.10$. H_a: The digits are not equally likely to occur. **(b)** Answers will vary.

10.13 **(a)**

	Extracurricular activities (hours per week) <2	2 to 12	>12
C or better	11/20 = 0.55	68/91 ≈ 0.747	3/8 = 0.375
D or F	9/20 = 0.45	23/91 ≈ 0.253	5/8 = 0.625

(b) Among the students who spend very little time in extracurricular activities, slightly more than half got a C or better. The group of students who spent *some* time in extracurricular activities had a high percent who received a grade of C or better (75%), while only 38% of students who spent *a lot of* time in extracurricular activities received a grade of C or better.

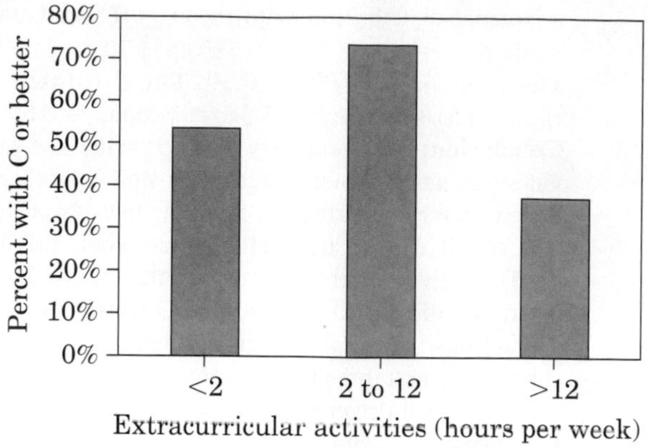

10.15 **(a)**

Alcohol consumption	Smoked	Did not smoke	Total
Yes	20	13	33
No	48	96	144
Total	68	109	177

(b) Among the drinkers, 60.6% smoked. Among the teetotalers, 33.3% smoked. Based on these data, people were more likely to resume smoking if they consumed alcohol.

(c)

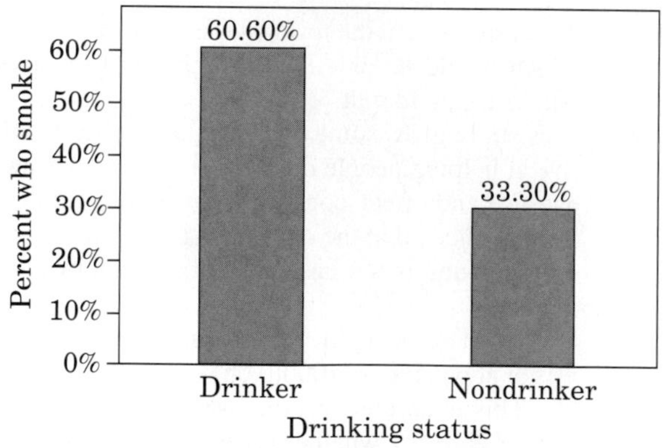

10.17 **(a)**

	Temperature: Cold	Neutral	Hot
Hatched	16	38	75
Did not hatch	11	18	29

(b) 59.3% of the cold eggs hatched, 67.9% of the neutral eggs hatched, and 72.1% of the hot eggs hatched. The researchers were mistaken in believing that the eggs would not hatch in cold water, though they don't hatch as well in cold water.

10.19 **(a)** $X^2 = 6.91$. D or F > 12. **(b)** df = 2; $0.01 < P < 0.05$. The chi-square statistic is significant at the 5% level but not at the 1% level. **(c)** We conclude that at the 5% level there is a statistically significant relationship between grade earned and number of hours spent in extracurricular activities.

10.21 **The hypotheses:** H_0: Water temperature and hatching rates are not related. H_a: Water temperature and hatching rates are related. **The sampling distribution:** The observed and expected values (O/E) are given in the following table:

	Temperature		
	Cold	**Neutral**	**Hot**
Hatched	16/18.63	38/38.63	75/71.74
Did not hatch	11/8.37	18/17.37	29/32.26

df = 2. **The data:** The observed cell counts are close to the expected cell counts, so it does not appear that water temperature and hatching rates are associated. **The *P*-value:** $X^2 = 1.703$. *P*-value > 0.25. **Conclusion:** These results are not significant. These data do not support the hypothesis that water temperature and hatching rates are related.

10.23 **(a)** The most effective treatment was desipramine, followed by lithium.

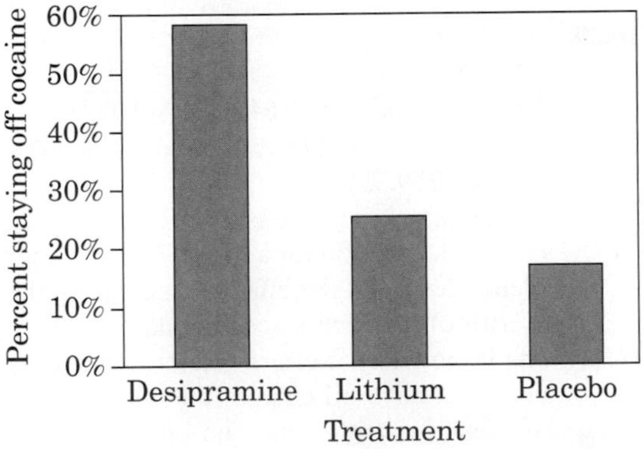

(b)

Treatment type	Stayed off cocaine	Did not stay off cocaine
Desipramine	14	10
Lithium	6	18
Placebo	4	20

10.25 **(a)** 0.01 < *P*-value < 0.05. **(b)** 0.25 < *P*-value. **(c)** 0.20 < *P*-value < 0.25.

10.27 The following table gives the counts and percents, by major, for males and females:

	Female		Male	
	Count	**Percent**	**Count**	**Percent**
Accounting	68	30.2%	56	34.8%
Administration	91	40.4%	40	24.8%
Economics	5	2.2%	6	3.7%
Finance	61	27.1%	59	36.6%
Total	225		161	

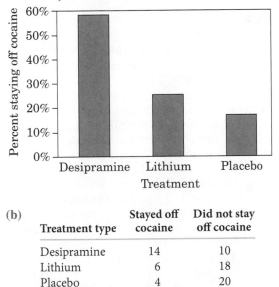

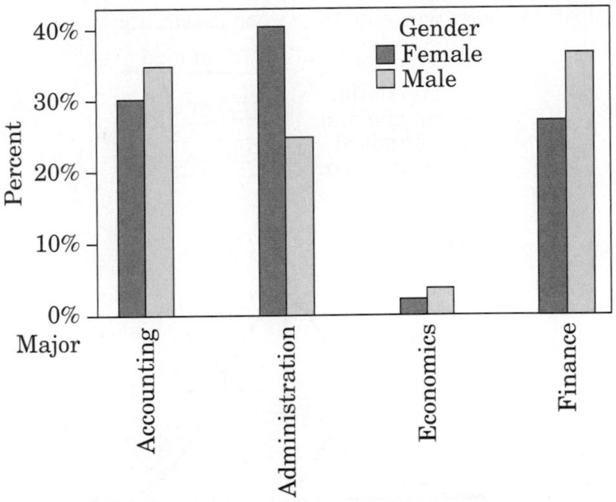

Administration is the most popular major for women. Finance, followed closely by accounting, is the most popular for men. The least popular major for men and women is economics.

10.29 **Hypotheses:** H_0 : Nuthatches do not prefer particular types of trees when they're searching for seeds and insects; that is $p_{\text{firs}} = 0.54$, $p_{\text{pines}} = 0.40$, $p_{\text{other}} = 0.06$. H_a : Nuthatches prefer particular types of trees when they're searching for seeds and insects. **The sampling distribution:** df = 2. **The data:** The expected counts are firs = 84.24, pines = 62.4, other = 9.36. **The *P*-value:** $X^2 = 7.4182$. $0.01 < P < 0.05$. **Conclusion:** We reject the null hypothesis at the 5% level but not at the 1% level. At the 5% level, we conclude that nuthatches exhibit a preference when searching for seeds and insects.

10.31 **(a) Hypotheses:** H_0 : The coin is fair; that is, $p_{\text{heads}} = 0.5$, $p_{\text{tails}} = 0.5$. H_a : The coin is not fair. **The sampling distribution:** df = 1. **The data:** The expected counts are heads = tails = 100. **The *P*-value:** $X^2 = 9.68$. $0.001 < P < 0.01$. **Conclusion:** We reject the null hypothesis at the 1% level and conclude that the coin is not fair. **(b) Hypotheses:** The parameter is the proportion of coin spins that resulted in "tails." $H_0: p = 0.5$. $H_a: p \neq 0.5$. **The sampling distribution:** If the null hypothesis is true, $\hat{p}$ follows a Normal distribution with mean = $p = 0.5$ and standard deviation = 0.0354. **The data:** $\hat{p} = 0.61$. **The *P*-value:** *P*-value = $2 \cdot P(\hat{p} > 0.61) = 0.0018$. **Conclusion:** There is enough evidence, at the 1% level, to conclude that the coin is unfair. **(c)** The results from the two tests are the same.

10.33 Answers will vary. Here's one possibility:

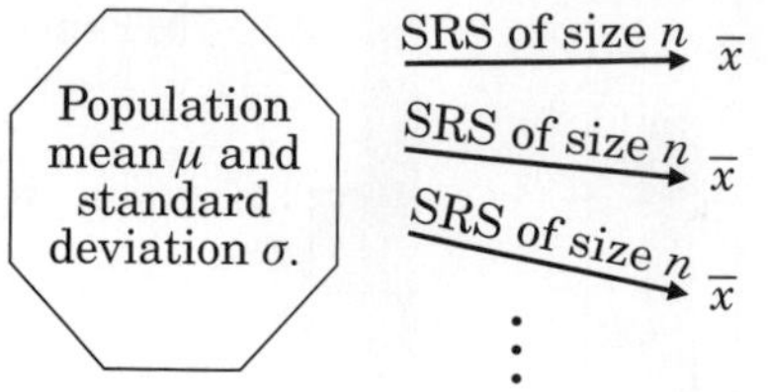

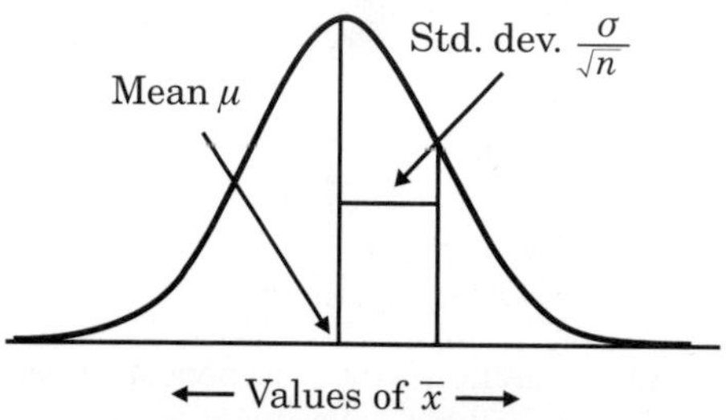

10.35 (a) The histogram appears to be more or less Normal (the bar centered at 0.54 makes it appear somewhat less so) and centered at the expected mean of 0.50.

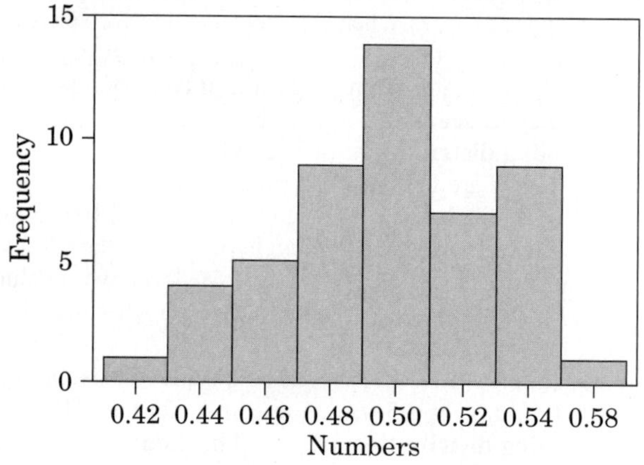

(b) The population mean has a theoretical value of 0.50.

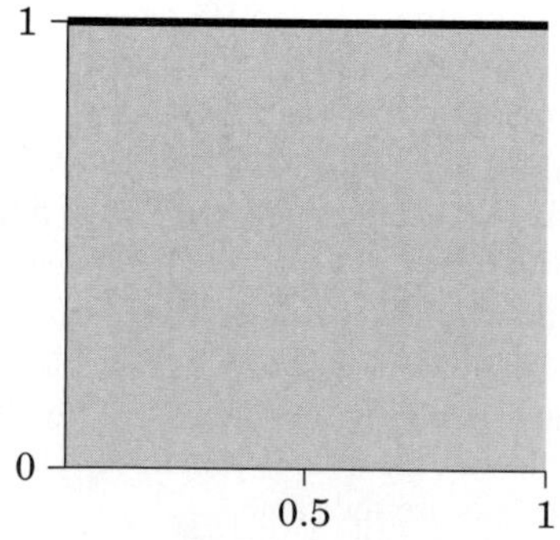

10.37 (a) $\bar{x}$ is approximately Normally distributed with mean 115 and standard deviation 6.

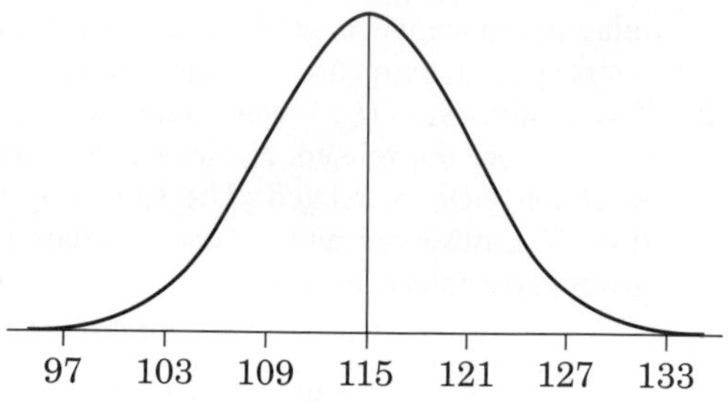

(b) and (c) If the null hypothesis were true, it would not be surprising to get a value as far removed from 115 as 118.6, but it would be fairly unusual to get one as far removed as 125.7.

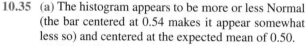

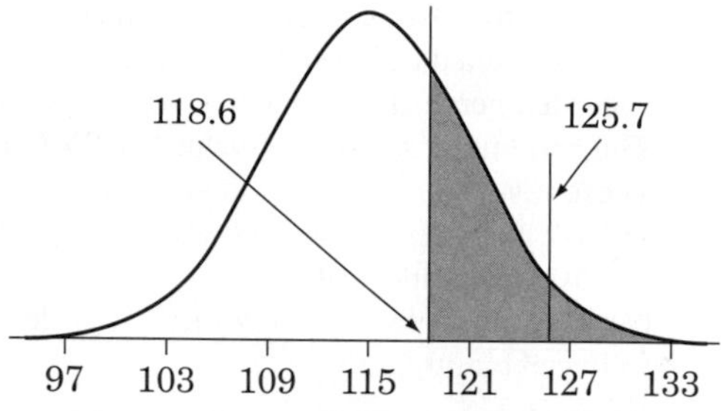

10.39 (a) 95% CI: $275 \pm 1.96(58/\sqrt{1077}) = (271.54, 278.46)$.
(b) 90% CI: $275 \pm 1.645(58/\sqrt{1077}) = (272.09, 277.91)$. 99% CI: $275 \pm 2.575\,(58/\sqrt{1077}) = (270.45, 279.55)$.
(c) The margin of error is 2.91 for a 90% CI, 3.46 for a 95% CI, and 4.55 for a 99% CI. Increasing the confidence level has the effect of increasing the margin of error of the confidence interval.

10.41 **The hypotheses:** Let μ = the mean increase, in dollars, of credit card charges this year compared with last year for customers who charge at least \$2400. $H_0: \mu = 0$. $H_a: \mu > 0$. **The sampling distribution:** If the null hypothesis is true, the sampling distribution of $\bar{x}$ for samples of size 200 would be approximately Normal with mean 0 and standard deviation 7.637. **The data:** The sample mean is 332. The corresponding standardized score is 43.47. **The *P*-value:** *P*-value ≈ 0. **Conclusion:** We reject the null hypothesis and conclude that it is likely that the mean amount charged has increased this year compared with last year.

10.43 (a) The sample size is 200 and the central limit theorem tells us that, for large samples, the sampling distribution of $\bar{x}$ will be approximately Normal regardless of the shape of the population from which the individual observations are drawn. (b) Answers will vary. One possibility is that the economy improved, so people spent more. Randomly divide the customers into two groups. One of these groups has their fees waived and the other group does not. Compare the two groups in terms of the increase in the amount charged from the prior year.

10.45 (a) $t^* = 3.106$. (b) $t^* = 1.699$. (c) $t^* = 2.110$.

10.47 95% CI: $7.364 \pm (2.228)(2.086) = (2.72, 12.01)$. The confidence interval gives us a range of plausible values for the difference in depression scores while taking a placebo versus taking a caffeine capsule.

10.49 **The hypotheses:** The parameter of interest is μ, the mean number of additional hours of sleep gained after taking the drug. $H_0: \mu = 0$. $H_a: \mu > 0$. **The sampling distribution:** We will use a t distribution with df $= 10 - 1 = 9$. **The data:** $\bar{x} = 2.33$ and $s = 2.0022$. The corresponding standard error is 0.6332, $t = 3.68$. **The *P*-value:** $0.0025 < P\text{-value} < 0.005$. **Conclusion.** We reject the null hypothesis. The data support the hypothesis that the drug is effective in increasing sleep time.

10.51 No, the student is not correct. The interval gives a range of plausible values for the mean NAEP quantitative score for men aged 21 to 25. It says nothing about the individual scores.

10.53 (a) Because the distribution of individual weights is not Normally distributed. (b) Since the sample size is large, and the data are not very non-Normal, the central limit theorem tells us that the distribution of the mean weight is approximately Normal. The standard deviation $= 6.39$. $P(\bar{x} > 200) = 0.0594$.

10.55 (a) Let μ be the average difference in speed of completing a task wearing an unscented mask minus wearing a scented mask. $H_0: \mu = 0$. $H_a: \mu > 0$. (b) $\bar{x} = 0.957$, $s = 12.55$. On average, there is about a one-second improvement, which is not enough to be considered practically important. (c) There aren't any outliers.

Stem-and-leaf of diff N = 21
Leaf unit = 1.0

−2	5
−2	
−1	7
−1	10
−0	876
−0	4430
0	24
0	789
1	24
1	69
2	
2	5

(d) If the null hypothesis is true, the sampling distribution of $\bar{x}$ has mean 0.957 and standard deviation 12.55. The standardized score is 0.35, so $P\text{-value} > 0.05$. This is not significant and does not provide evidence that wearing a scented mask reduces the time needed to complete a task.

10.57 **The hypotheses:** Let μ = average nightly study time for first-year students at a university. $H_0: \mu = 2.5$ hours, or 150 minutes. $H_a: \mu < 150$ minutes. **The sampling distribution:** If the null hypothesis is true, the sampling distribution of $\bar{x}$ for samples of size 269 would be approximately Normal with mean 150 minutes and standard deviation 3.963. **The data:** The sample mean is 137 minutes. The standardized score is -3.28. **The *P*-value:** $P\text{-value} = 0.0005$. **Conclusion:** We reject the null hypothesis and conclude that first-year students at the university study less than 2.5 hours per night on average.

10.59 (a) $\bar{x} = 7.5$ hours. $s = 64.536$ minutes. (b) His interpretation is incorrect. The confidence interval is a statement regarding the population mean lifetime of the batteries, not the individual lifetimes. (c) No. The confidence interval gives information about the value of the population mean. It tells us nothing about the value of another sample mean. (d) We are 95% confident that the mean lifetime of the company's AA batteries is between 430 and 470 minutes.

10.61 **The hypotheses:** Let μ = mean response time to all accidents involving life-threatening injuries this year in a particular city. $H_0: \mu = 6.7$ minutes. $H_a: \mu < 6.7$ minutes. **The sampling distribution:** If the null hypothesis is true, the sampling distribution of $\bar{x}$ for samples of size 400 would be approximately Normal with mean 6.7 minutes and standard deviation 0.1. **The data:** $\bar{x} = 6.48$ minutes. $z = -2.2$. **The *P*-value:** $P\text{-value} = 0.0139$. **Conclusion:** Using a 5% significance level, we reject the null hypothesis and conclude that the mean response time this year was less than 6.7 minutes, the previous year's average response time.

10.63 (a)

	Goal		
Gender	**Grades**	**Popular**	**Sports**
Female	130/51.8%	91/36.2%	30/12.0%
Male	117/51.5%	50/22%	60/26.4%

(b) The most popular response for both males and females was "make good grades." The second most popular response was "to be popular" for females and "to be good at sports" for males.

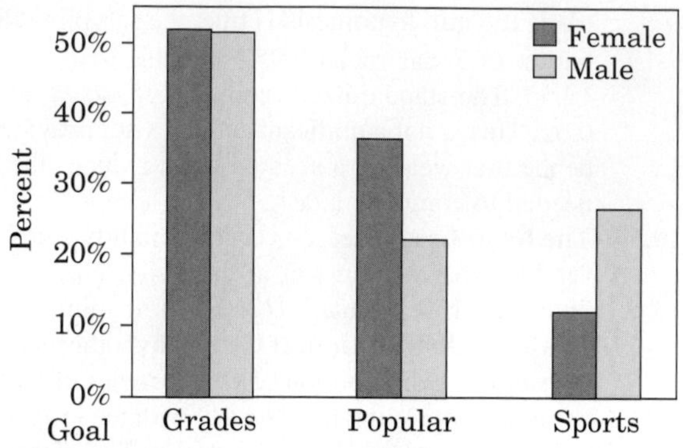

10.65 **(a)** Let μ = the mean decrease in systolic blood pressure for patients who eat dark chocolate. $H_0: \mu = 0$. $H_a: \mu > 0$. **(b)** The standard score for $\overline{x}$ is 2.20, which gives a P-value of 0.035. **(c)** The P-value of 0.035 is significant at the 5% level and provides good evidence that there has been a reduction in systolic blood pressure for patients who eat dark chocolate. **(d)** The sample size was very small. We would have more confidence in the results if the sample size were increased to control for variability.

10.67 **(a)** This was an experiment since a treatment was "applied" to the subjects. **(b)** **The hypotheses:** H_0: Treatment and pregnancy are unrelated. H_a: Treatment and pregnancy are related. **The sampling distribution:** The data (O/E) are

	Acupuncture	**Control**
Pregnant	34/27.5	21/27.5
Not pregnant	46/52.5	59/52.5

df = 1. **The data:** There are slightly more pregnancies in the acupuncture group and fewer in the control group than we would expect if the null hypothesis were true. **The P-value:** $X^2 = 4.68$. $0.01 < P < 0.05$. **Conclusion:** This result is significant at the 5% level but not at the 1% level. Using a 5% significance level, we reject the null hypothesis and conclude that pregnancy and treatment are related. **(c)** Since the subjects were aware of which treatment they were receiving, there was no placebo group. It's possible that the observed results are due in part to the placebo effect.

INDEX

Note: A page number in **boldface** indicates a definition; in *italics,* a figure; and followed by *t,* a table.

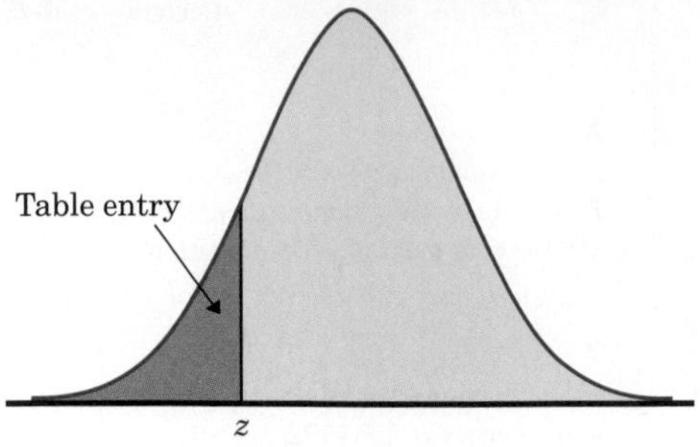

Table entry for z is the area under the standard Normal curve to the left of z.

Table A Standard Normal probabilities

z	.00	.01	.02	.03	.04	.05	.06	.07	.08	.09
−3.4	.0003	.0003	.0003	.0003	.0003	.0003	.0003	.0003	.0003	.0002
−3.3	.0005	.0005	.0005	.0004	.0004	.0004	.0004	.0004	.0004	.0003
−3.2	.0007	.0007	.0006	.0006	.0006	.0006	.0006	.0005	.0005	.0005
−3.1	.0010	.0009	.0009	.0009	.0008	.0008	.0008	.0008	.0007	.0007
−3.0	.0013	.0013	.0013	.0012	.0012	.0011	.0011	.0011	.0010	.0010
−2.9	.0019	.0018	.0018	.0017	.0016	.0016	.0015	.0015	.0014	.0014
−2.8	.0026	.0025	.0024	.0023	.0023	.0022	.0021	.0021	.0020	.0019
−2.7	.0035	.0034	.0033	.0032	.0031	.0030	.0029	.0028	.0027	.0026
−2.6	.0047	.0045	.0044	.0043	.0041	.0040	.0039	.0038	.0037	.0036
−2.5	.0062	.0060	.0059	.0057	.0055	.0054	.0052	.0051	.0049	.0048
−2.4	.0082	.0080	.0078	.0075	.0073	.0071	.0069	.0068	.0066	.0064
−2.3	.0107	.0104	.0102	.0099	.0096	.0094	.0091	.0089	.0087	.0084
−2.2	.0139	.0136	.0132	.0129	.0125	.0122	.0119	.0116	.0113	.0110
−2.1	.0179	.0174	.0170	.0166	.0162	.0158	.0154	.0150	.0146	.0143
−2.0	.0228	.0222	.0217	.0212	.0207	.0202	.0197	.0192	.0188	.0183
−1.9	.0287	.0281	.0274	.0268	.0262	.0256	.0250	.0244	.0239	.0233
−1.8	.0359	.0351	.0344	.0336	.0329	.0322	.0314	.0307	.0301	.0294
−1.7	.0446	.0436	.0427	.0418	.0409	.0401	.0392	.0384	.0375	.0367
−1.6	.0548	.0537	.0526	.0516	.0505	.0495	.0485	.0475	.0465	.0455
−1.5	.0668	.0655	.0643	.0630	.0618	.0606	.0594	.0582	.0571	.0559
−1.4	.0808	.0793	.0778	.0764	.0749	.0735	.0721	.0708	.0694	.0681
−1.3	.0968	.0951	.0934	.0918	.0901	.0885	.0869	.0853	.0838	.0823
−1.2	.1151	.1131	.1112	.1093	.1075	.1056	.1038	.1020	.1003	.0985
−1.1	.1357	.1335	.1314	.1292	.1271	.1251	.1230	.1210	.1190	.1170
−1.0	.1587	.1562	.1539	.1515	.1492	.1469	.1446	.1423	.1401	.1379
−0.9	.1841	.1814	.1788	.1762	.1736	.1711	.1685	.1660	.1635	.1611
−0.8	.2119	.2090	.2061	.2033	.2005	.1977	.1949	.1922	.1894	.1867
−0.7	.2420	.2389	.2358	.2327	.2296	.2266	.2236	.2206	.2177	.2148
−0.6	.2743	.2709	.2676	.2643	.2611	.2578	.2546	.2514	.2483	.2451
−0.5	.3085	.3050	.3015	.2981	.2946	.2912	.2877	.2843	.2810	.2776
−0.4	.3446	.3409	.3372	.3336	.3300	.3264	.3228	.3192	.3156	.3121
−0.3	.3821	.3783	.3745	.3707	.3669	.3632	.3594	.3557	.3520	.3483
−0.2	.4207	.4168	.4129	.4090	.4052	.4013	.3974	.3936	.3897	.3859
−0.1	.4602	.4562	.4522	.4483	.4443	.4404	.4364	.4325	.4286	.4247
−0.0	.5000	.4960	.4920	.4880	.4840	.4801	.4761	.4721	.4681	.4641

(continued)

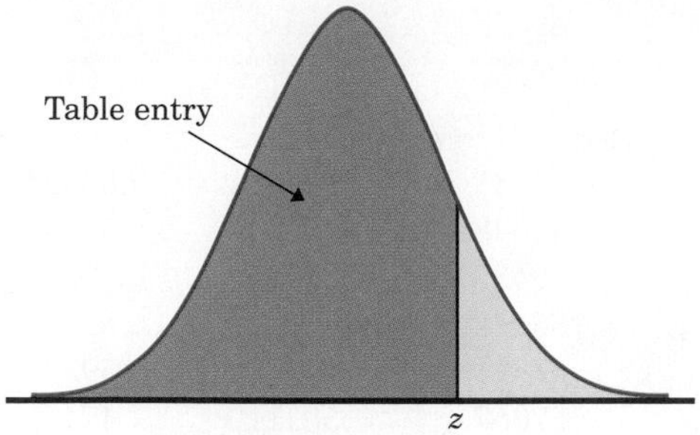

Table entry for z is the area under the standard Normal curve to the left of z.

Table A Standard Normal probabilities (*continued*)

z	.00	.01	.02	.03	.04	.05	.06	.07	.08	.09
0.0	.5000	.5040	.5080	.5120	.5160	.5199	.5239	.5279	.5319	.5359
0.1	.5398	.5438	.5478	.5517	.5557	.5596	.5636	.5675	.5714	.5753
0.2	.5793	.5832	.5871	.5910	.5948	.5987	.6026	.6064	.6103	.6141
0.3	.6179	.6217	.6255	.6293	.6331	.6368	.6406	.6443	.6480	.6517
0.4	.6554	.6591	.6628	.6664	.6700	.6736	.6772	.6808	.6844	.6879
0.5	.6915	.6950	.6985	.7019	.7054	.7088	.7123	.7157	.7190	.7224
0.6	.7257	.7291	.7324	.7357	.7389	.7422	.7454	.7486	.7517	.7549
0.7	.7580	.7611	.7642	.7673	.7704	.7734	.7764	.7794	.7823	.7852
0.8	.7881	.7910	.7939	.7967	.7995	.8023	.8051	.8078	.8106	.8133
0.9	.8159	.8186	.8212	.8238	.8264	.8289	.8315	.8340	.8365	.8389
1.0	.8413	.8438	.8461	.8485	.8508	.8531	.8554	.8577	.8599	.8621
1.1	.8643	.8665	.8686	.8708	.8729	.8749	.8770	.8790	.8810	.8830
1.2	.8849	.8869	.8888	.8907	.8925	.8944	.8962	.8980	.8997	.9015
1.3	.9032	.9049	.9066	.9082	.9099	.9115	.9131	.9147	.9162	.9177
1.4	.9192	.9207	.9222	.9236	.9251	.9265	.9279	.9292	.9306	.9319
1.5	.9332	.9345	.9357	.9370	.9382	.9394	.9406	.9418	.9429	.9441
1.6	.9452	.9463	.9474	.9484	.9495	.9505	.9515	.9525	.9535	.9545
1.7	.9554	.9564	.9573	.9582	.9591	.9599	.9608	.9616	.9625	.9633
1.8	.9641	.9649	.9656	.9664	.9671	.9678	.9686	.9693	.9699	.9706
1.9	.9713	.9719	.9726	.9732	.9738	.9744	.9750	.9756	.9761	.9767
2.0	.9772	.9778	.9783	.9788	.9793	.9798	.9803	.9808	.9812	.9817
2.1	.9821	.9826	.9830	.9834	.9838	.9842	.9846	.9850	.9854	.9857
2.2	.9861	.9864	.9868	.9871	.9875	.9878	.9881	.9884	.9887	.9890
2.3	.9893	.9896	.9898	.9901	.9904	.9906	.9909	.9911	.9913	.9916
2.4	.9918	.9920	.9922	.9925	.9927	.9929	.9931	.9932	.9934	.9936
2.5	.9938	.9940	.9941	.9943	.9945	.9946	.9948	.9949	.9951	.9952
2.6	.9953	.9955	.9956	.9957	.9959	.9960	.9961	.9962	.9963	.9964
2.7	.9965	.9966	.9967	.9968	.9969	.9970	.9971	.9972	.9973	.9974
2.8	.9974	.9975	.9976	.9977	.9977	.9978	.9979	.9979	.9980	.9981
2.9	.9981	.9982	.9982	.9983	.9984	.9984	.9985	.9985	.9986	.9986
3.0	.9987	.9987	.9987	.9988	.9988	.9989	.9989	.9989	.9990	.9990
3.1	.9990	.9991	.9991	.9991	.9992	.9992	.9992	.9992	.9993	.9993
3.2	.9993	.9993	.9994	.9994	.9994	.9994	.9994	.9995	.9995	.9995
3.3	.9995	.9995	.9995	.9996	.9996	.9996	.9996	.9996	.9996	.9997
3.4	.9997	.9997	.9997	.9997	.9997	.9997	.9997	.9997	.9997	.9998

Table B Random digits

Line								
101	19223	95034	05756	28713	96409	12531	42544	82853
102	73676	47150	99400	01927	27754	42648	82425	36290
103	45467	71709	77558	00095	32863	29485	82226	90056
104	52711	38889	93074	60227	40011	85848	48767	52573
105	95592	94007	69971	91481	60779	53791	17297	59335
106	68417	35013	15529	72765	85089	57067	50211	47487
107	82739	57890	20807	47511	81676	55300	94383	14893
108	60940	72024	17868	24943	61790	90656	87964	18883
109	36009	19365	15412	39638	85453	46816	83485	41979
110	38448	48789	18338	24697	39364	42006	76688	08708
111	81486	69487	60513	09297	00412	71238	27649	39950
112	59636	88804	04634	71197	19352	73089	84898	45785
113	62568	70206	40325	03699	71080	22553	11486	11776
114	45149	32992	75730	66280	03819	56202	02938	70915
115	61041	77684	94322	24709	73698	14526	31893	32592
116	14459	26056	31424	80371	65103	62253	50490	61181
117	38167	98532	62183	70632	23417	26185	41448	75532
118	73190	32533	04470	29669	84407	90785	65956	86382
119	95857	07118	87664	92099	58806	66979	98624	84826
120	35476	55972	39421	65850	04266	35435	43742	11937
121	71487	09984	29077	14863	61683	47052	62224	51025
122	13873	81598	95052	90908	73592	75186	87136	95761
123	54580	81507	27102	56027	55892	33063	41842	81868
124	71035	09001	43367	49497	72719	96758	27611	91596
125	96746	12149	37823	71868	18442	35119	62103	39244
126	96927	19931	36809	74192	77567	88741	48409	41903
127	43909	99477	25330	64359	40085	16925	85117	36071
128	15689	14227	06565	14374	13352	49367	81982	87209
129	36759	58984	68288	22913	18638	54303	00795	08727
130	69051	64817	87174	09517	84534	06489	87201	97245
131	05007	16632	81194	14873	04197	85576	45195	96565
132	68732	55259	84292	08796	43165	93739	31685	97150
133	45740	41807	65561	33302	07051	93623	18132	09547
134	27816	78416	18329	21337	35213	37741	04312	68508
135	66925	55658	39100	78458	11206	19876	87151	31260
136	08421	44753	77377	28744	75592	08563	79140	92454
137	53645	66812	61421	47836	12609	15373	98481	14592
138	66831	68908	40772	21558	47781	33586	79177	06928
139	55588	99404	70708	41098	43563	56934	48394	51719
140	12975	13258	13048	45144	72321	81940	00360	02428
141	96767	35964	23822	96012	94591	65194	50842	53372
142	72829	50232	97892	63408	77919	44575	24870	04178
143	88565	42628	17797	49376	61762	16953	88604	12724
144	62964	88145	83083	69453	46109	59505	69680	00900
145	19687	12633	57857	95806	09931	02150	43163	58636
146	37609	59057	66967	83401	60705	02384	90597	93600
147	54973	86278	88737	74351	47500	84552	19909	67181
148	00694	05977	19664	65441	20903	62371	22725	53340
149	71546	05233	53946	68743	72460	27601	45403	88692
150	07511	88915	41267	16853	84569	79367	32337	03316

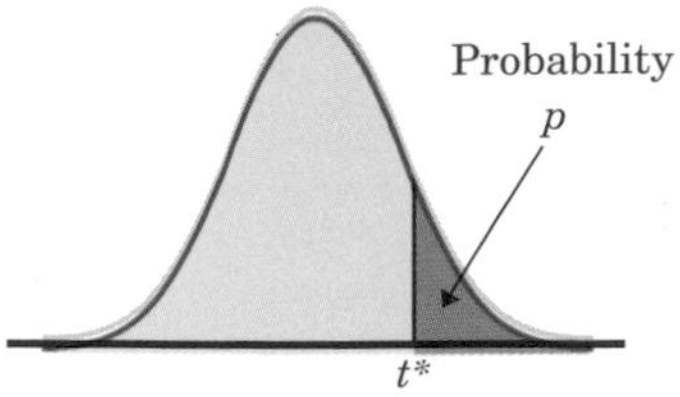

Table entry for p and C is the critical value t^* with probability p lying to its right and probability C lying between $-t^*$ and t^*.

Table C *t* distribution critical values

	Upper tail probability *p*											
df	**.25**	**.20**	**.15**	**.10**	**.05**	**.025**	**.02**	**.01**	**.005**	**.0025**	**.001**	**.0005**
1	1.000	1.376	1.963	3.078	6.314	12.71	15.89	31.82	63.66	127.3	318.3	636.6
2	0.816	1.061	1.386	1.886	2.920	4.303	4.849	6.965	9.925	14.09	22.33	31.60
3	0.765	0.978	1.250	1.638	2.353	3.182	3.482	4.541	5.841	7.453	10.21	12.92
4	0.741	0.941	1.190	1.533	2.132	2.776	2.999	3.747	4.604	5.598	7.173	8.610
5	0.727	0.920	1.156	1.476	2.015	2.571	2.757	3.365	4.032	4.773	5.893	6.869
6	0.718	0.906	1.134	1.440	1.943	2.447	2.612	3.143	3.707	4.317	5.208	5.959
7	0.711	0.896	1.119	1.415	1.895	2.365	2.517	2.998	3.499	4.029	4.785	5.408
8	0.706	0.889	1.108	1.397	1.860	2.306	2.449	2.896	3.355	3.833	4.501	5.041
9	0.703	0.883	1.100	1.383	1.833	2.262	2.398	2.821	3.250	3.690	4.297	4.781
10	0.700	0.879	1.093	1.372	1.812	2.228	2.359	2.764	3.169	3.581	4.144	4.587
11	0.697	0.876	1.088	1.363	1.796	2.201	2.328	2.718	3.106	3.497	4.025	4.437
12	0.695	0.873	1.083	1.356	1.782	2.179	2.303	2.681	3.055	3.428	3.930	4.318
13	0.694	0.870	1.079	1.350	1.771	2.160	2.282	2.650	3.012	3.372	3.852	4.221
14	0.692	0.868	1.076	1.345	1.761	2.145	2.264	2.624	2.977	3.326	3.787	4.140
15	0.691	0.866	1.074	1.341	1.753	2.131	2.249	2.602	2.947	3.286	3.733	4.073
16	0.690	0.865	1.071	1.337	1.746	2.120	2.235	2.583	2.921	3.252	3.686	4.015
17	0.689	0.863	1.069	1.333	1.740	2.110	2.224	2.567	2.898	3.222	3.646	3.965
18	0.688	0.862	1.067	1.330	1.734	2.101	2.214	2.552	2.878	3.197	3.611	3.922
19	0.688	0.861	1.066	1.328	1.729	2.093	2.205	2.539	2.861	3.174	3.579	3.883
20	0.687	0.860	1.064	1.325	1.725	2.086	2.197	2.528	2.845	3.153	3.552	3.850
21	0.686	0.859	1.063	1.323	1.721	2.080	2.189	2.518	2.831	3.135	3.527	3.819
22	0.686	0.858	1.061	1.321	1.717	2.074	2.183	2.508	2.819	3.119	3.505	3.792
23	0.685	0.858	1.060	1.319	1.714	2.069	2.177	2.500	2.807	3.104	3.485	3.768
24	0.685	0.857	1.059	1.318	1.711	2.064	2.172	2.492	2.797	3.091	3.467	3.745
25	0.684	0.856	1.058	1.316	1.708	2.060	2.167	2.485	2.787	3.078	3.450	3.725
26	0.684	0.856	1.058	1.315	1.706	2.056	2.162	2.479	2.779	3.067	3.435	3.707
27	0.684	0.855	1.057	1.314	1.703	2.052	2.158	2.473	2.771	3.057	3.421	3.690
28	0.683	0.855	1.056	1.313	1.701	2.048	2.154	2.467	2.763	3.047	3.408	3.674
29	0.683	0.854	1.055	1.311	1.699	2.045	2.150	2.462	2.756	3.038	3.396	3.659
30	0.683	0.854	1.055	1.310	1.697	2.042	2.147	2.457	2.750	3.030	3.385	3.646
40	0.681	0.851	1.050	1.303	1.684	2.021	2.123	2.423	2.704	2.971	3.307	3.551
50	0.679	0.849	1.047	1.299	1.676	2.009	2.109	2.403	2.678	2.937	3.261	3.496
60	0.679	0.848	1.045	1.296	1.671	2.000	2.099	2.390	2.660	2.915	3.232	3.460
80	0.678	0.846	1.043	1.292	1.664	1.990	2.088	2.374	2.639	2.887	3.195	3.416
100	0.677	0.845	1.042	1.290	1.660	1.984	2.081	2.364	2.626	2.871	3.174	3.390
1000	0.675	0.842	1.037	1.282	1.646	1.962	2.056	2.330	2.581	2.813	3.098	3.300
z^*	0.674	0.841	1.036	1.282	1.645	1.960	2.054	2.326	2.576	2.807	3.091	3.291
	50%	**60%**	**70%**	**80%**	**90%**	**95%**	**96%**	**98%**	**99%**	**99.5%**	**99.8%**	**99.9%**
	Confidence level *C*											

Table D Critical values of z^*

Confidence level C	Critical value z^*	Confidence level C	Critical value z^*
50%	0.67	90%	1.64
60%	0.84	95%	1.96
70%	1.04	99%	2.58
80%	1.28	99.9%	3.29

Table E Chi-square critical values

	Significance level α						
df	**0.25**	**0.20**	**0.15**	**0.10**	**0.05**	**0.01**	**0.001**
1	1.32	1.64	2.07	2.71	3.84	6.63	10.83
2	2.77	3.22	3.79	4.61	5.99	9.21	13.82
3	4.11	4.64	5.32	6.25	7.81	11.34	16.27
4	5.39	5.99	6.74	7.78	9.49	13.28	18.47
5	6.63	7.29	8.12	9.24	11.07	15.09	20.51
6	7.84	8.56	9.45	10.64	12.59	16.81	22.46
7	9.04	9.80	10.75	12.02	14.07	18.48	24.32
8	10.22	11.03	12.03	13.36	15.51	20.09	26.12
9	11.39	12.24	13.29	14.68	16.92	21.67	27.88